Forest Nursery Manual: Production of Bareroot Seedlings

Duryea, Mary L., and Thomas D. Landis (eds.). 1984. *Forest Nursery Manual: Production of Bareroot Seedlings.* Martinus Nijhoff/Dr W. Junk Publishers, The Hague/Boston/Lancaster, for Forest Research Laboratory, Oregon State University, Corvallis. 386 p.

The *Forest Nursery Manual* presents state-of-the-art information about current bareroot-nursery practices and research in the northwestern United States and Canada. Information on practices was gathered through a detailed nursery survey and incorporated into 30 chapters written by leading scientists and nursery managers. The *Manual* emphasizes the major stages of seedling production—selecting and developing the optimal site, assuring seed quality and vigorous early seedling growth, managing the soil and water, culturing bareroot seedlings, and harvesting and outplanting. A section on upgrading nursery practices is intended to stimulate thinking towards improving nursery management. A comprehensive glossary and tables of nursery conversion factors are included as appendices.

Forest Nursery Manual:
Production of Bareroot Seedlings

Edited by Mary L. Duryea and Thomas D. Landis

A cooperative project of

Nursery Technology Cooperative
Department of Forest Science
Oregon State University
Corvallis, Oregon 97331 U.S.A.

and

U.S.D.A. Forest Service
State and Private Forestry
Pacific Northwest Region
Portland, Oregon 97208 U.S.A.

1984 • Martinus Nijhoff/Dr W. Junk Publishers
The Hague/Boston/Lancaster

for Forest Research Laboratory, Oregon State University
Corvallis

Managing Editor: Carol R. Perry
Cover Design: Don Poole

Distributors

For the United States and Canada: Kluwer Boston, Inc., 190 Old Derby Street, Hingham, MA 02043 U.S.A.

For all other countries: Kluwer Academic Publishers Group, Distribution Center, P.O. Box 322, 3300 AH Dordrecht, The Netherlands.

Library of Congress Cataloging in Publication Data

Main entry under title:

Forest nursery manual.

(Forestry sciences)

"Cooperative project of Nursery Technology Cooperative, Department of Forest Science, Oregon State University . . . and USDA Forest Service, State and Private Forestry, Region 6."

Papers presented at the Bareroot Nursery Technology Workshop held at OSU in October 1982.

Includes index.

1. Trees—Seedlings, bareroot—Handbooks, manuals, etc. 2. Forest nurseries—Handbooks, manuals, etc. I. Duryea, Mary L. II. Landis, Thomas D. III. Oregon State University. Nursery Technology Cooperative. IV. United States. State and Private Forestry. Pacific Northwest Region. V. Bareroot Nursery Technology Workshop (1982: Oregon State University). VI. Series.

SD404.25.F67 1983 634.9'564 83-17347

ISBN 90-247-2913-0

ISBN 90-247-2914-9 (pbk.)

Second printing 1987

The material contained in this publication is supplied with the understanding that there is no intended endorsement of a specific product or practice. Users of pesticides should always consult the appropriate regulatory agencies for the latest information on registration and application. Individual authors are responsible for the opinions expressed in and content of their chapters.

Printed in the U.S.A.

Contents

Managing the Soil and Water (continued)

Culturing the Bareroot Seedling

Harvesting and Outplanting the Bareroot Seedling

Improving Nursery-Management Style: Selected Topics

Upgrading Nursery Practices

List of Contributors

Lawrence P. Abrahamson, Senior Research Associate, College of Environmental Science and Forestry, State University of New York, Syracuse, New York 13210.

Donald E. Boyer, retired U.S.D.A. Forest Service, Route 3, Box 299, Newberg, Oregon 97132.

A. Nigel Burdett, Research Scientist, British Columbia Ministry of Forests, Research Branch, 1450 Government Street, Victoria, B.C. V8W 3E7 Canada.

Robert K. Campbell, Principal Plant Geneticist, U.S.D.A. Forest Service, Pacific Northwest Forest and Range Experiment Station, Forestry Sciences Laboratory, 3200 Jefferson Way, Corvallis, Oregon 97331.

Charles B. Davey, Carl Alwin Schenck Professor of Forestry, Soil Science, and Plant Pathology, North Carolina State University, Raleigh, North Carolina 27650.

Robert J. Day, Professor, School of Forestry, Lakehead University, Thunder Bay, Ontario P7B 5E1 Canada.

Mary L. Duryea, Assistant Professor, Leader, Nursery Technology Cooperative, Department of Forest Science, Oregon State University, Corvallis, Oregon 97331.

James W. Edgren, Reforestation Specialist, U.S.D.A. Forest Service, Pacific Northwest Region, Post Office Box 3623, Portland, Oregon 97208.

Will B. Ellington, President, Lava Nursery, Inc., 5301 Culbertson Road, Post Office Box 370, Parkdale, Oregon 97041.

John C. Gordon, Dean, School of Forestry and Environmental Studies, Yale University, 205 Prospect Street, New Haven, Connecticut 06511.

Philip F. Hahn, Manager, Forest Research, Rex Timber, Inc., a subsidiary of Georgia-Pacific Corporation, Post Office Box 1618, Eugene, Oregon 97440.

Richard G. Hallman, Resource Staff Assistant, Missoula Equipment Development Center, Building 1, Fort Missoula, Missoula, Montana 59801.

David B. Hannaway, Assistant Professor, Department of Crop Science, Oregon State University, Corvallis, Oregon 97331.

Stephen M. Hee, Manager, Western Nurseries/Greenhouses, Land and Timber Production and Facilities, Weyerhaeuser Company, 7935 Highway 12 S.W., Rochester, Washington 98579.

Richard D. Iverson, Research Forester, Natchez Forest Research Center, Route 3, Box 312-B, Natchez, Mississippi 39120.

Thomas D. Landis, Western Nursery and Greenhouse Specialist, U.S.D.A. Forest Service, State and Private Forestry, Post Office Box 25127, Denver, Colorado 80225.

Denis P. Lavender, Professor, Department of Forest Science, Oregon State University, Corvallis, Oregon 97331.

Stephen E. McDonald, Research Forester, U.S.D.A. Forest Service, Timber Management Research, P.O. Box 2417, Washington, D.C. 20013.

William S. McGuire, Professor, Department of Crop Science, Oregon State University, Corvallis, Oregon 97331.

Randy Molina, Botanist, U.S.D.A. Forest Service, Pacific Northwest Forest and Range Experiment Station, Forestry Sciences Laboratory, 3200 Jefferson Way, Corvallis, Oregon 97331.

Frank E. Morby, Nursery Superintendent, U.S.D.A. Forest Service, J. Herbert Stone Nursery, 2606 Old Stage Road, Central Point, Oregon 97502.

Steven K. Omi, Research Assistant, Department of Forest Science, Oregon State University, Corvallis, Oregon 97331.

Peyton W. Owston, Principal Plant Physiologist, U.S.D.A. Forest Service, Pacific Northwest Forest and Range Experiment Station, Forestry Sciences Laboratory, 3200 Jefferson Way, Corvallis, Oregon 97331.

Gary A. Ritchie, Project Leader, Seedling Physiology, Weyerhaeuser Company, Western Forestry Research Center, 534 N. Tower Street, Centralia, Washington 98531.

Craig B. Royce, Nursery Manager, D. L. Phipps Oregon State Forest Nursery, Route 3, Box 193, Elkton, Oregon 97436.

David G. Simpson, Research Scientist, British Columbia Ministry of Forests, Kalamalka Research Station, 1450 Government Street, Victoria, B.C. V8W 3E7 Canada.

Frank C. Sorensen, Principal Plant Geneticist, U.S.D.A. Forest Service, Pacific Northwest Forest and Range Experiment Station, Forestry Sciences Laboratory, 3200 Jefferson Way, Corvallis, Oregon 97331.

Jack R. Sutherland, Research Scientist, Pacific Forest Research Centre, Canadian Forestry Service, 506 W. Burnside Road, Victoria, B.C. V8Z 1M5 Canada.

Yasuomi Tanaka, Forest Nursery Ecologist, Weyerhaeuser Company, Western Forestry Research Center, 534 N. Tower Street, Centralia, Washington 98531.

Barbara E. Thompson, Research Forester, International Paper Company, 34937 Tennessee Road, Lebanon, Oregon 97355.

James M. Trappe, Principal Mycologist, U.S.D.A. Forest Service, Pacific Northwest Forest and Range Experiment Station, Forestry Sciences Laboratory, 3200 Jefferson Way, Corvallis, Oregon 97331.

Robert van den Driessche, Senior Tree Physiologist, British Columbia Ministry of Forests, Research Branch, 1450 Government Street, Victoria, B.C. V8W 3E7 Canada.

Benno P. Warkentin, Professor and Head, Department of Soil Science, Oregon State University, Corvallis, Oregon 97331

Timothy L. White, Supervisor, Western Research, International Paper Company, Post Office Box 3860, Portland, Oregon 97208.

Chester T. Youngberg, Professor Emeritus, Department of Soil Science, Oregon State University, Corvallis, Oregon 97331.

Acknowledgments

Producing a manual such as this requires the coordinated efforts, diverse skills and expertise, and continuing energies and commitment of many people.

Bob Tarrant headed the task force that proposed the establishment of the Nursery Technology Cooperative and recognized the need for an update on nursery technology.

We are grateful to Tim White, who was a valued sounding board throughout—providing excellent reviews of many stages of production, including the original outline of the *Manual* and the questionnaires.

We appreciate the hospitality of all nurseries and seed-processing plants who participated in the Oregon State University (OSU) Nursery Survey, conducted in fall 1981, and the thoughtful responses of managers and staff to our questionnaires. David Simpson of the B.C. Ministry of Forests deserves special mention for his efforts in surveying the Canadian organizations.

The Bareroot Nursery Technology Workshop, held at OSU in October 1982, previewed the *Manual* for over 250 people. The success and "smooth sailing" of the Workshop are largely attributable to the organizational talents of Pam Henderson and the contributions of Helen Dufur, Steve Omi, and Allan Doerksen.

The secretarial and clerical skills of Char Singkofer were invaluable in preparing the nursery questionnaires, workshop programs, and related correspondence. Julie Cone helped by typing some of the *Manual* chapters on word processor.

Special thanks are extended to all those individuals who took time from busy schedules to carefully review and critique chapters.

We are grateful for the production assistance of: Alan Smith for indexing; Steve Omi for developing the glossary and conversion tables; Martha Brookes, Tawny Blinn, Susan Bell, Tom Brookes, Beth Marshall, and Martha Burdick for proofreading; Allan Doerksen for photographic work; Don Poole for the cover and brochure designs; Joan Barbour for drafting and layout; Wes Patterson and Don Ferguson for typesetting; George Shaw and Dwayne Downing for printing; and Carol Perry for substantive chapter editing.

The organizational and creative talents of our managing editor, Carol Perry, were instrumental in the development of the *Manual*. Her professional insights and ability to visualize the comprehensive nature of the final product were invaluable. We are grateful for her help in making this an enjoyable project.

John Gordon provided continual support and encouragement. His foresight and belief in the need for this *Manual* motivated and inspired us. For this, and for his creative input, we are thankful.

We acknowledge with appreciation the financial support of both the Nursery Technology Cooperative and the Department of Forest Science, Oregon State University. The Division of State and Private Forestry of the Pacific Northwest Region, U.S.D.A. Forest Service, also provided financial assistance.

Finally, we are indebted to our chapter authors—without whose tremendous dedication, patience, diligence, persistence, and promptness this *Manual* never could have been produced.

Introduction

Chapter 1
Development of the *Forest Nursery Manual*: A Synthesis of Current Practices and Research

M. L. Duryea and T. D. Landis

Abstract

The *Forest Nursery Manual*—a joint effort between the Department of Forest Science, Oregon State University, and the U.S.D.A. Forest Service, Division of State and Private Forestry—presents state-of-the-art information about current bareroot-nursery practices and research in the Northwest (northwestern United States and province of British Columbia, Canada). The *Manual* emphasizes all stages of seedling production, from nursery-site selection through outplanting. To gather information, 21 Northwest bareroot nurseries and eight seed-processing plants were surveyed with two in-depth questionnaires. Survey results were interpreted and incorporated into chapters written by leading scientists and nursery managers. Over 250 people previewed the *Manual* at the Bareroot Nursery Technology Workshop, held at Oregon State University in October 1982. The growing size and sophistication of the Northwest bareroot-nursery industry underscore the pressing need for this up-to-date manual.

1.1 Objective and Rationale

The *Forest Nursery Manual* is the result of a coordinated effort between the Department of Forest Science, Oregon State University (OSU) and the U.S.D.A. Forest Service, Division of State and Private Forestry. Current forest-nursery practices and research are reviewed and synthesized into a state-of-the-art presentation on bareroot nursery technology in the Northwest (northwestern United States and province of British Columbia, Canada).

In 1979 a task force was appointed by the Dean of the OSU School of Forestry and the State Forester of Oregon to study and report on the status of forest-nursery management in the Northwest. In addition to recommending the establishment of a Nursery Technology Cooperative at OSU, the task force also recognized the need for a detailed review of bareroot nursery technology.

At about the same time, nursery specialists of the U.S.D.A. Forest Service identified the need for a bareroot nursery manual. The Forest Service publication *How to Grow Tree Seedlings in Containers in Greenhouses* (see Appendix 1, this chapter)[1] has been well received worldwide. On that basis, a decision was made to initiate plans for a similar manual on bareroot nurseries.

Although many nursery handbooks have been published over the years (see Appendix 1), no up-to-date manual discusses current bareroot-nursery practices in depth. *Regenerating Oregon's Forests*, a recent publication, has a chapter on seedlings; the *Forest Nursery Handbook*, developed for bareroot nurseries in British Columbia, focuses largely on cultural practices. Other recent nursery publications (such as *Nursery Management: Administration and Culture*) deal primarily with ornamental nursery practices. None of these has comprehensively addressed operations, problems, and needs of forest-tree nurseries producing large numbers of bareroot seedlings for commercial use.

Bareroot nurseries are a sizable industry in the Northwest, annually producing about 278 million seedlings. New nurseries are being started and existing nurseries expanded to meet the increased seedling-production needs projected for the next decade. In addition, foresters are becoming more and more aware of the importance of high-quality seedlings to reforestation success. New nursery practices such as seedling vigor testing and the advent of new seedling stock types (e.g., plug+1 transplants; see chapter 16, this volume) are not covered in current nursery texts. For all these reasons, an up-to-date bareroot nursery manual is essential.

1.2 The OSU Nursery Survey

1.2.1 Survey participants

To meet our objectives of summarizing current bareroot-nursery practices, we conducted a survey (the OSU Nursery Survey) of 21 Northwest bareroot nurseries, 16 in the United States and 5 in Canada[2] (Fig. 1). To be selected, nurseries had to (1) produce more than 6 million seedlings per year and (2) be located in, or supply seedlings to, the Northwest. The 16 U.S. nurseries produced 229 million seedlings in 1980—approximately 95% of the bareroot seedlings grown in the northwest U.S. (Fig. 2). The five Canadian nurseries, located in British Columbia, produced 49 million seedlings—approximately 89% of the bareroot seedlings grown in the province.

The 21 nurseries ranged in elevation from 30 m (100 ft) to 1,282 m (4,206 ft). The oldest nursery, the U.S.D.A. Forest Service Wind River Nursery, was established in 1909 and the

[1] All nursery handbooks mentioned are fully cited in Appendix 1, Bareroot Nursery Handbooks, at the end of this chapter.

[2] The Canadian portion of the Survey was conducted by David Simpson, B.C. Ministry of Forests.

In Duryea, Mary L., and Thomas D. Landis (eds.). 1984. *Forest Nursery Manual: Production of Bareroot Seedlings*. Martinus Nijhoff/Dr W. Junk Publishers, The Hague/Boston/Lancaster, for Forest Research Laboratory, Oregon State University, Corvallis. 386 p.

newest, the Tyee Tree Nursery, in 1979. Douglas-fir [*Pseudotsuga menziesii* (Mirb.) Franco] is the most common species produced in the surveyed nurseries, and 2-year-old (2+0) seedlings are the most common stock type (Fig. 2).

Because not all seed collection and processing are done by nurseries, eight Northwest seed-processing plants also were surveyed (Fig. 3). These data are reported and discussed in chapter 4, this volume.

1.2.2 The questionnaires

With the help of chapter authors, we developed two in-depth questionnaires for the OSU Nursery Survey; copies are found in Appendices A and B at the end of this volume. Questionnaire #1, a series of tables, requested specific numerical information on topics such as fertilization, root culturing and equipment; because of its detailed nature, this questionnaire was mailed to nurseries prior to our visits, which were made in fall 1981. Questionnaire #2, in short-answer form, was administered during those visits and was completed in about [illegible] hours at each nursery.

Data collected from the Survey were later distributed to the authors for interpretation and use in their individual chapters.

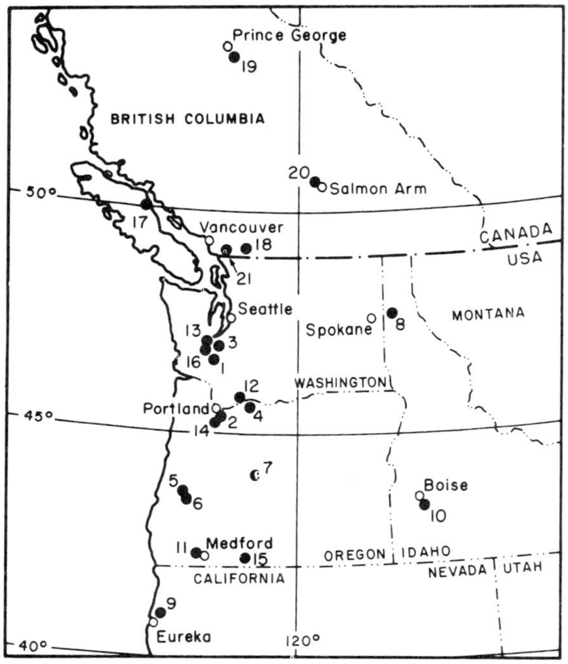

1. Industrial Forestry Association Toledo Nursery
2. Industrial Forestry Association Canby Nursery
3. Industrial Forestry Association Greeley Nursery
4. Lava Nursery, Inc.
5. State of Oregon D. L. Phipps Forest Nursery
6. Tyee Tree Nursery
7. U.S.D.A. Forest Service Bend Nursery
8. U.S.D.A. Forest Service Coeur d'Alene Nursery
9. U.S.D.A. Forest Service Humboldt Nursery
10. U.S.D.A. Forest Service Lucky Peak Nursery
11. U.S.D.A. Forest Service J. Herbert Stone Nursery
12. U.S.D.A. Forest Service Wind River Nursery
13. Washington Department of Natural Resources Lt. Mike Webster Nursery
14. Weyerhaeuser Company Aurora Nursery
15. Weyerhaeuser Company Klamath Falls Nursery
16. Weyerhaeuser Company Mima Nursery
17. B.C. Ministry of Forests Campbell River Nursery
18. B.C. Ministry of Forests Chilliwack River Nursery
19. B.C. Ministry of Forests Red Rock Nursery
20. B.C. Ministry of Forests Skimikin Nursery
21. B.C. Ministry of Forests Surrey Nursery

Figure 1. Listing and location of the 21 Northwest bareroot nurseries participating in the OSU Nursery Survey.

1.3 Preparing the *Manual*

Leading scientists with a practical knowledge of on-the-ground operations were invited to write chapters for the *Manual*. To present a balanced point of view, we chose authors from public agencies and institutions as well as private industry and also solicited chapters on selected topics in nursery management from practicing nursery managers in the Northwest.

As editors, we interacted regularly with authors during all stages in the writing process. First, we provided each author with a list of topics establishing the subject area to be covered in his or her chapter. Each author then developed a working outline and tentative title. Rough drafts were technically reviewed by us and by other scientists and nursery personnel; final drafts were technically edited at the Forest Research Laboratory, OSU, where the *Manual* was designed and produced.

Participants of the Bareroot Nursery Technology Workshop, held at OSU, previewed the *Manual* in preliminary form on October 26-28, 1982. Attendance of over 250 people at the

Total 1980 Production: 278 million seedlings

16 U.S. nurseries: 229 million

5 Canadian nurseries: 49 million

Species produced	% of total
Douglas-fir [*Pseudotsuga menziesii* (Mirb.) Franco]	6[illegible]
Ponderosa pine (*Pinus ponderosa* Dougl. ex Laws.)	1[illegible]
Lodgepole pine (*Pinus contorta* Dougl. ex Loud.)	[illegible]
Spruces (*Picea* spp.)	1[illegible]
True firs (*Abies* spp.)	[illegible]
Other	
	10[illegible]

Stock types produced	% of total
1+0	[illegible]
2+0	[illegible]
3+0	1
Bareroot transplant	7
Plug+1 transplant	2
	1[illegible]0

Figure 2. Total 1980 production of bareroot seedlings in the 21 nurseries participating in the OSU Nursery Survey, by seedling species and stock type.

B.C. Ministry of Forests, Duncan Seed Centre, Duncan, British Columbia
Brown Seed Company, Vancouver, Washington
Crown Zellerbach Nursery, Aurora, Oregon
Esses Tree Seed Company, Inc., Montesano, Washington
Pacific Forest Seeds, Medford, Oregon
Rex Timber, Inc., a subsidiary of Georgia-Pacific Corporation, Eugene, Oregon
Simpson Timber Company, Albany, Oregon
Weyerhaeuser Company Seed Plant, Rochester, Washington

Figure 3. The eight Northwest seed-processing plants responding to the seed section of the OSU Nursery Survey.

workshop reinforced our belief in the pressing need for state-of-the-art information on nursery practices. Stimulating discussions between authors and other participants resulted in important additions to many chapters; in particular, chapter 28, Designing Nursery Experiments, was proposed and subsequently added to the *Manual*.

The *Forest Nursery Manual* is organized into seven sections emphasizing the major stages of seedling production—from selecting a nursery site and starting seedlings through growing, harvesting, and outplanting. Where possible and appropriate, cross-referencing directs readers to other chapters that might provide additional useful information. The final section, Upgrading Nursery Practices, is intended to stimulate thinking toward improving nursery management. A comprehensive glossary and tables of common nursery conversion factors are also included as appendices.

Appendix 1—Bareroot Nursery Handbooks

Abrahamson, L. P., and D. H. Bickelhaupt (eds.). 1980. **Proceedings, North American forest tree nursery soils workshop.** State Univ. New York, Coll. Environ. Sci. and Forestry, Syracuse. 333 p.

Aldhous, J. R. 1972. **Nursery practice.** Her Majesty's Stationery Office, London. Forestry Comm. Bull. 43. 184 p.

Armson, K. A., and V. Sadreika. 1979. **Forest tree nursery soil management and related practices.** Ministry of Natural Resources, Ontario. 179 p.

Bluhm, W. L., and J. L. Green. 1979. **Growing nursery stock—is it for me?** Oregon State Univ. Ext. Serv., Corvallis. Ext. Manual 5. 56 p.

Cleary, B. D., R. D. Greaves, and R. K. Hermann (eds.). 1978. **Regenerating Oregon's forests.** Oregon State Univ. Ext. Serv., Corvallis. 286 p.

Davidson, H., and R. Mecklenburg. 1981. **Nursery management: administration and culture.** Prentice-Hall, Inc., Englewood Cliffs, New Jersey. 450 p.

Kummel, J. F., C. A. Rindt, and T. T. Munger. 1944. **Forest planting in the Douglas-fir region.** U.S.D.A Forest Serv., Washington, D.C. 154 p.

Schubert, G. H., and R. S. Adams. 1971. **Reforestation practices for conifers in California.** Dep. of Conservation, Div. of Forestry, State of California, Sacramento. 359 p.

Stoeckeler, J. H., and G. W. Jones. 1957. **Forest nursery practice in the Lake States.** U.S.D.A. Forest Serv., Washington, D.C. Agric. Handb. 110. 124 p.

Stoeckeler, J. H., and P. E. Slabaugh. 1965. **Conifer nursery practice in the Prairie-Plains.** U.S.D.A. Forest Serv., Washington, D.C. Agric. Handb. 279. 93 p.

Tillotson, C. R. 1917. **Nursery practice on the national forests.** U.S.D.A. Forest Serv., Washington, D.C. Bull. 479. 86 p.

Tinus, R. W., and S. E. McDonald. 1979. **How to grow tree seedlings in containers in greenhouses.** U.S.D.A. Forest Serv., Rocky Mountain Forest and Range Exp. Sta., Fort Collins, Colorado. Gen. Tech. Rep. RM-60. 256 p.

Toumey, J. W., and C. F. Korstian. 1942. **Seeding and planting in the practice of forestry.** 3rd ed. John Wiley and Sons, Inc., New York. 520 p.

van den Driessche R. 1969. **Forest nursery handbook.** B. C. Forest Service, Victoria. Res. Note 48. 44 p.

Wakeley, P. C. 1954. **Planting southern pines.** U.S.D.A. Forest Serv., Southern Forest Exp. Sta., New Orleans, Louisiana. Agric. Monograph 18. 233 p.

Williams, R. D., and S. H. Hanks. 1976. **Hardwood nurseryman's guide.** U.S.D.A Forest Serv., Washington, D.C. Agric. Handb. 473. 78 p.

Developing a Forest-Tree Nursery

Chapter 2
Nursery-Site Selection, Layout, and Development

F. E. Morby

Abstract

Because no potential nursery site is perfect, site selection inevitably requires compromise. A three to seven-member selection team should be given responsibility for establishing and ranking site-selection criteria; climate, soil, water, topography, previous land use, production potential, land availability and cost, and proximity to services are key considerations. Potential sites should be visited and evaluated and the best site chosen. A development team should then lay out the nursery, formulate an action plan, and document current development and possible future expansion in a comprehensive master plan. Careful site selection and planning plus proper management are essential to the economical production of high-quality nursery stock.

2.1 Introduction

A nursery site must be located with the realization that a perfect site does not exist and that choice of site will require compromise. However, careful attention to the selection of a permanent nursery site will amply repay all the effort expended. An unsatisfactory site will sooner or later (generally, sooner) increase the cost of operations and could lead to unnecessarily high seedling losses and poor stock production [1]. Such a situation will leave customers dissatisfied and may cause the nursery to fail.

Fifteen of the 21 Northwest nurseries questioned in the OSU Nursery Survey (see chapter 1, this volume) ranked site-selection characteristics (Table 1). The six most important considerations were: (1) soil workability and drainage, (2) soil texture, (3) water supply, (4) land cost, (5) climate, and (6) soil depth. On the basis of these and related concerns, this chapter provides guidelines for selecting the optimum nursery site. (See chapter 29, this volume, for more information on solving site problems.)

Table 1. Nursery-site characteristics ranked by 15 Northwest bareroot nurseries (OSU Nursery Survey).

	Ranking[1]					
Characteristics	1	2	3	4	5	Total
Climate	2	1	3	....	1	7
Elevation	....	....	....	....	2	2
Aesthetics	....	....	....	....	....	....
Proximity to markets	2	....	1	....	2	5
Water supply	....	4	1	3	2	10
Soil depth	1	1	1	2	1	6
Soil workability and drainage	2	4	3	3	....	12
Land cost	2	1	....	4	....	7
Proximity to work force	....	1	....	....	3	4
Soil fertility (including pH and cation exchange capacity)	2	....	1	1	1	5
Local topography	....	1	2	....	1	4
Politics	1	....	....	....	1	2
Previous land use	....	....	....	....	....	....
Freedom from weeds	....	....	....	....	....	....
Soil texture	3	1	3	2	1	10
Other—adequate acreage	....	1	....	....	....	1

[1] 1 (most important) to 5 (least important).

2.2 Selection

2.2.1 The team approach

First, a list of possible sites should be screened by the person or group wishing to establish a nursery in a given market or use area. Because selecting and establishing a permanent nursery requires a large capital investment [9], a team approach for final selection is probably best.

The team should be composed of at least three of the following:

- Experienced nursery manager
- Reforestation specialist, silviculturist, or other potential customer

In Duryea, Mary L., and Thomas D. Landis (eds.). 1984. *Forest Nursery Manual: Production of Bareroot Seedlings.* Martinus Nijhoff/Dr W. Junk Publishers, The Hague/Boston/Lancaster, for Forest Research Laboratory, Oregon State University, Corvallis. 386 p.

- Soils specialist
- Forest pathologist
- Civil engineer
- Soil Conservation Service representative
- Entomologist

Team members must be capable of blending their varied backgrounds and individual areas of expertise together, and all should have a physical sense, or "feel," for the land. Because team input is so diversified, it is far less likely that the site selected will be difficult or impossible to manage.

The selection team should first review criteria for all potential sites using a site checklist (Fig. 1); this form may be modified to ensure that all selection criteria are listed and properly emphasized. Then each site should be visited (Fig. 2) and its merits and drawbacks discussed. Finally, the selection team should meet after all site visits have been completed to make the final selection. The entire selection process should be carefully documented in a written report.

2.2.2 Site-selection criteria

2.2.2.1 Climate

Growing-season requirements will vary with stock type. A long growing season (150 days or more) provides an adequate period to produce 1+0, 2+0, 3+0 [for slow-growing species such as Pacific silver fir (*Abies amabilis* Dougl. ex Forbes) and some sources of western white pine (*Pinus monticola* Dougl. ex D. Don)], and transplant stock. A growing season of less than 150 days would reduce the chances of consistently growing shippable 1+0 seedlings but would be adequate for other age classes.

Temperature.—Possible nursery sites whose daily temperatures consistently exceed 105°F for extended periods (3 weeks or more) should be avoided; extremely hot periods reduce growth and may cause burning of foliage. Short periods of daytime temperatures of 110°F or more can tax irrigation systems, but properly designed irrigation systems can protect seedlings from burning during those periods (see chapters 11 and 12, this volume). Growth of most species is greatly impeded by ambient temperatures of 90°F and above.

Field-planting periods must be discussed with customers. Seedlings to be outplanted from December through early March cannot be lifted and processed when cold daytime temperatures keep soil frozen. Those seedlings to be outplanted from late March through June can endure frozen nursery soil from December to early March and be lifted while still dormant. However, from 120 to 150 cold nights (49°F or below) seem to be necessary before peak root-regeneration potential is reached [4]. Moderately low combined day and night temperatures during fall and early winter are necessary for bud cooling, preparing seedlings for optimum budbreak about 2 weeks after outplanting. About 300 bud-cooling hours are required in the temperature range 28 to 40°F ([7]; see also chapter 23, this volume).

Extremely low temperatures can be detrimental to seedlings not protected by snow or mulch. Extreme cold can drive frost deep into the soil, delaying lifting and seedling processing well into spring. If low temperatures recur annually, the species and basic seed sources that the nursery can produce will be limited because stock cannot be lifted and processed. If this fact is ignored, seedlings may not be available when needed for outplanting.

Precipitation.—Proposed nursery sites that have a record of frequent heavy snows persisting into the normal seedling-processing season should be avoided. Snow melting in late spring can radically reduce the time frame for processing seedlings, which can place undue stresses on workers and managers, facilities, equipment, and seedlings, and create dissatisfied customers because specified outplanting dates cannot be met.

POTENTIAL NURSERY SITE

NAME OF AREA ____________ DATE ________

LOCATION ____________

SOIL SURVEY TYPE ____________

TOP SOIL
1. TEXTURE ASSESSMENT ________
2. DEPTH ________
3. pH ________

SUBSOIL
1. TEXTURE ASSESSMENT ________
2. DEPTH ________
3. pH ________

DEPTH OF WATER TABLE ____________

DRAINAGE ____________

WATER SUPPLY:
1. ADJACENT CREEK ________
2. WELL ________
3. OTHER ________
4. RISE OR FALL (FEET) TO SUPPLY ________
5. pH ________

TOPOGRAPHY:
1. (a) LEVEL ________
(b) ROLLING ________
(c) IRREGULAR ________
2. SLOPE:
________TO________DEGREES
3. UNSUITABLE (i.e., more than 6° or very irregular) ________

COVER:
1. MAIN TREE SPECIES ________
2. COMMON MEMBERS OF GROUND FLORA ________

CLIMATIC DATA (USE NEAREST STATION IN DEPARTMENT OF TRANSPORT TABLES):
1. TOTAL ANNUAL PRECIPITATION ________
2. SPRING PRECIPITATION:
MARCH ____ APRIL ____ MAY ____
3. NUMBER OF FROST-FREE MONTHS ____
4. HIGHEST TEMPERATURE (AND MONTH RECORDED) LAST YEAR ________

APPROXIMATE ACREAGE ________

ACCESS ____________ OWNERSHIP ________

Figure 1. A nursery-site checklist such as this one is recommended for use by the site-selection team (adapted from [8]).

Figure 2. Selection team evaluating a possible nursery site.

High rainfall areas are best avoided. However, the season in which the precipitation occurs is important. Heavy spring rains can delay spring operations such as adding soil amendments, starting a cover or green manure crop, or sowing tree seed. Summer rains tend to be a problem only when they occur as cloudbursts and result in flooding, erosion, or seedling wash-out. Frequent summer rains may be detrimental, however, because rains may disrupt stock hardening processes already induced by withholding irrigation. Areas with heavy winter rains should be avoided; heavy rain saturates nursery soil to the point of hindering lifting, damaging soil structure, and causing flooding and erosion.

Wind.—Areas with frequent, long-lasting, high-velocity winds—particularly where humidity is low and winds are drying and from the east—should be avoided. Winds will affect irrigation application and uniformity and may result in soil movement. High winds can desiccate seedlings, and soil carried by winds can blast stems and foliage. Wind can restrict spraying of pesticides, cause tree-seed cover to be blown away, and displace or scatter seedbed mulches.

2.2.2.2 Soil

Perhaps the most important factor in establishing a nursery is the correct choice of soil (see chapter 6, this volume). Other site features, including fertility, moisture, and microclimate, can be manipulated by the nursery manager [2], but moving or significantly modifying large masses of soil is, at the very least, impractical and costly. An intensive soil survey, coupled with representative soil sampling, will help the selection team choose the site with the most suitable soil.

Texture.—Sandy loams or loamy sands with good drainage are excellent for nurseries. Light soils can be worked in weather conditions too wet for heavier soils—an important consideration in the Northwest. The content of clay and silt (particles <0.05 mm in diameter) in the soil should be within 15 to 25%.

Depth.—The top 4 feet of soil should be free of claypan, hardpan, shale, iron concretions, calcareous substrata, or mottled gley layers [9]. Without artificial drainage, this depth seems a reasonable minimum; where artificial drainage has been installed, however, a minimum clear soil depth of 2 feet is probably acceptable [8]. The top 18 inches of soil should be free of stones, which are expensive to remove, make the soil difficult to work, and interfere with nursery cultural practices.

Soil pH.—The optimum soil reaction, or pH, for most tree species is between pH 5.0 and 6.0 (see chapter 7, this volume). Soils of lower pH may have fewer available nutrients, whereas soils of higher pH encourage the invasion of fungus diseases [9]. Soil pH can be altered with soil additives such as sulfur or by injecting phosphoric or sulfuric acid into irrigation water.

2.2.2.3 Water

Securing an adequate supply of domestic and irrigation water can be a major problem. Water rights must be obtained for any water source. Therefore, special consideration must be given to a site where the quantity and quality of water are adequate for current and possible future requirements (see chapter 11, this volume).

All water needs and the timing of those needs must be considered. For example, in most nurseries, irrigation is necessary during the growing season and for frost protection. Restrictions on flow and on periods of delivery must be closely scrutinized. Is the water source reliable during drought years? Can breaks in canals, pipelines, and other delivery systems be expected? What are the time frames for repair? Are backup sources available in emergency situations? Is domestic water available through a city, village, or other municipality? Are there restrictions on quantity? Are costs high? Is it feasible to develop an on-site water source? Is water quality high? Are there any potential delivery problems? If no water is available near the site, can a transmission line be constructed?

Irrigation-water sources.—Lakes are a good source of irrigation water. Storage capacity, draw-down, other uses, and contaminants must be examined before any commitment is made. Screening may be necessary to remove water-borne debris.

Streams are sometimes used for nursery irrigation and must be checked for water rights, other uses, and quality. In addition, attention must be paid to intakes, diversions for pumping stations, protection during runoff periods, and maintenance of the stream channel to ensure maximum carrying capacity. Stream water may need to be screened to alleviate contamination by vegetation, weed seeds, frogs, fish, algae, and other water-borne debris.

Irrigation water delivered through open ditches is usually controlled by irrigation districts and is subject to specific short delivery periods. Such a source is not reliable unless storage is made available on site and therefore is not recommended.

Water drawn from wells is probably one of the best irrigation sources for most locations. Draw-down and pumping capacity must be checked to ensure that water is available in reliable quantities when it is required.

Domestic or irrigation pipelines are reliable. In many instances, clean water will be supplied with adequate pressure and volume to eliminate the need for pumping. The two types of pipelines are similar, both generally well designed and constructed, although domestic water lines usually have more connections creating a high demand for water and more concern for failure of the system. Systems must be reviewed to ensure that maintenance is adequate and repairs are timely.

Water quality.—Chemical contaminants may be introduced into an irrigation source through the soil or from precipitation or surface runoff. Contamination by minerals such as calcium or boron, for example, will usually be found in well water. However, because streams, lakes, and ditches also may have mineral contaminants, any potential site must have its water sources evaluated for mineral content and concentration.

Water originating from any open source (lake, stream, or ditch) is subject to contamination by weed seeds. High concentrations of these can lead to unwanted vegetation in seedbeds and cover crops—a major problem. Special, well-designed screening devices can alleviate this problem.

Water-borne diseases can infect root systems and foliage. If pathogens such as *Phytophthora*, a fungus causing root disease, are present, chemical water treatment may be necessary (see chapter 19, this volume).

2.2.2.4 Topography

The area for nursery beds should be level, or nearly so. A slight slope (2% maximum) is beneficial for better surface drainage, but slopes greater than 2% can cause erosion, necessitating expensive control measures, and may cause undesirable translocation of soluble fertilizer salts [9]. Furthermore, all mechanical equipment used in forest nurseries operates best on level ground. Moderate slopes and small rough areas may be leveled by terracing and grading, but these operations can be expensive and usually expose infertile subsoil that can cause future seedling-growth problems. Fertile topsoil should be removed and stockpiled before any major soil-moving operation is attempted and evenly redistributed afterward over the leveled area (see chapters 5 and 13, this volume).

The importance of aspect will depend on nursery-site latitude and altitude. In most localities in the temperate zone, eastern and southeastern aspects should be avoided because of greater frost danger, and southern and southwestern aspects because of excessive dryness during periods of drought. Where irrigation is available, southern aspects in northern latitudes at high elevations are best because of their greater warmth. For most sites, though, a northwestern aspect is best because vegetative growth starts later in spring and is not subjected to injury by frost. Water loss through evaporation from the soil surface is not so rapid on northwestern aspects.

Topographic undulations can cause water to accumulate. Standing water, no matter how little, causes complete destruction of nursery stock because of oxygen depletion or buildup of toxic gases. Irregular topography complicates installation of irrigation systems, causes irrigation-line leakage, and makes it difficult to operate nursery equipment. Damage from early fall or late spring frosts can be catastrophic to growing seedlings. Frost hollows, which occur wherever cold air can accumulate—in valley bottoms and large topographic depressions, especially where trees bar cold-air drainage—must be avoided.

As mentioned earlier in this chapter, heavy snows and frozen soils should be avoided. Even though these conditions can occur throughout the elevational zones, they occur more frequently at the middle to higher elevations. Elevation requires special attention; it is mandatory to choose an elevation that will ensure stock dormancy as well as lifting dates that meet customer requests.

2.2.2.5 Previous land use

Past use of the land may influence its value as a potential nursery site (see chapter 5, this volume). For example, past practices that have altered soil acidity or caused toxic chemicals to accumulate will be detrimental to growing seedlings. Has the site been altered? If so, when, and what was done? If the land has been leveled, were any problems associated with the leveling? If so, has time ameliorated them? An intensive survey of topsoil depth will reveal previous land leveling. Will additional leveling be required and, if so, are any problems anticipated?

Are there any areas where water accumulates from surface or subsurface flows? Has the land been drained through a subsurface system? If so, when was the drainage system installed? Did it solve the drainage problems? Is it still functional? Are there any conflicts with neighbors caused by runoff onto adjacent land? How has runoff been handled in the past?

What irrigation system was employed? Was it functional? Have there been any problems with the water source? Are water rights secure and transferable? Will the available irrigation be adequate for seedling growth? Are water quality and quantity acceptable?

What was the previous cropping schedule? Were disease problems associated with any particular crop? If so, what steps were taken to alleviate the problem, or does it still exist? Current vegetation in the area should be carefully examined for root diseases and foliar disorders by the forest pathologist on the site-selection team and recommendations made on suitability.

What pesticides were used? Were any spilled? If spills occurred, identify where, what the chemical was, and what was done to prevent soil contamination. Is there a written pesticide-use record?

Are there any known insect problems? If so, are they soil borne or foliage associated? Do adjacent lands or associated crops have insect problems? If so, what control techniques are employed? Are there any insect associations that may be a concern for the tree-seedling crop? The entomologist on the site-selection team should make a thorough evaluation.

Ideally, the new site should be relatively free from annual and perennial weeds and weed seeds. Any previous crop species that is difficult to eradicate can become a weed problem. Costs of weed control can be very high; therefore, obtaining a weed-free site and managing to keep it weed free will be cost effective. Vegetation on the site should be identified by the selection team and control measures evaluted for all species (see chapter 18, this volume).

2.2.2.6 Site production potential

To help determine the acreage needed for the seedling growing area, the selection team must estimate potential requests for seedlings. A rule of thumb is 500,000 seedlings/0.4 ha (1 acre), but this figure may vary with species or seedbed density. For this calculation, subtract all nonproductive areas—roads, streams, reservoirs, administrative site, and anywhere else that seedlings will not be grown—from the total nursery-site area.

Site growing potential can be derived with the following formula:

$$P = \frac{A \times [1 - (C+F)] \times U \times D \times (\text{m}^2/\text{ha or ft}^2/\text{acre})}{R}$$

where

- P = Annual production capacity, in 1000s of seedlings
- A = Production area (acres or hectares)
- C = Estimated cull factor
- F = Estimated overrun factor
- U = Actual seedbed area, %
- D = Density objective (number of seedlings to be grown per square foot); density desired at seedling harvest age
- R = Crop rotation

Many nursery sites have been selected and developed with little or no allowance made for future expansion. Regardless of how remote it may seem, expansion should be considered. To do so, the site-selection team must examine areas adjacent or close to the property.

2.2.2.7 Proximity to customers, labor, and services

Proximity of the nursery to seedling customers, work force, transportation, utilities, and facilities for people all should be evaluated by the site-selection team. Locating the site geographically close to seedling customers seems to be most judicious, although, with the advent of transportation systems and refrigerated trucks, this is not as paramount as it once was. Often, other criteria prevail.

Labor force.—The nursery should be within easy commuting distance—about 35 miles—of an adequate, dependable labor supply. The number of workers needed varies widely, depending on size of the nursery, extent of mechanization, amount of work contracted out, degree of chemical weed control, and type of stock grown. The peak period of employment occurs during the short 2- to 9-week processing season, when about 1 person is required for every 65,000 seedlings processed. About 10% of this work is supervisory and administrative. Over an entire year, about 1.6 fulltime equivalents (FTEs; 1 FTE is 2,080 person-hours) are required per million seedlings produced. Both male and female workers are needed; typically, 50 to 60% of the work force is women. Certain specialized positions such as tractor operator and irrigation specialist require some previous agricultural experience. Wages are usually higher than in other agricultural work, making it easier to recruit a reliable work force.

Transportation.—A good transportation network is essential. Rail, truck, bus, or plane can be used to transport seedlings, but refrigerated transportation equipment is mandatory. County or state roads that are well traveled, maintained, and connected to freeways will aid the transport of both seedlings and people; easy connections for seedling customers and nursery administrators expedite travel and reduce transportation and per diem costs. Motels, hotels, restaurants, and other facilities convenient for people in transit are a must, as are limousine, taxi, bus, and air transport to neighboring cities and states.

Refrigerated seedling storage.—Access to commercial tree-seedling storage is mandatory to ensure that stock can be stored without loss of vigor for up to 3 months. Potential storage may be found in the fruit or produce industry.

Utilities and fossil fuels.—Telephone, electric power, and other utilities required for nursery operation must be already available or easily secured. The history of these utilities as well as their current cost, supply, and reliability must be evaluated.

What is the commercial rate for electric power? Are the needed power types (single and 3 phase) and voltages (110, 208, 220, and 440 V) available? If an increased need develops, will power be available? Is there a potential for power failures? Are failures frequent and long term? Will backup on-site power be required? Is propane or natural gas available? What are the costs? Are fuel distributors available close-by? Will they provide service on short notice? What are the on-site storage requirements?

Other services.—Is a sewer and garbage disposal system available? What are the costs? What are the restrictions on materials that can be dumped in the system? Is there a discharge point for release of waste water from seedling processing and surface runoff? Are there restrictions on point discharge, and are permits obtainable?

What commercial repair, maintenance, and labor services are available? Are they able to respond with little or no advance warning? Are service contractors close—within 35 miles—or must they travel in excess of 50 miles?

Are electrical contractors equipped to handle main power, power panels, switches, automatic controls, warning systems for fire and burglar, refrigeration, and flow control for pumping stations? Are plumbing contractors and those that service refrigeration and heating systems available, reliable, and capable? Do all these services carry adequate parts inventories? If not, where must parts come from, and what are the turnaround times? For example, can parts be obtained and repairs made to refrigeration systems in from 48 to 72 hours?

Are contracting organizations available for seedling lifting and weeding? How flexible are these contractors? Will those available for tree planting, timber-stand management, and agricultural work also weed seedbeds and lift seedlings? Are janitorial contractors available? At what cost? Would in-house or contracted labor be more cost effective?

2.2.2.8 Land availability and cost

Are the sites under consideration actually for sale and within the price range given to the selection team? What are the owners' sale stipulations? Look at **total** developed cost. Unimproved land may initially cost less but require such large capital outlays for development that ultimate total cost may be more. Land that may initially cost more, on the other hand, may be developed to the point that few subsequent improvements are needed, and total cost may be less.

2.2.3 Evaluating criteria and selecting a site

All selection criteria are discussed by the team and the major ones listed and evaluated [5]:

Objective score = score x weighted value

where **score** reflects how well the site satisfies individual criteria (1, lowest, to 10, highest), and **weighted value** reflects the relative significance of each criterion (1, lowest, to 10, highest). The rationale behind the weighting of criteria and site assessment should be discussed for each site. Once sites have been rated and discussion is complete, the choice can be identified. For example, in Table 2, criteria have been ranked and weighted for three potential nursery sites. Site #3, with a composite score of 359, would be the preferred site.

Table 2. Evaluating criteria for three potential nursery sites with the Kepner-Tregoe [5] method.

Ranked criteria	Corresponding weighted value[1]	Weighted value (and corresponding objective score) for three sites		
		#1	#2	#3
1. Soils	10	8 (80)	6 (60)	10 (100)
2. Water	9	10 (90)	10 (90)	5 (45)
3. Climate	9	6 (54)	8 (72)	8 (72)
4. Topography	7	5 (35)	5(35)	10 (70)
5. Land availability and cost	8	10 (80)	8 (64)	9 (72)
Total objective score		(339)	(321)	(359)

[1] 1 (lowest) to 10 (highest).

2.3 Layout and Development

2.3.1 The team approach

Like site selection, layout and development benefit from the team approach. The development team should consist of the nursery manager; civil, electrical, and mechanical engineers; landscape and structural architects; and consultants for soils, irrigation, subsurface drainage, or other areas where on-site team expertise is weak or lacking.

Every effort must be made to visit similar facilities for comparison. Development of a new nursery requires a large initial monetary investment, and any new technology either already developed or under consideration must be evaluated for potential incorporation into development plans. New ideas always surface when other nurseries are visited and when both positive and negative sides of a particular site or procedure are discussed.

2.3.2 Access and traffic flow

The nursery should be as compact as possible—nearly square or regular in shape—to minimize the length of the boundary

Table 3. Comparative analysis on effects of access for four potential nursery-entry points.[1]

Entry	Cost, $	Security	Traffic	Site lines	School	Residential areas	Wildlife	Seedling growing area	Composite score
1	104,350	9	10	8	9	10	4	8	58
2	112,800	2	9	9	8	6	6	7	47
3	44,850	4	4	8	3	0	8	9	36
4	57,200	3	5	7	5	6	9	10	45

[1]0 (high impact) to 10 (low impact).

fence and reduce the time lost moving from one part of the nursery to another [1].

Roads provide access to the site (see 2.2.2.7) and to growing fields. When the site is developed, all access roads should be paved; they must be capable of taking heavy "semi" truck and tractor traffic in all kinds of weather. Parking areas must be evaluated and particular attention given to pedestrian and vehicle traffic flows. Possible conflicts with people, vehicles, buildings, and landscaping must be taken into consideration. The potential maximum number of future employees must be anticipated and allowances made for future parking if the need is identified.

When considering connecting points (entries and exits) to existing road systems, the development team should solicit input from the local community. A decision matrix such as that shown in Table 3 is extremely helpful. In that case, four entry points were rated from 0 (high impact) to 10 (low impact) in eight categories and their composite scores determined. Because of anticipated conflicts, entry #1 was chosen—though it was far from the least costly.

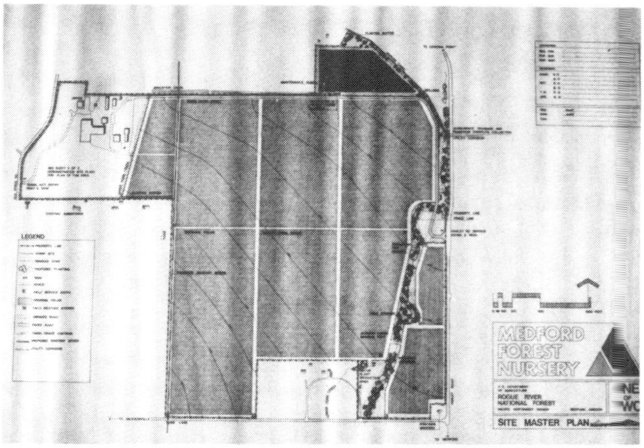

Figure 3. Master plan for a Northwest bareroot nursery.

2.3.3 Administrative site

The administrative site includes administrative offices; storage areas for equipment, trees, seed, pesticides, other chemicals, and fuels; shops; a fuel-dispensing station; an employee center; and seedling-processing facilities. The type, number, and location of required buildings can be determined with the team approach. Other administrative development could include employee-enrichment areas (in the form of parklike surroundings), holding areas for irrigation water or soil amendments, a culled-seedling disposal area, and an area for holding scrap material and used equipment until sale is possible (potential aesthetic conflicts with neighbors may arise in this last case).

Although possible future expansion must always be kept in mind, the administrative complex must optimize the use of space to avoid being spread out. The results of poor or inadequate planning will cause the manager and staff considerable anxiety in future years.

2.3.4 The master plan

Once agreement has been reached on placement of all structures and development begins, a master plan—a dynamic tool—must be made to document the team decision (Fig. 3). Once the development team has disbanded, this plan will stand as an illustrated document of site layout, indicating growing areas, roads, buildings, outdoor storage areas, reservoirs, streams, fences, neighbors, possible expansion areas for buildings, and other site development. The master plan is not cast in concrete, however, and can and must be updated as management needs change.

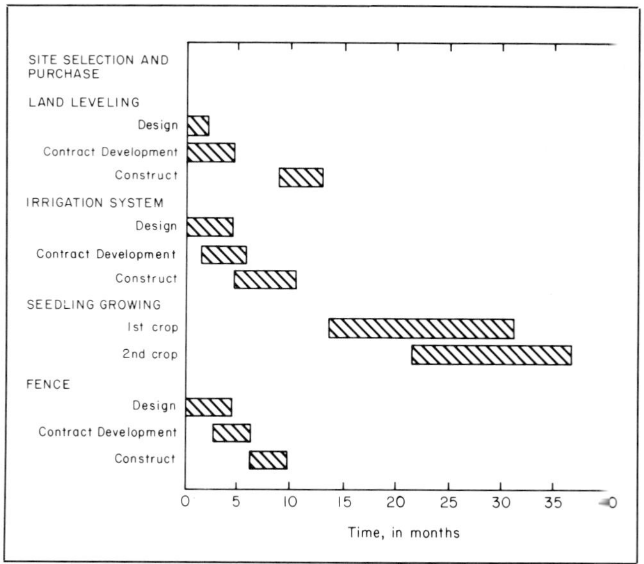

Figure 4. Partial nursery action plan, developed as a timeline.

2.3.5 Development program

To properly develop a site, an action plan must be prepared. One approach is to construct a critical-path chart that shows events and operations on a timeline (Fig. 4). Tree-production scheduling must be coordinated with site development. Structures that are needed first must be built first. For example, because seedlings need to be processed and stored in refrigeration rooms at the end of the second year, that complex must be ready for the first crop.

Throughout nursery development, the action plan is continuously reviewed—by an individual, a team, or a concerned outsider—and revised, as needed. Critical factors that may have been overlooked initially are identified and incorporated. It is important for everything to be viewed objectively and in proper perspective.

2.3.6 Budgeting and accountability

Budgeting is critical and must have highest priority in the development process. Budgets should be planned 2 to 3 years in advance to ensure that funding, people, and facilities will be available when needed. The budget and the action plan must be developed together. If shortages of funds or people are anticipated, construction may have to be delayed or other alternatives sought.

The process of "fixing accountability" identifies objectives and action steps [6] and the individuals responsible for their accomplishment in the outlined time frames. Responsibilities must be reasonable, however, and should be adjusted if necessary to ensure that the work can realistically be completed.

2.4 Conclusions and Recommendations

- A three- to seven-member team should be designated to select a nursery site. The team should develop site-selection criteria and establish priorities, then visit and evaluate possible sites on the basis of the chosen criteria, and finally select the best site.
- A development team should lay out the nursery, formulate an action plan, and then document nursery-site development in a flexible but clearly defined master plan.
- Possible future expansion of facilities and staff must always be considered.
- A perfect nursery site does not exist; tradeoffs are inevitable—but nursery soil should not be compromised.
- Wise planning and thoughtful decision making—plus proper management—are essential for the economical production of high-quality nursery stock for reforestation [3].

References

1. Aldhous, J. R. 1972. Nursery practice. Her Majesty's Stationery Office, London. Forestry Commun. Bull. 43. 184 p.
2. Chavasse, C. G. R. 1980. The means to excellence through plantation establishment: the New Zealand experience. Pages 119-139 *in* Proc., Forest plantations, the shape of the future. Weyerhaeuser Science symp., April 30-May 3, 1979. Weyerhaeuser Co., Tacoma, Washington.
3. Goor, A. Y., and C. W. Barney. 1976. Forest tree planting in arid zones. Ronald Press, New York. 409 p.
4. Krugman, S. L., and E. C. Stone. 1966. The effect of cold nights on the root-generating potential of ponderosa pine seedlings. Forest Sci. 12:451-459.
5. Kepner-Tregoe, Inc. 1973. Problem analysis and decision making. Princeton Research Press, Princeton, New Jersey. 99 p.
6. Morrisey, G. L. 1976. Management by objectives and results in the public sector. Addison-Wesley Publishing Co., Reading, Massachusetts. 278 p.
7. Ritchie, G. A. 1980. Seedling physiology. Silviculture Institute, Oregon State Univ., Corvallis. Unpublished rep.
8. van den Driessche, R. 1969. Forest nursery handbook. B. C. Forest Serv., Victoria. Res. Note 48. 44 p.
9. Wilde, S. A. 1958. Forest soils—their properties and relation to silviculture. Ronald Press, New York. 537 p.

For Further Reading

Cleary, B. D., R. D. Greaves, and R. K. Hermann (eds.). 1978. Regenerating Oregon's forests. Oregon State Univ. Ext. Serv., Corvallis. 286 p.

Schubert, G. H., and R. S. Adams. 1971. Reforestation practices for conifers in California. Dep. of Conservation, Div. of Forestry, State of California, Sacramento. 359 p.

Toumey, J. W., and C. F. Korstian. 1942. Seeding and planting in the practice of forestry. John Wiley and Sons, Inc., New York. 520 p.

Chapter 3
Equipment for Forest Nurseries

R. G. Hallman

Abstract

Modern machinery and equipment can increase efficiency and productivity in all phases of nursery operations without sacrificing quality or safety. Seed processing is facilitated with kilns, tumblers, separators, dewingers, scalpers, and grinders. Seedbeds are prepared with plows, harrows, rock rakes, packers, bed formers, and levelers; seed is then sown with drills or broadcast seeders. Fertilizers and soil amendments are applied with spreaders, seeders, or soil injectors; weeds are controlled by row or path cultivators. Sprinkler systems not only irrigate seedlings but also spray fertilizers, pesticides, or herbicides and provide frost protection. Saws, pruners, and mowers are used to trim seedling tops and roots. Transplanting and lifting machines allow seedlings to be moved from bed to bed or to processing areas, where they are packaged with balers or bundlers. Both seed and seedlings may be stored in trays, bins, crates, boxes, bags, drums, or tubs on pallets or racks in walk-in refrigerators, freezers, or sheds, or moved from place to place by forklifts, conveyors, carts, trucks, or tractors with attachments. Many machines are commercially available in a variety of types and sizes, but some must be custom built; some are highly specialized, but others, with adaptations, can serve common nursery functions.

3.1 Introduction

Nursery managers realize that maintaining trees as a renewable resource requires highly productive nurseries. Modern machinery and equipment can offer an efficient means of increasing nursery productivity without sacrificing seedling quality or employee safety. But staying abreast of developments in nursery equipment can be difficult.

This chapter informs nursery managers about current developments in nursery equipment and offers ideas about how custom-built and commercial nursery machines and equipment can serve common functions. For details on equipment types and specifications, the Missoula Equipment Development Center (MEDC), Missoula, Montana, has compiled a comprehensive catalog of nursery equipment, including descriptions of equipment typically used for common nursery functions and a list of supply sources; construction drawings of selected machinery are also available.[1]

3.2 Cone Storage and Handling

After harvesting, cones must be properly stored to dry (see chapter 4, this volume). If cones are first partially dried, kiln-drying time is reduced. Ventilating cones in storage also maintains seed fertility and keeps cones from molding.

Cone-storage methods vary with extractory types, volume of cones, and available facilities. Nurseries store cones in sheds, on floors or adjustable racks, on trays, or in ventilated containers like bins. Most nurseries store loose cones in burlap bags or on wooden trays, though aluminum, fiberglass, and plastic trays are also used. Wire-mesh or plastic bins or wooden crates store cones compactly with ventilation. Although storage bins and crates are stacked, air circulates freely around the cones.

Stored cones can be moved by hand or with various kinds of forklifts (Fig. 1) and conveyors, all of which are commercially available.

3.3 Seed Processing, Storage, and Handling

Nurseries use a variety of machines, including kilns, tumblers, scalpers, dewingers (Fig. 2), air-screen cleaners (Fig. 3), and air and gravity separators, to complete cone drying and seed processing (see chapter 4, this volume).

Kilns dry and open cones. Most nurseries use kilns with trays that hold cones dried by circulating hot air; however, some use rotating kilns that hold cones in a drum heated with circulating air. Although nursery kilns are often custom built, some, such as the International Seed Company kiln and the McPherson kiln, are commercially available.

Nurseries use tumblers, air or gravity separators, dewingers, and scalpers to extract seeds from cones, dewing seeds, and

[1] For more information about the catalog and drawings, contact the U.S.D.A. Forest Service, Equipment Development Center, Building #1, Fort Missoula, Missoula, MT 59801; phone 406-329-3157.

In Duryea, Mary L., and Thomas D. Landis (eds.). 1984. *Forest Nursery Manual: Production of Bareroot Seedlings*. Martinus Nijhoff/Dr W. Junk Publishers, The Hague/Boston/Lancaster, for Forest Research Laboratory, Oregon State University, Corvallis. 386 p.

remove cone bracts. Cone grinders, extractor tumblers, and fanning mill clippers are also commonly used, as are preheated bins and powered conveyors. Once extracted and dewinged, seeds may be sized before storage.

Extracted seeds are tested for moisture content and fertility. Some extractories dry seed in ovens to reduce moisture content before storage. Seed quality and fertility must be maintained because seeds are often stored for years. Therefore, most nurseries store seeds in walk-in freezers (Fig. 4), and many have refrigerator-freezers custom built.

Nurseries use various sizes and types of commercially available containers—cloth, paper, or burlap bags, cardboard or wooden boxes, and fiber or metal drums (Fig. 5)—to store seeds. Lining containers with plastic bags helps maintain proper seed moisture.

3.4 Soil Fumigation

Most U.S. nurseries apply methyl bromide gas, frequently using soil injectors with pressurized tanks, to control soil pathogens (see chapters 18 and 19, this volume). After fumigation, the soil must be covered with tarps or plastic sheets (Fig. 6) to help it retain the gas. However, many nurseries contract fumigation service because it can be dangerous and requires special equipment.

3.5 Ground Preparation

Nurseries prepare seedbeds with common farming equipment, including plows, tillers, harrows, rock pickers, rakes, and packers, all available commercially in a wide variety of sizes and types, and with specialized equipment, including bed formers and levelers (see chapter 5, this volume). The equipment chosen depends on nursery size, soil type, needed tillage depth, availability, and preferences of nursery personnel.

Plows are common primary tillers (Fig. 7). Straight blade plows have curved blades with a flat bottom, disk plows have circular blades, and chisel plows have straight, vertical shanks.

Harrows are common secondary tillers, breaking clods and smoothing plowed soil. Harrows are also available in different sizes with blade styles that range from vertical spiked teeth to disks.

Rock pickers and rakes remove rocks from tilled soil. Rock rakes have adjustable inclining teeth that skim the soil surface to catch rocks and lift them into holders.

Figure 1. Common forklift, useful for moving batches of stored cones as well as numerous other nursery containers.

Before sowing seedbeds, some nurseries form and roll them smooth with mechanical bed formers (Fig. 8) and rollers, most of which are custom built. Such seedbed formers combine light disk-shaped plowshares and rakes with roller packers and leveling bars. Separate soil levelers and packers that use adjustable blades and rakes on rubber-tired frames are also available.

3.6 Sowing

Sowing is a critical nursery operation in which seed density, planting depth, and timing must be carefully controlled (see

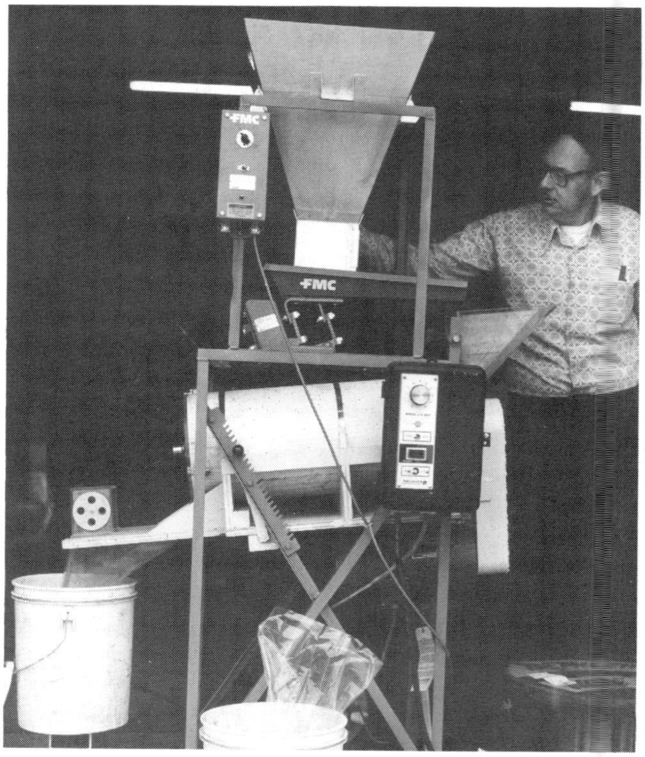

Figure 2. MEDC dewinger.

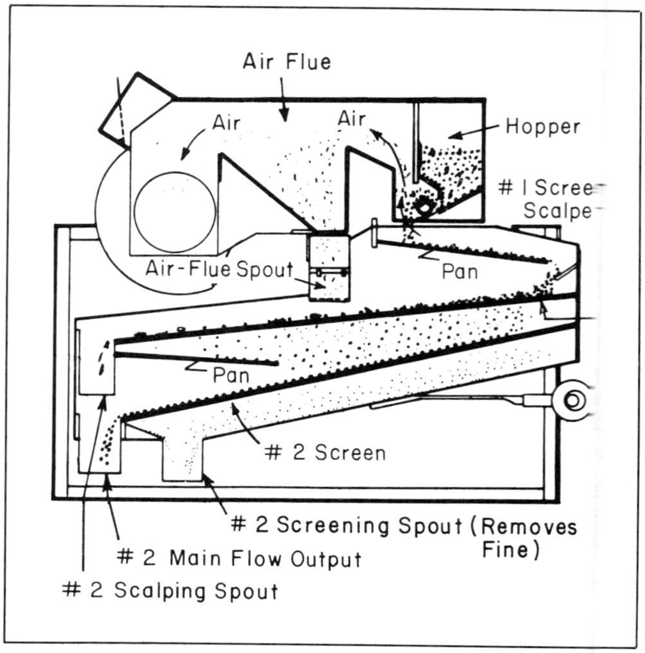

Figure 3. Schematic of air-screen cleaner [adapted from MEDC catalog; see text footnote 1].

chapters 5 and 15, this volume). Most nurseries use agricultural seed drills for sowing, though some prefer broadcast seeders. Seed drills commonly used include the Whitefish Nursery seeder, the Wind River drill seeder (Fig. 9a), and the Love-Øyjord seeder (Fig. 9b). However, precise seed placement remains a problem for many nurseries.

3.7 Irrigation

Most nurseries supply and control the water in seedbeds with commonly available agricultural irrigation systems that use impulse sprinklers attached to movable sections of pipe (Fig. 10) (see chapter 11, this volume). The pressure of water pumping through the pipes rotates sprinkler heads; water trajectory and patterns of rotation at each sprinkler head are adjustable. Some nurseries use injector pumps to apply fertilizers, pesticides, and herbicides through irrigation systems.

Figure 4. Nurseries commonly use walk-in freezers to store seeds.

Figure 5. Fiber drums are widely used for seed storage.

Figure 6. Plastic layer helps trap methyl bromide gas in newly fumigated soil.

Figure 7. Tractor-drawn plow used to prepare ground for seedbeds.

Figure 8. Rototiller bed former shapes and smooths plowed ground into raised beds.

3.8 Fertilization and Soil Amendments

Applied at nurseries to replenish soil nutrients, fertilizers may be either organic or chemical. Though both types are commercially available, specific crop and soil requirements will determine which type a nursery needs (see chapters 7 and 8, this volume).

Figure 9. Drill seeders like the (a) Wind River and (b) Love-Øyjörd are commonly used to sow nursery seed.

Figure 10. Impulse sprinkler systems (sprinkler head shown in inset) have adjustable water trajectories and rotation patterns.

Nurseries apply organic fertilizers—manure or mulch, for example—with commercial manure spreaders (Fig. 11) or mulchers. Chemical fertilizers may be either solid (usually granular) or liquid; granular chemicals are spread with granular applicators, broadcast seeders (Fig. 12), or spreaders and liquid chemicals by soil injectors, sprayers, or irrigation injector systems.

Soil amendments—commonly, sand, sawdust, or mulch—modify soil texture, add organic matter to the soil, and increase the soil's capacity for moisture (see chapters 9 and 10, this volume). Nurseries apply amendments with a variety of machines, all of which must adjust to control application density and width. Sand spreaders (Fig. 13) come in various sizes and types. Manure spreaders can effectively spread sawdust if modified to increase their holding capacity and decrease their spreading density. Manure spreaders or broadcast seeders also can apply mulch. Some nurseries mix water with mulch and spray it on newly sown seedbeds with hydromulchers (Fig. 14).

3.9 Seedbed Cultivation

Weeds rob seedlings of moisture and nutrients. Most nurseries periodically control weeds mechanically with various row and path cultivators, but some still weed by hand.

Many types and sizes of row cultivators, like the Buddingh wheel hoe (Fig. 15a), are commercially available. Path cultivators (Fig. 15b) include large row cultivators, rototillers, and custom-built weeders.

3.10 Root and Top Pruning

Root pruning reduces top growth and encourages full root development in nursery stock, although the timing and frequency of pruning depend on species, desired size and type of

Figure 11. Tractor-drawn manure spreader applies organic fertilizer to soil.

Figure 12. Broadcast seeder may be used to spread granular (solid) chemical fertilizer.

stock, and growth stage (see chapter 15, this volume). Two different models of mechanical pruners—either to cut tap roots or to trim lateral roots—are used. Reciprocating and fixed-bottom pruners (Fig. 16a), fixed and disk side pruners, and root wrenchers are available in both models. Most root pruners are tractor mounted.

Top pruning removes new top growth from seedlings. Seedlings are top-pruned regularly before lifting to produce short sturdy seedlings, obtain favorable root-to-top ratios, and reduce transpiration surface; these effects make seedlings hardier against drought. Nurseries use adjustable tractor-mounted sickle-bar, flail, or rotary mowers (Fig. 16b) to prune seedling tops.

3.11 Pesticide Spraying

Nurseries apply pesticides for weed, insect, and disease control with a variety of sprayers (see chapters 18 and 19, this volume). Because chemical treatment may leave toxic residue in soil, however, nursery personnel must consider the possible consequences of different compounds when choosing pesticides.

Most nurseries apply chemical pesticides with tractor-mounted boom sprayers (Fig. 17) or spraying kits mounted on tractor-drawn tilling equipment. Hand sprayers or portable mist blowers facilitate applications for small treatments. Many sizes and models of pesticide sprayers are commercially available.

Figure 13. Sand spreader useful for adding mulches to soil.

Figure 14. Hydromulcher spraying a mulch-water mixture on prepared beds.

Figure 15. (a) Buddingh wheel hoe row cultivator and (b) Coeur d'Alene path cultivator keep nursery beds weed free.

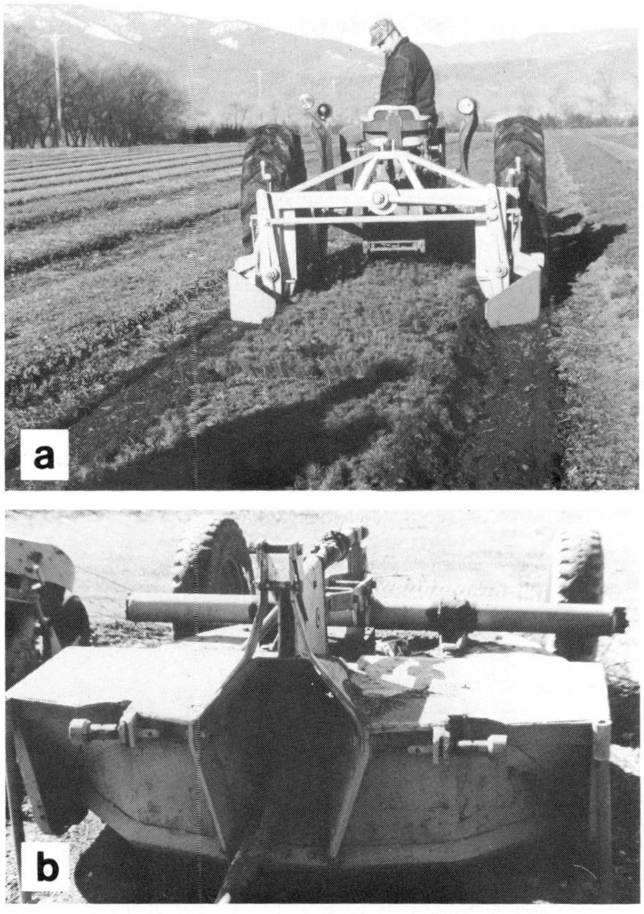

Figure 16. (a) MEDC reciprocating root pruner and (b) tractor-drawn rotary mower used for top mowing both control seedling growth.

3.12 Frost Protection

Seedlings must be protected against frost, which can damage immature seedlings at nurseries located in valley bottoms or where surface winds are restricted, until they harden and become dormant (see chapters 12, 14, and 15, this volume). Encouraging early dormancy by restricting water can reduce frost damage. Sprinkler systems are often used to protect seedlings that have not yet hardened from frost (see Fig. 10) by keeping frost from settling on young trees.

3.13 Transplanting

Nursery stock is often grown in one seedbed for 1 to 3 years and then transplanted, either manually or mechanically, to another seedbed. Transplanting generally produces large, sturdy seedlings, but the extra handling increases production costs.

Most nurseries use tractor-drawn transplanters (Fig. 18); some use hand-transplanting boards for small jobs. Although commercially available transplanters are often unsuited to nursery work, many can be modified to perform satisfactorily.

3.14 Field Lifting, Handling, and Transportation

Field lifting describes the removal of trees from nursery seedbeds (see chapters 21 and 22, this volume). Most nurseries use tractors with rigid undercutting blades and agitators (Fig. 19) to disturb seedlings and loosen seedbed soil so that seedlings can then be lifted manually. Manual field-lifting equipment includes pickup belts, conveyors, and forklifts.

Figure 17. Boom sprayers efficiently apply pesticides to large acreages of nursery beds.

Figure 18. Tractor-drawn transplanter.

In recent years, mechanical seedling lifters have been introduced and widely used. These tractor-drawn machines have hydraulic undercutting blades, conveyors, and spaces for seedling containers. Mechanical lifters like the Grayco Harvester (Fig. 20) are sometimes modified to accommodate individual nursery needs.

After trees have been lifted, they must be placed in containers before being moved to the packing area. Nurseries commonly use boxes, bins, and tubs as field containers; custom-made fabric slings, which are usually handled manually, also may be used. Mechanical seedling harvesters have racks for carrying boxes and tubs. Equipment needs for field handling depend on field lifting and transportation methods, nursery size, distance to packing area, and volume of seedlings handled.

Most nurseries move seedlings from fields to packing sheds with tractor-drawn trailers. But trailers are difficult to turn in the field, slow to load and unload, and too light for rigorous use. Nurseries also use flatbed trucks, pickups, tractor attachments, and forklifts. A wide variety of equipment suitable for transporting seedlings is commercially available.

Similarly, as nursery size and labor costs tend to increase it becomes important to move workers around the nursery as efficiently as possible. In recent years, a variety of homemade and commercial crew carriers has appeared. Scooters also are popular, sometimes replacing light trucks and buses; some nurseries even use bicycles.

3.15 Sorting, Grading, and Counting

After seedlings are lifted and moved to packing sheds, they must be sorted, graded, and counted. Even though these operations are manual, efficiency is increased with commercially available counters, scales, custom-built conveyors, and

Figure 19. Seedling lifters loosen soil so that seedlings can be manually extracted.

Figure 20. Grayco Harvester, which lifts seedlings mechanically rather than manually.

sorting tables. Most nurseries use moving belt systems for sorting, grading, and counting. MEDC has developed a new stacked three-belt system that is more efficient in grading and counting in less space than other systems (Fig. 21).

Packing-shed workers often trim seedling roots uniformly when sorting, commonly using various custom-built electric pruning saws (Fig. 22), fabric saws, and paper cutters. A variety of equipment that trims roots is commercially available.

3.16 Packaging, Storage, and Handling

Once seedlings have been sorted, graded, counted, and pruned, they are packaged for shipping or storage. Many nurseries use mechanical bundling machines, which wrap seedlings in burlap and other bags, to package seedlings; a packing medium like sphagnum moss or "shingle toe" (cellulose fiber) is often included to keep seedlings moist (see chapter 22, this volume). Most nurseries, however, pack seedlings manually in boxes or bags. Commercially available mechanical devices that package seedlings include balers, bag closers, staplers, and strapping equipment.

Packaged seedlings may be stored for months before shipping. Seedlings can be lifted and processed and then refrigerated at the nursery until needed. Most nurseries store seedlings in large, walk-in refrigerators, usually custom built, but others store them in sheds or on permanent racks or pallets. Because controlling the temperature and relative humidity of stored seedlings is crucial, nurseries often monitor these conditions with sensors that trigger alarm systems if damaging temperature fluctuations occur.

A variety of commercially available equipment, including forklifts, roller conveyors, skids, belt conveyors, and carts, is most commonly used to help employees move seedlings from storage to trucks (Fig. 23). In general, good scheduling and good equipment are the key components of designed flow patterns for all seed and seedling handling.

3.17 Shipping for Outplanting

Customers must receive seedlings promptly and in good condition. Therefore, nurseries often deliver seedlings in refrigerated trucks (Fig. 24) either owned by the nurseries themselves or contracted specifically for seedling hauling. Planned deliveries and refrigerated equipment allow nurseries to control the temperature and humidity of seedlings to ensure vigor.

Figure 21. Three-tiered belt system, which improves processing efficiency.

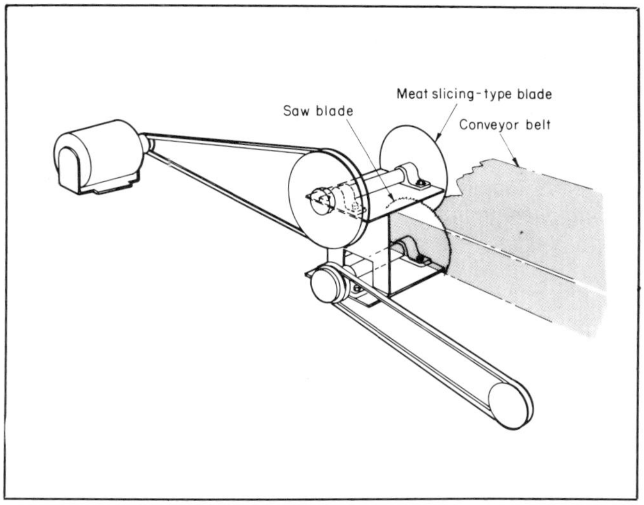

Figure 22. Schematic of root-pruning saw.

Figure 23. Trailer used to move containers of seedlings from one area of the nursery to another.

Figure 24. Seedlings are often transported to the field for outplanting in refrigerated trucks.

3.18 Conclusions

Proper equipment in good working condition is essential for high-quality nursery operations. Some machinery must be custom built and tailored to specific nursery needs, but other equipment is commercially available in a range of types and sizes and readily adaptable to nursery needs. Tractors, for example, are indispensable. Large tractors are used for lifting, plowing, and disking; small tractors for seeding, cultivating, and towing. All sizes accept attachments that increase versatility. Special features sometimes make particular tractors the best choice; for instance, hydrostatic-drive tractors are ideal for operations that require steady, slow speeds.

The right choice of equipment, in combination with manual operations, will facilitate all phases of seedling production, from cone storage and handling through outplanting.

Starting the Bareroot Seedling

Chapter 4
Assuring Seed Quality for Seedling Production: Cone Collection and Seed Processing, Testing, Storage, and Stratification

Y. Tanaka

Abstract

This chapter summarizes current technology concerning cone collection and seed processing, testing, storage, and stratification for the six major conifer species—Douglas-fir, ponderosa pine, lodgepole pine, noble fir, white fir, and western hemlock—produced as seedlings in Northwest bareroot nurseries. Though great advances have been made in the past 20 years, further refinements are deemed necessary to continue improving seedling-production technology, especially as use of valuable seed-orchard seed is favored over natural-stand seed. Suggested future refinements should include: (1) determining patterns of seed retrievability to capture maximum seed yield; (2) devising a method for separating nonviable and low-vigor seed from viable and high-vigor seed; (3) developing a method for improving the correlation between laboratory and field germination; (4) designing an effective long-term seed-storage method for true firs; and (5) developing a quick seed treatment for nursery sowing which shortens or eliminates stratification requirements.

4.1 Introduction

Seed quality has great impact on the quality of planting stock. For the last 20 years, the technology of producing seedlings has advanced greatly. Parallel to this advancement, seed quality also has improved dramatically. This chapter brings together information on cone collection and seed processing, testing, storage, and stratification drawn from the current literature and from questionnaires sent to 21 nurseries and eight seed-processing plants (extractories) in the Northwest (OSU Nursery Survey; see chapter 1, this volume). Discussions mainly focus on the six major coniferous species being produced by these nurseries: Douglas-fir [*Pseudotsuga menziesii* (Mirb.) Franco var. *menziesii*], ponderosa pine (*Pinus ponderosa* Dougl. ex Laws. var. *ponderosa*), lodgepole pine (*Pinus contorta* Dougl. ex Loud. var. *contorta*), noble fir (*Abies procera* Rehd.), white fir [*Abies concolor* (Gord. & Glend.) Lindl. ex Hildebr.], and western hemlock [*Tsuga heterophylla* (Raf.) Sarg.]. Where knowledge is lacking on these conifers, information on others is cited to illustrate important points.

4.2 Cone Collection

Careful attention to cone collection is critical to obtaining good quality forest-tree seed. Successful collection depends on understanding seed maturation and dispersal characteristics of each species, knowing local weather trends, and evaluating crop quality, harvesting procedures, and cone-storage methods.

In Duryea, Mary L., and Thomas D. Landis (eds.). 1984. *Forest Nursery Manual: Production of Bareroot Seedlings.* Martinus Nijhoff/Dr W. Junk Publishers, The Hague/Boston/Lancaster, for Forest Research Laboratory, Oregon State University, Corvallis. 386 p.

4.2.1 Seed maturation

Cone collection should begin only when seed is mature. Immature seed can bring about various problems including (1) slow and incomplete germination [4, 29, 97], (2) low-vigor seed, resulting in smaller seedlings [30, 107], (3) greater susceptibility to disease [20, 124], (4) reduced storage capability [63], and (5) increased incidence of abnormal seedlings [76]. In addition, extraction of immature seed is more difficult than that of mature seed [85, 116]. Various maturation indicators, reflecting visual, physical, biochemical, or climatic changes, can be used effectively to prevent harvest of immature seed.

4.2.1.1 General maturation indicators

Cone color [26, 57, 100], bract color [30], seed wing color [96], scale color [103], and color and firmness of embryo and megagametophyte [78, 96] can be **visual indicators** of seed maturity. These indicators, though indirect and subjective, have proved reasonably practical in many instances [131].

Cone moisture content [33, 86], cone specific gravity [53, 85, 96], and embryo development [30, 96, 103] can be **physical indicators** of seed maturity. Loss of cone and seed moisture is closely associated with seed ripening [46], and the decrease in cone moisture content and cone specific gravity has been used to indicate maturity. Of these two indicators, specific gravity (SG) is usually preferred because it can easily be determined in the field. This method has been successfully applied to various pines (*Pinus* spp.) and true firs (*Abies* spp.) using flotation liquids such as water (SG = 1.0) and various mixtures of kerosene (SG = 0.80), light motor oil (SG = 0.88), and linseed oil (SG = 0.93). The ratio of embryo length to embryo cavity length, which can be determined quickly in the field with a sharp knife and a 10X magnifying hand lens [41], also can be used to judge maturity [30].

Changes occurring within conifer seeds can be **biochemical indicators** of seed maturity. On the basis of observed correlation of reducing sugar content and germination, Rediske [106] recommended that Douglas-fir cone collection be initiated when reducing sugar content has fallen to 13 mg/g of seed weight. In a subsequent study, Rediske and Nicholson [108] found that, in noble fir, the increase in crude fat content is more closely related to seed maturation and recommended the threshold value of 250 mg/g of seed for beginning cone collection. Although measuring biochemical indicators is time consuming and requires special laboratory equipment, it is thought to be more reliable than methods based on visual observation.

Changes in temperature, particularly during the summer in which seeds mature, can strongly influence the rate of seed maturation and are used as **climatic indicators.** Consequently, degree-day summations should be potentially more reliable than calendar date, especially at high latitude or high altitude, where summer temperature may limit seed development. Tanaka and Cameron [135] reported that 1,310 degree-days are required for ponderosa pine seed to mature at high elevations in southeastern Oregon. Zasada [152] related cone and seed development in white spruce [*Picea glauca* (Moench) Voss] to summer heat-sum and found that 625 degree-days were required to produce cones that could be successfully after-ripened in Alaska. Heat-sums are not extensively used for cone-collection purposes, probably due to lack of sufficient information. However, together with other climatic parameters such as precipitation and radiation, heat-sums would be a useful tool for field collection of coniferous cones in the Northwest [46].

Information (as of 1974) on cone- and seed-maturation indicators for many coniferous species in the United States is available in *Seeds of Woody Plants in the United States* [138]. Edwards [46] also provides an extensive discussion on various types of maturation indicators.

4.2.1.2 Maturation indicators used in the Northwest

Maturation indicators for the six major coniferous species in the Northwest are summarized in Table 1. Those used by the one nursery and six seed-processing plants involved in cone collection (OSU Nursery Survey) are, in order of frequency: cone, wing, and scale color, firmness of embryo and megagametophyte, and embryo development. Somewhat surprisingly, seed moisture content and specific gravity are not currently used, probably indicating that visual observation of the above characteristics is preferred because it is less time consuming. One seed plant extensively relies on biochemical indicators, using crude fat for noble fir and ponderosa pine and reducing sugar for Douglas-fir; on the basis of past experience, these biochemical indicators seem highly reliable.

Table 1. Cone and seed maturation indicators for the six major conifers in the Northwest.

Species	Maturation indicators	Reference
Douglas-fir	Reducing sugar 13 mg/g or less	[106]
	Embryo:cavity length ratio greater than 90%	[30]
	Browning of cone bracts	[30]
	Cone moisture content lower than 50%	[104]
	Firm, nonmilky megagametophyte enclosing a yellowish-green embryo	[78]
	Main harvest period of squirrels	[30]
Ponderosa pine	Specific gravity 0.85 or less (central Idaho)	[85]
	Specific gravity 0.84 or less (California)	[51]
	Specific gravity 0.94 to 0.99 (South Dakota)	[39]
	Specific gravity 0.88 or less (Arizona and New Mexico)	[111]
	Heat-sum 1,000 to 1,110 degree-days	[135]
Lodgepole pine	Specific gravity 0.43 to 0.89	[77]
Noble fir	Specific gravity 0.90 or less	[53]
	Crude fat 0.25 g/g of seed	[108]
White fir	Specific gravity 0.96 or less, uniformly brown seed wing, embryos pale yellow-green, 94% of the embryos fully elongated	[96]
Western hemlock	Brown cones with red-brown tips	[56]
	Cones opening after drying	[46]

4.2.2 Seed dispersal

Although seed-maturation characteristics have been extensively studied, little is known about timing of seed dispersal in relation to cone characteristics or climatic variables. Most observations relate the timing of seed dispersal to calendar dates, but such correlations may be of little value in the field because of yearly variation in weather patterns.

We have found that ponderosa pine seed in southeastern Oregon starts disseminating when cone moisture content drops to approximately 120% on a dry-weight basis. Once the rate of moisture loss in early August has been determined, it has been possible to predict approximate dates of seed dispersal for this species. Together with knowledge of seed maturation rate, approximate seed-dispersal dates could be of practical importance to cone collectors. The field observations made by our laboratory also indicated that the earlier seed matured, the more quickly it started disseminating, probably due to faster drying of cones. Similar observations should be of value in capturing the maximum seed yield of other conifers that have a responsive reflex of cone scales.

4.2.3 Artificial ripening

It is important that cone collection be initiated after seed has attained full maturity. However, immature seed can be artificially ripened during cone storage in certain species. Artificial ripening has been successful on noble fir [108], grand fir [*Abies grandis* (Dougl. ex D. Don) Lindl.] [102], white fir [96], and Nordmann fir [*Abies nordmanniana* (Stev.) Spach] [95]. Because of this potential increase in germination during cone storage, true fir cones are usually stored longer than those of other conifers before seed extraction. Douglas-fir [125], several species of pines [13, 76], and white spruce [150, 152] have also shown increased germination during artificial ripening. However, despite the findings of various researchers and the potential benefits, artificial ripening has not been extensively used for conifers other than true firs in the Northwest—probably because there is more risk of poorer germination and reduced seed yield in other species.

4.2.4 Weather

Weather conditions significantly impact cone collection. Except for pines with serotinous cones or some cypresses (*Cupressus* spp.) or junipers (*Juniperus* spp.) for which year-round collection is possible, the optimum cone-collection period for most conifers at any given location is relatively short. This period occurs sometime between late summer and late fall but could vary by up to 2 to 3 weeks depending on weather conditions. For example, if the snow melts late at high elevations during a cool spring, flowering may be so late that seed maturation could be delayed significantly [131]. A hot, dry summer may shorten the optimum cone-collection period by causing early seed fall, whereas cool, rainy conditions may delay it. Seed generally ripens earlier at lower elevations and on south and west slopes and later at higher elevations and on north and east slopes [122].

In addition to general spring and summer weather trends that determine seed-maturation and dispersal patterns, weather conditions during the cone-collection period itself are also important for the cone-harvesting operation. High winds or rain may preclude tree climbing, disrupt access to collection areas, and reduce pickers' productivity. In many areas in the Northwest, a drying east wind during the fall collection period may cause seed to disseminate too quickly, thereby reducing seed yield. For these reasons, daily forecasts and 5-day outlooks are valuable aids to coordinating cone-collection activities [41].

4.2.5 Crop quality

Once seed maturity has been determined, the quality of cone crops to be harvested must be evaluated. This generally is done by estimating the number of good seeds present in several representative cones, sliced lengthwise with a sharp knife. Cones of Douglas-fir, western hemlock, and pines are sliced through the center; those of true firs are cut lengthwise ¼ to ½ inch to one side of center [43]. A variety of knife assemblies is available for slicing conifer cones [123, 148, 149].

Minimum acceptable seed-count requirements may vary from year to year according to supply and demand. Average good seed counts are 6 for Douglas-fir, 8 for western hemlock, 10 for ponderosa and lodgepole pine, and more than 50% of the seed (if seed has good appearance) for noble and white fir [43]. Lodgepole pine in certain areas produces cones that are very hard and, therefore, difficult to section. To extract seeds, such cones can be dipped in boiling water for 10 seconds, then placed in an oven at 65°C for 3 to 4 hours [41]. A minimum of 20 filled seeds per cone is required before a crop can be harvested. In addition to the filled-seed count, damage by biotic agents such as insects and disease, climatic extremes, or other abnormalities also should be assessed because these affect seed yield and are important factors in selecting areas from which to collect. Dobbs et al. [41] do not recommend collection if more than 50% of seeds are damaged. Several articles may be of help in identifying and assessing insect [59, 74] and disease [25, 62] damage.

4.2.6 Collecting methods

Cones are collected from western conifers: (1) by climbing standing trees, (2) from felled trees, and (3) from squirrel caches. Collecting cones from standing trees—the surest method to harvest seed of known origin, quality, and maturity—is often time consuming, expensive, and dangerous. Cones can be picked much more easily from felled trees in logged areas, but pickers should ascertain whether seeds were sufficiently mature when the trees were felled. Cones should be picked immediately after felling so as to minimize seed loss due to cone opening or mammal, bird, and insect damage. Squirrel-cached cones are easy to collect, but their use is sometimes questioned because the source and quality of the crop tree are not known. No evidence suggests, however, that seeds collected by squirrels are inferior to those collected by other means. All three of these methods are commonly used by cone collectors in the Northwest (OSU Nursery Survey).

Other methods less frequently mentioned in the Survey were helicopter collection and mechanical seed harvester. Helicopter collection has been experimentally tested in Canada by Dobbs et al. [42]. Mechanical tree shakers, regularly used on southern pines [27, 75, 137], have been tried only experimentally for western conifers. Although not easily adaptable to Northwest terrains for natural-stand collection, mechanized cone collection should play an important future role when western seed orchards are in full production.

4.2.7 Cone storage

Cones are stored (1) because processing equipment is not usually capable of extracting seeds from all harvested cones at once [81]; (2) to decrease cone moisture content, thereby reducing kiln drying time; and (3) to artificially ripen seeds of species such as true firs and improve seed-germination potential.

In large-scale cone collection, cones are usually placed in burlap bags, which are stored either temporarily near collection sites or in storage sheds at the extractory. However, great care should be exercised to maintain seed quality during cone storage. Burlap bags should not be filled to the tops, so that cone scales can fully expand upon drying; if scales cannot open sufficiently, seed extraction may be severely impaired [131]. Burlap bags should not be stacked up in large piles; this can lead to seed losses due to overheating or to insect and disease damage. Warm, moist environments can harm seed quality [81, 109]; hence, good ventilation should be provided. At a few seed-processing plants, cones of true firs and spruces are stored on ventilated mesh screens for artificial ripening (OSU Nursery Survey).

There has been an attempt to rank the different species according to relative ability to withstand prolonged cone storage [81]. At one seed-processing plant in the Northwest, cones of western hemlock are extracted first and those of true firs last. The ranking is primarily based on intuition and experience, but such information is valuable in scheduling cone processing.

OSU Survey respondents from most seed plants indicated that they store cones from 1 to 6 months, depending on species and size of cone crops. Several studies conducted with western species have confirmed the success of current cone-storage practices and have shown that, if cones are handled properly and storage conditions are optimum, seed could be safely stored in intact cones for up to 4 to 6 months [79, 82, 106, 109]. For cone storage beyond 4 months, it may be advantageous to install frost protection because subfreezing temperatures could significantly reduce seed germinability [133].

4.3 Seed Processing

After cones are harvested and stored, seeds are extracted and prepared for either immediate sowing or storage. This series of operations, called seed processing, includes kiln drying, cone tumbling, scalping, dewinging, and cleaning and sorting. (Seed-processing equipment is also discussed in chapter 3, this volume.)

4.3.1 Kiln drying

Given good drying conditions, cones of most conifers open readily. Under natural storage conditions, however, cones may not be thoroughly and uniformly dried, especially when weather is humid and cool. Cones should therefore be kiln dried to facilitate extraction.

Kilns are of two types: rotating and progressive [131]. In rotating kilns, a batch of cones is loaded into and dried within a drum where temperature and humidity are usually controlled. Such kilns, although not suitable for drying a large quantity of cones, can provide specific drying temperatures and relative humidities for small-lot processing. In progressive kilns, loaded trays are moved at certain time intervals to expose cones to increasingly warmer air as they dry. This type of kiln is more suitable for large-batch processing.

Kilns are generally operated at temperatures between 32 and 60°C [1]. Although studies have shown that the biologically lethal temperature of most tree seed is around 66°C [12, 113], the operational maximum temperature should not exceed 43°C [32, 54]. Because cones often have a high moisture content after storage, however, drying should be started at low temperatures that are progressively elevated. Drying cones with high moisture content immediately at high temperatures should be avoided because it could lead to case hardening and result in partial cone opening and incomplete seed extraction [78]. However, the problem of case hardening can, to some extent, be overcome by moistening scales or soaking cones in water.

Air humidity is as important a factor as temperature. Low humidity is the key to more complete drying. For example, cones can be successfully dried at the relatively low temperature of 32°C if relative humidity is below 30% [1]. Cones of most major conifers in the Northwest readily open upon drying. However, lodgepole pine cones from certain geographic areas are serotinous and require a short soak in hot water before kiln drying [112]. Additional soaking cycles with water have been reported to increase seed extraction by 20 to 84% [140, 144].

4.3.2 Cone tumbling

In rotating kilns, cones are dried and tumbled simultaneously, and seeds fall out as cones open. Generally, loose seeds drop through perforations in the drum. Cones dried in progressive kilns are subjected to shaking action by tumblers to extract seeds from cones. A tumbler is a rectangular or round wire-mesh container mounted horizontally on its long axis, which turns at a slow speed. Small quantities of cones may be tumbled in batches. In large-scale continuous operation, the tumbler axis is inclined so that rate of cone movement through the tumbler can be regulated [131].

4.3.3 Scalping

Seeds coming from the tumbler must be separated from a mixture of cone fragments, hardened pitch, foliage, dust, and other debris. This step, called scalping, is achieved by vibration, air movement, or screens, alone or in combination. The most commonly used equipment has several layers of vibrating screens of different-sized mesh. Coarse materials such as scales and twigs are retained on the uppermost screen and slide down to be collected in one bin, while fine particles are screened to be deposited in another bin; the seed is usually collected through an intermediate screen [47].

4.3.4 Dewinging

Once debris has been eliminated, wings must be removed from many conifer seeds. Although wings are often loosened during tumbling and scalping, dry or wet dewinging may also be required. Dry dewinging, a technique which employs a rubbing action to remove wings from dry seed, is generally used for Douglas-fir, pines, and true firs. Small lots can be dewinged in a cloth bag; lots of up to 5 kg are better handled in a Dybvig macerator [19]; and large lots are best dewinged with a brush-type dewinger, although auger-type dewingers have also been used successfully.

Because dry dewinging is the processing step that is most likely to cause seed damage, extra caution should be exercised to use proper equipment and to minimize unnecessary friction. In one study, for example, three cycles of brush-dewinging seeds of subalpine fir [*Abies lasiocarpa* (Hook.) Nutt.] destroyed 50% of the originally viable seeds [pers. commun., 49].

Because dry dewinging can mechanically damage seed, many seed workers prefer wet dewinging, especially for pines and spruces. The principle of wet dewinging is that wings are more hygroscopic than seed and, upon wetting, are released cleanly. The Kason Vibrator [47] and a rotating cement mixer with a soft brush [144] have been successfully used for wet dewinging. However, because seed absorbs moisture during wet dewinging, it must be redried sufficiently before storage. Germination tests verified that a 20- to 30-minute water soak, followed by wet dewinging and air drying for 16 hours at 26 to 30°C to 4 to 8% moisture content, did not adversely affect seed quality [144].

4.3.5 Cleaning and sorting

Empty seed, partially filled seed, and other foreign particles are removed from good seed in the final cleaning. Scalpers and fanning mills are often used for species that have few scales, such as Douglas-fir and pines, but vibratory gravity tables are best for true firs. Pneumatic seed cleaners have also been successfully used for various conifer species [45, 126, 151]. All this equipment, in combination, further improves sorting efficiency. Flotation sorting with water, alcohols, and other organic liquids has been used to clean red spruce (*Picea rubens* Sarg.) [8], true firs [pers. commun., 49], and several pines [10, 88, 143], although this method has only been tested experimentally with western species.

A noteworthy development in seed sorting is the IDS (incubation-drying-separation) method, developed by Simak [129], which can separate nonviable, as well as empty and partially filled, seed from viable seed. Fully imbibed seed is first incubated for a short time, then gradually dried, and finally separated by various specific-gravity methods. Because empty and nonviable seeds lose water more quickly during the drying phase, differences between nonviable and viable seeds are magnified, making subsequent separation by standard gravity methods more effective. Scots pine (*Pinus sylvestris* L.) seed of low germinability was successfully upgraded by this method experimentally [129].

4.3.6 Seed processing in the Northwest

Most processing work is done at seed plants in the Northwest. All eight seed-processing plants responding to the OSU Nursery Survey process their own seed as well as seed harvested by other organizations. However, four of the 16 nurseries responding do at least part of the processing at their own facilities. The remaining nurseries have private or state plants process their seeds.

Seven of the eight seed-processing plants and seven of the 16 nurseries set their own standards of purity for commercial seed transactions or nursery sowing (Table 2). The seed plants and nurseries replying to the Survey had a generally higher standard of purity than the Western Forest Tree Seed Council [130] recommendations for four of the six major coniferous species; the lower accepted purity standards of the true firs (see Table 2) may indicate possible difficulties in removing nonseed components without adversely affecting seed germination. Seeds of true firs are known to be especially sensitive to handling and mechanical damage [47].

Table 2. Minimum purity standards recommended by the Western Forest Tree Seed Council [130] and established by seed-processing plants and nurseries (OSU Nursery Survey) for the six major conifers in the Northwest.

Species	Tree Seed Council	Seed plants and nurseries
	%	
Douglas-fir	95	95-99
Ponderosa pine	95	95-99
Lodgepole pine	90	99
Noble fir	95	90-98
White fir	95	90-98
Western hemlock	90	95

4.4 Seed Testing

Seed testing evaluates seedlot quality and is essential for both seedling production and commercial seed transactions. Most tree-seed tests are conducted with methods based on rules of the Association of Official Seed Analysts (AOSA) [7] or the International Seed Testing Association (ISTA) [66]. Testing methods pertinent to western conifers are also available from the Western Forest Tree Seed Council [130].

4.4.1 Sampling

The first step in seed testing is to draw a sample that represents the entire seedlot. A seedlot is defined as a unit of seed of reasonably uniform quality from a particular location or elevation [21]. Seedlot size varies with testing rules and among laboratories. ISTA [66], for example, has determined that a seedlot should be less than 5,000 kg for seeds the size of beech (*Fagus* spp.) seed or larger, or 1,000 kg for seeds smaller than beech. The Western Forest Tree Seed Council [130] recommends that lots in excess of 227 kg be divided into equal smaller lots for sampling.

Loose seeds in containers should be sampled with seed-sampling probes long enough to reach all areas in the containers. The sample should be composed of equal portions taken from evenly distributed volumes of the lots to be sampled, each sample proportional to the size of the container. Samples should be subdivided in the testing laboratory with a mechanical divider until a subsample of the desired weight is obtained.

4.4.2 Physical characteristics

4.4.2.1 Purity

Purity tests measure the percent by weight of four major components: (1) pure seeds of the test species, (2) seeds of other crop species, (3) weed seeds, and (4) inert matter (leaves, cone scales, etc.). The purity test is usually the first test performed for a given lot and is especially important for commercial transactions, which are based on weight.

4.4.2.2 Moisture content

Seed moisture content is most often determined with the air-oven method [66]. Seed samples are heated in ovens; the weight loss that occurs during drying is considered to be seed moisture. ISTA rules prescribe oven drying at 105°C for 16 hours for all tree seeds except those of the genera *Abies*, *Cedrus*, *Fagus*, *Picea*, *Pinus*, and *Tsuga*. Seeds of those genera contain a significant amount of volatile oils and resins which may be lost at the above temperature. Therefore, their moisture content must be determined by toluene distillation [66]. Electronic moisture meters, though not as accurate as the above methods, are frequently used by various seed workers; they give rapid measurements desirable, for example, when checking moisture in a large number of seedlots being dried before storage.

Seed moisture content can be expressed as a percentage of water loss of either total **fresh** weight or corresponding **oven-dry** weight. Seed moisture content has been expressed on a dry-weight basis in some research [99], but international usage is exclusively on the fresh-weight basis. To avoid misunderstandings, the base should always be clearly specified.

4.4.2.3 Weight

Seed weight, required for calculating sowing rates in nursery sowing and direct seeding, is a function of seed size, moisture content, and proportion of full seed in a given lot. The commonly used unit is the weight of 1,000 pure seeds (1,000 seed weight). ISTA [66] specifies weighing eight random samples of 100 seeds each from the pure-seed component; however, some laboratories use two or more samples of 500 seeds each. When means of replicates vary more than 10%, additional samples should be weighed. All weights should be accurate to three significant digits.

4.4.3 Biological characteristics

4.4.3.1 Tests to estimate seed viability

Germination potential, perhaps the most important quality measurement in seed testing, is used to determine sowing rates as well as whether seed must be sown immediately or can be stored. Seeds of different species have different requirements for optimum germination. This potential can be (1) evaluated directly by germinating seeds under predetermined conditions or (2) estimated indirectly with biochemical staining, embryo excision, cutting tests, x-ray radiography, or hydrogen peroxide tests.

The most reliable method is germination in a controlled environment. At least 400 seeds, usually divided into four replicates of 100 seeds each, from the pure-seed component of the purity test [7] are normally prechilled for up to 28 days and germinated on suitable substrates (Table 3). Substrates should (1) be nontoxic, (2) be free of molds or other microorganisms, and (3) provide adequate aeration and moisture [71]; those recommended by AOSA [7] are blotter papers, paper towels, washed sand, vermiculite, perlite, and peat moss. Most (over 70%) of the coniferous species listed in the AOSA rules are germinated under alternating temperatures (30°C for

8 hours in the light, 20°C for 16 hours in the dark). An intensity of 750 to 1,500 (±250) lux [75 to 150 (±25) foot-candles] is recommended [71]. Seed is counted as germinated when all essential structures appear normal. Retests are necessary when an extremely high proportion of full, ungerminated seed is left at the end of the test, or when variation among test replicates exceeds the accepted tolerances [7].

Although controlled-environment germination tests are reliable, they are often time consuming, especially for dormant species requiring prechilling. Several rapid methods of estimating viability have been proposed, two of which—tetrazolium staining and embryo excision—are now recognized as official testing procedures.

The tetrazolium test is the most commonly practiced biochemical staining method [66, 94]. Seeds are immersed in 2,3,5-triphenyl tetrazolium chloride. Living cells stain red as tetrazolium is reduced by dehydrogenase enzymes to form a stable red triphenyl formazan, which is insoluble in water. The method is fast but lacks uniformity in staining [83]; therefore, results can be difficult to interpret. Other biochemical staining methods applied to seed testing with varying degrees of success include those using salts of selenium and tellurium [9] and Indigo Carmine [73, 93].

The excised embryo test is recommended for several species of pines including Coulter pine (*Pinus coulteri* D. Don), Jeffrey pine (*Pinus jeffreyi* Grev. & Balf.), and sugar pine (*Pinus lambertiana* Dougl.) [66]. Excised embryos are cultured on moist filter or blotter paper in covered dishes under light for 10 to 14 days at 18 to 20°C. Viable embryos remain firm and white and turn green, indicating growth, whereas dead ones turn dark or are covered with mold. This method is fast but requires skilled analysts.

Other quick methods include the cut, x-ray, and hydrogen peroxide tests [82]. In the cut test, seed is bisected and then rated visually; this is the simplest but most unreliable method because distinguishing seeds damaged during handling and storage is very difficult. The x-ray test is fast, especially when Polaroid film is used [44], and development of contrast techniques has greatly expanded x-ray test capabilities [127]. Disadvantages are difficulty in interpretation and relatively high equipment costs. The hydrogen peroxide (H_2O_2) test allows assessment of root growth in 1% H_2O_2 [32]. It is simpler to perform than the excised embryo test and is more objective and easier to interpret than x-ray. However, as with other quick tests, it tends to overestimate viability, compared with germination tests.

4.4.3.2 Seed vigor

Nursery bed germination is usually slower and less complete than laboratory germination. Therefore, various laboratories have attempted to define and determine seed vigor to improve prediction of nursery germination. Three major groups of expressions have been proposed: (1) mathematical values based on standard laboratory test results, (2) germination under stressful conditions, and (3) biochemical testing.

Mathematical expressions have been most widely tested. They include the number of days required to attain a certain proportion of total germination [18, 28], germination value [36], modified germination value [40], and the Weibull function [22, 114]. Germination under stressful conditions has been developed mainly for seeds of agricultural species; most widely used are the cold test for corn [31, 68] and the accelerated aging test for soybean [87, 136]. However, application of these tests or development of new procedures for tree seed has been rather limited. Biochemical tests have been tried to a limited extent; the few reported include tetrazolium staining [94] and the GADA (glutamic acid decarboxylase activity) test [21].

4.4.4 Seed testing in the Northwest

Over 70% of the nurseries and seed-processing plants conduct some type of seed-quality test at their own facilities (OSU Nursery Survey); the remaining organizations send all their samples to outside commercial laboratories. The most commonly used outside laboratory is the Oregon State University Seed Laboratory (Corvallis, Oregon). Samples are also sent to private laboratories, other state laboratories, and the National Tree Seed Laboratory (Macon, Georgia). Of the 17 organizations that conduct their own tests, three conduct all of their tests; the rest have certain types of tests done by outside laboratories—including checking their own test results. According to the OSU Survey, the tests most commonly conducted, in order of frequency, are seed moisture content (see 4.4.2.2), 1,000 seed-weight determination (see 4.4.2.3), purity test (see 4.4.2.1), and germination test (see 4.4.3.1). Cut, x-ray, and H_2O_2 tests are used less frequently. No organization indicated use of seed-vigor expressions, although a few have tried Czabator's [36] germination value.

4.5 Seed Storage

Irregular and often infrequent seed production by many of the major tree species necessitates seed storage—sometimes

Table 3. AOSA seed-testing procedures [7] for the six major conifers in the Northwest.[1]

Species	Temperature,[2] °C	Test duration, days	Additional directions
Douglas-fir	20-30	21	Light[3]; prechill 21 days at 3 to 5°C. Vermiculite recommended if top of blotter not used.
Ponderosa pine	20-30	21	Light; prechill 28 days at 3 to 5°C.
Lodgepole pine	20-30	28	Light; prechill 28 days at 3 to 5°C.
Noble fir	20-30	28	Light; prechill 14 days at 3 to 5°C. Vermiculite recommended if top of blotter not used.
			Dark; prechill 21 days at 3 to 5°C.
White fir	20-30	28	Light; many lots complete in 14 to 21 days; few sources from the coastal region may need prechill for 21 days at 3 to 5°C.
Western hemlock	20	28	Light.

[1]Substrates for all species were the tops of blotters and covered petri dishes with (a) two layers of blotters, or (b) one layer of absorbent cotton, or (c) five layers of paper toweling, or (d) three thicknesses of filter paper, or (e) top of sand or soil.

[2]Single numeral indicates constant temperature. Two numerals separated by a dash indicate an alteration of temperature, the test to be held at the first temperature for approximately 16 hours and at the second temperature for approximately 8 hours per day.

[3]Where prescribed, light should be provided by a cool-white fluorescent source. Illuminance for dormant seed should be 750 to 1,250 lux (75 to 125 foot-candles). Seeds should be illuminated for at least 8 hours of every 24 and, where temperatures alternate (see footnote 2), during the high-temperature period only.

for several years—to maintain supplies through years of poor seed production. Because of this, considerable research has been carried out on seed storage. Storage is one area of forest-tree seed technology for which sufficient information is available for most species of interest.

Successful seed storage requires knowledge of the seed characteristics of different trees as well as of the factors influencing storage capacity, such as seed quality before storage, seed moisture content, and storage temperature and method. These aspects have been reviewed by Baldwin [9], Barton [17], Holmes and Buszewicz [63], Jones [70], Magini [84], Wakeley [143], and Wang [145].

4.5.1 Seed longevity

The life span of seeds varies with species. Seeds are classified into three biological categories according to their life span under natural conditions: (1) microbiotic seeds (life span not exceeding 3 years), (2) mesobiotic seeds (life span from 3 to 15 years), and (3) macrobiotic seeds (life span from 15 to more than 100 years) [34]. Seeds of most conifers and hardwoods are microbiotic. Under regulated storage conditions, however, longevity of many tree seeds can be extended more than tenfold. For example, the viability of naturally dispersed seed of spruce and many pines extends only into the first growing season and, occasionally, into the second growing season. Under subfreezing storage, seed viability of these same species can easily be maintained at high levels for 10 years or longer [17]. Storage over 10 years is not usually required for seedling-production purposes but may become vital to future tree-breeding programs. Under optimum storage conditions, seed viability of certain trees might be maintained indefinitely, but the maximum potential for maintaining original seed viability has not yet been determined for most species [145].

4.5.2 Seed quality

Seed quality has a significant impact on storage capability. Factors affecting quality are seed maturity, cone handling, and seed extraction and processing. Immature seeds are not only poor in germinability and liable to be further damaged by seed processing but also are difficult to store successfully [2, 3, 30, 65]. Overheating during extraction [3] and damage caused by dewinging [3, 50, 72] also have been found to adversely affect seed quality. Injured seeds are not suitable even for short-term storage because they have a high rate of respiration, undergo spontaneous heating, and deteriorate rather quickly [63, 153].

4.5.3 Seed moisture content

Of all the factors influencing seed storage, moisture content may be the single most important one in maintaining germinability. Various researchers [17, 63, 70] have demonstrated the detrimental effect of high seed moisture on tree-seed viability; increased rates of respiration and changes in carbohydrates and fats presumably cause seeds to use their food reserves [84, 153]. Excessively low seed-moisture content also may reduce storage capability. Some species, including Douglas-fir, can tolerate drying to 0% moisture content [119]; however, overdrying can destroy the monomolecular layers that protect against oxidation [55]. Recommended seed-moisture content for storing Douglas-fir, ponderosa pine, lodgepole pine, and western hemlock is 6 to 9% (wet-weight basis); that for true firs is 9 to 12% [130]. However, Danielson and Grabe [38] showed that optimum moisture content for noble fir is also 6 to 9%.

4.5.4 Storage temperature

The effect of storage temperature on the retention of tree-seed viability has been thoroughly investigated [3, 17, 60, 63, 64, 70, 91, 120, 121]. The general relationship between storage temperature and moisture content was described by Barton [17] as follows: at a given moisture content, the higher the storage temperature, the faster the deterioration of seed viability; the lower the storage temperature, the greater the tolerance to high moisture content and the better the retention of viability. Some studies have shown that temperature slightly above or below freezing may be sufficient to prevent deterioration for short-term storage [14, 120], but the retention of viability has generally been better when seeds were stored at −18°C, particularly for longer periods [3, 13, 122]. Of the 24 organizations (nurseries and seed-processing plants) replying to the OSU Nursery Survey, 16 store seeds at their own facilities. Of these, 11 store them at −15 to −18°C (5 to 0°F) and four at −5 to −12°C (23 to 10°F); one stores them at 0.5°C (33°F), but only for short periods. For a seed moisture content of 6 to 9% (wet-weight basis), these storage temperatures generally seem within the safe range (Table 4) for storage up to 7 years. This length of storage time should maintain supplies through years of poor western-conifer seed production.

4.5.5 Storage method

Tree seeds can be stored either wet or dry. Large seeds of hardwood species require moist conditions and are usually kept wet for short-term storage, whereas small seeds of conifer species, including all major conifers in the Northwest, are stored dry. Seed moisture content is controlled by storing properly dried seed in tightly closed containers or by regulating humidity in the storage area (as for many agricultural seeds); in the Northwest, dried seed is generally placed in closed containers, although some facilities do use humidity-regulated storage rooms. These facilities, although costly, are effective in minimizing reabsorption of moisture by dried seed, especially in areas with humid climates.

The most frequently used storage containers are plastic bottles with screw tops, polyethylene bags, and fiberboard drums [145]. More than 80% of the organizations surveyed store their seeds in rigid drums lined with polyethylene bags (OSU Nursery Survey). This method is common in many areas because it is relatively inexpensive and effectively prevents uptake or loss of moisture by seed from the atmosphere. Of 16 organizations, nine use plastic bags as containers; five of those use 1- to 6-mil plastic, and the remaining four use 7- to 8-mil plastic. Thin bags are subject to ripping but easier to handle when cold. However, plastic containers are not completely impermeable to moisture [145]. Use of thicker materials may be desirable for seed requiring low seed-moisture content if external humidity is high and seeds are to be stored for a long period.

4.5.6 Retesting

Retesting is often recommended for seeds stored for a relatively long period (5 years or more). Even under ideal storage conditions, certain poor-quality seedlots rapidly lose their viability. Fourteen of 16 organizations conduct viability tests at 2- to 6-year intervals (OSU Nursery Survey). Seed moisture content is retested by only three of the 16, however, probably indicating that moisture content does not fluctuate significantly under current storage conditions.

Even if seed moisture changes only minimally in storage, additional moisture could be introduced when seeds are withdrawn. Therefore, it is particularly important that sealed containers removed from cold storage be permitted to reach ambient temperatures before being opened to avoid condensation of water within the container [145]. Sahlen and Bergsten [118] found that temperature was completely equalized in the center of a 28-liter container, with walls 2.5 mm thick, 36 hours after the container was moved from a −16°C storage room to 22°C ambient temperature. To minimize repeated opening and resealing and to reduce storage space, the use of small-sized containers (10- to 25-kg capacity) has been recommended [63].

4.6 Seed Stratification

Tree seeds, unlike agricultural seeds, are in many cases characterized by deep dormancy. This is true for most Northwest conifers. Seeds of different species or different geographical origins often require different pretreatments and conditions for optimum germination. The most commonly used pretreatment to break dormancy is stratification—which usually is moist cold treatment for up to several months. Stratification is generally known to bring about changes in anatomy or physiology, including embryo growth [105], and in metabolism [106, 117 Physiologically, breaking of dormancy has often been explained in terms of a shift in the inhibitor-stimulator balance. Presumably, although it may not directly affect the level of inhibitors [147[illegible], stratification could increase growth-stimulator levels, which would then counteract the effects of inhibitors in breaking dormancy [141, 142, 146].

Successful cold stratification requires: (1) proper moisture content, (2) low temperature, (3) adequate aeration, and (4) proper length of time. In practice, seed originally was stratified by placing it between moisture-holding media such as peat moss or sand in boxes, tanks, trays, and other suitable containers and maintaining it there under cold, moist conditions [2[illegible]].

Table 4. Effect of storage conditions and periods on seed viability of the six major conifers in the Northwest.

Species	Storage condition	Storage period, years	Effect on viability	References
Douglas-fir	Sealed, 5°C, 13.6% mc[1]	3	Reduced by 60%	[16[illegible]
	Sealed, 5°C, 5.8% mc	3	Maintained	
	Sealed, −18°C, 5.8% mc	3	Maintained	
	Sealed, −18°C, 13.6% mc	3	Maintained	
	Sealed; room temperature, 0 and −18°C; 6.5-9.5% mc	5-7	−18°C better than 0°C; substantial loss at room temperature after 2-3 years	[illegible]
	Sealed, −18°C, 6-9% mc	10-20	Maintained	[10[illegible]]
Ponderosa pine	Canvas bags, −4°C, 15% mc	3	Maintained	[1[illegible]]
	Canvas bags, −11°C, 17% mc	3	Reduced by 15%	
	Canvas bags, −18°C, 10% mc	3	Reduced by 9%	
	Sealed, 5°C, 5.1% mc	3	Reduced by 10%	[12[illegible]]
	Sealed, 0°C, 5.1% mc	3	Reduced by 8%	
	Sealed, −5°C, 5.1% mc	3	Reduced by 1%	
	Sealed, −18°C, 5.1% mc	3	Reduced by 10%	
	Sealed, room temperature, 8.1% mc	7	Reduced by 31%	[illegible]3]
	Sealed, 0°C, 8.1% mc	7	Maintained	
	Sealed, −18°C, 8.1% mc	7	Reduced by 9%	
	Airtight, 4.5°C	10	Maintained	[1[illegible]]
	Airtight, 0 and −18°C	14	Maintained	[[illegible]5]
	Airtight, cellar	14	Substantial loss	
Lodgepole pine	Airtight, 4.5°C	9+	Maintained	[[illegible]1]
	Airtight, 4.5°C	11-20	Substantial loss in some lots	[1[illegible]0]
	Sealed, 0°C, 8.8% mc	7	Maintained	[illegible]3]
	Sealed, 0°C	2	Maintained	[illegible]6]
Noble fir	Sealed; room temperature, 0 and −18°C; 9.0% mc	7	Reduced by 41, 11, and 10%, respectively	[illegible]3]
	Sealed; 8°C for 9 years and −4°C for an additional 7 years; 7, 8, 11, and 13% mc	16	Reduced by 6-16% after 9 years and 30-50% after 16 years	[[illegible]4]
	Sealed, 5°C	5	25%	[1[illegible]0]
	Sealed, −10°C	3-5	Maintained	[illegible]7]
	Sealed, room temperature	1	Total loss	
	Sealed; 20, 5, and −18°C; 4% mc	2	Maintained	[illegible]38]
	Sealed; 5 and −18°C; 6, 8, and 9% mc	2	Maintained	
	Sealed, −18°C, 12% mc	2	Reduced by 5-8%	
	Sealed; 20 and −18°C; 16 and 17% mc	2	Greatly reduced	
White fir	Sealed, room temperature, 6.3% mc	7	Complete loss	[3]
	Sealed, 0°C, 6.3% mc	7	Reduced by 17%	
	Sealed, −18°C, 6.3% mc	7	Reduced by 10%	
	Sealed, 5°C	5	4-53%	[[illegible]20]
	Sealed, 5°C	10	6%	
	Sealed, 5°C	20	8%	
Western hemlock	Airtight, 5°C	20	1 and 13%	[[illegible]20]
	Sealed; room temperature, 0 and −18°C	5-7	−18°C usually superior to 0°C; complete loss at room temperature	[3]
	Sealed, 5°C, 7.7% mc	2	Maintained	[16]
	Sealed, 5°C, 11.0% mc	2	Substantial loss	
	Sealed, −18°C, 7.7% mc	2	Maintained	
	Sealed, −18°C, 11.0% mc	2	Maintained	
	Canvas bags, −4°C, 8% mc	3	Complete loss	[15]
	Canvas bags, −11°C, 12% mc	3	Complete loss	
	Canvas bags, −18°C, 8% mc	3	Maintained	

[1]mc = moisture content.

Some nurseries use outdoor soil pits. More recently, stratification in polyethylene bags has become common at many nurseries and seed-processing plants. This method, called "naked stratification," requires no moisture-holding medium and less effort in preparing seed for subsequent sowing [6]. Seed is soaked in water in containers lined with plastic or mesh bags, drained of excess water, and kept at low temperatures for a predetermined period of time; bags are often loosely fastened to allow aeration. All nurseries and seed-processing plants responding to the OSU Survey use some type of naked stratification.

4.6.1 Water soaking

The rate of water absorption varies among species. Most conifers require 1 to 2 days of soaking to achieve full imbibition. It has been suggested that warm water can speed up water absorption by seed, and that running water and aeration can improve oxygen availability; however, this has not yet been substantiated experimentally for Northwest conifers. One study showed that running water was of no benefit to noble fir [unpubl. data, 134]. Twelve nurseries and five seed-processing plants responding to the OSU Survey stratify seed; nine of these soak seed for 24 hours, the other eight for 36 to 48 hours. During soaking, four organizations aerate water, whereas three use running water. These practices are probably beneficial, although the effectiveness should be determined for each species.

4.6.2 Temperature

After draining, seeds are stored in the fully imbibed state. A few species, such as yew (*Taxus* spp.) [61, 92] and yellow-cedar [*Chamaecyparis nootkatensis* (D. Don) Spach] [58], require storage at warm temperatures before cold storage; however, most coniferous species require low temperatures throughout. For loblolly pine (*Pinus taeda* L.), McLemore [89] found optimum stratification temperature to be 10°C, but Robinson et al. [115] reported that, for this same species, gradually increasing temperature over a 4-week period gave the best stratification results. Temperatures above 5°C are not desirable because they increase the risk of overheating and subsequent deterioration, although freezing temperatures also can damage seeds at high moisture content. Consequently, low temperatures of 2 to 5°C have been adopted as accepted operational practice in most cases. All except one organization use stratification temperatures between 1 and 5°C (OSU Nursery Survey); the exception uses 0°C. Even in this case, however, embryos would not experience freezing due to their osmotic potential, which is lower than that of water. Premature germination during prolonged stratification can be minimized if seeds are held at 2°C, rather than 5°C, for both Douglas-fir and ponderosa pine [39].

4.6.3 Aeration

Aeration during stratification is necessary to supply oxygen for seed respiration and to allow carbon dioxide and heat to escape [23]. Lack of aeration could therefore lead to deterioration of seed quality through buildup of toxic substances. The most commonly used technique is to leave a small air space at the neck of each bag in which seed is stratified and to massage the whole bag periodically. A few Northwest nurseries also use fine-meshed bags hung so that air may circulate (OSU Nursery Survey).

4.6.4 Duration

Optimum stratification length varies among species and seedlots [5]. In general, the longer the stratification period, the greater the rate of germination, especially under suboptimal germination temperatures [5, 52, 132]. For this reason, seed destined for colder environments must be stratified long enough for quick and complete germination. However, prolonged stratification can cause seeds of some species to germinate prematurely [5, 98, 121]; furthermore, vigor and total germination may be reduced if seeds are stratified for excessively long periods [5, 98].

In the Northwest, stratification periods vary from 28 to 90 days for noble fir and Douglas-fir and 28 to 45 days for ponderosa and lodgepole pine (OSU Nursery Survey). These variations probably reflect the nursery environment under which seed is to be sown and germinated.

Premature germination sometimes occurs during prolonged stratification. Although premature germination is a serious concern in nurseries because the fragile seeds can be damaged in handling or during mechanical sowing, redrying and storing of stratified seed are possible. Danielson and Tanaka [39] reported that ponderosa pine seed air-dried to 26% and Douglas-fir seed air-dried to 37% can be stored for 9 and 3 months respectively without losing the beneficial effect of stratification or having their viability adversely affected. Subsequently, Edwards [48] tested the efficacy of surface-drying true fir seed after 1 month of stratification at saturation moisture, followed by 3 months of storage at 35% moisture content; this treatment not only prevented premature germination but also improved total germination and germination rate.

The exact mechanism behind the benefit of surface drying is not completely understood. It may be related to improved gaseous exchange brought about by removing the water film from the seed surface, which increases oxygen availability to the seed and facilitates the release of any accumulated toxic gases. Although the surface-drying technique provides the option of storing stratified seed for prolonged periods without losing stratification effects, the lower limit of seed moisture content should be determined for each species. Seeds can be stored safely below certain thresholds but seem to then require restratification after storage [11, 90]. This induction of secondary dormancy suggests that seed moisture and dormancy are closely related.

4.6.5 Other treatments to improve germination

Although stratification is an effective method to break dormancy, it is often time consuming. Past research has shown that hydrogen peroxide [28], gibberellic acid [110], ethylene [24], microwave irradiation [69], or osmotic agents [128] can stimulate germination of conifer seed. However, these studies were usually conducted under optimum germination conditions and may not be effective under the suboptimal temperature conditions frequently encountered in the field in early spring when seed is sown at Northwest bareroot nurseries. Further work is required to develop a quick, effective method that would facilitate germination under a wide range of temperatures and that would either eliminate the need for stratification or shorten the stratification requirement.

4.7 Future Research Needs

Technology of seed procurement and utilization has advanced significantly in the past 20 years. Further refinements are deemed necessary, however, especially because we are now moving into a transition period in which more valuable seed from seed orchards will be preferred to seed from natural stands. Some suggestions for these refinements follow:

- **Cone collection:** Though a great deal is known about seed-maturation characteristics, relatively little is known about the timing of seed dissemination. To maximize the yield of high-quality seed in cone collection, a complete picture of the pattern of seed retrievability—which is influenced by both seed maturation and dissemination—is essential, especially in seed orchards where individual clones can be closely monitored.

- **Processing:** Currently available seed-cleaning procedures remove all of the empty seed and some of the partially developed seed. A method is needed for separating all the nonviable from viable seed, even including seed that looks fully developed but does not germinate. This is particularly important as precision sowing and uniform spacing of seedlings are introduced to maximize utilization of seedling-production areas. Each seed should have the potential of germinating, emerging through the soil surface, and forming a healthy seedling. Some effort is being made toward achieving this goal [129].
- **Testing:** Currently used seed-testing methods for western conifers provide information on germination potential of seed under optimum laboratory environments; however this often correlates poorly with nursery-bed emergence. A procedure should be developed by which germination potential in the nursery bed can be accurately assessed, to improve predictability of crop establishment.
- **Seed storage:** Some true fir species, such as noble fir, produce infrequent cone crops, with large crops occurring at intervals of 3 to 6 years depending on location. There has been some concern that the viability of true fir seed deteriorates during storage within a relatively short time; current storage procedures have shown inconsistent results [3, 14]. Seed condition before storage and storage environment need to be more closely examined. The National Seed Storage Laboratory is investigating the feasibility of using liquid nitrogen to store noble fir seed for periods up to 50 years. Such an approach may be necessary to maintain the viability of a large crop of true fir until the next crop is available.
- **Seed treatment:** Coniferous seeds are generally characterized by deep dormancy requiring prolonged stratification of 60 to 90 days. Unfortunately, this requirement reduces planning and scheduling flexibility of nursery crops. Developing a quick seed treatment that would shorten or eliminate stratification requirements would be most beneficial.

Acknowledgments

I thank the following persons for reviewing the manuscript: C. C. Boyd and G. A. Ritchie, Weyerhaeuser Western Forestry Research Center, Centralia, Washington; T. K. Smith, Weyerhaeuser Lands and Timber, Tacoma, Washington; M. L. Duryea, Department of Forest Science, Oregon State University, Corvallis; D. G. W. Edwards, Pacific Forest Research Centre, Victoria, B. C.; T. Landis, U.S.D.A. Forest Service, Denver, Colorado; and C. L. Leadem, B. C. Forest Service, Victoria. I also thank S. A. Godsey, D. M. Loucks, D. R. Park, V. J. Robinson, and F. M. Tanaka for their assistance in preparing the manuscript.

References

1. Aldous, J. R. 1972. Nursery practice. Her Majesty's Stationery Office, London. Forestry Commun. Bull. 43. 184 p.
2. Allen, G. S. 1956. The effect of date of cone collection upon the viability, germination behavior, and storage characteristics of western hemlock seed. Forestry Chronicle 32:262-263.
3. Allen, G. S. 1957. Storage behavior of conifer seeds in sealed containers held at 0°F., 32°F., and room temperature. J. Forestry 55:278-281.
4. Allen, G. S. 1958. Factors affecting the viability and germination behavior of coniferous seed. Part II. Cone and seed maturity, *Pseudotsuga menziesii* (Mirb.) Franco. Forestry Chronicle 34:275-282.
5. Allen, G. S. 1960. Factors affecting the viability and germination behavior of coniferous seed. Part IV. Stratification period and incubation temperature, *Pseudotsuga menziesii* (Mirb.) Franco. Forestry Chronicle 36:18-29.
6. Allen, G. S., and W. Bientjes. 1954. Studies on coniferous tree seed at the University of British Columbia. Forestry Chronicle 30:183-196.
7. Association of Official Seed Analysts. 1981. Rules for testing seeds. J. Seed Technology 6(2). 162 p.
8. Baldwin, H. I. 1932. Alcohol separation of empty seed and its effects on the germination of red spruce. American J. Botany 19:1-11.
9. Baldwin, H. I. 1942. Forest tree seed of the northern temperate regions with special reference to North America. Chronica Botanica Co., Waltham, Massachusetts. 240 p.
10. Barnett, J. P. 1970. Flotation in ethanol affects storability of spruce pine seeds. Tree Planters' Notes 21(4):18-19.
11. Barnett, J. P. 1972. Drying and storing stratified loblolly pine seeds reinduces dormancy. Tree Planters' Notes 23(3):10-11.
12. Barnett, J. P. 1976. Earlier collection dates for Southern pine cones. Pages 50-56 *in* Proc., Southeastern area nurserymen's conf. Cosponsored by Alabama Forestry Comm., South Carolina State Commission of Forestry, and U.S.D.A. Forest Serv. Southeastern Area, State and Private Forestry.
13. Barnett, J. P., and B. F. McLemore. 1970. Storing southern pine seeds. J. Forestry 68:24-27.
14. Barton, L. V. 1953. Seed storage and viability. Contributions from Boyce Thompson Institute 17:87-103.
15. Barton, L. V. 1954. Effect of subfreezing temperatures on viability of conifer seeds in storage. Contributions from Boyce Thompson Institute 18:21-24.
16. Barton, L. V. 1954. Storage and packeting of seeds of Douglas-fir and western hemlock. Contributions from Boyce Thompson Institute 18:25-37.
17. Barton, L. V. 1961. Seed preservation and longevity. Leonard Hill Books Ltd., London. 216 p.
18. Bates, C. G. 1913. The technique of seed testing. Proc., Society of American Foresters 8:127-138.
19. Belcher, E. W., Jr., and R. P. Karrfalt. 1978. The processing of conifer seed. Pages 9-18 *in* Proc., Small lot forest seed processing workshop. U.S.D.A. Forest Serv. Southeastern Area, State and Private Forestry.
20. Bloomberg, W. J. 1969. Disease of Douglas-fir seeds during cone storage. Forest Sci. 15:176-181.
21. Bonner, F. T. 1974. Seed testing. Pages 136-152 *in* Seeds of woody plants in the United States. U.S. Dep. Agric., Washington, D.C. Agric. Handb. 450.
22. Bonner, F. T., and T. R. Dell. 1976. The Weibull function: a new method of comparing seed vigor. J. Seed Technology 1:96-103.
23. Bonner, F. T., B. F. McLemore, and J. P. Barnett. 1974. Presowing treatment of seed to speed germination. Pages 126-135 *in* Seeds of woody plants in the United States. U.S. Dep. Agric., Washington, D.C. Agric. Handb. 450.
24. Borno, C., and I. E. P. Taylor. 1975. The effect of high concentrations of ethylene on seed germination of Douglas-fir [*Pseudotsuga menziesii* (Mirb.) Franco]. Can. J. Forest Res. 5:419-423.
25. Boyce, J. S. 1948. Forest pathology. 2nd ed. McGraw-Hill Book Co., Inc., New York. 550 p.
26. Buszewicz, G. M., and A. G. Gordon. 1971. *Pinus merkusii*. Rep. on Forest Res. Forestry Comm. (London):27-28.
27. Chappell, T. W. 1968. Harvesting pine cones with mechanical tree shakers. Pages 65, 68 *in* Proc., Forest engineering conf. American Society of Agric. Engineers, St. Joseph, Michigan.
28. Ching, T. M. 1959. Activation of germination in Douglas-fir seed by hydrogen peroxide. Plant Physiology 34:557-563.
29. Ching, T. M. 1960. Seed production from individual cones of grand fir (*Abies grandis* Lindl.). J. Forestry 58:959-961.
30. Ching, T. M., and K. K. Ching. 1962. Physical and physiological changes in maturing Douglas-fir cones and seeds. Forest Sci. 8:21-31.
31. Clark, B. E. 1954. Factors affecting the germination of sweet corn in low-temperature laboratory tests. New York Agric. Exp. Sta. Bull. 769. 24 p.
32. Cobb, H. C. 1959. Seed collection and processing. Pages 40-46 *in* Direct seeding in the South symp. Duke Univ., Durham, North Carolina.

43. Cram, W. H., and H. A. Worden. 1957. Maturity of white spruce cones and seed. Forest Sci. 3:263-269.

44. Crocker, W., and L. V. Barton. 1953. Physiology of seeds. Chronica Botanica Co., Waltham, Massachusetts. 267 p.

45. Curtis, J. D. 1955. Effects of origin and storage method on the germinative capacity of ponderosa pine seed. U.S.D.A. Forest Serv., Intermountain Forest and Range Exp. Sta., Ogden, Utah. Res. Note INT-26. 5 p.

46. Czabator, F. J. 1962. Germination value: an index combining speed and completeness of pine seed germination. Forest Sci. 8:386-396.

47. Danielson, H. R. 1972. Quick tests for determining viability of Douglas-fir seed. Pages 104-110 *in* Proc., Joint meeting, Western Forestry Nursery Council and Intermountain Forest Nurserymen's Assoc., Olympia, Washington, Aug. 8-10.

48. Danielson, H. R., and D. F. Grabe. 1973. Storage of noble fir seeds. Proc., Assoc. of Official Seed Analysts 63:161-165.

49. Danielson, H. R., and Y. Tanaka. 1978. Drying and storing stratified ponderosa pine and Douglas-fir seeds. Forest Sci. 24:11-16.

50. Djavanshir, K., and H. Pourbeik. 1976. Germination value—a new formula. Silvae Genetica 25:79-83.

51. Dobbs, R. C., D. G. W. Edwards, J. Konishi, and D. Wallinger. 1976. Guideline to collecting cones of B. C. conifers. B. C. Ministry of Forests/Can. Forestry Serv., Victoria. Joint Rep. 3. 99 p.

52. Dobbs, R. C., C. R. Silversides, and J. Walters. 1977. Development and evaluation of an aerial cone rake. Forest Management Institute, Ottawa. Inf. Rep. FMR-X-100. 19 p.

53. Douglas, B. S. 1969. Collecting forest seed cones in the Pacific Northwest. U.S.D.A. Forest Serv., Pacific NW Region, Portland, Oregon. 15 p.

54. Edwards, D. G. W. 1973. Polaroid film for rapid seed radiography. Proc., IUFRO International symp. on seed processing. Bergen, Norway. Vol. 1, Pap. 6. 8 p.

55. Edwards, D. G. W. 1979. An improved air seed-sorter for laboratory use. Can. Forestry Serv., Pacific Forest Res. Centre, Victoria. Rep. BC-X-188. 11 p.

56. Edwards, D. G. W. 1980. Maturity and quality of tree seeds—a state-of-the-art review. Seed Sci. and Technology 8:625-657.

57. Edwards, D. G. W. 1981. Cone collection and processing. Effects on seed quality and yield. Pages 12-37 *in* Proc., High-quality collection and production of conifer seed. Northern Forest Res. Centre, Edmonton. Inf. Rep. NOR-X-235.

58. Edwards, D. G. W. 1982. Storage of prechilled *Abies* seeds. Pages 195-203 *in* Proc., IUFRO International symp. on forest tree seed storage. Petawawa National Forestry Institute, Chalk River, Ontario.

59. Edwards, D. G. W. 1982. Personal communication, Pacific Forest Res. Centre, Victoria, B.C.

30. Eliason, E. J., and C. E. Heit. 1940. The results of laboratory tests as applied to large scale extraction of red pine seed. J. Forestry 38:426-429.

51. Fowells, H. A., and G. H. Schubert. 1956. Seed crops of forest trees in the pine region of California. U.S. Dep. Agric., Washington, D.C. Tech. Bull. 1150. 48 p.

52. Fowler, D. P. 1959. Rapid germination of white pine seed. Forestry Chronicle 35:203-211.

53. Franklin, J. F. 1965. An exploratory study of cone maturity in noble fir. U.S.D.A. Forest Serv., Pacific NW Forest and Range Exp Sta., Portland, Oregon. Res. Note PNW-21. 12 p.

54. Graci, A. 1973. New techniques and progress in processes of extraction of forest seeds. Proc., IUFRO International symp. on seed processing. Bergen, Norway. Vol. 1, Pap. 7. 14 p.

55. Harrington, J. F. 1972. Seed storage and longevity. Pages 145-245 *in* Seed biology (T. T. Kozlowski, ed.). Vol. III. Academic Press, New York and London.

56. Harris, A. S. 1969. Ripening and dispersal of a bumper western hemlock-Sitka spruce seed crop in southern Alaska. U.S.D.A. Forest Serv., Pacific NW Forest and Range Exp. Sta., Portland, Oregon. Res. Note PNW-105. 11 p.

57. Harris, A. S. 1971. Alaska-cedar. U.S.D.A. Forest Serv., Washington, D.C. American Woods FS-224. 7 p.

58. Harris, A. S. 1974. *Chamaecyparis* Spach. Pages 316-320 *in* Seeds of woody plants in the United States. U.S. Dep. Agric., Washington, D.C. Agric. Handb. 450.

59. Hedlin, A. F. 1974. Cone and seed insects of British Columbia. Can. Forestry Serv., Pacific Forest Res. Centre, Victoria. Inf. Rep. BC-X-90. 63 p.

60. Heit, C. E. 1967. Propagation from seed. Part 10. Storage methods for conifer seeds. American Nurseryman 126:14-15, 38-54.

61. Heit, C. E. 1969. Propagation from seed. Part 18. Testing and growing seed of popular *Taxus* forms. American Nurseryman 129:10-11, 118-128.

62. Hepting, G. H. 1971. Diseases of forest and shade trees of the United States. U.S. Dep. Agric., Washington, D.C. Agric. Handb. 386. 658 p.

63. Holmes, G. D., and G. Buszewicz. 1958. The storage of seed of temperate forest tree species. Forestry Abstracts 19:313-322, 455-476.

64. Huss, E. 1954. Studies on the importance of water content for the storage of conifer seed. Meddelanden fran Statens Skogsforskningsinstitut (Sweden) 44(7). 60 p. (Forestry Abstracts 16:1633.)

65. Huss, E. 1956. On the quality of conifer seed and other factors affecting the plant %. Meddelanden fran Skogsforskningsinstitut (Sweden) 46(9). 59 p. (Forestry Abstracts 18:2754.)

66. International Seed Testing Association. 1976. International rules for seed testing. Seed Sci. and Technology 4:4-180.

67. Isaac, L. A. 1934. Cold storage prolongs the life of noble fir seed and apparently increases germinative power. Ecology 15:216-217.

68. Isely, D. 1950. The cold test for corn. Proc., International Seed Testing Assoc. 16:299-311.

69. Jolly, J. A., and R. L. Tate. 1971. Douglas-fir tree seed germination enhancement using microwave energy. J. Microwave Power 6:125-130.

70. Jones, L. 1966. Storing pine seed: what are best moisture and temperature conditions? Georgia Forest Res. Council, Macon. Georgia Forest Res. Pap. 42. 8 p.

71. Justice, O. L. 1972. Essentials of seed testing. Pages 301-370 *in* Seed biology (T. T. Kozlowski, ed.). Vol. III. Academic Press, New York and London.

72. Kamra, S. K. 1967. Studies on storage of mechanically damaged seed of Scots pine (*Pinus sylvestris* L.). Studia Forestalia Suecica 42. 18 p.

73. Kamra, S. K. 1972. Comparative studies on germinability of *Pinus silvestris* and *Picea abies* seed by the indigo carmine and X-ray contrast methods. Studia Forestalia Suecica 99. 21 p.

74. Keen, F. P. 1958. Cone and seed insects of western forest trees. U.S. Dep. Agric., Washington, D.C. Tech. Bull. 1169. 168 p.

75. Kmecza, N. S. 1970. Using tree shakers for pine cone collection in Region 8. Tree Planters' Notes 21(1):9-11.

76. Krugman, S. L. 1966. Artificial ripening of sugar pine seeds. U.S.D.A. Forest Serv., Pacific SW Forest and Range Exp. Sta., Berkeley, California. Res. Pap. PSW-32. 7 p.

77. Krugman, S. L., and J. L. Jenkinson. 1974. *Pinus* L. Pages 598-638 *in* Seeds of woody plants in the United States. U.S. Dep. Agric., Washington, D.C. Agric. Handb. 450.

78. Kummel, J. F., C. A. Rindt, and T. T. Munger. 1944. Forest planting in the Douglas-fir region. U.S.D.A. Forest Serv., Portland, Oregon. 154 p.

79. Lavender, D. P. 1958. Viability of Douglas-fir seed after storage in the cones. Oregon Forest Lands Res. Center, Corvallis. Res. Note 31. 8 p.

80. Lavender, D. P., and W. H. Engstrom. 1956. Viability of seeds from squirrel cut Douglas-fir cones. Oregon State Board of Forestry, Salem. Res. Note 27. 19 p.

81. Leadem, C. L. 1980. Seed viability of *Abies*, *Picea*, and *Tsuga* after storage in the cones. Pages 57-67 *in* Proc., IUFRO International symp. on forest tree seed storage. Petawawa National Forestry Institute, Chalk River, Ontario.

82. Leadem, C. L. 1981. Quick methods for determining seed quality in tree seeds. Pages 64-72 *in* Proc., High-quality collection and production of conifer seed. Northern Forest Res. Centre, Edmonton. Inf. Rep. NOR-X-235.

83. Mackay, D. B. 1972. The measurement of viability. Pages 172-208 *in* Viability of seeds (E. H. Roberts, ed.). Syracuse Univ. Press, Syracuse, New York.

84. Magini, E. 1962. Forest seed handling, equipment and procedures. II. Seed treatments, storage, testing, and transport. Unasylva 16:20-35.

85. Maki, T. E. 1940. Significance and applicability of seed maturity indices for ponderosa pine. J. Forestry 38:55-60.
86. Mátyás, V. C. 1972. Möglichkeiten der Frühernte in Weisskiefer-Samenplantagen. (The possibility of early harvesting in Scots pine seed orchards.) Silvae Genetica 21:191-193.
87. McDonald, M. B., Jr. 1977. The influence of seed moisture on the accelerated aging seed vigor test. J. Seed Technology 2:18-28.
88. McLemore, B. F. 1965. Pentane floatation for separating full and empty longleaf pine seeds. Forest Sci. 11:242-243.
89. McLemore, B. F. 1966. Temperature effects on dormancy and germination of loblolly pine seed. Forest Sci. 12:284-289.
90. McLemore, B. F., and J. P. Barnett. 1968. Moisture content influences dormancy of stored loblolly pine seed. Forest Sci. 14:219-221.
91. Mirov, N. T. 1946. Viability of pine seed after prolonged cold storage. J. Forestry 44:193-195.
92. Mitiska, L. J. 1954. The propagation of *Taxus* by seeds. Proc., International Plant Propagators' Society 4:69-75.
93. Moore, R. P. 1969. History supporting tetrazolium seed testing. Proc., International Seed Testing Assoc. 34:242.
94. Moore, R. P. 1971. Tetrazolium evaluation of tree and shrub seeds. 16th International Seed Testing Assoc. Congress, Washington, D.C. Preprint 69. 7 p.
95. Muller, C. 1971. La postmaturation des graines d' *Abies nordmanniana*. (After-ripening of seeds of *Abies normanniana*.) Revue Forestiere Francaise 23:436-439.
96. Oliver, W. W. 1974. Seed maturity in white fir and red fir. U.S.D.A. Forest Serv., Pacific SW Forest and Range Exp. Sta., Berkeley, California. Res. Pap. PSW-99. 12 p.
97. Olson, D. L., and R. R. Silen. 1975. Influence of date of cone collection on Douglas-fir seed processing and germination: a case history. U.S.D.A. Forest Serv., Pacific NW Forest and Range Exp. Sta., Portland, Oregon. Res. Pap. PNW-190. 10 p.
98. Olson, J. S., F. W. Stearns, and H. Nienstaedt. 1959. Eastern hemlock seeds and seedlings: response to photoperiod and temperature. Connecticut Agric. Exp. Sta., New Haven. Bull. 620. 70 p.
99. Owen, E. B. 1956. The storage of seeds for maintenance of viability. Commonwealth Agric. Bureaux, Farnham Royal, Bucks, England. Commonwealth Bur. of Pastures and Field Crops Bull. 43. 81 p.
100. Owens, J. N. 1975. Guide for the collection of yellow cedar cones. B. C. Forest Serv., Res. Div., Victoria. 7 p.
101. Owston, P. W., and W. I. Stein. 1974. *Pseudotsuga* Carr. Pages 674-683 *in* Seeds of woody plants in the United States. U.S. Dep. Agric., Washington, D.C. Agric. Handb. 450.
102. Pfister, R. D. 1966. Artificial ripening of grand fir cones. Northwest Sci. 40:103-112.
103. Pfister, R. D. 1967. Maturity indices for grand fir cones. U.S.D.A. Forest Serv., Intermountain Forest and Range Exp. Sta., Ogden, Utah. Res. Note INT-58. 7 p.
104. Pogoda, G. 1962. The influence of harvesting time on the germinability of Douglas-fir seed. Allgemeine Forstzeitschrift 17:283. (Forestry Absracts 23:5061.)
105. Pollock, B. M., and H. O. Olney. 1959. Studies of the rest period. I. Growth, translocation, and respiratory changes in the embryonic organs of the after-ripening cherry seed. Plant Physiology 34:131-142.
106. Rediske, J. H. 1961. Maturation of Douglas-fir seed: biochemical study. Forest Sci. 7:204-213.
107. Rediske, J. H. 1969. Effects of cone picking date on Douglas-fir seed quality. Forest Sci. 15:404-410.
108. Rediske, J. H., and D. C. Nicholson. 1965. Maturation of noble fir seed—a biochemical study. Weyerhaeuser Co., Centralia, Washington. Weyerhaeuser Forestry Pap. 2. 15 p.
109. Rediske, J. H., and K. R. Shea. 1965. Loss of Douglas-fir seed viability during cone storage. Forest Sci. 11:463-472.
110. Richardson, S. D. 1959. Germination of Douglas-fir seed as affected by light, temperature, and gibberellic acid. Forest Sci. 5:174-181.
111. Rietveld, W. J. 1978. Forecasting seed crops and determining cone ripeness in southwestern ponderosa pine. U.S.D.A. Forest Serv., Rocky Mountain Forest and Range Exp. Sta., Fort Collins, Colorado. Gen. Tech. Rep. RM-50. 12 p.
112. Rietz, R. C. 1941. Kiln design and development of schedules for extracting seed from cones. U.S. Dep. Agric., Washington, D.C. Tech. Bull. 773. 70 p.
113. Rietz, R. C., and O. W. Torgeson. 1937. Kiln temperatures for northern white pine cones. J. Forestry 35:836-839.
114. Rink, G., T. R. Dell, G. Switzer, and F. T. Bonner. 1979. Use of the Weibull function to quantify sweetgum germination data. Silvae Genetica 28:9-12.
115. Robinson, D. W., D. L. Weeks, and C. E. Posey. 1972. The effect of stratification temperature on the germination of two southern pines. Oklahoma State Univ., Stillwater. Agric. Bull. B-698. 15 p.
116. Roe, E. I. 1960. August-collected cones yield poor red pine seed. U.S.D.A. Forest Serv., Lake States Forest Exp. Sta., St. Paul, Minnesota. Tech. Note 575. 2 p.
117. Ross, S. D. 1969. Gross metabolic activity accompanying the after-ripening of dormant Douglas-fir seeds. Botanical Gazette 130:271-275.
118. Sahlen, K., and U. Bergsten. 1982. Temperature and moisture content changes in *Pinus silvestris* seed stored in a container during freezing and warming. Pages 17-27 *in* Proc., IUFRO International symp. on forest tree seed storage. Petawawa National Forestry Institute, Chalk River, Ontario.
119. Schönborn. A. von. 1964. The storage of forest tree seed. BLV Verlagsgesellschaft, Munich. 158 p. (Forestry Abstracts 25:4957.)
120. Schubert, G. H. 1954. Viability of various coniferous seeds after cold storage. J. Forestry 52:446-447.
121. Schubert, G. H. 1955. Effect of storage temperature on viability of sugar, Jeffrey and ponderosa pine seed. U.S.D.A. Forest Serv., California Forest and Range Exp. Sta., Berkeley. Forest Res. Notes 100. 3 p.
122. Schubert, G. H., and R. S. Adams. 1971. Reforestation practices for conifers in California. Dep. of Conservation, Div. of Forestry, State of California, Sacramento. 359 p.
123. Seal, D. T., J. D. Matthews, and R. T. Wheeler. 1965. Collection of cones from standing trees. Forestry Commission, London. Forest Record 39. 48 p.
124. Shearer, R. C. 1977. Maturation of western larch cones and seeds. U.S.D.A. Forest Serv., Intermountain Forest and Range Exp. Sta., Ogden, Utah. Res. Pap. INT-189. 15 p.
125. Silen, R. R. 1958. Artificial ripening of Douglas-fir cones. J. Forestry 56:410-413.
126. Silen, R. R. 1964. A laboratory seed separator. Forest Sci. 10:222-223.
127. Simak, M. 1957. The X-ray contrast method for seed testing Scots pine (*Pinus silvestris*). Meddelanden fran Statens Skogsforskningsinstitut (Sweden) 47. 22 p.
128. Simak, M. 1976. Germination improvement of Scots pine seeds from circumpolar regions using polyethylene glycol. Pages 145-153 *in* Proc., IUFRO 2nd International symp. on physiology of seed germination. Fuji, Japan.
129. Simak, M. 1981. Bortsortering av matat-dott fro ur ett fröparti. (Removal of filled-dead seeds from a seedlot.) Sveriges Skogsvardsfoerbunds Tidskrift 5:31-36.
130. Stein, W. I., (ed.). 1966. Sampling and service testing western conifer seeds. Western Forest Tree Seed Council, Western Forestry and Conservation Assoc., Portland. 36 p.
131. Stein, W. I., P. E. Slabaugh, and A. P. Plummer. 1974. Harvesting, processing, and storage of fruits and seeds. Pages 98-125 *in* Seeds of woody plants in the United States. U.S. Dep. Agric., Washington, D.C. Agric. Handb. 450.
132. Tanaka, Y. 1976. Stratification and other pre-treatments of Douglas-fir seed for nursery bed germination. Pages 163-173 *in* Proc., IUFRO 2nd International symp. on physiology of seed germination. Fuji, Japan.
133. Tanaka, Y. 1982. Effect of freezing temperatures on noble fir seed during cone storage. Pages 68-74 in Proc., IUFRO International symp. on forest tree seed storage. Petawawa National Forestry Institute, Chalk River, Ontario.
134. Tanaka, Y. 1982. Unpublished data, Western Forestry Res. Center, Weyerhaeuser Co., Centralia, Washington.
135. Tanaka, Y., and P. C. Cameron. 1979. Maturation of ponderosa pine seeds in southern Oregon. Pages 218-225 *in* Proc., U.S.D.A. Forest Serv./IUFRO/Mississippi State Univ. International symp. on flowering and seed development in trees. U.S.D.A. Forest Serv., Southern Forest Exp. Sta., Starkville, Mississippi.
136. Tekrony, D. M., and D. B. Egli. 1977. Relationship between laboratory indices of soybean seed vigor and field emergence. Crop Sci. 17:573-577.

137. Tietz, J. G. 1971. Evaluation of cone collection equipment. U.S.D.A. Forest Serv., Equipment Development Center, Missoula, Montana. Project Record ED&T 1553. 33 p.
138. U.S. Department of Agriculture. 1974. Seeds of woody plants in the United States. U.S. Government Printing Office, Washington, D.C. Agric. Handb. 450. 883 p.
139. Van Deusen, J. L., and L. D. Beagle. 1973. Judging ripeness of seeds in Black Hills ponderosa pine cones. U.S.D.A. Forest Serv., Rocky Mountain Forest and Range Exp. Sta., Fort Collins, Colorado. Res. Note RM-235. 4 p.
140. Van Haverbeke, D. F. 1976. Soaking and retumbling controlled-pollinated Scots pine cones increases seed yields. Tree Planters' Notes 27(4):8-9, 33.
141. Villiers, T. A. 1961. Dormancy in tree seeds: a brief review of recent work. Proc., International Seed Testing Assoc. (Wageningen) 26:515-536.
142. Villiers, T. A., and P. F. Wareing. 1960. Interaction of growth inhibitor and a natural germination stimulator in the dormancy of *Fraxinus excelsior* L. Nature 185:112-114.
143. Wakeley, P. C. 1954. Planting the southern pines. U.S. Dep. Agric., Washington, D.C. Agric. Monograph 18. 233 p.
144. Wang, B. S. P. 1973. Collecting, processing and storing tree seed for research use. Proc., IUFRO International symp. on seed processing. Bergen, Norway. Vol. 1, Pap. 17. 12 p.
145. Wang, B. S. P. 1974. Tree-seed storage. Can. Forestry Serv., Dep. Environ., Ottawa. Publ. 1335. 32 p.
146. Wareing, P. F. 1961. Dormancy of woody plants. Pages 1216-1219 *in* Recent advances in botany. Vol. II. IX International Botanical Congress, 1959.
147. Wareing, P. F. 1965. Endogenous inhibitors in seed germination and dormancy. Encyclopedia of Plant Physiology 15:909-924.
148. Wilson, B. C. 1968. A cutter for sampling cone seed quality. Tree Planters' Notes 19(2):8-9.
149. Winjum, J. K., and N. E. Johnson. 1960. A modified-knife cone cutter for Douglas-fir seed studies. J. Forestry 58:487-488.
150. Winston, D. A., and B. D. Haddon. 1981. Effects of early cone collection and artificial ripening on white spruce and red pine germination. Can. J. Forest Res. 11:817-826.
151. Woollard, R. F., and R. R. Silen. 1973. All-pneumatic laboratory seed cleaner successful. Tree Planters' Notes 24(4):15-17.
152. Zasada, J. C. 1973. Effect of cone storage method and collection date on Alaskan white spruce (*Picea glauca*) seed quality. Proc., IUFRO International symp. on seed processing. Bergen, Norway. Vol. 1, Pap. 19. 10 p.
153. Zeleny, L. 1954. Chemical, physical and nutritive changes during storage. Pages 46-76 *in* Storage of cereal grains and their products (J. A. Anderson and A. W. Alcock, eds.). American Assoc. Cereal Chem., St. Paul, Minnesota.

Chapter 5
Establishing a Vigorous Nursery Crop: Bed Preparation, Seed Sowing, and Early Seedling Growth

B. E. Thompson

Abstract

Many aspects of preparing a production nursery bed, such as correcting drainage problems, eliminating disease potential, and maintaining fertility and pH, are specific to site history and soil properties. Efficiency of field use is a major concern in designing a nursery and should be considered when aligning beds and installing irrigation systems. Sowing should be done in spring, commonly in April or May, after soil temperatures at the 10-cm (4-in.) depth reach 10°C. Though seeders commonly used tend to produce clumpy distributions, adverse effects on seedling quality and quantity are reduced at low densities. Both seedling quality and cost are affected by seedbed density; therefore, great care must be exercised in prescribing densities. Sowing formulas must consider the desired seedling density as well as expected yields and various aspects of seed quality and quantity. Expected tree, yield, and damage percents, derived from experience, should be reevaluated annually. Proper care and tending after sowing are critical for obtaining high tree percents. Diseases, birds, and weather are the most common causes of loss, and preventive measures should be taken whenever possible.

5.1 Introduction

As with most endeavors, getting off on the right foot is important in growing a quality nursery crop. Care taken in ground preparation, sowing, and early seedbed monitoring will result in better drainage and fewer disease problems as well as more and better seedlings at harvest. Early and thorough planning will provide the flexibility needed for nursery personnel to cope with changing conditions yet maintain quality.

This chapter presents procedures for bed preparation, seed sowing, and early seedling maintenance which are based on current practice in the Northwest and on the available literature. Universal practices are simply mentioned. Those that are controversial or that vary greatly from nursery to nursery are examined more closely. Alternatives are discussed and, if warranted, recommendations made. Theory and practice in many important areas such as fertilization, weed control, and seed properties as they relate to sowing and early growth are not covered extensively because these topics are fully addressed by others in chapters 7, 18, and 4, respectively, of this volume. Further information on many subjects addressed here also is discussed by Armson and Sadreika [2] and Aldhous [1].

Each step in establishing a vigorous crop has a number of alternative approaches. When possible, several alternatives and their pros and cons are described. The lists, however, are not exhaustive and other possibilities exist. A good rule of thumb when weighing alternatives is to ask questions such as:

- How much will a given treatment cost?
- What is the expected outcome of an alternative treatment?
- What extra cost or loss in seedling quality can be expected from **not** following a prescribed treatment or preferred alternative?
- Is this cost or loss acceptable?

Alternatives and combinations of alternatives are many. Each nursery manager must ask the above questions and reach decisions based on his or her own situation. Adherence to sound principles and proper timing will result in rapid establishment and growth of seedlings.

5.2 Production-Area Development

Past land use and present condition are among the most important factors to consider when establishing a new nursery and preparing it for sowing.

5.2.1 Previous land use

First, ascertain what was growing on the site during the last 5 years. If the land was in agricultural use, determine whether any of the previous crops could have become infested with diseases or insects that also attack crop-tree species or whether weeds present are difficult to control. If diseases or pests are suspected, identify the problem by soil assay or insect trap-

In Duryea, Mary L., and Thomas D. Landis (eds.). 1984. *Forest Nursery Manual: Production of Bareroot Seedlings.* Martinus Nijhoff/Dr W. Junk Publishers, The Hague/Boston/Lancaster, for Forest Research Laboratory, Oregon State University, Corvallis. 386 p.

ping and take appropriate management measures before a nursery crop is sown. If previous crops are unknown or assays impractical, presowing soil fumigation is good insurance because it will remove most pathogenic fungi and weeds. It may, however, negatively affect beneficial soil organisms such as mycorrhizae and bacteria [6], and it is very expensive ($1,000 to $1,500/acre) (see chapters 19 and 20, this volume). Once pests have been identified, specific treatments can be applied that cost less and maintain beneficial soil microorganisms.

Hard-coated weeds such as clover and vetch are not killed by fumigation. Summer fallow with tillage and irrigation will remove these weeds best. When seeds sprout, allow them to grow rapidly until the first true leaves appear; then till in. After each tilling, a new crop will appear, so the sequence must be repeated. Irrigation helps ensure rapid growth during July and August so that as many "crops" of weeds as possible can be grown and killed during a single summer. If time does not permit a summer fallow, an extensive herbicide schedule, hand weeding, or both will control the problem, but some seedlings will be damaged.

Former forest sites pose some additional problems. Clearing, removing debris, and leveling can be expensive and time consuming. Tree roots and, in many cases, rocks complicate the task: seeds sown on a rock or root have little chance of developing into acceptable seedlings. However, seedling diseases usually are less prevalent, weed problems generally are fewer, and beneficial mycorrhizae-forming fungi often are endemic.

5.2.2 Leveling land and orienting beds

Once cleared, a field must be leveled and sloped. If it is already reasonably level, with a slope of 2 to 3% [10], adequate finish leveling may be accomplished with a land plane. However, where drainage is restricted or frost pockets are formed by natural contours, more severe measures are needed (see chapters 6 and 13, this volume).

Depending on depth of the cuts and fills, removing topsoil and storing it for return after leveling may be the first step. As with all major projects requiring heavy equipment, land leveling should be done when the soil is driest to minimize the chance of massive compaction [23]. If possible and practical, fields should be leveled and sloped so that the least amount of soil is moved. Careful attention to this aspect will save money and minimize compaction. Desired slope also depends on soil characteristics. Sandy soils require less slope than heavier soils. Increased slope favors bed erosion in sandy soil but aids surface drainage in finer textured soil [14]. Fields can be tiled for supplemental drainage before final leveling and sloping.

Bed orientation, important to ease of operation and seedling growth, should be considered **before** final slope and road positions are planned. Where no additional sloping is necessary on a newly acquired field, orient beds to run perpendicular to the land contour for maximum drainage. Where contours must be drastically changed, consider the possible advantages and disadvantages of east-west or north-south orientation. In an area where lifting is restricted due to frozen ground, orienting beds in a north-south direction will facilitate early thawing by the morning sun [1]—and thereby lifting. East-west orientation increases the possibility of sunscald in the summer and will cause growth differentials across the bed due to shading [unpubl. data, 22]; where frozen soil is not a problem, east-west orientation can produce acceptable results.

5.2.3 Laying out irrigation and road systems

Once bed orientation has been determined, irrigation and road systems must be designed and installed. Bed lengths in Northwest nurseries range from 76.25 to 152.5 m (250 to 500 ft) (OSU Nursery Survey; see chapter 1, this volume) under normal circumstances because a pressure drop can occur in longer irrigation lines. At the Forest Service Nursery in Medford, Oregon, beds are 244 m (800 ft) long, but the possible pressure drop is compensated for by starting the lateral pipes at the top of the slope, adding the downhill rush of the water to the line pressure.

Nozzle size, sprinkler pressure, and spacing of lateral lines and of sprinklers on laterals all are important to uniform delivery and application of water. For uniform water application in winds up to 5 mph, sprinklers should be placed so that the longest of the two distances in rectangular spacing (the distance between sprinklers either on adjacent laterals or on the same lateral) is 60% of the no-wind diameter (see manufacturer's specification) of the sprinkler used, and so that the sum of the two spacings is not more than 105% of that diameter [16, 18]. For example (Fig. 1), if sprinklers have a no-wind diameter (C) of 20.4 m (67 ft), the largest spacing in one direction could be 60% of 20.4 m (67 ft), or 12.2 m (40.2 ft). Therefore, the distance between sprinklers on the adjacent laterals (B) is 12.2 m. Then, solving the formula for A:

$$105\% \, C = A + B$$
$$1.05(67) = A + 40$$
$$A = \sim 9 \text{ m} \; (\sim 30 \text{ ft})$$

Sprinkler-pattern diameter should be increased or spacing along the lateral decreased where higher wind speeds are expected. No additional uniformity is achieved with triangular, rather than rectangular, spacing [18]. Water-droplet size, a result of nozzle size and pressure, can greatly influence crusting of the soil surface. Therefore, during germination, a smaller nozzle size may be desirable to reduce crusting [16]. (For more information on irrigation and nursery-site layout, see chapters 11 and 2 respectively.)

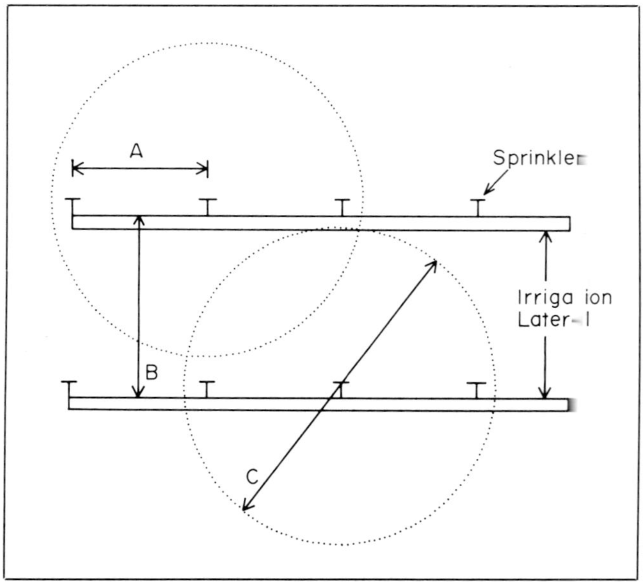

Figure 1. Correct sprinkler distance for uniform water distribution in a typical nursery section. A is the distance between sprinklers along the same irrigation line, B the distance between adjacent irrigation lines, and C the sprinkler-pattern diameter. If C = 67 ft, then A or B, whichever is larger, should not exceed 0.6 × 67, or 40 ft, and A + B should not exceed 1.05 × 67, or 70 ft.

5.2.4 Field efficiency

Field efficiency (the amount of area growing trees divided by the amount of area cultivated in a given field) is primarily determined by irrigation-system design. Distance between irrigation laterals is a major component affecting efficient land use within the production field. Once a fixed irrigation system is installed, it is difficult, if not impossible, to improve field efficiency because the area saved in narrower paths will result in wider unused strips along irrigation lines. In the Northwest, various combinations of bed and tractor-path widths, numbers of seedling rows per bed, and distances between irrigation lines are used (Table 1), resulting in field efficiencies ranging from 55 to 71%; ⅔ of the nurseries report efficiencies of 58 to 63%.

Table 1. Factors affecting field efficiency (OSU Nursery Survey).

	Highest	Lowest	Most common
Bed width, m (in.)	1.27 (50)	1.07 (42)	1.22 (48)
Path width, m (in.)	0.76 (30)	0.53 (21)	0.61 (24)
Rows of seedlings/bed[1]	8	7	8
Beds between irrigation lines	9	5	6
Distance between irrigation lines, m (ft)	15.9 (52)	9.15 (30)	12.2 (40)

[1] For 2+0 beds.

Width of tractor paths also is important for calculating efficiency. As path width increases from a theoretical minimum of one tractor tire width (38 cm, or 15 in.) to the practical maximum of two tractor-tire widths (76 cm, or 30 in.), field efficiency drops, and the required distance between irrigation lines increases. Because tractor operations can be difficult when paths are too narrow, most nurseries compromise with path widths of 61 cm (24 in.). A typical nursery section (that distance between two irrigation laterals) (Fig. 2) would include six beds, each 122 cm (48 in.) wide, with 61-cm (24-in.) tractor paths and irrigation lines 12.2 m (40 ft) apart (OSU Nursery Survey). Such an arrangement would result in 60% field efficiency.

5.3 Field Preparation

5.3.1 Preparing the soil

After the field has been sloped and leveled and once road and irrigation systems have been installed, there is little difference in the steps necessary to cultivate and prepare beds in a new nursery field or one used many times before. In both situations, examine the physical and chemical properties of the soil.

Compaction is the most commonly cited soil physical problem (see chapters 6 and 29, this volume). If compaction is known or suspected, deep subsoiling during the late fall is recommended to improve drainage. A green manure crop that has a fibrous root system can improve soil physical properties and may increase organic matter (see chapters 9 and 10, this volume).

Fertility and pH are important soil chemical properties (see chapters 7 and 8, this volume). Soil tests for phosphorus, potassium, calcium, and magnesium provide a basic inventory of mineral nutrients. If these elements are significantly deficient, the appropriate fertilizer should be added before bed forming. For optimum conifer growth, pH should be kept between 5 and 6. The pH is best adjusted during a fall fallow period so that the reaction of amendments with the soil will have been effected before sowing. Sulfur is used to lower soil pH and lime to raise it.

Soil is most often fumigated to remove weeds or pathogens in fall (OSU Nursery Survey). Methyl bromide/chloropicrin (67%/33%) is preferred in the U.S. because of its proven effectiveness. Law prohibits its use in Canada (see chapter 19, this volume). Note that methyl bromide reacts with sulfur; therefore, if sulfur is being used to reduce pH, it should be applied after fumigation [pers. commun., 24].

Figure 2. Typical nursery beds within sections. Note raised beds and tractor-path width.

Following fumigation, the field is left fallow over winter. In spring, when the soil is dry enough to work, cultivation for sowing should begin. What implements are used and in what order depends mainly on soil texture. In any event, the objective is to prepare the field so that a bed former can be used to produce level, even seedbeds. Care should be taken not to mix unfumigated soil from below or from edge areas into the fumigated field. Nurseries with sandier soil usually require minimal presowing soil preparation. Before final presowing cultivation, fertilizer can be spread and incorporated into the soil (see chapter 7, this volume). Final preparation should leave the soil fluffed and mixed, ready for bed shaping.

5.3.2 Marking and forming seedbeds

Some nurseries, especially those with inexperienced tractor drivers, have difficulty getting beds formed straight and at the proper intervals between irrigation lines. Most nurseries have adopted some homemade equipment to remedy this situation. Many possibilities exist. With an experienced tractor driver and relatively short beds, "eyeballing" the beds works surprisingly well. Some nurseries run string lines to get straight beds; others rely on tractor-mounted bed markers.

A method that uses a bed marker and the irrigation line as a reference point can be adapted wherever a fixed irrigation system exists. First, irrigation lines are positioned in the field. Their proper placement is important and should be done with surveying equipment so that the distance between any two lines is constant. To form the first two beds along the outside of the nursery section, next to the laterals, a chain is suspended from a bar attached to the front of the tractor at the proper distance from the outside wheel; while driving, the tractor operator keeps the chain over the irrigation line and thereby creates a straight bed. A bed marker, which need consist only of a wheel on a pipe, is mounted on the side of the tractor opposite to that with the irrigation tracing chain; it marks where the tractor wheel should run for the next bed, usually about 1.8 m (6 ft) away, depending on the path width desired. Many other systems that accomplish the same results exist. An accurate but expensive method employs a tractor equipped with a laser-tracking device and a laser-emitting target; however, this kind of precision is not necessary for growing quality seedlings.

Two general types of bed formers are commonly in use in the Northwest. One type simply moves soil from the tractor path onto the raised bed and levels off the bed surface. This type, often homemade, can be mounted behind or under the tractor; Whitfield Manufacturing Co. (Mableton, Georgia) makes one commercially that works on this principle. A second type combines final tilling and bed forming by attaching a bed shaper to the back of a 1.8-m (6-ft) rototiller or roterra. This second type requires less field preparation before bed shaping; but the rototiller may produce a soil layer that is compacted, restricting drainage, and should therefore be avoided on heavier soils (see chapter 6, this volume).

Beds, however shaped and formed, are generally raised between 7.5 to 15 cm (3 to 6 in.) above the tractor paths to increase drainage and promote warming of the seedbed (OSU Nursery Survey). Both an increased germination rate and more rapid root growth can be expected. Beware of deep tractor paths, however—they can cause problems during the second growing season. Implements such as individual-bed fertilizer spreaders and even tractor bellies can scrape the tops of tall 2+0 seedlings. The resultant damage can reduce the number of shippable seedlings (seedlings that pass all standards for both size and form) and may promote disease development.

5.4 Sowing

Once the previous sequence of steps has been followed, the field is ready for sowing. Successful sowing depends on type of seeder, sowing date and depth, bed density, and sowing formula.

5.4.1 Seeders

In Northwest nurseries, the two most common seeders are the Øyjörd seeder (manufactured by Love Co., Garfield, Washington) and the Wind River seed drill (OSU Nursery Survey). According to the Survey, neither is wholly satisfactory—the largest complaint is clumpy seed distribution, a problem that no commercially available seeder has as yet solved. A comparison of ease of operation of these seeders and the Stan Hay found the Øyjörd to be slightly superior (Table 2) [13].

Table 2. Summary of operational seeder characteristics [adapted from 13].

	Øyjörd	Stan Hay	Wind River
Ease of adjustment	Excellent	Poor	Fair
Range of adjustment	Adequate	Adequate	Adequate
Ease of calibration	Excellent	Poor	Fair
Clean out	Excellent	Poor	Fair
Seed damage	Low	Moderate	Low
Range of travel speeds	Adequate	Limited	Adequate
Variation with speed	Low	High	Low
Number of hoppers	One	One/row	One with one pocket/row
Depth control	Good	Good	Good
Seed covering	Good	Good	Good
Construction	Good	Good	Good

5.4.2 Sowing date

Regardless of seeder chosen, deciding when to sow is important and depends on several factors, some of which are related to seedling growth and some of which are wholly operational. Sorensen [19] found that Douglas-fir [*Pseudotsuga menziesii* (Mirb.) Franco] sown as germinants set bud earlier and had a longer shoot-elongation period when sown earlier. Each day of earlier sowing between April 23 and May 12 increased height of 1+0 seedlings by about 0.5 mm. Differences in budset and height persisted through the second year, indicating that early sowing can increase seedling size.

To attain germinants in the nursery by April 23, seed must be sown in the cold, wet nursery soil April 1 or before. This is often impractical because the soil is unworkable. Furthermore, germination is slower in cold, wet soil, creating a situation where preemergence damping-off can reduce total germination [20]. Data from the OSU Nursery Survey indicate that most sowing in the Northwest is done between mid-April and early June. Sowing is best done as early as possible after average soil temperature at the 10-cm (4-in.) depth exceeds 10°C. In Oregon's Willamette Valley, this usually occurs in early to mid-April.

Although spring sowing is the norm, fall sowing can provide natural stratification and has been shown to produce excellent seedlings. The advantages are outweighed, however, by the disadvantages. Fall-sown beds must be protected from rodents and birds, should be mulched to prevent frost heaving, and may be lost to early spring frosts. Bed preparation and nursery-space availability also may be problems [15].

5.4.3 Sowing depth

Sowing depth is crucial to producing a uniform bed of seedlings. As with many other seeding parameters, the accuracy with which a selected depth can be achieved depends on uniformity of the prepared seedbed and soil properties. If a bed lacks a level, flat top, then sowing depth will vary greatly from row to row and even along a given row. This variation can lessen germination, retard growth, and reduce crop uniformity. Experiments have shown how germination can vary with sowing depth for slash pine (*Pinus elliottii* Engelm.) (Fig. 3) [17]. Similar results can be expected with Northwest species.

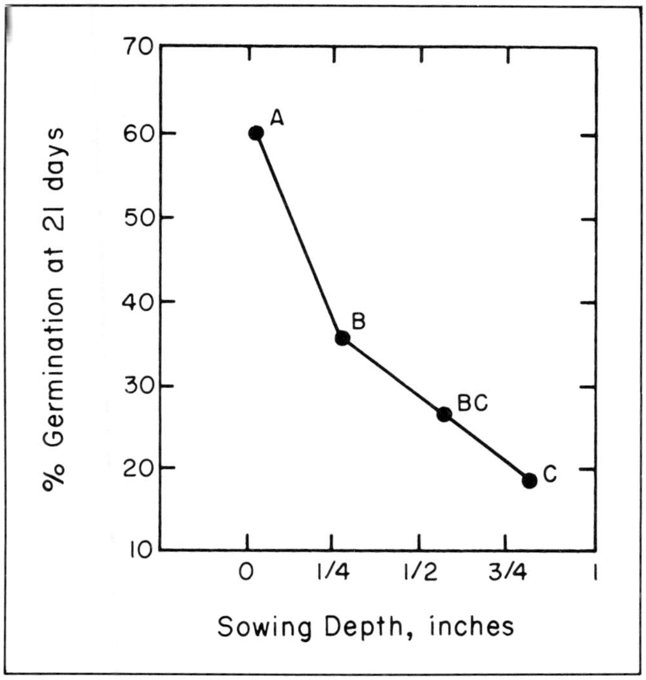

Figure 3. Effect of sowing depth on speed of germination of slash pine seedlings. Points followed by the same letter do not differ significantly at the 95% confidence level, according to Duncan's new multiple range test (adapted from [17]).

Best germination is obtained by sowing seed only as deep as necessary to cover it and prevent erosion or birds from removing it. Reported sowing depths for Douglas-fir and ponderosa pine (*Pinus ponderosa* Dougl. ex Laws.) vary from "just covered" to 1.25 cm (½ in.); the most commonly used depths range from 0.31 to 0.62 cm (⅛ to ¼ in.) (OSU Nursery Survey). In the Canadian nurseries surveyed, seed was surface planted and then covered with a ¼ inch of sand. Seedbeds are not mulched with water-holding material at the time of sowing in the Northwest (OSU Nursery Survey).

5.4 4 Seedbed density

In recent years, nurseries have responded to the research results on seedbed density and spacing by decreasing the number of seed sown per square meter. The density at which a nursery chooses to sow depends on the seedling characteristics desired at harvest and the economics of seedbed use.

Experiments have shown that increasing bed density is closely correlated with decreasing stem diameter and dry weights in Douglas-fir [9, 11, 25] and three western pine species [4]; it has been negatively correlated with height growth in some experiments [9] but not in others [11]. When outplanting results from seedlings grown at various densities were compared, survival did not differ significantly, but seedlings initially grown at lower densities (108 to 215 seedlings/m², or 10 to 20 seedlings/ft²) grew taller than those grown at higher densities [11] (see also chapter 15, this volume).

Logic dictates that seedling spacing within a row should be as uniform as possible for optimum growth. However, many seeders produce a clumpy distribution. At the lower densities commonly used (108 to 215 seedlings/m², or 10 to 20/ft²), the range of spacing created with a Wind River drill did not affect either stem diameter or shoot:root ratio of seedlings [3]. Similar research comparing morphology of seedlings grown in hand-thinned, uniformly spaced beds vs. those grown in operationally sown beds of the same average density showed that spacing did not affect seedling caliper, whereas density did [unpubl. data, 22].

To determine sowing density (the number of viable seed sown per square meter to achieve a given density at lifting), the nursery manager must know the intended diameter specification of a shippable tree at lifting and the diameter distribution of seedlings grown at various densities in the nursery. The manager must estimate tree percent (the number of trees in a nursery bed at lifting relative to the number of viable seed sown) and then must designate an acceptable yield percent (the percentage of trees meeting a specific size criterion, regardless of form). On the basis of yield percent, a density is chosen which will give the maximum number of shippable seedlings per square meter at the lowest cost. Yield percent is an important economic factor not only because seed is becoming more expensive, but because lifting and handling many more seedlings than are shipped can also be expensive.

Bunting [5] gives an excellent account of the dilemma faced when choosing a bed density:

> "The choice of which density to grow your seedlings at is always a compromise. It would be nice to be able to grow the beds thin enough so that there would be no built in cull factor, on the other hand, the thinner you grow your beds the fewer shippable trees you produce per acre and hence production costs go up. When you increase densities to try and maximize the number of shippable trees you will produce per acre you increase your cull percentage and your lifting costs. Since lifting and shipping costs can be greater than the cost of producing the stock, this is no small consideration."

A good way to choose the correct growing density (density in a bed at lifting, usually for 2+0 seedlings; sowing density x tree percent) for a specific diameter limit and a given nursery is to construct a set of curves similar to those in Figure 4 for 2+0 Douglas-fir [unpubl. data, 22]. It is important to remember that a yield percent of 75 does not mean that 75% of the seedlings will be shippable, only that 75% will meet the diameter standard. Of those meeting the diameter standard, a certain percentage (the damage percent) will exhibit some defect such as root damage, multiple tops, or disease that will make them undesirable. The percentage of the crop that is shippable

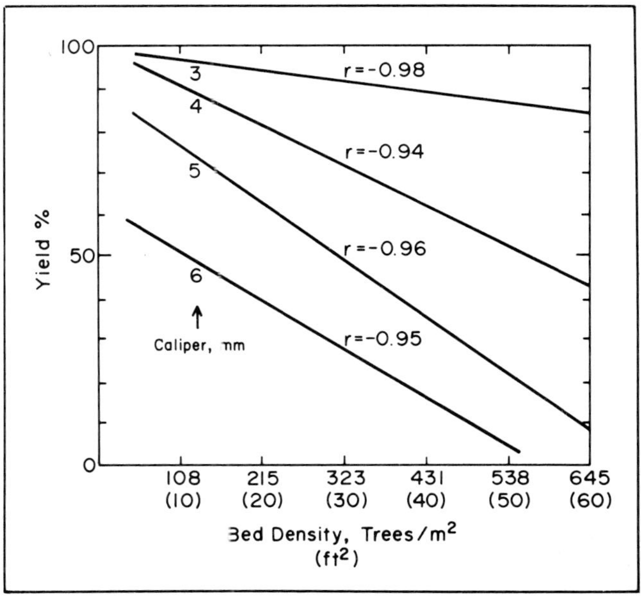

Figure 4. Yield percent for various caliper standards and bed densities for 2+0 seedlings [unpubl. data, 22].

(shippable percent) is thus calculated as yield percent x (1 − damage percent). Barring a large loss to disease or insects, a damage estimate of 5 to 10% is probably conservative when calculating sowing formulas. When possible, this factor should be based on past experience with the nursery in question.

A bed density is chosen based on seedling specification (e.g., 4-mm caliper) and on an acceptable compromise between maximum yield percent and maximum number of seedlings per square meter. In the Northwest, growing densities vary by stock type and range as follows (OSU Nursery Survey):

Stock type	Seedlings/m² (ft²)
2+0	161-323 (15-30)
2+0 for 2+1	376-538 (35-50)
1+0 for 1+1	538-753 (50-70)

5.4.5 Sowing formula

Before seed can be sown, a formula must be employed that calculates the amount of seed necessary to sow to produce the desired number and size of seedlings. In general, all sowing formulas have the same basic form; they differ radically, however, in the number and refinement of the factors used in their calculation.

Some nurseries use a "nursery factor" that is a combination of all possible causes of seedling loss between germination and shipping. However, as a composite, it tells the grower little about what specific problem is causing the loss. Other nurseries have been compiling data for many years and have factors for correcting sowing formulas based on nursery field, species, and sometimes even field-by-species interaction. For a new nursery or one that does not want or cannot afford to compile very specific data, the seed and nursery factors discussed in the following two sections (5.4.5.1 and 5.4.5.2) should allow nursery personnel to track seedling survival and yields with minimum effort and to pinpoint problem areas for improvement.

5.4.5.1 Seed factors

Before any sowing formula can be used, the nursery manager must know the quality of the seedlot intended for sowing. Most nurseries will obtain the data necessary from an independent seed-testing laboratory, although some prefer to perform the tests themselves (see chapter 4, this volume).

In any case, seedlot quality is determined on the basis of three major factors. Seed purity percent is the percentage of the seedlot, by weight, that is seed, not debris. A high percentage (25%) of debris can cause seeders to plug and reduce seeding accuracy. Seed germination percent is the percentage of the total number of seed tested that germinate after a standard treatment and set period of time in the laboratory. This factor can vary greatly from one lot of a given species to another. Estimates from past experience can be very inaccurate and should only be used as a last resort. Furthermore, results from laboratory and field germination often do not agree, probably due to variation in stratification length and, of course, germination conditions. If no better data are available, however, laboratory germination is probably a reasonable approximation of field germination. A 1,000 seed weight is determined by weighing 1,000 seeds plus the accompanying debris. The pure, live seed (the expected number of germinants to be produced) per kilogram can then be calculated. The difference between laboratory and field germination, if known, can be used to modify the pure, live seed calculation.

Seed is wet stratified before sowing for periods varying from 1 week to 4 months depending on species (see chapter 4, this volume) and surface dried just before sowing. The seed is spread on screens to air dry until surface moisture no longer holds seeds together and they do not stick to the hand. Drying allows the seed to flow smoothly through the drill and reduces clumping. No seed in the Northwest is treated with bird repellent or fungicide before sowing (OSU Nursery Survey).

5.4.5.2 Nursery factors

Nursery factors affecting the sowing formula include tree, yield, and shippable percents, bed density, and number of seedlings ordered.

Tree percent depends heavily on cultural practices and environmental influences. Refinement can be made by calculating this factor at each inventory (e.g., 1+0 tree percent, summer 2+0 tree percent, and fall 2+0 tree percent) and using these numbers to project final 2+0 bed density and tree percent. For example, a higher than expected 1+0 tree percent may indicate that trees should be thinned to obtain the lower density needed for the expected yield percent. Calculating intermediate tree percents may also allow the nursery manager to pinpoint the time of loss and, thereby, the cause. Cultural practices can then be changed to minimize losses.

Any tree percent used in the sowing formula is, at best an average of many years of experience and, at worst, a conservative educated guess. Owston and Stein [15] reported a range of 25 to 77% for Douglas-fir (unweighted average for 18 nurseries, 51%); similar tree percents (48 to 80) are reported for ponderosa pine [12]. Most Northwest nurseries use a tree percent of 60 to 80 for Douglas-fir (OSU Nursery Survey).

However, as previously noted, certain size and form criteria must be met for seedlings to be considered shippable. Seedlings present in the tree percent that meet the size criterion make up the yield percent; of these, some will be lost to the damage percent. The remainder are the shippable percent. As discussed earlier, yield percent varies with bed density. Selecting an economically feasible yield percent for a given size criterion also sets the best bed density for that seedlot (Fig. 4).

To complete the sowing calculation, only the number of seedlings ordered by the customer is needed.

5.4.5.3 Calculating the sowing formula

Once seed and nursery factors have been determined, calculating the sowing formula is easy:

(1) Pure, live seed per kilogram (PLS/kg) =
 (germination %/100) x (purity %/100) x (1,000 g/1,000 seed weight)

(2) Number of shippable seedlings produced per kilogram of seed =
 PLS/kg x (tree %/100) x (yield %/100) x (shippable %/100)

(3) Kilograms of seed required =
 (Number of seedlings ordered)/(number of seedlings/kg)

(4) Meters of nursery space required =
 [(Kilograms of seed required) x (PLS/kg) x (tree %/100)]/bed density

For example, let's assume we received an order for 1,000,000 Douglas-fir seedlings. Seed test results are 83% germination, 98% purity, and 1,000 seed weight of 11.6 g. Pure, live seed per kilogram can then be calculated:

(83/100) x (98/100) x (1,000/11.6)= 70,120

From past records, we know our tree percent is 80 and our shippable percent 86. Seedlings must have a 4-mm caliper to meet the size criterion, and an economic analysis indicates we need a yield percent of 87.5% to balance packing and bed-area costs. Number of seedlings produced per kilogram of seed is calculated:

70,120 x (80/100) x (87.5/100) x (86/100) = 42,212

Thus, the kilograms of seed required would be

1,000,000/42,212 = 23.69

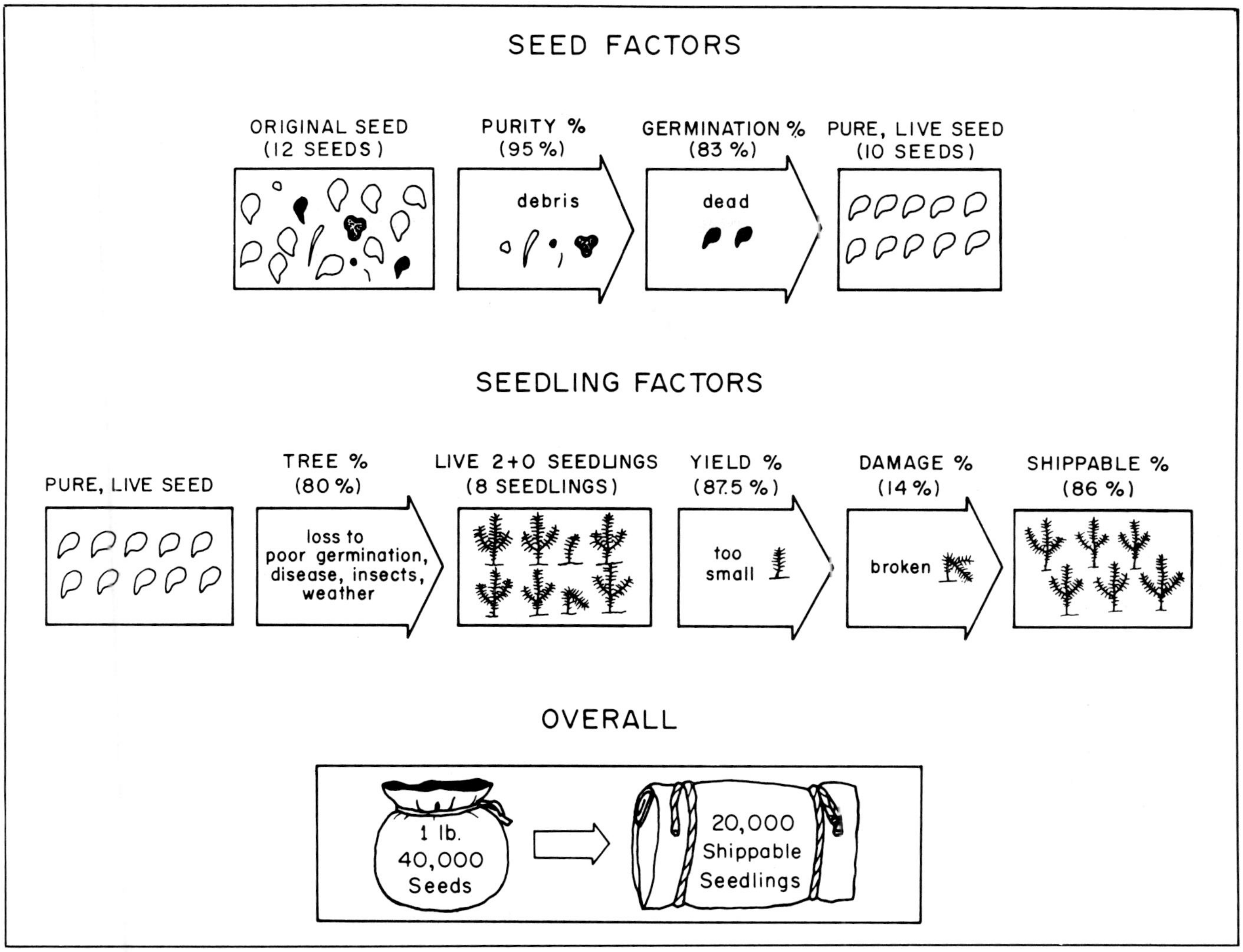

Figure 5. Fate of seed and seedlings in the nursery.

From Figure 4, we determine that the best density at which to grow our seedlings would be 250 seedlings/m^2. In that case, bed space required (in square meters) can be calculated:

$$[(23.7) \times (70{,}120) \times (80/100)]/250 = 5{,}318$$

Figure 5 demonstrates how seed and seedling factors determine the ultimate number of seedlings produced.

Each year, as practices change, actual tree, yield, and damage percents should be calculated and any large changes from past years investigated as to cause. These new percents should be used as part of the updated data base for new seed- and seedling-requirement calculations.

5.5 Care of the Seedbed during Germination and Early Growth

A critical nursery period follows sowing. Seeds and seedlings are vulnerable to the environment and predators, and major losses can occur if seedbeds are not protected. If conditions are not right or care is not taken, viable seed either will not germinate or will be lost to predators or disease. Even after seedlings have emerged, predation and disease continue to be problems. In addition, new seedlings may have to survive hail storms, scorching sun, and high temperatures. As spring progresses, weeds, if not checked, can reduce growth and even kill young seedlings by usurping water and sunlight. However, certain steps can be taken to nurture seedlings through this period and ensure their vigor.

5.5.1 Irrigation

After sowing, the soil surface should not be allowed to dry out. If soil becomes dry, the seed may dry too much, all advantages of stratification may be lost, and slow, spotty germination may result. Seed of species such as western hemlock [*Tsuga heterophylla* (Raf.) Sarg.], which are small and surface sown, are often covered with burlap to hold in moisture and prevent seed from being washed away. Careful watch must be kept daily because more than one watering per day may be required to maintain proper moisture status on spring days with high evaporative demand. As a rule, finer textured soils require more frequent watering than coarser textured ones (see chapter 12, this volume).

Too much water during germination, however, is not desirable. Excessive water promotes preemergence damping-off and growth of seed-borne fungi by decreasing soil temperature and increasing soil moisture [20]. A correct balance is important. If emergence is much slower than expected and examination of seed reveals that it is "rotting in the ground," decreasing water and spraying with an appropriate fungicide to prevent further damage can be beneficial.

After germination in the nursery has peaked, the watering regime should be changed. Frequent, shallow irrigation should be superseded by longer periods of irrigation. Soil should be kept between −0.1 and −0.75 bar at a depth of 15 cm (6 in.) for optimum growth and irrigated only when it approaches −0.75 bar [8]. This regime is followed until hardening-off treatments are begun in midsummer (mid-July to mid-August). Predawn plant moisture stress is then allowed to increase to between 12 and 15 bars before rewatering to promote budset [7] (see chapter 15).

During the period from germination to dormancy, irrigation water is applied to cool and shade young (1+0), tender seedlings. In the exposed surface of a nursery bed, soil-surface temperatures can rapidly rise to over 45°C (112°F) on a warm, sunny day. This can literally "cook" the root-collar area and kill the seedling. To prevent damage, the soil surface can be cooled by irrigation.

Critical soil temperatures used for cooling vary with seedling age and species. Damage is most apt to occur in younger seedlings of species adapted to cool, moist climate; for instance, Douglas-fir and western hemlock are less tolerant to heat damage than most pines. Some nurseries use air temperature as a guide for determining need for cooling, but the majority use soil-surface temperature, usually measured 0.5 to 1 cm below the surface. A typical guideline might be: irrigate when soil temperature exceeds 32°C (90°F) before July 1; 35°C (95°F) before August 1; and 38°C (100°F) until winter (composite from OSU Nursery Survey).

How long and how often a nursery applies water when critical temperatures are reached varies greatly and depends at least partially on soil type. Some nurseries irrigate 5 to 10 minutes during every hour the temperature is above that considered critical; others water for an hour; and still others water until the soil temperature drops below a fixed, safe temperature [e.g., 25°C (77°F)]. One nursery reported that due to its soil properties, irrigation does not reduce soil temperature but merely prevents further increase. Research data on critical temperatures and how they vary with the season for various species are not available. Therefore, nursery managers should adopt a reasonable schedule of cooling and adhere to it until sound research evidence produces a better one.

5.5.2 Diseases and insects

After emergence, damping-off can continue to be a problem (see chapter 19, this volume). Fumigation is the most common method used in the U.S. to rid the soil of damping-off fungi. In Canadian nurseries, where fumigation is not used, seed is covered with 0.5 cm of sand. This decreases the moisture around the seed and increases soil-surface temperature. No matter what is done to reduce occurrence of damping-off, it can still be a problem in certain years. Careful and frequent monitoring of the crop for any sign of disease and rapid treatment with a fungicide drench at the first sign can make a significant difference in tree percent.

Although not generally found in the Northwest (OSU Nursery Survey), insects can be a problem in the nursery. Cutworms probably pose the greatest threat. Patches of seedlings clipped just above the ground line soon after emergence can be a sign of this pest. Control can be achieved with a number of insecticides. Recently, another yet unidentified insect pest has caused extensive damage at nurseries in Oregon's Willamette Valley. Buds and stems of 1+0 and 2+0 seedlings have been damaged late in the growing season [pers. commun., 21].

5.5.3 Birds and rodents

As soon as the seed is sown and until the seedcoat is shed, birds can create a serious problem in the nursery. They seem particularly fond of pine seed and can destroy a crop in a very short time. In southern pine nurseries, seed is commonly coated with bird repellent. Northwest nurseries use either scare tactics or screening (OSU Nursery Survey). Where the problem is prevalent, owls painted on balloons, hawk decoys on posts, loud noises, and shotguns have all been tried, with varying degrees of success.

Screening is the most expensive and labor-intensive method, but also the most effective. A frame of wood as wide as the bed and 2 or 3 m long, covered with fiberglass screen, can be placed over the bed until seedcoats are shed. These frames are relatively inexpensive, reusable, and effective. Where possible, to reduce damage, nurseries can wait to sow until certain migratory birds have passed through their area. Rodents are generally not as great a problem but, where encountered, are usually controlled by poison bait and traps.

5.5.4 Weeds

The last pests to be considered, but the ones found most frequently, are weeds (see chapter 18, this volume). Recent developments in herbicides have made weed control easier and cheaper than ever before. As discussed previously, weed control begins when a field is selected for nursery production. As many weeds as possible, both in and around the field, are eliminated at that time. Where practiced, fumigation eliminates many residual seeds. Yet even with all this presowing care, weed seeds still blow into the nursery from surrounding fields and germinate if additional steps are not taken.

Most nurseries control weeds by applying a selective herbicide within a week after sowing. The most commonly used chemicals in Douglas-fir nurseries are oxyfluorfen (Goal 2E®) and bifenox (Modown®). Weed control usually begins to falter about midsummer and can be reinforced by a second application of herbicide or supplemental hand or machine weeding. Recent evidence that oxyfluorfen may limit germination of some conifer species and accumulates over time in some soils may cause reevaluation of its widespread use [unpubl. data, 2].

5.6 Conclusions and Recommendations

In seedbed preparation and sowing, planning is all-important. Many of the potential problems such as poor drainage, weeds, and disease can be eliminated or greatly ameliorated by careful planning and field pretreatment. The difference between a successful, economically run nursery and one that is continually trying to solve preventable problems at undue expense is planning.

Before preparing a new field for sowing, review the past history of the area; decide if any potential problems exist and how to eliminate them or at least alleviate their impact on the nursery crop. Determine an appropriate slope and orientation for the field. Before irrigation lines, roads, and beds are established, considerations of field efficiency should be balanced with those of ease of operation. Once irrigation lines are established, it is almost impossible to improve field efficiency.

Beds should be raised 6 to 12 cm above ground level and have smooth, even surfaces; this allows more uniform sowing. Seed should be sown as early as possible in spring, after the soil temperature at the 10-cm depth exceeds 10°C. Most commercially available seeders do an adequate job of distributing seed, although the Øyjörd is generally more versatile. Sowing depth should be uniform for uniform germination. A sowing depth of "just covered" to 0.5 cm is recommended for Douglas-fir.

Sowing density and subsequent growing density have a pronounced effect on the resulting size of the seedling and its production cost. Managers should consider growth characteristics in the nursery and costs of various nursery operations when determining growing density. A growing density of 160 to 325 seedlings/m^2 (15 to 30 seedlings/ft^2) is recommended for 2+0 Douglas-fir.

In addition to planning, crop monitoring and recordkeeping are necessary to obtain the numbers needed for a viable sowing formula. Accurate tree, yield, and damage percents are essential to determining the correct amount of seed and bed space to grow a given seedlot.

Proper care and tending after sowing ensure the early success of the crop. The secret is to be aware of the possible problems and keep a constant vigil. Diseases, birds, weather, and weeds are the most common causes of loss.

Although a vigorously growing 1+0 crop free from weeds and pests is the goal of all nurseries, many factors may combine to make this difficult to achieve. Bareroot nurseries are always at the mercy of nature. Hail, excessive rain, and cloudy or hot weather all take their toll. It is important, therefore, that nursery managers understand and optimize all those factors over which they exert a measure of control. Good seedbed preparation and sowing practices and early seedling care all increase the likelihood of success. Having accepted the challenge of growing seedlings and succeeded, nursery managers and personnel may find nothing more beautiful than looking over their fields of lush green, uniform, healthy seedlings.

References

1. Aldhous, J. R. 1972. Nursery practice. Her Majesty's Stationery Office London. Forestry Commun. Bull. 43. 184 p.
2. Armson, K. A., and V. Sadreika. 1979. Forest tree nursery soil management and related practices. Ontario Ministry of Natural Resources, Toronto. 179 p.
3. Arnold, E. W. 1972. Effect of sowing accuracy on seedling-size 1-0 and 2-0 Douglas-fir. Weyerhaeuser Washington Nursery. Pages 20-29 *in* Proc. (H. W. Anderson, J. A. Bryan, and R. P. Eide, eds.). Western Forest Nursery Council and Intermountain Forest Nurserymen's Assoc., Olympia, Washington.
4. Baron, F. J., and G. H. Schubert. 1963. Seedbed density and pine seedling grades in California nurseries. U.S.D.A. Forest Serv., Pacific SW Forest and Range Exp. Sta., Berkeley, California. Res. Note PSW-31. 14 p.
5. Bunting, W. R. 1980. Seedling quality: growth and development—soil relationships, seedling growth and development, density control relationships. Pages 21-42 *in* Proc., North American forest tree nursery soils workshop (L. P. Abrahamson and D. H. Bickelhaupt, eds.). State Univ. New York, Coll. Environ. Sci. and Forestry, Syracuse.
6. Campagna, J. P., and D. P. White. 1973. Nursery soil fumigation affects growth and phosphorus nutrition of pine and spruce seedlings. Forestry Chronicle 49:219-223.
7. Cleary, B. D., and J. B. Zaerr. 1980. Pressure chamber techniques for monitoring and evaluating seedling water status. New Zealand J. Forestry Sci. 10(1):133-141.
8. Day, R. J. 1980. Effective nursery irrigation depends on regulation of soil moisture and aeration. Pages 52-71 *in* Proc., North American forest tree nursery soils workshop (L. P. Abrahamson and D. H. Bickelhaupt, eds.). State Univ. New York, Coll. Environ. Sci. and Forestry, Syracuse.
9. Edgren, J. W 1976. Seedbed density, diameter limit culling and 2-0 Douglas-fir seedling production. Proc., Western Forest Nursery Council and Intermountain Forest Nurserymen's Assoc., Richmond, B.C. 9 p.
10. Hart, J. B., Jr. 1980. Basic nursery soil physical properties. Pages 44-51 *in* Proc., North American forest tree nursery soils workshop (L. P. Abrahamson and D. H. Bickelhaupt, eds.). State Univ. New York, Coll. Environ. Sci. and Forestry, Syracuse.
11. Iverson, R. D. 1981. Low bed densities enhance Douglas-fir seedling size and performance. International Paper Co., Western Forest Res. Center, Lebanon, Oregon. Tech. Note 60. 6 p.
12. Krugman, S. L., and J. L. Jenkinson. 1974. *Pinus* L. Pine. Pages 598-638 *in* Seeds of woody plants in the United States. U.S. Dep. Agric., Washington, D.C. Agric. Handb. 450.
13. Lott, J. R., and B. J. Lowman. 1976. Evaluating precision seeders for tree seedling nurseries. Report ED&T 2522. U.S.D.A. Forest Serv. Equipment Development Center, Missoula, Montana. 49 p.
14. May, J. T. 1982. Seed preparation. *In* Southern pine nursery handbook (C. Lantz, ed.). (In press.)
15. Owston, P. W., and W. I. Stein. 1974. *Pseudotsuga* Carr. Douglas-fir. Pages 674-683 *in* Seeds of woody plants in the United States. U.S. Dep. Agric., Washington, D.C. Agric. Handb. 450.
16. Pair, C. H., W. W. Hinz, C. Reid, and K. R. Frost. 1975. Sprinkler irrigation. Irrigation Assoc., Silver Spring, Maryland. 615 p.
17. Rowan, S. J. 1980. Planting depth and seedbed mulch affect germination of slash pine seed. U.S.D.A. Forest Serv., Southeastern Forest Exp. Sta., Asheville, North Carolina. Res. Note SE-292. 3 p.
18. Shearer, M. N. 1980. Requirements for quality irrigation. Pages 107-112 *in* Proc., Joint meeting, Intermountain Forest Nurserymen's Assoc. and Western Forest Nursery Council, Boise, Idaho.
19. Sorensen, F. C. 1978. Date of sowing and nursery growth of provenance of *Pseudotsuga menziesii* given two fertilizer regimes. J. Applied Ecology 15:273-280.
20. Sutherland, J. R. and R. L. Anderson. 1980. Seedling disease and insect problems related to nursery soil conditions in North America. Pages 182-190 *in* North American forest tree nursery soils workshop (L. P. Abrahamson and D. H. Bickelhaupt, eds.). State Univ. New York, Coll. Environ. Sci. and Forestry, Syracuse.
21. Tanaka, Y. 1982. Personal communication, Weyerhaeuser Co., Centralia, Washington.
22. Thompson, B. 1982. Unpublished data, International Paper Co., Lebanon, Oregon.
23. van den Driessche, R. 1979. Soil management in Douglas-fir nurseries. Pages 278-292 *in* Forest soils of the Douglas-fir region (P. E. Heilman, H. W. Anderson, and D. M. Baumgartner, eds.). Washington State Univ. Ext. Serv., Pullman.
24. Whalen, M. 1981. Personal communication, Weedco, Portland, Oregon.
25. Wilson, B. C., and R. K. Campbell. 1972. Seedbed density influences height, diameter and dry weight of 3-0 Douglas-fir. Tree Planters' Notes 23(2):1-4.

Managing the Soil and Water

Chapter 6
Physical Properties of Forest-Nursery Soils: Relation to Seedling Growth

B. P. Warkentin

Abstract

The physical properties desirable in forest-nursery soils are those that provide the optimum environment for root growth and function. Because these properties are difficult to alter, they should be used to determine which sites are chosen for forest nurseries. Solid grains, about half the soil volume, provide the framework for stable soil pores and an anchor for plant roots. The large pores, about half the pore space, allow for necessary gas exchange; the small pores store water for plant use. Soil cultivation and drainage change the relative proportions of solids and pores. Tillage can improve soil tilth, but excess tillage usually results in undesirable changes in the soil. Soil organic matter content is important to tilth as well as to stability of structure. Soil compaction, including crusting, represents a special problem, which can be both caused and ameliorated by tillage. Increased soil resistance to compaction and adequate organic matter maintenance should be major objectives in nursery soil management for Northwest forest nurseries.

6.1 Introduction

The vigor of a shoot depends upon the health of the root. But though the shoot is always in sight, the root is not. The root must proliferate in the soil to supply the shoot with the necessary water, oxygen, nutrients, and structural support. Any hindrance to root growth restricts root functions and thus influences shoot growth. Yet such hindrances may not be obvious until the soil is examined to see how the root has developed.

Our objective in managing nursery soils is to manipulate them to provide an optimum physical environment for root growth. In this chapter, physical properties of soils are examined to this end **from a root's perspective.** What are the root's requirements? What are the impediments in the soil to meeting those requirements? How can soil management remove or alleviate those impediments to improve the physical environment, so roots—and shoots—can thrive?

Specific nursery practices for applying the principles are not thoroughly discussed here. This is due in part to my lack of familiarity with all nursery soil problems and the management options available to cope with them, but also in part to a conviction that forest-nursery managers are best able to design practices to apply the principles.

6.2 Physical Environment for Root Growth

The physical characteristics desired in nursery soil are:

- Optimal proportions of air and water in soil pores after natural drainage
- Rapid drainage of excess water from soil
- Adequate infiltration rate for rainfall or irrigation water
- High resistance to compaction
- Low shear strength for easy harvest of seedlings
- Low adhesion of soil to seedling roots
- Absence of frost heaving, erosion, and soil splash onto seedlings

These characteristics are optimized in loamy sands or sandy loams whose silt plus clay contents are 10 to 25% and whose organic matter contents can adequately stabilize soil structure to maintain pore size and continuity.

Adequate "root room," a concept that has been used to evaluate the root's environment [6], requires adequate depth of freely draining soil. Root proliferation also requires low soil resistance to root growth. Compacted layers, poor drainage, and root pruning decrease root room. Seedbed density also affects root room, as do various management practices to change shoot:root ratios. The mass of roots is approximately half the mass of shoots, although this ratio can vary from 0.3 to 0.8, depending upon management factors [10]. A good root environment, with adequate aeration and low resistance to root penetration, favors high root mass.

Root growth depends upon movement, or fluxes, of materials in the soil. Fluxes of water as it redistributes in the soil, of oxygen as it diffuses to the root and of carbon dioxide and other gases as they diffuse away from the root, and of nutrients as they either diffuse or flow to the root are all necessary

In Duryea, Mary L., and Thomas D. Landis (eds.). 1984. *Forest Nursery Manual: Production of Bareroot Seedlings.* Martinus Nijhoff/Dr W. Junk Publishers, The Hague/Boston/Lancaster, for Forest Research Laboratory, Oregon State University, Corvallis. 386 p.

for root function. In addition, the root itself moves as it extends into the soil, and any resistance that it meets influences its function. Root hairs, mycorrhizae, and enhanced biological activity in the rhizosphere all involve fluxes. Most root growth occurs in early spring and again in autumn, when shoot growth is not great, although it can also occur during summer, depending upon species and weather [10].

Soil physical characteristics are important in nursery-site selection because they are hard to change. Soil fertility, for example, can be changed quite readily by adding fertilizer and lime, but it is usually impractical to mix in enough off-site soil to change soil texture or depth. The easiest way to modify physical characteristics is to add organic matter from mulches such as sawdust or cover crops (see chapters 9 and 10, this volume). It is no wonder, then, that in considering soil physical factors for site selection, Northwest nursery managers listed good soil workability and drainage as the dominant desirable characteristics (OSU Nursery Survey; see chapter 1, this volume) (Table 1).

Table 1. Important soil factors considered by Northwest nurseries in selecting sites (OSU Nursery Survey).

Soil factor	% of time: Considered	% of time: Listed 1st or 2nd in importance
Workability and drainage	67	38
Texture	57	28
Depth	43	19
Fertility	33	14

6.3 Soil Structure

Soil structure refers to the size, shape, and arrangement of soil grains and pores. The relative sizes of the different physical and biological components are shown in Figure 1. The solid portion of a soil is composed of soil grains, or particles, of different sizes (Table 2). Different proportions of these sizes produce the texture classes (e.g., sandy loam, silt loam, etc.). The pore space of a soil—that portion between the grains or grain aggregates—is occupied by air and water. Porosity (volume of pores divided by total soil volume) of soils in desirable physical condition is around 0.5.

Table 2. Grain-size fractions.

Name	Size, mm
Clay	<0.002
Silt	0.002-0.05
Very fine sand	0.05-0.1
Fine sand	0.1-0.25
Medium sand	0.25-0.5
Coarse sand	0.5-1.0
Very coarse sand	1.0-2.0
Gravel	>2.0

Aeration, which occurs only in pores not filled with water, consists both of oxygen diffusing through soil to the root and of carbon dioxide and gases such as ethylene diffusing from the root to the soil surface. Diffusion rates are 1,000 times slower in water than in air; therefore, water-filled pores do not contribute to aeration. In a well-drained soil, most diffusion occurs in transmission pores larger than 0.05 mm (Table 3). For good aeration, a minimum of 20% of the pores should be filled with air [4]; when only about 10% are, diffusion of gases falls to zero. The pores constituting this 10% porosity are isolated by water; their continuity is limited, and diffusion cannot readily occur.

Optimum bulk density (mass of dry soil divided by soil volume) varies with grain size and nature of the soil. A clay or clay loam should have a bulk density of 1.0 to 1.1 g/cm³ because a large part of its porosity is in the very small residual pores (Table 3). Soils developed on volcanic parent materials have optimum bulk densities of 0.9 to 1.1 g/cm³. Some very sandy soils can provide an adequate root environment at a bulk density of 1.45 g/cm³ because the pores are mostly storage and transmission pores.

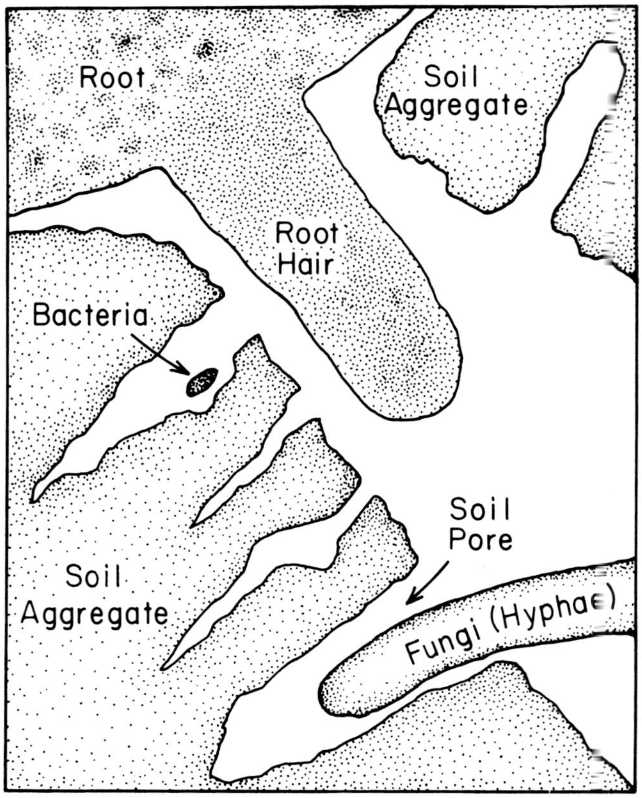

AVERAGE SIZES, μm	
Soil Bacteria	2.0 x 0.05
Fungi (Hyphae)	5
Root Hair	10
Root	200
Soil Pores	
Empty at wilting	0.1
Empty at field capacity	15
Soil Aggregate	5,000

Figure 1. Architecture of the soil, showing relative sizes of biological and physical components (adapted from [12]). Root hairs can penetrate only the largest soil pores holding water available to plants; even bacteria cannot penetrate the smaller pores.

Table 3. Pore-size classifications in soils (adapted from [5]).

Classification	Size, μm (mm)	Function
Fissures	500-5,000 (0.5-5)	Allow rapid drainage
Transmission pores	50-500 (0.05-0.5)	Allow flow of water, diffusion of gases
Storage pores	0.05-50	Hold water available to plants
Residual pores	0.005-0.05	Hold water not available to plants

An ideal sandy loam soil, then, from a root's perspective, could have a bulk density of 1.3 g/cm³, an air-filled porosity of 20%, and a water content of 23% at field capacity (see 6.4). On a volume basis, this soil would have 50% solids, 30% water, and 20% air.

6.4 Soil Water

Water infiltrates the soil at the surface and redistributes below. The force of gravity pulls water down. Soil water potential (Ψ_{soil}) (the negative pressure that must be applied to prevent pure water from moving into soil) pulls water into drier soil and decreases as soil water content decreases. In dry soils, soil water potential is much smaller than the force of gravity. A small amount of added water will, therefore, be distributed in the surface layers of dry soil. Soil water potential depends upon pore size and is lowest in the smallest pores; thus, water will be sucked first into these. The force of gravity can remove water from larger (>0.05 mm) pores; thus, drainage will empty those.

The water retention curve (Fig. 2) relates soil water potential with water content. Varying water potentials can be applied to a soil sample in the laboratory, the resulting water content measured, and a water retention curve produced. The shape of the curve for a particular soil gives information on the amount of water released to plants at different potentials. Plant roots will first absorb the water held at the highest potential in the largest pores (see chapter 12, this volume).

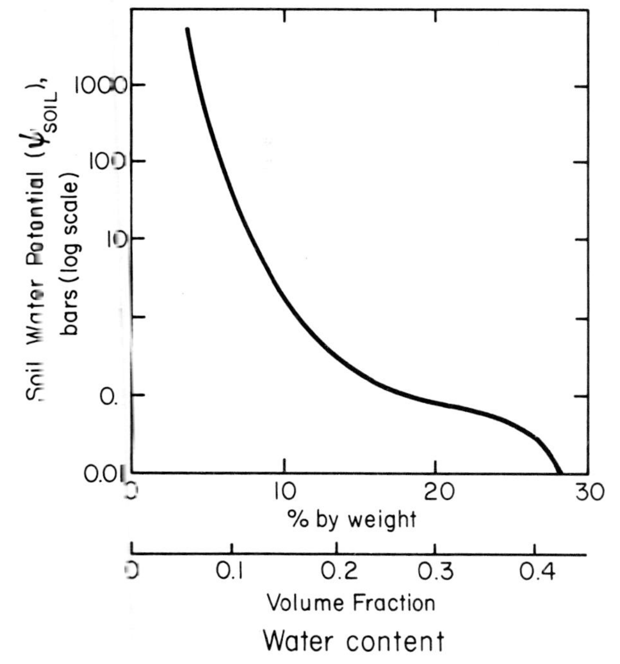

Figure 2. Water retention curve for a typical sandy loam with bulk density of 1.5 g/cm³.

Water content may be expressed by weight or volume. Water content by weight, usually a percentage, is the mass of water divided by the mass of oven-dry soil. Water content by volume is the volume of water divided by total soil volume and can be calculated as water content by weight multiplied by bulk density. This volumetric water content is a useful number because it is directly related to centimeters (inches) of water in a soil. A value of 0.10 means 0.10 cm of water per centimeter of soil, or 10 cm of water per meter of soil (1.2 inches of water per foot of soil).

A saturated soil without growing roots will drain to field capacity (water content of a soil saturated by rainfall or irrigation and then allowed to drain for 24 to 48 hours). The water remaining in the soil at field capacity, held at lower soil water potentials, will drain only very slowly. Any water added in irrigation beyond that required to increase the water content to field capacity is wasted. Excess water will drain away before an appreciable part of it can be used by plants, leaching fertilizer and preventing good aeration (see chapter 13, this volume, for more detailed discussion of land drainage).

Clearly, water control is a very important part of managing soils for intensive use such as forest nurseries. Nursery managers should know the field capacity of their soils, which can be readily measured: (1) after the soil is saturated by irrigation or rainfall, a small area is covered with a plastic sheet to prevent evaporation; (2) after 48 hours, soil samples are taken at 6- and 12-inch depths, and water content is determined.

The total "available water" stored in the soil is defined as the difference between field capacity and wilting percentage (water content at which a plant wilts and will recover turgor only if more water is added). Figure 3 shows the water retention curve from Figure 2 replotted on the basis of available water instead of water content. For this soil, half of the available water is released between −0.1- and −0.3-bar soil water potential and less than a quarter is left at −1 bar.

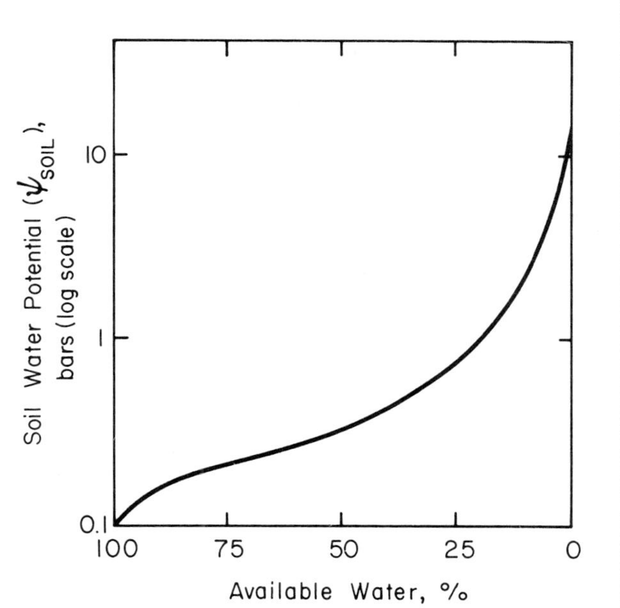

Figure 3. Water retention curve for typical sandy loam (Fig. 2) replotted on the basis of available water.

Growth is not constant over the water-content range from field capacity to wilting percentage (Fig. 4). Atmospheric conditions, which determine atmospheric demand and the amount of water transpired, shape the growth vs. available water curve. In many situations, growth is measurably reduced once about half of the available water has been used. For maximum growth, the soil should be irrigated when it has dried to that point.

It is often assumed that the effects of water stress are temporary, restricted to the period of actual stress, and that maximum growth resumes once plants are irrigated. However, water stress has longer term effects. These may be due to changes in roots on drying which make roots less able to absorb water. New root growth must then occur before maximum shoot growth can.

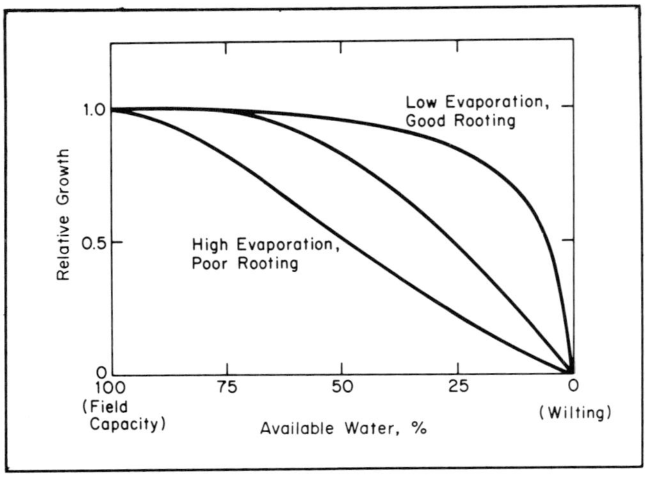

Figure 4. Generalized relation between plant growth and available soil water.

6.5 Soil Temperature

The temperature of the soil, as well as that of the soil relative to the air, affects root growth and function. The root zone has a gradient in soil temperature within which temperatures change diurnally and with depth. Both low and high soil temperatures can be a problem in forest nurseries.

Soil temperature can be controlled through change in either water content of the soil or absorbing and evaporating properties of the soil surface. The amount of heat required to raise the temperature of a volume of water is twice as high as that required for an equal volume of soil. Drainage to decrease water content will increase soil temperature. Mulches can control soil temperature by decreasing or increasing absorption of insolation and by reducing evaporation (see chapter 12).

6.6 Tilth and Tillage

Soil tilth is hard to define but easy to recognize from feel and from kicking the soil with your toe. Tilth is the physical condition of a soil related to ease of tillage, suitability as a seedbed, and impedance to seedling emergence and root growth. A soil in good tilth has aggregates that are rounded and porous, in a range of sizes, that crush easily in the hand. From the standpoint of the plant root or the germinating seed, tilth is the system of pores of different sizes that supply the roots' needs. That system of pores—and its stability—is the concern of tilth and tillage.

Tillage can be used to increase the volume of transmission pores and of the larger (>10 μm) storage pores (Table 3), though it does not influence the smaller storage pores and residual pores. These small pores are determined by physical and chemical interaction between soil grains, such as swelling and shrinking, and by intergrain movement caused by biological activity in the soil. In sum, tillage directly affects the larger pores and indirectly affects the smaller pores by changing the environment for biological and physical-chemical action.

6.6.1 Seedbed preparation

The main physical effect of plowing is a loosening of the soil. How long this increase in porosity lasts depends upon the stability of soil structure (Fig. 5). A soil with stable structure will retain a high porosity for many months; this is typical of soils that have been in grass sod. Other soils will revert to their original porosity within weeks; this undesirable condition is characteristic of soils low in organic matter or those under continuous cultivation.

Tillage operations that prepare a fine seedbed from a plowed soil decrease porosity by making smaller aggregates that can fit more closely together, eliminating the larger pores and fissures. The close fit is assumed to be required to assure that the seed contacts the soil adequately to imbibe moisture for germination. However, the decreased porosity created in a fine seedbed is not optimum for later plant growth. Therefore, soil in seedbed preparation should be manipulated only to the degree of fineness needed for adequate seed germination. It may be better to accept a lower germination percentage to ensure a favorable environment for subsequent growth.

Many studies have been carried out to determine optimum size of aggregates for seed germination and plant growth. The results depend upon the watering regime—with fine aggregates, aeration is limiting; with coarse aggregates, water is limiting. Generally, growth is optimum for aggregates of 0.5 to 2 mm. This observation indicates that rototillers create aggregates that are too small.

Some tillage operations are used for weed control (see chapter 18, this volume). Where weed-control alternatives are available, the benefits and hazards of the extra tillage need to be evaluated.

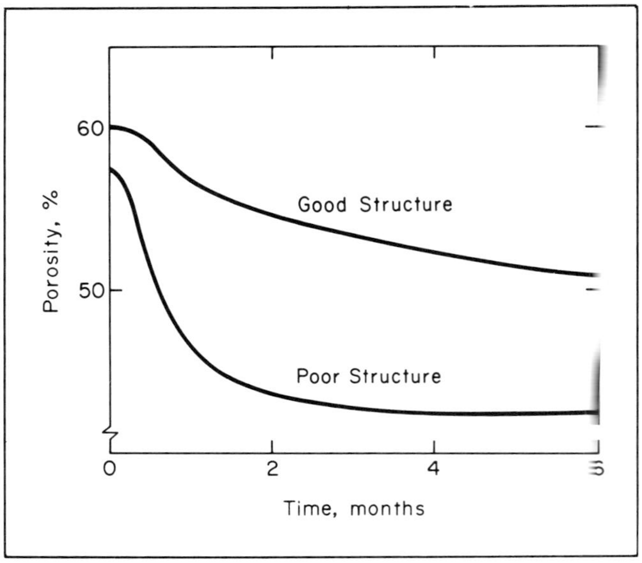

Figure 5. Total porosity of soil after plowing.

6.6.2 Organic matter

Organic matter is the energy source for biological activity in the soil. Molecules produced during decomposition stabilize soil aggregates, thus maintaining good soil structure. This process, like others previously described, is dynamic: because decomposition is continuous, a supply of fresh organic matter is always needed for stable soil structure (see chapter [illegible], this volume).

In most agricultural crops, the root and some of the shoot are left in the soil as a source of fresh organic matter. The root can be 40% of the weight of the shoot. In a forest nursery, however, root and shoot are both removed. Therefore, organic matter must be either brought in from somewhere else and added to the soil or grown on site as a green manure crop (see chapter 10, this volume). A green manure crop grown alternately with nursery seedling crops helps to maintain organic matter, and its prolific roots stabilize soil structure.

6.6.3 Response to tillage: soil workability

Soil workability depends primarily upon soil cohesion, the forces holding soil grains and aggregates together, and soil adhesion, the forces holding soil to tillage implements. Both of these properties vary with water content. Figure 6 shows this variation for sand and clay, which manifest the extremes in cohesion and adhesion. Clays develop the highest cohesive and adhesive forces because of the small grain size and large surface area. In contrast, sands develop cohesion or adhesion only through water films—when the soil is saturated or dry, these properties are not displayed. Agricultural soils fall between the extremes of sand and clays.

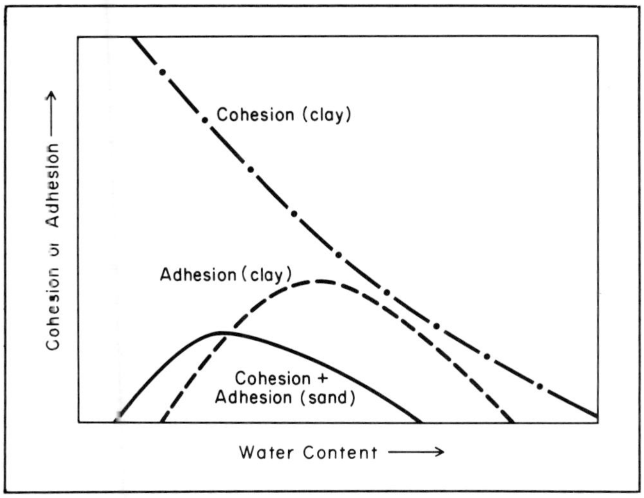

Figure 6. Cohesion and adhesion of clay and sand as a function of water content.

Achieving tilth requires breaking down large aggregates against the force of cohesion. The energy needed increases as soil becomes drier. However, good tilth cannot be achieved at high moisture contents because the broken aggregates will not remain as separate units. Tillage is, therefore, most effective at intermediate water contents, the correct water content depending upon texture and kind of clay, and is best determined by experience with a particular soil.

The main practical way to alter soil workability is by adding organic matter. Organic matter increases cohesion of sandy soils and decreases cohesion of clays; in either case, the result is better tilth or range of pore sizes, for optimum root growth. Organic matter also decreases adhesion of most soils.

6.7 Soil Compaction

Soil compaction is the rearrangement of soil aggregates into a position of higher bulk density, hence lower porosity, as a load is applied to the soil (Fig. 7). Because rearrangement is easiest into large pores, most of the loss of porosity is in the large pores where water flow and gas exchange occur. Rearrangement also increases soil strength and resistance to root penetration. As a result, a small amount of compaction can have a large influence on root growth.

Soils differ in inherent compactability due to grain-size composition and organic matter content. Soils having predominantly one grain size are not easily compacted. However, when a range of sizes is present, small grains can be moved into pores between larger grains, increasing compaction. Organic matter stabilizes aggregates, increases their strength, and decreases compaction.

Aggregates can be rearranged to a higher bulk density from both applied pressure and shear. A tractor tire without slip would apply only pressure to a soil; however, if slip is present, which is the usual case, both pressure and shear are applied. A tractor tire or track produces equal pressure lines that form bulb shapes below it (Fig. 8), but maximum pressure drops off

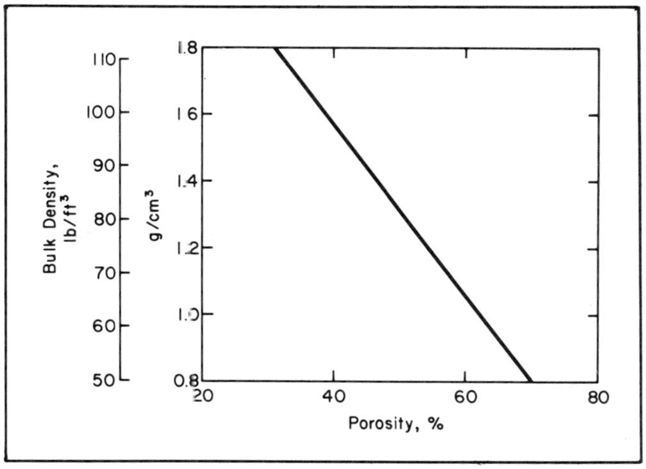

Figure 7. Calculated relationship between porosity and bulk density.

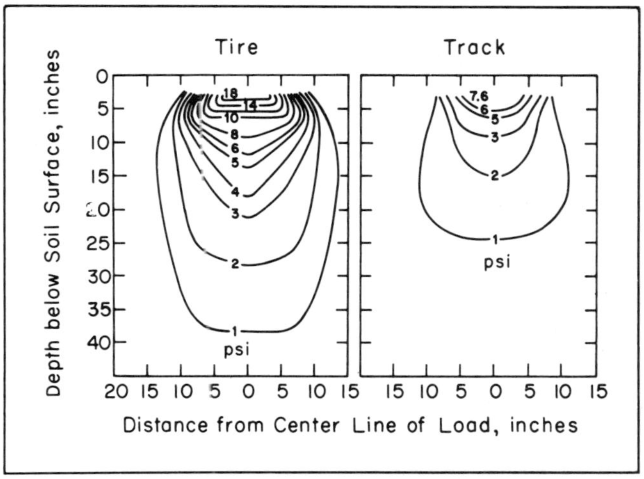

Figure 8. Mean normal stress (psi) under a wheel with a 13-38 tire and a 12-inch track (adapted from [8]).

quickly with depth and with lateral distance from the tire or track. Because track pressures are lower than tire pressures, vehicles with tracks may be preferred to minimize compaction.

The main soil variable determining compaction is water content. Increasing water content to a certain point makes the soil easier to compact; for example, the loose sandy loam soil in Figure 9 is more compactable at 11% water content than at 6%. Above an optimum water content, compaction decreases and soil puddling increases. Over the range of water contents below field capacity, where soils are usually tilled, compaction increases with increasing water content.

6.7.1 Recognizing soil compaction

Compacted layers in soils become apparent when they interfere with water movement or root growth. Various methods are available to measure severity and depth of compaction.

Soil penetrometers measure the force required to push a probe through a soil, giving a quantitative comparison among different fields in the nursery and from year to year. This is a relative measure, however, because no probe pushed rapidly into soil can adequately reproduce the process by which a root grows. Penetrometer readings show whether compacted layers are present and where they occur. Because soil resistance to the penetrometer probe increases with decreasing water content, a comparison of readings is valid only if water content is the same.

Mechanical resistance and poor aeration, the two main factors limiting growth in compacted soils, often work together and are hard to separate. Compaction increases mechanical resistance by increasing soil strength and decreases the volume of large pores in which gas exchange occurs. For example, root growth of maize seedlings (Fig. 10) is shown as a function of these variables; applied pressure formed the mechanical resistance, and different oxygen concentrations simulated aeration. In Figure 11, root elongation of pea seedlings is shown as a function of bulk density and water content of a soil. The mechanical resistance due to these two variables was measured as penetrometer resistance. If only mechanical resistance was present, root elongation would have followed the line segments labeled "b." Decreased root elongation shown by the line segments "a" resulted from decreased aeration in the wetter soil samples.

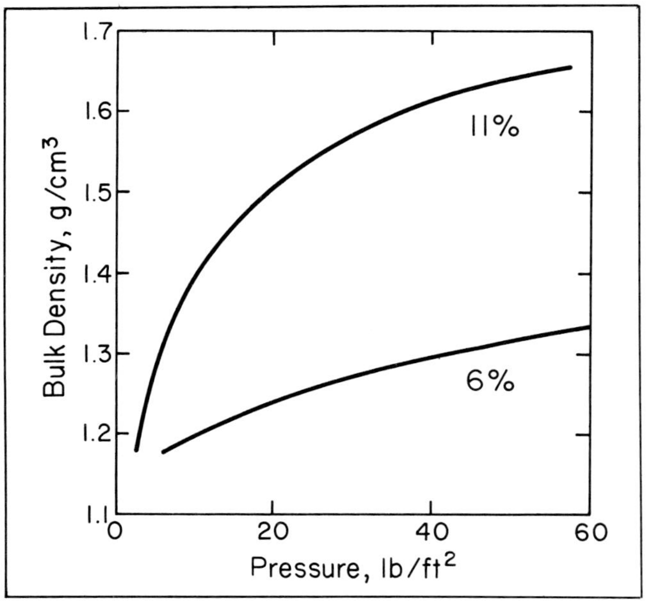

Figure 9. Compaction of a sandy loam soil at two different water contents (adapted from [3]).

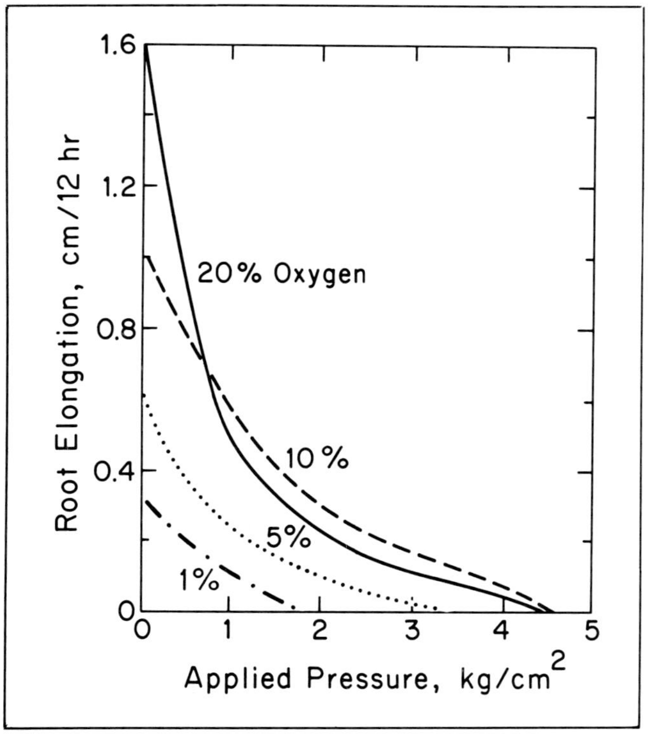

Figure 10. Effect of interaction of aeration and mechanical resistance on growth of maize roots (adapted from [2]).

Many measurements and observations have shown that roots can exert large forces to grow; for example, roots can crack pavement. However, growth rates under those conditions are very slow and are inadequate to support active shoot growth. Figures 10 and 11 show that root growth decreases very quickly as soil resistance increases.

Bulk density can be readily measured, but its measurement requires some care. Soil can be cored at different depths and then dried, and its bulk density calculated by dividing its oven-dry weight by the volume of the core. Bulk density also can be measured nondestructively. Nuclear density probes, largely research tools, can be placed on or into soil for this purpose. However, this equipment is expensive, and special safety precautions must be taken with its radioactive source. Air permeameters also are used in research work to evaluate compacted layers which restrict air movement.

Bulk density values for sandy loam soils normally fall in the range of 1.3 to 1.5 g/cm³. High (above 6%) organic matter contents result in lower bulk density. Soils with high contents of halloysite clay, amorphous clay minerals, or pumice grains have low bulk densities, around 1.0 g/cm³; these soils also usually have good drainage, good workability when wet, and lower susceptibility to compaction.

6.7.2 Minimizing soil compaction

On the basis of the compaction process described above, the options for minimizing soil compaction fall into four groups:

(1) Decrease the amount of pressure applied to soil and the number of times it is applied. Use tractors and equipment with low ground pressure (see Fig. 8); use equipment as few times as possible; and dedicate certain soil areas for tractor wheels.

(2) Increase the resistance of the soil to compaction by adding organic matter to increase soil aggregate stability and by draining soils to decrease water content.

(3) Till the soil at lower water contents, where it is more resistant to compaction.

(4) Use tillage equipment that has the least compacting influence. Avoid creating tillage pans; till to different depths at different times; do not use vibrating tools.

6.7.3 Improving compacted zones

Though avoiding compaction is most desirable in principle, it is often not possible in practice. Seedbed preparation requires certain tillage operations, and seedlings often must be lifted when the soil is wet. Compacted soil zones do occur—and amelioration is necessary.

Reversing compaction requires moving soil aggregates into an arrangement of greater porosity and then stabilizing that new arrangement in some way. The most common practice for overcoming compacted subsoils is ripping. Ripping shanks, at 40- to 80-cm (15- to 30-in.) spacing, are pulled through the soil 40 to 80 cm (15 to 30 in.) deep. A second pass at right angles to the first is common.

Although most nursery managers claim good results with ripping, soil scientists have theoretical reasons for believing that changes in the soil due to ripping may be only transitory. If the same forces, whether natural or due to soil manipulation such as tillage, continue to act on the soil, then soil particles would again settle into the original bulk density. To prevent this, the soil would have to be modified in some way. Many measurements of porosity and rooting have failed to detect differences between ripped and unripped fields [unpubl. data, 11].

What are the requirements for effective ripping? To increase porosity, the soil volume must be increased (Fig. 12). If ripping does not increase height of soil at the surface, which is a function of design, angle, and depth of the implement used, it cannot increase soil porosity. There is a critical depth for each type and shape of ripping tine [9].

If the soil is dry, ripping will shatter it, producing the desired effect. If the soil is wet, aggregates can readily flow back into a structure with the original porosity. However, even under ideal conditions of loosening and shattering, ripping may produce small soil clods, loosely arranged, which have large pore spaces between them but low porosity within. Ripping cannot change the internal porosity of 1- to 5-mm clods. Though the ripped soil has some large fissures, which aid aeration, root penetration, and water drainage, root hairs cannot easily extract water and nutrients from clods. The increased drainage due to large fissures probably accounts for much of the benefit obtained from ripping.

The effects of ripping can probably best be evaluated visually. Expose a soil face to below the depth of the ripping tine. The amount of shattering and lifting and any compaction from the shaft or tine can then be observed (Fig. 12).

6.7.4 Soil crusting

Soil crusts are thin layers at the surface that are either rigid enough to decrease seedling emergence and damage stems of growing plants, or impermeable enough to decrease infiltration of water into soil or gas exchange between soil pores and the atmosphere. The crusts, commonly 1 to 5 mm thick, result from movement and bonding of soil grains into a new, more dense arrangement due to falling water drops from either rainfall or irrigation. Crusting is most common in soils with a high content of fine sand or coarse silt. Stable aggregates are difficult to maintain in these soils, and special management practices often are necessary to overcome effects of crusts.

Some soils contain plate-shaped grains such as unweathered mica in their sand and silt fractions. If the aggregate stability of these soils is low, the grains will disperse on wetting. These plate-shaped grains then settle out with a preferred orientation along their horizontal axis. On drying, clay grains will bond the adjacent plates. The effect is analogous to sheets of paper allowed to fall and then glued in spots. A hard and impermeable crust forms.

Another common mechanism for crust formation is for dispersed grains of fine sand and coarse silt to flow into pores between aggregates or large grains. The silt grains seal the pores in a thin layer of soil at the surface. When the soil dries, a hard and impervious crust forms.

Because crusts result from grain movement and rearrangement, wetting alone will not remove them although it decreases their strength. Light cultivation will obliterate the effect of crusts by bringing uncrusted soil to the surface; however, a crust will form at the next wetting.

Preventing crusts requires increasing aggregate strength to prevent dispersion of grains, protecting the surface from the energy of water striking it, and decreasing the bonding between grains on drying. Mulches protect the surface from direct impact of falling water drops. Organic matter commonly is added to increase aggregate stability and decrease bonding between soil grains on drying; decomposing organic molecules coat mineral grains and interfere with strong grain-to-grain bonds. Type of cultivation also influences aggregate stability; for example, rototillers destroy aggregates, thereby increasing the chance of crusting.

Various materials such as phosphoric acid and vermiculite have been added to soils as anticrusting agents; the effect is usually to decrease bonding. Chemicals can be added in a band above the seed; vermiculite can be covered with a thin

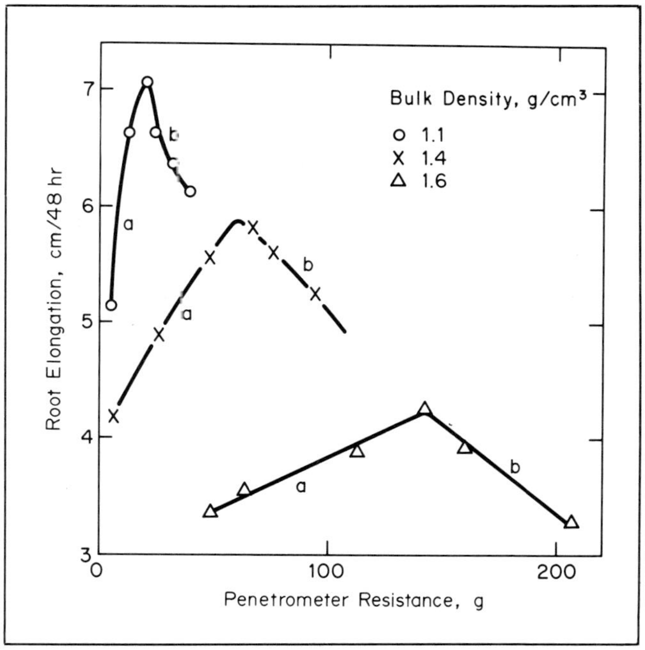

Figure 11. Effects of aeration (a) and mechanical resistance (b) on growth of pea roots (adapted from [1]).

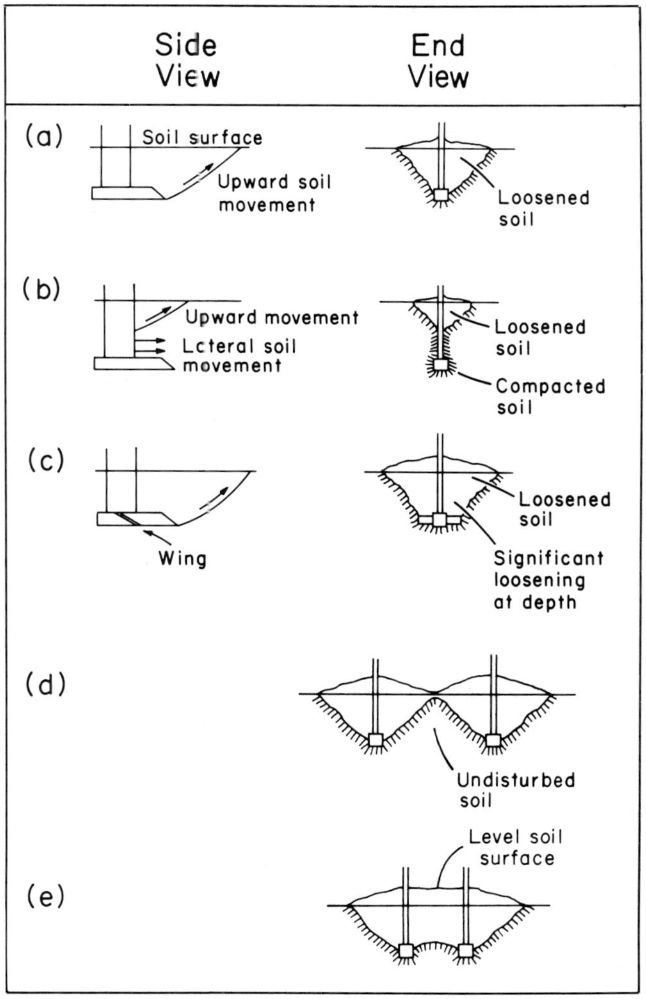

Figure 12. Effects of various subsoil tines: subsoiler (a) above critical depth and (b) below critical depth; (c) winged subsoiler above critical depth; soil disturbance (d) at wide tine spacing and (e) at narrow tine spacing (adapted from [9]).

layer of soil to prevent it from being blown away by wind or washed away by water. Hemphill [7] found vermiculite more effective than phosphoric acid (Table 4) in promoting emergence of vegetable seeds.

Table 4. Evaluation of anticrusting materials for vegetable crops (adapted from [7]).

Treatment	Number of seedlings emerged per meter of row at 14 days		
	Carrot	Lettuce	Onion
Control	4.7	4.7	5.2
Phosphoric acid	8.8	9.0	8.3
Vermiculite	15.7	27.3	17.3

6.8 Northwest Nurseries: Assessment of Soil Physical Properties

Of the soils identified in the OSU Nursery Survey (see chapter 1, this volume), 54% were sandy loams, 18% loamy sands, 12% loams, 8% silt loams, and 8% clay loams.

The sandy loams had 2 to 3% organic matter. Most managers prefer an organic matter content of at least 5% and are actively adding organic matter to their soils. Most tillage equipment can be used on sandy loam soils, which are relatively easy to till; tilth improves as organic matter is added. Good tilth in sandy loams depends predominantly on maintaining high levels of actively decomposing organic matter.

The loams to clay loams, which are finer grained, store more water and can be tilled effectively only over a narrow range of water contents. Seedbed preparation, which consists of breaking the soil into fine aggregates, is more difficult, as is separating soil from roots during lifting. However, these characteristics depend upon the type of clay. For example, the clay fraction of soils formed on volcanic materials has relatively low cohesion and adhesion.

Rototillers are used in almost all nurseries. Some managers realize that this equipment destroys soil structure by pulverizing the soil too much and leaving a compacted layer just below depth of working; nevertheless, rototilling is often the easiest way to get the fine seedbed desired for uniform germination of small seeds. Although rototillers are often singled out, any tillage operation has the potential to destroy soil structure and create tillage pans in the soil. Therefore, the best general management guide is to decrease the number of tillage operations to the minimum necessary for seedbed preparation.

Of the soil-related problems identified by nursery managers (Table 5), compaction and organic matter maintenance were the major concerns. Preventing compaction requires that soil not be worked when it is at a water content near field capacity. Because this is often unavoidable, compacted zones have to be ameliorated by ripping. The large emphasis on maintaining organic matter seems justified, based on its many benefits, although a few soils have inherently good physical properties at low organic matter levels. Certain amorphous or oxide clay minerals can impart good structure. For most soils, however, organic matter is essential to good soil structure.

Table 5. Soil-related problems in Northwest nurseries (OSU Nursery Survey).

Problem	% of time	
	Considered a problem	Listed 1st or 2nd in importance
Compaction	62	24
Organic matter maintenance	62	19
Poor drainage	43	19
Wind abrasion	34	19
Too much variation	29	14
Too "heavy"	29	14
Uneven topography	24	5

Poor drainage was identified as a problem, although most managers rated their soils as having good drainage. The poor drainage may result from unevenness in the fields, identified as a problem by a quarter of the managers. Inadequate drainage may be more of a problem than managers realize, however, because its effects are subtle. Decreased root growth, decreased efficiency in nutrient uptake, and plant changes due to decreased aeration will reduce seedling growth uniformly over an area; therefore, the amount of the decrease may not be apparent. Intensive management of soils usually requires artificial drainage.

Soil variation within a field makes it difficult to manage the field uniformly. Variation in physical properties is hard to correct; often, the variation occurs over such a small scale that different units cannot be separated as different fields. Land with excessive variation over a small area should be avoided for use as nursery sites.

Soil splash, often a first step in crust formation, has been identified as a problem in some nurseries. Soil splash will be controlled by the same preventative measures used to control crust formation.

Success in maintaining a good physical environment for growth of seedlings depends upon a wise choice of site and wise manipulation of the soil. The objective of this manipulation is to maintain a stable soil structure with a sufficient volume of pores of different sizes to allow for the important fluxes of air, water, and nutrients and for water storage. The soil must resist compaction, puddling, and crusting to maintain this pore assemblage. No "magic" substance can be added to soil to achieve a stable structure—this is truly a management concern. Maintenance of organic matter, adequate subsurface drainage, use of soil-building crops in a rotation, and judicious tillage are all parts of a successful soil-management program.

Technical Paper 6536, Oregon Agricultural Experiment Station Oregon State University, Corvallis.

References

1. Eavis, B. W. 1972. Soil physical conditions affecting seedling root growth. Plant and Soil 36:613-622.
2. Gill, W. R., and R. D. Miller. 1956. A method for study of the influence of mechanical impedance and aeration on the growth of seedling roots. Soil Sci. Society of America Proc. 20:154-157.
3. Gill, W. R., and G. E. Vanden Berg. 1967. Soil dynamics in tillage and traction. U.S. Dep. Agric., Washington, D.C. Agric. Handb. 316. 511 p.
4. Grable, A. R. 1971. Effects of compaction on content and transmission of air in soils. Pages 154-164 *in* Compaction of agricultural soils (K. K. Barnes, W. M. Carleton, H. M. Taylor, R. I. Throckmorton, and G. E. Vanden Berg, eds.). American Society of Agric. Engineers, St. Joseph, Michigan.
5. Greenland, D. J. 1979. Structural organization of soils and crop production. Pages 47-56 *in* Soil physical properties and crop production in the tropics (R. Lal and D. J. Greenland, eds. Wiley and Sons, New York.
6. Hardy, F. 1974. Root room. Tropical Agric. 51:272-278.
7. Hemphill, D. D. 1982. Anticrustant effects on soil mechanical resistance and seedling emergence. Horticultural Sci. 17:39-393.
8. Reaves, C. A., and A. W. Cooper. 1960. Stress distribution in soils under tractor loads. Agric. Engineering 41:20-21, 31.
9. Spoor, G. 1982. Deep loosening and mole drainage on clay soils. Tropical Agric. 59:97-102.
10. van den Driessche, R. 1969. Forest nursery handbook. B.C. Forest Serv., Victoria. Res. Note 48. 44 p.
11. Vomocil, J. A. 1982. Unpublished data, Dep. of Soil Science, Oregon State Univ., Corvallis.
12. Warkentin, B. P. 1982. Clay soil structure related to soil management. Tropical Agric. 59:82-91.

For Further Reading

Arkin, G. F., and H. M. Taylor (eds.). 1981. Modifying the root environment to reduce crop stress. American Society of Agric. Engineers, St. Joseph, Michigan. Monograph 4. 407 p.

Barnes, K. K., W. M. Carleton, H. M. Taylor, R. I. Throckmorton, and G. E. Vanden Berg (eds.). 1971. Compaction of agricultural soils. American Society of Agric. Engineers, St. Joseph, Michigan. Monograph. 471 p.

Brady, N. C. 1974. The nature and properties of soils. 8th ed. Macmillan, New York. 639 p.

Flegmann, A. W., and R. A. T. George. 1977. Soils and other growth media. Avi Publishing Company, Inc., Westport, Connecticut. 170 p.

Chapter 7
Soil Fertility in Forest Nurseries

R. van den Driessche

Abstract

The soil cation-exchange complex serves as a reservoir of nutrients which are released into the soil solution, where they are accessible to seedlings. Although macronutrients are most readily available in soils of pH 6 to 7, micronutrients are most available in more acid soils; therefore, pH values of 5.0 to 6.0 are recommended for forest nurseries. Under such conditions, available nitrogen is primarily in the ammonium form, and phosphorus can form insoluble iron and aluminum compounds. Sulfur, potassium, calcium, magnesium, and micronutrients are seldom deficient in forest nurseries because sufficient fertilizer is added as "maintenance" dressings, or supplies from native minerals are adequate. Recommended fertilizer applications for a 2-year nursery rotation range from 112 to 285 kg of nitrogen, 67 to 200 kg of phosphorus, and 75 to 150 kg of potassium per ha. Recommended nutrient levels in both soils and seedlings are tabulated and some effects of nutrients on seedling growth and physiology mentioned.

7.1 Introduction

The primary purpose of forest nurseries is to produce trees to form new forests. Therefore, maintaining adequate fertility in bareroot nursery soils is important to assure production of high-quality planting stock. Gathering the appropriate information on maintaining adequate nursery soil fertility into a single publication has been attempted many times previously [e.g., 1, 9, 58, 71, 77] and undoubtedly will be necessary again as conditions change and new information becomes available. In this chapter, the main factors affecting soil fertility are outlined and the management measures that can alter or maintain fertility described. Particular attention is devoted to fertilizers and their use, and some effects of nutrients on seedling growth and physiology also are included.

7.2 Soil Cation-Exchange Capacity

Soils are derived primarily from minerals but also contain organic matter. The colloidal fractions (<0.002 mm in diameter) of both mineral soil and soil organic matter are the chemically active portions. The colloidal mineral fraction is constituted of clays consisting of particles (micelles) of silicate and alumina arranged in crystal lattice structures [e.g., 20]. These micelles carry an overall negative charge and so can attract and adsorb positively charged particles (cations) such as hydrogen (H^+) or positively charged metallic ions such as ammonium (NH_4^+), potassium (K^+), calcium (Ca^{++}), and magnesium (Mg^{++}). Colloidal organic matter (humus) also carries negative charges and behaves as micelles do except that it carries many times more negative charges for the same amount of dry weight.

Cations adsorbed to clay micelles and organic matter can be displaced by other cations that are more positively charged or as a result of mass action. The quantity of cations which can be adsorbed or displaced is a measure of the cation exchange capacity (CEC) of the soil. This measurement is important for soil fertility because nutrient cations held on the soil CEC are not leached but are available for plant growth. Although CEC is normally measured at pH 7, nursery soils may have lower pH values and high organic matter contents such that the effective CEC may be lower than the measured CEC.

Nutrients are released from the CEC complex into the soil solution in the form of ions (Table 1), which are absorbed by plants. The proportion of the CEC occupied by bases is referred to as the percent base saturation. The remaining CEC is assumed to be occupied by H^+ ions which confer an acid reaction to the soil; therefore, soils with low percent base saturation tend to be acidic. Low percent base saturation also implies that the supply of nutrient cations for plant growth is low.

CEC is measured in milliequivalents (meq), which relate the combining capacity of soil and nutrient cations. For example, a soil with a CEC of 20 meq could adsorb 20 meq Ca, which

In Duryea, Mary L., and Thomas D. Landis (eds.). 1984. *Forest Nursery Manual: Production of Bareroot Seedlings*. Martinus Nijhoff/Dr W. Junk Publishers, The Hague/Boston/Lancaster, for Forest Research Laboratory, Oregon State University, Corvallis. 386 p.

equals 400 mg Ca (20 x equivalent wt. of Ca, 20.0 g; see Table 1), 20 meq K (782 mg K), or 20 meq Mg (243 mg Mg).

The meq values are normally expressed on the basis of 100 g dry soil. For practical purposes, 1 mg of nutrient/100 g of soil is equivalent to 22 kg/ha (20 lb/acre) furrow slice. Thus, 0.4 meq K/100 g soil would represent

0.4 x 39.1 = 15.64 mg K/100 g
15.64 x 22 = 344 kg K/ha (312 lb K/acre) furrow slice

where 39.1 is the equivalent weight of K (Table 1).

CEC values for a number of nurseries in the Pacific region of Canada and the United States ranged from 8 to 30 meq/100 g soil, and base saturation was usually less than 50% [71]. The OSU Nursery Survey (see chapter 1, this volume) showed the mean CEC of 16 nurseries to be 12.5 meq/100 g (range, 6 to 29 meq/100 g).

Table 1. Ionic forms of macronutrients and their equivalent weights.

Nutrient	Ionic form	Equivalent weight[1]
Potassium	K^{+}	39.1
Calcium	Ca^{++}	20.0
Magnesium	Mg^{++}	12.2
Nitrogen (nitrate)	NO_3^{-}	62.0
Nitrogen (ammonium)	NH_4^{+}	18.0
Sulfur (sulfate)	SO_4^{--}	48.0
Phosphorus (phosphate)	PO_4^{---}	31.7

[1]Equivalent weight is the weight that will combine with (for unlike charges) or replace (for like charges) the weight of some other element. Thus, 39 g of K^{+} will replace 20 g of Ca^{++} or combine with 48 g of SO_4^{--}.

7.3 Soil pH

Soil pH, or reaction, is described by the pH scale in which 7 is neutral for soils measured in water. Soils measuring from 5.5 to 6.5 are generally regarded as slightly acid, those from 4.5 to 5.5 as acid, and those less than 4.5 as strongly acid. Values lower than 3.5 are rare. Alkaline soils have pH values above 7.5.

The pH is normally measured in a mixture of 1 part soil to 1 part distilled water; but other ratios (e.g., 1:5) may be used, or 0.1 molar calcium chloride may be used instead of distilled water. However, different methods result in different values. In particular, the calcium chloride method indicates pH values about 0.5 units lower than those obtained in water.

Soil pH affects availability of nutrients to plants and influences the composition of soil flora and fauna, including some crop pathogens. The macronutients nitrogen (N), K, Ca, and Mg are most readily available at soil pH values above 6, but maximum availability of P is restricted to between pH 6 and 7 [20]. The micronutrient metals iron (Fe), manganese (Mn), zinc (Zn), copper (Cu), and cobalt (Co) are most available in soils with pH values below about 5.5. Most conifers tend to become chlorotic on soils of neutral or alkaline pH because of their inability to obtain adequate Fe and Mn. However, extremely acid soils (pH < 4.5) are infertile because they do not retain nutrient cations such as NH_4^{+}, K^{+}, and Ca^{++} to any extent. Incidence of damping-off is reduced when nursery soil pH is maintained in the region of 4.5 to 6.0 [62], and weed problems also are reduced on acid soils (see chapters 18 and 19, this volume).

Ideal values for conifer nursery soils are pH 5 to 6 and those for hardwood nursery soils pH 6 to 7 [65]. Aldhous [1] warns against allowing nursery soil pH to become too high and recommends pH 5 for conifer nurseries, pH 5.5 for hardwoods, and pH 6 for poplars. Growth of several Northwest conifer species is optimal between pH 5 and 5.5 [13].

7.4 Nutrient Form and Availability

7.4.1 Nitrogen

Three forms of N occur in soil: (1) organic N associated with the soil humus, (2) ammonium N (NH_4-N) fixed within the lattice of clays such as vermiculite, and (3) soluble inorganic ammonium and nitrate compounds [20]. As soil organisms slowly break down organic matter, ammonium is released into the soil solution. In the presence of adequate bases, ammonium is nitrified to produce nitrate ions (NO_3^{-}). Nitrification is probably slow in forest nursery soils because most, but not all, are low in bases. In any case, most conifer seedlings grow well with a predominantly NH_4-N source [43, 49, 69]; Douglas-fir [*Pseudotsuga menziesii* (Mirb.) Franco] can be grown exclusively on NH_4-N [75]. Although the NH_4-N fixed within the clay lattice is relatively unavailable to plants, the NH_4^{+} adsorbed within the measurable CEC is available to plants and relatively resistant to leaching; NO_3^{-}, on the other hand, is readily leached from soil.

The common Kjeldahl chemical analysis for N in soils determines all the N present other than nitrate, which is normally excluded. Although this value generally indicates soil N status, it is frequently a poor guide to N fertilizer requirements, which may be better judged from measuring of mineral N in light nursery soils. Effects of fertilizing a sandy loam soil with ammonium nitrate were readily detected by measuring extractable mineral N [74].

7.4.2 Phosphorus and sulfur

Occurring primarily in the earth's surface as insoluble apatite [$Ca_5(PO_4)_3F$ or $Ca_5(PO_4)_3OH$], P is present in the soil mainly as inorganic phosphates (50 to 70%) and organic P (30 to 50%), which together compose the solid phase. A very small amount of P, in proportion to the solid phase, is present in the liquid phase as orthophosphate ions ($H_2PO_4^{-}$, HPO_4^{--}, and PO_4^{---}). Though the proportion of orthophosphate ions is influenced by pH, it is thought that plants can take up any one of these ions. These soluble phosphates react with Fe and aluminum (Al) under acid soil conditions to form insoluble $FePO_4 \cdot 2H_2O$ and $AlPO_4 \cdot 2H_2O$ and with Ca to form apatites in neutral and calcareous soils. The P thus becomes fixed in a form that is unavailable to plants.

Phosphorus fertilizers dissolve in water to release orthophosphates which, if not absorbed by plants, are steadily rendered unavailable by the fixation process just described. Soils with more clay tend to fix more of the P supplied by fertilizer than those with less clay, but soils with more organic matter tend to fix less. Organic matter improves P availability because it (1) competes with phosphate ions for binding sites on soil particles, (2) produces organic anions which chelate (form organic compounds with nutrients available to plants) Al, Fe, and Ca, and (3) slowly releases P during decomposition.

In forest soil, trees appear to rely, at least partly, on mycorrhizae for obtaining soil P [30] (see chapter 20, this volume). Even in the nursery, P deficiency has been detected in white spruce [*Picea glauca* (Moench) Voss] after soil fumigation [24]. This lack was attributed to destruction of mycorrhizal fungi essential to P uptake but could be rectified by using adequate P fertilizer.

Heavy use of superphosphate fertilizer has been reported to greatly accentuate Cu deficiency symptoms in Sitka spruce [*Picea sitchensis* (Bong.) Carr.] [13].

Analysis of total soil P is relatively uninformative because only a small fraction of P is available for plant growth. Consequently, several methods for determining available P have been devised which employ a variety of extracting agents (such as sodium bicarbonate and various dilute acids) and give a variety of results, seldom comparable. One suitable method for acid nursery soils is the dilute acid-fluoride procedure (Bray and Kurtz No. 1 solution; see [33], p. 159) which is used in Ontario [7] and British Columbia [71]. A sodium bicarbonate extraction solution (Olsen's method; see [33], p. 164) is widely used for calcareous soils of high pH.

Sulfur (S) occurs mainly in soil organic matter and is absorbed by plants as sulfate. Unlike P, it is not rendered unavailable to plants by reaction with other soil components, and it is usually present in adequate amounts in fertilized nursery soils because many fertilizers contain it in substantial amounts. For example, ammonium sulfate contains 24% S, potassium sulfate 13% S, and calcium superphosphate 12% S. Nevertheless, S can be deficient in nurseries [18]. In fact, incidence of S deficiency may be increasing due to use of fertilizers with low S content as well as reduced industrial SO_2 emissions. The amount of S in a conifer is closely related to the amount of N; the ratio is 1 part S to 14 parts N by weight [66]. Knight [35] recommends ensuring that 1/15 as much S as N is applied to light soils to safeguard against deficiency.

7.4.3 Potassium, calcium, and magnesium

The cations K^+, Ca^{++}, and Mg^{++} are adsorbed on soil cation-exchange sites where they are available to plants. Potassium-containing minerals, which are widespread, weather to release K. Exchangeable Ca and Mg also are derived from weathering of soil minerals, but these are not so ubiquitous. Ca is more readily displaced than K and can become depleted in acid nursery soils. However, even K can be rapidly depleted by leaching in sandy nursery soils with a pH of 5 or less [37].

7.4.4 Micronutrients

The metallic micronutrients Fe, Mn, Cu, and Zn occur as cations in the soil solution at low pH. At high pH they are converted to insoluble oxides and hydroxides which are not available to plants; for example, Douglas-fir seedlings growing in an artificial container soil (pH 6) to which too much Ca (50 meq/100 g) had been added showed Mn deficiency symptoms and contained no detectable foliar Mn. Healthy seedlings growing in a medium with a similar Ca level but at pH 5.5 contained 2 ppm foliar Mn [unpubl. data, 76]. Availability of micronutrients is generally reduced by increasing soil pH, as occurs with liming, although availability of molybdenum (Mo) is increased by raising pH. Boron (B) is generally more available under acid conditions; however, its concentration usually decreases down the soil profile. Therefore, deficiencies may be accentuated in dry weather because surface root activity is curtailed by lack of water [20]. Organic matter can complex metallic cations and render them unavailable to plants or can also produce molecules that chelate micronutrients.

7.5 Recommended Nutrient Levels

Soil and plant analysis can help the nursery manager maintain adequate nutrient levels for satisfactory plant growth (see chapter 8, this volume). These analyses are not always easy to interpret, however, and the quality and growth of stock over the previous few years provide equally important guidance for changes in nursery fertility.

7.5.1 Soils

The soil nutrient levels to be expected in a fertile nursery on the Pacific Coast growing Douglas-fir seedlings (Table 2) can probably be taken as a guide for most nurseries in the Northwest, although the range of values may be large [71]. Analyses from nurseries in other regions [e.g., 73] show levels similar to or somewhat lower than those in Table 2. Note that percentage of Kjeldahl N is largely a function of soil organic matter content and seldom indicates the N available to plants.

Table 2. Expected range in analytical values for Douglas-fir nursery soils (adapted from [71]).

	Range	Analytical method
pH	4.8-5.5	1 soil:1 water paste
Organic matter, %	3-5	Wet oxidation
N, %	0.20-0.25	Kjeldahl
P, ppm	100-150	Bray and Kurtz No. 1, dilute acid-fluoride
K, meq/100 g	0.20-0.30	
ppm	78-117	
Ca, meq/100 g	3.0-8.0	1 N ammonium acetate leachate at pH 7
ppm	600-1,600	
Mg, meq/100 g	0.7-2.0	
ppm	170-486	
CEC, meq/100 g	10-20	

Soil is most conveniently sampled when it is in fallow or after cover cropping so that recommendations for adjusting fertilizer schedules can be prepared before sowing the new crop (see chapter 8, this volume). One of the best ways of using soil analysis is to maintain records for each management unit and interpret them for the effects of different management procedures.

7.5.2 Seedlings

Because nutrient concentrations in conifers vary with season, it is conventional to sample seedlings in late autumn or early winter when nutrient levels are relatively stable. Whole shoots or entire plants commonly are analyzed for 1 + 0 seedlings, but only needles usually are removed and tested in older stock. Nutrient concentrations in needles are higher than those in stems and roots, but concentrations in 1 + 0 seedlings are higher than those in 2 + 0 seedlings. It is convenient to sample whole 1 + 0 seedlings in mid-October for chemical analysis so that inadequacies in plant nutrient concentrations can be rectified by fertilizing during the second year of growth (see chapter 8, this volume).

Nutrient concentrations vary with growing conditions of the crop, and from year to year, but certain ranges can be expected (Table 3). Seedlings with concentrations below the lower limit of the 50% range may be inadequately supplied with nutrients, and those with levels close to the minima almost certainly require appropriate fertilizing. Micronutrient levels in whole 1 + 0 Douglas-fir seedlings whose roots had been thoroughly washed ranged from 30 to 101 ppm B, 108 to 180 ppm Mn, and 47 to 66 ppm Zn [71]. In 1 + 0 white spruce, Mn ranged from 328 to 1,456 ppm [70]. The higher values were associated with applying chelated micronutrients to the seedlings.

Nutrient levels expected in needles of adequately supplied 2 + 0 Douglas-fir (Table 4) are probably a guide to levels that can be expected in most other conifer species grown in Northwest nurseries. The deficiency level of K in white spruce needles was 0.13 to 0.21% [31]. The Mn level ranged from 636 to 2,852 ppm in 2 + 0 white spruce foliage in a nursery experiment where Mn chelate was used [70]. The deficiency level of S for Sitka spruce needles was about 0.08% and the sufficiency level about 0.16% [18]. The deficiency level of Cu in Sitka spruce was 2.5 ppm and in Douglas-fir 4 ppm [60].

Interpreting foliar nutrient concentrations can be complicated by effects of environmental factors and interactions between nutrients. However, in healthy plants the ratios between the different nutrients are fairly constant. Work with Douglas-fir, Sitka spruce, and western hemlock [*Tsuga heterophylla* (Raf.) Sarg.] seedlings shows that if the percentage of N is set at 100, the proportions of the other nutrients are, approximately, 16 P, 60 K, 5 Ca, 5 Mg, 9 S, and 0.7 Fe under favorable nutrient conditions [32]. These proportions can serve as a guide to nutrient imbalance within the seedling.

Table 3. Means and ranges of morphological and nutrient-concentration values for samples of 40 (80 white spruce) 1+0 seedlings of five species collected on October 15, 1968 to 1978.

Measurement	Mean	Minimum	Maximum	50% range[1]
Coastal Douglas-fir (233 observations)				
Seedling dry wt., g	0.48	0.08	1.36	....
Shoot length, cm	9.93	3.60	19.22	....
Shoot: root ratio	0.50	0.29	1.05	....
N, %	1.61	0.74	2.55	1.41-1.81
P, %	0.20	0.11	0.45	0.17-0.23
K, %	0.85	0.43	1.32	0.74-0.95
Ca, %	0.30	0.01	0.61	0.22-0.38
Mg, %	0.11	0.06	0.22	0.10-0.13
Interior Douglas-fir (70 observations)				
Seedling dry wt., g	0.29	0.07	0.81	....
Shoot length, cm	6.34	3.50	12.80	....
Shoot: root ratio	0.56	0.30	1.15	....
N, %	1.93	1.34	2.79	1.73-2.13
P, %	0.25	0.18	0.41	0.22-0.28
K, %	0.84	0.63	1.44	0.74-0.95
Ca, %	0.31	0.05	0.67	0.22-0.40
Mg, %	0.12	0.07	0.17	0.10-0.13
Sitka spruce (44 observations)				
Seedling dry wt., g	0.23	0.08	0.47	....
Shoot length, cm	5.68	3.40	11.90	....
Shoot: root ratio	0.51	0.25	1.03	....
N, %	2.03	1.06	2.62	1.80-2.26
P, %	0.25	0.16	0.37	0.18-0.33
K, %	1.15	0.74	1.42	1.05-1.25
Ca, %	0.51	0.32	0.71	0.45-0.57
Mg, %	0.16	0.11	0.25	0.14-0.18
White spruce (234 observations)				
Seedling dry wt., g	0.18	0.03	0.60	....
Shoot length, cm	3.90	1.70	9.20	....
Shoot: root ratio	0.57	0.24	0.97	....
N, %	2.59	0.24	3.50	2.28-2.91
P,%	0.32	0.22	0.42	0.30-0.35
K, %	0.90	0.52	1.26	0.83-0.98
Ca, %	0.49	0.12	0.87	0.39-0.59
Mg, %	0.15	0.10	0.22	0.14-0.17
Lodgepole pine (53 observations)				
Seedling dry wt., g	0.58	0.15	1.58	....
Shoot length, cm	6.55	2.70	14.60	....
Shoot: root ratio	0.42	0.30	0.62	....
N, %	1.99	1.38	2.66	1.76-2.22
P,%	0.25	0.17	0.33	0.22-0.28
K, %	0.95	0.73	1.24	0.86-1.03
Ca, %	0.32	0.19	0.52	0.27-0.37
Mg, %	0.13	0.10	0.17	0.12-0.14

[1]Range in mineral nutrient concentrations for 50% of observations; range calculated as mean ±0.68 standard deviation.

7.5.3 Deficiency symptoms

Inadequate mineral nutrition usually results in reduced seedling growth before any characteristic deficiency symptoms become evident. Visual symptoms of macronutrient deficiencies have been described by Purnell [52], Sucoff [61], Stone [59], Benzian [13], Baule and Fricker [11], and Armson and Sadreika [9]. Morrison's [46] summary (Table 5) shows that, in many instances, symptoms are rather similar for deficiencies of different nutrients. Thus, determining the particular nutrient causing the deficiency is seldom possible without supporting evidence, such as tissue analysis or alleviation of symptoms by nutrient addition.

Table 5. Visual deficiency symptoms in conifers (adapted from [46]).

Nutrient	Deficiency symptoms
N	General chlorosis and stunting of needles increasing with severity of deficiency; in most severe cases, needles short, stiff, yellow-green to yellow; in some cases, purple tipping followed by necrosis of needles at end of growing season.
P	Youngest needles green or yellow-green; older needles distinctly purple-tinged; purple deepens with severity of deficiency; in very severe cases in seedlings, all needles purple.
K	Symptoms vary: usually needles short, chlorotic, with some green near base; in some severe cases, purpling and necrosis with top dieback, or little or no chlorosis of needles but purpling, browning, or necrosis.
Ca	General chlorosis followed by necrosis of needles, especially at branch tips; in severe cases, death of terminal bud and top dieback; resin exudation.
Mg	Yellow tipping of current needles followed in severe cases by tip necrosis.
S	General chlorosis of foliage followed in severe cases by necrosis.
Fe	More or less diffuse chlorosis confined in milder cases to new needles; in more severe cases, bright yellow discoloration with no bud development.
Mn	Needles slightly chlorotic; in severe cases, some necrosis of needles.
B	Tip dieback late in growing season with associated chlorotic-to-necrotic foliage, intergrading to dieback of leading shoot with characteristic crooking.
Zn	Extreme stunting of trees with shortening of branches; needles yellow, short, crowded together on twig, sometimes bronze-tipped; older needles shed early, with resultant tufting of foliage; in severe cases, trees rosetted with top dieback.
Cu	Needles twisted spirally, yellowed or bronzed; "tip-burn" or necrosis of needle tips evident; in severe cases, young shoots twisted or bent.
Mo	Chlorosis of leaves followed by necrosis of tissue, beginning at tip and eventually covering whole leaf.

Table 4. Nutrient concentrations[1] expected in dry needles of 2 + 0 Douglas-fir in October.

Level	Nutrient concentrations											
	N	P	K	Ca	Mg	S	SO_4	Fe	Mn	B	Cu	Zn
	%						ppm					
Adequate	1.8	0.18	0.8	0.20	0.12	0.18	80	......	390-1,294	9-39	5.1-7.7	17-53
Low	1.2	0.14	......	......	......	......	......	39-51	......	......	......	......
Very low	1.0	0.09	......	......	......	......	......	......	......	5	2.4-5.1	......

[1]A variety of sources has been used; these are cited in [72]. P concentrations have been revised downward from that paper. Micronutrient values are from [59].

7.6 Fertility Management

Managing nursery soils is something of an art because specifications for many soil and crop characteristics cannot be precise. Location, climate, soil, weather, and many other factors make each nursery unique. Though guidelines and limits can be provided, much depends upon the individual nursery manager, who can keep adequate records of cultural treatments, particularly soil amendments, and conduct regular analyses of soils and crops. By being aware of how soil fertility factors are changing and how stock is growing in the nursery and performing after outplanting, the nursery manager can develop prescriptions to maintain adequate soil fertility.

7.6.1 Controlling soil pH

Raising soil pH can be relatively easy, but reducing it is much more difficult. Thus, any attempts to increase nursery soil pH should be careful and conservative. Fertilizers modify soil pH, with most of the commonly used N and P sources tending to acidify the soil. This effect is usually small and can readily be offset by an occasional small amendment of dolomitic limestone. The considerable ability of soils to resist change in pH is due to the buffer capacity resulting mainly from the reserve acidity of the cation exchange complex. The pH is detected in the soil solution, but this is in equilibrium with the cation exchange complex. Greater proportions of clay or organic matter in the soil provide a larger cation exchange complex and so buffer the soil solution against pH change. High buffer capacity implies stable soil pH.

Soil pH can be reduced with S, aluminum sulfate [$Al_2(SO_4)_3$], and sulfuric acid (H_2SO_4). But these substances are toxic to conifer seedlings at high concentrations and should therefore be applied as long before sowing as possible. Adding more than 1,680 kg/ha (1,500 lb/acre) of S to Ontario nurseries reduced survival of red pine (*Pinus resinosa* Ait.) in the seedbed, though average seedling dry weight increased up to at least 2,520 kg/ha (2,250 lb/acre) of S [47]. Experience in Ontario nurseries has shown that 560 kg/ha (500 lb/acre) of S reduces soil pH 0.5 units over the range pH 5.5 to 7.0 [9].

Sulfur and slaked lime [$Ca(OH)_2$] were applied to silt loam soils at two coastal nurseries in spring to obtain plots with different soil pH [68]. Douglas-fir sown during the same spring showed little adverse effect from these S applications. Sulfur applied at 4,480 kg/ha (4,000 lb/acre) decreased pH by about 1 unit from the control in June and 1.4 units in September (Table 6). Slaked lime applied at 4,480 kg/ha (4,000 lb/acre) increased pH by about 1.1 units in June and about 1.0 in September. Between September and June of the following year, no further major changes in pH occurred.

Organic materials are safe acidifying agents whose effects often become appreciable only after several years of continuing application (see chapter 9, this volume). Hop waste (from a brewery) was effective in reducing pH of a calcareous silt loam (pH 7.8 to 8.0) at a nursery in the East Kootenay region of British Columbia. A dressing 2.5 cm (1 in.) thick, worked into the soil, reduced pH 1.2 units after 2 years; a similar dressing of commercial peat decreased pH 1.0 unit.

Table 6. Average changes in soil pH obtained with S and slaked lime [$Ca(OH)_2$] at two coastal nurseries.

Treatment	March application, kg/ha	Soil pH: Year 1 June	Soil pH: Year 1 Sept.	Soil pH: Year 2 June
Control	0	5.6	5.4	5.5
S	1,680	4.9	4.5	4.7
	4,480	4.5	4.0	3.9
$Ca(OH)_2$	1,680	6.3	5.9	5.9
	4,480	6.7	6.3	6.3

Ground limestone or dolomitic limestone (which contains Mg as well as Ca) is equally good for raising soil pH. The effectiveness of a unit quantity of limestone in changing pH is influenced by both soil texture and soil organic matter content (Table 7). Both clay and organic matter increase the soil's buffer capacity, making it more difficult to either raise or lower the existing pH.

Most nurseries in the Northwest irrigate heavily, and the pH and dissolved salt content of irrigation water can influence soil pH. Water with high pH, containing cations and especially bicarbonates, tends to raise soil pH. Acid injection into the irrigation system is possible [8] when water sufficiently low in bicarbonates is not available.

Table 7. Ground limestone (1,000 kg/ha) required to raise existing soil pH to one of three chosen pH values in soils of different textures (adapted from [1]).

Existing pH by soil analysis	Sands, loamy sands 5.0	5.5	6.0	Sandy loams 5.0	5.5	6.0	Silty loams, silt loams, loams, sandy clay loams 5.0	5.5	6.0	Clay loams, silty clay loams, clay 5.0	5.5	6.0	Soils in previous class, but high in organic matter[1] 5.0	5.5	6.0
3.0	5.0[2]	6.3	7.5	6.0	7.7	9.0	8.2	10.2	12.2	10.0	12.5	15.0	11.9	14.9	18.0
3.2	4.5	5.8	7.0	5.5	7.0	8.5	7.3	9.4	11.4	9.0	11.5	14.0	10.8	13.7	17.0
3.4	4.0	5.3	6.5	4.9	6.4	7.9	6.5	8.5	10.5	8.0	10.5	13.0	9.5	12.6	15.6
3.6	3.5	4.8	6.0	4.3	5.8	7.3	5.6	7.8	9.8	7.0	9.5	12.0	8.4	11.3	14.3
3.8	3.0	4.3	5.5	3.6	5.1	6.7	4.9	6.9	8.9	6.0	8.5	11.0	7.2	10.2	13.2
4.0	2.5	3.8	5.0	3.0	4.5	6.0	4.1	6.2	8.2	5.0	7.5	10.0	6.0	8.9	11.9
4.2	2.0	3.3	4.5	2.5	4.0	5.5	3.3	5.3	7.3	4.0	6.5	9.0	4.8	7.8	10.8
4.4	1.5	2.8	4.0	1.9	3.4	4.9	2.5	4.5	6.5	3.0	5.5	8.0	3.6	6.5	9.5
4.6	1.0	2.2	3.5	1.3	2.8	4.3	1.6	3.6	5.6	2.0	4.5	7.0	2.4	5.4	8.4
4.8	0.5	1.8	3.0	0.6	2.1	3.6	0.9	2.9	4.9	1.0	3.5	6.0	1.3	4.1	7.2
5.0	......	1.3	2.5	......	1.5	3.0	......	2.0	4.1	......	2.5	5.0	......	3.0	6.0
5.2	......	0.8	2.0	......	1.0	2.5	......	1.3	3.3	......	1.5	4.0	......	1.8	4.8
5.4	......	0.3	1.5	......	0.4	1.9	......	0.4	2.5	......	0.5	3.0	......	0.6	3.6
5.6	......	......	1.0	......	......	1.3	......	......	1.6	......	......	2.0	......	......	2.4
5.8	......	......	0.5	......	......	0.6	......	......	0.9	......	......	1.0	......	......	1.3

Column heads: Intended pH for soil-texture class.

[1]If the soil contains more than 10% organic matter, use the next higher soil-texture class.
[2]To convert to tons/acre, multiply by 0.398.

7.6.2 Organic matter

Organic matter consists of three principal components—(1) plant, animal, and microbial residues in various stages of decomposition, (2) humus, and (3) live microorganisms [27]—and affects soil in various ways (see chapter 9, this volume). It increases the CEC, buffer capacity, and water retention; provides a substrate for microbial activity, which can influence soil crumb structure; supplies some nutrients; appears to play an important part in P nutrition, both by supplying P and by rendering other sources of P more available to plants; and interacts with micronutrients to increase their availability.

Organic materials added to nursery soils are decomposed by microorganisms that respire. Their respiration causes a large portion of the organic matter to be lost by oxidation. Thus, fresh organic matter must continually be added if a particular level of soil organic matter is to be maintained.

In the past, large additions of peat, forest duff, or compost were considered essential for maintaining nursery soil fertility [53, 58]. Long-term experiments comparing the relative merits of organic composts and mineral fertilizers conducted over 15 years in two English nurseries [17] and 20 years in one Scottish nursery [40] generally showed that mineral fertilizers alone produced as much or more seedling growth as added organic matter. Related studies of mycorrhizal development showed no consistent differences between Sitka spruce and Scots pine (*Pinus sylvestris* L.) treated with organic compost and mineral fertilizer [41]. The English nurseries had sandy loam soils, and the soil organic carbon (C) decreased in one of them from 0.7 to 0.6% with no organic matter added. The Scottish nursery apparently started with 17% organic C and still contained about 7% C after 20 years with no organic matter added; even this latter level is high by comparison with Northwest nurseries. But interestingly, the low organic matter content in the English nurseries did not prevent production of satisfactory crops. Apparently, high organic matter content of nursery soils is not essential if adequate, and sufficiently frequent, applications of inorganic fertilizers are made. It would be unwise, however, to allow the level to fall too low because of the other benefits of adequate soil organic matter.

Economics also must be considered. In the Northwest, it should be possible to maintain an organic matter content of about 4% in nursery soil [27]; costs outweigh advantages somewhere near 5%, but advantages make the expenditure worthwhile at 2 to 3%. The organic matter content of 21 Northwest nurseries varied from 2 to 6% (average 3.7%), and the range for different management areas within nurseries was greater (0.9 to 12.0%) (OSU Nursery Survey). Thus, an organic matter level of about 4% is a practical goal for most nurseries in this region.

Various forms of organic amendments are added to nursery soils, but the additional C may increase the C:N ratio sufficiently to reduce the amount of N available to the crop. For example, dry softwood sawdust can immobilize about 6 kg (12 lb) of N per ton and dry hardwood sawdust about 12 kg (25 lb) of N per ton [2]. Thus, supplemental N fertilization may be necessary when certain organic amendments are made.

Cover cropping, which seldom or never increases the level of soil organic matter or soil N [27, 54], may, however, benefit the soil by conserving nutrients otherwise lost by leaching and improving soil physical and biological properties (see chapter 10, this volume).

7.6.3 Fertilization

Fertilizers can be organic (such as compost or manure) or inorganic (the various salts of nutrient elements now widely used in forest nurseries). The concentrations of nutrient elements in organic fertilizers are usually low; for example, composts may contain 2 to 4% N and 0.2 to 1.8% P [13] and farmyard manure about 1.1 to 1.5% N [4]. Inorganic fertilizers, on the other hand, are manufactured to definite nutrient specifications, referred to as "the analysis," and may contain high nutrient concentrations. For example, urea fertilizer contains 45% N. The properties and behaviors of fertilizers are described in various publications [e.g., 9, 23, 44].

According to an ancient but awkward convention, concentration in the analysis appears as **percentage** of N but as **percentage of the oxide** of P, K, Ca, and Mg; this means that a fertilizer specified as 21-0-0 contains 21% N, but one specified as 0-20-0 contains 20% P_2O_5. Although the fertilizer analysis usually states only the percentages of N, P_2O_5, and K_2O, other nutrient elements also may be present. For instance, calcium superphosphate (0-20-0) contains Ca and S as well as P (Table 8). Because nearly all work is done in terms of nutrients and not nutrient oxides, it is convenient to convert nutrient oxide values to nutrient values. To convert (1) P_2O_5 to P, multiply by 0.437; (2) K_2O to K, multiply by 0.830; (3) CaO to Ca, multiply by 0.714; and (4) MgO to Mg, multiply by 0.60.

Increased absorption of P occurs in the presence of N. In fact, the greatest stimulation of absorption takes place when N is intimately mixed with P [45]. Thus, ammonium phosphate fertilizers are very effective sources of P, particularly when banded below the seed before sowing. Because chloride damage to conifers can result from applying potassium chloride [[illegible]4], potassium sulfate may be a safer K source, particularly for spruce.

Soil pH can be changed by adding mineral fertilizers. Ammonium and urea salts, and even ammonia solutions, make the soil more acid. Ammonium sulfate is particularly effective in reducing soil pH. Nitrate fertilizers containing a base [KNO_3 or $Ca(NO_3)_2$] increase soil pH. Phosphate fertilizers either have no effect on soil pH or increase it, unless they contain ammonium, in which case they reduce it. Potassium sulfate and chloride have negligible effects on soil pH.

7.6.3.1 Nutrient elements

Fertilization of nursery soils is necessary to replace lost nutrients. Conifer seedlings, removed complete with root systems and, often, soil, contain substantial quantities of mineral nutrients when they leave the nursery. Weeding nursery beds by hand represents a further loss of nutrients. By contrast, in agriculture, frequently only the seed or part of the root is removed; the remainder of the plant is left to decompose and return its nutrients to the soil. About ⅓ of the nursery field is uncropped paths and headlands, however, and the crop is usually removed once every 2 years, not annually.

Amounts of nutrients removed vary from 50 to 200 kg N, 4 to 35 kg P, and 25 to 105 kg K in 2 + 0 conifer crops [[illegible]3]. However, simply replacing these amounts of nutrients in the form of inorganic fertilizers is inadequate because fertilizer recovery is relatively low. Measurements made on 1 + 0 Sitka spruce crops show that only 13 to 16% N, 2 to 4% P, 10 to 22% K, and 2 to 4% Mg were recovered from added fertilizers [[illegible]6]. Although recoveries by larger 2 + 0 seedlings may be greater, they are unlikely to exceed 50%. Thus, amounts of nutrients applied in fertilizers during a rotation tend to be much in excess of the quantities removed in the crop, as is evident from a summary of fertilizer recommendations from North America, Britain, and Germany [73]. The average quantities of nutrients applied per hectare and per rotation in 19 nurseries in the Northwest were 224 kg N, 126 kg P, 103 kg K, 9 kg Mg, 136 kg S, and 557 kg of ground limestone (OSU Nursery Survey), but the quantities applied at individual nurseries varied immensely (Table 9). In most cases, the total amounts shown as top dressings were applied as several smaller doses during the growing season.

Nitrogen.—Most conifers respond rapidly to N fertilizer. In general, the earlier N is applied, the better. Early May-sown seedlings usually benefit from a top dressing of N fertilizer in late June, and rising 2 + 0 stock can be fertilized from March onwards. Seedlings sown in sandy soils of low N status may benefit from ammonium phosphate (11-55-0) banded into the soil before sowing. On the other hand, on heavier soils and where damping-off occurs, N fertilization may not be advisable during the first year of growth; excessive use of N fertilizer on 1 + 0 Douglas-fir seedlings almost invariably accentuates damping-off [56]. Mortality in 1 + 0 Douglas-fir seedbeds was found to be lower with ammonium nitrate than with ammonium sulfate [68].

Several applications of 22 to 44 kg/ha (20 to 40 lb/acre) of N should be made during the second growing season as a top dressing (Table 10). Crops should be watered immediately after dry, soluble, N fertilizers have been applied to wash fertilizer off to prevent foliage damage (fertilizer "burn"). The key to efficient N fertilization of conifers seems to be little and often. In general, N should not be applied to seedlings after July; otherwise dormancy may be delayed. During the second or subsequent years of growth, N fertilizer can be applied after buds are set and there is no chance of inducing flushing; this would normally be in late September or October.

Phosphorus.—Growth responses to P fertilizers are seldom detected if maintenance dressings of 67 to 135 kg/ha (60 to 120 lb/acre) of P are applied each rotation. However, such responses may be evident [56], particularly when a new nursery is being developed on previously unfertilized land.

Phosphorus fertilizers must be incorporated into the soil close to seedling roots. Thus, they are applied and cultivated

Table 8. Some common fertilizers, their analysis, and factors for determining nutrient amounts for application.

Fertilizer	Analysis	Nutrient	%	Factor[1]	Nutrient	%	Factor	Nutrient	%	Factor
Ammonium sulfate $(NH_4)_2SO_4$	21-0-0	N	21	4.76	S	24	4.17	......	......	......
Ammonium nitrate NH_4NO_3	33-0-0	N	33	3.03	......	......	......	......	......	......
Urea $CO(NH_2)_2$	45-0-0	N	45	2.22	......	......	......	......	......	......
Sulfur-coated urea	32-0-0	N	32	3.13	S	22	4.55	......	......	......
Calcium nitrate $Ca(NO_3)_2$	16-0-0	N	16	6.25	Ca	24	4.17	......	......	......
Ammonium phosphate $NH_4H_2PO_4$	11-55-0	N	11	9.09	P	24	4.17	......	......	......
Diammonium phosphate $(NH_4)_2PO_4$	21-55-0	N	21	4.76	P	24	4.17	......	......	......
Calcium superphosphate $CaH_4(PO_4) + 2CaSP_4 \cdot 2H_2O$	0-20-0	P	8.7	11.5	Ca	20	5	S	11	9.09
Triple superphosphate $Ca(H_2PO_4)_2$	0-45-0	P	19.6	5.1	Ca	14	7.15	......	......	......
Phosphoric acid H_3PO_4	0-52-0	P	22.7	4.4	......	......	......	......	......	......
Potassium sulfate K_2SO_4	0-0-50	K	41	2.44	S	17	5.89	......	......	......
Potassium chloride KCl	0-0-62	K	51	1.96	Cl	46	2.17	......	......	......
Sul-Po-Mag®	0-0-22	K	18	5.6	Mg	11	9.09	S	11	9.09

[1]The factor may be used to determine the actual weight of a nutrient in a fertilizer. For example, to supply 50 kg of N as ammonium sulfate (21-0-0), multiply 50 by the N factor: 50 x 4.76 = 238 kg ammonium sulfate.

Table 9. Nutrient elements applied during one crop rotation for 19 nurseries in the Northwest (data from OSU Nursery Survey).

	N	P	K	Lime	Mg	S
Presowing treatment						
Nurseries applying nutrient, %	21	84	58	37	16	53
Application rates, kg/ha[1]						
Average	42	46	73	1,500	38	61
Range	22-56	5-87	18-148	750-2,240	22-50	14-185
Median	45	45	55	1,000	40	55
Year 1 top dressing						
Nurseries applying nutrient, %	84	68	37	0	0	42
Application rates, kg/ha						
Average	103	69	43	......	......	79
Range	36-152	10-139	23-62	......	......	9-130
Median	110	75	45	......	......	85
Year 2 top dressing						
Nurseries applying nutrient, %	84	68	42	0	11	53
Application rates, kg/ha						
Average	152	78	107	......	30	134
Range	53-306	20-140	23-208	......	28-33	33-248
Median	160	75	110	......	30	150

[1]To convert to lb/acre, multiply by 0.89.

into the soil before bed formation (Table 10). Where seed is drill sown, P fertilizer can be banded into the soil 3 to 5 cm below the drill. Banding ammonium phosphate fertilizer (e.g., 11-55-0) below drill-sown white spruce and Engelmann spruce [*Picea engelmannii* Parry ex Engelm.] substantially improves growth and is standard practice in many nurseries.

Top dressings of P fertilizers are relatively ineffective except on very sandy soils. Only relatively soluble P fertilizers such as ammonium phosphate (11-55-0 or 21-55-0) should be used for top dressings if they are to be applied at all, and this should be done early in the year, for example, March or April of the second year for 2 + 0 seedlings.

Calcium superphosphate (0-20-0) is a good P fertilizer for acid, sandy soils because it supplies Ca and S as well as P (Table 8) and tends to reduce soil acidity. Triple superphosphate (0-45-0) can be used if soil pH is too high or if soil Ca level is already high. Ammonium phosphate (11-55-0) should be used if the superphosphates do not provide adequate P nutrition.

In agricultural practice, heavy applications of P fertilizer have sometimes caused micronutrient deficiencies [50]. Zn and Cu are the elements most frequently affected by high P levels, and calcium superphosphate can accentuate Cu deficiency in conifers [13].

Potassium.—Positive growth responses of conifer seedlings to K fertilizers are seldom detected in Northwest nurseries, probably because maintenance dressings of K fertilizer prevent decline in soil K levels. However, a small increase in Douglas-fir root dry weight due to K fertilization was detected in a sandy loam nursery soil containing 0.25 meq K/100 g [71]. Both quantity and frequency of application seemed to affect growth. Evidence of K deficiency (yellowing and necrosis of apical needles and 0.3% foliar K) also has been noted in white spruce 2 + 0 seedlings and transplants from this nursery. More frequent top dressing of K fertilizer throughout the second growing season seems to have remedied the problem. There also is evidence that excessive K fertilization can result in undesirably high soil K levels, which reduce seedling growth [67]. Douglas-fir crops growing on soils containing more than 0.45 meq K/100 g should not be fertilized with K.

Table 10. Recommended yearly total applications of N, P, and K and typical fertilizer schedules for three crop age classes.

Age class	Application method	kg/ha[1]	Fertilizer	kg fertilizer/ha
		Nitrogen		
		0 to 120 kg/ha		
1 + 0	Band at sowing	30	11-55-0	280
	Top dress as at least 4 separate doses	22	21-0-0 or	106 (x4)
			33-0-0	67 (x4)
		112 to 165 kg/ha		
2 + 0	Top dress in early March	30	11-55-0	280
	Top dress as at least 6 separate doses	22	21-0-0 or	106 (x6)
			33-0-0	67 (x6)
		90 to 180 kg/ha		
Transplants	Top dress as at least 4 separate doses	45	21-0-0 or	210 (x4)
			33-0-0	134 (x4)
		Phosphorus		
		67 to 134 kg/ha		
1 + 0	Work in before sowing	67	0-20-0 or	770
			0-45-0	340
	Band at sowing	67	11-55-0	280
		0 to 67 kg/ha		
2 + 0	Top dress in early March	67	11-55-0	280
		67 to 134 kg/ha		
Transplants	Work in before planting	67	0-20-0 or	770
			0-45-0	340
		Potassium		
		50 to 75 kg/ha		
1 + 0	Work in before sowing	50	0-0-53	112
	Top dress in July	25	0-0-53	56
		25 to 75 kg/ha		
2 + 0	Top dress in April	25	0-0-53	56
	Top dress in June	25	0-0-53	56
	Top dress in August	25	0-0-53	56
		50 to 100 kg/ha		
Transplants	Work in before planting	50	0-0-53	112
	Top dress in June	25	0-0-53	56
	Top dress in August	25	0-0-53	56

[1]To convert to lb/acre, multiply by 0.89.

Potassium fertilizers can be worked into the soil before sowing, but some top dressings are strongly advised from July onwards during the first growing season and throughout the second (Table 10).

Calcium.—A high level of Ca is undesirable in conifer nurseries because it raises soil pH and tends to promote growth of pathogenic fungi. Adding ground limestone may be justified if soil pH is much below 5; the quantity required can be determined from Table 7. Low soil Ca level (less than 3.0 meq Ca/100 g soil) or low seedling Ca content (less than 0.1% Ca in 1 + 0 shoots) also may necessitate addition of ground limestone. A single dressing of not more than 2,240 kg/ha (2,000 lb/acre) should be applied and well worked into the soil. Dolomitic limestone is commonly preferred in forest nurseries because it provides Mg as well as Ca. Because excessive liming can impair the nursery's ability to produce good-quality stock, considerable caution and expert advice are recommended.

Magnesium.—Nursery conifer crops seldom seem to require Mg fertilizers; nevertheless, Mg is applied in dolomitic limestone and occasionally in other fertilizers. Magnesium sulfate is a common fertilizer which occurs in two forms. One (Epsom salts) is very hydrated, requiring 10 kg (22 lb) of salt to provide 1 kg (2.2 lb) Mg; the other, a less hydrated form (Kieserite), requires 5 kg (11 lb) of salt to obtain 1 kg Mg. At least one manufactured fertilizer (Sul-Po-Mag®) containing K, Mg, and S also is available (Table 8).

Sulfur and micronutrients.—Because many fertilizers contain S (Table 8), nurseries are unlikely to show S deficiencies. Should the fertilizer schedule contain inadequate S, the simplest remedy is to switch to fertilizers containing S, such as calcium superphosphate, potassium sulfate, and ammonium sulfate. Flowers of sulfur and calcium sulfate (gypsum) can be used when only S is required. S should be applied at about 30 to 60 kg/ha (26 to 52 lb/acre).

So far as is known, no micronutrient deficiencies have been detected in bareroot nurseries of the Northwest. Iron chlorosis is apparently fairly common in conifer nurseries where soil pH is high and Ca is abundant [9], but it can be corrected by spraying with ferrous sulfate [36]. Should micronutrients be found deficient, various soluble fertilizers supplying the nutrients are available. Chelated micronutrients, although more expensive, may be more effective in correcting deficiencies.

7.6.3.2 Application methods

Fertilizers can be applied to conifer nurseries in several ways, depending on time of treatment during the rotation and nutrient being applied. For example, N is usually required in greater quantities during the second year of a 2-year rotation and so must be top dressed; but P must be placed as close to the roots as possible and so is mainly incorporated into the soil before sowing.

Fertilizers can be broadcast with many types of agricultural spreaders. When P or Ca fertilizers are applied before bed shaping, broadcast spreaders with rotary flingers, which cover an 8- to 10-m swath, can conveniently be used; these presowing fertilizers are normally disked into the soil. These same spreaders can be used to broadcast top dressings of N, K, or other fertilizers, treating four or five 1.2-m-wide beds in a single swath; but they also apply fertilizer to the path, where it is largely wasted. Broadcast spreaders that use a worm-gear-driven bar to meter the fertilizer, which then falls by gravity, can apply fertilizer to individual beds with little waste. Thorough watering is essential after top dressing to wash fertilizer off the crop to prevent fertilizer burn.

In some nurseries, P fertilizers are banded below the drill at a depth of about 3 cm immediately before sowing with a modified wheat drill. The relative insolubility and immobility of P make banding a very efficient method of applying this fertilizer.

Fertilizer can be applied to foliage through overhead irrigation systems towards the end of an irrigation period, but distribution may be uneven. Pesticide spray equipment of the high-pressure, low-volume type is more satisfactory for foliar feeding [23]. Pressure and correct nozzle selection are important because droplet size can affect crop response.

7.6.4. Foliar feeding

Many types of crops such as vegetables and fruit trees are treated with foliar nutrient sprays [80]. Nutrients are also applied to conifer seedlings in container nurseries through overhead sprays. But as far as is known, foliar nutrients are not applied in bareroot nurseries in the Northwest.

Nitrogen, in the form of urea, is the most common foliar-applied nutrient, although all macronutrients and micronutrients apparently have been used on various crops. Foliar applications tend to give rapid responses, and deficiencies of immobile elements, such as Ca, can often be more easily

Table 11. Fertilizer solutions used as foliar sprays on Monterey pine seedlings in New Zealand nurseries (adapted from [35]).

Element	Chemical source	Formula	Percentage of Element	Percentage of Sulfur	Solution,[1] % wt./vol. (compound)	Nutrient applied,[2] kg/ha (element)
N	Urea[3]	NH_2CONH_2	46	0	5	11.5
Mg	Epsom salts[4]	$MgSO_4 7H_2O$	10	13	5	2.5
Fe	Ferrous (iron) sulfate[4]	$FeSO_4 7H_2O$	20	11.5	5	5.0
B	Borax	$Na_2B_4O_7 10H_2O$	11.3	......	0.2-0.5	0.11-0.28
	Solubor	$Na_2B_8O_{13} 4H_2O$	20.5	......	0.2-0.5	0.20-0.51
Cu	Copper sulfate[4,5]	$CuSO_4 7H_2O$	25	12	0.5	0.62
Mn	Manganous sulfate[4]	$MnSO_4 4H_2O$	24	14	1	2.4
Zn	Zinc sulfate[4]	$ZnSO_4 7H_2O$	23	11	1	2.3

[1] Safe concentration (% wt./vol.) of single-salt solution. Where two or more compounds are combined in the same spray solution, concentration of each should be substantially reduced; e.g., to supply N, Mg, and Fe together, compound concentrations of 2% wt./vol. each would be more appropriate.

[2] Rate as kg/ha for element concerned when solution applied at standard rate of 500 liters/ha.

[3] Solution strength that can be safely tolerated depends on stage of growth and climate. If frosts are likely, concentration should not normally exceed 2% wt./vol. (equivalent to 1.15 kg/ha N), even for frost-free conditions.

[4] Mg, Fe, Cu, Mn and Zn can alternatively be supplied in chelated form. Generally, a concentration of 0.05% wt./vol. (compound) applied in 500 liters will be suitable. EDTA chelates contain 9.8% Cu, 9.8% Mn, 6% Mg, 14% Zn, and 14% Fe (element) while EDDHAFe supplies 6% Fe.

[5] Burning can be avoided by adding 1.25 kg of sodium carbonate for each kilogram of copper sulfate in the spray solution.

corrected with this method. However, quantity of spray and droplet size, as well as nutrient concentration, must all be controlled to prevent foliage burn. Urea is often applied in solutions containing 400 to 800 g/100 liters (4 to 6 lb/100 gal.).

Fertilizers have been applied to foliage in New Zealand forest nurseries [35]. Stage of crop development and weather conditions have been found to affect results. Applications should not be made in cold weather. If more than one nutrient is to be applied in the same solution, the concentrations of each must be reduced so that the overall concentration of salts in solution remains about the same. Adding wetting agents to the solution may actually increase likelihood of damage by urea sprays. The 5% concentration (weight/volume, i.e., 5 kg/100 liters) recommended for Monterey pine (*Pinus radiata* D. Don) (Table 11) might be more concentrated than desirable for smaller, slower growing seedlings such as spruce.

7.7 Seedling Responses to Nutrients in the Nursery

7.7.1 Growth

Newly germinated conifer seedlings contain adequate nutrients and show little response to different levels of external nutrient supply for up to 6 weeks after germination [29]. However, nutrient uptake by a nursery crop increases continuously, if somewhat irregularly, throughout the remainder of the season [e.g., 5]. This is supported by the observation that greatest growth is achieved when frequent top dressings of soluble fertilizers are made once seedlings are past the cotyledon stage. P uptake by a 2 + 0 Douglas-fir crop may reach a maximum in September, and K may actually be lost from the soil through leaching in fall [55]. Although nutrients are required continuously throughout the season, rate of P uptake, measured as milligrams of P per gram of seedling per unit time, varies considerably in white spruce [5] and to a smaller extent in Douglas-fir [6]. In white spruce, P uptake rate is high in early summer, drops in August, and increases again in fall. By contrast, N and K uptake rates are highest at the beginning of the growing season and then decrease steadily.

Seedling growth can be reduced and sometimes modified by withholding particular nutrients. Withholding N or P tends to restrict shoot growth more than root growth [22]. This was found to be true in Monterey pine seedlings, where reducing the N supply also reduced stem diameter in relation to height and decreased the number and length of branches [79].

Because undercutting and wrenching procedures tend to remove or damage part of the root system, intensively wrenched seedlings may require additional fertilization to compensate for their reduced root systems. Additional fertilization also is sometimes necessary to offset effects of wet or cold weather when fertilizers are leached or uptake is reduced by low temperature. However, dressings of ammonium nitrate applied to 2 + 0 pine seedlings growing on wet, cool soil resulted in reduced growth and disease symptoms attributed to nitrate accumulation [38].

7.7.2 Drought stress

Under drought conditions, high N levels have generally been found detrimental to growth and survival, intermediate N levels have either been beneficial or have had no effect, and low N levels have had the least effect on tree growth [51]. High levels of N tend to promote shoot growth, and seedlings with large shoots transpire more water than those with small shoots. Even when this was taken into consideration, high N supply reduced lodgepole pine (*Pinus contorta* Dougl. ex Loud.) recovery from drought stress [28].

High levels of foliar K are associated with reduced transpiration rates in trees. In an experiment with Sitka spruce, where water-use efficiency was calculated as grams of water used per gram of new shoot dry weight, seedlings with 1.0% foliar K used 188 g water/g new shoot, whereas seedlings with 1.9% foliar K used only 156 g water/g new shoot [19]. Increased K concentration was also shown to increase drought survival of Scots pine but not Norway spruce (*Picea abies* L.) seedlings [25]. Adequate K nutrition has been shown to increase drought avoidance of young dormant Douglas-fir seedlings in frozen soil [39].

7.7.3 Cold hardiness

Nutrition can influence seasonal growth pattern, which in turn can alter seedling susceptibility to low temperature. For example, fertilization may either prolong growth in the fall or cause earlier bud flushing in spring. Heavy N fertilization is often found to delay dormancy and result in fall frost damage to nursery seedlings, but the same effect can also be achieved with heavy P fertilization [42]. The time of fertilizer application clearly influences the outcome. For example, applying N and K to nursery beds so late in the season that growth was unaffected substantially reduced frost damage to Sitka spruce and western hemlock seedlings [14].

Low B levels have been implicated in frost damage to tree species [21, 26], but whether this is a symptom of nutrient imbalance or due to the failure of a function performed by B alone is unknown. Internal nutrient balance may be important; Timmis [64] found that, after a hardening period, young Douglas-fir seedlings with an internal K:N ratio of about 0.6 were hardier than those with a ratio of 1.3. This result also makes it unlikely that K level alone is important in promoting cold hardiness. Other evidence now shows that high levels of K do not directly increase cold hardiness in trees [10, 25].

7.8 Nutrient Effects on Stock Performance after Outplanting

How seedling nutrient status affects performance after outplanting is not clearcut. General biological principles indicate that small chlorotic seedlings will not survive as well or grow as fast as large green seedlings after planting out [5]. In several experiments, however, the benefit of nursery fertilization to survival after planting could not be demonstrated [12, 34, 48, 63]. In an experiment with red pine and Scots pine, the 10% higher survival shown by stock fertilized in the nursery was not significant [78]. Yet nursery fertilization of jack pine (*Pinus banksiana* Lamb.), red pine, and white pine (*Pinus strobus* L.) in the Lake States gave a slight but consistent gain in field survival [58].

Fertilization generally increases seedling size, which could be advantageous on planting sites where competition with other species occurs (see chapter 24, this volume). Survival and height growth of outplanted Douglas-fir were increased by nursery fertilization, which also increased seedling size [57]. Similar results were obtained in another study with Douglas-fir [74], although competition on the planting site was minimal. This latter study suggested that an optimal foliar N concentration (close to 2%) for survival exists and that 2 + 0 seedlings with needle nutrient concentrations above or below that value did not survive as well.

In maritime climates, fertilizing nursery stock in the fall after budset can increase nutrient reserves, with subsequently increased growth after planting. Two-year-old Douglas-fir seedlings were fertilized with 56 kg/ha (50 lb/acre) of N in September and outplanted the following spring [3]. Fertilized trees were still 13% taller than unfertilized trees 5 years after planting. Similar results have also been reported for Sitka spruce fertilized in the nursery after budset [15].

It seems likely that seedlings with high internal nutrient concentrations will often survive better and usually grow more than seedlings with low nutrient concentrations (see chapter 15, this volume). However, the relationship between nutrient status and performance after outplanting may be subtle, influenced by factors such as cold-storage conditions and moisture relationships at the planting site.

7.9 Conclusions

Developing and maintaining a high level of fertility in bareroot nurseries are essential for producing good-quality nursery stock. However, soil fertility is only one of a number of factors influencing stock quality: fertile nursery soil does not compensate for poor practices such as overdense sowing, unseasonal lifting, or inadequate undercutting and wrenching.

Achieving an optimal supply of nutrients to conifer seedlings growing in nursery soil over a 2-year rotation requires skill and attention to detail. Soil features such as drainage and texture, which usually vary throughout the nursery, and changes in weather must continually be taken into account. These demands undoubtedly contribute to the steady increase in popularity of container-grown seedlings, for which fertility and climate can be reasonably well controlled. Though container nurseries can provide very favorable growing conditions, they are equally less forgiving, and mistakes in technique are more disastrous than in bareroot nurseries. Yet there is little reason why the nutrition furnished to bareroot seedlings should not be comparable to that attained in container systems. Many factors contributing to nursery soil fertility can be measured and at least partly controlled, and ensuring a reasonable level of health and vigor in nursery stock should be possible by soil and seedling analysis.

Correct timing and sufficient frequency of fertilization may still be lacking in many bareroot nurseries. These points should be further investigated, as should the possible benefit of slow-release fertilizers for maintaining a steady nutrient supply in seedbeds.

References

1. Aldhous, J. R. 1972. Nursery practice. Her Majesty's Stationery Office, London. Forestry Commun. Bull. 43. 184 p.
2. Allison, F. E. 1965. Decomposition of wood and bark sawdust in soil, nitrogen requirements, and effects on plants. U.S. Dep. Agric., Washington, D.C. Tech. Bull. 1332. 29 p.
3. Anderson, H. W., and S. P. Gessel. 1966. Effects of nursery fertilization on outplanted Douglas-fir. J. Forestry 64:109-112.
4. Armson, K. A. 1959. The use of farmyard manure in forest tree nurseries. Forestry Chronicle 35(2):100-103.
5. Armson, K. A. 1960. White spruce seedlings: the growth and seasonal absorption of nitrogen, phosphorus and potassium. Univ. Toronto Forestry Bull. 6. 37 p.
6. Armson, K. A. 1965. Seasonal patterns of nutrient absorption by forest trees. Pages 65-75 *in* Forest-soil relationships in North America (C. T. Youngberg, ed.). Oregon State Univ. Press, Corvallis.
7. Armson, K. A. 1977. Forest soils: properties and processes. Univ. Toronto Press. 390 p.
8. Armson, K. A., and R. D. Carman. 1961. An acid injection system for nursery irrigation water. Tree Planters' Notes 46:11-13.
9. Armson, K. A., and V. Sadreika. 1979. Forest tree nursery soil management and related practices. Ontario Ministry of Natural Resources, Toronto. 179 p.
10. Aronsson, A. 1980. Frost hardiness in Scots pine (*Pinus sylvestris* L.). II. Hardiness during winter and spring in young trees of different mineral nutrient status. Studia Forestalia Suecica 155:1-27.
11. Baule, H., and C. Fricker. 1970. The fertilizer treatment of forest trees. Translated by C. L. Whittles. BLV Verlags, München. 259 p.
12. Bell, T. I. W. 1968. Effect of fertilizer and density pretreatment on spruce seedling survival and growth. Her Majesty's Stationery Office, London. Forestry Comm. For. Res. 67. 67 p.
13. Benzian, B. 1965. Experiments on nutrition problems in forest nurseries. Her Majesty's Stationery Office, London. Forestry Commun. Bull. 37. 251 p.
14. Benzian, B. 1966. Manuring young conifers: experiments in some English nurseries. Proc. Fertilizer Society of London 94:3-35.
15. Benzian, B., R. M. Brown, and S. C. R. Freeman. 1974. Effect of late-season top-dressings of N (and K) applied to conifer transplants in the nursery on their survival and growth on British forest sites. Forestry 47:153-184.
16. Benzian, B., and S. C. R. Freeman. 1973. Reference experiments on young conifers at Woburn experimental farm, 1961-69. Rep. Rothamsted Exp. Sta. for 1973, Part 2:152-171.
17. Benzian, B., S C.R. Freeman, and H. D. Patterson. 1972. Comparison of crop rotations, and of fertilizer with compost, in long term experiments with Sitka spruce (*Picea sitchensis*) in two English nurseries. Forestry 45:145-176.
18. Bolton, J., and B. Benzian. 1970. Sulfur as a nutrient for Sitka spruce (*Picea sitchensis*) seedlings and radish (*Raphanus sativus*) grown on a sandy podzol in England. J. Agric. Sci. Cambridge 74:501-504.
19. Bradbury, I. K., and D. C. Malcolm. 1977. The effect of phosphorus and potassium on transpiration, leaf diffusive resistance and water-use efficiency in Sitka spruce (*Picea sitchensis*) seedlings. J. Applied Ecology 14:631-641.
20. Brady, N. C. 1974. The nature and properties of soils. 8th ed. MacMillan Publishing Co., Inc., New York. 639 p.
21. Braekke, F. H. 1979. Boron deficiency in forest plantations on peatland in Norway. Meddelelser fra det Norske Skogforsoeksvesen 35:213-236.
22. Brouwer, R. 1962. Nutritive influences on the distribution of dry matter in the plant. Netherlands J. Agric. Sci. 10:399-408.
23. California Fertilizer Association, Soil Improvement Committee. 1980. Western fertilizer handbook. The Interstate Printers and Publishers, Inc., Danville, Illinois. 269 p.
24. Campagna, J. P., and D. P. White. 1973. Nursery soil fumigation affects growth and phosphorus nutrition of pine and spruce seedlings. Forestry Chronicle 49:219-223.
25. Christersson, L. 1976. The effect of inorganic nutrients on water economy and hardiness of conifers. II. The effect of varying potassium and calcium contents on water status and drought hardiness of pot-grown *Pinus sylvestris* L. and *Picea abies* (L.) Karst. seedlings. Studia Forestalia Suecica 136:1-23.
26. Cooling, E. N. 1967. Frost resistance in *Eucalyptus grandis* following the application of fertilizer borate. Rhodesia, Zambia and Malawi J. Agric. Res. 5:97-100.
27. Davey, C. B., and H. H. Krause. 1980. Functions and maintenance of organic matter in forest nursery soils. Pages 130-165 *in* Proc., North American forest tree nursery soils workshop (L. P. Abrahamson and D. H Bickelhaupt, eds.). State Univ. New York, Coll. Environ. Sci. and Forestry, Syracuse.
28. Etter, H. M. 1969 Growth, metabolic components and drought survival of lodgepole pine seedlings at three nitrate levels. Can. J. Plant Sci. 49:393-402.
29. Etter, H. M. 1971. Nitrogen and phosphorus requirements during the early growth of white spruce seedlings. Can. J. Plant Sci. 51:61-63.
30. Harley, J. L. 1969. The biology of mycorrhiza. 2nd ed. Leonard Hill Books, London.
31. Heiberg, S. O., and D. P. White. 1951. Potassium deficiency of reforested pine and spruce stands in northern New York. Soil Sci. Society of America Proc. 15:369-376.
32. Ingestad, T. 1979. Mineral nutrient requirements of *Pinus silvestris* and *Picea abies* seedlings. Physiologia Plantarum 45:373-380.
33. Jackson, M. L. 1958. Soil chemical analysis. Constable and Co. Ltd., London. 498 p.
34. Knight, H. 1957. Growth and survival of experimental plantations of Douglas-fir. B. C. Forest Serv., Victoria. Res. Note 33. 22 p.
35. Knight, P. J. 1978. Fertilizer practice in New Zealand forest nurseries. New Zealand J. Forestry Sci. 8:27-53.
36. Korstian, C. F., C. Hartley, L. F. Watts, and G. G. Hahn. 1921. A chlorosis of conifers corrected by spraying with ferrous sulphate. J. Agric. Res. 21:153-171.
37. Krause, H. H. 1965. Effect of pH on leaching losses of potassium applied to forest nursery soils. Soil Sci. Society of America Proc. 29:613-615.
38. Landis, T. D. 1976. Nitrogen fertilizer injures pine seedlings in Rocky Mountain nursery. Tree Planters' Notes 27(4):29-32,35.
39. Larsen, J. B. 1978. Untersuchungen über die Bedeutung der Kalium und Stickstoffversorgung für die Austrocknungsresistenz der Douglasie (*Pseudotsuga menziesii*) im Winter. Flora 167:197-207.

40. Low, A. J., and A. L. Sharpe. 1973. The long term effects of organic and inorganic fertilizer regimes at Teindland nursery. Scottish Forestry 27:287-295.
41. Levisohn, I. 1965. Mycorrhizal investigations. Her Majesty's Stationery Office, London. Forestry Comm. Bull. 37:228-235.
42. Malcolm, D. C., and B. C. Y. Freezaillah. 1975. Early frost damage on Sitka spruce seedlings and the influence of phosphorus nutrition. Forestry 48:139-145.
43. McFee, W. W., and E. L. Stone. 1968. Ammonium and nitrate as nitrogen sources for *Pinus radiata* and *Picea glauca*. Soil Sci. Society of America Proc. 32:879-884.
44. McVickar, M. H., G. L. Bridger, and L. B. Nelson. 1963. Fertilizer technology and usage. Soil Sci. Society of America, Madison, Wisconsin. 464 p.
45. Miller, M. H. 1971. Effect of nitrogen on phosphorus absorption by plants. Pages 643-668 *in* The plant root and its environment (E. W. Carson, ed.). The Univ. Press of Virginia.
46. Morrison, I. K. 1974. Mineral nutrition of conifers with special reference to nutrient status interpretation: a review of the literature. Can. Forestry Serv., Dep. Environ. Publ. 1343. 74 p.
47. Mullin, R. E. 1969. Soil acidification with sulfur in a forest tree nursery. Sulfur Institute J. 5:2-3.
48. Mullin, R. E., and L. Bowdery. 1978. Effects of nursery seedbed density and top dressing fertilization on survival and growth of 3 + 0 red pine. Can. J. Forest Res. 8:30-35.
49. Nelson, L. E., and R. Selby. 1974. The effect of nitrogen sources and iron levels on the growth and composition of Sitka spruce and Scots pine. Plant and Soil 41:573-588.
50. Olsen, S. R. 1972. Micronutrient interactions. Pages 243-264 *in* Micronutrients in agriculture (J. J. Mortvedt, P. M. Giordano, and W. L. Lindsay, eds.). Soil Sci. Society of America, Madison, Wisconsin.
51. Pharis, R. P., and P. J. Kramer. 1964. The effect of nitrogen and drought on loblolly pine seedlings. Forest Sci. 10:143-150.
52. Purnell, H. M. 1958. Nutritional studies of *Pinus radiata* Don. I. Symptoms due to deficiency of some major elements. Australian Forestry 22:82-87.
53. Rayner, M. C., and W. Neilson-Jones. 1944. Problems in tree nutrition. Faber and Faber, London.
54. Rogers, T. H., and J. E. Giddens. 1957. Green manure and cover crops. Pages 252-257 *in* Soil, the yearbook of agriculture 1957.U.S. Dep. Agric., Washington, D.C.
55. Schaedle, M. 1959. A study of the growth of Douglas fir (*Pseudotsuga menziesii*) seedlings. M.S. thesis, Univ. British Columbia, Vancouver. 98 p.
56. Sinclair, W. A., D. P. Cowles, and S. M. Hee. 1975. Fusarium root rot of Douglas fir seedlings: suppression by soil fumigation, fertility management, and inoculation with spores of the fungal symbiont *Laccaria laccata*. Forest Sci. 21:390-399.
57. Smith, J. H. G., A. Kozak, O. Sziklai, and J. Walters. 1966. Relative importance of seedbed fertilization, morphological grade, site, provenance, and parentage to juvenile growth and survival of Douglas fir. Forestry Chronicle 42:83-86.
58. Stoeckeler, J. H., and G. W. Jones. 1957. Forest nursery practice in the Lake States. U.S.D.A. Forest Serv., Washington, D.C. Agric. Handb. 110. 124 p.
59. Stone, E. L. 1968. Micronutrition of forest trees: a review. Pages 132-175 *in* Forest fertilization. Tennessee Valley Authority, National Fertilizer Development Center, Muscle Shoals, Alabama.
60. Strullu, D. G., and M. Bonneau. 1978. Contribution a l'étude des carences en cuivre chez les Abietacées. Can. J. Botany 56:2648-2659.
61. Sucoff, E. I. 1961. Potassium, magnesium and calcium deficiency symptoms of loblolly and Virginia pine seedlings. U.S.D.A. Forest Serv., NE Forest Exp. Sta., Upper Darby, Pennsylvania. Sta. Pap. 164. 18 p.
62. Sutherland, J. R., and E. van Eerden. 1980. Diseases and insect pests in British Columbia forest nurseries. B.C. Ministry of Forests/Can. Forestry Serv. Rep. 12. 55 p.
63. Switzer, G. L., and L. E. Nelson. 1963. Effects of nursery fertility and density on seedling characteristics, yield, and field performance of loblolly pine (*Pinus taeda* L.). Soil Sci. Society of America Proc. 27:461-464.
64. Timmis, R. 1974. Effect of nutrient stress on growth, bud set, and hardiness in Douglas fir seedlings. Pages 187-193 *in* Proc., North American containerized forest tree seedling symp. (R. W. Tinus, W. I. Stein, and W. E. Balmer, eds.). Great Plains Agric. Council Publ. 68.
65. Tinus, R. W. 1980. Nature and management of soil pH and salinity. Pages 72-86 *in* Proc., North American forest tree nursery soils workshop (L. P. Abrahamson and D. H. Bickelhaupt, eds.). State Univ. New York, Coll. Environ. Sci. and Forestry, Syracuse.
66. Turner, J., and M. J. Lambert. 1978. Sulfur nutrition of conifers in relation to response to fertilizer nitrogen, to fungal infections and soil parent material. Pages 546-564 *in* Proc., Forest soils and land use (C. T. Youngberg, ed.).5th North American Forest Soils conf., Colorado State Univ., Ft. Collins.
67. van den Driessche, R. 1963. Nursery experiments with Douglas fir. Commonwealth Forestry Review 42(3):242-254.
68. van den Driessche, R. 1969. Forest nursery handbook. B.C. Forest Serv., Victoria. Res. Note 48. 44 p.
69. van den Driessche, R. 1971. Response of conifer seedlings to nitrate and ammonium sources of nitrogen. Plant and Soil 34:421-439.
70. van den Driessche, R. 1977. Fertilizer experiments in conifer nurseries of British Columbia. B. C. Ministry of Forests, Victoria. Res. Note 79. 32 p.
71. van den Driessche, R. 1979. Soil management in Douglas fir nurseries. Pages 278-292 *in* Forest soils of the Douglas fir region (P. E. Heilman, H. W. Anderson, and D. M. Baumgartner, eds.). Washington State Univ. Ext. Serv., Pullman.
72. van den Driessche, R. 1979. Estimating potential response to fertilizer based on tree tissue and litter analysis. Pages 214-220 *in* Forest fertilization conf. (S. P. Gessel, R. M. Kennedy, and W. A. Atkinson, eds.). Univ. Washington, Seattle. Institute of Forest Resources Contribution 40.
73. van den Driessche, R. 1980. Health, vigour and quality of conifer seedlings in relation to nursery soil fertility. Pages 100-120 *in* Proc., North American forest tree nursery soils workshop (L. P. Abrahamson and D. H. Bickelhaupt, eds.). State Univ. New York, Coll. Environ. Sci. and Forestry, Syracuse.
74. van den Driessche, R. 1980. Effects of nitrogen and phosphorus fertilization on Douglas fir nursery growth and survival after outplanting. Can. J. Forest Res. 10:65-70.
75. van den Driessche, R., and J. Dangerfield. 1975. Response of Douglas fir seedlings to nitrate and ammonium nitrogen sources under various environmental conditions. Plant and Soil 42:685-702.
76. van den Driessche, R. 1982. Unpublished data, B.C. Ministry of Forests, Victoria.
77. Wilde, S. A. 1958. Forest soils. The Ronald Press Co., New York. 537 p.
78. Wilde, S. A., R. Wittenkamp, E. L. Stone, and H. M. Galloway. 1940. Effect of high rate fertilizer treatments of nursery stock upon its survival and growth in the field. J. Forestry 38:805-809.
79. Will, G. M. 1977. The influence of nitrogen supply on the growth form of *Pinus radiata* seedlings. Forest Sci. 23:64-68.
80. Wittwer, S. H., M. J. Bukovac, and H. B. Tukey. 1963. Advances in foliar feeding of plant nutrients. Pages 429-455 *in* Fertilizer usage and technology (M. H. McVickar, G. L. Bridger, and L. B. Nelson, eds.). Soil Sci. Society of America, Madison, Wisconsin.

Chapter 8
Soil and Tissue Analysis: Tools for Maintaining Soil Fertility

C. T. Youngberg

Abstract

Systematic monitoring through soil and plant analysis is essential for understanding and managing soil systems in forest nurseries. Analysis services are offered by Oregon State University, University of Idaho, and seven commercial laboratories in the U.S. Northwest, as well as the British Columbia Ministry of Forests. Suggested target fertility levels for raising Douglas-fir in Northwest nurseries are: pH of 5.0 to 6.0, total nitrogen (N) of 0.18 to 0.23%, available phosphorus (P) of 25 to 50 ppm, available potassium (K) of 80 to 120 ppm, exchangeable calcium (Ca) of 2 to 4 meq/100 g, and exchangeable magnesium (Mg) of 1 to 2 meq/100 g. Suggested ranges in macronutrient concentrations in Douglas-fir needle tissue are: 1.2 to 2% N, 0.1 to 0.2% P, 0.3 to 0.8% K, 0.2 to 0.5% Ca, 0.10 to 0.15% Mg, and 0.1 to 0.2% sulfur (S). The lower levels indicate deficiencies and the higher levels adequacy. Success of the fertility monitoring program depends on careful sampling and handling, consistency in laboratory services used, and meticulous recordkeeping.

8.1 Introduction

In view of the trends in reforestation research and resulting reforestation programs, the goals and objectives of a forest nursery are closely related to, if not dictated by, the goals and objectives of a given reforestation program. The nursery manager is expected to produce seedlings "tailor made" for specific planting sites. This may result in very complex management systems of which soil-fertility management is only one.

Although certain basic principles of soil management may apply to all forest nurseries, a sound soil-management program must be based upon a thorough understanding of the soil system of each individual nursery so that a monitoring program can be established to fit existing soil conditions. Knowledge of both physical and chemical conditions of the soil is important because these influence interpretation of analysis data (see chapters 6 and 7, this volume). For example, poor physical conditions such as compaction may result in poor drainage and aeration, which in turn will impact nutrient uptake.

A systematic sampling program must be the base upon which a sound soil-management program is developed. Benefits will accrue only if the data generated are accurate, interpreted correctly, and put to use and if the results are then evaluated. However, data are only as good as the samples analyzed. Consistent quality control in the sampling program, analytical procedures, and recordkeeping is essential so that valid trends may be distinguished from anomalies.

In this chapter, soil analysis and tissue analysis are discussed as valuable tools for monitoring soil fertility. Suggested target nutrient levels for Douglas-fir [*Pseudotsuga menziesii* (Mirb.) Franco] and for species grown in intermountain nurseries are recommended; however, interpretation of those levels will be influenced by soil conditions at a given nursery. Examples are drawn from the Oregon State University Soil Testing Laboratory in the Department of Soil Science (OSU Lab) because it is the one with which I am most familiar, but other Northwest facilities are named which provide similar valuable services.

8.2 Available Laboratories

In addition to the services offered at the OSU Lab, one other state-owned laboratory, one Canadian laboratory, and seven private laboratories in the Northwest offer soil and plant analysis services:

Agri-Check, Inc., Umatilla, Oregon
British Columbia Ministry of Forests, Victoria
Century Testing Laboratories, Inc., Bend, Oregon
Chinook Research Laboratories, Inc., Corvallis, Oregon
HR Consulting Services, Umatilla, Oregon
Marr Wadoups and Associates, Kennewick, Washington
Soil and Plant Lab (office in Bellevue, Washington; lab in Santa Clara, California)
United States Testing Co., Inc., Richland, Washington
University of Idaho, Moscow

The OSU Nursery Survey (see chapter 1, this volume) indicates that three nurseries (under single management) use Agri-Check, two use Soil and Plant Lab, five use the B.C. Ministry of Forests Lab, and nine use the OSU Lab. In addition, 15 nurseries not included in the Survey use the OSU Lab.

The analytical methods used by the above-listed laboratories are generally the same as those of the OSU Lab. At present, however, the results of nursery soil analysis from

In Duryea, Mary L., and Thomas D. Landis (eds.). 1984. *Forest Nursery Manual: Production of Bareroot Seedlings.* Martinus Nijhoff/Dr W. Junk Publishers, The Hague/Boston/Lancaster, for Forest Research Laboratory, Oregon State University, Corvallis. 386 p.

these labs cannot be compared with those of the OSU Lab because nurseries are not submitting the duplicate samples necessary for comparison.

The methods used by the OSU Lab for pH, exchangeable potassium (K), calcium (Ca), and magnesium (Mg), cation exchange capacity (CEC), organic matter, and total nitrogen (N) are essentially the same as those used by the State University of New York at Syracuse (SUNY Lab) and the labs servicing nurseries in the southern and southeastern U.S. At an ad hoc meeting in Detroit in 1980, persons involved in forest-nursery soil testing agreed to aim at standardizing analytical methods for all tests frequently used, except for phosphorus (P) extraction techniques, so that comparisons of nurseries from different regions could be more meaningful. Such comparisons have limitations, however. For example, data for soil samples from the same sample areas in the Wind River Nursery (Carson, Washington)—analyzed by Wilde and Associates (Madison, Wisconsin), the SUNY Lab, and the OSU Lab—were compared. Absolute values for individual samples varied, but trends among samples were similar.

8.3 Soil Analysis

8.3.1 Sampling and handling

Soils should be routinely sampled at the end of the seedling crop rotation so that changes in nutrient levels can be monitored and fertilizer and lime added before establishment of a cover crop or new seedling crop. This is especially important in the case of the macronutrients P, K, Ca, and Mg, which do not readily move into the soil when surface applied.

The first step in the sampling procedure is to stratify the area on the basis of obvious soil differences, e.g., wet areas, areas having striking textural differences, or areas where topsoil has been removed as a result of land leveling. Most nurseries already have sampling patterns (e.g., predetermined lines or zigzag patterns) established within compartments or seedling blocks. The usual technique is to obtain a composite soil sample of each area according to the sampling pattern by coring soil to a depth of 15 cm (6 in.). The most efficient tool is a sampling tube having a 2-cm (¾-in.) diameter. A minimum of 30 cores per sample unit are placed in a clean (free of fertilizer or other chemicals) plastic pail and thoroughly mixed. A 225-g (½-lb) subsample sufficient for routine analysis is placed in a container and labeled. If particle-size analyses are desired, the sample should be split and placed in two containers. Samples are shipped to the soil-testing laboratory with information regarding tests desired. Samples may be air dried to reduce shipping weight.

An alternative, but more costly, method is random sampling. Randomly distributed samples are collected within each sample area so that an estimated mean value for each parameter measured can be calculated. If 20 samples are required to estimate the mean value of each parameter, the cost becomes prohibitive. This particular sampling method is used primarily for research purposes.

8.3.2 Testing

The basic tests available at the OSU Lab[1] for assessing soil nutrient levels are given in Table 1. Tests for mineralizable N and calcium carbonate ($CaCO_3$) equivalent also are available. Mineralizable N is determined with an anaerobic incubation technique [11] to provide an estimate of N availability. The $CaCO_3$-equivalent test determines the amount of acid or sulfur (S) required to lower the pH of alkaline soils and is used primarily in intermountain nurseries. Soil test #15 is designed for sodic soils [pH 8.5 to 10, >15% exchangeable sodium (Na)] but will generally not be needed because such soils are avoided in selecting nursery sites.

In P analysis, the dilute acid-fluoride method of Bray and Kurtz [2] is used for acid soils and the sodium bicarbonate method of Olsen et al. [5] for alkaline soils.

Ammonium N (NH_4-N) and nitrate N (NO_3-N) tests are not common in nursery soil analysis, but they might be used to determine the amount of available soil N at the beginning of the growing season or the time and rate of early-season N fertilization.

Soil tests are useful within limits. Perhaps the most serious limitation is the arbitrariness of extraction procedures. Chemical extracting solutions do not necessarily remove the same amount of a nutrient element that a plant can. CEC measurements, which indicate the buffer capacity of the soil and its resistance to rapid change in pH as cations are added or leached, are adjusted to a standard pH for convenience, whereas exchange capacities are strongly pH-dependent in many soils.

A related and serious limitation is the lack of data correlating seedling growth response, quality, and performance after outplanting with soil-test values and fertilizer additions. Comparisons must be made for each species produced at a given nursery. Thus, soil-test values are, at best, only a starting point and must be related to overall soil-management practices and seedling performance. Furthermore, it cannot be overemphasized that the benefits derived from a soil-testing program depend on meticulous recordkeeping for soil-test data, soil-management practices, and seedling performance.

Table 1. Soil tests available at the OSU Lab.

Test #	Item tested
1	pH, P, K, Ca, Mg
2	pH, P, K, Ca, Mg, and boron (B)
3	Cation exchange capacity (CEC)
4	Organic matter (OM)
5	Total nitrogen (TN)
6	Ammonium nitrogen (NH_4-N)
7	Nitrate nitrogen (NO_3-N)
8	Ammonium and nitrate nitrogen
9	Sulfate sulfur (SO_4-S)
10	pH
11	B
12	Zinc (Zn)
13	Manganese (Mn)
14	Zn and Mn
15	pH, P, K, Ca, Mg, and soluble salts (SS)—Na if pH > 7.4
16	$CaCO_3$ equivalent
17	SS

8.3.3 Interpretation

Before his death in 1981, S. A. Wilde had undoubtedly used soil-test data more often for making fertilizer recommendations for forest nurseries than any other person in North America. The basis for his recommendations was soil fertility standards developed for northern conifers [12] and northern hardwoods [14], as well as many years of accumulated experience. Using a similar approach, Youngberg and Austin [17] developed fertility standards for Douglas-fir. With some modification, these are presented in Table 2. It should be emphasized that these standards are only targets and are subject to revision as experience is gained.

Similar values are presented by van den Driessche in chapter 7, this volume. The levels of soil-test P recommended in that chapter are higher than those presented in Table 2; the range in total N is slightly higher. Because the method of P analysis referred to in chapter 7 is the same as that used by the

[1] The methods used by the OSU Lab are summarized in Berg and Gardner [1]; this report is available on request from the Department of Soil Science, Oregon State University, Corvallis.

OSU Lab, the differences in recommended levels are probably due to soil differences. The British Columbia nurseries generally have acid, sandy soils [pers. commun., 9], whereas soil pH in nurseries in the U.S. Northwest and northern California ranges from 5.0 to over 6.0. In soils with a pH range below 5.0, added P is strongly fixed by aluminum (Al) and iron (Fe). Landis [pers. commun., 4] has developed soil-fertility targets for intermountain nurseries (Table 3.).

Table 2. Soil-fertility levels recommended for Douglas-fir.[1]

pH	Total N	Available P	Available K	Exchangeable Ca	Exchangeable Mg
	%	ppm	ppm	meq/100 g	meq/100 g
5.0-6.0	0.18-0.23	25-50	80-120	2.0-4.0	0.8-1.5

[1] Based on OSU Lab values.

Table 3. Soil-fertility targets recommended for intermountain nurseries [pers. commun., 4].

	Range
pH	
Most conifers	5.5-6.5
Hardwoods and junipers	6.5-7.5
Electrical conductivity, mmhos/cm	
Conifers	<2.0
Hardwoods	<4.0
Organic matter, %[1]	2.0-5.0
CEC, meq/100 g	7.0-12
$CaCO_3$ equivalent, %	0
Total N, %	0.10-0.20
P, ppm[2]	30-60
lb P_2O_5/acre	175-350
K, ppm	100-200
lb K_2O/acre	300-600
Ca, ppm	500-1,000
meq/100 g	2.5-5.0
Mg, ppm	120-240
meq/100 g	1-2

[1] Determined by Walkley-Black [10] method.
[2] Determined by Olsen et al. [5] sodium bicarbonate method.

Fertilizer added to make up the difference between the soil-test level and the desired level may not necessarily be adequate to supply seedling needs. As mentioned, these values are only targets. The actual amounts needed to meet crop requirements may vary considerably among nurseries due to differences in soil properties such as texture, structure, drainage, aeration, acidity, and clay mineralogy. Amounts of fertilizer required to supply needed levels of nutrients will vary even among soils having similar soil-test values. For example, nursery A (sandy-textured soil) and nursery B (sandy loam soil strongly influenced by volcanic ash) may both have test values of 10 ppm P. To raise the level to 50 ppm, adding the phosphate fertilizer equivalent of 40 ppm might be sufficient for soil in nursery A, but the allophanic or amorphous colloids weathered from the volcanic ash impart phosphate-fixing properties to the soil in nursery B; therefore, more phosphate fertilizer would be required for nursery B than A to attain the desired level. In some soils, clay minerals impart K-fixing properties to the soil, influencing the availability of added potash fertilizers. Even within a given nursery, the amount of fertilizer needed to supply the desired level may vary over time due to changes in physical conditions caused by cultural practices. For example, poor aeration resulting from these practices can depress the uptake of K; increase the availability of Fe, causing P fixation; and increase the availability of manganese (Mn), causing Mn toxicity. Recommendations for, or decisions made concerning, fertilizer additions generally assume good soil physical condition (see chapter 6, this volume). If these conditions do not exist, soil-test values may not accurately indicate nutrient availability.

For most fertilizer recommendations, it is probably better to aim too high rather than too low. In the case of N, however, overfertilization will result in poor shoot:root ratio and will delay hardening off (see chapter 15). In the case of liming, only sufficient lime to raise the pH to the desired level should be added; overliming can increase the incidence of damping-off and root rot [13].

For alkaline soils, CEC can be used to determine the amount of S or acid needed to acidify the soil; for acidic soils, it can be used to determine the amount of lime needed to raise pH to the desired level. The data from Table 4 illustrate the use of CEC and other soil-test values for making a decision on liming as well as increasing Mg levels. Dolomitic limestone is often used for liming because it supplies Mg as well as Ca. In the example in Table 4, the Mg level should be increased in both nurseries. Nursery A (pH 6.0) has 7 milliequivalents (meq) of exchangeable acid [CEC − (K + Ca + Mg)]. One ton of dolomite/acre would add approximately 1 meq of Ca and 1 meq of Mg. In this case, the desired increase in Mg could be effected without causing an excess of bases. On the other hand, Nursery B (pH 6.7) has only 1.4 meq of exchangeable acid. Adding 1 ton of dolomite/acre would result in an excess of bases, making soil alkaline (pH > 7.0). Some other means of increasing Mg—such as the more costly addition of $MgSO_4$ (Epsom salts)—would be called for.

Table 4. Soil-test data used for liming and Mg fertilization recommendations.

Nursery	Exchangeable K	Exchangeable Ca	Exchangeable Mg	CEC	pH
	meq/100 g	meq/100 g	meq/100 g	meq/100 g	
A	0.32	5.0	0.37	12.7	6.0
B	0.42	6.0	0.35	8.2	6.7

CEC, a function of the contents of clay and organic matter, is a fairly stable parameter. Therefore, it should not be necessary to redetermine CEC every time a soil from a given area is tested. If organic matter content decreases over time, so probably will CEC. This relationship could be used to determine the advisability of obtaining a CEC analysis.

Nitrogen tests are the most difficult to interpret. Total N data provide information on the total amount of N in the seedling root zone, but nothing about its availability. Ammonium and nitrate N tests show **how much** of these forms of N are present in the soil when sampled, although this is partly a function of time of sampling and stage of seedling growth. Levels are usually low during periods of rapid growth but tend to build during the dormant season. Even if levels are high in the fall and early winter, winter rainfall will leach nitrate N to depths below the seedling root zone. Because ammonium N is held on the exchange complex, it is less subject to leaching losses; therefore, testing for this form some time before seeding might be a good indicator of the need for N fertilization. However, the demand for N by newly germinating seedlings is so small that N fertilization before seeding is probably a waste of money. Total N and organic matter data are used primarily for monitoring levels from one rotation to the next and to indicate the need for building up organic levels.

8.3.4 Monitoring soil fertility: an example

The changes in soil fertility over time in three nurseries are shown in Table 5. Data are mean values for all blocks at the Bend and Humboldt Nurseries but only for a single block at the Lava Nursery. The 1961 data for the Humboldt Nursery and the 1975 data for the Lava Nursery are from samples analyzed before these nurseries were established. Baseline data are being determined for each block at the Lava Nursery.

Over 28 years, management practices at the Bend Nursery have resulted in a wider range in pH values, increases in P, K, and Ca, and an increase in organic matter. The soil is a coarse pumice sand. The nursery is in a low rainfall area (high desert), and native soil organic matter is naturally low. Irrigation and organic amendments have increased organic matter levels.

Over 20 years, pH values and P levels have increased slightly at the Humboldt Nursery. The increase in P has probably resulted from residual buildup from phosphate fertilizer applications. Potassium levels decreased during the first 10 years but are now at or above initial levels. The increases in exchangeable Ca and Mg, as well as pH, are the result of adding dolomitic limestone. Organic matter and total N have decreased over the 20-year period as a result of frequent cultivation.

Over 6 years, pH has not changed significantly, P and Ca have decreased slightly, and K and Mg have increased at the Lava Nursery. Initially, the soils were low in Mg, so $MgSO_4$ and dolomitic limestone were added. Organic matter has also decreased, probably due to cultivation.

Remember, however, that the data in Table 5 are mean values. For careful monitoring, data for individual sample areas should be used for comparison. But this will require more detailed recordkeeping (see chapter 27, this volume). Computer printouts such as those from the OSU Lab (Figs. 1 and 2) give the kinds of specific breakdowns essential for thorough analysis. To facilitate more detailed interpretations of cultural practices, more frequent sampling and analysis would be required.

Table 5. Mean soil-test values at different times at three Northwest forest nurseries.

		Available		Exchangeable			Organic
Year	pH range	P	K	Ca	Mg	Total N	matter
		---- ppm ----		-- meq/100 g --		----- % -----	
			Bend Nursery				
1954	6.4-6.7	21	337	4.3	.....	0.06	1.3
1968	6.3-6.4	12	449	5.9	2.8	.....	2.5
1982	5.8-7.4	41	466	5.5	2.7	0.09	2.2
			Humboldt Nursery				
1961	5.1-5.3	5	100	1.5	0.65	0.31	8.0
1971	5.4-5.7	17	60	1.8	0.46	0.26	7.1
1982	5.4-6.2	13	120	3.2	1.4	0.22[1]	6.3
			Lava Nursery[2]				
1975	6.4-6.8	12	147	6.5	0.43	0.17	5.2
1981	6.3-6.6	7	221	5.5	1.4	0.16	3.7

[1] 1981 data.
[2] Means for one block only.

8.4 Tissue Analysis

The nutrient concentration of seedling tissue is a measure of the soil's ability to provide nutrients to a seedling crop. Because tissue analysis does not rely so heavily on arbitrary extraction procedures, it can be very useful for calibrating soil-test values.

Most tissue sampling is done in the fall (October-November), when seedlings are generally dormant and nutrient levels somewhat stabilized. However, if the objective is to evaluate the efficiency of fertilizer uptake, periodic sampling during the growing season should be scheduled. The use made of tissue analysis will determine the time of sampling and kinds of samples taken.

8.4.1 Sampling and handling

For analyzing 1+0 seedlings, the whole seedling is sampled. For analyzing 2+0 seedlings or transplants, only the needles (usually the current year's needles) are sampled. If, however, fertilizer-uptake efficiency or total nutrient uptake is to be evaluated, the whole seedling should be sampled. Samples submitted to the lab should represent soil conditions that are not too diverse. For more details, see Solan [6].

Tissue samples should be washed to remove soil and dust, especially if Fe analysis is desired, and sent as quickly as possible to the laboratory. If a drying oven is available, samples can be dried at 65 to 70°C for 24 hours; 10 g of dry plant tissue is adequate for lab analysis. If fresh seedlings cannot be sent to the lab soon after sampling, they should be stored in a refrigerator until ready for shipping; upon receipt, they are dried, if necessary, and ground in a Wiley mill to pass a 20-mesh screen in preparation for analysis.

8.4.2 Testing

The OSU Lab can analyze N, P, K, Ca, Mg, and S as individual elements or as a combined package. Additional analyses are available for boron (B), copper (Cu), Fe, Mn, Na, and zinc (Zn).

PAGE 1 OF 1

OREGON STATE UNIVERSITY
SOIL TESTING LABORATORY
CORVALLIS, OREGON 97331

Forest Nursery Soil Testing Service

OSU Forest Nursery Technology Center

NAME: EVERGREEN FOREST NURSERY
JOHN DOE
Address: RT 2 BOX 257
GILCHRIST OR 97737

Date Sampled: 7/20
Date Received: 7/25
Date Completed: 8/ 2/82

Sample From: East of Cascades XX
West of Cascades

Comments: ACID SOILS - USE BRAY P TEST. SEND RESULTS TO DR. YOUNGBERG.

Sample No.	Lab No.	pH	Bray P ppm	K ppm	Ca m/100g	Mg m/100g	CEC m/100g	OM %	TN %
4-PP	66111	6.0	38	304	4.70	1.90	7.60	0.91	0.04
4-LP	66112	5.9	41	276	4.20	1.60	7.40	1.10	0.05
5-2	66113	6.8	52	319	7.50	2.30	10.2	1.80	0.05
5-5	66114	6.5	54	280	5.00	1.30	9.40	1.20	0.06
5-8	66115	6.4	63	401	7.30	2.00	10.1	1.90	0.07
5-11	66116	6.7	48	331	6.20	1.60	8.30	1.80	0.05
5-13	66117	6.8	45	319	6.00	1.60	8.90	1.60	0.05
6-2	66118	6.5	41	253	5.90	1.90	8.10	1.70	0.05
6-5	66119	9.0	39	245	5.70	1.90	9.30	1.40	0.05
6-8	66120	6.4	39	218	4.90	1.60	7.60	1.60	0.05
6-11	66121	6.3	43	222	5.20	1.70	8.50	1.30	0.06
6-14	66122	6.2	45	234	4.70	1.70	7.50	1.30	0.04
6-17	66123	6.4	45	273	5.70	1.80	10.1	1.30	0.05
7-2	66124	6.2	32	265	5.20	1.80	9.30	1.30	0.05
7-5	66125	5.9	40	280	5.20	1.90	9.00	1.70	0.07
7-8	66126	6.2	33	335	5.90	2.40	10.2	1.90	0.05
7-11	66127	6.3	35	343	6.00	2.70	9.60	1.80	0.06
12	66128	5.9	27	187	4.09	2.09	8.80	1.50	0.05

Figure 1. Computer printout reporting results of soil analysis for a typical forest nursery in the Northwest.

Forest Nursery Soil Testing Service

OREGON STATE UNIVERSITY
SOIL TESTING LABORATORY
CORVALLIS, OREGON 97331

OSU Forest Nursery Technology Center

NURSERY SOIL FERTILITY MONITORING FORM

NAME: EVERGREEN FOREST NURSERY
JOHN DOE
Address: RT 2 BOX 257
GILCHREST OR 97737

Sample No. 4-PP

8/02/82

Date	pH	Bray P ppm	K ppm	Ca m/100g	Mg m/100g	CEC m/100g	OM %	TN %
09/80	6.2	50	400	4.85	2.10	8.00	0.82	0.03
07/81	5.9	44	325	4.68	1.50	7.10	1.02	0.05
08/82	6.0	38	304	4.70	1.90	7.60	0.91	0.04

Figure 2. Computer printout reporting monitored nutrient levels in a typical nursery sample over time.

The sample size (0.5 to 1.0 g) used for digestion and analysis depends on the number of elements to be determined and the approximate elemental concentration in the tissue. A Kjeldahl digest is used for N and P. All cations including K, Ca, Mg, Fe, Mn, Cu, molybdenum (Mo), and Zn are digested with a nitric-perchloric acid mix. S and B are dry ashed. Elemental determinations are made using standard methods.[2]

8.4.3 Interpretation

The range in concentration of macronutrients in 2+0 Douglas-fir needle tissue collected in the dormant season (fall-early winter) is given in Table 6 ([unpubl. data, 16]; see also chapter 7, this volume). Concentrations below the low values indicate probable deficiencies, and those above the high values suggest possible luxury consumption.

Table 6. Range in nutrient concentrations in needle tissue of 2+0 Douglas-fir seedlings.

Level	N	P	K	Ca	Mg	S
	%					
Low	1.2	0.1	0.3	0.2	0.1	0.1
High	2.0	0.2	0.8	0.5	0.15	0.2

Micronutrient data for nursery-grown seedlings are scarce; most of the data available are for larger trees [7]. Availability of micronutrients is strongly influenced by pH. For example, Fe deficiency (chlorosis) is often observed on seedlings in nurseries with strongly alkaline soils. Toxicity problems may be caused by strongly acid soils. In 1972, pronounced Mn toxicity symptoms were observed on 2+0 Douglas-fir seedlings in a poorly drained area with strongly acid soil (pH 4.5) at the Wind River Nursery.

Micronutrient problems often occur on old, strongly weathered soil material. Fortunately, however, most of the forest nurseries in the Northwest are sited on young, relatively nutrient-rich soils. The levels of available nutrients in Northwest nurseries, even those on sandy glacial soils, are considerably higher than those in nurseries on strongly weathered soils in the southeastern U.S. Because most Northwest soils are only slightly to moderately acid, micronutrient problems will likely be minimal. Sewage sludge and other "exotic" amendments, which may cause toxicity problems, should not be used without first analyzing them for micronutrients.

[2] All procedures and methods used by the plant analysis laboratory are on file with the Department of Soil Science, Oregon State University, and are available on request.

The tissue analysis done by the OSU Lab for forest nurseries thus far has shown that for all species analyzed, elemental concentrations are generally within the ranges given in Table 6. In a few instances, concentrations of P and Mg have been low, but not deficient; those for K and Ca have varied from midrange to above the high levels in Table 6; and those for total N have ranged from low to very high, with most in the high range. Data on N concentration in seedling tissue from four Northwest nurseries (Table 7) seem to indicate that more N is being added to soils than is needed; concentrations much over 2% suggest overfertilization. Concentrations of the other macronutrients in seedling tissue from these four nurseries (Table 8) indicate that the nutritional status of the seedlings is satisfactory.

Table 7. Foliar N concentration of 2+0 Douglas-fir seedlings from four Northwest nurseries.

Nursery	Percent N Mean	Range	Remarks
1	1.59	1.24-2.03	7 of 15 samples < 1.6
2[1]	1.95	1.78-2.03	7 samples
3	1.78	1.26-2.67	32 of 37 samples < 2.0
4	2.29	1.92-2.57	1 of 13 samples < 2.0

[1] Total soil N = 0.22%; similar data unavailable for the other three nurseries.

Turner and Lambert [8] and Knight [3] have emphasized the importance of S in conifer seedling nutrition; Knight recommends adding 1 part of S for every 15 parts of N added as fertilizer. Foliar analysis for total N and total S is a valuable way of assessing this aspect of fertility management; a ratio at or below 15 N:1 S is suggested for adequate S nutrition and protein synthesis.

Foliar S data were available from three of the four nurseries discussed in Tables 7 and 8; their N:S ratios ranged from 7:1 to 23:1. Seedlings from nursery 2 had foliar S concentrations ranging from 0.19 (adequate) to 0.24% (high) and N:S ratios of

Table 8. Mean soil and foliar levels of four macronutrients for 2+0 Douglas-fir seedlings in four Northwest nurseries.

Nursery	P Soil	P Foliar	K Soil	K Foliar	Ca Soil	Ca Foliar	Mg Soil	Mg Foliar
	ppm	%	ppm	%	ppm	%	ppm	%
1	79	0.17	93	0.70	1.5	0.33	0.61	0.19
2	13	0.21	120	0.60	3.2	0.46	1.4	0.30
3	....	0.23	....	0.78	.....	.....	.....	.....
4	18	0.15	79	0.50	7.3	0.55	1.8	0.19

8:1 to 9:1; those from nursery 3 had foliar S concentrations ranging from 0.12 (low) to 0.19% (adequate) and N:S ratios of 7:1 to 15:1; and those from nursery 4 had foliar S concentrations ranging from 0.14 (midrange) to 2.24% (high) and N:S ratios of 9:1 to 23:1. Seedlings with N:S ratios greater than 15:1 had high foliar N concentrations and S levels in the midrange. Sulfur deficiencies are known to exist in some Northwest soils [15], and color in Christmas trees has been observed to improve after addition of S. The use of fertilizers containing S should adequately supply that element to Northwest soils.

8.5 Combined Soil and Tissue Analysis

Either soil analysis data **or** plant analysis data can form the basis for fertilizer recommendations. From time to time, however, it is advantageous to have **both** types of analysis to verify the validity of management recommendations.

Soil analysis data also were available for three of the four nurseries examined in Tables 7 and 8. As might be expected, the correlations between foliar and total soil N were not consistent. Furthermore, foliar N was more responsive to fertilizer N than were foliar P, K, Ca, and Mg to fertilizer additions containing those elements.

In nursery 1, soil-test levels for P were well above the minimum recommended value, and those for K were within the recommended range (Table 2). Foliar P was midrange to high, and K was adequate (Table 6). Exchangeable Ca and Mg were both low (Table 2); however, foliar Ca was midrange and Mg high.

Soil-test values for P in nursery 2 (Table 8) were below those suggested in Table 2 and well below those recommended in chapter 7, this volume. However, foliar P concentrations were above the high levels suggested in Table 6 and in chapter 7. Correlation was good between foliar and soil K, Ca, and Mg. Total soil N was adequate (Table 2); in this case, the correlation between foliar and soil N was good. Because information was not available on N fertilization regimes, its influence could not be evaluated.

Soil-test levels for P in nursery 4 (Table 8) were less than the low values recommended in Table 2 and in chapter 7. Foliar P concentrations were in the midrange. Foliar and soil K, Ca, and Mg correlated reasonably well.

It should be emphasized that this discussion concerning the use of combined soil and plant analysis is based on general comparisons of data from a limited number of nurseries. The values cited for both soil and foliar levels are means. Sufficient data were not available to detect any nutrient interactions or dilution effects, although a comparison of foliar N and P data for nurseries 2 and 4 (Tables 7 and 8) suggests that there may be a slight dilution effect from high N on foliar P in nursery 4. Some correlations were good and others poor. Only careful sampling can assure that both soil and tissue samples come from the same area. Moreover, information on fertilizers and their rates of application are necessary for adequate interpretation of any analysis. Careful recordkeeping is therefore essential. Obviously, the use of combined soil and tissue analysis is an area requiring concentrated research.

8.6 Conclusions and Recommendations

Soil fertility is only one important factor among the many necessary for producing high-quality nursery stock. Soil and plant analysis are readily available tools that enable forest-nursery managers to monitor the fertility status of their soils. The success of the monitoring program depends on careful sampling—which requires sampling the same area each time, careful handling of samples, and consistency in laboratory services.

Suggested target nutrient levels for Douglas-fir in Northwest nurseries are: pH of 5.0 to 6.0, total N of 0.18 to 0.25%, available P of 25 to 50 ppm, available K of 80 to 120 ppm, exchangeable Ca of 2 to 4 meq/100 g, and exchangeable Mg of 0.8 to 1.5 meq/100 g. Suggested levels for conifers and hardwoods in intermountain nurseries are: pH of 5.5 to 6.5 for most conifers (6.5 to 7.5 for hardwoods and junipers), total N of 0.1 to 0.2%, available P of 30 to 60 ppm, available K of 100 to 200 ppm, exchangeable Ca of 2.5 to 4 meq/100 g, and exchangeable Mg of 1 to 2 meq/100 g. Because amounts of fertilizer added to achieve desired levels will vary with soil type, tissue analysis is a useful cross-check for assessing the success of fertilizer-management regimes. Suggested ranges in macronutrient concentrations in Douglas-fir needle tissue are: 1.2 to 2% N, 0.1 to 0.2% P, 0.3 to 0.8% K, 0.2 to 0.5% Ca, 0.10 to 0.15% Mg, and 0.1 to 0.2% S.

Seedling nutrient status is assumed to influence performance after planting. Researchers and nursery personnel should seek to uncover the relationships between nutrient status and outplanting, bearing in mind, however, that many factors in nursery culture other than seedling nutrient status profoundly affect survival and growth of outplanted trees.

References

1. Berg, M. G., and E. H. Gardner. 1978. Methods of soil analysis used in the soil testing laboratory at Oregon State University. Oregon Agric. Exp. Sta., Corvallis. Special Rep. 321. 44 p.
2. Bray, R. H., and L. T. Kurtz. 1945. Determination of total, organic and available forms of phosphorus in soils. Soil Sci. 59:39-45.
3. Knight, P. J. 1978. Fertilizer practice in New Zealand forest nurseries. New Zealand J. Forestry Sci. 8:27-53.
4. Landis, T. 1982. Personal communication, U.S.D.A. Forest Serv., Denver, Colorado.
5. Olsen, S. R., D. V. Cole, F. S. Watanabe, and L. A. Dean. 1954. Estimation of available phosphorus in soils by extraction with sodium bicarbonate. U.S. Dep. Agric., Washington, D.C. Circular 939. 19 p.
6. Solan, F. M. 1980. Soil and plant tissue techniques for tree nurseries. Pages 228-235 *in* Proc., North American forest tree nursery soils workshop (L. P. Abrahamson and D. H. Bickelhaupt, eds.). State Univ. New York, Coll. Environ. Sci. and Forestry, Syracuse.
7. Stone, E. L. 1968. Microelement nutrition of forest trees: a review. Pages 132-175 *in* Forest fertilization: theory and practice. Tennessee Valley National Fertilizer Development Center, Muscle Shoals, Alabama.
8. Turner, J., and M. J. Lambert. 1978. Sulphur nutrition of conifers in relation to fertilizer nitrogen, to fungal infections, and soil parent material. Pages 546-564 *in* Proc., Forest soils and land use (C. T. Youngberg, ed.). 5th North American Forest Soils conf., Colorado State Univ., Ft. Collins.
9. van den Driessche, R. 1982. Personal communication, B. C. Ministry of Forests, Victoria.
10. Walkley, A., and I. A. Black. 1934. An examination of the Degtjareff method for determining soil organic matter and a proposed modification of the chronic acid titration method. Soil Sci. 37:29-38.
11. Waring, S. A., and J. M. Bremner. 1964. Ammonium production in soil under waterlogged conditions as an index of nitrogen availability. Nature 201(4922):951-952.
12. Wilde, S. A. 1938. Soil fertility standards for growing northern conifers in forest nurseries. J. Agric. Res. 57:945-952.
13. Wilde, S. A. 1958. Forest soils. Ronald Press, New York. 537 p.
14. Wilde, S. A., and W. E. Patzer. 1940. Soil fertility standards for growing northern hardwoods in forest nurseries. J. Agric. Res. 61:215-222.
15. Will, G. M., and C. T. Youngberg. 1978. Sulfur status of some central Oregon pumice soils. Soil Sci. Society J. 42:132-134.
16. Youngberg, C. T. 1982. Unpublished data, Dep. of Soil Science, Oregon State Univ., Corvallis.
17. Youngberg, C. T., and R. C. Austin. 1954. Fertility standards for raising Douglas-fir in forest nurseries. J. Forestry 52:4-6.

Chapter 9
Nursery Soil Organic Matter: Management and Importance

C. B. Davey

Abstract

Organic matter is important in nursery management because of its favorable effects on the physical, chemical, and biological properties of the soil. Organic matter may be added by incorporating into the soil either cover or green manure crops grown on the site or organic amendments brought from elsewhere. Some constituents of organic matter decompose very quickly and others much more slowly, but both types are important in maintaining favorable soil conditions and productivity.

9.1 Introduction

"Now *here*, you see, it takes all the running *you* can do to keep in the same place. If you want to get somewhere else, you must run at least twice as fast as that."

—The Queen to Alice in *Through the Looking Glass* [7], Chapter 2, by Lewis Carroll (1832-1898)

To paraphrase the Queen, we could easily state: "Now *here*, in the nursery, you see, it takes all the running *you* can do to maintain your soil organic matter level. If you want to increase it, you must run at least twice as fast as that."

With very few exceptions, nursery managers are faced with a constant struggle in keeping their soil organic matter content at an appropriate level. In this regard, they differ from many other tillers of the soil. Basically, forest-nursery management is a mining operation with respect to organic matter. Most management activities accelerate the decomposition of organic matter; further, during harvest (lifting), the entire plant, including roots with adhering soil and organic matter, is removed. It is no wonder, then, that the soil must be replenished frequently.

There are really only two fundamental ways in which organic material can be added to the soil. The first way is to grow a crop on the land and incorporate it into the soil. Such crops may be referred to as **catch crops** if they are grown principally for catching and holding nutrients on the site, **cover crops** if they are grown principally for erosion control, or **green manure crops** if they are grown principally as organic amendments for the soil [13] (see also chapter 10, this volume). In this chapter, I discuss mainly green manure crops. The second way is to transport organic matter from another place to the nursery and incorporate that into the soil. Both of these approaches are employed frequently in the Northwest, as shown clearly by the OSU Nursery Survey.

Following a brief look at the current status of soil organic matter levels and management in Northwest nurseries, we will explore organic matter dynamics, types, and sources. The importance of organic matter to the physical, chemical, and biological properties of the soil will be stressed, and these properties will be related to the growth and harvesting of forest-tree seedlings and transplants.

9.2 Nursery Survey Results

The results of the OSU Nursery Survey (see chapter 1, this volume) showed that most managers (86%) felt that their soil organic matter level was not as high as it should be. They reported current levels ranging from 1 to 7% (average 3.6%) but estimated that levels should range from 2 to 10% (average 5.0%). When asked to list their five major nursery-management problems in order of importance, 62% included soil organic matter maintenance among their top five, and 14% regarded it as their greatest problem.

Of the eight nurseries not including organic matter maintenance among their top five problems, three reported adding sawdust, manure, or both and growing green manure crops. Two felt that their organic matter levels were near optimum and used green manure crops to maintain those levels. One nursery reported being so pushed for production that it had no opportunity to include a green manure crop in the rotation, though both manure and sawdust were applied, and one was so new that it had not evolved to the point of needing to enhance its organic matter. Finally, one had so many other major problems that organic matter maintenance did not make the top five; however, in attempts to deal with some of its other problems (e.g., soil compaction, poor drainage, and crusting), that nursery stressed the use of organic amendments and green manure crops.

Of the 17 nurseries (81%) that reported adding organic matter other than green manure crops, 12 added sawdust or

In Duryea, Mary L., and Thomas D. Landis (eds.). 1984. *Forest Nursery Manual: Production of Bareroot Seedlings*. Martinus Nijhoff/Dr. W. Junk Publishers, The Hague/Boston/Lancaster, for Forest Research Laboratory, Oregon State University, Corvallis. 386 p.

bark, six added peat, two added manure, and one added sludge. Obviously, several reported using more than one source. The really disturbing fact is that 90% of the managers said that they foresaw a shortage of such materials in the future. Those fortunate managers that did not expect shortages reported ample local supplies of either peat or manure.

Although composting offers several advantages—weed and pest control as well as stabilization of organic matter through lowering of the carbon to nitrogen (C:N) ratio and elimination of toxic decomposition products—it also is an inconvenience. Thus, despite the fact that several managers said they were interested in using composts, only 14% said they actually did.

Cover or green manure crops were included in the management of 76% of the nurseries, and 71% used them in every rotation. Peas [some cowpeas, *Vigna sinensis* (L.) Endl.; some field peas, *Pisum sativum* var. *arvense* (L.) Poir.; some not specified] were listed most frequently. They were followed in order of decreasing frequency by oats (*Avena sativa* L.), sudangrass [*Sorghum bicolor* (L.) Moench], and lupines (*Lupinus* L. spp.). Several other crops were reported used, each in a separate nursery. Exactly half (8 of 16) of the nursery managers said they used cover or green manure crops to increase the soil organic matter level; two were less optimistic, hoping only to maintain the present level; and others cited different reasons, including improving soil structure, controlling weeds, preventing erosion, and conserving nutrients.

9.3 Organic Matter Dynamics

Soil organic matter has been variously defined. This is not surprising, however, because it results from a complex and dynamic system comprising three principal segments: (1) organic residues (plant, animal, and microbial) in various stages of decomposition, (2) true humic materials, and (3) live organisms, principally microbes. The turnover rate in this mostly biochemical system is determined by the nature of the organic residues added to the soil and the physical and chemical nature of the soil. These all affect the microbial species populations and their rates of biological activity.

In forest nurseries, most of the organic residues added to the soil are of plant origin. They are composed principally of carbohydrates, primarily cellulose, and lignin but include varying amounts of other constituents. Though microorganisms degrade these materials at various rates, eventually, most of the carbon is either returned to the atmosphere as respiratory carbon dioxide (CO_2) or resynthesized into the bodies of microbes. The final true humus is dark colored, predominantly aromatic material of high molecular weight. Its rate of decomposition is very slow.

The early stages of the breakdown of organic residues are quite rapid, especially if the residue is an immature green manure crop. The process is slower when materials such as sawdust or bark are added to the soil. Peat represents material which has already undergone the initial stages of humification; thus, its breakdown is slow. Any plant residue incorporated with the soil undergoes continuous decomposition, the rate of which decreases with time as the remaining compounds become increasingly resistant to decay. Much of any added material will be gone in weeks or months, and nearly all of it will be gone in a few years. But some will remain, in a highly altered form, even after several centuries.

Summarizing the studies of several workers in a new text on humus chemistry, Stevenson [22] reported trends in organic matter system dynamics. Long-term crop rotations have generally resulted in a slow decrease in organic matter content, leading to a steady state in 50 to 100 years. Yet in a study where barley (*Hordeum vulgare* L.) was grown continuously and manure applied annually, the soil organic matter content kept increasing and still had not reached equilibrium when the experiment was terminated after 94 years.

Though the exact age of soil organic matter cannot be determined, its mean residence time (MRT) can be by ^{14}C dating. In his summary, Stevenson [22] found that MRT varied from 250 to 1,900 years. In a virgin prairie soil, MRT was almost 1,200 years. But continuous clean cultivation had resulted in the accelerated loss of the younger fractions such that its MRT had increased to 1,900 years. Conversely, where considerable manure had been added, MRT had decreased to less than 900 years.

9.3.1 Optimum soil organic matter levels

The question can very legitimately be asked by any nursery manager: "What is the optimum soil organic matter level?" Unfortunately, there is no simple, single answer. However the factors affecting the answer are sufficiently well understood that an answer can be given for a specific site.

The organic matter level represents a dynamic equilibrium among those factors favoring organic matter accumulation and those dedicated to its decomposition. The most important soil variables are moisture, temperature, fertility, and texture. A long, moist, **cool** growing season and fertile soil favor the accumulation of organic matter through the growth of vegetation such as a green manure crop. However, a long, moist, **warm** growing season and fertile soil favor the action of saprophytic microbes that decompose organic matter. Clay and humified organic matter tend to be closely associated in the soil, which reduces the surface area available for attack by saprophytic organisms. Thus, fine-textured soils tend to have higher organic matter contents than coarse-textured soils

Assessing all of the above variables together, we can arrive at some reasonable ranges for desired soil organic matter content in forest nurseries. Areas where Sitka spruce [*Picea sitchensis* (Bong.) Carr.] predominates and soils that have a moderate amount of clay provide the ultimate mix of climatic and soil conditions favoring organic matter accumulation. In those soils, 7 to 10% organic matter would be desirable. Where coastal Douglas-fir [*Pseudotsuga menziesii* (Mirb.) Franco var. *menziesii*] predominates, summers are slightly warmer and drier; in those soils, 5 to 8% organic matter would be desirable. At higher elevations and in inland areas well beyond any coastal influence, growing seasons tend to be decidedly shorter, hotter, and drier; in those soils, 3 to 6% organic matter would be desirable. In contrast, in the typical sandy nursery soil of the U.S. Southeast coastal plain—with its long, hot, moist summers and moist, mild winters—saprophytes are active nearly year round and decomposition dominates. In those soils, managers must be content with 1 to 2% organic matter.

9.3.2 Methods for determining soil organic matter content

Basically, three different methods can be used to determine soil organic matter content: loss on ignition, wet oxidation by acid hydrolysis, and wet oxidation by alkaline hydrolysis.

Loss on ignition might seem the simplest, but it tends to yield values that are too high. It removes all of the organic matter, including charcoal and other inert, nonreactive materials. Additionally, at certain temperatures, it actually begins to destroy some of the mineral matter. Thus, except in the case of peats, mucks, and other soils that are mostly organic matter, this method is seldom used.

Wet oxidation by acid hydrolysis, by far the most common method, is intended to determine the "active" soil organic matter only. Chromic acid is used to oxidize organic matter so that the amount of reactive carbon in the soil can be determined. The resulting value is then multiplied by a factor used to convert carbon content to organic matter content. (This method is assumed in all of the equilibrium organic matter contents given in 9.3.1.)

Despite its popularity, wet oxidation by acid hydrolysis has certain problems. First, some laboratories heat the flask in which the reaction takes place, which results in elevated values. Second, not all laboratories use the same multiplication factor to convert carbon to organic matter, which has a variable effect on the results. Third, a significant amount of chromium leaves the laboratory in its wastewater; for this reason, state environmental protection agencies have required some soil-testing laboratories to cease using this method.

A few labs, especially those that have been required to abandon the chromic acid method, have adopted wet oxidation by alkaline hydrolysis. Sodium hydroxide is used to determine organic matter that is partly humified. Because this method detects neither inert carbonaceous materials such as charcoal nor fresh nonhumified organic matter, it tends to give lower values than the chromic acid method. Experience with nursery soils that frequently receive fresh sawdust or green manure crops has indicated that organic matter contents determined with alkaline hydrolysis range from ½ to ⅔ of those determined with the chromic acid method.

One final word on methodology is warranted. Essentially all soil-testing laboratories sieve soil samples before they are analyzed. Thus, any large fragments of residue such as green manure crop stems or roots or fresh sawdust or bark will be removed and not included in the analysis. As a result, some soil amendments may take more than a year to appear in the organic matter test. This is much more likely with sawdust or bark than it is with more easily decomposable or finely divided materials.

9.4 The Two Organic Matters

Fresh organic material added to soil begins a continuum of reactions that are of ever-decreasing rates and that eventually terminate only when the last atom of carbon reenters the atmosphere as CO_2 many years later. Nonetheless, it is convenient to divide organic matter into two basic types: (1) highly reactive, recently added fresh organic materials and (2) much less reactive, more nearly stable materials of the later stages of humification. Both types serve important functions in the soil, and each deserves separate discussion.

9.4.1 Highly reactive organic materials

Fresh organic residues, especially green manure crops, contain a wide variety of compounds. These vary from water-soluble substances (such as sugars, amino acids, and some starches) to less soluble materials (such as pectins, proteins, and more complex starches) to insoluble celluloses and lignin. In addition, there are varying amounts of fats, oils, waxes, and extractives such as resins and terpenes.

Most of the water-soluble substances are immediately available to microbes and are metabolized quickly unless they are physically inaccessible because of location within large fragments. The pectins, proteins, and starches also are readily metabolized. The hemicelluloses, alphacelluloses, and lignins are increasingly difficult to decompose. Although fats, waxes, oils, resins, and terpenes were once thought to be quite resistant to decomposition, we now know that most of them are readily metabolized by specific microbes.

During the rapid decomposition phase, several important functions are performed, principally by bacteria and some "sugar fungi" [15]. The bacteria produce many polysaccharide gums that improve soil structure. The metabolic rate in the soil results in the suppression of various pathogens through (1) nonspecific reactions, such as the production of a high level of CO_2 in the soil atmosphere which is fungistatic to *Rhizoctonia solani* Kühn [18]; (2) competition for specific nutrients; (3) parasitism by some facultative organisms of pathogens; and (4) production of specific antibiotics and other antimetabolites in the soil. Small amounts of nitrogen may be fixed by free-living bacteria, principally in the genera *Clostridium* and *Bacillus*, and nutrients may be mineralized during this phase. The major carbon-containing constituents, cellulose and lignin, are attacked only slightly. Thus, the C:N ratio is narrowed only a small amount.

The period of very high microbial activity lasts from 1 to a few weeks. Most of the organisms involved can respond quickly to the presence of readily available food, grow rapidly, and produce resting structures, principally spores. These microbes are generally poor competitors and have low tolerance of antibiotics. Thus, their chief advantage is speed.

During the very early stages of rapid decomposition, soil oxygen often becomes temporarily limiting. This results in the incomplete metabolism of some constituents and the production of certain volatile organic compounds and some organic acids with low molecular weight which are toxic to germinating seeds and young plants. However, this plant-toxic period is short lived, nearly always less than 2 weeks and usually less than 1 [17], and toxic substances are easily metabolized as soon as the oxygen level permits. The important point for nursery managers is that it is frequently unwise to plant seeds of trees or green manure crops in soil less than a week after crop residue or other easily decomposed material has been added to the soil. This is not, however, a problem with compost or peat because both have passed this stage of decomposition [13]. Fresh sawdust or bark, at high rates of application, may cause some minor toxicity problems, but weathered bark or sawdust generally does not.

9.4.2 Less reactive organic materials

As the initial burst of microbial activity begins to diminish, a different group of microbes, composed principally of fungi and some actinomycetes, becomes dominant. This group can utilize the celluloses and eventually the lignins. The period of their activity is measured in months or years.

Again, several important changes—chiefly physical and chemical ones—take place in the soil. As organic matter particles break down, the macropore volume of the soil increases, which increases water infiltration and gaseous exchange. In fine-textured soils, surface crusting is reduced.

During this more leisurely phase of decomposition, the nitrogen in the system is passed efficiently from one crop of microorganisms to the next, as a significant amount of carbon is lost through respiration as CO_2. The result is a lowering of the C:N ratio. The general course of biological activity is to transform organic debris from identifiable particles composed of identifiable compounds to humified materials whose origin is impossible to detect and whose chemical composition is highly altered from the original. Early concepts of humus formation held that humus represented biochemically altered lignin [24]. But we now know that humic substances are resynthesized by microbes and are not simply degraded lignin [22].

Humic substances affect the soil in many ways:

- They are brown or black, which facilitates soil warming.
- They readily retain water and thus are particularly important in sandy soils.
- They combine with clays to stabilize soil structure.
- They are highly buffered and so help stabilize the soil reaction (pH).
- They have very high cation exchange capacities (CEC) —some exceed 1,000 milliequivalents per 100 grams of soil—and thus increase the soil's CEC and hence its ability to hold cations against leaching.

- They slowly mineralize and provide plant-available sources of nitrogen, phosphorus, and sulfur.
- They readily combine with many organic molecules such as pesticides and thereby affect the application rate needed for effective pest control.

The turnover rate of organic matter may seem difficult to determine because some fractions are metabolized in a few hours whereas others may require centuries. However, in many soils, added organic matter decomposes at rates which metabolize about 2/3 of the material in 1 year and 4/5 of it in 2 years.

One possible complication can arise in assessing the effect of added organic residues on soil organic matter. Once soil microorganisms become stimulated ("primed") because of the added organic matter, they are likely to metabolize any organic material in the soil, including native organic matter. This was demonstrated clearly with a soil-incorporated sudangrass cover crop that was ^{13}C-labeled [5]. The fraction of ^{13}C-label in the respiratory CO_2 was related to the breakdown rate of both the sudangrass and the native organic matter in the soil. After the sudangrass was incorporated, the oxidation rate of the native organic matter more than tripled. Indeed, because of the priming action of the microorganisms, it is possible to add easily decomposable organic material to the soil and, following its normal decomposition, reach a soil organic matter level actually lower than if the material had not been added. This condition is restricted almost exclusively to easily decomposable materials such as immature green manure crops. It does, however, help us understand why increasing soil organic matter content with such crops is, at best, difficult.

One recent, interesting study [23] suggested that soil organic matter in Georgia might be maintained between 1.4 and 1.6% with green manure crops. The organic matter level increased soon after the various tested crops were turned under at the end of the first summer. However, this level did not change again, despite cover crops the succeeding winter and summer. A crop of 1 + 0 pine (*Pinus* spp.) was then grown in the nursery and soil organic matter content decreased to the pretreatment level during that growing season [pers. commun., 21]. Thus, the net flux for the 3-year rotation of green manure crops and tree seedlings was zero. The green manure crops had served a valuable purpose in the soil, but the gain in soil organic matter content lasted less than 1 year.

9.5 Sources of Organic Matter

As stated earlier, only two general sources of organic matter are available for nurseries: that which is grown as a cover, catch, or green manure crop on the soil into which it will be incorporated, and that which is brought to the site from elsewhere and incorporated into the soil (organic soil amendment).

9.5.1 Cover and green manure crops

Cover and green manure crops are important in the soil physically, chemically, and biologically and should be included in all nursery rotations (see chapter 10, this volume, for details). However, their potentially positive or negative influence on the succeeding tree crop needs more investigation.

For example, a periodically recurring outbreak of a *Fusarium* root rot in the Saratoga Nursery, New York, was traced to the use of buckwheat (*Fagopyrum sagittatum* Gilib.) as a green manure crop preceding the tree crop [8]. The problem was corrected by eliminating buckwheat as a green manure crop in that nursery. Green manure and cover crops also can be associated with beneficial microorganisms. Preliminary results indicate that the preceding green manure crop stimulated both mycorrhiza formation and seedling growth on endomycorrhizal hardwood seedlings in Virginia [unpubl. data, 20].

9.5.2 Organic soil amendments

Traditional organic soil amendments include sawdust, bark, peat, and manure. Because some of these are locally scarce or because other organic residues are locally abundant, a wide variety of other materials has been used on nursery soils with varying levels of success.

Briefly, this list includes: hammermilled cones from seed extractories; leaves collected from city streets; spoiled hay, straw, and other agricultural wastes; organic sludges including sewage, paper mill, fish, and mint; commercially processed and dried sewage sludge; brewery and cannery wastes; and spent mushroom compost. The *Western Fertilizer Handbook* [6] provides the nitrogen, phosphate, potash, and organic matter contents of a range of such materials. On an absolute scale, these amendments carry only limited amounts of mineral nutrients, but on a relative scale they vary greatly (Table 1).

It is difficult to generalize about such a diverse group of substances, but many have high C:N ratios and thus require that extra nitrogen be added to the soil to avoid immobilization of soil nitrogen during the early stages of decomposition. The amount of nitrogen needed depends more on the ease of decomposition of the material than on its C:N ratio. For example, sawdust from red alder (*Alnus rubra* Bong.) has a C:N ratio of 134, that from Douglas-fir a ratio of 623, and that from western hemlock [*Tsuga heterophylla* (Raf.) Sarg.] a ratio of 1,244 [2]. Even though it contains the most nitrogen per unit of carbon, red alder decomposes so readily that it required more supplementary nitrogen when added to soil than either of the other two species [3, 4]. As a general guideline for using Douglas-fir sawdust, Bollen and Lu [4] recommended applying from 5 to 10 pounds of actual nitrogen per ton of sawdust applied in the first year, with amounts halved in each of the following 2 years.

Table 1. Average analysis of organic materials (adapted from [6]).

Material	N	P_2O_5	K_2O	Organic matter
	- - - - - -	- - - - - - %	- - - - - -	- - - - - -
Bulky organics				
Pine sawdust	0.1	0.01	0.05	[illegible]8
Steer manure[1]	2.0	0.54	1.92	50
Horse manure	0.7	0.34	0.52	50
Hog manure	1.0	0.75	0.85	30
Sheep manure	2.0	1.00	2.50	50
Poultry manure	1.6	1.25	0.90	50
Poultry droppings[2]	4.0	3.20	1.90	74
Seaweed (kelp)	0.2	0.10	0.60	30
Alfalfa hay	2.5	0.50	2.10	35
Grain straw	0.6	0.20	1.10	30
Organic concentrates				
Dried blood	13.0	1.50	....	80
Fish meal	10.4	5.90	....	80
Sewage sludge				
Digested	2.0	3.01	....	50
Activated	6.5	3.40	0.30	80
Castor pomace	6.0	2.75	0.50	80

[1]All manures include some bedding material.
[2]Droppings are bedding-free.

9.5.2.1 Sludge

The several sludges that are available vary widely in their composition. Thus, before any sludge is used, its chemical composition should be determined. This is especially true of sewage sludges. For example, in a study of the sludges produced in the Tualatin Basin of Oregon [19], Portland sludge was found to contain a higher concentration of lead, cadmium, nickel, and zinc than sludges from Forest Grove, Hillsboro, Oregon City, or Aloha, whereas Hillsboro sludge contained the most copper.

One difficulty in using sludges is knowing how much of any of these potentially toxic elements the soil can tolerate. A useful guide has been provided by Hausenbuiller [16] who stated that the tolerance level varies with the soil CEC. For each milliequivalent of CEC per 100 grams of soil, no more than 100 kg/ha (or lb/acre) of lead, 50 of zinc, 25 of copper, 10 of nickel, or 1 of cadmium should be allowed to accumulate in soil.

Most sludges have very high (close to 99%) water contents and consequently may be quite expensive to transport and spread. Even dewatered sludge cake, which must be handled and spread as a solid, is still at least 50% water.

Several cities thoroughly dry their digested sewage sludge and market it as a soil amendment. Such products are generally excellent soil amendments but are rather expensive to use. As a group, they tend to contain ample nitrogen and phosphorus but very little potassium. Some 25 U.S. cities are now using a new method of composting sewage sludge which provides a solid, easily handled, nearly odorless, weed-and-disease-free end product which may be particularly useful in forest nurseries [1].

9.5.2.2 Composts

Various composts can be used advantageously in nurseries. Generally, they are prepared by combining carbonaceous wastes such as sawdust with nitrogen and other nutrients contained in manures [14], sewage sludges [1], or chemicals [9, 11]. Such materials are placed in a suitable physical environment for several weeks. The end product, stabilized and pest-free, can be applied to soil immediately ahead of a seedling crop. The advantages of using composts are obvious. The disadvantages are that their preparation requires much prior planning and considerable handling of bulky materials. Costs vary tremendously, depending on the availability of the constituents.

Some unexpected benefits have resulted from using sawdust composts. Applying 40 yd³/acre to a sandy soil increased the number of mycorrhizal roots of pine seedlings 50%, even though the compost itself contained no detectable mycorrhizal inoculum [unpubl. data, 10]. However, using spent mushroom compost as an organic amendment was disappointing. At reasonable application rates, excessive amounts of soluble salts can become a problem [unpubl. data, 12].

9.5.2.3 Application rates

Application rates of organic soil amendments are difficult to express. Volume (cubic yards per acre, cubic meters per hectare, or depth) is frequently mentioned, but compaction can vary greatly. Thus, the actual amount applied at the same apparent rate can vary considerably. Likewise, expressions of weight are quite arbitrary because of great variation in water content, especially for peat, which can easily hold several times its own weight in water. The only completely nonambiguous rate is oven-dry weight per unit of area; though fine for research, measuring oven-dry weight is seldom operationally practical. Application rates should probably be expressed in terms of volume, with the organic amendments moist but not saturated and not compacted.

In the OSU Nursery Survey, five groups of nonaqueous materials were listed: (1) sawdust and bark, (2) peat, (3) compost, (4) manure, and (5) sludge. Application rates were given principally in terms of volume (cubic yards per acre or cubic meters per hectare). Cubic yards per acre doubled approximately equals cubic meters per hectare. Thus, some comparisons can be made with all rates converted to the cubic yards per acre equivalent (Table 2).

Table 2. Application rates of organic soil amendments.

Amendment	Application rate	
	Range	Typical
	yd³/acre	
Sawdust and bark	90-270	100
Peat	100-270	100
Compost	40-270	50
Manure	30-100	50
Sludge	80-100	80

Suppose, for example, we apply Douglas-fir sawdust at 100 yd³/acre per rotation. In addition to assumptions noted earlier, we will assume that the oven-dry weight of a cubic yard of the sawdust is 500 lb and that the soil down to the bottom of the rooting depth weighs 2,000,000 lb/acre. Finally, because of the length of a typical rotation in Northwest nurseries, approximately 90% of the added sawdust will be gone by the time we are ready to add more at the beginning of the next rotation. We can then calculate the immediate effect on soil organic matter content, the effect at the end of 1 year, and the effect at the end of the rotation:

Immediate effect

100 yd³ of sawdust at 500 lb dry weight/yd³ = 50,000 lb/acre of sawdust; 50,000 lb of sawdust added to 2,000,000 lb/acre of soil represents an immediate increase of 2.5% in soil organic matter content.

At end of 1 year

Research has shown that, generally, ⅔ of the sawdust will decompose during the first year and ⅓ remain in the soil; ⅓ of the immediate 2.5% gain represents a 0.8% gain in soil organic matter content at the end of the first year.

At end of rotation

Though 90% of the sawdust decomposes during the rotation, 10% remains in the soil; 10% of the 2.5% immediate gain represents a 0.25% gain in the nearly stable fraction of the soil organic matter content.

This may seem like a small victory, but it is solid progress nonetheless and is much better than can be done with cover or green manure crops.

9.6 Conclusions and Future Outlook

In sum, organic matter is good for the trees, good for the soil, and perhaps even good for the soul. Some sources of organic matter, like green manure crops, can be home grown; others, like sawdust, peat, or manure, must be imported. Some organic matter is dynamic and highly reactive. It decomposes rapidly but favorably affects soil biology and, to some extent, physical and chemical soil properties. Some is more stable and less reactive. It decomposes slowly, its rate decreasing with time. But eventually, it is fully resynthesized into true humic materials and only many years later is totally metabolized and eliminated from the soil. During all this time, perhaps over centuries, it favorably affects physical and chemical soil properties and, to some extent, soil biology.

Nearly all nursery managers see soil organic matter maintenance as an important step in the production of quality trees. Actual management strategy among nurseries varies considerably, though, primarily because of soil properties, climate, and local availability of organic amendments.

The best information currently available suggests that a shortage of the traditional organic amendments is likely in the foreseeable future. Therefore, we need to devote some time and resources to studies of catch, cover, and green manure crops; to alternative sources of organic amendments; and to changes in rotation schedules. Certainly, we now grow fewer transplants than we did only a few years ago. The proportion of 2 + 0 and plug + 1 stock has increased considerably. Each of these changes offers the opportunity to use more green manure crops and add more organic amendments to the soil over any number of years.

We can be cautiously optimistic about the future. Organic matter is as essential as nutrients and water to producing quality trees. Fortunately, we have more time to make appropriate additions to the soil. Unfortunately, this has lulled some nursery managers into a false sense of security, and they have let their soils become depleted. A whole litany of physical, chemical, and biological problems can then ensue, and it may not be immediately apparent that loss of organic matter is the cause of them all.

The road back to soil and tree health is long and difficult. Thus, prudent nursery managers will not neglect soil organic matter. Rather, they will determine the appropriate level for the existing soil and climate and design a steady, deliberate management program to reach and maintain that level.

Acknowledgment

While preparing this paper, I was graciously provided a draft copy of "Organic Amendments in Forest Nursery Management" by Susan Blumenthal, Soil Scientist with the U.S.D.A. Forest Service, Rogue River National Forest. This comprehensive paper promises to be a significant synthesis of a great deal of information on this important subject. I commend it to your attention when it becomes available.

References

1. Anonymous. 1982. Sludge may help nurseries. J. Forestry 80:34.
2. Bollen, W. B. 1953. Mulches and soil conditioners: carbon and nitrogen in farm and forest products. Agric. and Food Chemistry 1:379.
3. Bollen, W. B. 1969. Properties of tree barks in relation to their agricultural utilization. U.S.D.A. Forest Service, Pacific NW Forest and Range Exp. Sta., Portland, Oregon. Res. Pap. PNW-77. 36 p.
4. Bollen, W. B., and K. C. Lu. 1957. Effect of Douglas-fir sawdust mulches and incorporations on soil microbial activities and plant growth. Soil Sci. Society of America Proc. 21:35-41.
5. Broadbent, F. E. 1948. Nitrogen release and carbon loss from soil organic matter during decomposition of added plant residues. Soil Sci. Society of America Proc. 12:246-249.
6. California Fertilizer Association. 1975. Western fertilizer handbook. Interstate Printers and Publishers, Inc., Danville, Illinois. 250 p.
7. Carroll, L. 1923. Through the looking-glass. Macmillan, New York. 224 p.
8. Davey, C. B. 1950. New York State Conservation Dep., Albany. Unpublished rep. 6 p.
9. Davey, C. B. 1953. Sawdust composts: their preparation and effect on plant growth. Soil Sci. Society of America Proc. 17:59-60.
10. Davey, C. B. 1955. Unpublished data, Univ. of Wisconsin, Madison.
11. Davey, C. B. 1955. Transformation of sawdust in the course of its decomposition under the influence of *Coprinus ephemerus*. Soil Sci. Society of America Proc. 19:376-377.
12. Davey, C. B., and R. Eide. 1969. Unpublished data, Oregon State Univ., Corvallis.
13. Davey, C. B., and H. H. Krause. 1980. Functions and maintenance of organic matter in forest nursery soils. Pages 130-165 *in* Proc., North American forest tree nursery soils workshop (L. P. Abrahamson and D. H. Bickelhaupt, eds.). State Univ. New York, Coll. Environ. Sci. and Forestry, Syracuse.
14. Galler, W. S., C. B. Davey, W. L. Meyer, and D. S. Airan. 1978. Animal waste composting with carbonaceous material. U.S. Environmental Protection Agency, Washington, D.C. EPA-600/2-78-154. 96 p.
15. Garrett, S. D. 1981. Soil fungi and soil fertility. 2nd ed. Pergamon Press, Inc., New York. 150 p.
16. Hausenbuiller, R. L. 1978. Soil science, principles and practice. 2nd ed. Wm. C. Brown Co., Dubuque, Iowa. 611 p.
17. Papavizas, G. C., and C. B. Davey. 1960. Rhizoctonia disease of bean as affected by decomposing green plant materials and associated microfloras. Phytopathology 50:516-522.
18. Papavizas, G. C., and C. B. Davey. 1962. Activity of Rhizoctonia in soil as affected by carbon dioxide. Phytopathology 52:759-766.
19. Shotzko, R. T., C. Allison, V. V. Volk, and A. G. Nelson. 1977. Projecting farm income effects of sewage sludge utilization in the Tualatin Basin of Oregon. Agric. Exp. Sta., Oregon State Univ., Corvallis. Special Rep. 498. 67 p.
20. Snyder, C. S. 1982. Unpublished data, North Carolina State Univ., Raleigh.
21. South, D. 1982. Personal communication, Auburn Univ., Auburn, Alabama.
22. Stevenson, F. J. 1982. Humus chemistry. John Wiley and Sons, New York. 443 p.
23. Sumner, M. E., and J. H. Bouton. 1981. Organic matter maintenance in forest nurseries. Georgia Forestry Commission, Macon. Res. Pap. 24. 5 p.
24. Waksman, S. A. 1936. Humus. Williams and Wilkins Co., Baltimore. 526 p.

Chapter 10
Cover and Green Manure Crops for Northwest Nurseries

W. S. McGuire and D. B. Hannaway

Abstract

Green manure crops are planted in Northwest bareroot nurseries primarily to produce organic matter to enrich soil. Available species include grasses, legumes, and brassicas. Legumes are deep rooting, decompose quickly, and can increase soil nitrogen by fixation but require good drainage and well-fertilized soils and can increase soil-borne pathogens. The most suitable species for short-term rotations are annual grasses, including small grains, certain legumes, and spring-sown brassicas. For longer rotations, tall fescue is well adapted to varied soil conditions, and birdsfoot trefoil, a legume, produces well on acid soils. The pros and cons of other species for various rotation lengths are discussed.

10.1 Introduction

The terms **cover crops** and **green manure crops** are frequently used interchangeably. Certain distinctions between the two, however, can be made. Cover crops are grown primarily for soil cover to help prevent various forms of erosion. Green manure crops are grown primarily to produce organic matter and are usually incorporated into the soil for the benefit of succeeding crops. Although both objectives may be accomplished with the same crop in some cases, the grower should determine the prime objective before selecting and planting the crop.

Results of the OSU Nursery Survey (see chapter 1, this volume) show that 75% of Northwest nurseries use cover crops and do so to produce organic matter. These crops generally are planted in spring and plowed under in late summer or early autumn; because this is a period of low erosion potential in the Northwest, they can be considered green manure crops.

This chapter will familiarize nursery managers with the most suitable plant species for organic matter production in the Northwest.

10.2 Benefits of a Cover or Green Manure Crop

The benefit of a cover crop for soil protection is well established. The kinetic energy of falling raindrops dislodges soil particles. This "splash effect" results in the breakdown of soil aggregates, forming a less pervious surface layer, which in turn creates a "puddling effect." The result is decreased infiltration rate and increased runoff. The splash effect can also cause some downslope movement of particles, but most sediment is lost by surface-water runoff. Cover crops protect the soil from the splash effect by intercepting the raindrops and absorbing the kinetic energy. When a cover crop is used to control water or wind erosion, rapidity of establishment and increased seeding rate should be considered in selecting an appropriate species.

Some effects of soil organic matter provided by green manure and cover crops also are discussed in chapter 9, this volume. Organic matter increases soil aggregation and structure, water-holding capacity, and aeration. The high cation-exchange capacity (CEC) of organic matter helps soil retain cationic nutrients and buffers against changes in soil pH. Plant nutrients are released when soil organic matter is mineralized (decomposed). Reactive forms of organic matter chelate (form available compounds with micronutrients) iron and aluminum, preventing the formation of insoluble metal phosphates in acid soils. These physical and chemical advantages, as well as biological effects of pesticide interaction and energy sources for organisms, are fully discussed by Davey and Krause [2].

Although the benefits of organic matter just mentioned may be derived from both cover crops and other organic materials such as manure or compost, which are brought in, some benefits are peculiar to the organic matter produced by the growing crop. Deep-rooted cover crop species can reclaim nutrients that would otherwise be lost from lower soil levels. Cover crops can also catch and hold nutrients for later use (catch crops).

Cover crops can aid weed control. Although clean fallow is effective in weed control, use of a cover crop by no means

In Duryea, Mary L. and Thomas D. Landis (eds.). 1984. *Forest Nursery Manual: Production of Bareroot Seedlings*. Martinus Nijhoff/Dr W. Junk Publishers, The Hague/Boston/Lancaster, for Forest Research Laboratory, Oregon State University, Corvallis. 386 p.

indicates no weed control. Competition from a thickly sown and vigorous cover crop will discourage weed invasion. Later seed production of weeds can be prevented by mowing or turning under before seed matures. Use of a straight grass cover allows broadleaf weed control by herbicides; similarly, use of straight legume cover allows weedy grass control.

10.3 Species Longevity and Nursery Rotation

Available green manure crops include winter annuals and spring and summer annuals, biennials, and perennials.

Winter annuals require some amount of vernalization (low temperature) before proceeding to good vegetative growth and, later, reproduction. They are normally fall sown in areas with mild winters. Examples are crimson clover (*Trifolium incarnatum* L.), hairy vetch (*Vicia villosa* Roth), and the winter-type (fall-sown) small grains such as wheat (*Triticum aestivum* L.) and oats (*Avena sativa* L.). If cold requirement is insufficient (i.e., if seeds are sown in April-May), these species are likely to stay in the rosette stage and produce little organic matter until the following spring, thus behaving as biennials. However, some annual species, although more productive when fall sown, will produce fairly well if spring sown; examples are rye (*Secale cereale* L.) and annual ryegrass (*Lolium multiflorum* Lam.).

Other annuals, such as spring or common vetch (*Vicia sativa* L.), lupine (*Lupinus* spp.), peas (*Pisum sativum* subsp. *arvense* L.), and the spring-type small grains, have less cold requirement and can be sown successfully in March-April. Summer annuals include sudangrass (*Sorghum bicolor* L.) and corn (*Zea mays* L.), both of which need warm soil temperatures to germinate, and annual sweetclover (*Melilotus alba* var. *annua* Coe).

True biennials do not flower until the second growing season. Sweetclover (*Melilotus alba* Desr.) and the *Brassica* spp., including rape and kale, are green manure crops in this group. Sown in later summer or autumn, the brassicas are grown as winter annuals for seed production and winter feed for livestock; but when they are spring sown, vegetative production is rapid and continues through the summer with irrigation. New hybrid cultivars of kale are available.

Perennial species of green manure crops produce over a period of 3 years or longer. Typically, perennial grasses and legumes establish less rapidly and produce less the first year than annuals and biennials. The longevity of the green manure crop is, therefore, important in regard to the length of the desired rotation.

10.3.1 Short-term rotations

March-April to August-September is the most common rotation for seedling nurseries. Therefore, the most suitable species are the spring and summer annuals (Table 1). Where seasons are longer, two crops may be grown, as in the Weyerhaeuser Co. nursery at Aurora, Oregon; a spring crop of Austrian peas is followed by sudangrass in late May (OSU Nursery Survey). Also, if the full spring-summer season is utilized, the same crop may be grown twice. More production probably could be obtained by cutting the first crop and allowing regrowth. Crops suitable for regrowth are rape, kale, annual ryegrass, and sudangrass, although regrowth of sudangrass would extend well into September, possibly too late for turning under. Regrowth is more successful when higher stubble is left, especially for rape and kale.

10.3.2 Longer term rotations

If land is available over the winter season (fall, winter, spring), winter annuals can be planted in milder areas to provide good production the following spring. Biennial species also can be used. But if land is available for 2 or more years, it can be sown repeatedly to annuals or biennials—or to perennials (Table 1).

Legume species grown for 3 or more years include alfalfa (*Medicago sativa* L.), if soil pH, drainage, and nutrients are adequate, and birdsfoot trefoil (*Lotus corniculatus* L.), which tolerates poor soil drainage, low pH, and lower soil fertility. Among the grass species, tall fescue (*Festuca arundinacea* Schreb.) and meadow foxtail (*Alopecurus pratensis* L.) have wide tolerance to poor drainage and soil pH 5.0 to 8.0; these grasses can be sown singly or together to maintain a relatively weed-free stand for several years. Tall fescue has a strong, deep, fibrous root system that is very effective in promoting improved soil structure and tilth. Other locally adapted grasses may be useful in a long-term rotation, including orchardgrass (*Dactylis glomerata* L.) and timothy (*Phleum pratense* L.).

Certain species should be avoided in nurseries. Subterranean clover (*Trifolium subterraneum* L.), a winter annual, produces viable seed (including hard seed) at or below the soil surface before a good quantity of organic material is produced. White clover (*Trifolium repens* L.) reproduces through the summer and can become a weed problem. Rescuegrass (*Bromus carthartcus* Vahl.), although a very productive grass for up to 3 years, produces seed sporadically in late summer and could also become a weed problem in subsequent years. Species with rhizomes or stolons should be avoided because these are more difficult to eradicate.

10.4 Using Legumes to Produce Organic Matter

10.4.1 Advantages

Several advantages accrue from using legumes, as compared with grasses, for cover or green manure crops:

10.4.1.1 Deep rooting

Cultivated legumes typically have deep taproots. Deeper root penetration and channel development in soil horizons improve drainage, decrease the need for frequent irrigation, and allow plants to reclaim nutrients from lower soil levels.

Rooting depth depends on species and longevity. Though nursery managers are interested mostly in annuals, some land may be out of production long enough that perennials may be used to take advantage of deep root development. Alfalfa is the deepest rooting perennial legume, with roots penetrating 8 to 12 m if subsoil conditions are suitable. Red clover (*Trifolium pratense* L.), trefoil (*Lotus* spp.), and biennial sweetclover have relatively strong taproots 2 to 3 m deep. In contrast, white clover has a branching taproot that is normally restricted to the top 60 cm of soil. Annual legumes with rooting depth up to 1 m include vetches, crimson clover, and arrowleaf clover (*Trifolium vesiculosum* Savi). Deeper rooting annual legumes such as lupines and annual sweetclover may have roots that penetrate to a depth of 2 m or more.

10.4.1.2 Faster decomposition

Decomposition rate of leguminous green manure crops is greater than that of grass species, particularly for the herbaceous legumes, because of a more favorable carbon to nitrogen (C:N) ratio. However, this rapid decomposition rate could be a disadvantage if the objective were to increase soil organic matter over a long period.

10.4.1.3 Increased soil nitrogen

Cultivated legumes likely to be used in nursery management fix atmospheric N in root nodules following infection with effective strains of *Rhizobium* bacteria. The amount of N fixed

Table 1. Recommended green manure species for various rotation lengths.

Species	Sowing rate, kg/ha	Sowing time	Dry matter production,[1] metric ton/ha	Remarks
		Short term (March-April to August-September)		
Legumes				
Spring vetch	75	March-April	4-6	Often sown with 100 kg/ha spring grain
Lupines	100	March-April	6	Will produce on acid soil
Annual sweetclover	14	March-April	6	Use Hubam cultivar; needs soil pH of about 6.0
Grasses				
Spring grains (wheat, oats, barley, rye)	150	March-April	6	Use local cultivars
Annual ryegrass	30	March-April	6-8	Will regrow after cutting for 2 to 3 crops
Sudangrass	35-40	May-June	8-10	Increase sowing rate 50% with sorghum hybrid
Corn	20	Late April-June	12-14	Sowing rate is for 70,000 plants/ha and depends on seed size
Other				
Brassicas (rape, kale)	6	March-May	8	Can grow 2 crops if planted very early
		Longer term		
		Fall sown (plow under in spring or summer)		
Legumes				
Crimson clover[2]	20	Sept.-early Oct.	4-6	Plow under when flowering
Arrowleaf clover[2]	15	Sept.-early Oct.	6	Flowers 1 to 2 months later than crimson clover
Hairy vetch[2]	50	Sept.-early Oct.	4	Often sown with winter grain
Grasses				
Winter grain (oats, rye, wheat, barley)	120	October	6	Use local cultivars
Annual ryegrass[2]	30	Sept.-early Oct.	6-8	Mow or plow in
Other				
Brassicas[2] (rape, kale)	6	October	6	Plow in when flowering in spring
		Spring sown (for 2-year use)		
Legumes (sweetclover)	14	April-June	8	For low-rainfall, nonirrigated areas, use yellow flowered
Grasses (tetraploid ryegrasses)	30	April-June	8	Oregon annual will produce through second year
Other (rape, kale[2])	6	April-June		Will flower in early spring; plow in and replant
		Perennial (3 years or more)		
Legumes				
Alfalfa	12-15	April-June	12-16	Use local cultivars; needs soil pH 6.0 or more
Birdsfoot trefoil	8	April-June	9-10	Use Granger or Cascade cultivars
Grasses				
Tall fescue	25	April-June	10-12	Use Alta or Fawn cultivars
Meadow foxtail	20	April-June	8-12	Light, fluffy seeds, difficult to drill
Timothy	10	April-June	8-10	Regrowth is less than for other grasses
Orchardgrass	18	April-June	10-12	Use local cultivars

[1]Based on production under irrigation at Corvallis, Oregon.
[2]Not winter hardy at higher elevations or east of Cascades.

depends on the legume species and the length of the growing season; estimates are about 85 kg/ha for vetches and 140 kg/ha for annual clovers for a full season of growth (fall sown, maturing the next summer). Where productive, alfalfa may fix over 500 kg/ha N in a season.

Efficient inoculation of legume seed just before sowing is essential to ensure high rates of N fixation. Available commercial peat inoculum is specifically labeled for each legume species and should be used by the date shown on the container. The inoculum should be applied liberally to seeds moistened with skim milk, sugar solution, or a weak solution of gum arabic. Application to water-moistened seeds may be sufficient if seeds are protected from the sun and drying and if planting occurs within a few hours of inoculation.

According to OSU Nursery Survey results, many Northwest nursery soils are acidic, which can reduce survival of the nodule bacteria of some legume species. Alfalfa, sweetclover, and related medics and burr clovers (*Medicago* spp.) have little tolerance to acidity and require a soil above pH 6.0 for nodulation and adequate growth. The true clovers (*Trifolium* spp.), however, are relatively acid tolerant and can be grown successfully at pH 5.5. Vetches and lupines, and their *Rhizobium* strains, can tolerate even greater acidity. Where soil pH values are below 5.5, effective nodulation is increased by using lime-pelleted seed [7]. Lime coating protects *Rhizobium* from desiccation when seeds are surface sown or drilled into dry soil and from acidity when seeds are sown into acid soils or when inoculated seeds are sown in contact with acid fertilizers.

10.4.2 Disadvantages

Several disadvantages temper the use of legumes as green manure crops in nurseries:

10.4.2.1 Hard seed

Most legume species have relatively hard seedcoats which are impervious to water and can retard germination, possibly carrying over to grow as weeds in the following seedling crop. This problem can be reduced by scarifying the seed (scratching the seedcoat against abrasive material), a process available at most seed-cleaning plants. In addition, the legume crop should be turned under before seed matures.

10.4.2.2 Need for good drainage

Inadequate soil drainage may prevent good legume growth. Alfalfa and sweetclover require good drainage; vetches, lupines, and annual legumes require moderate drainage; and white clover and the trefoils will tolerate poor drainage. However, soil drainage for legumes is not a significant problem in the Northwest with spring-summer, one-season crops. The legumes most tolerant to waterlogging or flooding are alsike clover (*Trifolium hybridum* L.), white clover, and big trefoil (*Lotus pendunculatus* Car.) and are seldom used as nursery cover or green manure crops.

10.4.2.3 Need for high fertility levels

Legume production may require additional nutrient application. Some areas of western Oregon are deficient in molybdenum (Mo), which is essential in the N-fixation process. Although Mo can be deficient in soils with high pH values, it is more often deficient in acid soils due to decreased availability. Correcting this deficiency by liming or by adding very small amounts (0.25 kg/ha) of Mo to the soil often results in much higher production. Molybdenum can be added to inoculated seeds or to the coating of pelleted seeds if the seeds are drilled soon after into moist soil; delay may adversely affect *Rhizobium* survival.

10.4.2.4 Increased disease problems

The use of legumes as green manure crops may increase disease problems of succeeding tree crops. This potential problem is discussed later in this chapter (10.6).

10.5 Cultural Practices for Green Manure Crops

10.5.1 Seedbed preparation

Seedbeds for cover crops are prepared by plowing, followed by disking and harrowing or by rototilling. The main purpose of plowing is to eliminate vegetative material, dead or alive, that may interfere with sowing. Often, a seedbed may be prepared with only a disk and harrow. Most green manure or cover crops have small seeds, and a firm seedbed aids in controlling the depth of planting, which improves seedling emergence.

10.5.2 Sowing

Amount of seed sown depends very much on seed size. In addition, increased rate of sowing may compensate to some extent for the shorter productive life in the short-term rotation. The suggested sowing rates (Table 1) are somewhat higher than those for the same species when used in pasture or forage production. Because of competition, mainly for light, in the irrigated, well-fertilized field, high sowing rates generally do not increase yield.

One exception to the above is corn production for green manure. It produces few if any tillers. The corn crop likely would be turned under in August or September, before grain matures, or even before ears form. Plant population should be increased from the usual 60,000 to 70,000 plants/ha to nearly double that number. Seeds may be drilled in closer rows or even sown with a grain drill.

Drilling is the most efficient method of sowing because it controls planting depth and seed distribution. Broadcast seeding can be a satisfactory method if seeds are covered by harrowing or with a corrugated roller.

Proper seeding depth depends on seed size. The small seed of clovers and trefoils should be sown about 1 cm deep. Vetches and lupines can be planted deeper, up to 3 cm. Grasses are sown approximately 2 cm deep and sudangrass, corn, and small grains somewhat deeper, up to 4 to 5 cm.

The seedbed may be rolled or cultipacked before sowing seed to provide the firmness necessary to control depth of seed placement. Rolling after sowing is beneficial because it firms the soil around seed to ensure better moisture contact for germination and better moisture conservation. Rolling may not be required, however, if the drill has press wheels, if a Brillion-type seeder is used, or if rain or irrigation follows shortly after sowing.

10.5.3 Planting and plow-down times

Planting time depends on species longevity and local climatic conditions. Winter annuals are sown from late September to early October in mild-winter areas. Small grains may be sown into November. East of the Cascade Mountains and at higher elevations, annual clovers and vetches are not hardy enough for winter planting and should be sown in early spring.

Spring-sown annuals may be planted as soon as conditions permit, from March through April; however, summer annuals need warm soil to germinate. In western Oregon, for example, corn may be planted from late April through June and sudangrass from mid-May through June. Brassicas for short-term (spring-summer) use generally are sown from April to May, but also can be fall sown in mild-winter areas, if land is available at that time.

Annual sweetclover is sown in spring. Biennial sweetclover (for longer rotation) may be sown either in spring or summer. Perennial species may be sown in the spring, or in the summer if irrigation is available.

Annual and biennial green manure crops should be plowed under and perennial crops mowed or chopped when plants are in the flowering stage, before seeds are produced.

10.6 Green Manure Crops and Seedling Pests

Although adding green plant material to nursery soil improves soil and nutrient conditions, green manure crops sometimes may increase pathogens (fungi and nematodes), insects, and other soil organisms (see chapter 19, this volume). This is one reason why green manure crops are not currently used in 25% of Northwest nurseries (OSU Nursery Survey).

Green manure crops may affect soil nematode populations. McElroy [6] tested 31 plant species for host suitability of the corky root pathogen (*Xiphinema bakeri* Williams) in the Fraser Valley of British Columbia and found that the nematode increased with rye, orchardgrass, and several species of weeds but decreased with brassicas. Where corky root occurs, rape or kale might be a suitable green manure crop. Corky root also can be controlled by keeping soil dry and working it frequently from August through September [9]. Other species which may be planted in corky root areas are annual ryegrass and spring-type grains. They are sufficiently vigorous to be plowed under in August, followed by soil drying and working.

Other pathogens that have a wide range of hosts may increase on green manure crops. Number of microsclerotia of the root-rot fungus *Cylindrocladium scoparium* Max. decreased with corn but increased with soybean (*Glycine max* L. Merr.) after a 5- to 6-month decomposition period [10]. Legumes, particularly alfalfa and red clover, are more susceptible to root rot than grasses [1]. Incorporating flax (*Linum usitatissimum* L.) or sorghum-sudangrass cover crops into a sandy nursery soil for 2 successive years significantly reduced the number of *C. floridanum* Sobers and Seymour propagules [3]; after four growing seasons, both species of cover crops reduced root-rot potential to innocuous levels. Menge and French [8] reported that cover crops influence soil fungi, even if they are nonhosts, because root-infecting fungi can grow in the rhizosphere of nonsusceptible plants.

Cover crops also can influence the incidence of *Phytophthora* fungi, whose hosts include Douglas-fir [*Pseudotsuga menziesii* (Mirb.) Franco], alfalfa, and arrowleaf clover [4]. *Phytophthora* can be particularly severe on alfalfa and clover in the poorly drained soils of the Northwest. Several alfalfa cultivars with moderate to high resistance to *Phytophthora* are available for areas where this fungus is a potential problem.

Sorghum and sorghum-sudangrass residues contain toxins which may temporarily inhibit growth of the following crop of some conifer seedlings [5]. These crops, apparently beneficial for controlling root rot in Wisconsin, were found to damage seedlings, largely by eradicating mycorrhizal fungi. Damage decreased with early plowing, watering, and delayed fall seeding of the seedling crop. (See also chapter 20, this volume, for other examples of the relationship between green manure crops and mycorrhizae.)

10.7 Conclusions and Recommendations

Relatively few species of legumes, grasses, or other plant families are suited for use as green manure crops in bareroot nurseries. A species must be adapted to local conditions and must be sufficiently vigorous in its establishment and growth to make a significant contribution in organic matter during a relatively short period. The selected species also must favor the desirable soil fungi and inhibit at least some soil-borne pathogens.

Leguminous green manure crops have the advantage of deep rooting and can fix N but often increase the number of nematodes and soil-borne disease organisms. In addition, according to the OSU Nursery Survey, most nursery soils in the Northwest, certainly those west of the Cascade Mountains, are too acid for the production of certain legumes. Yet, it is desirable to maintain acid soils for suitable seedling production and to discourage pathogens, particularly damping-off organisms (those which cause rotting of seeds and succulent seedlings).

In the Northwest, the most suitable species for short-term green manure crops are annual grasses, including small grains, particularly oats and rye, corn, annual ryegrass, and sudangrass; among the legumes, spring vetch, peas, lupines, and annual sweetclover (on less acid soil); and spring-sown rape or kale. For a long-term rotation, tall fescue is widely adapted to varied soil conditions and is usually very productive. Birdsfoot trefoil, a legume, will fare well on acid soils, either planted alone or in combination with grass.

References

1. Bugbee, W. M., and N. A. Anderson. 1963. Host range and distribution of *Cylindrocladium scoparium* in the North Central states. Plant Disease Reporter 47:512-515.
2. Davey, C. B., and H. H. Krause. 1980. Functions and maintenance of organic matter in forest nursery soils. Pages 130-165 *in* Proc., North American forest tree nursery soils workshop (L. P. Abrahamson and D. H. Bickelhaupt, eds.). State Univ. New York, Coll. Environ. Sci. and Forestry, Syracuse.
3. Hadi, S. 1974. Epidemiology and control of *Cylindrocladium* stem canker and root rot of conifers in nurseries. Dissertation Abstracts International B 35(4):1473-1474.
4. Hansen, E. M., and P. B. Hamm. 1983. Morphological differentiation of host-specialized groups of *Phytophthora megasperma*. Phytopathology 73:129-134.
5. Iyer, J. G. 1980. Sorghum-sudan green manure: its effect on nursery stock. Plant and Soil 54:159-162.
6. McElroy, F. D. 1972. Studies on the host range of *Xiphinema bakeri* and its pathogenicity to raspberry. J. Nematology 4:16-22.
7. McGuire, W. S., and D. B. Hannaway. 1981. Making lime-pelleted seeds. Oregon State Univ. Ext. Serv., Corvallis. Ext. Circ. 1084. 3 p.
8. Menge, J. A., and D. W. French. 1976. Effect of plant residue amendments and chemical treatments upon the inoculum potential of *Cylindrocladium floridanum* in soil. Phytopathology 66:1085-1089.
9. Sluggett, L. J. 1972. Corky root disease of Douglas fir nursery seedlings. Can. Forestry Serv., Pacific Forest Res. Centre, Victoria, B.C. Pest Leaflet 53. 5 p.
10. Thies, W. G., and R. F. Patton. 1970. The biology of *Cylindrocladium scoparium* in Wisconsin forest tree nurseries. Phytopathology 60:1662-1668.

Chapter 11
Water Management

R. J. Day

Abstract

Water is managed in forest-tree nurseries to (1) control available soil moisture and foster germination, growth, and specific physiological responses, (2) provide solutions for transporting and infiltrating water-soluble fertilizers and leaching excessive salt concentrations, (3) protect crops from extreme drought, soil heating, freezing, or frost heaving, (4) promote germination of weed seed on fallow land for herbicide-free weed control and regulate growth of cover crops, (5) minimize potentially polluting losses of fertilizers and biocides, and (6) limit the amount of water to the optimum needed for crop production. A nursery's water requirements for all water-management purposes must be determined either with tables or by calculation from climatic data. Water quality (salinity level) must be acceptable for adequate crop growth. The quantity of the water supply or of stored water must be sufficient for all needs, even in the driest years. Although "ditch and flood" type methods have been used for irrigation over the years, most modern nurseries rely on sprinkler irrigation; a fully permanent, semipermanent, or solid set rotating sprinkler system is recommended. System design must be carefully tailored to water resource, soil depth and type, irrigation need, pressure head, friction loss, sprinkler layout, land elevation, and local winds. Nursery managers should consult an irrigation engineer to plan the best irrigation system their budgets will allow, recognizing the likely need for future expansion.

11.1 Introduction

Water management may be defined as "the scientific regulation of water for the production, conditioning, and protection of bareroot nursery crops." The term **water management** is to be preferred over others such as "irrigation" or "supply with water" [9] because it embraces many water-management techniques in addition to irrigation.

In forest-tree bareroot nurseries, water may be managed [15]:

- To control available soil moisture to promote the germination, establishment, and growth of the crop—or to slow or stop that growth, if necessary—and to foster root-regeneration potential, bud formation, frost hardiness, or other physiological responses.
- To provide solutions for transporting and infiltrating water-soluble fertilizers and leaching excessive salt concentrations.
- To protect the crop from excessive atmospheric drought, soil heating, freezing, or frost heaving.
- To promote the germination of weed seed on fallow land before cultivation as a herbicide-free weed control measure.
- To promote the germination and establishment and regulate the growth of cover crops on fallow land and noncrop areas.
- To minimize potentially polluting losses of fertilizer and biocides.
- To limit the amount applied to the optimum needed for crop production.

This chapter acquaints nursery personnel with the principles, practices, and methods of modern water management. Because forest-tree nurseries are most efficiently irrigated by sprinklers, water-management systems associated with sprinkler irrigation will be stressed.

11.2 History of Water Management

Water management is an ancient procedure. The earliest evidence of irrigation is a water storage dam in Egypt dating to 5,000 B.C. [23, 30]. Perhaps the earliest water managers were

In Duryea, Mary L., and Thomas D. Landis (eds.). 1984. *Forest Nursery Manual: Production of Bareroot Seedlings.* Martinus Nijhoff/Dr W. Junk Publishers, The Hague/Boston/Lancaster, for Forest Research Laboratory, Oregon State University, Corvallis. 386 p.

the priests of ancient Egypt, who built conduits connecting the temples of the river gods to the Nile and used "Nilometers" (graduated pillars) to forecast the floodcrest and the success of each season's irrigation and deposit of fertile silt. In Pliny's time (1st century B.C.), floodcrests marking the Nilometers at 12, 13, 15, and 16 cubits (18.0, 19.5, 22.5, and 24.0 ft) were taken to indicate famine, scarcity, safety, and plenty, respectively.

The history and development of early man seem closely related to the development of water-management technology not only for the Egyptians of the Nile valley but also for civilizations in four other principal river valleys: the Mesopotamian of the Tigris-Euphrates valley, the Indian of the Indus valley, the Chinese of the Yellow River valley, and the Andean of the coastal river valleys of Peru [23]. Because the principles of irrigation technology are basic and few, their worldwide coevolution should not be surprising—regardless of whether or not contact was made between the early irrigators [4]. Common features of ancient and modern irrigation systems are types of water sources (streams, rivers, and wells), storages (storage dams and cisterns), diversion dams, and methods for lifting and transporting water and applying it to the land.

11.3 Planning A Water-Management Program

11.3.1 Determining nursery water requirements

Before a new nursery is established, it is essential to estimate or calculate the **water requirements** for all potential water-management purposes. This can be done either by consulting tables showing the average water requirements in various regions or by obtaining climatic data from a station at or near the proposed nursery site and computing seasonal and annual water requirements.

11.3.1.1 Water requirements from tables

Because the average amount of water needed to produce forest-tree nursery crops in any region is approximately similar to that needed for agricultural crops, tables giving the average annual irrigation requirements of agricultural crops may be obtained from local agricultural extension agencies and the information applied to nursery production. Table 1 compares the average annual water requirements for agricultural crops in broad regions of the United States [17].

11.3.1.2 Water requirements from climatic data

Computing the water requirements of a bareroot nursery from climatic data recorded at or near the nursery site is preferable to relying on annual tables used for agricultural crops (see 11.3.1.1). To do this, it is best to obtain the monthly means of precipitation and temperature for as many previous years as possible, or at least for mean and extremely dry years.

The **irrigation need**—the principal component of the water requirement—is the amount of water required to maintain nursery soil within an optimum range of available soil-moisture levels throughout the growing season each year [2]. The irrigation need for a given bareroot nursery can readily be estimated by computing water balances for all (or just the mean and extremely dry) years in the past by the Thornthwaite method [27, 28]. Although methods described by Blaney and Criddle [3] or Penman [18] could be used for the same purpose, in this chapter all examples of estimating irrigation need will be by the Thornthwaite method.

For example, Figure 1a shows the monthly Thornthwaite water balance for the Ontario Ministry of Natural Resources Thunder Bay Forest Station, based on climatic data recorded from 1947 to 1978 (a 32-year average). In an "average" year, potential evapotranspiration (PET) exceeds precipitation (P) in May, June, July, and August; thus, these months will require irrigation. In such a year, PET – P = 0.62 inches in May, 0.81 inches in June, 2.03 inches in July, and 0.59 inches in August, for a total of 4.05 inches (15.7, 20.6, 51.6, and 15.0 mm, for a total of 102.9 mm). Therefore, the total amount of irrigation water needed to maintain soil moisture close to field capacity would approximately equal 4.05 inches (102.9 mm). Water needed for purposes other than irrigation will of course increase this amount.

The problem with using estimates of irrigation need based on an average year is that an average year never occurs. For example, the monthly Thornthwaite water balance for the Thunder Bay Forest Station in the extremely dry year of 1975 (Fig. 1b) showed that PET exceeded P from May through September, 1 month more than in an average year; PET – P = 3.27 inches in May, 0.47 inches in June, 2.58 inches in July, 2.73 inches in August, and 0.04 inches in September, for a total of 9.09 inches (83.1, 11.9, 65.5, 69.4, and 1.0 mm, for a total of 230.9 mm). The irrigation need in the dry year of 1975 was more than double that of the average year. Obviously, nursery staff must be ready to provide for all irrigation needs and water requirements of the crop in the driest years if an effective water-management program is to be developed. To

Table 1. Average annual irrigation-water requirements by region in the United States (adapted from [17]).

Region	(a) Net required by crop, in.[1]	(b) Application efficiency, %	(c) a x 100/b, in.	(d) Storage delivery efficiency, %	(e) c x 100/d, in.	(f) Estimated recovery of losses, %	(g) Total water requirements [e – (c x f)/100], in.
Eastern U.S.							
Moistest region	2.43	60.0	4.05	60.0	6.75	20.0	5.94
Driest region	5.25	60.0	8.75	65.0	13.46	20.0	11.71
Mean	4.07	60.0	6.78	62.0	10.94	20.0	9.58
Western U.S.							
Moistest region	5.62	50.0	10.52	60.0	17.53	55.0	11.74
Driest region	8.59	45.0	19.10	55.0	34.71	55.0	24.21
Mean	8.45	46.0	18.37	52.0	35.33	56.0	25.04
Addenda							
Pacific Northwest	4.04	60.0	6.73	60.0	11.22	20.0	9.87
Western Lake States[2]	6.88	17.2	60.00	17.2	28.67	60.0	18.35

[1]To convert in. to cm, multiply by 2.54.
[2]Western Lake States are Minnesota and Wisconsin.

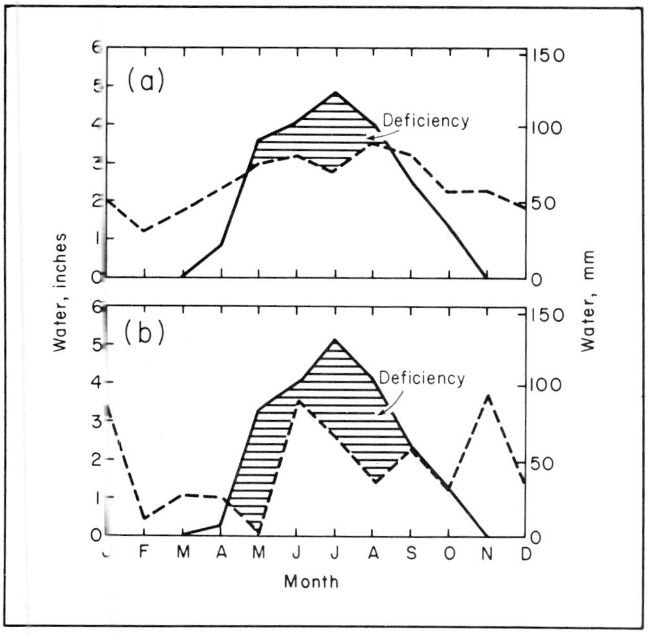

Figure 1. Precipitation (P, dashed line) and potential evapotranspiration (PET, solid line) for the Thunder Bay Forest Station for (a) an average year [mean deficiency, or PET − P = 4.05 inches (~103 mm)] and (b) an extremely dry year [acute deficiency, or PET − P = 9.09 inches (~231 mm)].

evaluate the water requirements for a bareroot nursery in detail, it is also essential for nursery staff to estimate the irrigation need for dry periods not indicated by monthly means and to be sure that adequate water supplies will be available for all water-management purposes at such times. The severity of dry periods can best be identified and assessed by computing a daily water balance for past years by the Thornthwaite method [5].

Calculating the percentage probability of monthly or periodic irrigation-water requirements during the time in which irrigation is needed each year is also very useful. This tells the nursery staff how often they are likely to have to supply specific quantities of irrigation water. For example, the percentage probability of monthly irrigation-water requirements at the Thunder Bay Forest Station was computed by the Thornthwaite method from climatic data recorded over the 32-year period noted in Figure 1 (Table 2). At Thunder Bay, the probabilities of an irrigation-water requirement of more than 1 inch (25.4 mm) in May, June, July, August, September, and October are 12, 35, 63, 38, 16, and 6%, respectively. In the hot summer months of July and August, the probabilities of an irrigation need of more than 2.5 inches (65.3 mm) are 34 and 25%, respectively.

Once the water requirements of a nursery have been computed, the amount of water to be applied per acre (or per hectare) can be rapidly determined. Table 3 gives the volumes of water necessary to satisfy monthly water requirements ranging from light (0.25 inches, or 6.35 mm) to heavy (2.5 inches, or 63.5 mm).

Table 2. Percentage probability of monthly irrigation-water requirements at the Thunder Bay Forest Station, computed by the Thornthwaite method.

Water requirements (basis 1947-1978),		Probability of monthly water requirements, %					
in.	mm	May	June	July	Aug.	Sept.	Oct.
<0.1	<2.5	59	31	19	28	68	63
0.1-0.5	2.5-12.7	19	31	6	15	10	25
0.6-1.0	12.8-25.4	10	3	12	19	6	6
1.1-1.5	25.5-38.1	3	10	19	0	10	3
1.6-2.0	38.2-50.8	3	10	10	10	0	3
2.1-2.5	50.9-65.3	3	3	0	3	0	0
>2.5	>65.3	3	12	34	25	6	0

Table 3. Volume of water per unit area needed for various monthly irrigation-water requirements.

Water requirement,		Volume of water			
		Per acre,		Per hectare,	
in.	mm	ft³	U.S. gal.	m³	L (x 10³)
0.25	6.35	908	6,789	635	635
0.50	12.70	1,815	13,578	1,270	1,270
1.00	25.40	3,630	27,156	2,450	2,450
1.50	38.10	5,445	40,734	3,810	3,810
2.00	50.80	7,260	54,312	5,080	5,080
2.50	63.50	9,075	67,318	6,350	6,350

11.3.2 Water quality

Water quality is defined in terms of the elemental composition and concentration of salts dissolved in the irrigation water [1,7,8]. As the ratio of precipitation to potential evapotranspiration varies seasonally in arid and semi-arid climates, so do the salinity and resultant quality of irrigation water. Salinity is a common problem in poorly managed container nurseries when fertilizer salts are allowed to build up in the medium without adequate leaching, but is not usually a problem in bareroot nurseries except in the prairie regions of Canada and the United States. In these regions, where potential evapotranspiration exceeds precipitation, the quality of irrigation water may be more critical than that of the soil for growing healthy nursery crops [14].

Bareroot conifer crops are readily damaged by an excess of salts in the soil solution. The damage initially takes the form of brilliant reddening of needle tips; this is followed by progressive browning of the foliage and may be accompanied by resin bleeding from the roots. Salts injure bareroot stock in four ways: (1) by increasing the osmotic pressure of the soil solution, causing stress and drought; (2) by decreasing soil permeability owing to loss of soil structure and aggregation caused by the deflocculation of soil colloids (particularly in clays); (3) by direct ion toxicity from sodium, chloride, borate, and other ions; and (4) by change in nutrient availability owing to changes in pH and associated solubility and to antagonisms between ions.

The best method of evaluating water quality is to determine: (1) conductivity in micromhos/centimeter of total salts, (2) pH, and (3) concentrations of the specific ions sodium (Na^+), potassium (K^+), calcium (Ca^{++}), magnesium (Mg^{++}), carbonate (CO_3^{--}), bicarbonate (HCO_3^-), sulfate (SO_4^{--}), chloride (Cl^-), nitrate (NO_3^-), and boron (B), measured in milliequivalents/liter (meq/L) [14].

11.3.2.1 Osmotic stress

The following values are often used to assess the effects of salts on growth [14]:

Salt hazard	Conductivity, micromhos/cm
Low	<250
Medium	250-750
High	751-2,250
Very high	>2,250

11.3.2.2 Reduced soil permeability

The effects of salts on soil permeability are generally determined from two indexes, the Adjusted Sodium Adsorption

Ratio (ASAR) and the Residual Sodium Carbonate (RSC). ASAR is the relative proportion of deleterious ions (i.e., Na^{+}, CO_3^{--}, and HCO_3^{-}) to beneficial ions (i.e., Ca^{++} and Mg^{++}). Although sodium ions are usually the principal offenders, carbonate ions are included in the ASAR because they can dissolve beneficial calcium and magnesium ions [1]. RSC reflects the harmful effects of salts in deflocculating clays and dissolving soil organic matter. RSC values are computed by subtracting the sum of the carbonate and bicarbonate ions from the sum of the calcium and magnesium ions. The following index values are often used to judge the effects of salt on soil permeability:

Type of value	Effect on soil permeability		
	Good	Marginal	Poor
ASAR	<6.00	6.00-9.00	>9.00
RSC	<1.25	1.25-2.50	>2.50

11.3.2.3 Direct ion toxicity

Na^{+}, Cl^{-}, and B ions injure plant tissues directly. Na^{+} and Cl^{-} can be absorbed through the foliage, causing plasmolysis (osmotic dehydration of cell protoplasm) and tissue death. The following values are given by Ayers [1] and Landis [14]:

Absorption site and ion	Effect of ion toxicity, meq/L		
	Good	Marginal	Poor
Foliage			
Sodium	<3.0	>3.0	....
Chloride	<3.0	>3.0	....
Root			
Sodium (ASAR)	<3.0	3.0-9.0	> 9.0
Chloride	<4.0	4.0-10.0	>10.0
Borate	<0.5	0.5-2.0	> 2.0

11.3.2.4 Changes in nutrient availability

Changes in nutrient availability can only be defined by determining the effects of salts and of pH changes caused by salts on foliar levels of absorbed nutrients. Prime examples are absorption of phosphate ions (PO_4^{---}) when pH is high or when Ca^{++} or Mg^{++} ions are in excess, or of ferrous (Fe^{++}) or ferric (Fe^{+++}) ions when CO_3^{--} or HCO_3^{-} ions are in excess (i.e., "lime-induced chlorosis").

11.3.3 Water quantity

A dependable, abundant source of water adequate to meet all water-management needs is an indispensable component of any nursery or prospective nursery site (see chapter 2, this volume). Yield of a proposed source must be determined before a nursery site is approved. This source must be able to meet the demand for all water requirements regardless of the aridity of the season, the severity of the irrigation need, and the need for water for other purposes.

Water sources for bareroot nurseries are usually streams, rivers, or wells on or near the nursery property. In the Northwest, where many bareroot nurseries have to depend on wells for water, managers must be sure that the amount of water needed in the driest seasons can be supplied either directly or from storage in cisterns or reservoirs. The details of well construction specifically for use in irrigation programs can be found in Israelsen and Hansen [13]. For many bareroot nurseries, a supply stream can be dammed and a storage reservoir created. Such reservoirs are designed both to provide adequate water supplies in periods of peak water requirements and to buffer fluctuations in downstream flow [13, 16, 17, 29].

11.4 Irrigation Systems

As in the past, most agricultural water is still applied by the ditch with flood, border, or furrow methods [4, 16, 23, 29, 30]. With the development of efficient mechanical pumps and turbines in the 19th century, sprinkler irrigation systems came into common use for agriculture in the developed countries and are now almost universally used in nursery practice. In this chapter, the ditch with flood, border, or furrow methods will only be briefly described; sprinkler irrigation systems, which are more important in nursery practice, will be discussed in detail (see 11.4.2 and, especially, 11.5).

11.4.1 Ditch with flood, border, or furrow systems

These systems run water onto the surface of the land from a nearby irrigation ditch. As a result, they can only be employed on relatively flat terrain. None of these systems are particularly satisfactory for most of the water-management objectives listed in this chapter's introduction (see 11.1): transporting and infiltrating fertilizers; leaching excessive salt concentrations; protecting the crop from atmospheric drought, soil heating, freezing, or frost heaving; promoting the germination of weed seed and regulating the growth of cover crops on fallow land; restricting the losses of fertilizer and biocides; and limiting the amount of water to the optimum needed for crop production.

11.4.1.1 Flood irrigation

Flood irrigation systems can only be operated on land with very gentle topography because large areas must be uniformly supplied with water from irrigation ditches. Fields are typically subdivided by dikes or border ridges into strips or basins 300 to 1,200 ft (91 to 365 m) wide. The water is usually transported from a supply canal to secondary ditches that parallel the strips or basins. By opening gaps or setting up short siphons at 30- to 60-ft (9- to 18-m) intervals along the secondary ditches, the strips are flooded each time irrigation is required. As soon as the water covers the entire strip and has remained long enough for infiltration, irrigation is complete. The openings in the secondary ditches are then closed or the siphon tubes removed to shut off the water.

11.4.1.2 Border irrigation

Border (or strip) irrigation is similar in many respects to flood irrigation except that each strip, 300 to 1,200 ft (91 to 365 m) wide, is subdivided into substrips 30 to 60 ft (9 to 18 m) wide. Water is then allowed to enter each substrip in succession until the main strip is completely irrigated. Border irrigation has the advantage of limiting the area under irrigation at any one time, providing superior depth control and permitting more uniform, less wasteful applications even when the ground is gently sloping.

11.4.1.3 Furrow irrigation

Furrow (or corrugation) irrigation differs from both the flood and border systems because it is limited to row crops. The water is run into furrows which are made either by cultivating the rows for the crop or by digging special shallow ditches called rills between the crop rows. The furrows or rills are run down a gradual slope. However, use of furrow irrigation must be restricted to suitable soils (usually loams): water in pervious soils (i.e., sands and sandy loams) may sink beneath the irrigation furrows or rills before it is absorbed by the crop lands and that in impervious soils (i.e., clays) may be so slowly absorbed that it never reaches the crop's root system.

11.4.2 Sprinkler systems

Sprinkler irrigation is completely different from the ditch methods described in 11.4.1 in that mechanical turbines or pumps are used to pressurize a pipeline system that delivers water at a specific head pressure to sprinklers spaced so that their spray simulates rainfall. Sprinkler irrigation is álmost always used on bareroot nurseries because, unlike the ditch methods, it satisfies all the purposes of water management listed in this chapter's introduction (see 11.1).

Sprinkler systems have considerable advantage over all other irrigation systems because they can be calibrated for use on almost any soil type, can deliver the exact amount of water needed for any water-management purpose, and can be used on uneven or gently rolling terrain. Their main disadvantage is that their water distribution is readily affected by winds over 7 mph (11 kph); other disadvantages are their high initial capital cost and higher maintenance costs and power requirements than other systems. A minor disadvantage of sprinkler systems with overland supply pipes is that the branch lines feeding the laterals block access to one end of each nursery compartment; to overcome this problem, some nurseries have installed fully permanent systems with subterranean branch and even lateral lines.

Two principal types of sprinkler systems—oscillating nozzle line and rotating sprinkler—are used in bareroot nurseries. Although nozzle line systems were favored in the 1950s at some locations, rotating sprinkler systems are used almost exclusively in modern forest-tree nurseries because of cost advantages, simplicity of installation and maintenance, and superior rates of application.

11.4.2.1 Oscillating nozzle line

Oscillating nozzle line systems consist of galvanized steel pipelines, tapped and fitted at regular intervals with aligned nozzles, spaced at uniform intervals in a parallel pattern across each compartment to be irrigated. Each nozzle line is attached at the header supply line to a water-powered motor which causes it to oscillate in rowlock-shaped supports so that water is sprayed upwards through a 120 to 165° arc. Oscillating nozzle lines produce a uniform overlapping spray pattern that covers their length and sweep.

Because nozzle lines must be supported on semipermanent posts set above the height of mechanical equipment, these irrigation systems tend to lack mobility and versatility. They also tend to be more expensive to install and maintain than rotating sprinkler systems. Water motors are complex, subject to breakdown, and costly to buy, repair, and maintain. In addition, application rates generally are low; maximum is ¼ inch (0.64 cm) of water/hour, or, with special nozzles, ⅓ inch (0.85 cm) of water/hour. Furthermore, because the nozzles in the lines have small orifices, nozzle clogging may be a serious maintenance problem unless the irrigation water is either very clean or filtered. Thus, although many older nurseries have oscillating nozzle line systems, new nurseries rarely install them.

11.4.2.2 Rotating sprinkler

Rotating sprinkler irrigation systems generally consist of portable aluminum pipes joined by couplings fitted with risers (just high enough to clear the crop) and rotating sprinkler heads [19-22]. In some nursery installations, the main, branch, lateral feeder, and lateral lines are permanently buried beneath the ground to permit superior machine access. Usually, several lateral pipelines fitted with sprinkler heads at 30-ft intervals are spaced at 30, 40 50, or 60 ft (9.14, 12.19, 15.24, or 18.29 m) across each compartment, so that the water radiating from the sprinkler heads is sprayed in a circular overlapping pattern. The distances between sprinkler heads in lateral lines should be approximately 40 to 60% of the area covered by their circular spray patterns (see chapter 5, this volume). It is essential to select sprinkler heads with orifices that can achieve this objective at available head pressure. Because rotating sprinklers are made that can operate from 3 psi (21 kPa), with a discharge of as little as 1 U.S. gallon (3.78 liters)/minute, to 100 psi (690 kPa), with a discharge of 110 U.S. gallons (416 liters)/minute, almost any rate of application or arrangement of sprinklers is possible. Thus, rotating sprinklers have many advantages over oscillating nozzle lines—including considerably lower cost. One problem of rotating sprinklers is the difficulty in obtaining a uniform irrigation pattern because the spray is circular; however, this can be overcome by designing sprinkler-head layouts with effective overlapping spray patterns.

In most modern bareroot nurseries, rotating sprinkler systems with lateral lines in place throughout the water-management period are universally preferred. Such systems are essential where water may be required at any time to implement any of the objectives listed in this chapter's introduction (see 11.1).

11.5 Rotating Sprinkler System Design

The designs of all sprinkler irrigation systems are similar in principle yet infinitely variable in layout and degree of mobility. In this section, sprinkler-system components, types, and design factors are discussed.

11.5.1 System components

The following components are common features of all sprinkler systems:

- **Pumps or turbines** located at the water source, supplying and pressurizing the distribution pipelines (Fig. 2). These may be static or mobile and are usually either diesel or electric. Because the pumps are often located at the lowest point in the system, a check valve usually must be installed to prevent uncontrollable back pressures when pumps stop or are shut down.
- **Main line or lines** connecting the pumps to the branch supply lines. These are the largest diameter supply lines and are fitted with pressure-regulating valves (PRVs) just before they connect with the branch lines (Fig. 3). PRVs ensure an acceptably uniform head pressure in the branch lines regardless of variation in pump pressure.
- **Branch and connector lines** linking the main lines to compartment header lines. The branch lines are usually smaller in diameter than the main lines and are often arranged in loops extended from the PRV at the end or ends of a main line or lines. Looping branch lines ensure a reasonably uniform pressure at any point on the line.
- **Compartment header lines** connecting the branch lines to the lateral supply lines. These are usually considerably smaller in diameter than the branch lines because they are only required to supply water to the individual compartments.
- **Lateral lines** delivering the irrigation water from the compartment header lines to the sprinkler heads. These are the smallest diameter pipelines in the irrigation system and may even be stepped down to a smaller diameter at half their length. The laterals are fitted with sprinkler heads at regular intervals and are laid out in parallel lines across each nursery compartment so that uniform amounts of irrigation water may be applied to the soil or crop.

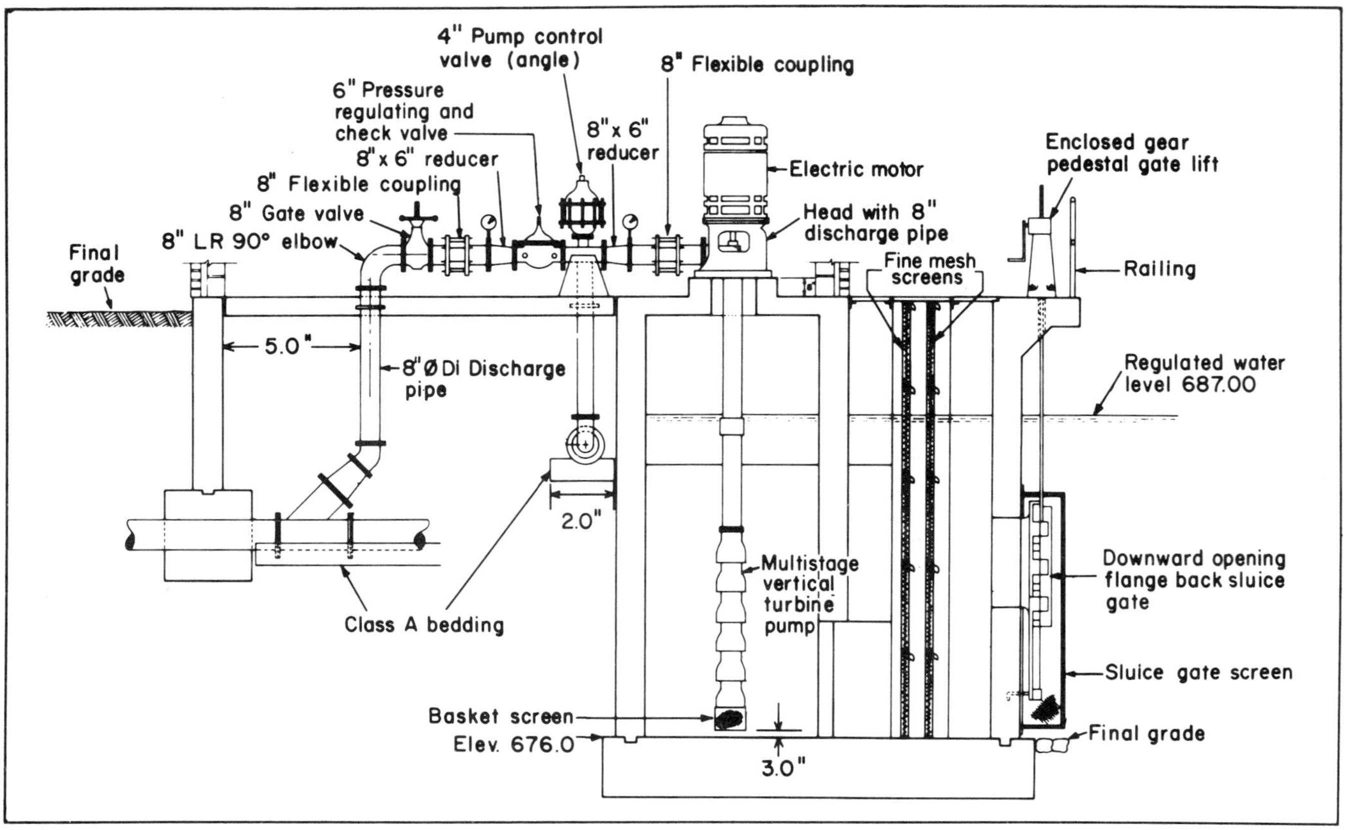

Figure 2. Profile of a modern bareroot-nursery pumping station.

11.5.2 System types

The types of rotating sprinkler systems used on bareroot nurseries are: (1) fully permanent, (2) solid set, (3) semipermanent, (4) fully portable with manually moved laterals, and (5) fully portable with mechanically moved laterals. Solid set and semipermanent systems are most commonly chosen (Table 4). Both these systems have semipermanent laterals that remain in position on the nursery compartments throughout the entire water-management period but are removed at the end of the period to permit cultivation and for overwintering. Fully permanent systems have been installed in recent years at some nurseries (e.g., the Ontario Ministry of Natural Resources Dryden Tree Nursery), but they are not common because of the additional cost of burying and then servicing the lateral lines. The comparative advantages of fully permanent, solid set, and semipermanent sprinkler systems will stimulate argument among nursery managers for years to come because all these systems work well. Their advantages and disadvantages may be summarized as follows:

Fully Permanent

- **Advantages**

  —Vehicular and machinery access is optimal because the branch, compartment header, and lateral lines are buried.

  —Lateral lines stay in the design pattern all the time, and risers and sprinklers threaded into them remain vertical, ensuring a uniform irrigation pattern.

- **Disadvantages**

  —Permanent risers fitted with sprinkler heads prevent complete soil cultivation or ripping close to or across the buried lateral lines. Hand or chemical weed control near the risers is usually necessary.

  —Machinery damage to the risers is not uncommon because most nursery machines must work near them. Repair of ruptured lateral lines or damaged risers is slower and more expensive because excavation is required.

  —Lateral and other pipelines not installed below frost level must be drained and blown out with compressed air before winter.

Solid Set

- **Advantages**

  —Vehicular and machinery access is fair because the main and branch lines are buried.

  —Lateral and compartment header lines can be removed at the end of the irrigation season and at the end of each crop rotation, clearing the compartment for complete cultivation and ripping.

- **Disadvantages**

  —Compartment header lines feeding the laterals block vehicular and machinery access to one end of each compartment.

  —Sprinkler risers (and heads) tilt unless staked or guyed, disrupting the uniformity of the irrigation pattern.

Semipermanent

- **Advantages**

  —None over those listed for solid set.

- **Disadvantages**

  —All those of solid set. In addition to compartment header and lateral lines, main and branch lines block ready access to many parts of the nursery. To overcome this problem at road intersections, main and branch lines may have to be set in underpasses or bridged.

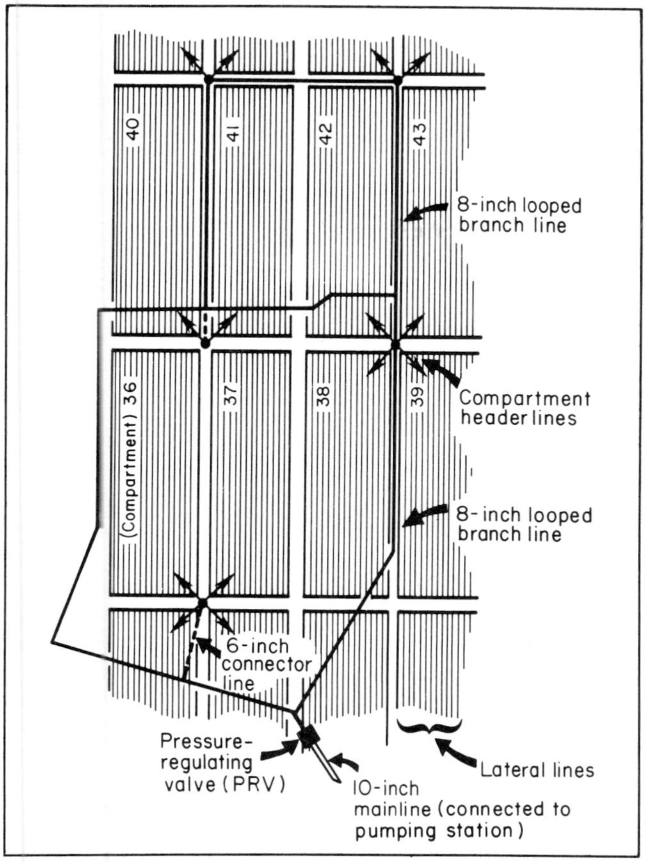

Figure 3. Portion of a nursery sprinkler-system layout.

Fully portable systems with manually moved laterals are sometimes used on undercapitalized bareroot nurseries. However, such systems are inadequate to supply irrigation to meet all the requirements of a fully developed water-management program and should be upgraded to a semipermanent or solid set system as soon as funding permits. Fully portable systems with mechanically moved laterals are rarely found on bareroot nurseries for similar reasons.

11.5.3 Design factors

Sprinkler-system design for a bareroot nursery depends on the following components: (1) water resource, (2) soil depth and type, (3) irrigation need, (4) pressure head, (5) relative land elevation, (6) system friction loss, (7) sprinkler layout, and (8) local wind characteristics.

11.5.3.1 Water resource

Because an adequate water supply is an important constraint in establishing a bareroot nursery, the water resource should always be of sufficient quality and quantity to meet nursery water requirements (see 11.3.2 and 11.3.3). Because many bareroot nurseries are established adjacent to watercourses or wells with delivery rates that are lower than the rates of use during irrigation periods, water storages such as dams and cisterns often must be built to create a reservoir of water. For a nursery obtaining its water from a stream, impoundment minimizes fluctuations in streamflow which may affect others during periods of heavy water use. Water storage also allows water to warm before it is applied to the crop or soil.

11.5.3.2 Soil type and depth

Bareroot nursery stock is generally grown in soils plowed at least 7 inches (18 cm) deep. Because seedling roots are almost always confined to the plow layer, water should be applied to keep this layer optimally moist—that is, in an optimum range of soil water potential (between −0.1 and at most −0.75 bar [6]). To maintain the plow layer in the optimal range, soil-moisture retention curves—the percentage of total soil-moisture content by weight [%TSMC (wt)] plotted over soil moisture tension (SMT)—must be derived for the range of soil types on each nursery (see 11.5.3.3 and also chapters 6 and 12, this volume).

Principal factors affecting water relations of cultivated bareroot-nursery soils are texture and organic matter content (see chapters 6 and 9, this volume). Light soils (sands and sandy loams) have a lower water-holding capacity than medium soils (loamy sands and loams); heavy soils (clay loams and clays) have the highest water-holding capacity. Ideally, bareroot nurseries should be located on light soils so that seedlings may be harvested with minimal root loss; thus, organic matter is usually added to improve soil water-holding and cation-exchange capacities.

Data presented in Table 5, drawn from a typical irrigation handbook [10], suggest that "available moisture" is the soil water between field capacity (−0.1bar SMT) and wilting coefficient (−15.0 bars SMT). Although soil water in this range is technically "available," growth of seedling crops in sandy soils is severely limited at SMTs greater than −0.75 bar [6]. Table 5 clearly shows that the light soils preferred for bareroot nurseries hold the least water yet have the highest percentage of available water and, conversely, that heavy soils hold the most water yet have the least available. Thus, the soils most desirable for bareroot nurseries are also those that tend to need regular irrigation and a well-planned program of water management.

Table 4. Comparison of five basic types of rotating sprinkler systems.

	Type of sprinkler system				
System component	Fully permanent	Solid set	Semipermanent	Fully portable, manually moved laterals	Fully portable, mechanically moved laterals
Water source	Single	Single	Single	Single or several	Single or several
Pumping plant	Static	Static	Static or semimobile	Mobile	Mobile
Main lines	Static buried	Static buried or surface	Static buried or surface	Portable surface	Portable surface
Branch lines (Loops)	Static buried	Static buried or surface	Static buried or surface	Portable surface	Portable surface
Lateral lines	Static buried	Static surface	Portable surface	Portable surface	Mechanically moved, surface
Use at bareroot nurseries	Rarely	Commonly	Commonly	Rarely	Very rarely

Table 5. **General water relations of various soil types in inches of water/foot of soil (in./ft)[1] and percent total soil-moisture content (%) (adapted from [10]).**

Soil type	Moisture-holding capacity				Available moisture (field capacity – wilting coefficient),	
	At field capacity,		At wilting coefficient,			
	in./ft	%	in./ft	%	in./ft	%
Light (sands to loamy sands)	1.25	100	0.25	20	1.00	80
Medium (loamy sands to loams)	2.25	100	0.56	25	1.69	75
Heavy (clay loams to clays)	3.67	100	1.28	35	2.39	65

[1]To convert in./ft to cm/m, multiply by 8.3332.

11.5.3.3 Irrigation need

Soil texture and depth, depth of root development, and soil moisture content govern the amount of water to be applied at any irrigation. To irrigate a nursery crop scientifically, nursery staff need the following information to compute the amount of water to be applied [6]:

- The **average depth of the plow layer** in centimeters (or inches)—conventionally, this has been 18 cm (7 in.). During germination or for 1+0 crops, it may be desirable to reduce the soil depth to that exploited by the seedling roots.
- The **bulk density (BD)**, or dry weight of the soil per unit volume in grams per cubic centimeter. The BD of most bareroot nursery soils averages 1.3 g/cm^3, ranging from 0.9 g/cm^3 in sands to 1.6 g/cm^3 in clay loams.
- The **soil-moisture retention curve**—%TSMC(wt)/SMT—with exact values for %TSMC(wt) at field capacity (−0.1 bar) and at the upper limit of dryness (normally between −0.5 and −0.75 bar [6]).

From this information, the amount of irrigation required to maintain the soil within the optimum moisture range for growth may be calculated:

(1) Compute %TSMC by volume [%TSMC(vol)] at field capacity before irrigation:

%TSMC(vol) = %TSMC(wt) x BD/1

where 1 = the density of water at 20°C.

This computation is essential for scientific application of water: the volume of water (in centimeters) held in the soil before irrigation and that which must be applied by irrigation must be determined to return the soil to an optimally moist condition at field capacity.

(2) Compute, in centimeters:

(a) The amount of water in the soil at field capacity, W(fc):

$$W(fc) = \frac{\%TSMC(vol) \text{ at field capacity} \times \text{soil depth}}{100}$$

(b) The amount of water in the soil before irrigation, W(bi):

$$W(bi) = \frac{\%TSMC(vol) \text{ before irrigation} \times \text{soil depth}}{100}$$

(c) The amount of irrigation water to be applied, W(i):

W(i) = W(fc) − W(bi)

(3) Compute the volume of irrigation water to be applied per unit area:

(a) In liters/hectare (L/ha):

$$L/ha = \frac{W(i) \text{ in cm} \times 10^6}{10.0}$$

(b) In U.S. gallons/acre (gal./ac):

$$gal./ac = \frac{W(i) \text{ in cm} \times 10^6}{10 \times 3.7853 \times 2.471} = W(i) \text{ in cm} \times 10{,}691$$

where 1 U.S. gallon = 3.7853 liters and 1 hectare = 2.471 acres.

In practice, it is necessary to know both the amount of water to be applied and the rate of infiltration of the soil between field capacity and the upper limit of dryness (i.e., between −0.1 and −0.75 bar [6]). The irrigation system can then be designed so that the rate of application is slightly lower than the rate of infiltration capacity of the soil, to avoid flooding [2]. If the sprinkler system is not designed in this way, the scientific application of irrigation water becomes a more complex and expensive procedure because water will have to be applied several times to return the soil to field capacity and to avoid flooding. In practice, it is always best to determine the infiltration rate and capacity of the soil in the range to be maintained by irrigation before the sprinkler system is designed. General relationships between soil moisture-holding capacities and infiltration rates of various soil types are given in Table 6 [11].

Table 6. Generalized relationship between soil moisture-holding capacity and approximate infiltration rate in various soil types (adapted from [11]).[1]

Soil type	Moisture-holding capacity	Infiltration rate/hour
	in. (cm)	in. (cm)
Light (sands to loamy sands)	0.75 (1.9)	1.5 (3.8)
Medium (loamy sands to loams)	1.30 (3.3)	0.75 (1.9)
Heavy (clay loams to clays)	2.15 (5.5)	0.5 (1.3)

[1]For a 7-inch (18-cm) plow layer.

11.5.3.4 Pressure head

The pressure head required to operate a sprinkler system is usually the sum of the following:

- **Sprinkler head pressure (SHP)**, the pressure required to operate the sprinkler heads with an appropriate overlap.
- **Frictional loss pressure (FLP)**, the pressure drop caused by frictional losses in the main, branch, compartment header, and lateral lines between the pumps and the highest sprinkler head in the system.
- **Lift pressure (LP)**, the pressure required to lift irrigation water from its source to the highest sprinkler head in the system.

Pressure head (PH) may thus be calculated:

PH = SHP + FLP + LP

If the water source is above rather than below the sprinkler system, so that head pressure is added to the system, LP should be subtracted, rather than added, in the above equation.

11.5.3.5 Relative land elevation

For an effective water-management program, it is best that slope vary less than 5% within any one nursery compartment or block of compartments operating from a single or looped branch line. To ensure that the amount of water prescribed is uniformly applied, the nursery should be designed so that blocks of compartments are located on level land and are served and circumscribed by looped branch lines fed by main lines fitted with PRVs (see Fig. 3). Thus, a bareroot nursery may be located on sloping ground provided that the areas to be irrigated from any looped branch line are level or are terraced to less than 5% slope.

Because the water source is usually located below the compartments in bareroot nurseries, blocks of compartments may be established at one or more levels above the pumping station provided that each block is fitted with a separate PRV. The PRVs will then reduce head pressure to the optimum for irrigation at each level.

11.5.3.6 System friction loss

Friction losses occur in the main, branch, compartment header, and lateral lines of all sprinkler systems. Losses in the main, branch, and compartment header lines should not exceed 10 psi (70 kPa), or the cost of operating the pumps will rapidly rise above that of fitting larger supply pipelines [11]. Losses in the lateral lines should be kept below 20% of the operating pressure to ensure uniform water application. A 20% pressure variation in the lateral lines causes approximately a 10% variation in discharge from the sprinkler heads from the point at which each lateral tees onto the compartment header line to its distal end. Because variations in discharge from the sprinkler heads are virtually impossible to eliminate, most sprinkler systems attempt to keep the variation in sprinkler-head discharge to within 10%.

Friction losses vary with the type of pipe used and the pressure (or rate of flow) applied to the line. They are least in plastic pipes and progressively increase in cement asbestos, aluminum, and steel pipe, especially aging and corroded steel pipe. Friction losses have been estimated for the above types of pipe and may readily be determined by reference to North Plains [16] or Gray [11].

11.5.3.7 Sprinkler layout

Because the fully permanent, solid set, or semipermanent sprinkler systems used at bareroot nurseries are costly, it is wise to design the best sprinkler layout for the crops to be grown before calculating pipeline diameters or considering pumping requirements. Each layout must be tailored to the soil, wind, and crop conditions at each nursery. Generally, the design process begins with selection of the type and spacing of sprinkler heads that will be optimum for the infiltration rate of the soil to be irrigated. The amount of water to be applied to the whole compartment per unit time is then calculated so that the size of the lateral, compartment header, branch, and main lines may be determined.

For example, Figures 4a-e show recently designed sprinkler layouts for seedbed and transplant compartments at the Dryden and Thunder Bay Forest Stations [12, 24-26]. Figure 4a depicts a typical, fully permanent, square-spaced sprinkler layout, with all pipes buried, in a single compartment at Dryden; all compartments at Dryden, whether seedbed or transplant, are so equipped [25]. Triangular offset and square sprinkler-head layouts in the solid set system are compared for seedbed (Figs. 4b, c) and transplant (Figs. 4d, e) compartments at Thunder Bay; these layouts can be interchanged to accommodate specific crop types. All of these are used here to illustrate how sprinkler layouts may be designed.

Compartments at Dryden are 200 ft (60 m) wide and 450 ft (137 m) long and are serviced by four 3-inch (7.6-cm) polyvinyl chloride (PVC) lateral lines spaced 52 ft (15.8 m) apart. The two outer lateral lines are fitted with 11 #25 Rain Bird® impact sprinklers [21, 22], spaced 40 ft (12.2 m) apart, each delivering 4.03 U.S. gallons (15.3 liters) of water/minute at 50 psi (345 kPa) over a 20-ft (6.09-m) radius; nozzles are 3/64 inch (3.6 mm). The two inner lateral lines are fitted with 11 #30 Rain Bird® impact sprinklers spaced 40 ft (12.2 m) apart, each delivering 6.84 U.S. gallons (25.9 liters)/minute at 50 psi (345 kPa) over a 48-ft (14.6-m) radius; nozzles are 5/32 by 3/32 inch (4.0 by 2.4 mm). Thus, the rate of application over the whole compartment averages 239.14 U.S. gallons (905.2 liters)/minute [(22 x 4.03) + (22 x 6.84)].

The rate/minute/unit area applied to a Dryden compartment (Fig. 4a) can be more usefully expressed:

(a) $\text{U.S. gallons/minute/acre} = \dfrac{43{,}560 \text{ ft}^2 \times 239.14 \text{ gal.}}{200 \text{ ft} \times 450 \text{ ft}} = 115.7$

(b) $\text{Liters/minute/hectare} = \dfrac{10{,}000 \text{ m}^2 \times 905.2 \text{ L}}{61 \text{ m} \times 137 \text{ m}} = 1{,}083.2$

where 43,560 ft^2 = 1 acre and 10,000 m^2 = 1 hectare.

(c) $\text{Inches/hour/acre} = \dfrac{43{,}560 \text{ ft}^2 \times 60 \text{ min} \times 239.14 \text{ gal.}}{200 \text{ ft} \times 450 \text{ ft} \times 27{,}154} = 0.256$

(d) Centimeters/hour/hectare =

$$\frac{10{,}000 \text{ m}^2 \times 60 \text{ min} \times 905.2 \text{ L}}{61 \text{ m} \times 137 \text{ m} \times 100{,}000} = 0.65$$

where 27,154 is the factor used to convert U.S. gallons/acre to inches, and 100,000 is the factor used to convert liters/hectare to centimeters.

The lateral header lines supplying the Dryden compartments are made of 4-inch (10.2-cm) PVC pipe. These are connected to 6-inch (15.2-cm) cement asbestos compartment header lines, which in turn are connected to 8-inch (20.3-cm) ductile iron looped main lines and a PRV. Finally, the PRV is connected to the 10-inch (25.4-cm) ductile iron main line and the pumps. Two electric 75-hp vertical turbines (see Fig. 2) able to deliver 1,700 U.S. gallons (6,435.0 liters)/minute are used, permitting approximately seven 2.0-acre (1.2-ha) compartments (1,700/239 gallons, or 6,435/905 liters) to be irrigated simultaneously.

The sprinklers in all the Dryden compartments are arranged as those shown in Figure 4a, regardless of whether the compartments are used for seedbeds or transplant beds. The distance between sprinklers is small enough to provide sufficient overlap, but the overlap clearly will be greater in the center of the compartment than at the edges; in practice, this tends to cause a wet strip in the center and, possibly, dry edges.

To irrigate seedbeds, Thunder Bay sets out five laterals spaced 40 ft (12.2 m) apart. Each lateral is equipped with 12 #30 Rain Bird® impact sprinklers [21, 22] at 30-ft (9.1-m) intervals, arranged in a triangular offset pattern (Fig. 4b). Each sprinkler delivers 10.9 U.S. gallons (41.25 liters)/minute at 55 psi (379 kPa) over a 48.5-ft (14.8-m) radius; nozzles are 3/16 by 1/8 inch (4.8 by 3.2 mm). Figure 4c shows a similar layout except that the sprinklers are arranged in a square pattern. Although it may be argued that triangular offset spacing provides more uniform irrigation than square spacing, comparing Figures 4b and 4c suggests this is not so.

To irrigate transplant beds, Thunder Bay sets out three laterals spaced 60 ft (18.2 m) apart. Each lateral is equipped with eight #70 Rain Bird® impact sprinklers at 70-ft (21.3-m) intervals, arranged in a triangular offset pattern (Fig. 4d). Each sprinkler delivers 13.7 U.S. gallons (51.85 liters)/minute at 55 psi (379 kPa) over a 60.5-ft (18.4-m) radius; nozzles are 7/32 by

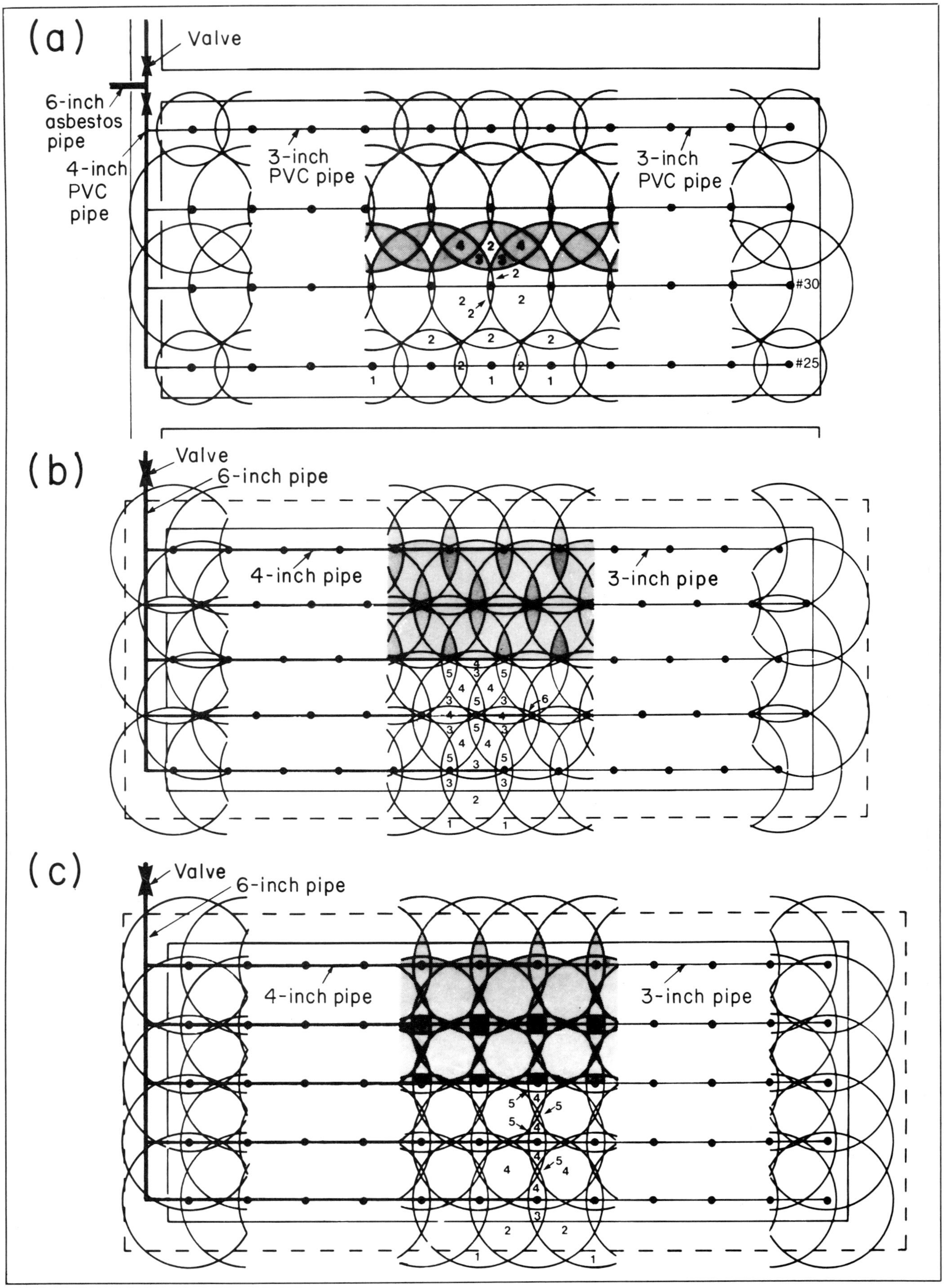
(a)
Valve
6-inch asbestos pipe
4-inch PVC pipe
3-inch PVC pipe
3-inch PVC pipe
#30
#25
(b)
Valve
6-inch pipe
4-inch pipe
3-inch pipe
(c)
Valve
6-inch pipe
4-inch pipe
3-inch pipe

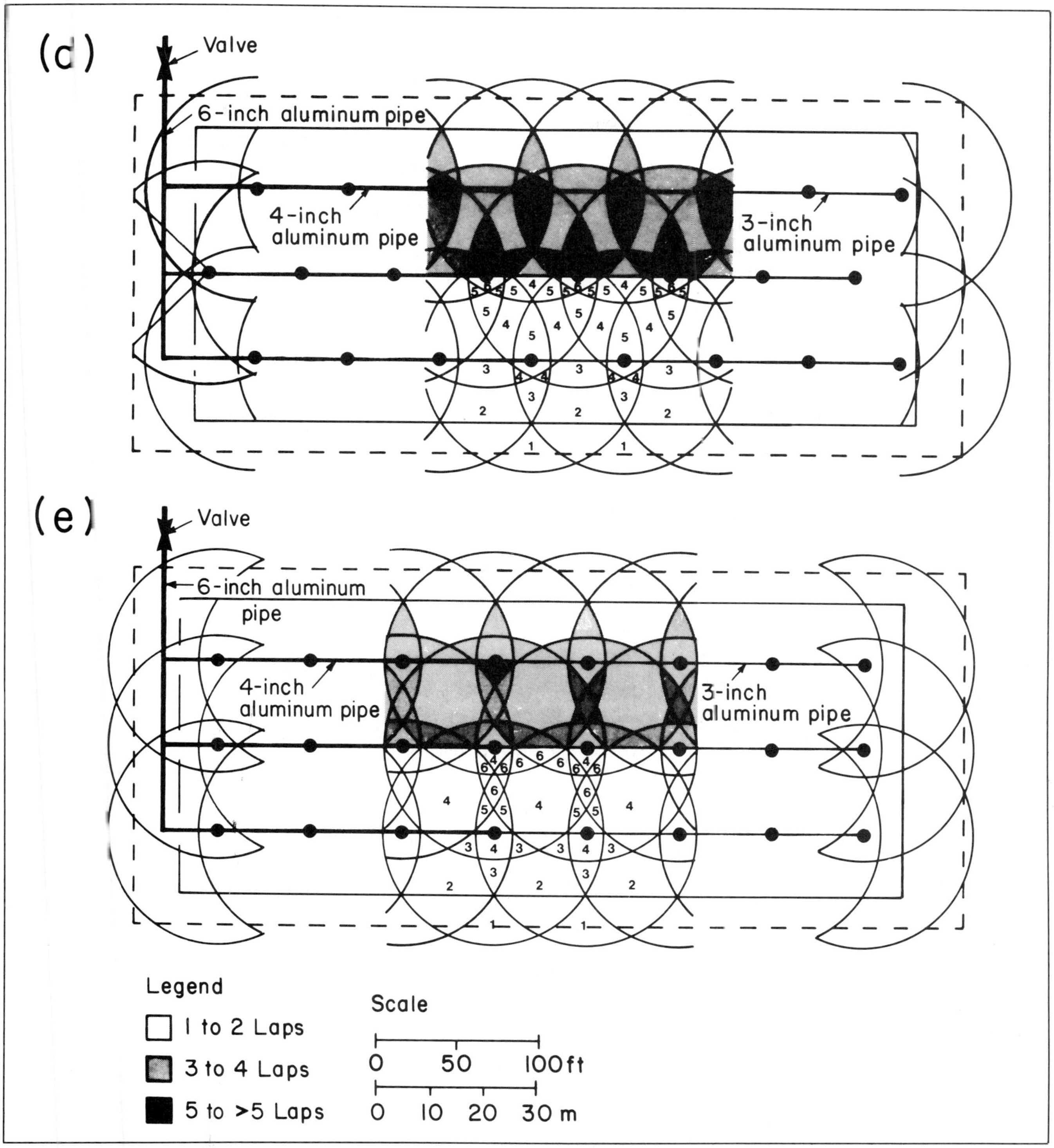

Figure 4. Sprinkler irrigation layouts at the Dryden and Thunder Bay Forest Stations: (a) fully permanent, square-spaced system for seedbeds or transplant beds at Dryden; (b,c) solid set system for seedbeds at Thunder Bay for (b) triangular offset and (c) square patterns; (d e) solid set system for transplant beds at Thunder Bay for (d) triangular offset and (e) square patterns. See text for all system specifications.

11/64 inch (5.6 by 4.4 mm). Figure 4e shows a similar layout except that the sprinklers are arranged in a square pattern. Again, comparing Figures 4d and 4e shows that differences in overlap and uniformity between triangular and square arrangements are minimal.

In addition to versatility in arrangement of lateral lines and sprinkler heads, solid set and semipermanent irrigation systems permit changes in riser height to accommodate crop growth. When the crop is young (e.g., 1+0 seedbeds), the risers can be set close to the ground surface to minimize wind effects; as the crop matures (e.g., 2+2), taller risers can be used.

11.5.3.8 Local wind characteristics

Local wind conditions may affect sprinkler-system design. Tests of sprinkler heads and determination of their distribution curves have shown that the maximum distance between sprinklers under normal wind conditions—winds less than 6 mph (10 kph)—should not be more than 60% of the diameter of the area covered by the sprinkler head. The following percentages are recommended for sprinkler heads:

	Mean wind speed, mph	kph	Spacing, % of irrigated diameter
Nil	0	0	65
Up to	6	10	69
Up to	8	13	50
Above	8	13	≤30

On nurseries that experience high winds, a large number of small sprinkler heads are preferable to a small number of large ones. For winds in excess of 8 mph (13 kph), the distance between sprinklers should be reduced to 20 ft (6.1 m) to ensure adequate overlap.

11.6 Conclusions and Recommendations

The facts presented in this chapter should make nursery staff more aware of the need to plan a proper water-management program and to match it with an irrigation system that can implement it.

If planning such a program indicates a need for a new or modified irrigation system, nursery managers should solicit the assistance of an irrigation engineer in designing or modifying the system. Comprehensive knowledge of the water requirements for irrigation and all other water-management purposes and a basic understanding of sprinkler-system design are essential for all nursery managers, especially if they are to cooperate with the consulting irrigation engineer.

The following steps are recommended for nursery managers planning a new irrigation system or renovating an inadequate one:

- Determine the nursery's water requirements for all water-management purposes by using tables or, better, by calculation with the Thornthwaite method. Be sure to use the water requirements of the driest years on record as the basis for determining the maximum amount of water that may be needed.
- Determine the maximum quantity of water that will be needed for all water-management purposes on the nursery, even in the driest years.
- Determine the quality of the water supply to be sure that it meets the standards for acceptable crop growth; be sure to check quality of stored or impounded water, especially in the driest season.
- Determine the infiltration rate of water into the nursery soil over the optimal soil-moisture range for growing seedling crops.
- Select the type of sprinkler system most suited to the individual nursery; a fully permanent, solid set, or semi-permanent rotating sprinkler system is recommended.
- Select sprinkler heads and design sprinkler layouts that will be optimal for supplying irrigation water to each compartment at rates slightly lower than the infiltration rate. Be sure that the sprinkler-head layouts and riser heights are (1) matched to crop types, (2) can operate satisfactorily in the winds that prevail on the nursery, and (3) are optimal for water-management purposes *in addition to* irrigation.
- Determine how many compartments are to be irrigated at one time and design the pipelines and pumping system needed to supply them. Be sure to allow for future nursery expansion—it almost always occurs.
- Consult and cooperate with an irrigation engineer in planning the best bareroot-nursery irrigation system your budget will permit.

References

1. Ayers, R. S. 1977. Quality of water for irrigation. Pages 134-1[illegible]4 *in* Proc., American Society of Civil Engineers. J. Irrigation and Drainage Division. Vol. 103, IR2.
2. Baver, L. D. 1956. Soil physics. John Wiley and Sons, Inc., New York. 489 p.
3. Blaney, H. F., and W. G. Criddle. 1950. Determining water requirements in irrigated areas from climatic data. U.S. Dep. Agric., Soil Conservation Serv., Washington, D.C. 96 p.
4. Cantor, L. M. 1970. A world geography of irrigation. Oliver and Boyd, Edinburgh. 252 p.
5. Chong, C. Y., and R. J. Day. 1974. Soil moisture and irrigation need on the Thunder Bay Forest Station. School of Forestry, Lakehead Univ., Thunder Bay. Unpubl. rep. 20 p.
6. Day, R. J. 1980. Effective nursery irrigation depends on regulation of soil moisture and aeration. Pages 52-71 *in* Proc., North American forest tree nursery soils workshop (L. P. Abrahamson and D. H. Bickelhaupt, eds.). State Univ. New York, Coll. Environ. Sci. and Forestry, Syracuse.
7. Doneen, L. D. 1975. Water quality for irrigated agriculture. Pages 56-76 *in* Plants in saline environments (A. Poljakoff-Mayber and J. Gale, eds.). Springer-Verlag, Berlin.
8. Eaton, F. M. 1966. Total salt and water quality appraisal. Pages 501-509 *in* Diagnostic criteria in plants and soils (H. Chapman, ed.). Univ. of California, Division of Agric. Sci., Citrus Res. Center, Riverside, California.
9. Fowler, H. W., and F. G. Fowler. 1976. The concise Oxford dictionary. Oxford Univ. Press, Oxford. 1,368 p.
10. Fry, A. W., and A. S. Gray. 1971. Sprinkler irrigation handbook. 10th ed. Rain Bird Manufacturing Co., Glendora, California. [illegible]3 p.
11. Gray, A. S. 1952. Sprinkler irrigation handbook. 3rd ed. Rain Bird Manufacturing Co., Glendora, California. 40 p.
12. Hrdlica, V. K. 1982. Ministry of Natural Resources Tree Nursery irrigation system at Dryden, Ontario. Alterations and design report. Ontario Ministry Government Services on Project NR-74780. Unpub. rep. 11 p.
13. Israelsen, O. W., and V. E. Hansen. 1962. Irrigation principles and practices. 3rd ed. John Wiley and Sons, Inc., New York. 447 p.
14. Landis, T. D. 1982. Irrigation water quality in tree nurseries in the inland west. Pages 60-67 *in* Proc., 1981 Intermountain Nurserymen's Assoc. meeting, Edmonton, Alberta, Aug. 11-1[illegible]. Can. Dep. of Environ. Inf. Rep. NOR-X-241. 120 p.
15. Morby, F. E. 1981. Irrigation regimes in a bareroot nursery. Pages 55-59 *in* Proc., 1981 Intermountain Nurserymen's Assoc. meeting, Edmonton, Alberta, Aug. 11-13. Can. Dep. of Environ. Inf. Rep. NOR-X-241. 120 p.
16. North Plains. (Not dated.) Irrigation handbook and directory. North Plains Press, Aberdeen, South Dakota. 327 p.
17. Oliver, H. 1972. Irrigation and water resources engineering. Edward Arnold, London. 124 p.
18. Penman, H. L. 1963. Vegetation and hydrology. Commonwealth Agric. Bureau, Commonwealth Bureau of Soils, Harpenden Technical Commun. 53. 124 p.
19. Rain Bird. 1977. Irrigation equipment. Rain Bird (Canada), Inc., Guelph, Ontario. 53 p.
20. Rain Bird. 1978. General maintenance manual—irrigation system equipment. Rain Bird Sprinkler Manufacturing Co., Glendora, California. 120 p.
21. Rain Bird. 1982. Agriculture irrigation handbook. Rain Bird Inc., Glendora, California. 48 p.
22. Rain Bird. 1982. Turf irrigation equipment. Rain Bird Sales, Turf Division, Glendora, California. 77 p.
23. Smith, H. 1976. Man and water—a history of hydro-technology. Peter Davies, London. 239 p.
24. Sorokoski, R. 1977. Report on irrigation water supply system for Ministry of Natural Resources Tree Nursery at Thunder Bay, Ontario. Rep. Ontario Ministry Government Services. 3[illegible] p.

25. Sorokoski, R. 1978. Ministry of Natural Resources Dryden Tree Nursery irrigation system. Systems operation. Project NR-74613. Rep. Ontario Ministry Government Services. 16 p.
26. Sorokoski, R. 1981. Ministry of Natural Resources Tree Nursery irrigation system. Thunder Bay. Project NA-74619. Rep. Ontario Ministry Government Services. (Not paginated.)
27. Thornthwaite, C. W., and J. R. Mather. 1955. The water balance. Drexel Institute of Technology, Lab. of Climatology, Centerton, New Jersey. VIII(1). 86 p.
28. Thornthwaite, C. W., and J. R. Mather. 1957. Instructions and tables for computing potential evapotranspiration and the water balance. Drexel Institute of Technology, Lab. of Climatology, Centerton, New Jersey. X(3). 311 p.
29. Turner, J. H., and C. L. Anderson. 1980. Planning for an irrigation system. American Assoc. for Vocational Materials, Engineering Center, Athens Georgia. 120 p.
30. Zimmerman, J. D. 1966. Irrigation. John Wiley and Sons, Inc., New York. 516 p.

Chapter 12
Irrigation in Forest-Tree Nurseries: Monitoring and Effects on Seedling Growth

S. E. McDonald

Abstract

The main objective of nursery irrigation is to avoid unwanted seedling moisture stress and its negative consequences for seedlings. Soil water potential, best measured by the tensiometer, decreases as soil water content drops; this relationship changes with soil texture. The secret of effective nursery irrigation is to keep soil pores filled with the proper balance of water and air to minimize moisture stress. Plant water potential, best measured by the pressure chamber, is the single most useful indicator of seedling moisture stress; predawn readings are the most stable, midday readings the second most stable. Soil-moisture retention curves, soil- and plant-moisture monitoring procedures, and careful observation together form the best approach for properly monitoring and controlling irrigation, assuming the irrigation system is a good one. However, because crop responses vary due to environmental modification, nursery climate, tree species, and seed source, managers need phenological information to fully anticipate seedling response to cultural operations. Seedlings must be protected from the damaging effects of frost and heat; overhead irrigation sprinkling is the most common, effective method for accomplishing both. Top dormancy of seedlings in late summer should be encouraged so that trees can become hardy long before the first frost; proper irrigation scheduling assures the desired seedling growth early in the season and induces dormancy (by imposing a *moderate* moisture-stress level) later on, thereby enhancing frost hardiness. In sum, knowing when and when *not* to irrigate should help nursery managers implement the most effective irrigation monitoring and application programs possible.

12.1 Introduction

The distribution of vegetation over the earth's surface is controlled more by the availability of water than by any other single factor [25]. The ecological importance of water thus reflects the physiological importance of water in plant processes.

Growth—and all related physiological and metabolic functions [19, 20, 23, 26, 50]—is the first process to be retarded when sufficient water is lacking. The effect of the resulting seedling moisture stress on growth of nursery stock can be profound; elongation of roots and shoots and growth in volume and dry weight are usually severely limited. Yet, although continued severe stress will either damage or kill stock, moderate stress can benefit seedlings, for example, by inducing dormancy ([58]; see also chapter 15, this volume). In addition, water applied at the proper time and in the proper way can prevent heat damage or help make seedlings frost hardy in fall. Therefore, applying or withholding water to benefit plants requires a thorough knowledge not only of species and site but also of seedling physiology (see chapter 14, this volume).

Water in forest-tree nurseries is best regulated through carefully designed irrigation systems and practices (see chapter 11, this volume). This chapter should help nursery managers plan and implement the most effective irrigation monitoring and application programs possible.

In Duryea, Mary L., and Thomas D. Landis (eds.). 1984. *Forest Nursery Manual: Production of Bareroot Seedlings*. Martinus Nijhoff/Dr W. Junk Publishers, The Hague/Boston/Lancaster, for Forest Research Laboratory, Oregon State University, Corvallis. 386 p.

12.2 Basic Water Relations

The main objective of irrigation—the artificial application of water to plants—is to avoid unwanted moisture stress in plants. In simple terms, moisture stress occurs whenever the rate of transpiration (loss of water from plants as vapor) exceeds the rate of absorption, leaving plant cells and tissues less than fully turgid [25]. Such stress can vary in degree from a small decrease in water potential (see 12.2.1), detectable only by instruments, to transient midday wilting, to death by desiccation.

Plant water transport can be viewed as a simple input-output system: soil water is the input and plant transpiration to the atmosphere the output [39]. Under optimal conditions, water transpired to the atmosphere is replenished by water absorbed by plant roots, although some lag between transpiration and root uptake is normal [24]. Midday water deficits occur in most plant species because absorption tends to lag behind transpiration (Fig. 1). This lag results because water flowing through plants meets resistance and because rates of water absorption and transpiration are controlled by different sets of factors. Transpiration rate is controlled by (1) leaf area and structure, (2) stomatal opening, and (3) those factors affecting the steepness of the vapor-pressure gradient from plant to air. Absorption rate, on the other hand, is controlled by (1) rate of water loss, (2) extent and efficiency of the root system, and (3) water potential and hydraulic conductivity of the soil. It is not surprising that processes controlled by different sets of factors are not perfectly synchronized, even though they are partly interdependent and linked together by the continuous water columns extending from roots to leaves.

Atmospheric evaporative demand is generated primarily by increasing air temperature and decreasing humidity, although radiation intensity and windspeed contribute indirectly. At constant relative humidity, evaporative demand increases exponentially, not linearly, with air temperature because the atmosphere can hold more water vapor as air temperature rises (Fig. 2) [25]. Consequently, under conditions of increasing temperature and low humidity, plants can suffer significant short-term moisture stress even in well-watered soil. The nursery manager can moderate evaporative demand by shading (see 12.5.2.2) and midday overhead sprinkling (see 12.5.2.1), depending on nursery location and tree species being grown [39]. Low soil water content can also induce seedling moisture stress, but this can easily be controlled by the nursery manager (see 12.3).

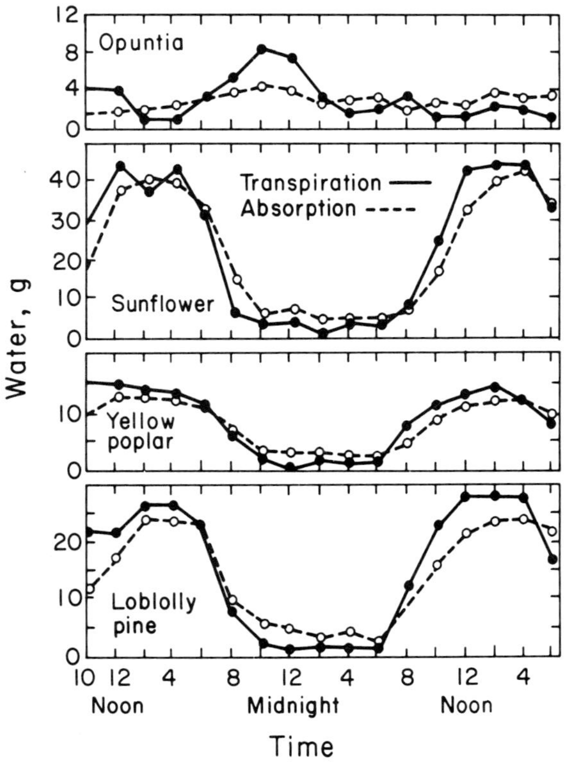

Figure 1. During the day, absorption lags behind transpiration in each of these four plant species (adapted from [24]).

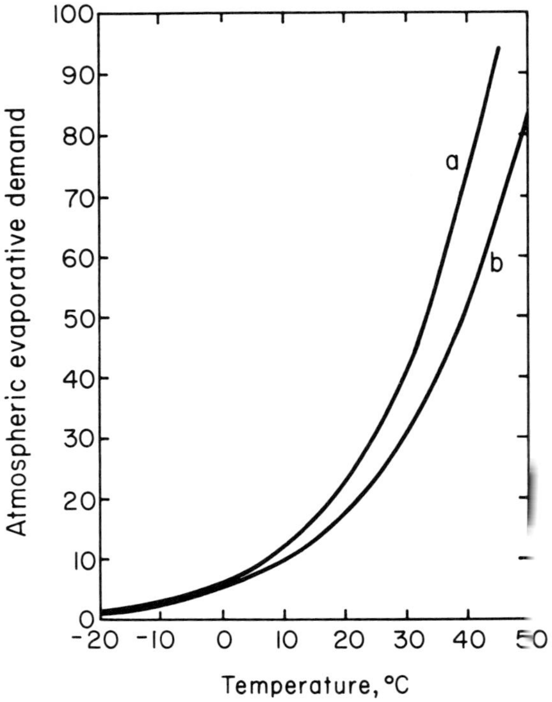

Figure 2. Effect of air temperature on the ability of the atmosphere to hold water vapor, expressed as (a) saturation vapor pressure, in millibars, and (b) saturation vapor density, in grams/cubic centimeter of air (x 10^6) (adapted from [25]).

12.2.1 Water potential

Water relations in plants and soil were once discussed in terms such as "suction" and "diffusion pressure deficits" but are now considered in thermodynamic terms—that is, as **potentials** ([25]; see also chapter 23, this volume).

Total water potential (Ψ) is a physical-chemical parameter whose components quantify soil particle-water attractions, salt-solution influences, plant-xylem tensions, and cell-turgor effects. For our purposes here, it is probably sufficient to say that Ψ is a measure of the capacity of water to do work, expressed in dynes/square centimeter (dynes/cm^2) or ergs/cubic centimeter (ergs/cm^3), but more commonly in atmospheres (atm), bars, or megaPascals (MPa).[1]

The water potential in any system is decreased by:

- Matric forces (surface and microcapillary forces in soils, cell walls, protoplasm, and other substances that adsorb or bind water)
- Addition of solutes
- Negative pressures (tensions), such as those in the xylem of transpiring plants

[1] 1 MPa = 10 bars ~ 10 atm ~ 150 psi ~ 10^6 dynes/cm^2 (or ergs/cm^3).

12.2.2 Soil water potential

The chemical potential of soil water (soil water potential, Ψ_{soil}) is of considerable importance in soil-water relations. The principal forces contributing to Ψ_{soil} are those associated with the soil matrix, with the osmotic characteristics of the soil solution, and with the total pressure on the soil water [25]. Adsorption and capillarity—the two mechanisms by which water is retained in shrinking and nonshrinking soils—are associated with the structure and characteristics of the soil matrix and are termed matric forces; together, these constitute the **matric potential, Ψ_m**. The osmotic forces associated with the soil solution constitute the **osmotic potential, Ψ_π**; the pressure forces generate the **pressure potential, Ψ_p**. Thus, soil water potential comprises three main component potentials:

$$\Psi_{soil} = \Psi_m + \Psi_\pi + \Psi_p$$

Soil water potential is also affected by external force fields such as gravity, which constitutes a **gravitational potential, Ψ_g**. The equation can then be written:

$$\Psi_{soil} = \Psi_m + \Psi_\pi + \Psi_p + \Psi_g$$

Gravitational and matric potentials are the most important components of nursery Ψ_{soil} and are the ones nursery managers are normally concerned with.

Apart from these thermodynamic terms, several other terms have been used in soil and plant science to describe soil-water characteristics significant to plant growth. The two most important of these are **field capacity** and **permanent wilting percentage** (see also chapter 6, this volume). The field capacity of a soil—its water content after gravitational drainage has slowed such that water content is relatively stable [25]—has been widely used to refer to the upper limit of soil water stored for plant growth. Field capacity is not a true equilibrium value, but a condition of such slow water movement that moisture content does not change appreciably between measurements; it is usually reached 1 to 3 days after soil has been thoroughly wetted by rain or irrigation. Permanent wilting percentage—the soil water content at which plants remain permanently wilted unless water is added to the soil [25]—has been widely used to refer to the lower limit of soil water available to plants.

Soil water potential decreases nonlinearly as soil water content drops (Fig. 3) [39]. Furthermore, the relationship changes with soil texture [18]. For example, in coarse-textured nursery soils (e.g., sands and sandy loams), the soil pores are large, and the amount of water retained by adsorptive and capillary forces is small compared to that retained in fine soils (clays).

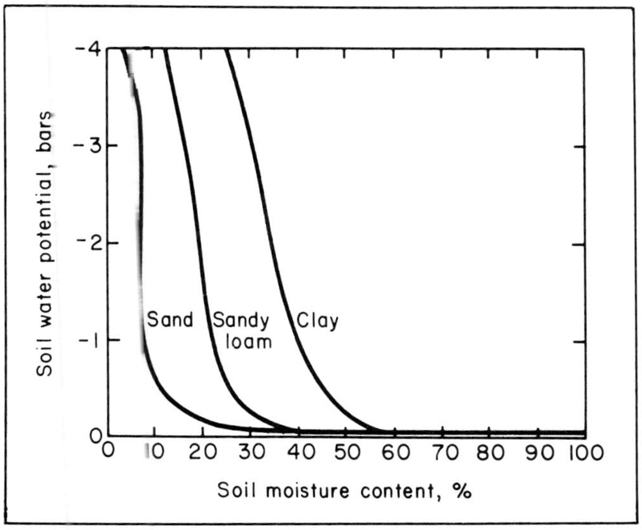

Figure 3. Typical soil-moisture retention curves for three different soil types (adapted from [39]).

Adsorptive forces at the surface of soil particles cause the water in the pores to adhere to the particle walls and form menisci (curved surfaces) at all water-air boundaries. Because pore size directly affects the curvature of such menisci, it therefore affects the matric potential. Thus, a coarse-textured soil with a few large pores partially filled with water would have low angles of curvature of the menisci and low matric potential. Conversely, a fine-textured soil at the same moisture content containing numerous, smaller pores would have high angles of curvature of the menisci and high matric potential (Fig. 4) [12].

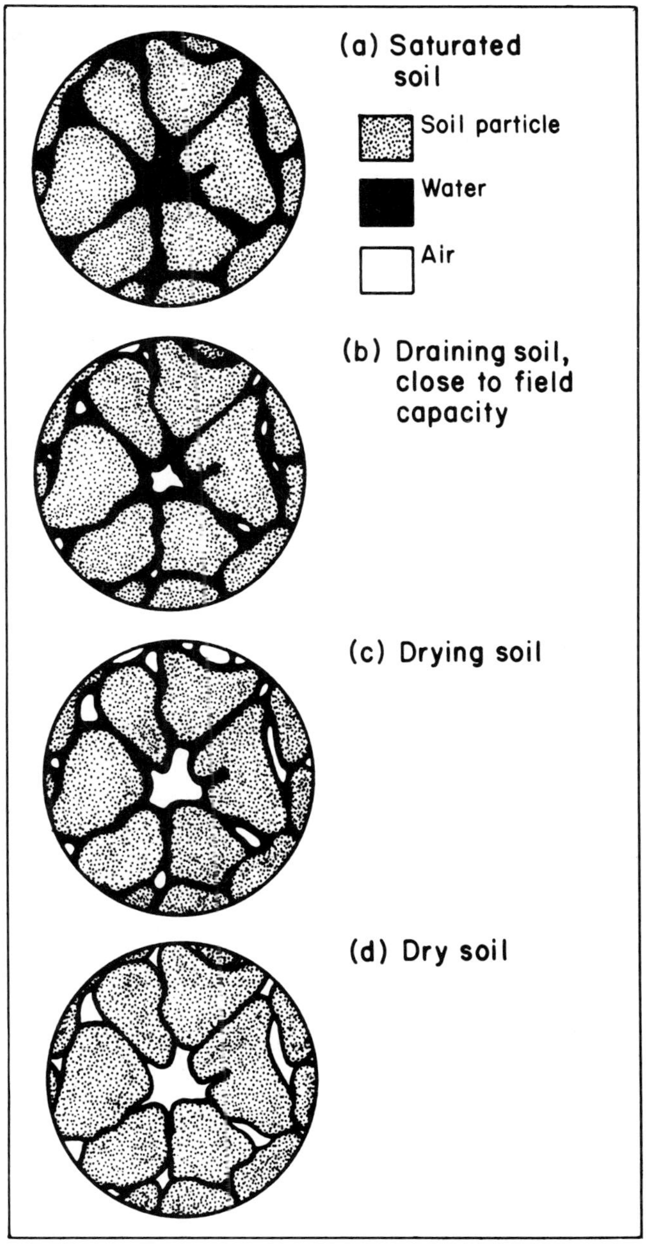

Figure 4. The changing status of water and air in soil pores (adapted from [12]). (a) All pores filled with water; soil at very high matric potential; photosynthesis and transpiration negligible; growth impossible due to lack of oxygen. (b) Large pores now partly filled with air; matric potential still high; excellent growth possible due to good supply of air and water. (c) All large pores and some medium pores filled with air; matric potential decreasing; satisfactory growth possible unless high atmospheric evaporative demand causes daytime moisture stress. (d) All large and medium pores filled with air; matric potential very low; root extension negligible and top growth severely limited due to long periods of moisture stress.

As the water held in either fine- or coarse-textured soils is depleted by drainage, drying, or absorption by seedling roots, the water held in the large pores at high matric potential is withdrawn first, followed by that at successively lower matric potentials. Thus, as water contained in soil pores is used up, matric potential decreases, and the space initially filled with water becomes filled with air [12].

The secret of effective nursery irrigation is to keep soil pores filled with *both* water (at high matric potential—i.e., −0.1 to −0.5 bar) *and* air to minimize seedling moisture stress [12].

12.2.3 Plant water potential

Traditionally, irrigation has been controlled by measuring and adjusting soil moisture content and inferring the resultant effects on seedling moisture stress. More recently, internal plant water potential (Ψ_{plant}) has also been measured directly. As a dynamic indicator and an integrator of effects of soil water potential and atmospheric evaporative demand with plant response [39], Ψ_{plant} is the single most useful measure of moisture stress in plants and will be used throughout this chapter. Plant moisture stress (PMS) is the absolute value of Ψ_{plant}.

Predawn Ψ_{plant} readings, which indirectly measure Ψ_{soil}, are the most stable measures of Ψ_{plant} available. For greatest accuracy, predawn readings should be taken while it is still completely dark [39]. In summer, Ψ_{plant} usually drops to a plateau by midmorning and remains roughly at that level until late afternoon (Fig. 5). These midday readings are the second most stable measures of Ψ_{plant} [39]. However, interpreting midday measurements is more difficult because they reflect Ψ_{soil}, atmospheric evaporative demand, and physiological plant response through stomatal closure. Even when soils are well watered, Ψ_{plant} may remain in the range of −7 to −9 bars on a cool, humid day and may drop to −9 to −12 bars on a hot, dry day. As soil water is depleted, the midday plateau may drop to −15 bars or lower (Fig. 5); without irrigation, the reading will continue to fall to as low as −50 bars—until the plant dies [39]. As Ψ_{soil} decreases, first midday and then predawn Ψ_{plant} decreases (Fig. 6) [50]. Generally, when midday Ψ_{plant} drops to −12 to −15 bars, moisture stress probably begins to impair growth [39]. The time required to reach this level depends on soil water available to each seedling, evaporative demand, and seedling characteristics.

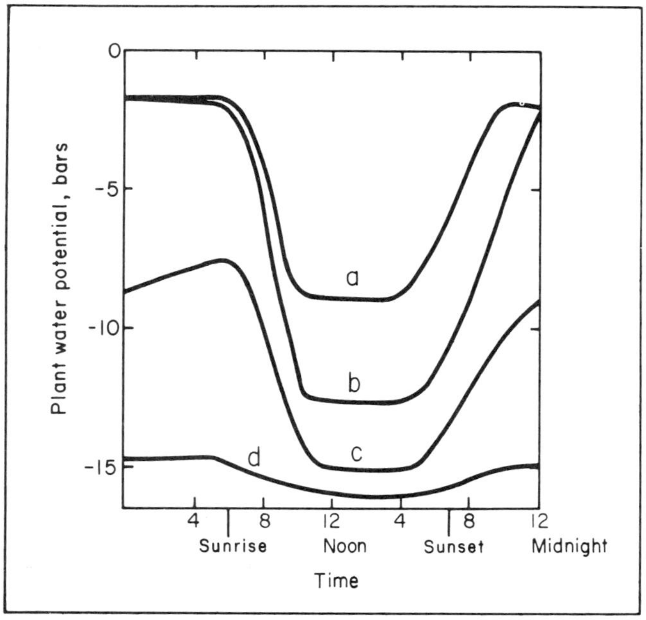

Figure 5. Patterns of plant water potential for a nursery seedling at (a) high soil-water potential and low atmospheric evaporative demand, (b) high soil-water potential and high evaporative demand, (c) low soil-water potential and high evaporative demand, and (d) extremely low plant water potential (severe seedling moisture stress) (adapted from [39]).

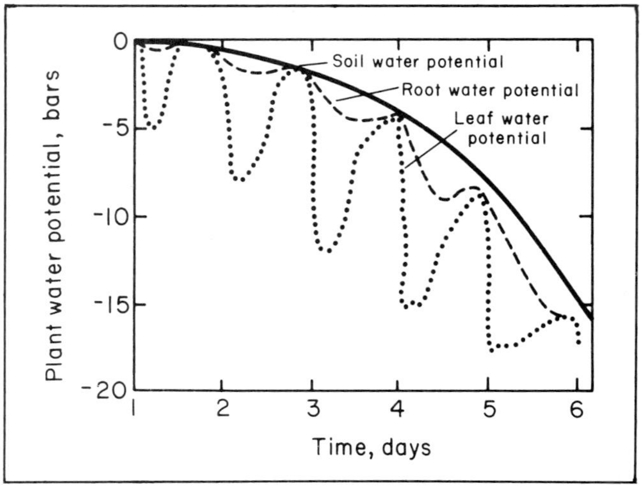

Figure 6. As soil water potential decreases, plant water potential also decreases (seedling moisture stress rises) (adapted from [50]).

Variability in Ψ_{plant} is significantly greater at midday than before dawn. Therefore, when Ψ_{plant} measurements are used for monitoring moisture stress at outdoor nurseries, the effect of changing evaporative demand on midday Ψ_{plant} must be accounted for. Seedlings in a greenhouse experience less climatic variability and should produce more consistent readings. Besides atmospheric influences, Ψ_{plant} can vary with tree species, age, phenological stage, and other factors. Consequently, although some indication of expected values is given here, each nursery should conduct tests under its own conditions [39].

12.3 Monitoring Irrigation

To monitor irrigation completely and professionally at a nursery, managers must rely on:

- **Soil-moisture retention curves** for individual nursery soils
- An effective procedure for rapidly assessing **soil moisture status**
- An accurate means of monitoring **seedling moisture stress**
- An understanding of **seedling response to irrigation**

12.3.1 Soil-moisture retention curves

Soil-moisture retention curves (Figs. 3 and 7), which illustrate the relationship between the percentage of soil moisture by weight and matric potential, provide nursery managers a means for monitoring the matric potential of a soil. These curves must be developed for each soil type at the nursery and can be obtained from most soil laboratories. Soil samples should be collected from the plow layer of the nursery field; ideally, these samples should be undisturbed cores although, in practice, samples from cultivated soils have been satisfactory [3].

Figure 7 illustrates how a soil-moisture retention curve should be used in nursery irrigation scheduling. Irrigation should usually be initiated at a matric potential of approximately −1 bar, although the exact point at which seedling growth is affected is not known. Armson and Sadreika [3] and Glerum and Pierpoint [17] indicate that the top growth of coniferous nursery stock is curtailed well before the wilting point (Fig. 7, d) is reached. Day and MacGillivray [13] and Day et al. [15] have shown that the roots of coniferous nursery stock would not develop in sandy loam soils at matric potentials of less than −0.6 to −1.5 bars. McDonald and Running [39] recommended that irrigation be initiated at approximately −0.5 to −0.8 bar for western U.S. nurseries. Because the matric potential that restricts either top or root growth is not known, irrigation should be applied to maintain the plow layer between field capacity (Fig. 7, a) and the point on the soil-moisture retention curve at which the matric potential begins to drop towards the wilting point (Fig. 7, c) [12].

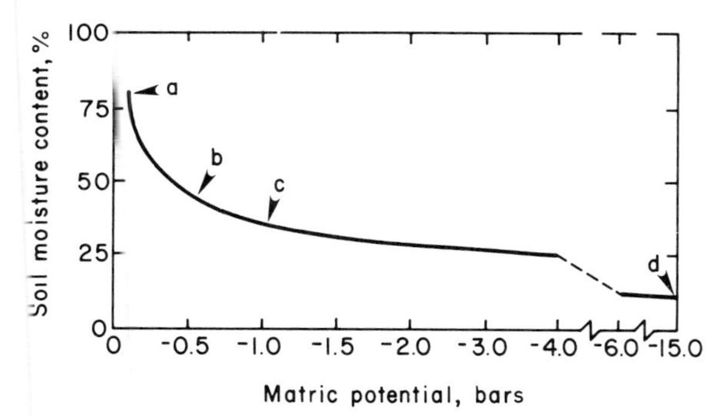

Figure 7. Soil-moisture retention curves for heavily peated nursery soil at Dryden, Ontario (adapted from [14]). (a) At 80% soil moisture content and −0.1 bar matric potential, the soil is at field capacity; just drained, it is moist and well aerated, ideal for crop growth. (b) At 45% soil moisture content and −0.5 bar matric potential, the soil is at the upper limit for irrigation but still ideal for crop growth. (c) At 37% soil moisture content and −0.1 bar matric potential, matric potential begins to drop steeply; further soil moisture losses cause large decreases in matric potential. (d) At 12% soil moisture content and −15.0 bars matric potential, the soil is at the wilting point; crops may survive in this very dry soil, but growth is impossible and physiological quality will decline.

12.3.2 Monitoring soil moisture status

Moisture in nursery soils is monitored by estimating either soil moisture content or matric potential [12]. Because several excellent reviews are available [11, 22, 35, 55, 56], only the merits of the more practical methods for Northwest nurseries—gravimetric, neutron probe, visual and tactile, tensiometer, and electrical resistance—are described here.

It is essential to remember that soil moisture content usually varies both horizontally and vertically in the plow layer [3, 43]. Soil moisture varies horizontally in nursery soils due to changes in soil type or irregular irrigation patterns. Soil moisture varies vertically after irrigation or rainfall because water applied to the soil surface moves downward into the plow layer with a definite wetting front over 48 hours [12], leaving the zone behind the wetting front at field capacity. Vertical arrangement of water in a nursery soil is important because inadequate irrigation will not permit the wetting front to advance to the bottom of the plow layer; this is a common cause of poor root development in nurseries [2, 37].

12.3.2.1 Gravimetric method

This most basic of all methods is generally used for calibrating other methods [12]. Soil samples are randomly collected from the plow layer with a soil auger or sampling tube; samples are then placed in preweighed metal containers, weighed, and dried to constant weight in an oven at 105°C for 12 hours. Samples are reweighed after drying, and the percentage of total soil moisture content by weight [%TSMC(wt)] is computed as follows:

$$\%TSMC(wt) = \frac{\text{wet wt} - \text{dry wt}}{\text{dry wt}} \times 100$$

If the bulk density (BD) is known, %TSMC by volume [%TSMC(vol)] may also be computed:

$$\%TSMC(vol) = \%TSMC(wt) \times \frac{BD}{\text{density of water}}$$

Usually, from 10 to 25 samples are needed to estimate mean %TMSC for a nursery field.

The gravimetric method works well in most nursery soils because of their homogeneity; however, it normally takes too long for everyday use.

12.3.2.2 Neutron probe

The neutron probe can indirectly measure %TSMC(wt) if bulk density is known. High-energy neutrons are emitted into the soil from a radioactive source contained in the probe. The neutrons are slowed down and thermalized by elastic collisions with other nuclei [12]. Because the slowing down or moderating of the neutrons is caused almost entirely by hydrogen nuclei in soil water, the number of thermal neutrons detected per unit time is directly proportional to the %TSMC(wt). Though accurate, neutron probes are expensive and cumbersome and require highly trained, licensed personnel for their operation.

12.3.2.3 Visual and tactile approach

Most often, soil water content is assessed at the surface and (or) root zone by eye and by touch to evaluate irrigation need [39]. Of 99 U.S. nurseries, nearly all determined irrigation schedules by visually observing soil dryness [1].

There are, however, problems with the visual and tactile approach. First, it is entirely subjective; observations are not based on quantifiable procedures. Second, water status may not be assessed at the proper sampling depth; for example, seeing and feeling the surface soil are useless when the irrigation need is at the bottom of the plow layer. Third, guidelines compiled for use by irrigators are, at best, imprecise; even the most experienced and methodical personnel could reach different conclusions due to differences in interpretation or judgment.

In its favor, this method requires no mechanical equipment that can fail and forces the irrigator to closely examine the plants and soil [39].

For best results, the visual and tactile approach should be accompanied by some mechanical method—for example, gravimetric sampling (see 12.3.2.1) or tensiometer readings (see 12.3.2.4)—for calibration, continuity, and quantification [39].

12.3.2.4 Tensiometer

Tensiometers, used in about 10% of western bareroot nurseries [38], can measure matric potential directly in the field [39]. But they can only provide indirect inferences about internal seedling moisture stress.

A tensiometer is basically a porous cup filled with water which is buried in the soil and connected to a vacuum gauge. This gauge registers the pressure drop on the water in the cup, which is in equilibrium with the matric potential of the soil water. Because tensiometers operate well in the 0 to −0.8 bar range, they are ideal for monitoring irrigation in forest-tree nurseries [12]; however, when matric potential drops below −0.8 bar, air begins to enter the porous cup, and the tensiometer becomes inoperative [25].

A minimum of 10 to 12 tensiometer readings is needed to determine mean matric potential. Therefore, one portable tensiometer may be more practical than a large number of static ones. If mean matric potential is used to monitor irrigation, soil-moisture retention curves are required to determine the equivalent %TSMC(wt) when computing the irrigation need [12].

12.3.2.5 Electrical resistance

Sensors containing a pair of electrodes, usually in "blocks" or "sandwiches" made of gypsum, plaster of paris, or fiberglass cloth, are planted in the soil at specified depths to indirectly measure matric potential. As the water content of these blocks changes with soil moisture content, so does the electrical resistance between the two electrodes. Readings on a resistance meter connected to these electrodes are converted to an index of soil moisture content. Resistance blocks are sensitive over a −0.5 to −15 bar range of matric potential and are therefore effective in dry soils.

The electrical resistance is calibrated against actual soil moisture content by taking readings at various soil moistures, finding the soil moisture content by the gravimetric method (see 12.3.2.1), and plotting a curve relating the true soil moisture content to the electrical resistance reading. However, calibration can be a problem because resistance is decreased by salt in the soil and by increasing temperature [39]. Furthermore, seedling roots must be near blocks for readings to be meaningful. The major problem is that electrical resistance does not usually calibrate well with matric potential, %TSMC, or available soil moisture content [12].

Only if equipment can be satisfactorily calibrated can electrical resistance methods be useful for monitoring soil moisture. However, the problems are great, and better methods are available [12].

12.3.3. Measuring seedling moisture stress

The principal methods of monitoring the internal water status, or water potential, of plants include thermocouple or thermistor psychrometer, gravimetric vapor exchange, dye, freezing point, and pressure chamber ([12, 45]; see also chapter 23, this volume). Of these, only the pressure chamber method developed by Scholander et al. in 1965 [47; also 10, 16] is sufficiently practical for nursery use and probably is the simplest, most rapid, and most accurate method suitable for the field [6, 38].

A small twig or needle is cut from the seedling and placed in a steel chamber with the cut end protruding from the lid (Fig. 8). Think of the water column in a seedling as a rubber band [39]. As moisture stress increases in the seedling, this rubber band is stretched. When the twig or needle is cut from the seedling, the tension on that rubber band water column is released, so that water shrinks back from the cut surface. By slowly applying pressure to the cut twig in the chamber, water will be forced back to the cut surface by a pressure equal to the tension originally on that water column. Ψ_{plant} is determined by reading the pressure gauge the instant water appears at the cut surface and recording the pressure in negative bars [39]. Step-by-step instructions for using the pressure chamber are given in Day and Walsh [16].

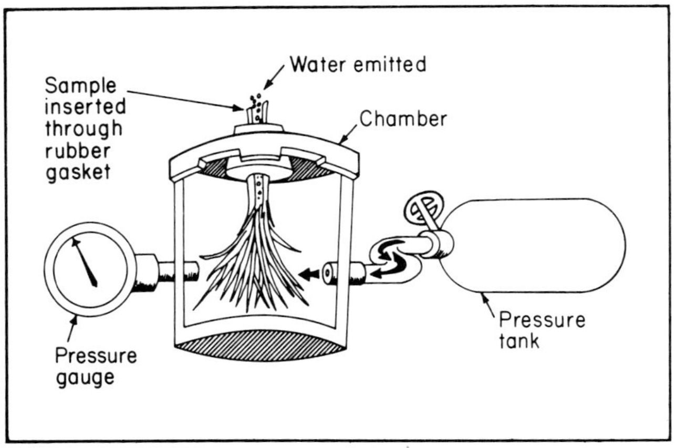

Figure 8. Pressure chamber for monitoring plant water potential (adapted from [10]).

Even though the measurement of Ψ_{plant} cannot be used to monitor soil moisture or to compute the irrigation need, it tells nursery managers whether the seedlings are adequately irrigated or under stress. As such, pressure-chamber measurement is the final test of any system of monitoring soil moisture supply and applying irrigation [12].

Seedling moisture stress depends on: (1) the supply of soil moisture and the ability of seedlings to absorb it, (2) the atmospheric evaporative demand, which is related to the temperature, relative humidity, or vapor-pressure saturation deficit, and (3) the ability of stock to control moisture loss by closing stomata [12]. For example, charting the midday pressure-chamber readings for a group of tree seedlings as they dry out produces a curve relating available soil water to midday Ψ_{plant} (Fig. 9) [39]. Note that at high soil-moisture availabilities, Ψ_{plant} is highly influenced by evaporative demand of the air at the leaf surface. The "adequate" segment (Fig. 9) reflects moisture stress developed solely by evaporative demand. The "stressed" segment reflects increased stress (lower Ψ_{plant}) as soil moisture is depleted and as the influence of available soil water gradually dominates the effect of evaporative demand. The "dangerous" segment indicates increasing stress despite total stomatal closure; if trees are not irrigated, damage will be irreversible.

However, a primary problem in interpreting midday Ψ_{plant} occurs along lines a-b and b-c in Figure 9 [39]. Note that from a to b soil water content is the same, but Ψ_{plant} differs as a result of differing evaporative demand. Conversely, from b to c Ψ_{plant} is the same, but soil water content differs again because of differing evaporative demand. Each nursery manager will have to determine the midday Ψ_{plant} curve for levels of evaporative demand at the nursery. This problem is simplified in greenhouses, where evaporative demand is more uniform [39]. Where internal seedling moisture stress is intentionally induced to stimulate root growth or apical dormancy (see chapter 15, this volume), the midday Ψ_{plant} curve can provide quantitative guides to the degree of stress that is safe.

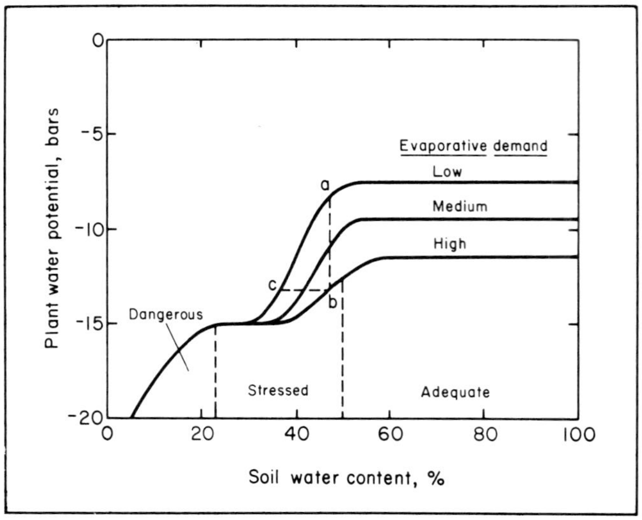

Figure 9. Changes in soil water content and atmospheric evaporative demand affect midday plant water potential (adapted from [39]).

12.3.4 Seedling response to irrigation

12.3.4.1 Seed germination and emergence

The first critical period of irrigation begins immediately after seed is sown and lasts until the seedling is established in the soil and is autotrophic (self-nourishing). A seed must imbibe (absorb) water additional to that present in its "dry" state. It is this first water, entering the seed either during stratification or in the seedbed (if seed is not stratified), that initiates biochemical and physiological processes leading to germination. Practically all seed can absorb enough water to germinate from soil at field capacity, and some seed can germinate in relatively dry soil (i.e., matric potential of −1 to −5 bars). However, germination on dry substrates inhibits root growth [4, 26, 27].

The frequency and rate of water application during germination are influenced by amount of rainfall, soil texture, type of mulch, temperature, sun, and wind. However, a moist germination period is preferred over a wet one because excessive moisture during that period can kill germinating seed [36]. The respiration that fuels metabolic processes during germination requires oxygen, and too much water can limit oxygen supply, especially after the seed coat begins to crack [27]; the embryo may not survive if conditions are too wet. Furthermore, overly wet conditions encourage certain diseases (damping-off) that may damage seedlings. May [36] indicated that the amount of water needed to keep the soil moist may vary from ¼ to ½ inch (6.4 to 12.7 mm) daily on sandy soils to ½ inch or more at 2- to 3-day intervals on heavy soils. Because some mulches, such as hydromulch, absorb moisture during watering and lose water rapidly by evaporation, special care is needed to provide enough moisture for both mulch and soil.

Irrigation during germination should be frequent and in small amounts. Amounts should be increased over time to wet the entire rooting depth as seedlings develop [36]; if managers continue to apply only small amounts of water once seedlings have emerged, shallow rooting will result. However, frequency of irrigation should be gradually decreased so that seedbeds are kept moist but not wet.

In many nurseries, irrigation may be needed not only to enhance germination but also to prevent hardening of the soil surface into a crust the emerging seedling cannot penetrate. This condition occurs in some interior western tree nurseries that have alkaline soils or water [36]. At some locations frequent sprinkling may also be needed once seedlings have emerged to keep the surface soil cooled by evaporation (see 12.5.2).

12.3.4.2 Seedling growth

Once the new seedling is established, it enters a true growth stage characterized by three primary phases [44, 49]. In the first (logarithmic) phase, the growth rate is initially slow (Fig. 10), apparently because the germinating seed has fewer cells capable of growth, but the rate continuously increases as more cells are formed. In the second (linear) phase, size continues to increase at a constant (usually maximum) rate for some time. We do not understand exactly why the growth rate should be constant in this phase, but one reason might be that stems and roots grow by meristems, which produce cells that grow mainly in length [46]. The final (senescence) phase is characterized by a decreasing growth rate (note drop in rate curve in Fig. 10b) as the plant reaches maturity. Although the curves in Figure 10 are generalized representations of many plant species, measured growth curves of trees often only approximate these generalized curves.

Trees commonly cease height growth temporarily in late summer, when temperatures are still warm and days are long [26]. Growth sometimes resumes again before true winter

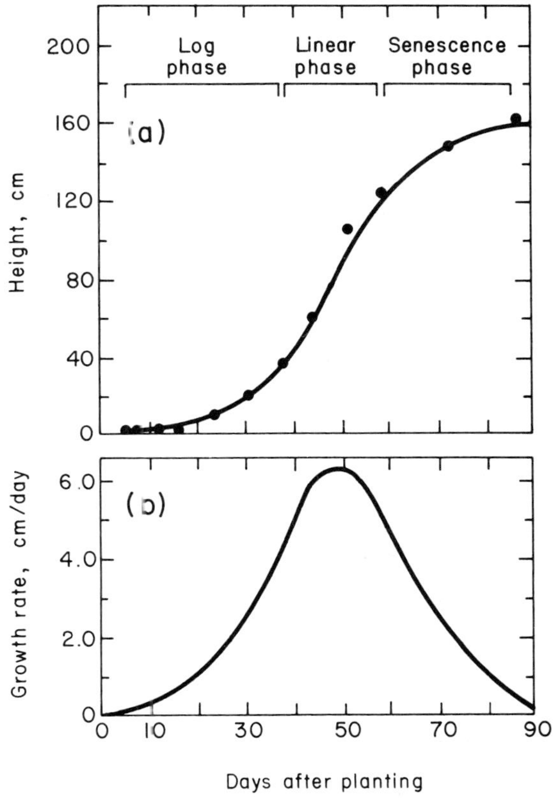

Figure 10. Idealized curves indicating seedling (a) height and (b) growth rate. The rate curve of (b) was obtained by taking the first derivative of (a).

dormancy, a deeper dormancy that results in part from increasing night lengths and in part from low fall temperatures ([46]; see also chapter 14 this volume). Stem diameter continues to grow at a decreasing rate (due to expansion of cells produced by the vascular cambium) until well after height growth stops, and photosynthesis continues until leaves become senescent and yellow (in deciduous trees) or until temperatures become too cold (in evergreens). Root growth can persist as long as water and nutrients are available and soil temperatures remain high enough.

12.3.4.3 Sensitive periods

All crops have moisture-sensitive periods—times during which a water deficit depresses growth much more than would typically be expected. These critical periods are not well defined for forest-tree seedlings. Plants are probably especially sensitive to moisture stress during budbreak and the early part of the linear growth phase (see Fig. 10) in spring and during budset and hardening in fall. Specific seedling reactions would depend, however, on interactions with the environment at a given nursery and, within a species, on the responses and tendencies of given varieties.

12.3.5 Irrigation-monitoring recommendations

Although there is no substitute for personal observation and judgment based on nursery experience, every nursery should use some kind of quantifiable method to indicate available soil water or internal seedling moisture stress [39]. Tensiometers are best for measuring soil moisture at forest-tree nurseries because they cover the critical 0 to −0.8 bar range, and pressure chambers are best for directly measuring Ψ_{plant}. Nursery managers using soil-moisture retention curves and soil- and plant-moisture monitoring procedures that can be correlated with both observed and measured crop responses have all the necessary tools for properly monitoring and controlling irrigation—assuming their irrigation system is a good one. However, because crop responses to any environmental modification—be it irrigation, fertilization, spacing, or shading—differ depending on nursery climate, tree species, and seed source, managers need phenological information to fully anticipate seedling response to cultural operations.

The OSU Nursery Survey (see chapter 1, this volume) provided a profile of irrigation monitoring at 21 Northwest nurseries. Though many nurseries use more than one moisture-monitoring method, all use the visual and tactile approach. Two-thirds use a pressure chamber; half use tensiometers, most of which are fixed installations; and only one out of five uses electrical resistance blocks. Only about 30% have soil-moisture retention curves for their soils, and several having them said they did not use them. Most nurseries alter irrigation practices and monitoring as the crop grows older and the season passes. About 75% feel the need for better equipment and guides for irrigation monitoring and control.

Obviously, considerable work is needed at most nurseries to refine irrigation monitoring and control, not only to improve water use and plant growth but also to maximize the benefits of other cultural practices.

12.4 Frost Protection with Irrigation

12.4.1 Damaging effects of low temperatures and frost

Plants vary greatly in their ability to tolerate cold. When properly conditioned, trees are among the most cold-hardy plants. Many temperate-zone trees will tolerate −40°C [21], and boreal forest species, like paper birch (*Betula papyrifera* Marsh.) and jack pine (*Pinus banksiana* Lamb.), will tolerate winter temperatures as low as −196°C [32]. On the other hand, newly emerged seedlings and lush new foliage are not conditioned to withstand cold weather and may be severely damaged by subfreezing temperatures.

Even for the same plant, the frost killing temperature may vary widely, depending on the manner of temperature change, the season, and the physiological state of the plant [32]. The killing temperatures of even the most frost-sensitive plants are slightly below the freezing point of free water, due to freezing-point depression of cell solutions by the solutes in the cell sap [46]. When intracellular freezing of water occurs, it is always fatal to the cell [33]. Plants may be killed at higher temperatures if freezing is rapid rather than gradual, and greater injury to plants may occur after long-continued freezing than after short freezing periods at the same temperature. However, plants that survive one freezing without damage may be injured after two or more freezings at that same temperature.

It is generally during spring that nursery managers are concerned about frost protection; some plants that survive the cold in winter may be killed by a very slight freezing during spring [28]. In some locations, early fall frosting of recently grown foliage also is a problem, though this problem may be avoided by discouraging late-summer top growth (see 12.6). Whether spring or fall, the frost-protection steps a manager can take are the same.

In woody plants, the freezing process can begin on leaf surfaces, in xylem elements, or both [7]. Initial formation of ice crystals on leaf surfaces near the freezing point of water is abetted by so-called "ice-nucleating bacteria," which serve as focal points for the crystallization process [34]. Ice that forms on leaf surfaces around bacteria or other nucleation centers (e.g., dust) can migrate and spread into leaves via stomata and into cells, killing them. The number of ice-nucleating bacteria on leaves has been experimentally reduced with bactericides such as streptomycin and cupric hydroxide; with competing bacteria that are not ice-nucleating types; and with chemicals such as chloroform, which inactivate nucleating bacteria [34]. Although such practices may ultimately have some application in forest-tree nurseries, especially those rearing broadleaved trees, there is little current indication that ice-nucleating bacteria significantly affect the freezing point of conifer leaves.

12.4.2 Other (nonfrost) winter damage

Several types of damage attributable to cold weather, but not directly to frost [7], can affect plants:

- **Frost heaving:** Repeated freezing (expansion) and thawing (contraction) of surface layers of soil can work small or recently transplanted seedlings up and out of the ground. This is common between the first and second growing seasons but can be avoided through mulches of straw or needles.
- **Frost smothering:** When saturated soil freezes so that little or no oxygen can reach tree roots, seedlings may die unless the soil thaws. Damage usually occurs if the situation persists over 48 hours.
- **Frost cracking:** The bark and outer layers of a tree trunk can crack because of forces in the bole caused by differentials of expansion when (1) the exterior thaws or (2) the interior xylem freezes.
- **Winter desiccation:** When roots are frozen and cannot absorb water, but the stem, branches, or needles are not frozen, dry winter air can desiccate foliage.
- **Winter burn:** The sun can raise foliage temperatures above freezing on the south side of plants in winter, even when air temperature is below freezing. At sunset the thawed foliage refreezes very rapidly. The rapidity of the freezing causes winter burn.
- **Winter scald:** Winter scald is like winter burn but affects the bark, not the foliage, of a woody plant. Often, trees' lower trunks are painted white to lessen this type of damage.

12.4.3 Frost types and conditions

Frosts can be divided into two types: (1) radiation frosts and (2) wind and advection frosts [9]. The two types may occur simultaneously; moreover, in some instances, a radiation frost may intensify a wind frost.

Radiation frosts occur on cloudless nights with little or no wind, when excessive amounts of heat energy in the soil and plants are lost to the sky as long-wave radiation, and leaves, air near the ground, and the soil surface may fall below freezing. Advection frosts occur when cold air flows from a higher location to a lower one, displacing lighter, less dense warm air; such flows will "pool" and concentrate if they encounter an obstruction, causing a "frost pocket." Wind in excess of 4 mph from cold regions causes a wind frost [9], which can occur at any time of the day or night and is not necessarily related to topography.

12.4.4 Protecting seedlings from frost damage

All agricultural frost-protection measures are meant to prevent crop freezing to the point that intracellular water crystallizes to ice and cell membranes rupture. Generally, crops are protected by preventing the air around them from becoming too cold or by placing insulating barriers between them and the cold air. However, only some agricultural methods are appropriate for forest-tree nurseries.

The major frost-protection methods used in agriculture are overhead sprinkling, heating, and wind machines. Less common but also sometimes employed are brushing, sanding, and windbreaks. Managers should be aware, however, that most frost-protection schemes can raise the temperature by only a few degrees, and some are effective only against radiation frost.

12.4.4.1 Overhead sprinkling

Overhead sprinkling—the most commonly used frost-protection method in tree nurseries—is effective against radiation or advection frosts or any combination of the two unless high winds cause poor sprinkler coverage [7]. Water droplets suspended in the air help check the flow of outgoing long-wave radiation. Sprinkling prevents frost by increasing the thermal conductivity and heat capacity of the ground and by releasing latent heat when water freezes [9]. The temperature of the plant will not fall below the freezing point as long as the change of state from water to ice is taking place.

In most tree nurseries, the new succulent growth is most vulnerable to frost damage; older wax-covered needles are usually more frost hardy. Overhead sprinkling is commonly begun in nurseries as the temperature drops to near 32°F and is continued until the temperature again rises above 32°F. Plant temperature declines immediately if sprinkling stops [9]. However, prolonged sprinkling should be avoided because the ice formed on plants can cause damage due to its weight and because overland flows resulting from extended sprinkling can become excessive; where sprinkling is commonly employed for frost protection, drainage ditches should be constructed to carry excess water away. Irrigation water may require a wetting agent to create a more uniform protective water film on hydrophobic surfaces like conifer needles [9].

The success of sprinkling largely depends on the amount and frequency of water application. Ideally, sprinkler rotation speed should be such that all water has just turned to ice at the next pass of the sprinkler. Thus, rotation speed should generally increase with the severity of frost. However, overhead sprinkling may have limited effectiveness or even increase frost damage if administered improperly. Many observers have noted that, under *light* radiation-frost conditions, minimal damage occurs to trees in unsprinkled areas whereas serious damage occurs to trees sprinkled with an insufficient amount of water. Businger [8] suggests two possible explanations for increased damage due to inadequate sprinkling: (1) when the air is dry, temperature of the sprinkled leaves will approach the wet-bulb temperature which may be significantly lower than the dry-bulb temperature; (2) the small amount of ice that forms on a leaf will prevent the undercooling of the cell solution and also may dilute the solution, thereby raising the freezing temperature. Another potentially dangerous situation is when air humidity is low (20% or less), a strong inversion exists, and temperature is barely subfreezing. If sprinklers are not operated until the air temperature rises *well above* the freezing point, the wet-bulb temperature of the leaves can fall back below freezing when sprinklers are shut off [7]. The principal lesson, then, is that frost damage can occur if sprinkler irrigation is improperly applied; to avoid such damage, continue to irrigate until all frost danger has passed.

Seventeen (80%) of 21 Northwest nursery managers use sprinkler irrigation to avoid frost damage (OSU Nursery Survey). Application varies with tree species, season (spring vs. fall), and nursery location. One nursery protects against frost only in spring and three only in fall, but most protect against frost in both seasons. Over 75% of the nurseries begin to irrigate when the air temperature is at or slightly above freezing (32 to 33°F); the rest wait until the temperature drops to as low as 25 to 28°F. Two nursery managers commented that their seedlings had experienced frost injury after irrigation for frost protection but admitted that they may have turned off the water too soon.

Nearly all nursery managers stipulated that irrigation must continue until all ice has disappeared if frost damage is to be avoided. Several said that the treatment was effective into the low 20s (°F) and that spring frosts were of more concern than fall frosts. No frost-protection method other than sprinkling was mentioned; however, the OSU Nursery Survey did not specifically solicit such a response.

12.4.4.2 Other potential frost-protection methods

Heating.—Heaters were utilized in the first successful attempts to prevent frost injury in California in the late 1800s [9] and have since been acknowledged as the best agricultural frost-protection measure. The idea is to warm the cold air in the lower layers of an inversion.

Heating is most effective on a night with a strong temperature inversion. An ordinary inversion ceiling is typically between 10 and 15 m above the ground; therefore, the depth of air to be heated is rather shallow [9]. In the absence of a temperature inversion, however, the principal value of direct heating is to radiate heat to trees and ground surfaces and to produce a pall of humid smoke, which forms a moderating screen that diminishes net radiation loss from the ground [9]. In the United States, the use of heaters generating large amounts of smoke has recently been declared illegal because of environmental degradation.

In general, a large number of small heaters are more effective than a few larger heaters [9] because the latter may set up a current that actually punctures the inversion layer and destroys the valuable warm ceiling. To protect trees against radiation frost, heaters should be evenly distributed so trees can absorb infrared radiation equally. However, to protect trees against advection frost, heaters should be placed in heavier concentrations along the upwind border. In hilly country, heaters should be concentrated mainly in the valleys so that the heat produced will move upward along the slope [9].

Oil heaters have been widely used in the United States; in Germany, coal, briquettes, and wood are the principal fuels. Oil is by far the most efficient because solid fuels cannot often be ignited quickly enough to avert frost damage [9]. In most areas, liquefied petroleum (LP) gas heaters are now legal to use. Because of its high cost, however, heating for frost protection is used only for a few high-priced crops, such as citrus fruits.

Wind machines.—Wind machines break up a nighttime temperature inversion by mechanically mixing air, their effectiveness increasing with the strength of the inversion [9]. Though several machines are needed to do any good, their combined effect is instantaneous.

Wind machines have been used increasingly, largely because their operating cost is only about 20% of that of heaters; however, they cannot protect trees in a freeze with cold daytime conditions, on cold soils, or with relatively warm air overhead on a clear, cold night. In steep valleys, the elevation of the temperature inversion may be too high for wind machines to be useful. Even when inversions are strong, the gain in ground-surface temperature with wind machines is rather small, usually less than 3 to 5°C. Therefore, it is common practice to install both wind machines and heaters in the field [9].

Brushing.—Brushing is a frost-protection scheme extensively used for vegetable crops in California [9]. Shields of brown kraft paper are attached to stakes on the north side of east-west rows, leaning over the plants. No plants are located on the shaded side, which is used for irrigation ditches. During the day, the shields deflect radiation to the plant and soil and also act as a windbreak; at night, the shields reduce radiation loss to the sky.

Brushing is more effective in protecting against radiation frost than wind frost. It is also more effective for small plants that have not outgrown the height of the kraft paper. This procedure may have a future in the culture and protection of tree-seedling crops grown from high-value genetically improved seed.

Sanding.—A sandy surface warms up easily and cools only slowly by radiation. Sand also minimizes evaporation, because of its low water content. Sanding can raise the temperature of loam and clay by several degrees and that of organic soils by even more, thus diminishing frost hazard [9]. In nurseries with sandy soils, this measure would be useless; but in nurseries with heavier soils and recurrent problems with spring frosts, it may have merit.

Windbreaks.—By excluding or diminishing the inflow of cold air and by shielding the field somewhat from the night sky, windbreaks can protect against frost [7]. However, windbreaks that are too dense can produce radiation frost in the lee or allow frost pockets to develop. Obviously, the merits of a shelterbelt like a windbreak for frost protection must be analyzed for specific instances. Where a nursery is subject to regular advection frosts, a windbreak to deflect the flow of cold air could be useful.

12.5 Controlling Heat in Seedbeds with Irrigation

12.5.1 Effects of heat

The number of seedlings that die outright because of heat stress is unknown. Even more uncertain is how many seedlings die from longer term heat stress and resultant indirect damage. Direct heat injury is due to cellular membrane injury, cell-component decomposition, or both; the effects are immediate and obvious. Indirect heat injury—due to metabolic disturbances—is more subtle and varies from minor reversible damage to death. If heat stress is moderate or short lived, the effect may be negligible; but if stress is severe or long term, the effect may be major.

The physiological complexity of plants precludes precise determination of cardinal temperatures because different processes and species have different temperature requirements [42]. Such temperatures also depend on the plant's state of development. General cardinal temperatures for cool- and warm-season plants are [9]:

	Temperature, C°	
Cardinal points	Cool-season plants	Warm-season plants
Lowest temperature for survival (killing point)	1 to very low	−1 to 10
Lowest temperature for growth	−1 to 5	15 to 18
Optimum temperature for growth	25 to 31	31 to 37
Highest temperature for growth	31 to 37	41 to 50
Highest temperature for survival (killing point)	40 to 45	50 to 52

12.5.1.1 Young seedlings

Very young trees are susceptible to heat damage because they are physically ill-equipped to deal with heat. The most apparent injury is direct, irreversible damage to tender seedling tissues, which can lead to death. Such damage, which often occurs just above the soil surface on the south and west sides of the stem, may be seen as depressed areas of necrotic tissue called "heat lesions."

Seedbed surfaces can become very hot in spring and summer. Evidently, the proximity of the seedling's tender cortical stem tissues to the hot soil surface is the key factor—because that is where the damage occurs (Fig. 11). Stem temperature reaches the killing point (about 40°C) there first because of energy concentrated at the hot soil surface, infrared radiation reflected from the soil surface to the stem, and lack of conduction and convection of heat away from the stem by moving air. The amount of direct insolation absorbed by the seedling seems less important than the amount absorbed by the soil, and air temperature is only indirectly related. Transpirational cooling, quite low in young trees, probably is inconsequential. Stems of young, heat-injured seedlings shrivel and become pale. At first, there is a definite boundary between the healthy and shriveled parts; then the healthy parts slowly decay. Drought-stricken seedlings, on the other hand, wilt along the entire length of the stem, which sometimes curves before shriveling or rotting at any one point; digging may show that soil is dry well below seedling roots [54].

The nursery manager's job is to keep the temperature of susceptible plant tissues below the lethal level. In the case of very young tree seedlings, the soil surface must be kept cool to prevent seedling damage. Where the nursery has high insolation rates, heavy soils, dark soils or mulches, or poor air circulation, sensitive species will have to be protected from heat damage most of the first growing season. Most true firs (*Abies* spp.), spruces (*Picea* spp.), coastal Douglas-fir [*Pseudotsuga menziesii* (Mirb.) Franco var. *menziesii*], white pines (*Pinus* spp.), hemlocks (*Tsuga* spp.), Northwest cedars (*Thuja* and *Chamaecyparis* spp.), and redwood [*Sequoia sempervirens* (D. Don) Endl.] are susceptible to heat damage as young seedlings. However, the degree of protection is contingent on the nursery environment and must be determined for each species at each nursery.

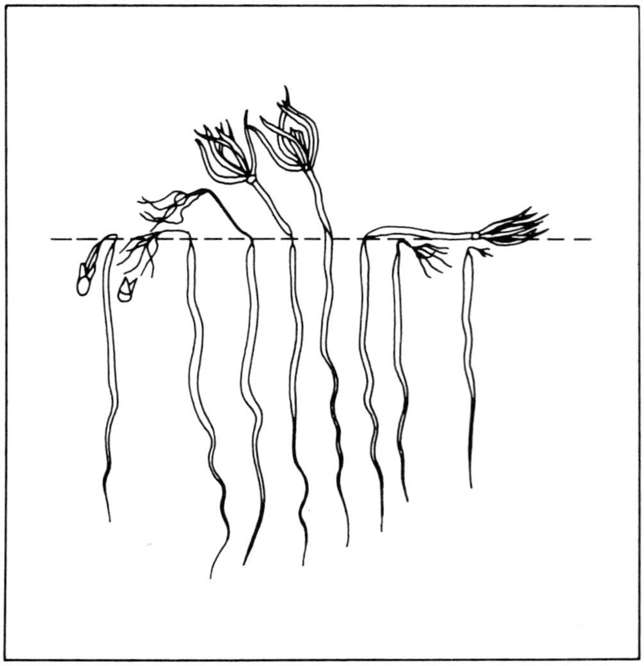

Figure 11. Pine seedling killed due to overheating at the soil surface (adapted from [41]).

12.5.1.2 Older seedlings

Older seedlings have developed a lignified outer stem that insulates sensitive tissues from hot soil—which is why older tree seedlings are more heat resistant than younger ones. But heat lesions may still develop on older seedlings, either on only one side (usually the south) or all around the stem. Fresh lesions are typically pale but sharply defined, at or just above the soil surface; older lesions on larger seedlings may be surrounded by slight swellings [54].

Indirect heat injury such as growth loss is more of a concern with older seedlings than direct heat injury. Although growth loss due to heat damage is hard to diagnose, it may be sizable, especially in low-elevation nurseries growing heat-sensitive species like true firs and spruce. Heat sensitivity can be expressed in growth curves from seedlings reared in growth chambers at various temperatures [53] (Fig. 12). Note that seedling growth is optimal within a certain temperature range, depending on species; growth of coastal redwood, which is particularly sensitive to heat, drops off dramatically outside a narrow optimum range (Fig. 12c).

12.5.2 Keeping seedlings cool

Sprinkling seedlings is the most common way to keep them cool, although other methods, like shading or mulching, also are employed.

Generally, soil temperature can be modified in two ways. First, the incoming or outgoing energy can be altered by: (1) placing an insulating layer, such as paper, mulch, screen, or glass, on or near the ground surface; (2) changing the absorptivity of the ground; or (3) varying the air temperature with a wind machine or shelterbelt. Second, the thermal properties of the ground can be modified by: (1) increasing or decreasing the absorptivity of the ground; (2) changing the thermal conductivity by rolling the seedbed before sowing or by cultivation and irrigation; (3) altering the heat capacity by adding or draining water; or (4) varying the rate of evaporation by removing weeds, regulating soil moisture, and placing mulch, screen, or sand on the soil surface [9].

12.5.2.1 Irrigation

Reducing soil-surface temperature with irrigation is the principal cooling method available to the forest-nursery manager. Managers need to know the soil-surface temperatures that cause irreversible damage to their crop at a given stage of development—and must prevent the soil from becoming that hot.

Irrigation is normally applied regularly after seed is sown and until it germinates to keep the soil surface moist. This not only prevents the development of a surface crust which the emerging seedlings may have difficulty penetrating but also assures adequate available moisture for seedling development. Additionally, irrigation cools seedbed surfaces as water evaporates. Evaporation is a powerful cooling process. Some 540 calories are required to convert 1 g of water at the boiling point (100°C) to vapor. When 1 g of water evaporates at 20°C, it absorbs 586 calories; at 30°C, it absorbs 580 calories. Irrigation during the heat of the day can reduce soil-surface temperatures by as much as 20°F (11°C) [36]. Once germination is complete, the intervals between irrigations gradually lengthen as seedlings develop deeper roots and are physiologically better equipped to endure normal moisture stresses. However, during this same period, the seedling is most susceptible to heat damage, and with some species in some locations, frequent, brief irrigations are needed to cool the soil surface.

An easy, reliable, and accurate method for determining soil-surface temperature is essential. Temperature is usually measured by a thermometer placed immediately below the soil surface. Dial thermometers can be read without disturbing the probe. Thermocouples or other sensing devices also can be used. An infrared thermometer would be an ideal device because no surface contact is needed, but it must be calibrated frequently. In any case, thermometers should be evenly distributed over the area monitored to account for variation in soil texture, slope, tilth, and soil color and to compensate for any instrument that may be defective.

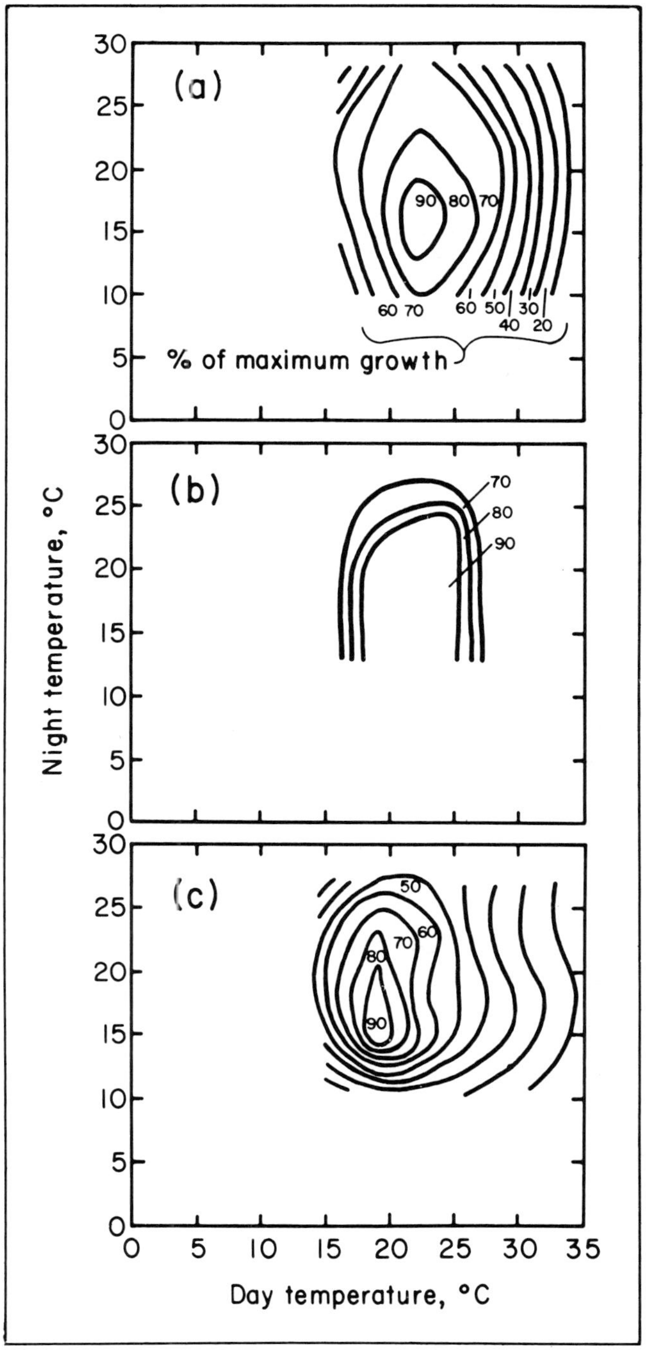

Figure 12. Curves indicating the effect of heat sensitivity on growth of (a) white spruce, (b) Douglas-fir, and (c) redwood (adapted from [53]). Values within curves represent the percentage of maximum possible growth.

In addition to cooling the soil surface, sprinkler irrigation during the hottest period of the day can drop the air temperature 10 to 15°F (5 to 8°C) or more around trees [36], cooling seedling foliage and reducing overall plant heat stress. The benefits of such cooling to seedling growth—in terms of increased photosynthesis, less photo-oxidation, or respirative energy loss—are hard to determine, but effects on growth could be estimated from the growth curves in Figure 12. Whether sprinkling in the heat of the afternoon is justified to enhance growth and development is debatable but probably deserves further study.

Guidelines for cooling seedlings with irrigation must be developed which consider seedling species, soil type, and climate. For example, the cooling regime for the U.S.D.A. Forest Service J. Herbert Stone Nursery in Medford, Oregon, begins with irrigation just after germination, in early June, when the soil-surface temperature is about 35°C, and continues through late August to the end of the summer growing season, when soil-surface temperature may reach 46°C [40]. Critical soil-surface temperatures are raised gradually through the growing season as seedlings develop greater heat tolerance; at the Stone Nursery, critical temperatures are increased by 1 to 2°C every 2 weeks. The cooling irrigation period is 30 minutes long when wind speed is 6 mph or less, 45 to 60 minutes long when wind speed exceeds 6 mph. Normally, this regime also provides all the water needed for normal seedling growth.

At the Coeur d'Alene (Idaho) Nursery, soil-surface temperature is never allowed to exceed 90°F during germination of Engelmann spruce (*Picea engelmannii* Parry ex Engelm.) and grand fir [*Abies grandis* (Dougl. ex D. Don) Lindl.] and for the first 2 weeks afterward, 95°F for the first 2 months, and 100°F for the first summer. Midday irrigation is applied as frequently as necessary to adhere to these rules and only long enough to reduce soil temperature to a stabilized lower level.

Overhead irrigation of pine seedlings in North Dakota, begun when soil temperature reached 120°F, reduced that temperature to 100°F after ½ hour of watering; the temperature reduction lasted for 4 hours or more (Fig. 13) [52].

About ⅓ of Northwest nursery managers irrigate to cool air temperature (OSU Nursery Survey), primarily to reduce seedling moisture stress and air temperature around plants and to enhance growth and development. This type of irrigation is not intended to prevent direct heat damage.

Over ¾ of Northwest nurseries surveyed use irrigation to reduce soil-surface temperatures. In most instances, irrigation to cool the soil surface was relatively brief, beginning in the 85 to 90°F range, with the critical temperature increasing gradually as seedlings became older and more heat tolerant. Differences among species were recognized; generally, spruces and true firs were regarded as more heat sensitive than Douglas-fir and pines. In sum, the variation in guidelines from one nursery to another suggests that nursery managers should experimentally determine what works best for their sites and the particular species grown there.

12.5.2.2 Shading

Shading prevents the buildup of high soil-surface temperatures by intercepting solar radiation and, in effect, insulating seedlings from the heat source. Though this technique is quite effective, the materials are expensive and their installation and removal labor intensive. Because shading simulates the early growth environment of many "shade-tolerant" species, "half-shade" (50% light transmission) has been recommended for many heat-sensitive species such as spruces, firs, and hemlocks in the Lake States [51] and the Great Plains [52].

Shading can be effected by suspending wooden snowfence on wires or boards over the sown seedbeds. However, snowfence is expensive and must be removed for cultural operations such as hand weeding; detailed information on this practice is given in Stoeckeler and Jones [51]. Woven polypropylene "shadecloth" can be used instead of snowfence; although somewhat more efficient and less cumbersome, it is still expensive. Wakeley [54] thoroughly tested seedbed shading in nurseries in the southern U.S. and found it to be costly and unnecessary for southern pines.

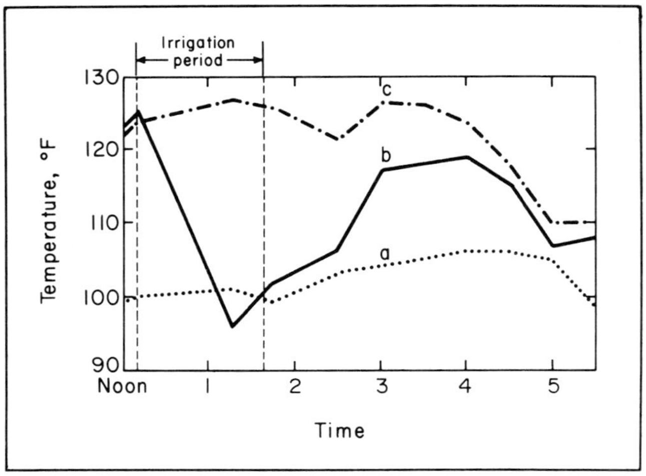

Figure 13. Effect of watering on soil-surface temperatures at Towner Nursery, North Dakota: (a) air temperature in shade; (b) soil-surface temperature in sun with 1½ hours of wetting; (c) soil-surface temperature in sun with no wetting (adapted from [52]).

"High shade" is employed at some nurseries that rear very heat-sensitive tree species. In such cases, shadecloth is suspended on cables between utility poles high enough to clear sprinkler irrigation and nursery equipment. However, most tree nurseries have abandoned costly shading in favor of irrigation cooling. As the value of tree crops rises due to increased seed values and nursery capital investments, shading—especially mechanized versions or "high shade"—may become economically viable at some nurseries, especially where irrigation water is scarce or where very sensitive or high-value species are grown.

12.5.2.3 Mulching

Different soils vary considerably in their ability to conduct and dissipate heat (Fig. 14). Sandy soils are very porous, so the sun's energy is concentrated in a thin surface layer which can

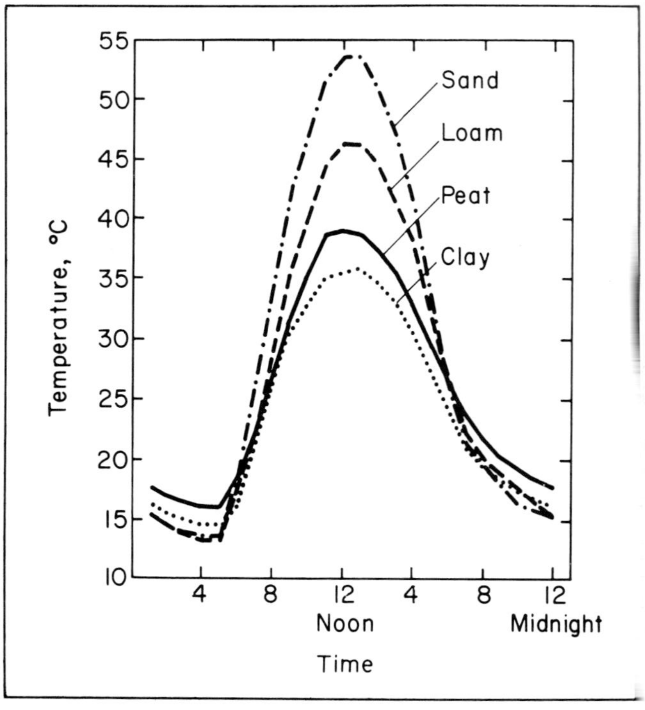

Figure 14. Daily pattern of surface temperatures of different soil types at Sapporo, Japan (adapted from [57]).

reach temperatures exceeding 50°C (122°F). Finer textured soils such as clays contain less air space and can therefore conduct heat away from the surface.

Mulches can either foster or hinder temperature buildup in soil surfaces. A thin, light-colored layer of fibrous or particulate organic matter can help hold down surface temperatures in newly sown seedbeds. A light-colored mulch has a higher heat capacity, greater surface area, and higher albedo (reflectivity) than most nursery soils and can reflect or dissipate incoming solar radiation. A dark porous mulch, on the other hand, absorbs more sunlight and can allow soil-surface temperatures to rise high enough to damage young seedlings.

12.6 Controlling Seedling Dormancy with Irrigation

12.6.1 Importance of top dormancy

Northwest nursery managers realize the importance of encouraging seedling dormancy in late summer so that the trees can become frost hardy before the advent of freezing weather and be adequately hardy for lifting and storage (see chapters 14, 15, and 21, this volume).

Frost hardiness develops in two stages. The first stage prepares seedlings for temperatures down to −1 to −2°C [7, 31]; in this stage, physiological activity actually increases [48] and moderate moisture stress is apparently tolerated without negative effect, although severe stress is disruptive [7]. Though genetically controlled, the first stage is triggered by photoperiod and temperature. The second stage of hardening is actually triggered by frost and, once invoked, proceeds rapidly. No photoperiodic or translocatable factors are involved, and the numerous physiological changes occurring at that time make protoplasm more resilient [31]. But frost hardiness cannot develop in actively growing seedlings.

For many temperate-zone conifers, the relatively long, cool nights of late summer and early fall provide the required stimulus for inducing top dormancy. However, Lavender [29] found that moisture stress inhibited top growth and induced dormancy in conifer species such as Douglas-fir, white fir [*Abies concolor* (Gord. & Glend.) Lindl. ex Hildebr.], and blue spruce (*Picea pungens* Engelm.). This is of particular interest to nursery managers: because irrigation can extend active shoot growth of Douglas-fir and ponderosa pine well into the fall [30], seedlings will be subject to frost injury unless nurseries develop irrigation schedules that include a period of late-summer moisture stress to induce seedling dormancy before the first fall frosts.

12.6.2 Scheduling irrigation to induce dormancy

Obviously, no single irrigation schedule will work at all nurseries because of differences in climate, cultural regimes, species, and genotypes (see chapter 15, this volume). One of the few actual examples of an irrigation regime designed to induce seedling dormancy was developed over a 5-year period at the D. L. Phipps Oregon State Nursery at Elkton, Oregon [58]. These researchers used a pressure chamber to monitor predawn Ψ_{plant} in 2+0 Douglas-fir seedlings for three different watering regimes: wet (−5 bars), medium (−8 bars), and dry (−15 bars). Intermediate tests revealed that wet-regime seedlings were not sufficiently dormant in fall but that dry-regime seedlings were too small to meet minimum size standards. After additional testing, an intermediate water regime was found to produce a balance between seedling size and dormancy. The final irrigation schedules emphasize the importance of seedling size class and timing of the various irrigation stress treatments (adapted from [58]):

Seedling class	Predawn Ψ_{plant}, bars			
	Until July 9	After July 9	About Aug 3	About Aug 20
1+0	−5	−10	−12	−15
	Until June 1	June 1-15	After June 15	
2+0	−5	−8 to −10	−15[1]	
	Until July 1	July 1-Aug 1	After Aug 1	
2+1	−7	−10	−15	

[1]In seedlings held for 2+1, keep predawn Ψ_{plant} between −10 and −15 bars during this period.

Blake et al. [5] also studied the effect of moisture on seedling dormancy and found that lower Ψ_{plant} between mid-July and late August consistently halted shoot growth and increased frost hardiness of Douglas-fir seedlings. Of particular interest is their observation that delaying the moisture stress treatment until late August interfered with hardiness development; they concluded that this stress must be relieved in the fall if frost hardiness is to develop properly. Supplemental irrigation may even be needed to complete the first stage of hardening before the weather becomes too cold [30].

Zaerr et al. [58] emphasize that any irrigation schedule is only a guide and must take seedling growth rates and nursery climate into account. They recommend the following principles to regulate seedling growth, induce dormancy, and enhance frost hardiness:

- Monitor predawn Ψ_{plant} to schedule irrigation
- Promote shoot growth early in the season
- Induce dormancy in late summer by gradually increasing moisture-stress levels
- Deepen dormancy by relieving stress in early fall
- Tailor irrigation schedules to accommodate the soil, climate, species, seedling class, and cultural practices of the particular nursery

Twenty of 21 (95%) Northwest nurseries reduce watering to harden seedlings in fall (OSU Nursery Survey). Of those 20, approximately half monitor Ψ_{plant} to schedule irrigation, and half cease irrigation after seedlings reach a certain size or on a certain date.

There was no consensus regarding regimes among nursery managers who monitor Ψ_{plant} to induce dormancy—which reflects both site variation among nurseries and a difference in opinion about the required stress levels. The various regimes can be summarized as follows:

- 46% of the nurseries allow predawn Ψ_{plant} to decrease to the −10 to −15 bar range in June or early July and maintain this stress level until the fall rains begin
- 27% use the calendar date of August 1 to allow *midday* Ψ_{plant} to fall to −20 bars
- 27% allow predawn PMS to fall to −8 to −10 bars by July 15 and then usually begin to irrigate again after budset in August or September

Nursery managers using calendar date as a criterion for irrigation scheduling felt that irrigation should be reduced after root pruning in late June or early July and stopped altogether in September.

Although all nursery personnel clearly were aware of the importance of inducing seedling dormancy with moisture stress, there were obvious differences in the timing and magnitude of stress levels. More research in this area is needed.

12.7 Conclusions and Recommendations

- The main objective of irrigation is to avoid unwanted seedling moisture stress—which can vary from a small decrease in water potential to death by desiccation. Knowing when and when *not* to water should help nursery managers implement the most effective irrigation monitoring and application programs possible.
- Soil water potential—a combination of matric, osmotic, pressure, and gravitational potentials—as well as field capacity and permanent wilting percentage are soil-water characteristics significant to plant growth. As water held in soil pores is depleted, matric potential decreases, and soil pores initially filled with water become filled with air. The secret of effective nursery irrigation is to keep soil pores filled with the proper balance of both water and air to minimize seedling moisture stress.
- Plant water potential is the single most useful measure of moisture stress in plants. Predawn readings, which indirectly measure soil water potential, are the most stable; midday readings are the second most stable, but are more difficult to interpret because they reflect not only soil water potential but also atmospheric evaporative demand and physiological response through stomatal closure.
- Every nursery should use a quantifiable method to monitor available soil water or internal seedling moisture stress. Tensiometers are best for measuring soil water in forest-tree nurseries because they cover the critical 0 to −0.8 bar range, and pressure chambers are best for directly measuring plant water potential. Soil-moisture retention curves (which relate matric potential to percentage of soil moisture by weight), soil- and plant-moisture monitoring procedures, and careful observation together form the best approach for properly monitoring and controlling irrigation—assuming the irrigation system is a good one. However, because crop responses vary due to environmental modification, nursery climate, tree species, and seed source, managers need phenological information to fully anticipate seedling response to cultural operations.
- Seedlings must be protected from the damaging effects of frost resulting from intracellular water crystallizing to ice and rupturing cell membranes. Overhead irrigation sprinkling is the most common frost-protection method and is the most effective, unless high winds cause poor sprinkler coverage. The success of sprinkling largely depends on the amount and frequency of water application; to avoid the damage that can occur from improper application, continue to irrigate until the temperature rises above 0°C. Heaters, wind machines, brushing, sanding, and windbreaks are used for agricultural frost protection at varying costs with varying degrees of success.
- Heat injury to seedlings can be direct (due to cell-membrane injury or decomposition) or indirect (due to metabolic disturbances). The soil surface around young seedlings must be kept cool to prevent heat damage, especially where the nursery has high insolation rates; heavy, dark soils or mulches; or poor air circulation. Older seedlings, whose lignified outer stem helps insulate tissues from hot soil, may suffer indirect injury such as growth loss and therefore also warrant attention. Sprinkling seedlings with irrigation water is the most common, effective method for keeping soil-surface temperatures down, although shading and mulching also have been used; in addition, sprinkling helps lower air temperature, reducing overall heat stress. However, guidelines for cooling seedlings with irrigation must take into account species, soil type, and climate.
- Nursery managers should encourage top dormancy of seedlings in late summer so that trees can become frost hardy long before the first fall frost. Frost hardiness, which progresses in two stages, cannot develop in actively growing seedlings. Scheduling irrigation to produce a *moderate* level of moisture stress allows managers to aid dormancy induction. Monitoring predawn plant water potential to schedule irrigation, promoting shoot growth early in the season, inducing dormancy in late summer by gradually increasing moisture-stress levels, deepening dormancy in early fall by relieving moisture stress (through irrigation, if necessary), and tailoring irrigation schedules to soil and stock types, climate, and cultural practices should assure the desired seedling growth and enhance frost hardiness.

References

1. Abbott, H. G., and D. S. Fitch. 1977. Forest nursery practices in the United States. J. Forestry 3:141-145.
2. Armson, K. A. 1972. Distribution of conifer roots in a nursery soil. Forestry Chronicle 48:141-143.
3. Armson, K. A., and V. Sadreika. 1979. Forest tree nursery soil management and related practices. (Metric ed.) Ontario Ministry of Natural Resources, Toronto. 179 p.
4. Belcher, E. W. 1975. Influence of substrate moisture level on the germination of seed of selected *Pinus* species. Seed Sci. and Technology 3:597-604.
5. Blake, J., J. Zaerr, and S. Hee. 1979. Controlled moisture stress to improve cold hardiness and morphology of Douglas-fir seedlings. Forest Sci. 25:576-582.
6. Boyer, J. S. 1969. Measurement of water status in plants. Annual Review of Physiology 20:351-364.
7. Burke, M. J. 1978. Comments on frost protection and heat injury. From lectures on environmental requirements of crop and horticultural plants. Horticulture 675. Dep. of Horticulture, Colorado State Univ., Fort Collins.
8. Businger, J. A. 1965. Frost protection with irrigation. Meteorological Monographs 6(28):74-80.
9. Chang, J. 1968. Climate and agriculture: an ecological survey. Aldine Publishing Co., Chicago, Illinois. 296 p.
10. Cleary, B., and J. Zaerr. 1980. Pressure chamber techniques for monitoring and evaluating seedling water status. New Zealand J. Forestry Sci. 10(1):133-141.
11. Cope, F., and E. Trickett. 1965. Measuring soil moisture. Soils and Fertilizers 28:201-208.
12. Day, R. J. 1980. Effective nursery irrigation depends on regulation of soil moisture and aeration. Pages 52-69 *in* Proc., North American forest tree nursery soils workshop (L. P. Abrahamson and D. H. Bickelhaupt, eds.). State Univ. New York, Coll. Environ. Sci. and Forestry, Syracuse.
13. Day, R. J., and G. R. MacGillivray. 1975. Root regeneration of fall-lifted white spruce nursery stock in relation to soil moisture content. Forestry Chronicle 51:196-199.
14. Day, R. J., and J. Paisley. 1977. Soil moisture tension curve for Dryden Tree Nursery. Lakehead Univ., Thunder Bay, Ontario. School of Forestry Silviculture Rep. 1977-3. 4 p.
15. Day, R. J., J. T. Stupendick, and J. M. Butler. 1976. Root periodicity and root regeneration potential are keys to successful plantation establishment. *In* Proc., Ontario Ministry of Natural Resources—Great Lakes Forest Res. Centre, Plantation establishment symp. Lakehead Univ., Thunder Bay, Ontario.
16. Day, R. J., and S. J. Walsh. 1980. A manual for using the pressure chamber in nurseries and plantations. Lakehead Univ., Thunder Bay, Ontario. School of Forestry Silviculture Rep. 1980-2. 49 p.
17. Glerum, C., and G. Pierpoint. 1968. The influence of soil moisture deficits on seedling growth of three coniferous species. Forestry Chronicle 44:76-79.
18. Hausenbuiller, R. L. 1972. Soil science, principles and practices. Wm. C. Brown Co., Dubuque, Iowa. 483 p.
19. Hsiao, T. C. 1973. Plant responses to water stress. Annual Review of Plant Physiology 24:519-570.
20. Hsiao, T. C., E. Acevedo, E. Fereres, and D. W. Henderson. 1976. Stress metabolism: water stress, growth, and osmotic adjustment. Philosophical Trans. of the Royal Society of London, Series B 273:479-500.

21. Kaku, S., and M. Iwaya. 1978. Low temperature exotherms in xylems of evergreen and deciduous broadleaved trees in Japan with references to freezing resistance and distribution range. Pages 227-240 *in* Plant cold hardiness and freezing stress (P. H. Li and A. Sakai, eds.). Academic Press, New York. 405 p.

22. King, K. M. 1967. Soil-moisture instrumentation, measurement and general principles of network design. Pages 269-285 *in* Proc., Soil moisture. Can. Dep. of Energy, Mines, and Resources, Ottawa. Hydrology symp. 6.

23. Kozlowski, T. T. (ed.). 1968. Water deficits and plant growth. Vol. 1. Academic Press, New York. 333 p.

24. Kramer, P. J. 1937. The relation between rate of transpiration and rate of absorption of water in plants. American J. Botany 24:10-15.

25. Kramer, P. J. 1969. Plant and soil water relationships: a modern synthesis. McGraw-Hill Book Co., New York. 390 p.

26. Kramer, P. J., and T. T. Kozlowski. 1960. Physiology of trees. McGraw-Hill Book Co., New York. 642 p.

27. Krugman, S. L., W. I. Stein, and D. M. Schmitt. 1974. Seed biology. Pages 5-40 *in* Seeds of woody plants in the United States (C. S. Schopmeyer, ed.). U.S. Dep. Agric., Washington, D.C. Agric. Handb. 450.

28. Larcher, W. 1969. Zunahme des frost abhartungsvermogen von *Quercus ilex* im laufe der individualent wicklung. Planta 88:130-135.

29. Lavender, D. P. 1981. Environment and shoot growth of woody plants. Forest Res. Lab., Oregon State Univ., Corvallis. Res. Pap. 45. 47 p.

30. Lavender, D. P., and B. D. Cleary. 1974. Conifer seedling production techniques to improve seedling establishment. Pages 177-180 *in* Proc., North American containerized forest tree seedling symp. (R. W. Tinus, W. I. Stein, and W. E. Balmer, eds.). Great Plains Agric. Council Publ. 68.

31. Leopold, A. C., and P. E. Kriedemann. 1975. Plant growth and development. 2nd ed. McGraw-Hill Book Co., New York. 545 p.

32. Levitt, J. 1956. The hardiness of plants. Academic Press, New York. 278 p.

33. Levitt, J. 1972. Responses of plants to environmental stresses. Academic Press, New York. 665 p.

34. Lindow, S. E., D. C. Arny, C. D. Upper, and W. R. Barchet. 1978. The role of bacterial ice nuclei in frost injury to frost sensitive plants. Pages 249-266 *in* Plant cold hardiness and freezing stress (P. H. Li and A. Sakai, eds.). Academic Press, New York.

35. Lull, H. W., and K. G. Reinhart. 1955. Soil moisture measurement. U.S.D.A. Forest Serv., Southern Forest Exp. Sta., New Orleans, Louisiana. Occasional Pap. 140. 56 p.

36. May, J. T. 1984. Soil moisture. *In* Southern pine nursery handbook (C. W. Lantz, ed.). U.S.D.A. Forest Serv., Southern Region, Atlanta, Georgia. (In press.)

37. McClain, K. M. 1973. Growth response of some conifer seedlings to regimes of soil moisture and fertility under greenhouse and field conditions. Master's thesis, Faculty of Forestry, Univ. of Toronto.

38. McDonald, S. E. 1978. Irrigation monitoring in western forest tree nurseries. Pages B-16 to B-49 *in* Proc., Western Forest Nursery Council and Intermountain Nurserymen's Assoc. combined nurserymen's conf. and seed-processing workshop, Eureka, California, Aug. 7-11. U.S.D.A. Forest Serv., State and Private Forestry, San Francisco.

39. McDonald, S. E., and S. W. Running. 1979. Monitoring irrigation in western forest tree nurseries. U.S.D.A. Forest Serv., Rocky Mountain Forest and Range Exp. Sta., Fort Collins, Colorado. Gen. Tech. Rep. RM-61. 8 p.

40. Morby, F. E. 1982. Irrigation regimes in a bare-root nursery. Pages 55-59 *in* Proc., Intermountain Nurserymen's Assoc. meeting (R. F. Huber, ed.). Can. Forestry Serv., Edmonton, Alberta. Inf. Rep. NOR-X-241.

41. Munch, E. 1913. Hitzeschaden an Waldpflanzen. Naturwissenschaftliche Zeitschrift fuer Forst.- und Landwirtschaft 11:557-562.

42. Parker, N. W. 1946. Environment factors and their control in plant environment. Soil Sci. 62:109-119.

43. Phillip, J. R. 1957. The physical principles of soil water movement during the irrigation cycle. Proc., International Congress on Irrigation Drainage 8:124-154.

44. Richards, F. J. 1969. The quantitative analysis of growth. Pages 3-76 *in* Plant physiology. Vol. 5A. Analysis of growth: behavior of plants and their organs (F. C. Steward, ed.). Academic Press, New York.

45. Ritchie, G. A., and T. M. Hinckley. 1975. The pressure chamber as an instrument in ecological research. Advances in Ecological Res. 9:165-254.

46. Salisbury, F. B., and C. W. Ross. 1978. Plant physiology. 2nd ed. Wadsworth Publ. Co., Belmont, California. 422 p.

47. Scholander, P. F., H. T. Hammel, E. D. Bradstreet, and E. A. Hemmingsen. 1965. Sap pressure in vascular plants. Science 148:339-346.

48. Siminovich, D., C. M. Wilson, and D. R. Briggs. 1953. Studies on the chemistry of living bark in relation to frost hardiness, VII. Plant Physiology 29:331-337.

49. Sinnott, E. W. 1960. Plant morphogenesis. McGraw-Hill Book Co., New York.

50. Slatyer, R. O. 1967. Plant-water relationships. Academic Press, New York. 366 p.

51. Stoeckeler, J. H., and G. W. Jones. 1957. Forest nursery practice in the Lake States. U.S. Dep. Agric., Washington, D.C. Agric. Handb. 110. 124 p.

52. Stoeckeler, J. H., and P. E. Slabaugh. 1965. Conifer nursery practice in the Prairie-Plains. U.S. Dep. Agric., Washington, D.C. Agric. Handb. 279. 93 p.

53. Tinus, R. W., and S. E. McDonald. 1979. How to grow tree seedlings in containers in greenhouses. U.S.D.A. Forest Serv., Rocky Mountain Forest and Range Exp. Sta., Fort Collins, Colorado. Gen. Tech. Rep. RM-60. 256 p.

54. Wakeley, P. C. 1954. Planting the southern pines. U.S. Dep. Agric., Washington, D.C. Agric. Monograph 18. 233 p.

55. Weibe, H. H., G. S. Campbell, W. H. Gardner, S. L. Rawlins, J. W. Cory, and R. W. Brown. 1971. Measurement of plant and soil water status. Utah Univ., Logan. Agric. Exp. Sta. Bull. 484. 71 p.

56. Wilson, R. G. 1971. Methods of measuring soil moisture. International Field Year for the Great Lakes Tech. Manual Series. No. 1. 20 p.

57. Yakuwa, R. 1946. Uber die boden temperaturen in den verschiedenen bodenarten in Hokkaido. Geophysical Magazine (Tokyo) 14:1-12.

58. Zaerr, J. B., B. D. Cleary, and J. L. Jenkinson. 1981. Scheduling irrigation to induce seedling dormancy. Pages 74-79 *in* Proc., Joint meeting, Intermountain Nurserymen's Assoc. and Western Forest Nursery Council, Boise, Idaho, Aug. 12-14, 1980. U.S.D.A. Forest Serv., Intermountain Forest and Range Exp. Sta., Ogden, Utah. Gen. Tech. Rep. INT-109. 148 p.

Chapter 13
Land Drainage

D. E. Boyer

Abstract

Under certain nursery soil conditions, land-drainage improvements and surface-water control can enhance seedling quality and production and provide greater operational flexibility. Good drainage minimizes soil compaction, allows maintenance of proper soil chemical and biological conditions, reduces surface runoff and erosion, decreases nutrient losses, and diminishes occurrence of soil pathogens. However, water management can be complex because sources of water include surface water from adjacent lands, paved areas, and wheel tracks; subsurface flows; and seasonal or perennial water tables. Good drainage-system design and installation rely on careful and complete analysis using all available resources, including aerial photography, topographic maps, field-by-field observations, and expert technical assistance.

13.1 Introduction

Due to the typical location of conifer nurseries and the nature of their operations, adequate surface and subsurface water drainage is a common concern (OSU Nursery Survey; see chapter 1, this volume). The percentage of land within the nursery that can receive, store, and transmit water for plant growth is usually out of balance with those areas causing rapid runoff or ponding such as roads, wheel tracks, and buildings. This frequently results in surface erosion, sediment production, and wet spots. Surface erosion and sediment production are often subtle expressions that are easily overlooked or accepted as part of the cost of doing business. However, wet spots impair trafficability and productivity because machinery can become mired down and operations disrupted, or seedling growth can be retarded.

Nursery soil-management techniques should take into account the source, quality, quantity, and timing of all waters added to the site. The natural or altered physical properties of the soil and its ability to utilize those waters are of equal importance. Because water quality is seldom a major problem in the Northwest, this discussion will be confined to other factors.

In this chapter, the impacts of poor land drainage and surface-water control are discussed and some typical drainage problems presented so that nursery managers may better combat the conditions they are likely to encounter and understand the management implications of land-drainage improvements.

13.2 Drainage Impacts

The physical properties of soils largely dictate their internal drainage characteristics. Two important properties—infiltration rate (the ability of a soil to absorb water) and percolation rate (downward movement of water through a soil)—can be affected by soil-forming processes and may be altered through heavy use (see chapter 6, this volume).

Under typical irrigation and cultivation practices, the most significant changes in physical properties take place in the wheel tracks between seedbeds and at the base of cultivation depths within seedbeds. When percolation rates are slow or become reduced, especially within seedbeds, the surface soil layer tends to stay moist for longer than normal periods. This condition affects biological and chemical properties within the seedbed; furthermore, the excess water can move laterally into the wheel-track zone, causing a loss of bearing strength necessary to support machinery. But maintenance or improvement of existing physical characteristics, along with careful surface-water management, can protect or enhance productivity and operations.

In Duryea, Mary L., and Thomas D. Landis (eds.). 1984. *Forest Nursery Manual: Production of Bareroot Seedlings.* Martinus Nijhoff/Dr W. Junk Publishers, The Hague/Boston/Lancaster, for Forest Research Laboratory, Oregon State University, Corvallis. 386 p.

13.2.1 Detriments

- **Greater soil compaction:** When subjected to continuous cultivation or frequent machinery traffic at optimum moisture contents, even the most resistant soils can become densified to the extent that internal drainage rates are diminished. Densified layers can form at the soil surface, where machinery wheels contact soil, and extend downward, or they can form at the base of the tillage operating plane. For example, rototilling at optimum moisture levels but repetitious depths can create a densified layer, commonly referred to as a "traffic pan" or "plow sole," at that depth; this layer can extend downward nearly twice that depth. Under such conditions, soil water contents can reach a saturation level above the densified layer whereby most of the pore space is filled with water and soil air is excluded.

- **Impaired chemical and biological conditions:** Detrimental effects of saturated soils are: (1) low pH levels and excess soluble manganese, which can become toxic to plants; (2) retardation of organic matter decomposition and mineralization of organically bound nutrients; (3) release of organic sulfur as toxic hydrogen sulfide; (4) denitrification, which converts nitrates to volatile forms of nitrogen (N) that are lost from the soil; and (5) promotion of pathogens [4].

- **Increased runoff and erosion:** When infiltration rates are reduced, the opportunity for surface runoff increases dramatically. Erosion may not be significant within the wheel-track zone, where this condition would be expected to occur, but runoff waters can either inundate adjacent areas or provide the energy to cause erosion on downslope areas. In seedbeds where percolation rates are diminished, soils become increasingly wetter, lose their resistance to detachability, and increase their susceptibility to transport.

13.2.2 Benefits

- **Enhanced operational efficiency:** Well-drained soil profiles can permit considerable flexibility in tractor access. Installing a shallow drainage system with closely spaced pipelines, along with soil-management practices such as subsoiling, can increase downward water movement. For example, one nursery manager reported that after the installation of his drainage system, he could enter his fields within 24 hours after a heavy rain—as opposed to a week previously [pers. commun., 3]. This particular system was installed 36 inches deep, on a 20-ft spacing, in sandy, flood-plain soils whose water table was associated with a rise in streamflow levels.

- **Warmer soil temperatures:** Properly drained soils warm earlier in the spring, permitting earlier sowing. Wet soils warm more slowly because water requires 4 to 5 times more heat to raise a unit weight 1° than is needed for the same weight of mineral soil. Plant growth and all chemical reactions are slowed approximately 25% for each 10°F that temperature drops [2]. Lyon et al. [6] reported that inadequately drained surface soil may be from 5 to 15°F cooler than contiguous areas relieved of excess water.

- **More uniform soil moisture:** Proper drainage allows soil moisture to be distributed more evenly over the entire field, eliminating wet spots. This permits earlier, more predictable, and more efficient tilling [2]. Northwest soils are inherently variable in their water-transmission characteristics. Layered soils with different textures and structure can temporarily restrict water movement. However, drainage installations with closely spaced pipelines favor the disruption of these layers and increase downward water movement. The net result is a soil profile uniform in moisture content.

- **Decreased soil-N losses:** Saturated soils create anaerobic (lacking oxygen) conditions which favor denitrification. Although some N is lost through the drainage systems themselves, most of those losses are not nearly as significant as the ones attributed to the combined effects of denitrification and lack of oxygen in wet soils.

- **Fewer soil pathogens:** Some diseases are particularly enhanced by excessive soil moisture or an irregular water supply—for example, *Pythium* and other damping-off fungi [8]. Well-drained soils tend to favor a balanced mixture of biotic populations, rather than to promote a few species.

- **Reduced surface erosion:** The loss of topsoil and the effect of that loss on productivity are difficult to assess. At present, the U.S.D.A. Soil Conservation Service has proposed soil-erosion tolerance levels for agricultural lands which allow from 3 to 5 tons/acre/year of topsoil to be lost. Considering the value of conifer nursery land and the fraction of an inch that it takes to produce 5 tons of topsoil, that level of loss may be more than the nursery manager is willing to accept. Erosion can be reduced on a well-drained soil by increasing the soil's capacity to hold water, thereby reducing runoff.

13.3 Common Drainage Problems and their Remedies

The solution to an excess water problem usually defies a "cookbook" approach. Most solutions are site specific. In addition to the technical aspects of drainage design and installation, there are often political or legal considerations, especially where the contributing lands and the receiving lands are not under one ownership.

Various problems and their possible solutions are offered here as examples. Although each problem requires a different investigation and design approach, all have this in common: the source of the excess water must be located and the feasibility of diverting or relieving it determined.

13.3.1 Surface water

The main objective in handling surface waters is to move them off the nursery site as quickly as possible without impairing the water quality of any receiving streams.

13.3.1.1 From adjacent lands

Nurseries located immediately downslope from steep uplands often experience surface-water runoff, as do lands subjected to some cultural practice such as grazing. Runoff may be from confined water flows, such as small creeks or drainageways, or from unconfined flows, such as slope wash or seepage.

This condition usually calls for a diversion system or physical structures to carry water through or around nursery beds. In one instance, surface water originating from an upslope watershed was confined to a channel that discharged water, at peak flow, onto the nursery; at low flows, it disappeared into the channel and emerged downslope as seepage. The recommended solution was to capture and contain the water in a concrete ditch, diverting it to an off-site discharge point. This corrective measure was expensive but appeared to be the most efficient and effective [1].

13.3.1.2 From paved areas, wheel tracks, and buildings

These sources of runoff, which also may be from confined or unconfined flows, probably account for most of the drainage problems within Northwest nurseries. As such, they offer the greatest opportunity for imaginative solutions.

Water collected on roads or in "permanent" wheel tracks has to be discharged somewhere but, unfortunately, often ends up in nursery beds or pools on roads (Fig. 1). In some situations, roads can be constructed so that they slope toward the center line, collecting and transporting the water off site. If the volumes of water are significant, a sediment pond also may be needed before the water is discharged into a stream. In some nurseries, surface erosion accompanies excess water accumulations. Remedies might include grassed waterways or temporary dams made of plastic-tied straw bales. Sandbagging has also been used at the end of seedling beds but can interfere with traffic movement; the bags should be small enough so they can be moved out of the way when necessary. In one nursery, water from seedling beds discharged onto perimeter roads and caused wet spots. It was recommended that the perimeter roads have constructed rock dips in the travel way to allow passage of water while maintaining a running surface for vehicles [1].

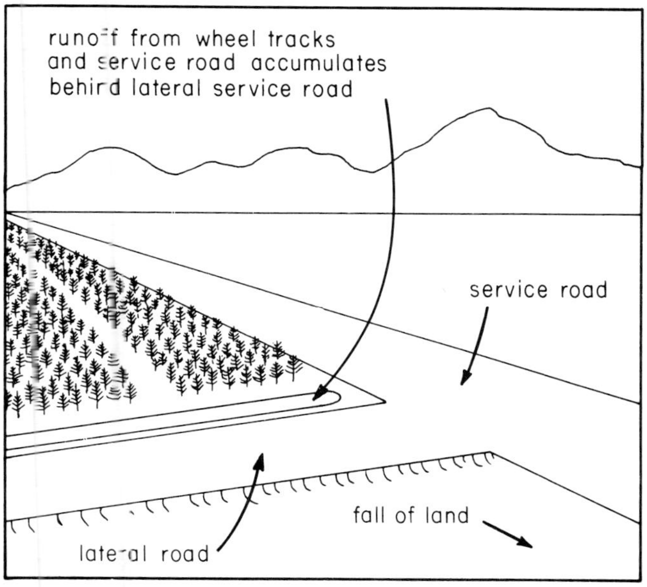

Figure 1. Surface-water collection against a service road.

13.3.2 Topographic depressions

Because land-forming processes seldom produce perfectly shaped landscapes, depressions are a common occurrence. Included in this category are man-made depressions, or catchments.

Although extensive land leveling is generally not recommended due to its negative impacts on soil nutrients, it is often advantageous to grade and smooth land to achieve uniform water application and eliminate the possibilities of surface-water accumulation. Occasionally, however, extensive leveling may be necessary; when the nursery manager is faced with this situation, the topsoil should be removed, stockpiled, and reapplied. Even though this is an expensive practice, it will pay in the long run.

Smoothing is simply the elimination of minor ridges and depressions in the field without altering the general topography. Smoothing usually requires 2 years to complete; land is graded and smoothed the first year and then graded and smoothed again the second year, after the soil has settled. The quality of the smoothing is best viewed immediately after a storm [2]. If puddles persist, the smoothing is not adequate and should be redone, provided that irregularities in topography—and not compaction—are the culprit. At any rate, any existing depressions should be investigated to see if soil is draining adequately; if it is not, a subsurface drainage system may also be required.

13.3.3 Subsurface water from adjacent lands

Some of the landforms large enough to support a nursery are also known for having unpredictable subsurface water flows; alluvial-colluvial fans and toeslopes are the most common landforms encountered (Fig. 2). Subsurface flows include seepage from ditches, reservoirs, and high-gradient streams. Usually, these water-table conditions are localized and have a gradient or hydraulic head behind the water flow.

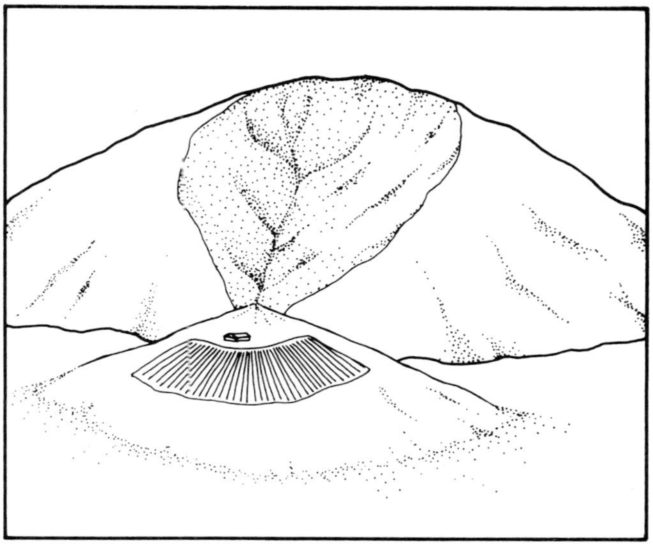

Figure 2. Nursery located on an alluvial-colluvial fan at the base of a watershed.

The normal intent of subsurface drainage is to lower the groundwater level in the soil or to prevent waterlogging from seepage. Two principal types of systems are used: open and closed. Open systems consist of one or more ditches that border or transect the land being drained; they are continuous and have a disposal system that carries the water to a natural drainageway. Closed systems, for the most part, consist of interconnected pipelines located below the water-table level, where they can collect and transport water to an open drainageway [4]. Heavy-textured soils, composed of expanding clays, can only be drained by open systems. However, because most Northwest nurseries are not located on clayey soils, this discussion will concentrate on closed systems (OSU Nursery Survey). Further, the maintenance problems associated with open ditches (e.g., weeds and rodents) are so numerous that such systems are usually discouraged.

There are two main types of closed systems: interception and relief. Interception systems are common for handling subsurface water from adjacent lands but are usually confined to small land areas. They may consist of a simple line of pipes located at the seepage source or the lower edge of an off-site slope, or they may have complex grids. Interception drains are similar in effect to surface drainage systems in that they remove water before it enters the soil. In contrast, relief systems drain already saturated soils. They may consist of a single pipeline or a complex grid, depending upon the size of the affected area, and can be installed in isolated depressions or soft spots that might be remedied by land smoothing. Whether for interception or relief systems, complex grids may have parallel, gridiron, or herringbone patterns (Fig. 3). The lay of the land, direction of subsurface water flow, and irregularities of subsoil characteristics dictate the layout to be used.

Drainage spacing and depth are functions of soil characteristics, chiefly percolation, and the type of drainage problem encountered. Pipeline size is a function of inflow rate and should also be determined on site.

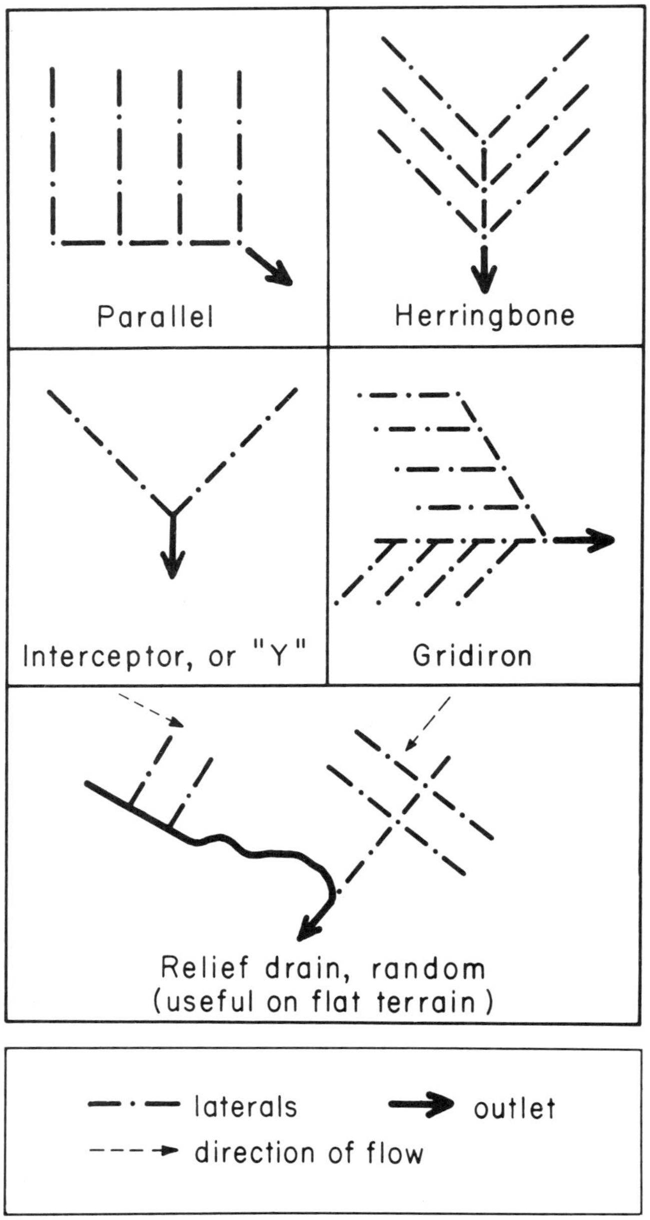

Figure 3. Typical drainage patterns (adapted from [4]).

13.3.4 Seasonal or perennial water table within the plant influence zone

The plant influence zone is usually considered to be the upper 5 feet of soil. In the Northwest, wetting of this top 5 feet is not uncommon because of abundant rainfall and irrigation. Seasonal or perennial water tables within the plant influence zone are most prevalent on flat-lying terrain adjacent to a body of water, such as is found in valley-fill positions (Fig. 4). The water flow is of a low gradient and usually fluctuates in response to changes in streamflow or lake level.

Intercepting water that moves in response to changes in streamflow or large expanses of water-table intrusion are considerably more complicated than other drainage problems because the required drainage system must have a discharge outlet lower than the system itself. On flat-lying terrain, this is often difficult to achieve. In one case, the elevation for an outlet was not suitable, and a sump basin with a pump was recommended. For these conditions, an open ditch also may be an alternative.

13.3.5 Impeded or slow water movement within the soil profile

This particular condition is common to "perched" water tables but may also affect soils that are naturally or artificially compacted (Fig. 5). Where downward drainage is blocked by an impermeable layer, a perched water table may form. If a perched water table exists, a complex grid with closely spaced pipelines may be the best drainage solution. If the soil conditions are not as severe as in the perched water-table situation but seasonal saturation occurs, there are new advances in drainage design that offer "controlled" water levels. Pipelines may be installed according to a variety of patterns (see Fig. 3) at shallow depths, depending on the crops to be grown and the cultural practices to be used. These shallow-depth drainage systems with closely spaced pipelines offer some potential in nursery fields where the timing of operations is critical. Furthermore, an elaborate field investigation seldom is needed for designing such systems, although several factors, such as slope available to the drainage lines, total acreage to be drained, and desired water-removal rates, need to be determined. Water-removal rates (i.e., drainage coefficients) of ½, ⅜, and ¼ inch in 24 hours—the basis for agricultural lands—would be sufficient for nurseries. For example, for a 40-acre field on a 2 ft/1,000 ft slope, the ¼ inch/24-hour rate would require an 8-inch main drain line, and the ½ inch/24-hour rate a 10-inch main drain line.

Drainage of seasonally saturated soils may be expensive but can provide benefits that include early sowing and flexibility in late-season entry. Successful systems have been installed in nurseries and filbert orchards on deep, moderately well-drained Willamette Valley soils that did not truly have perched water tables.

In any drainage-system installation, it is important to recognize that the soil around the pipeline must be saturated before water will flow into the pipeline. If soil saturation occurs for only brief periods, deep ripping may be a better solution than a drainage system.

13.4 Identifying Drainage Problems

If a drainage problem is suspected, a field investigation should be initiated to determine which one or combination of possible soil-water or environmental conditions is causing the difficulty (see 13.3). To gain an overview of the entire situation, both small- and large-scale aerial photographs should be obtained. In addition, topographic maps with the most usable

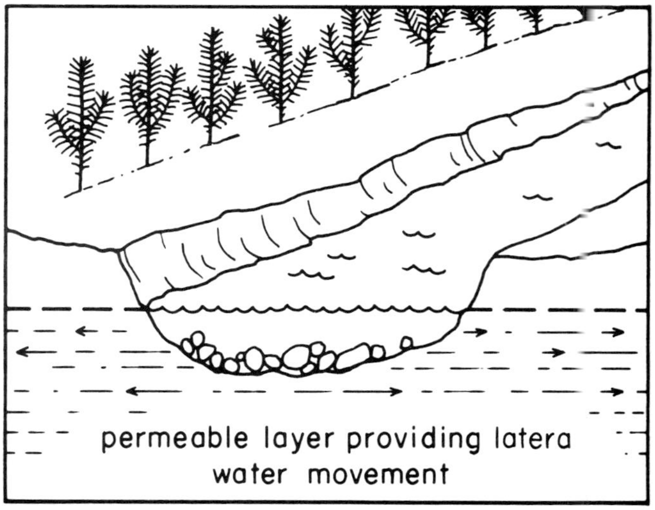

Figure 4. Streamflow within a valley creates lateral flow of subsurface water through permeable strata during high-runoff periods.

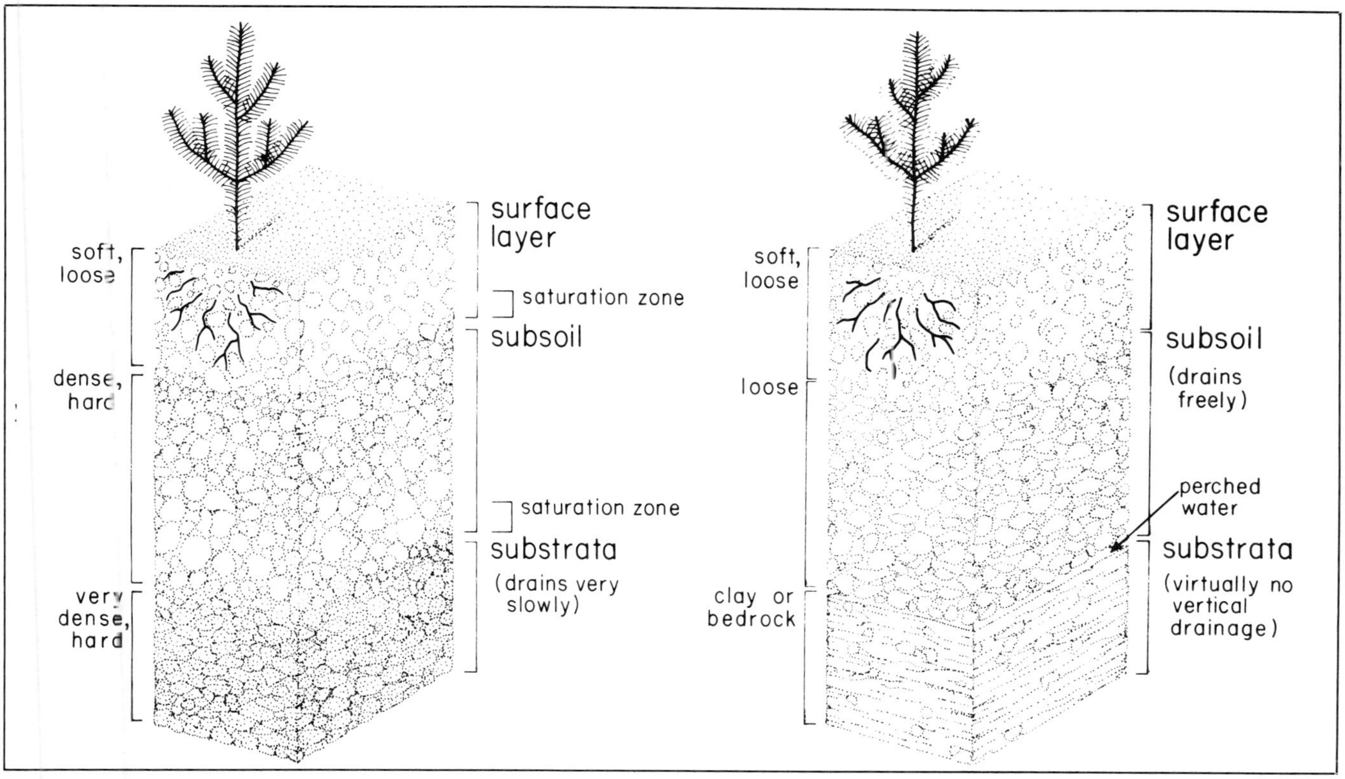

Figure 5. Physical properties of two typical Northwest soil profiles and their effects on water movement.

scale available are a valuable supplemental source of information in heavily wooded areas such as those west of the Cascade Mountains.

13.4.1 Off-site effects

Occasionally, problems at a particular site can be related to distant, off-site conditions, especially in mountainous terrain. Utilizing small-scale (1:60,000 or 70,000), high-altitude, aerial photography, the nursery manager can speculate on the contributions of runoff waters from lands adjacent to the nursery site (see 13.3.1.1). Stereoscopic viewing at this scale can aid in identification of possible sources of excess water. For example, in one nursery, it was determined that runoff from roads located upslope and ¾ mile from the affected fields was collected and discharged through permeable subsoil layers and surfaced midway downslope—in the nursery. This source of runoff could have easily been overlooked had the manager studied on-site conditions only. Similarly, the transmission of waters from distant or upslope watersheds into alluvial-colluvial fans or toeslopes (see Fig. 2) lacking well-defined channels can be a hidden source of drainage problems. Water will follow old, obscure channels and emerge as seepage [1].

Managers of nurseries located in valley-fill positions can also profit from studying aerial photos, which reveal changes in stream gradients, streamflow configuration, and old meanders. If the stream gradient is nearly level above or adjacent to the nursery site, subsurface flow into the nursery can be anticipated; but if the stream gradient slopes away from the nursery, the opportunity for subsurface flow is not as great. Streamflow configuration can indicate possible water movement from upstream or sidestream channels. Sinuous or meandering streams frequently have higher water-table levels within the adjacent lands than do "straight" streams. Old meander channels are often hard to detect on the ground but are usually observed in tonal differences as well as topographic depressions on aerial photographs.

13.4.2 On-site areas

Topographic maps, in combination with large-scale (1:15,840) aerial photography, are useful for identifying depressions where water is likely to collect. A closer examination may also be needed to determine how well a depression drains (see 13.4.3). Depending on the size of the nursery, it could be a worthwhile investment to have a topographic map constructed on a 100-ft grid. A well-prepared topographic map can also be used for other purposes, such as field and irrigation-system layout or traffic patterns.

Service-road and wheel-track runoff, as well as excess water from buildings and parking lots, should be inventoried to identify the disposition of all surface waters on the site. Obviously, the best time to evaluate this would be during a storm. However, sediment accumulations can be observed several days after a storm, provided that sediments are not cultivated or disturbed. Observations by field workers are also an important source of information.

13.4.3 Subsoil investigations

Subsoil and substrata conditions are evaluated by traversing the land and boring holes in the soil. Borings can be made randomly or according to an established grid spacing. Large-scale aerial photographs may be the basis for selecting where the random borings should be made according to changes in topography or tonal expressions on the photos. If borings expose saturated soil conditions or free water within the top 5 feet, then a problem exists. Depending on their texture, good nursery soils should be capable of draining from a saturated condition to field capacity within a couple of days. Abrupt changes in soil texture and structure—which can be natural or induced by traffic—often indicate pronounced reduction in soil-water movement (see chapter 6, this volume).

Examining subsoil colors can be revealing. Blues and grays predominate in saturated soils in which insufficient oxygen causes soil minerals to be chemically reduced. Seasonally saturated soils usually show alternating streaks of oxidized and reduced materials, principally yellows to rusty reds. These are normally referred to as "mottled" colors and may extend upward into the surface layers.

Careful examination of the changes in texture and structure and of color differences can help managers approximate the elevation of a water-table level or fluctuation. To plot these, record the water level (as depth below ground line) when soil borings are made. Then take a second reading 24 hours later to see if the water table is stabilized or rising. A rise in levels could indicate a substantial hydraulic head behind the flow, as opposed to a static "pool." A hydraulic head might be alleviated by an interception system, whereas a static pool might be remedied by a relief system.

13.4.4 Determining the source or direction of flow

By preparing a ground-surface topographic map and overlaying this with a water contour map constructed in the same manner, a nursery manager can pinpoint the source of water or its direction of flow. This technique, particularly useful on sloping lands, requires that the field be staked on a 100-ft grid. Elevations at each grid point are determined and plotted; borings are then made at each grid point, and the elevations of the stabilized water table are determined. Both land and water-table contours are plotted in different colors. The result provides the information necessary for designing the point of interception, depth, and spacing of drain lines (Fig. 6). If the borings are carefully made and logged, such characteristics as dense subsoil layers, gravel lenses, or aquifers and bedrock can be noted. This information can also be plotted in cross section, which would be useful to the drainage-system designer and installer.

13.5 Subsurface Drainage Installation

Because most nurseries produce 2+0 seedlings on ⅓ of the nursery site, it may take several years to complete a drainage installation. Careful planning is essential so that nursery operations are not disrupted.

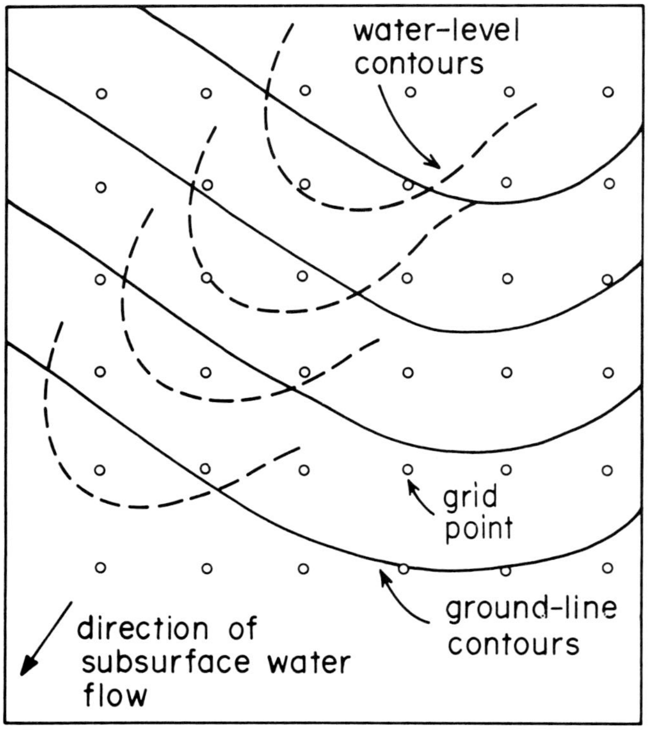

Figure 6. Surface and water-table contours plotted on a 100-ft grid.

13.5.1 Pipelines

Clay pipelines are still popular in some areas and work well as long as settling is not a problem. A clay or concrete pipeline may function satisfactorily for a century or more if properly planned, adequately constructed, and carefully maintained. Both clay and concrete are acceptable except in instances of strongly acid soils because the acids can dissolve the concrete or bonding.

A more recent innovation is plastic pipeline, or tubing. Tubing is corrugated for strength and perforated for infiltration efficiency. It can be quickly and accurately installed, especially by laser-guided machines—a single operator can easily lay over 4 miles of tubing daily with an alignment accuracy of within ⅛ inch. The problem of misalignment, so common with clay or concrete pipe, is eliminated because the tubing is continuous. Although laser-guided laying equipment is expensive, installation is rapid, particularly in large fields, and the final costs are competitive when compared with those of conventional trenching equipment.

Strength and durability of the selected pipeline are major considerations. Loads created by backfilling and, in some cases, surface loading must be considered. The U.S.D.A. Soil Conservation Service requires a minimum of 35 psi crush strength before fracture. The crushing resistance of plastic tubing is directly related to the cross-sectional design at the corrugations, and stretching adversely affects this structural integrity. Latest specifications by the Soil Conservation Service and American Society for Testing and Materials allow minimum installation stretch of 5% on plastic tubing [5].

13.5.2 Excavators

Regardless of the system and the purpose, two methods of excavator installation are popular: "plows" and trenchers. The plow is a long shank, much like a subsoiler tooth, attached to the rear of a crawler tractor. Tubing is fed automatically behind the shank, causing very little surface disturbance and minimal soil settling. The trencher, an excavator on a continuous belt, leaves a trench that requires backfilling. As with the plow, tubing is fed into the trench automatically; after installation, however, the trench may settle, creating traffic problems for a year or two. Most drainage contractors in the Northwest are equipped with laser-guided systems on both plows and trenchers.

Backhoes also are sometimes used in excavator installations. Though convenient for small jobs, they are slower and less accurate in laying a grade and, therefore, usually more costly.

13.5.3 Filters

Most nursery soils in the Northwest need filters—either gravel envelopes or nylon fabric—to reduce pipeline siltation. Sandy clays, clay loams, loams, sandy clay loams, loamy sands, and sands require removal or blockage of particles that would otherwise deposit in the drain line and cause plugging; silt loams and silts require sand and gravel filters (nylon fabric is unacceptable). Clays, silty clays, and silty clay loams, however, do not need filters [7].

Minimum trench width for a conventional filter envelope of graded sand and gravel usually is the width of the pipe (outside diameter) plus 8 inches. Plastic tubing wrapped with bonded nylon fabric, however, requires only the trench width necessary for the size of the tubing used (Fig. 7).

13.5.4 Outlet protection

Drainage outlets must be designed to prevent entry by small animals. A grill or flap gate is good insurance. Where plastic tubing is used and where the outlet discharges into a ditch that is frequently burned or cleaned, a short section of metal pipe should be installed at the outlet to protect the tubing.

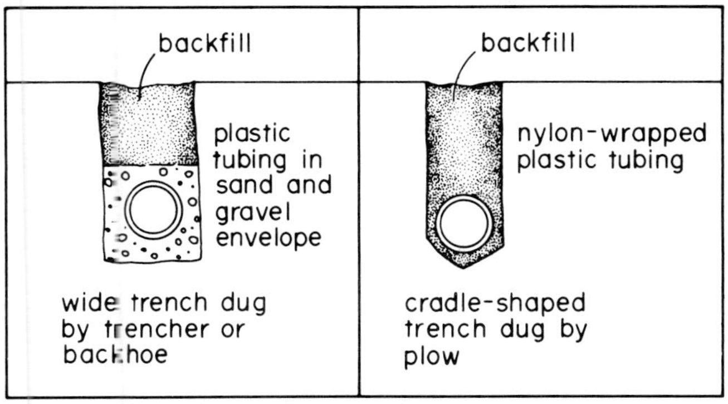

Figure 7. Typical filter systems (adapted from [5]). Note that nylon-wrapped plastic tubing allows a much narrower trench than tubing or pipe within a gravel and sand envelope.

13.6 Drainage Systems and Nursery Operations

13.6.1 Recycling drainage waters

The question often arises as to whether discharge waters should be collected and returned through the irrigation system. This practice sometimes is acceptable, but caution should be exercised. The water should be thoroughly analyzed to ensure that salt buildup does not create new problems for the nursery manager. Unfortunately, water quality is unknown until after a drainage system is installed.

Provisions could be made, however, to capture and recycle the water, if so desired. Salt concentrations can be diluted with normal irrigation water, or drainage water can be chemically desalinized. At this time, there is no known practical method of extracting pesticides from water. Therefore, recycling of water should not be relied upon until potential problems are identified and measures taken to correct them.

13.6.2 Ripping or subsoiling

A drainage system intended to intercept or relieve a water-table problem is usually installed at a depth of 3 to 5 feet. A system to correct temporary saturation conditions can be installed at 2 to 4 feet. However, installing a drainage system at these shallow depths could interfere with deep ripping or subsoiling. Moreover, if a nursery is suffering from a zone of densified or compacted soil, the shallow drainage system alone may provide only temporary relief.

The ideal soil-management program would consist of deep (24 to 30 inches) ripping to alleviate the compacted layers; then installation of drain lines at a depth (24 to 36 inches) more responsive to water movement; and, finally, shallow (12 to 16 inches) subsoiling when the land is lying fallow or is planted in cover crops. Development of a new compacted layer must be avoided if the shallow drainage system is to survive and function properly. A program to monitor compaction development could be initiated to alert the manager if the compacted zone approaches the subsoiler's capacity to perform. A shovel or probe would serve as the monitoring device.

13.6.3 Recommended management approach

Because a nursery may experience drainage difficulties for numerous reasons, the nursery manager must be able to recognize the severity and extent of any problem when planning and budgeting for its remedy. Managers are urged to analyze drainage problems using all available resources, initially with an overview approach, then with field-by-field observations.

Fortunately, free assistance is often available. Soil scientists and agricultural engineers from the U.S.D.A. Cooperative Extension Service, U.S.D.A. Soil Conservation Service, and U.S.D.A. Forest Service can provide investigative data and design criteria. Many Northwest drainage contractors are well qualified or employ agricultural engineers who are willing and capable of designing and installing a system that will perform as expected.

Nursery management is a complicated form of farming. So many factors are involved—from cone collection to outplanting. But nursery soils that are well drained can eliminate some of the negative or troublesome aspects of conifer production.

References

1. Boyer, D. 1977. Wind River Nursery land treatment plan. U.S.D.A. Forest Serv., Portland, Oregon. 44 p.
2. Donahue, R. L., R. W. Miller, and J. C. Shickluna. 1977. Soils, an introduction to soils and plant growth. 4th ed. Prentice-Hall Inc., Englewood Cliffs, New Jersey. 626 p.
3. Gonzales, O. 1982. Personal communication, U.S.D.A. Forest Serv., Grants Pass, Oregon.
4. Hausenbuiller, R. L. 1978. Soil science, principles and practices. 2nd ed. Wm. C. Brown Publishers, Dubuque, Iowa. 611 p.
5. Jeffords, O. 1978. Drainage handbook. Advanced Drainage Systems Inc., Columbus, Ohio. 47 p.
6. Lyon, T. L., H. O. Buckman, and N. C. Brady. 1952. The nature and properties of soils. 5th ed. The MacMillan Co., New York. 591 p.
7. U.S.D.A. Soil Conservation Service. 1976. Washington, D.C. Advisory Memo ENG. WA-6.
8. Walker, J. C. 1950. Plant pathology. 1st ed. McGraw-Hill Book Co., Inc., New York. 699 p.

Culturing the Bareroot Seedling

Chapter 14
Plant Physiology and Nursery Environment: Interactions Affecting Seedling Growth

D. P. Lavender

Abstract

Environmental factors (such as light, moisture, nutrients, density, and temperature) and plant physiological factors (such as carbohydrate reserves, hormone levels, frost hardiness, and dormancy) interact to shape growth and survival of coniferous seedlings in nursery fields and after outplanting. Nursery managers can manipulate moisture, nutrients, and density to achieve desired seedling morphology and vigor. However, the annual growth cycle of perennial plants has evolved in response to environmental pressures. When the environment is modified, as with heavy irrigation in a nursery, to permit growth at a time when natural seedlings are dormant, the ensuing phases of the growth cycle will not be properly synchronized with their environments. Seedlings so cultivated lack vigor after outplanting. Nursery managers should aim at keying their cultivation schedules to both environmental conditions and endogenous seedling physiology to ensure production of high-quality seedlings.

14.1 Introduction

The annual growth cycle of most temperate-zone plants seems regulated by endogenous, or internal, rhythms. But these rhythms may be overridden by exogenous, or environmental, factors which can, either collectively or individually, strongly limit or stimulate active growth [38]. Because the details of endogenous activity or of response to exogenous stresses or stimulation vary widely among temperate-zone plant species, botanists, horticulturists, foresters, and nursery personnel should be thoroughly familiar with the physiology of their plant populations and the environmental sequences necessary to produce plants of uniformly high vigor.

Cultivation according to physiological guidelines is essential to produce plants with maximum survival and growth potential. Such cultivation includes proper manipulation of seeds to assure a stand of well-spaced young seedlings by early June of the first year, irrigation schedules designed to promote growth in the spring and early summer and dormancy thereafter, and fertilizer applications which will provide the proper balance of the essential nutrients for optimum seedling growth and vigor.

14.2 Seedling Growth

14.2.1 The shoot

The first-year coniferous seedling commonly has an indeterminant growth habit; that is, shoot elongation results from production of cells by the apical meristem during the growth season. The significance of this habit to the nursery manager is that seedlings will often continue to grow as long as their environment favors growth [10, 42]. For example, it is not at all uncommon to observe first-year Douglas-fir [*Pseudotsuga menziesii* (Mirb.) Franco] seedlings in a nursery actively growing in October.

Research on the annual growth cycle of coastal Douglas-fir (var. *menziesii*) [5, 40, 41] suggests that, in nature, seedlings germinate in early spring and complete shoot elongation by midsummer, when increasing drought stimulates dormancy. Hermann and Lavender [25] demonstrated that seedlings grown from high-elevation seed sources entered dormancy earlier than those from low-elevation sources, whereas Rehfeldt [64] and Lavender and Overton [42] showed that Douglas-fir seedlings of the Rocky Mountain form (var. *glauca*) enter dormancy without appreciable environmental stress. Other western conifers have not been investigated as thoroughly as Douglas-fir. Nonetheless, the available data suggest that western hemlock [*Tsuga heterophylla* (Raf.) Sarg.] [36, 56, 57], true firs (*Abies* spp.) [39], and ponderosa pine (*Pinus ponderosa* Dougl. ex Laws) [37, 79] all have annual growth cycles comparable to that of the associated Douglas-fir and that these species respond to environmental stimuli in a similar manner.

After their first growing season, temperate-zone conifers generally demonstrate a determinant growth habit; that is, shoot elongation results from expansion of primordia laid

In Duryea, Mary L., and Thomas D. Landis (eds.). 1984. *Forest Nursery Manual: Production of Bareroot Seedlings*. Martinus Nijhoff/Dr W. Junk Publishers, The Hague/Boston/Lancaster, for Forest Research Laboratory, Oregon State University, Corvallis. 386 p.

down in buds produced the previous growing season. Under natural conditions, this period of shoot elongation is typically brief [38], seldom more than 3 months. Normally, most western species complete their season's height growth in a single flush; but if the environment in midsummer favors growth (particularly after a heavy rain), these species can produce lammas shoots (one or more additional flushes of terminal growth on the terminal shoot) or prolepsis growth (elongation of lateral buds at the base of the terminal bud or on lateral shoots) [37, 69]. Such growth is undesirable in western nurseries because it generally does not develop frost hardiness early and hence may be killed by fall frosts. More importantly, it indicates that the seedling is not proceeding through the dormancy sequence properly (see 14.4.4) and therefore will not have high resistance to the stresses inherent in the harvest-storage-outplanting sequence and will not grow vigorously after outplanting.

14.2.2 The root

The roots and, generally, the fungi which form symbiotic structures with the roots or rhizomes of higher plants absorb most of a plant's nutrient and moisture requirements and support the aerial part of the plant. The obvious importance of the woody plant root system has stimulated research for at least a century, but work has suffered from the following weaknesses. First, the opaque nature of soil has made direct observation of root growth almost impossible; unfortunately, the glass-soil interface of glass-fronted boxes or underground containers creates an atypical rooting environment that is exacerbated by the light used for observation. Second, growth and physiology of individual roots are extremely variable; the erratic growth rhythms of temperate-zone plant roots are more similar to the uncoordinated shoot-growth patterns of tropical plants than they are to the more regulated shoot-growth patterns of temperate-zone plants. Third, attempts to use environmental controls to study root growth and physiology have largely been frustrated by the extreme difficulty of maintaining or manipulating endogenous moisture and temperature gradients in soil isolated in discrete containers.

Notwithstanding the above, Sutton [80] reviewed extensive research demonstrating that a number of environmental factors may affect the growth, form, and physiology of roots. Such factors, however, seem to more heavily influence growth of second-year and older seedlings; the root form of first-year seedlings is often controlled more by genetics than environment [84].

Although Stone and his colleagues have repeatedly stressed the importance of the nursery environment in producing seedlings with a high potential for early root growth after outplanting [35, 76, 77, 78], there is no such consensus on the effects of planting techniques upon survival and growth of properly conditioned seedlings in plantations [22, 45, 60, 61, 62, 68, 91, 93, 94]. However, the results obtained with 1+0 Monterey pine (*Pinus radiata* D. Don) seedlings in New Zealand [86]—where careful nursery procedures preserved an intact root system and permitted twice the growth after outplanting of project-harvested seedlings—and the vigorous growth reported for 1+0 bareroot and container-grown Douglas-fir [21, 31] seem to support Toumey's [84] observation that roots of older seedlings are more affected by the environment (i.e., planting technique) and that the adverse impact of present planting techniques is reflected in the postplanting growth of older seedlings. Perhaps if planting techniques more compatible with maintaining seedling vigor were developed, the effects of nursery practice upon growth of second-year and older planting stock would be more evident. Conflicting evidence suggests both that well-developed root systems are associated with high seedling survival [23, 47] and that plantation growth and root form are not correlated [46]; certainly, until such conflict is resolved, the more subtle effects of nursery practice upon seedling growth will be difficult to assess.

14.2.3 Shoot:root ratio

Wakeley [93] concluded that measurements of seedling morphology were poor indicators of future field performance. However, subsequent reports equivocate on this point. Some workers [6, 7, 18, 47, 75, 92] suggest that seedling shoot:root ratios at the time of planting do predict seedling performance—a low shoot:root ratio would indicate good survival and growth potential—but others [1, 2, 53, 54, 55, 70, 99] disagree.

Several probable reasons underlie this sharp divergence of opinion. First, shoot:root ratio may vary as a result of:

- **Seedling age or size:** Older, larger plants generally have higher shoot:root ratios than smaller, younger ones [32, 87].
- **Seedling genetics:** Plants grown from seed collected in dry regions have lower shoot:root ratios than similar plants grown from seed collected in moist areas [42].
- **Environment:** Plants grown with high levels of water, nutrients, or both or with less than full sunlight often have higher shoot:root ratios than similar plants grown with relatively limiting levels of water and nutrients under full light [106].
- **Cultural practices:** Root or shoot pruning or wrenching, for example, may stimulate either high or low shoot:root ratios, but this effect is generally transitory [51]. Wareing [96] has shown, for example, that shoot growth may be quantified in terms of root growth according to the formula $S = cR^k$, where S is shoot growth, R is root growth, and c and k are constants specific for a given species and environment. Ledig and Perry [44] suggest that the constants are stable over a range of environments. Obviously, unless both c and k equal 1, the shoot:root ratio will change with time.

Second, stresses present at the planting site vary widely with climate and vegetation type. A plantation established on a relatively dry site in eastern Oregon and Washington, for example, may well have higher survival if the seedlings have a low shoot:root ratio. But the major stress on a typical Oregon Coast Range plantation will be competition for light [29], in which case seedling survival is more heavily dependent on shoot size than on shoot:root ratio.

Third, there is no standardized methodology for determining shoot:root ratios. Some workers use the dry weights of roots and shoots [32, 44, 47]; others use the relative volumes of these seedling parts [53, 54]; and still others use the relationship of foliage weight to root-surface area [18] as a "drought resistance index." Edgren and Iyer [18] note that shoot:root ratios calculated by the volumetric technique may be transformed to the drought index by dividing by 0.04.

Finally, reports frequently neither cite the probable cause for shoot:root differences nor demonstrate the probable effects of planting-site environment on the physiological parameters determining shoot:root ratio.

14.3 Exogenous Factors Affecting Growth

14.3.1 Light

Light profoundly affects the growth and development of temperate-zone plants in two ways. First, it is the energy source that drives photosynthesis, the process by which plants create the organic substrates necessary for growth. Second, light—or, more properly, the absence of light—regulates seedling development through a phenomenon termed photoperiodism; that is, daily dark periods of less than 10 hours stimulate active shoot elongation, whereas daily uninterrupted dark periods longer than 14 hours stimulate dormancy.

In spite of the extreme importance of light, however, the bareroot nursery manager can affect the light environment of seedlings by (1) reducing light intensity with shading materials; (2) manipulating density of both crop and weed species; and (3) controlling the photoperiod by installing either artificial light sources or blackout devices designed to shorten seedlings' daily exposure to light. Shading seedlings or manipulating their density, which may significantly affect morphology and carbohydrate reserves, will be discussed more fully in 14.3.4. Controlling the photoperiod, which has been done occasionally in eastern U.S. nurseries and in research trials, is not a technique used by Northwest nurseries and therefore will not be discussed in this chapter.

14.3.2 Moisture

Like light, moisture influences seedling growth and development by its presence or absence. The rate of photosynthesis, one major key to total seedling growth, may be sharply reduced by soil moisture deficits that are relatively small (−1 to −3 bars) [105]; but it may also be slowed by saturated soils, which produce an anaerobic environment [102]. In addition, excess moisture may promote growth of plant pathogens such as *Phytophthora*, *Pythium*, and *Fusarium* [19].

The regulatory role of moisture in the annual growth cycle of Northwest conifers, especially in initiating dormancy (see 14.4.4), reflects the region's climate, which is characterized by dry summers and wet winters. Such a precipitation pattern is similar to that of California and the Mediterranean area, but is sharply different from that of most land areas, which receive the majority of their annual precipitation during summer. Dormancy in perennial, temperate-zone plants indigenous to areas with moist summers is initiated primarily by shortening photoperiods in late summer and only secondarily by plant moisture stress [38]. Therefore, timing and intensity of irrigation in eastern U.S. nurseries do not impact the annual growth cycle. But in the Northwest, most species grown in coniferous forest nurseries have evolved to initiate dormancy primarily in response to midsummer drought [9].[1]

Nursery personnel can effectively manipulate plant moisture. For example, they can help protect seedlings from moisture stress by carefully noting both seedling and environmental conditions during nursery operations. Not infrequently, weather during the lifting and packing period may be sufficiently desiccating to cause severe moisture stress. Seedlings should be moistened thoroughly when dry days occur during harvest because even brief periods of moisture stress at that time will reduce subsequent seedling growth [15]. Furthermore, seedlings that are stressed when packed must endure many days in storage before such stress can be alleviated [15](see chapters 21 and 22, this volume). Conversely, however, irrigating at the wrong times—physiologically—can do damage. Frequent irrigation of nursery stock to relieve moisture stress due to late-summer drought can cause dormancy to be initiated too late to permit the sequence of physiological changes necessary for vigorous seedling growth [41](see chapter 15).

14.3.3 Nutrients

Seventeen elements have been shown to be essential to plant growth. Three of these—carbon, hydrogen, and oxygen—are absorbed from the atmosphere or from water. The remaining 14 are taken up from the soil. Several of these—nitrogen, phosphorus, potassium, calcium, sulfur, and magnesium—are termed macronutrients because harvesting an acre of coniferous seedlings commonly removes from 1 to more than 100 pounds of each of these elements [90]. The remaining elements—boron, chlorine, copper, zinc, iron, manganese, molybdenum, and cobalt—are required in much smaller quantities and, hence, are termed micronutrients. A healthy seedling, however, must be well supplied with **all** nutrients in proper proportions [28]. Any environmental or cultural factor that affects growth will, of course, affect seedling nutrient requirements. Though it is not possible to specify absolute soil-fertility standards, ranges within which vigorous seedlings may be grown can be specified (see chapters 7 and 8, this volume).

If a given nutrient is deficient, seedlings may compensate to some extent by increasing their capacity to take up the deficient ion [26]. More commonly, such stress is reflected by reduced growth and by distinct changes in the plant's habit. Plants require nitrogen, for example, to synthesize chlorophyll; nitrogen-deficient plants, therefore, often appear chlorotic. Low levels of phosphorus, which is essential to seedling metabolism, result in reddish-purple foliage. Boron is required for lignification; deficiency causes terminal dieback and necrotic buds. Other symptoms characteristic of malfunctioning physiology are exhibited by seedlings deficient in other nutrients [43].

van den Driessche [89, 90] reviewed reports that indicate both positive and negative effects of nursery fertilizer applications on subsequent seedling growth and survival. Both van den Driessche's trials with Douglas-fir [89] and those of Smith et al. [73] showed positive growth responses after outplanting for Douglas-fir seedlings fertilized with various levels of nitrogen in the nursery. Radwan et al. [63], however, suggest that the form of nitrogen fertilizer strongly affects response; in their trials, nitrate and urea fertilizers produced greater seedling response than did ammonia salts.

Several reports have suggested that cold hardiness in conifer seedlings may be affected by adding mineral nutrients to the nursery seedbed in late summer [90]. For example, potassium has been shown to increase drought resistance when soils are frozen in winter, and both potassium and nitrogen, applied too late to affect the dormancy cycle, have increased seedling frost hardiness in both Sitka spruce [*Picea sitchensis* (Bong.) Carr.] and western hemlock seedlings.

The above studies as well as others not mentioned here suffer from lack of positive control of nursery environmental factors other than nutrients and from lack of uniformity and control of physical and biological factors in the outplanting area. Although the results of such research may provide empirical guidelines for the moment, they fail to elucidate the physiological role of nutrients in seedling vigor. A range of carefully controlled, designed studies—such as those conducted by Ingestad [27], wherein all environmental factors including nutrients are fully controlled—is needed to answer questions about species, quantities, and timing of nursery nutrient applications.

Given the above caveats, the following points, discussed by van den Driessche [90] in his comprehensive review of nursery soil fertility, are valid, useful guides:

- Nutrient availability may be affected by soil pH and organic matter content.
- Harvest of 2+0 seedlings removes significant quantities of nutrients. Continual cropping of nursery soils, then, requires adding nutrients to maintain fertility.
- Adding nutrients, especially nitrogen or phosphorus, affects the growth of soil microorganisms and may stimulate pathogens.

[1] With the probable exception of ecotypes or species native to either the fog belt or to the *Abies amabilis* and *Tsuga mertensiana* zones in the Cascade Mountains [20].

- Frequent, light additions of nutrients will provide more constant levels of seedling nutrition than less frequent, heavy applications.
- Evaluating the effects of individual nutrients on certain aspects of seedling physiology is difficult because of the possible interactions of those particular nutrients with other aspects of seedling physiology. A comprehensive study relating nursery fertilization practices with seedling growth and survival after outplanting has not yet been made.

14.3.4 Density

Reports on a wide range of forest types throughout the world suggest that the density of coniferous seedlings in seedbeds dramatically affects seedling development [4](see chapters 5 and 15, this volume). A study from New Zealand [8] suggests that the optimum spacing between 1+0 Monterey pine seedlings is about ⅛ of their height. However, age of planting stock at the time of harvest and variations in seeding method and densities make it impossible to generalize about an optimum density for all nurseries.

For example, Mullin and Bowdery [52] demonstrated that white pine (*Pinus strobus* L.) and red pine (*Pinus resinosa* Ait.) seedlings grown at 15 seedlings/ft² survive and grow better than similar plants grown at 30 seedlings/ft². However, Shoulders [72] and Shipman [71] reported that loblolly pine (*Pinus taeda* L.) and slash pine (*Pinus elliottii* Engelm.) seedlings generally have equivalent survival whether grown at 20 or 40 seedlings/ft². The greater size and growth of the lower density stock in Shipman's [71] trials find support from the unpublished data of Mexal [50] who argues that 18 seedlings/ft² is the optimum density for 1-year-old loblolly pine.

In the Northwest, unpublished data from Weyerhaeuser Company reforestation projects [30] suggest that in the state of Washington, 25 2+0 Douglas-fir seedlings/ft² is the optimum spacing when costs of both nursery cultivation and plantation establishment are related to seedling survival. In British Columbia, Revel [65] proposes that 30 to 50 seedlings/ft² will produce the highest yield of plantable 2+0 Douglas-fir, but Edgren [17] argues that a spacing of no more than 20 seedlings/ft² is necessary to produce 2+0 seedlings with a 4-mm caliper. The foregoing data demonstrate that spacing in the seedbed affects seedling caliper. Lopushinsky and Beebe [47] note that seedling stem caliper is correlated with root development and that seedling survival on droughty sites is improved if plants have well-developed roots. Very probably, wide spacing in the seedbed permits increased photosynthesis, hence the increased food reserves necessary for vigorous growth after cold storage. It should be emphasized, however, that the densities referenced in this and the preceding paragraphs are **means,** which can vary significantly within treatments.

In summary, density itself affects seedlings indirectly—by impacting available light, moisture, and nutrients. Generally, wide spacing (lower density) promotes greater root development and higher levels of carbohydrate reserves, which are essential for development of cold hardiness [97], and reduces losses to insects and disease [74].

14.3.5 Temperature

Temperature is a measure of the heat energy available to plants. Higher plants, under normal growth conditions, are poikilothermic—that is, they assume the temperature of their environment. Further, the rates of most metabolic processes are strongly regulated by temperature; for example, a 10°C increase in temperature may cause a plant's respiratory rate to double [98]. The temperature at which maximum plant growth occurs is not necessarily that which permits maximum gross photosynthesis—it is the temperature at which the rates of the plant's synthetic processes exceed those of its catabolic processes by the greatest margin.

Controlled-environment trials with Douglas-fir seedlings suggest that the optimum temperature for growth of this species is 24°C [42]; similar results are reported for seedlings of other coniferous species [83]. But findings from trials with other plants suggest that the optimum temperature for growth may drop slightly as plants increase in size and age ([13] for Monterey pine, [95] for agricultural crops).

The preceding data are largely concerned with the effect of air temperature, primarily during the day, on active plant growth. However, Lavender and Overton [42] demonstrated that warm, not cool, nights stimulated Douglas-fir seedling dormancy under short photoperiods, and Lavender [38] reviewed data which indicate that soil, as well as air, temperatures may greatly influence the growth of plant shoots.

The optimum growth period for nursery stock in the Northwest seems to occur—not during the hot days of summer—but during the relatively mild days of spring. However, when daytime temperatures exceed 20°C, a decided moisture stress, which will limit photosynthesis, may develop by 10 a.m. even in seedlings growing in moist soil. Seedling growth in spring may be maximized by applying intermittent, light irrigation during bright spring days to reduce seedling moisture stress and subsequent stomatal closure [105].

Although nursery managers can do little to regulate the temperature of nursery seedbeds, they can produce superior seedlings by scheduling annual growth cycles so that seed and seedling physiology is compatible with environmental conditions. For example: (1) seed germination in the relatively cool soils of April or early May is facilitated by prescribing stratification periods of 3 months; (2) seedling quiescence (summer dormancy), initiated by midsummer drought, is associated with hot summer days; (3) early rest (winter dormancy) is stimulated by the mild temperatures of early fall. Failure to match seedling physiological states with the temperature regimes occurring naturally during those states may have a profound negative impact on seedling quality [38].

Temperature extremes may damage seedlings (see chapter 12, this volume). However, effects of high temperature may be minimized by proper seedling spacing and cultural regimes supplemented by occasional, light, cooling irrigation on hot days. Frost damage may be avoided by initiating dormancy in midsummer and by seedling spacing which permits maximum photosynthesis and production of carbohydrate reserves.

14.4 Physiological Factors Affecting Growth

14.4.1 Carbohydrates

The heterogenous group of compounds termed carbohydrates provides the principal substrates for producing the energy necessary for plant metabolism. Simple sugars may be converted to amino acids, the basic compounds of the proteins essential to cell structure.

Although carbohydrate levels in plant tissues have been studied for decades, the literature contains little really definitive data for several reasons. First, "carbohydrate" is an imprecise term. It includes monosaccharides, oligosaccharides, and polysaccharides and should include sugar derivatives such as alcohols, cyclitols, their methylated derivatives, and even such compounds as gluconomic acid. Second, carbohydrate levels may change after sample harvesting as seedling metabolism continues until tissue is killed; enzyme activity may interconvert various carbohydrate species. Probably the best harvest procedure is immediately placing sample tissues into liquid nitrogen, followed by freeze-drying and dry storage at about

0°F (−17°C) [unpubl. data, 101]. Third, before the development of sophisticated gas-liquid chromatography [12], the methodology used to analyze carbohydrates was not sufficiently precise to provide accurate estimates of many species of interest.

Carbohydrate levels have been related to: (1) development of cold hardiness [97], based on the hypothesis that relatively high levels of substrate are necessary if a plant is to cold-harden fully; (2) nursery cultural practice, in which the effects of box pruning and wrenching on carbohydrate content of Monterey pine seedlings were shown to increase the level of substrate [11]; (3) growth of seedling roots [103]; and (4) duration of cold storage [66]. In the last case, carbohydrate reserves of Douglas-fir seedlings decreased as length of storage increased; concurrently, root-growth potential declined for storage periods longer than 6 months. But Krueger and Trappe [34] reported little correlation between root activity and seedling carbohydrate reserves.

14.4.2 Hormones

A hormone, or plant-growth regulator, is a substance synthesized (usually in minute quantities) in one location (i.e., the plant root) but transferred to another location (i.e., the plant leaf), where it exerts an effect upon growth and differentiation. This concept is not without controversy [85] because, unfortunately, methodologies for isolating and identifying hormones have lacked the precision necessary for obtaining unequivocal data. Nevertheless, a substantial volume of literature has appeared in the past 50 years relating plant hormones to such a bewildering array of metabolic and differentiation processes that even a summary is beyond the scope of this chapter. The following are the major, accepted hormones and the growth parameters most characteristic of each [81]:

Auxins: Stimulate cell enlargement, rooting of cuttings, and apical dominance; inhibit abscission of leaves, fruits, and root elongation.

Gibberellins: Stimulate cell division, seed germination, and reproductive growth.

Cytokinins: Retard senescence; promote bud growth as well as cell division, expansion, and differentiation.

Ethylene: Stimulates fruit ripening, breaking of dormancy, and epinasty (downward twisting of leaves or other organs); inhibits elongation of shoots and roots.

Inhibitors (e.g., abscisic acid): Reduce growth; may inhibit seed germination; may control stomatal physiology.

Plant growth and differentiation are generally believed to be controlled by interactions of the above compounds, in the manner suggested by Khan [33] for seeds, such that high concentrations of promoters favor germination, high concentrations of inhibitors favor dormancy, and cytokinins facilitate the action of promoters.

Reviewing a wide range of literature investigating the role of plant-growth regulators in woody-plant seedling physiology, Zaerr and Lavender [104] concluded that limitations in analytical methods have prohibited satisfactory understanding of the relationship, if any, between levels and species of plant-growth regulators and seedling vigor. Current work at Oregon State University, Corvallis, is concerned with developing more efficient, definitive analysis procedures for plant regulatory compounds, especially gibberellins and cytokinins, and with describing the role of such substances in the growth of coniferous seedlings [3, 14, 49].

14.4.3 Frost hardiness

Frost or cold hardiness is the ability of plant cells to withstand temperatures below freezing without suffering irreversible physical damage. The nature of the changes that occur in plant cells during the hardening process is not fully known or understood, but the hardening process apparently involves changes (1) in cell membranes, to allow movement of water to extracellular ice crystals, and (2) in the protoplasm, to resist effects of desiccation [97].

Significant frost hardiness is developed in coniferous seedlings only if the plants have an adequate carbohydrate reserve and if active growth has ceased [97]. The weather sequence that best promotes frost hardiness is warm, dry days and nights to favor growth cessation; mild, short days and mild nights to initiate hardening; cool, short days and cool nights to develop moderate hardiness; and, finally, cool days and freezing nights to develop maximum cold hardiness [82]. If this sequence is begun in mid-July and completed by late November, seedlings should be frost hardy to from −20 to −30°C by early December.

Frost hardiness in plants is quite labile. A few days of mild temperatures during winter may greatly reduce a seedling's cold resistance so that at least part of the foregoing sequence must be repeated before maximum cold hardiness is restored. However, frost damage to buds or foliage (at least after mid-November) does not affect seedling survival significantly [24].

14.4.4 Dormancy

The growth habit of perennial, temperate-zone plants is generally characterized by a relatively short period (about 3 months) of active shoot elongation followed by a lengthy "dormancy." Dormancy is a general term for all instances in which a tissue predisposed to elongate (or grow in some other manner) does not do so (after [16]). Romberger [67, p. 74] describes the nomenclature of dormancy, which still tends to be vague and confusing. Although dormancy is an adaption to permit plant survival during periods of stress (e.g., drought or frost), a plant is not equally resistant to all environmental factors during the entire dormant period, nor are the phases of dormancy normally defined in relation to stress resistance.

14.4.4.1 Growth patterns during dormancy

Only the apical meristems demonstrate true, endogenous dormancy. This is in sharp contrast to the phenomenon of cold hardiness, a parallel associated physiological state which affects, at least in some degree, the **entire** plant.

Lateral meristems of Douglas-fir seedlings grow from about budbreak until midfall [34]. Root meristems of Douglas-fir seedlings grow mainly during two peak periods. The first and larger peak extends from late winter until shortly after budbreak; the second and smaller peak occurs from late summer until midfall. During the rest of the year, either adverse environment or competition with the shoot for substrates results in relatively little root activity [34].

Lyr and Hoffman [48] present data generally confirming the above root-growth pattern for other temperate-zone woody perennials, whereas Sutton [80] suggests that root growth is controlled by environment rather than endogenous rhythms; he notes that both dry soils in summer and cold soils in winter may strongly limit root growth. Given the above general patterns, transplanting seedlings in late summer allows plants so handled to develop strong root systems by utilizing the period of root growth in both fall and early spring.

Even the buds of dormant seedlings are not inactive for the entire dormant period. Initials that will develop into the following year's shoots are laid down from July until November, the rate decreasing with time [58, 59].

14.4.4.2 Phases of dormancy

Although the sequence of physiological changes occurring during dormancy is not clearly understood, recent data [9, 38, 41] describe the environments necessary to permit proper

Table 1. Dormancy sequence in Douglas-fir (adapted from [41]).

Phase of dormancy	Period of year	Physiology	Environment
1: Initiation of dormancy	July-late September	Cessation of growth, increased desiccation, lignification of tissues	Mild to hot days, shortening photoperiod, mild to strong moisture stress
2: Deep dormancy	Late September-early December	Accumulation of growth inhibitors, increased cold resistance	Mild temperatures, shortening days
3: Dormancy lifting	Early December-late February	Breakdown of inhibitors, virtual cessation of metabolic activity	Short days, low temperatures
4: Postdormancy	Late February-budbreak	Accumulation of growth promoters (gibberellins, cytokinins, auxins), gradual conversion of carbohydrate substrates	Lengthening days, mild temperatures, low moisture stress

progression through dormancy from early budset in summer until budbreak the following spring. It cannot be emphasized too strongly that any major deviations from the endogenous pattern of dormancy will greatly diminish seedling vigor and reduce the survival potential of affected seedlings when outplanted. The environments necessary to permit normal development of dormancy in Douglas-fir are presented in Table 1; the phases of dormancy—and the consequences of deviation from the proper progression—are described in detail below, by phase. Bear in mind, however, that the dates, conditions, and processes in Table 1 are approximations—all these can vary from one year to the next—and that the transitions between phases are gradual rather than sharp.

Phase 1—Initiation of dormancy: Shortening photoperiods stimulate many temperate-zone plants to initiate dormancy during late August and September. However, the Northwest is relatively unique in that most of its annual precipitation falls during winter rather than summer, as it does for the majority of agricultural regions throughout the world. Accordingly, the prime impetus for initiation of dormancy in Douglas-fir is drought.

In nature, seedlings commonly set a resting terminal bud no later than mid-July. However, because nurseries can irrigate seedbeds during the entire summer, the natural chronology can be altered. Seeds are commonly sown in May or early June, resulting in germination and early seedling growth no earlier than early June (as opposed to early April under most natural environments). Nurseries irrigate seedlings until mid-August to achieve the growth most foresters want. As a result, plants are actively growing in late August, when fall rains start, and continue to grow until late September or early October before initiating a bud. Obviously, the environment in October is not that of July. Bud development is slowed, and the seedling remains out of phase with the environment through the winter and following spring, with a corresponding reduction in field-survival potential.

Phase 2—Deep dormancy: This is the critical phase for nursery operations. If resting buds are not well formed by mid-August, the requirements of buds for shortening days and mild temperatures, which occur during September and October, will not be met. A seedling that sets bud in late September will experience the cold temperatures of late October and subsequent months before its physiology has progressed sufficiently to benefit from the chilling, and phase 3 of dormancy will not be completed satisfactorily. As a result, the seedling will have a delayed budbreak the following spring and lower field-survival potential.

Phase 3—Dormancy lifting: Virtually all perennial, temperate-zone plants have a strong requirement for exposure to temperatures between 0 and 5°C during winter. Some horticultural varieties are characterized by the number of hours of such chilling they require. Douglas-fir has been shown to require from 8 to 12 weeks of chilling at temperatures around 5°C [88, 100]. However, these data are based on laboratory trials in which the temperature was continually maintained at that level.

In nature, warm periods during winter are frequent. During those times, the chilling process is disrupted, and the warmth actually reverses part of what the previous cold had accomplished. (It is generally believed that low temperatures facilitate destruction of the hormones that inhibit plant growth.) Winter weather normally just satisfies the seedling's requirement for low temperatures. However, if the environments during phases 1 and 2 were not conducive to bud formation and development, the seedling will require a much longer period of chilling to complete phase 3 satisfactorily. Seedlings lacking the necessary chilling will begin to grow later than normal in the spring, and their field-survival potential will be correspondingly reduced.

Phase 4—Postdormancy: If seedlings have progressed properly through the first three phases of dormancy, they should enter phase 4 no later than early March. In this phase, the plant is ready to grow and remains dormant only so long as temperatures are unfavorable for growth. If seedlings have not progressed through the first three phases of dormancy properly, they will fail to grow in response to the warming temperatures of early spring. Lack of root growth will greatly reduce the plant's ability to take up necessary moisture from the soil and it will probably die of drought before mid-June.

By definition, dormancy is related to the ability of the apical meristem to grow. However, the concepts in Table 1 are based more on the resistance of seedlings to the stress inherent in the reforestation process than they are on the classical definition of dormancy. For example, the period from early October until early November usually corresponds with the time when the apical meristem is least likely to resume growth under favorable conditions. However, the period from late September until early December corresponds with the time when seedlings are most easily injured by the transplanting process. Accordingly, that period (late September to early December) has been identified as phase 2 so that nursery personnel can better interpret seedling physiology in terms of nursery operations.

To complicate the role of dormancy in seedling physiology still further, Owens and Molder [59] demonstrated that there is no strong correlation between the phases of dormancy and initiation and development of primordia in buds. It is clear, then, that the phases outlined in Table 1 cannot be identified with anatomical or morphological changes in seedlings but must result from changing hormonal levels. Until analytical techniques are sufficiently precise to accurately determine species and quantities of these compounds, the true nature of the physiology of dormancy will remain unknown.

14.5 Conclusions

This chapter has described the effects of a range of environmental factors and cultural treatments upon the physiology of coniferous seedlings. Most of this discussion has been based upon empirical trials, which generally suffer in that they are not sufficiently precise to permit uncritical extrapolation.

The nursery manager, then, should use the relationships presented here as general guides, realizing that specific nursery environments and specific genetic stock may produce results which deviate, at least in detail, from those outlined in this chapter. A thorough knowledge of the meteorological and edaphic characteristics of the nursery and of the genetic composition of the major stock types is necessary if nurseries are to consistently produce high-quality seedlings.

References

1. Anstey, C. 1971. Survival and growth of 1/0 radiata pine seedlings. New Zealand J. Forestry 16:77-81.
2. Baron, F. J., and G. H. Schubert. 1963. Seedbed density and pine seedling grades in California nurseries. U.S.D.A. Forest Serv., Pacific SW Forest and Range Exp. Sta., Berkeley, California. Res. Note PSW-31. 13 p.
3. Bell, T. E. 1983. A procedure for the purification and quantitation of abscisic acid and its application to a study of abscisic acid concentrations in needles of cold stored Douglas-fir [*Pseudotsuga menziesii* (Mirb.) Franco] seedlings. M.S. thesis, Oregon State Univ., Corvallis. 80 p.
4. Benson, A.D., and K. R. Shepherd. 1976. Effect of nursery practice on *Pinus radiata* seedling characteristics and field performance. I. Nursery seedbed density. New Zealand J. Forestry 6(1):19-26.
5. Blake, L, J. Zaerr, and S. Hee. 1979. Controlled moisture stress to improve cold hardiness and morphology of Douglas-fir seedlings. Forest Sci. 25(4):576-582.
6. Carlson, L. W. 1972. Survival of 2-0 and 3-0 jack pine seedling outplantings in southeastern Manitoba. Can. Forestry Serv. Bi-Mon. Res. Notes 28:25-26.
7. Carmean, W. H. 1971. Large, well balanced stock and control of grass competition needed for red pine plantings on sandy soils. Tree Planters' Notes 22(1):8-10.
8. Chavasse, C. G. R., and G. P. Bowles. 1976. The growth of precision-sown 1.5/0 radiata pine on a benign site. New Zealand Forest Serv., Forest Res. Institute, Rotorua. Forest Establishment Rep. 96 (unpubl.).
9. Cleary, B. D., R. D. Greaves, and R. K. Hermann (eds.). 1978. Regenerating Oregon's forests. Oregon State Univ. Ext. Serv., Corvallis. 287 p.
10. Copes, D., F. Sorensen, and R. Silen. 1969. Douglas-fir seedling grows 8 feet tall in two seasons. J. Forestry 67:174-175.
11. Cranswick, A. M. 1977. Sugar levels in box pruned and wrenched radiata pine seedlings. New Zealand Forest Serv., Forest Res Institute, Tree Physiology, Rotorua. Internal Rep. 8 (unpubl.).
12. Cranswick, A. M., and J. A. Zabkiewicz. 1979. Quantitative analysis of monosaccharides, cyclitols, sucrose, quinic and shikimic acids in P. *radiata* extracts on a glass SCOT capillary column by automated gas chromatography. J. Chromatography 171:233-242.
13. Cremer, K. W. 1972. Seasonal patterns of shoot development in *Pinus radiata* near Canberra. Australian Forest Res. 6:31-52.
14. Crozier, A., J. B. Zaerr, and R. O. Morris. 1980. High-performance steric exclusion chromatography of plant hormones. J. Chromatography 198:57-63.
15. Daniels, T. G. 1979. The effects of winter plant moisture stress on survival and growth of 2+0 Douglas-fir seedlings. M.S. thesis, Oregon State Univ., Corvallis. 86 p.
16. Doorenbos, J. 1953. Review of the literature on dormancy in buds of woody plants. Mededelingen Lanbouwhogeschool Wageningen 53:1-23.
17. Edgren, J. W. 1977. Seedbed density, diameter limit culling, and 2+0 Douglas-fir seedling production. Proc., Meeting of Western Forest Nursery Council and Intermountain Nurserymen's Assoc., Richmond B. C., Aug. 10-12, 1976.
18. Edgren, J. W., and J. G. Iyer. 1979. Immersion determination of seedling root-top ratio and its correlation with a drought resistance index. Tree Planters' Notes 30(1):8-10.
19. Filer, T. H., Jr., and G. W. Peterson. 1975. Damping-off. Pages 6-8 *in* Forest nursery diseases in the United States (G. W. Peterson and R. S. Smith, Jr., eds.). U.S.D.A. Forest Serv., Washington, D.C. Agric. Handb. 470.
20. Franklin, J. F., and C. T. Dyrness. 1973. Natural vegetation of Oregon and Washington. U.S.D.A. Forest Serv., Pacific NW Forest and Range Exp. Sta., Portland, Oregon. Gen. Tech. Rep. PNW-8. 417 p.
21. Hahn, P. F. 1983. Containerized seedling production in a shelterhouse system. Pages 110-128 *in* Proc., 1982 Western Nurserymen's conf., Medford, Oregon, Aug. 10-12. Southern Oregon State Coll., Ashland.
22. Hay, R. L., and F. W. Woods. 1974. Shape of root system influences survival and growth of loblolly seedlings. Tree Planters' Notes 25(3):1-2.
23. Hermann, R. K. 1964. Importance of top-root ratios for survival of Douglas-fir seedlings. Tree Planters' Notes 64:7-11.
24. Hermann, R. K. 1974. Frost damage in the nursery and its effect on field performance of Douglas-fir seedlings. Pages 117a-d *in* Proc., Western Forest Nursery Council meeting, Portland, Oregon, Aug. 5-7.
25. Hermann, R. K., and D. P. Lavender. 1968. Early growth of Douglas-fir from various altitudes and aspects in southern Oregon. Silvae Genetica 17(4):143-151.
26. Ingestad, T. 1974. Towards optimum fertilization. Ambio 3(2):49-54.
27. Ingestad, T. 1977. Nitrogen and plant growth: maximum efficiency of nitrogen fertilizers. Ambio 6(2-3):146-151.
28. Ingestad, T. 1979. Mineral nutrient requirements of *Pinus silvestris* and *Picea abies* seedlings. Physiologia Plantarum 45:373-380.
29. Iverson, R. D., and M. Newton. 1980. Large Douglas-fir seedlings perform best on Oregon coastal sites. International Paper Co., Wood Products and Resources Group, Lebanon, Oregon. Tech. Note 55. 9 p.
30. Jeffers, N. 1978. Unpublished data, Weyerhaeuser Co.
31. Jenkinson, J. L., and J. A. Nelson. 1983. 1-0 Douglas-fir: a bare root planting option. Pages 63-76 *in* Proc., 1982 Western Nurserymen's conf., Medford, Oregon, Aug. 10-12. Southern Oregon State Coll., Ashland.
32. Jones, E. W. 1968. A note on the dimensions of shoots and roots of planting stock. Forestry 41(2):199-206.
33. Khan, A. A. 1975. Primary, preventive and permissive roles of hormones in plant systems. The Botanical Review 41:391-420.
34. Krueger, K. W., and J. M. Trappe. 1967. Food reserves and seasonal growth of Douglas-fir seedlings. Forest Sci. 13(2):192-202.
35. Krugman, S. L., and E. C. Stone. 1966. The effect of cold nights on the root regeneration potential of ponderosa pine seedlings. Forest Sci. 12:451-459.
36. Kuser, J. 1980. Provenance variation in western hemlock [*Tsuga heterophylla* (Raf.) Sarg.] seedlings. Ph.D. thesis, Oregon State Univ., Corvallis. 88 p.
37. Lanner, R. M. 1976. Patterns of shoot development in *Pinus* and their relationship to growth potential. Pages 223-243 *in* Tree physiology and yield improvement (M. G. R. Cannell and F. T. Last, eds.). Academic Press, New York.
38. Lavender, D. P. 1981. Environment and shoot growth of woody plants. Oregon State Univ., Forest Res. Lab., Corvallis. Res. Pap. 45. 43 p.
39. Lavender, D. P. 1982. Seedling dormancy and frost hardiness. Pages 93-96 *in* Proc., Biology and management of true fir in the Pacific Northwest symp. (C. D. Oliver and R. M. Kenady, eds.). Coll. Forest Resources, Institute of Forest Resources, Univ. of Washington, Seattle. Contribution No. 45.
40. Lavender, D. P., K. K. Ching, and R. K. Hermann. 1968. Effect of environment on development of dormancy and growth of Douglas-fir seedlings. Botanical Gazette 129:70-83.
41. Lavender, D. P., and B. D. Cleary. 1974. Coniferous seedling production techniques to improve seedling establishment. Pages 177-180 *in* Proc., North American containerized forest tree seedling symp. (R. W. Tinus, W. I. Stein, and W. E. Balmer, eds.). Great Plains Agric. Res. Publ. 68.
42. Lavender, D. P., and W. S. Overton. 1972. Thermoperiods and soil temperatures as they affect growth and dormancy of Douglas-fir seedlings of different geographic origin. Oregon State Univ., Forest Res. Lab., Corvallis. Res. Pap. 13. 26 p.

43. Lavender, D. P., and R. B. Walker. 1979. Nitrogen and related elements in nutrition of forest trees. Pages 15-22 *in* Proc., Forest fertilization conf. (S. P. Gessel, R. M. Kenady, and W. A. Atkinson, eds.). Coll. Forest Resources, Institute of Forest Resources, Univ. of Washington, Seattle. Contribution No. 40. 275 p.

44. Ledig, F. T., and T. O. Perry. 1965. Physiological genetics of the shoot-root ratio. Pages 39-43 *in* Proc., Society of American Foresters, Division of Silviculture, Detroit, Michigan.

45. Little, S., and H. A. Somes. 1964. Root systems of direct-seeded and variously planted loblolly, shortleaf, and pitch pines. U.S.D.A. Forest Serv., NE Forest Exp. Sta., Upper Darby, Pennsylvania. Res. Pap. NE-26. 13 p.

46. Long, J. N. 1978. Root system form and its relationship to growth in young planted conifers. Pages 222-234 *in* Proc., Root form of planted trees symp. (E. Van Eerden and M. Kinghorn, eds.). B. C. Forest Serv./Canadian Forestry Serv., Victoria. Joint Rep. 8. 357 p.

47. Lopushinsky, W., and T. Beebe. 1976. Relationship of shoot-root ratio to survival and growth of outplanted Douglas-fir and ponderosa pine seedlings. U.S. Forest Serv., Pacific NW Forest and Range Exp. Sta., Portland, Oregon. Res. Note PNW-274. 7 p.

48. Lyr, H., and G. Hoffmann. 1967. Growth rates and growth periodicity of tree roots. International Review of Forestry Res. 1:181-236.

49. MacDonald, E. M. S., D. E. Akiyoshi, and R. O. Morris. 1981. Combined high performance liquid chromatography—radio immunoassay for cytokinins. J. Chromatography 214:101-109.

50. Mexal, J. G. 1979. Unpublished data, Weyerhaeuser Co., Hot Springs, Arkansas.

51. Morris, J. W. 1979. Growth and moisture relations of western hemlock seedlings as affected by root or shoot disturbance. M.S. thesis, Oregon State Univ., Corvallis. 78 p.

52. Mullin, R. E., and L. Bowdery. 1978. Effects of nursery seedbed density and top-dressing fertilization on survival and growth of 3+0 red pine. Can. J. Forest Res. 8(1):30-35.

53. Mullin, R. E., and C. Christl. 1981. Morphological grading of white spruce nursery stock. Forestry Chronicle 57(3):126-130.

54. Mullin, R. E., and C. Christl. 1982. Morphological grading of white pine nursery stock. Forestry Chronicle 58(1):40-43.

55. Nelson, E. A., and H. W. Anderson. 1966. Handling large nursery stock. Pages 33-41 *in* Proc., 10th Biennial Western Forest and Conservation Assoc., Placerville and Davis, California.

56. Nelson, E. A., and D. P. Lavender. 1976. Dormancy of western hemlock seedlings. Pages 103-107 *in* Western hemlock management (W. A. Atkinson and R. J. Zasoski, eds.). Coll. Forest Resources, Univ. of Washington, Seattle. 317 p.

57. Nelson, E. A., and D. P. Lavender. 1979. The chilling requirement of western hemlock seedlings. Forest Sci. 25(3):485-490.

58. Owens, J. N., and M. Molder. 1973. A study of DNA and mitotic activity in the vegetative apex of Douglas-fir during the annual growth cycle. Can. J. Botany 51(7):1395-1409.

59. Owens, J. N., and M. Molder. 1973. Bud development in western hemlock. I. Annual growth cycle of vegetative buds. Can. J. Botany 51(11):2223-2231.

60. Pierpoint, G., J. M. Paterson, J.G. Boufford, and C. Glerum. 1977. Irregular growth of outplanted red pine. II. The influence of handling and planting on first-year performance. Ontario Ministry of Natural Resources. Forest Res. Note 6. 4 p.

61. Pierpoint, G., J. M. Paterson, J. G. Boufford, and C. Glerum. 1978. Irregular growth of outplanted red pine. III. The influence of handling and planting on second-year performance. Ontario Ministry of Natural Resources. Forest Res. Note 14. 4 p.

62. Pierpoint, G., J. M. Paterson, J. G. Boufford, and C. Glerum. 1981. Irregular growth of outplanted red pine. VI. The influence of handling and planting on fifth-year performance. Ontario Ministry of Natural Resources. Forest Res. Note 29. 4 p.

63. Radwan, M. A., G. L. Crouch, and H. S. Ward. 1971. Nursery fertilization of Douglas-fir seedlings with different forms of nitrogen. U.S.D.A. Forest Serv., Pacific NW Forest and Range Exp. Sta., Portland, Oregon. Res. Pap. PNW-113. 8 p.

64. Rehfeldt, G. E. 1979. Patterns of first-year growth in populations of Douglas-fir (*Pseudotsuga menziesii* var. *glauca*). U.S.D.A. Forest Serv., Intermountain Forest and Range Exp. Sta., Ogden, Utah. Res. Note INT-255. 7 p.

65. Revel, J. 1965. Effect of method and density of sowing on Douglas-fir seedlings. Tree Planters' Notes 72:13-15.

66. Ritchie, G. A., and J. R. Dunlap. 1980. Root growth potential: its development and expression in forest tree seedlings. New Zealand J. Forestry Sci. 10(1):218-248.

67. Romberger, J. A. 1963. Meristems, growth and development in woody plants. U.S.D.A. Forest Serv., Washington, D.C. Tech. Bull. 1293. 214 p.

68. Rudolph, P. O. 1939. Why forest plantations fail. J. Forestry 37:377-383.

69. Rudolph, T. D. 1964. Lammas growth and prolepsis in jack pine in the Lake States. Forest Sci. Monograph 6. 70 p.

70. Schubert, G. H., and R. S. Adams. 1971. Reforestation practices for conifers in California. State of California, Div. of Forestry, Sacramento. 359 p.

71. Shipman, R. D. 1966. Low seedbed densities can improve early height growth of planted slash and loblolly pine seedlings. Tree Planters' Notes 76:24-29.

72. Shoulders, E. 1961. Effect of nursery bed density on loblolly and slash pine seedlings. J. Forestry 59:576-579.

73. Smith, J. H. G., A. Kozak, O. Sziklai, and J. Walters. 1966. Relative importance of seedbed fertilization, morphological grade, site, provenance, and parentage to juvenile growth and survival of Douglas-fir. Forestry Chronicle 42(1):83-86.

74. Smith, R. S., Jr. 1975. Grey mold of giant sequoia. Pages 47-49 *in* Forest nursery diseases in the United States (G. W. Peterson and R. S. Smith, Jr., eds.). U.S.D.A. Forest Serv., Washington, D.C. Agric. Handb. 470.

75. Stoeckeler, J. H., and G. A. Linstrom. 1950. Reforestation research findings in northern Wisconsin and upper Michigan. U.S.D.A. Forest Serv., Lake States Sta., Minneapolis, Minnesota. Pap. 23. 34 p.

76. Stone, E. C. 1970. Variation in the root-growth capacity of ponderosa pine transplants. Pages 40-46 *in* Regeneration of ponderosa pine (R. K. Hermann, ed.). School of Forestry, Oregon State Univ., Corvallis. 125 p.

77. Stone, E. C., J. L. Jenkinson, and S. L. Krugman. 1962. Root-regenerating potential of Douglas-fir seedlings lifted at different times of the year. Forest Sci. 8:288-297.

78. Stone, E. C., and E. A. Norberg. 1979. Use of root growth capacity in developing propagation regimes, storage criteria, and nursery stock certification. A symposium on regeneration and management of young true fir stands. Redding, California. 35 p.

79. Stone, E. C., and G. H. Schubert. 1959. Root regeneration of ponderosa pine seedlings lifted at different times of the year. Forest Sci. 5:322-332.

80. Sutton, R. F. 1969. Form and development of conifer root systems. Commonwealth Forestry Bureau, Oxford, England. Tech. Commun. 7. 131 p.

81. Thimann, K. V. 1972. The natural plant hormones. *In* Plant physiology, a treatise. Vol. VIB. Physiology of development: the hormones. Academic Press, New York. 365 p.

82. Timmis, R. 1978. Frost hardening of containerized conifer seedlings under constant and sequenced temperatures. Weyerhaeuser Co. Forestry Res. Tech. Rep. 042-2201/78/37. 21 p.

83. Tinus, R. W., and S. E. McDonald. 1979. How to grow tree seedlings in containers in greenhouses. U.S.D.A. Forest Serv., Rocky Mountain Forest and Range Exp. Sta., Fort Collins, Colorado. Gen. Tech. Rep. RM-60. 256 p.

84. Toumey, J. W. 1929. Initial root habit in American trees and its bearing in regeneration. Proc., International Congress of Plant Science, Ithaca, New York, 1926. 1:713-728.

85. Trewavas, A. 1981. How do plant growth substances work? Plant, Cell and Environ. 4:203-228.

86. Trewin, A. R. D. 1979. Poor handling can upset forest plans. Forest Industries Review (July):10.

87. van den Driessche, R. 1969. Forest nursery handbook. B. C. Forest Serv., Victoria. Res. Note 48. 44 p.

88. van den Driessche, R. 1975. Flushing response of Douglas-fir buds to chilling and to different air temperatures after chilling. B.C. Forest Serv., Research Division, Victoria. Res. Note 71. 22 p.

89. van den Driessche, R. 1980. Effects of nitrogen and phosphorus fertilization on Douglas-fir nursery growth and survival after outplanting. Can. J. Forest Res. 10:65-70.

90. van den Driessche, R. 1980. Health, vigour and quality of conifer seedlings in relation to nursery soil fertility. Pages 100-120 *in* Proc., North American forest tree nursery soils workshop (L. P. Abrahamson and D. H. Bickelhaupt, eds.). State Univ. New York, Coll. Environ. Sci. and Forestry, Syracuse.

91. Van Eerden, E., and J. M. Kinghorn. 1978. Proc., Root form of planted trees symp. B. C. Forest Serv./Canandian Forestry Serv., Victoria. Joint Rep. 8. 357 p.

92. Wahlenberg, W. G. 1928. Experiments with classes of stock suitable for forest planting in the northern Rocky Mountains. J. Agric. Res. 36:977-1000.

93. Wakeley, P. C. 1954. Planting the southern pines. U.S.D.A. Forest Serv., Washington, D.C. Agric. Monograph 18. 233 p.

94. Wakeley, P. C. 1969. Results of southern pine planting experiments established in the middle twenties. J. Forestry 67:237-241.

95. Wang, J. W. 1960. A critique of the heat unit approach to plant response studies. Ecology 41:785-790.

96. Wareing, P. F. 1950. Growth studies in woody species. I. Photoperiodism in first-year seedlings of *Pinus silvestris*. Physiologia Plantarum 3:258-275.

97. Weiser, C. J. 1970. Cold resistance and injury in woody plants. Science 169:1269-1277.

98. White, A., P. Handler, E. L. Smith, and D. Stetten, Jr. 1954. Principles of biochemistry. McGraw-Hill, New York. 1117 p.

99. Wilkinson, G. B. 1969. Some establishment problems in Southland and Otago. New Zealand J. Forestry 14:170-177.

100. Wommack, D. E. 1964. Temperature effects on the growth of Douglas-fir seedlings. Ph.D. dissertation, School of Forestry, Oregon State Univ., Corvallis. 176 p.

101. Zabkiewicz, J. A. 1979. Unpublished data, New Zealand Forest Res. Institute, Rotorua.

102. Zaerr, J. B. 1983. Short-term flooding and net photosynthesis in seedlings of three conifers. Forest Sci. 29(1):71-78.

103. Zaerr, J. B., and D. P. Lavender. 1974. The effects of certain cultural and environmental treatments upon the growth of roots of Douglas-fir [*Pseudotsuga menziesii* (Mirb.) Franco] seedlings. Pages 27-32 *in* 2nd International symp. on ecology and physiology of root growth. Academie-Verlag, Berlin.

104. Zaerr, J. B., and D. P. Lavender. 1980. Analysis of plant growth substances in relation to seedling and plant growth. New Zealand J. Forestry Sci. 10(1):186-195.

105. Zahner, R. 1968. Water deficits and growth of trees. Pages 191-254 *in* Water deficits and plant growth (T. T. Kozlowski, ed.). Academic Press, New York. 333 p.

106. Zimmerman, M. H., and C. L. Brown. 1971. Tree structure and function. Springer-Verlag, New York. 336 p.

Chapter 15
Nursery Cultural Practices: Impacts on Seedling Quality

M. L. Duryea

Abstract

Better understanding and implementation of nursery cultural practices to improve seedling quality will enable better matching of seedlings to forest sites, reducing the chance of regeneration delay and improving future growth of forest stands. This chapter reviews a number of important cultural practices and the ways in which they affect indicators of seedling quality (morphology and physiology) and, ultimately, field performance (growth and survival). Early spring sowing produces larger seedlings that can complete their growth and be hardened by midsummer. Lowering seedbed density results in more seedlings from a given amount of seed and can improve field survival and growth. A nursery irrigation schedule that imposes moderate stress on seedlings in midsummer induces earlier budset and seedling dormancy and increases field-survival

In Duryea, Mary L., and Thomas D. Landis (eds.). 1984. *Forest Nursery Manual: Production of Bareroot Seedlings.* Martinus Nijhoff/Dr W. Junk Publishers, The Hague/Boston/Lancaster, for Forest Research Laboratory, Oregon State University, Corvallis. 386 p.

potential. Most often, field survival and growth are improved with growing-season fertilization in the nursery; fall fertilization also may increase field growth of Northwest species. Root wrenching in dry soil and (or) hot, dry weather without immediate irrigation can greatly stress Douglas-fir seedlings and should be avoided because of increased chance of seedling mortality in nursery beds or reduced growth later in the field. Wrenching to mildly stress seedlings can induce budset and hardening and may benefit field growth and survival. Top pruning, to control shoot height and achieve crop-size uniformity, should be done during the period of active seedling growth in early summer to ensure proper development of terminal buds. Transplant seedlings have more fibrous root systems, larger stem diameters, and lower shoot:root ratios than seedlings of comparable age grown at a standard density; seedlings are most commonly transplanted in spring and are outplanted as 1+1s or 2+1s. It is important for nursery managers to be aware of interactions among the various nursery practices they employ; if a current practice is altered or discontinued or a new practice added, careful attention should be given to the effect of this change on other cultural practices in the nursery.

15.1 Introduction

A seedling is considered of high quality if it meets the expectations or standards of performance on a particular planting site. The first and most obvious performance standard is survival—without adequate survival a site must be replanted or interplanted. The second performance standard is rapid seedling growth. Levels of survival and growth which are considered adequate must be defined for each individual site. Failure to meet these specified levels means an increase in the time until a particular forest stand reaches merchantable size and may be harvested. This regeneration delay, caused by either a replant of the site or slow initial growth, results in a loss of value and volume yield for that forest site [19]. Better understanding and implementation of cultural practices to improve seedling quality should enable better matching of seedlings to forest sites, reducing the chance of regeneration delay and improving future growth of forest stands.

My objective in this chapter is to review a number of important cultural practices and the ways in which they affect indicators of seedling quality (morphology and physiology) and field performance (growth and survival). Three practices—root culturing, top pruning, and transplanting—are presented in more detail because they are not substantially covered elsewhere in the *Manual*.

15.2 Seedling Quality Criteria

In attempts to set standards for seedling quality, three types of criteria have been used: (1) stock-type description, (2) morphological characteristics, and (3) physiological condition. The role that each plays in describing seedling quality is discussed in this section.

15.2.1 Stock-type description

Stock is described by seedling age and growing location. A 1+0 is grown for 1 year in a seedbed and 0 years in a transplant bed; a 2+1 is grown for 2 years in a seedbed and 1 year in a transplant bed. Although studies to determine which stock type survives and grows best on a particular site have been common (see chapter 24, this volume), contradictory results from such comparisons suggest that variability in seedling morphology and physiology must play an important role. Foresters who formerly requested seedlings by stock-type description now realize that more information is needed to describe a seedling and predict its field performance. Some nurseries have already changed the seedling descriptions given to their customers to include average height, stem diameter, and shoot:root ratio in addition to the standard stock-type designation [97].

15.2.2 Morphological characteristics

Morphological characteristics are the physical or visually determinable attributes of a seedling. The major morphological criteria used to describe seedling quality—shoot height, stem diameter, root mass, and shoot:root ratio—are the basis for grading seedlings at the nursery; seedlings thought to have low survival and growth potential (culls) are eliminated. Some studies attempting to show how these morphological criteria are important to successful field performance are discussed in the next sections.

15.2.2.1 Shoot height

Seedling height at the time of outplanting can greatly influence growth rate in the field. Height increment of Douglas-fir [*Pseudotsuga menziesii* (Mirb.) Franco] planted as 4-year-old seedlings was strongly correlated with height at the time of planting [75]; at 5 and 10 years of age, height increment of the tallest seedlings was more than twice that of the smallest.

Survival of Monterey pine (*Pinus radiata* D. Don) in Australia was the same regardless of seedling height at the time of field planting, but growth rate during the early years in the field was strongly influenced by initial stock size [72]. Where seedlings were segregated into a large and a small stand, the two stands showed equal growth after 10 years in the field; however, where seedling sizes were mixed, most of the initially small seedlings remained smaller than the larger stock after 8½ years and productivity per acre was correspondingly lower, according to the proportion of small seedlings planted in the stand.

15.2.2.2 Stem diameter

Generally, seedlings with larger root-collar diameters (which tend to be larger stock) have better outplanting success [80]. Anstey [5] found stem diameter alone to be a valuable measurement of 1+0 Monterey pine seedling quality. Growth after three seasons in the field for seedlings 5 mm or more in diameter was twice that of seedlings with 2-mm diameter. On a harsh site, survival increased from 72% for seedlings with a 2-mm diameter to 89% for 4 mm, to 98% for 6 mm. Chavasse [23] found that root-collar diameter of Monterey pine and Douglas-fir was a better indicator of seedling quality than shoot height.

15.2.2.3 Root system

Root mass (including dry weight and overall fibrosity) has recently been recognized as one of the most important factors critical to field performance. Survival of Douglas-fir seedlings with poor root systems was significantly lower than that of seedlings with good root systems regardless of shoot-height class [40]; Hermann concluded that a high shoot:root ratio does not necessarily mean low survival if seedlings have a well-developed root system and that root development is a reliable criterion for predicting seedling survival.

In a more recent study, 2+0 ponderosa pine (*Pinus ponderosa* Dougl. ex Laws.) and Douglas-fir seedlings of three different top heights were separated according to root size (large and small) and then planted in north-central Washington [55]. Survival was 22 to 26% greater for Douglas-fir seedlings with large roots and 5 to 15% greater for ponderosa pine seedlings with large roots. Height growth for trees of both species with large roots was 1.2 to 1.7 times that of those with small roots

15.2.2.4 Morphological grades

In many studies, seedlings have been graded according to morphological characteristics, and then field performance of those ranked morphological grades has been tested. Wakeley [111] established three grades for southern pine nursery stock based on observable and measurable seedling characteristics. Each species had its own specifications for each of the three grades; grades 1 and 2 were considered plantable, and grade 3 was culled. Slash pine (*Pinus elliottii* Engelm.) seedlings from four nurseries and loblolly pine (*Pinus taeda* L.) seedlings from one nursery were separated into three grades similar to Wakeley's [111] and measured after 13 growing seasons in the field [17]. Rust infection and disease were no different among seedling grades. However, grades 1 and 2 generally survived and grew better in the field than grade 3 (Fig. 1), though some exceptions suggested that these grades are not always reliable for ranking subsequent survival and growth [17].

When white spruce [*Picea glauca* (Moench) Voss] seedlings were graded and then measured after 5 years in the field, shoot height, stem diameter, root volume, and shoot:root ratio were all highly significant predictors of subsequent growth, with larger seedlings performing best [69]. Growth and survival of white pine (*Pinus monticola* Dougl. ex D. Don) were predicted by shoot height, stem diameter, and root length [70].

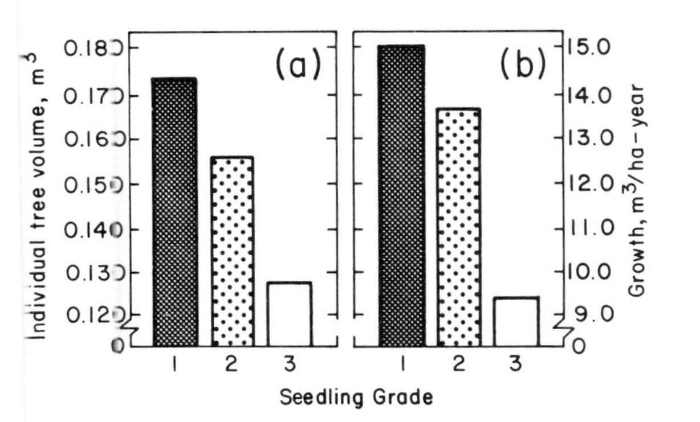

Figure 1. (a) Individual tree volume and (b) volume growth/ha for three seedling grades of slash pine after 13 growing seasons in the field (adapted from [17]).

15.2.3 Physiological condition

The variable results of using only stock-type description and (or) morphological characteristics to predict survival and growth have caused much dissatisfaction. Many authors point out that outplanting performance depends not only on seedling appearance, but on its preconditioning and resultant physiological state [23]. Others have mentioned a need for physiological grades for assessing seedling quality [46, 79, 82]. Chapter 23 (this volume) reviews various techniques for evaluating seedling quality.

Physiological condition of seedlings can influence field performance either independently or in conjunction with morphological characteristics. For example:

- 1+0 Monterey pine seedlings grown at lower density had better root-growth capacity after lifting than those grown at higher density. The seedlings with better root-growth capacity subsequently had better height growth and survival after 2 years in the field [10].
- Earlier dormancy induction due to moderate moisture stress resulted in greater cold hardiness of Douglas-fir seedlings and better growth-room survival [18].
- Root-growth capacity predicted white spruce survival independent of stock type and seedling size [62].
- Fall fertilization of Douglas-fir improved survival after 2 years in the field and growth for 5 years. Presumably, the seedlings that were fertilized had higher concentrations of nitrogen (N) than the unfertilized seedlings [4].
- Root wrenching of Monterey pine seedlings increased the proportion of total carbohydrates translocated to roots, compared to shoots. Roots then grew at the expense of shoots. When outplanted, wrenched seedlings had superior survival and growth, compared to unwrenched seedlings [77, 108].
- Root wrenching of Monterey pine and Douglas-fir many times during the growing season without adequate fertilization decreased seedling nutrient concentrations; seedlings in turn stagnated in the field [14, 77, 108].

How nursery practices influence seedling physiological condition—and, ultimately, field performance—is detailed for each cultural practice in the following sections.

15.3 Sowing

15.3.1 Seed quality

Seed quality is important for growing high-quality seedlings. Seed purity, weight, germination potential, and vigor must be accurately assessed so that the correct sowing rate can be calculated and an evenly spaced seedbed attained (see chapters 4 and 5, this volume). The need for stratification and the treatment time should be carefully determined for seed of different species and geographic origin because it can affect germination rate, vigor, and amount and, therefore, seedbed uniformity.

15.3.2 Sowing depth

Sowing depth can influence germination rate and amount and, thus, the final number of seedlings in the seedbed ([86]; also see chapter 5, this volume). Sowing depth of Douglas-fir seed at nurseries in the Northwest ranges from 1/16 to ½ inch (0.16 to 1.27 cm); most nurseries sow seed at ¼ inch (0.64 cm) (OSU Nursery Survey; see chapter 1, this volume). The recommended sowing depth for optimum germination of Douglas-fir ranges from ⅛ to ¼ inch (0.32 to 0.64 cm) [86, 103]. Sowing depth of other Northwest species varies from nursery to nursery: ponderosa pine—1/16 to ½ inch (0.16 to 1.27 cm), noble fir (*Abies procera* Rehd.)—⅛ to ¼ inch (0.32 to 0.64 cm), lodgepole pine (*Pinus contorta* var. *latifolia* Engelm.)—⅛ to ½ inch (0.32 to 1.27 cm), and spruce (*Picea* spp.)—3/16 to ¼ inch (0.48 to 0.64 cm) (OSU Nursery Survey). Recommendations for ponderosa pine vary from ¼ to ½ inch (0.64 to 1.27 cm) [86, 101, 103].

To ensure good growth and crop uniformity, it is important to choose a sowing depth proper for the tree species to be planted, to prepare a level seedbed, and to ensure consistent seed depth throughout the bed. A person continuously walking behind the seeder can check to make sure that proper depth control is maintained. Most nurseries sow on the shallow side, allowing an occasional seed to remain uncovered.

15.3.3 Sowing date

Seeds can be sown in fall or spring. Fall-sown seeds are planted dry, are naturally stratified in the seedbed over winter, and germinate earlier in the spring than spring-sown seed, producing larger 1+0s [86, 90, 103, 106]. In British Columbia, van den Driessche [102, 106] found shoot length and shoot and root dry weights to be larger for fall-sown than spring-sown Douglas-fir and, in another study, seedling dry weights of three of the four species tested to be greater for fall-sown than spring-sown stock (Fig. 2). However, fall sowing has some important disadvantages: (1) seed loss is often extensive during winter due to heavy rains, birds and rodents, or fungi, resulting in poor spacing and stocking of seedlings; (2) when seeds germinate too early in spring, young seedlings can be killed by frost unless protected; (3) natural stratification may be inadequate where nurseries are located in warm climates; and (4) irrigation may be needed to prevent drying of seed in an early spring drought. For these reasons, most sowing in the Northwest occurs in spring. A nursery manager who chooses to sow in fall is taking a great risk that yields will not be adequate.

Spring sowing, if done early enough, can produce 1+0 seedlings as large as those sown in fall. Sorensen [89] found final height of 1+0 Douglas-fir seedlings to be larger by 0.5 mm for each day of earlier sowing in spring; furthermore, these earlier sown seedlings set bud 1 month earlier than those sown later. The height difference was still evident in these seedlings as 2+0s and in the final crop (Fig. 3). In British Columbia, March-sown Douglas-fir seedlings were twice the height of June-sown seedlings and also had greater root length and root and shoot dry weight [102]. Early spring sowing at Webster Nursery (Olympia, Washington) resulted in larger roots and shoots in the fall of the 1+0 year [pers. commun., 3]; these early-sown seedlings were still larger when harvested as 2+0s. At a northern California nursery which had snow until May 16, early sowing (May 16) versus June 15 sowing of ponderosa pine, Jeffrey pine (*Pinus jeffreyi* Grev. & Balf.), Douglas-fir, white fir [*Abies concolor* (Gord. & Glend.) Lindl. ex Hildebr.], and incense-cedar (*Libocedrus decurrens* Torr.) resulted in (1) more rapid and complete germination, (2) more uniform density, (3) a larger number of seedlings, and (4) a greater number of superior 1+0s and, after transplanting, 1+1s and, again after transplanting, 1+1+1s [86].

Ten percent of the Northwest nurseries begin to sow in March, 50% in April, and 40% in May. Almost all nurseries are still sowing on May 15, and 40% are still sowing in June. Although poor weather conditions and wet soils limit access to seedbeds for sowing, there are often short periods of time—even a few days in spring—when nurseries can take advantage of dry weather to sow. If more than one seeder were available at a nursery, seeding could be completed during such favorable sowing "windows."

15.3.4 Conclusions

The importance of early sowing cannot be overemphasized. Early-sown 1+0 seedlings benefit from having the entire growing season and are large enough for hardening by July and August. The final 2+0 crop is larger the next year and again ready for hardening by midsummer. The result of early sowing means better crop control for the nursery manager, reducing the risk of growing seedlings that are too small and that must be "pushed" for additional growth in late summer. Earlier sowing often results in increased yield of high-quality seedlings.

15.4 Seedling Spacing and Seedbed Density

Seedbed density is the number of seedlings growing in an area of seedbed, expressed either on an area basis (seedlings per square meter or foot) or on a lineal basis (seedlings per lineal meter or foot). The spacing between seedlings can vary according to either the distance between drill rows or the distance between each seedling within a drill row. In the Northwest, 2+0 seedbeds have drill rows 6 inches (15.24 cm) apart (OSU Nursery Survey); spacing is usually varied within the drill row. For example, if seedlings are 1 inch (2.54 cm)

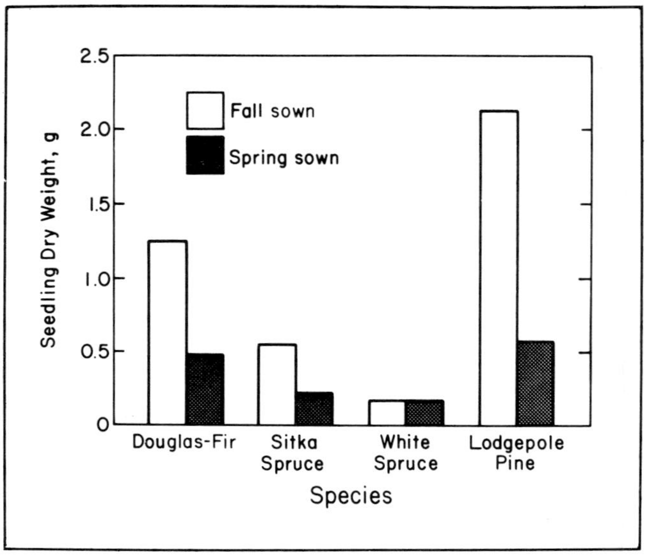

Figure 2. Dry weights of 1+0 seedlings grown from fall- and spring-sown seed for four species (adapted from [106]).

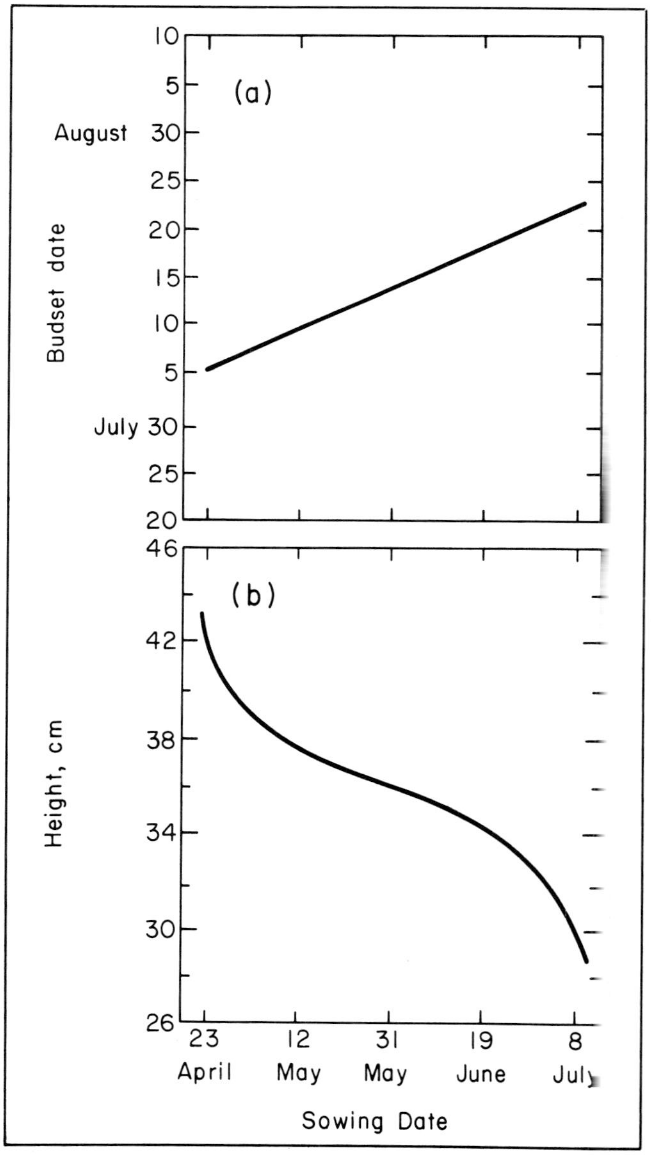

Figure 3. General effect of sowing date on (a) date of 2+0 seedling budset and (b) seedling height (adapted from [89]).

apart, a row would contain approximately 24 seedlings/ft² (258 seedlings/m²); if they are 2 inches (5.08 cm) apart, a row would contain approximately 12 seedlings/ft² (129 seedlings/m²). For a list of seedbed densities used in the Northwest, see chapter 5 this volume.

Uniform spacing between seedlings within a drill row is extremely important to seedbed density. Any local variation caused when seedlings clump in the seedbed results in lack of uniformity in growing conditions and therefore in greater variation in seedling quality. When specific seedbed densities are discussed in the following section, spacings are assumed even, permitting similar growing conditions for all seedlings within that density. However, lack of uniform spacing and inability to control the final growing density are some of the most important problems in Northwest nurseries. Almost all other nursery practices interact with seedbed density—thus density dictates how the crop will respond to practices such as fertilization, root wrenching, or irrigation.

15.4.1 Number of acceptable seedlings

Though lower seedbed densities increase yield percents and reduce the number of culls per lineal foot (meter) of seedbed [31, 84], determining the number of acceptable seedlings—those available for field planting—produced at these lower densities is the "bottom line."

Edgren [33] found that, when using diameter for culling, for a minimum acceptable stem diameter of 4 mm, 40 seedlings were acceptable and none were culls when grown at a seedbed density of 40 seedlings/lineal foot, whereas 48 were acceptable and 232 were culls when grown at a seedbed density of 280 seedlings/lineal foot (Table 1). Several studies have examined the quantity of seedlings produced within each of Wakeley's [111] morphological seedling grades (1, 2, and 3, where 3 = cull) at different seedbed densities [22, 84]. The proportion of large, morphologically high-grade seedlings was usually found to increase as seedbed density decreased [22]. Studies with Monterey pine have used the shoot height:stem diameter ratio as the basis for dividing seedlings into four grades to determine the number of seedlings produced at each seedbed density [10, 64].

These studies all attempt to identify the optimum seedbed density for producing the highest number of plantable seedlings. It should be emphasized, however, that this optimum may change with different nurseries, seed sources, cultural practices, and seedling-quality specifications.

15.4.2 Seedling morphology

In general, lowering seedbed density produces seedlings with larger stem diameters and heavier shoots and roots (dry weight). Seedling heights and shoot:root ratios are only sometimes affected by seedbed density.

Three+0 white spruce grown at 15 seedlings/ft² were larger and heavier and had a lower shoot:root ratio than those grown at 30 seedlings/ft² (Table 2) [67]. In another study, heights of 2+0 white spruce seedlings were greater when densities were reduced from 80 to 10 seedlings/ft² [7].

Lowering Douglas-fir 1+0 seedbed density increased seedling dry weight and stem diameter but did not affect shoot height [103]. More recently, van den Driessche [107] reported that lower seedbed density increased seedling dry weight, root-collar diameter, and, in this case, height of coastal (var. *menziesii*) and interior (var. *glauca*) Douglas-fir, Sitka spruce [*Picea sitchensis* (Bong.) Carr.], and lodgepole pine. Growing 1+0 and 2+0 ponderosa pine at lower densities at two California nurseries increased stem diameter and fresh weight [11]. At the Bend Nursery, stem diameter of 3+0 ponderosa pine grown at 10 seedlings/ft² was 7.1 mm, at 30/ft² 5.2 mm, and at 70/ft² 4.6 mm; however, height and shoot:root ratio were unaffected by seedbed density [31].

Monterey pine seedlings grown at lower densities were found to have larger root-collar diameters, shoot heights, and root and shoot dry weights [13]; in this study, spacing within the drill row affected seedling size more than distance between rows. Furthermore, variation in seedling size decreased as density decreased—an important point in this and other studies.

In summary, diameter is more affected by seedbed density than height [113], except possibly for white spruce. Decreasing seedbed density (increasing the growing space) for each seedling results in larger stem diameters, increased dry weights, and more uniform crop size for most species.

Table 2. Morphological characteristics of 3+0 white spruce grown at two seedbed densities and two nurseries (adapted from [67]).

Nursery	Seedbed density, seedlings/ft²	Shoot height, cm	Root length, cm	Stem diameter, mm	Total oven-dry weight, g	Shoot: root ratio
Midhurst	15	26.9	60.9	6.5	14.4	3.08
	30	24.9	52.0	5.3	9.8	3.35
		**	*	***	***	**
Orono	15	28.9	46.8	6.4	13.5	3.43
	30	27.4	44.2	5.7	10.8	3.60
		NS	NS	**	**	NS

NS = not significant
* = significant at the 5% level
** = significant at the 1% level
*** = significant at the 0.1% level

15.4.3 Seedling physiology

Very few studies have investigated differences in physiological condition of seedlings grown at varying seedbed densities. One study in New Zealand measured root-growth capacity of 1+0 Monterey pine grown at various spacings (distances between seedlings within the drill row) [10]. Seedlings were transplanted to pots and grown for 14 and 28 days; both number and total length of white rootlets increased as seedbed density decreased (Table 3). These lower-density-grown seedlings with

Table 1. Number of acceptable and cull 2+0 Douglas-fir seedlings (based on stem diameter as the sole grading criterion) grown at Humboldt Nursery (adapted from [33]).

Seedbed density		Minimum acceptable diameter, mm									
		2		3		4		5		6	
Seedlings/ ft²	Seedlings/ lineal ft	Accept	Cull	Accept	Cull	Accept	Cull	Accept	Cull	Accept	Cull
10	40	40	0	40	0	40	0	36	4	22	18
20	80	80	0	80	0	72	8	44	36	14	66
30	120	120	0	116	4	78	42	28	92	6	114
40	160	160	0	152	8	90	70	22	138	8	152
70	280	280	0	204	76	48	232	11	269	0	280

Table 3. Effect of spacing on root-growth capacity of 1+0 Monterey pine seedlings (adapted from [10]).

Spacing, cm apart within drill row	Number of white rootlets		Total length of white rootlets, mm	
	14 days	28 days	14 days	28 days
2	5	6	24	88
4	10	7	57	141
7	11	9	73	166
10	11	15	76	329

better root-growth capacity also had better growth and survival after 2 years in the field.

We can also speculate that larger seedlings grown at lower seedbed densities have more stored food reserves, which will promote better growth in the field. In addition, their needle surface area is greater, affording them greater photosynthetic capacity when outplanted, which could increase height growth.

15.4.4 Growth and survival

Seedlings grown at lower seedbed densities have altered morphological and, perhaps, physiological characteristics. However, once seedlings are planted in the field, their survival varies regardless of the density at which they originally were grown. But, most often, field growth of seedlings grown at lower density is superior for a number of growing seasons after planting.

In the southern United States, 1+0 slash and loblolly pine grown at 20, 30, 40, 50, and 60 seedlings/ft² survived the same in the field in a year with above-average rainfall; but after 2 years, field growth of seedlings grown at lower densities was superior to that of those grown at higher densities [83]. Shoulders [84] found that, in moderately dry years, loblolly and slash pine survived best when grown at lower densities in the nursery but that when rainfall after outplanting was adequate, seedbed density did not affect field survival. When slash and loblolly pine seedlings were graded according to size, the morphologically high-grade seedlings (from all densities) survived and grew better than the low grades after 5 years in the field. The proportion of high- to low-grade seedlings increased as bed density decreased—that is, low seedbed densities produced a greater number of larger seedlings which performed better in the field [22].

Five years after field planting of 1+0 Monterey pine grown at different densities, survival was similarly high for all density classes, but tree height and diameter at breast height were significantly greater for trees grown at lower densities (Table 4) [13]; stem volume was 70% larger on plots planted with seedlings grown at low density than on plots with seedlings grown at high density. This is one of the many examples in which initial seedling height differences became more pronounced with each year after field planting; the slightly larger seedlings grown at lower densities grew faster in the field, over time

Table 4. Effect of density on tree height, diameter at breast height, and stem volume 5 years after planting with 1+0 Monterey pine seedlings (adapted from [13]).

Seedbed density, seedlings/m²		Height, m	Diameter, cm	Volume, m³/ha
Low:	101	2.52[1]a	3.8a	4.77[2]
Medium:	231	2.34ab	3.3b	3.65
	205	2.44bc	3.4bc	3.45
High:	420	2.28c	3.1c	2.80

[1]Means followed by the same letter within a column are not significantly different at the 5% level.
[2]Not analyzed statistically.

increasing the difference between themselves and the trees grown at higher densities.

Field survival of 2+0 ponderosa pine seedlings from four seed zones was improved if seedlings were grown at lower densities [11]. Survival increased from 62 to 71 to 78 to 83% as growing densities decreased from 50 to 40 to 30 to 20 seedlings/ft², respectively. The shorter seedlings grown at higher bed densities remained smaller after the first field-growing season.

Three+0 white spruce seedlings grown at two densities 15 and 30 seedlings/ft²) had equal survival at four field sites [67]; however, tree height after 5 years in the field differed on many of these sites, with the trees grown at lower densities consistently taller (Table 5).

Table 5. Survival and height 5 years after field planting of 3+0 white spruce grown at two seedbed densities and outplanted on four sites (adapted from [67]).

Nursery	Seedbed density, seed-lings/ft²	Height, cm					Mean survival, %
		Site 1	Site 2	Site 3	Site 4	Mean	
Midhurst	15	69.6	69.0	67.2	67.6	68.4	88.2
	30	64.8	57.7	62.0	59.0	60.9	89.4
		NS	***	*	**		
Orono	15	75.7	67.4	67.7	63.0	68.4	89.3
	30	69.2	58.6	70.6	63.2	65.4	9[illegible]1
		*	*	NS	NS		

NS = not significant
* = significant at the 5% level
** = significant at the 1% level
*** = significant at the 0.1% level

Similarly, 2+0 Douglas-fir seedlings grown at lower densities were larger when outplanted and produced the best height growth during the first field-growing season under four different planting-site conditions [32] (Table 6). In this study, both stem diameter at time of lifting and field height growth were consistently higher as seedbed density decreased. In another study, coastal and interior Douglas-fir and Sitka spruce seedlings grown at wider spacings had 53 to 83% greater new shoot growth after one growing season and had better survival in the field after three growing seasons [107].

Table 6. First-year height growth of 2+0 Douglas-fir seedlings grown at five seedbed densities at the Wind River Nursery and outplanted on sites with different ground cover (adapted from 32]).

Seedbed density, seed-lings/ft²	Height growth, cm				
	Ground-cover type				
	Vegetation	No vegetation	Debris	No debris	Mean
10	5.7	5.2	5.1	4.7	5[illegible]
20	5.4	4.9	4.4	4.2	4[illegible]
30	4.7	4.2	4.1	4.2	4[illegible]
40	4.1	3.6	3.5	3.5	3[illegible]7
70	4.0	3.7	3.4	3.5	3[illegible]7

15.4.5 Conclusions

Some advantages of growing seedlings at lower seedbed densities are:

- Because the cull percent decreases with lower seedbed density, a larger number of seedlings may be obtained from a given amount of seed [11, 103]. As use of improved seed becomes more common, nursery managers will not want to waste it on culls.

- The higher cull percent at higher densities means a greater chance of directional selection, which could change the genotypic mix of seedlings produced (see chapter 17, this volume, for genetic implications).
- An increased number of culls means more time is spent grading. This increased time could result in increased stress on seedlings from more handling and exposure and definitely raises seedling production costs. Lowering seedbed densities and reducing the number of culls could perhaps eliminate the need for grading altogether [10].
- Lower seedbed density may shorten the time required to grow an acceptable seedling [31], i.e., a 2+0 seedling grown at low density may meet the same size specifications as a 2+1 seedling.
- On some field sites, survival might be improved by planting with seedlings grown at lower densities; on many sites height growth certainly can be improved, increasing stand volumes and possibly reducing future rotation lengths.
- Size of planting stock may be more uniform. Stock size varies more for seedlings grown at high than at low densities—a difference that is still evident after several years in the field. If young stands are highly variable in size, tree competition, growth, and eventually canopy closure could be delayed or uneven [13].

However, all these benefits of growing seedlings at lower densities must in turn be weighed against the costs of using more land to produce the same number of seedlings.

15.5 Irrigation

Irrigation guidelines are established on the basis of [65]:

- Tree species
- Present crop size in relation to seedling specifications
- Stage of crop development
- Weather conditions
- Soil characteristics
- Scheduling of other cultural practices
- Seedbed density

Because these factors vary from nursery to nursery, the best irrigation regime for one nursery may not suit another. However, there are times common to all nurseries when having an irrigation regime ready is critical: (1) to water freshly sown or germinated seed, (2) to maintain proper temperature and moisture control for young seedlings, (3) to promote plant growth, (4) to protect seedlings against frost, (5) to augment other cultural practices such as fertilization, root culturing, lifting, and transplanting, (6) to control moisture stress and harden seedlings, and (7) to help seedlings enter dormancy.

Methods for monitoring plant, soil, and air to determine irrigation needs are discussed in detail in McDonald and Running [60], Day [25], and Morby [65] and in chapters 11 and 12, this volume. This section focuses on one important use of irrigation—controlling moisture stress to promote onset of dormancy—and its possible effects on seedling quality.

15.5.1 Water in the forest environment

Conifers growing under natural conditions in the Northwest complete their height growth in late spring and early summer when adequate soil moisture is available from seasonal precipitation or snow melt. Trees then set bud and height growth ceases during the summer drought, which usually is a time of high evaporative demand, high air temperature, and low soil moisture. Resultant plant moisture stress (PMS) prevents second flushing, and trees enter the dormancy cycle (become hardened) (see chapter 14, this volume, for more information on dormancy). Trees typically have firm winter buds by late summer and will not resume growth or flush again even with early fall rains; in fact, these fall rains help deepen dormancy [115].

15.5.2 Water in the nursery environment

Irrigation in spring and early summer promotes growth of both new germinants and recently flushed second- and third-year seedlings. It is important to pay attention to unusually hot or dry periods in late spring, which could stress seedlings and hamper their growth. While seedlings are actively growing, frequent irrigation generally increases their height and dry weight [20, 39, 59]. However, it is crucial to closely monitor PMS throughout the growing season because too much or too little water can harm seedling quality and subsequent field performance.

15.5.2.1 Too much water

Unrestricted watering throughout the summer promotes growth. Seedlings will continue to grow, and if they do set bud, a second flush in the late summer or early fall is very likely. Although the increased plant size may seem favorable, delayed budset or second flushing is most often harmful to plant vigor because (1) the new, recently grown plant tissue is not hardy and is therefore susceptible to frost damage, and (2) delayed budset inhibits completion of the subsequent phases of dormancy, which may be necessary for seedlings to successfully tolerate nursery processing after lifting [65] and to ensure vigorous field growth the next spring [52, 115]. For example, at two hypothetical nurseries with different irrigation regimes (Fig. 4), seedlings grown with a restricted watering regime (R) completed their second-year height growth by mid-July, but those watered throughout the summer (U) continued to grow taller [52]. The potential field survival of seedlings grown with no imposed moisture stress (U) is low because they did not set bud until fall and were unable to adequately complete their dormancy cycle.

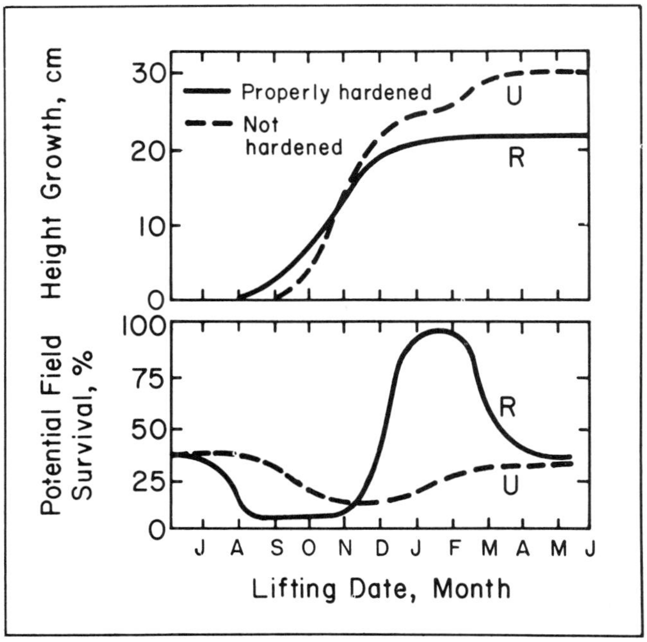

Figure 4. Cumulative nursery height growth and potential field survival of 2+0 Douglas-fir seedlings grown under restricted (R) and unrestricted (U) watering regimes throughout the summer (adapted from [52]).

15.5.2.2 Too little water

High moisture-stress regimes in the nursery can also negatively affect seedling morphology and physiology. Seedlings grown under a dry regime—watered beginning in spring only when predawn PMS reached 15 bars—set bud early but were too small to meet minimum size standards for plantable seedlings [115]. High PMS was found to inhibit budset of Douglas-fir grown under three different moisture-stress regimes [39]. In another study, although moderate stresses improved cold hardiness, higher stresses (10 to 15 bars) reduced cold hardiness of Douglas-fir seedlings [18]. Root growth as well as shoot growth may also be inhibited at these high stress levels [25]. High moisture stress results in smaller seedlings with decreased height and shoot and root dry weights, or low seedling vigor, or mortality [20, 39, 65, 115].

15.5.2.3 Moderate moisture stress

A nursery irrigation schedule that imposes moderate stress on seedlings in midsummer may result in:

- Earlier budset [39]
- Earlier induction of seedling dormancy [51, 115]
- Increased cold hardiness [18]
- Greater tolerance to exposure during lifting, storage, and handling [115]
- Smaller seedlings [81]
- No delay of budburst the following spring [52]
- Increased field-survival potential [18, 52, 115]

The date of moisture-stress induction also is important; seedlings undergoing earlier initiation of moisture stress were shorter and had higher root dry weights, resulting in lower shoot:root ratios (Table 7). In addition, cold hardiness of Douglas-fir seedlings decreased as initiation date of moisture stress was delayed [18].

In summary, most reports agree that moderate stress is favorable and that extremely low or high stress can be harmful to seedlings. The difficult point for nursery managers is defining what a moderate stress level should be.

Table 7. Effects of moisture-stress induction date on morphological characteristics of 2+0 Douglas-fir seedlings. Seedlings were subjected to moisture stress for 30 days, then well watered and lifted in mid-October (adapted from [18]).

Seedling characteristic	Moisture-stress induction date			
	July 15	Aug 1	Aug 15	Sept 1
Seedling height,[1] cm	21.15[2]a	23.20b	25.35c	25.15c
Root dry weight,[1] g	1.19a	0.94b	0.97b	0.81c
Seedling caliper, mm	4.02a	3.80a	4.07a	3.89a
Shoot:root ratio[1]	2.00a	2.60b	2.90c	3.12d

[1]Significantly affected by induction date at the 1% level.
[2]Means followed by the same letter within a row do not differ significantly at the 5% level.

15.5.3 Irrigation schedules in the Northwest

Many nurseries have developed irrigation schedules for inducing dormancy (e.g., [65, 115]). Of the 21 Northwest nurseries surveyed, 95% reduce watering in midsummer to harden seedlings (OSU Nursery Survey), though the date of initiation, the growth year (first growing season, second growing season, or both) in which stressing occurs, and the levels of stress employed vary widely. In the first year of producing a 2+0 crop, ⅓ of the nurseries do not reduce watering to stress seedlings; others water only to cool seedlings; some reduce water only if the crop has reached a particular size; and a few let seedlings reach predetermined stress levels (e.g., 10 bars predawn PMS after mid-August, 20 bars midday PMS by September 1, 15 bars predawn PMS by August 1). In the second year, most nurseries stop watering regularly in July to let seedlings reach predetermined PMS levels, then rewater when seedlings attain them; levels range from 8 to 15 bars predawn PMS, with most around 12. A few nurseries do not reduce irrigation until September. Not all nurseries use a PMS measurement to indicate when to water (see chapter 12, this volume, for other monitoring methods).

15.5.4 Irrigation regime and growth and survival

Most studies to determine irrigation levels for restricted watering have measured how restricted watering affects bud set date (as an indication of onset of dormancy) and morphology—but not how it affects growth and survival in the field. In one study [18], groups of 2+0 seedlings which had received three different stress treatments in the nursery (0 to 4, 4 to 6, and 6 to 8 bars predawn PMS) were lifted and stored for 30 days, then potted and placed in a growth room for 6 weeks. Survival was 78, 85, and 94%, respectively. The authors concluded that imposing *moderate* moisture stress (4 to 8 bars) on seedlings enhanced onset of dormancy.

Others noted that seedlings that had second-flushed in the nursery and were therefore not conditioned properly for winter chilling had less vigorous root growth, delayed budburst, and reduced survival potential in the field [52]. However, no published data are available on the effects of different nursery irrigation regimes on *field* survival and height growth, and some negative effects are possible if these watering levels are too high or too low. Seedlings overstressed in the nursery may lay down fewer needle primordia (in buds that set at the end of the second growing season), which can result in less field growth [37, 74], or they may have decreased food reserves available for growth the next spring [54]. Douglas-fir seedlings stressed by root wrenching in the nursery had impaired field growth up to 3 years after planting [29].

15.5.5 Conclusions

At this point we know that either a wet irrigation regime or a high stress regime may adversely affect seedling survival and growth, but we have not defined the optimum level of irrigation which will promote survival and growth. A general recommendation is to begin moderate stress (8 to 12 bars predawn PMS) as soon as the crop has reached its proper height and caliper. But because the optimum moderate stress level probably varies according to soil type, climatic conditions, seedling species, and so forth, exact levels will have to be defined for each nursery. Tailoring irrigation schedules and determining their effect on seedling quality at each nursery site are essential before proper prescriptions can be made.

15.6 Fertilization

One important goal of nurseries today is maintaining an adequate level of soil fertility to produce high-quality seedlings (see chapter 7, this volume). Long-term nursery productivity can only be assured by careful management of those factors affecting soil fertility—such as cation exchange capacity, pH, and organic matter content—and by proper fertilization (see chapters 6 through 10).

The signs of poor seedling nutrition are (1) a decrease in or cessation of growth and (2) under extreme conditions, visually recognizable deficiency symptoms. In contrast, seedlings with adequate nutrition grow to a specified size early in the summer, allowing ample time for hardening. Whereas it may be possible to improve seedling quality by altering the timing and level of fertilization in the nursery or by monitoring the nutritional status of seedlings during or after active growth, little is known yet about the optimum nutritional status or needs of outplanted seedlings, especially when comparing good and poor sites [6]. For example, should trees destined for a poor site be grown under nutrient-deficient conditions in the nursery, or should they be well supplied with nutrients when planted? Do trees grown under optimum nutrient conditions at the nursery grow best in the field? Is there danger in overfertilizing seedlings? Many of these questions remain unanswered, and some have different answers according to the species being grown, nursery soil, timing of application, or cultural practices used. Changing irrigation regime or seedbed density, for example, can also alter seedling response to fertilization [8]. This complex relationship among fertilization, site conditions, and other cultural practices makes fertilization decisions some of the most difficult in nursery management.

15.6.1 Seedling morphology

15.6.1.1 Growing-season fertilization

In general, fertilization—and especially N fertilization—during the first and second growing seasons produces 2+0 seedlings that are taller and heavier and have larger shoot diameters [4, 68, 88, 95]. Most often, seedlings have greater shoot:root ratios with fertilization [8, 68, 104] and may have greater root mass [4].

van den Driessche [104] found that shoot height, root and shoot dry weights, and shoot:root ratio increased in both 1+0 and 2+0 Douglas-fir seedlings in the nursery as more N was applied (Table 8). The rise in seedling dry weight also was correlated with increased foliar N levels, and maximum dry weights of both roots and shoots were obtained at 2.0 to 2.1% N concentration. Often, as in this study, when fertilizing during the growing season, an application level exists above which adding more fertilizer will not further increase seedling size; for example, increasing second-year N rates above 100 kg/ha did not further increase shoot height or caliper (Table 8) [104]. If the optimum level could be determined for each nursery site and species, unnecessary applications of costly fertilizers could be eliminated. In the same study, adding phosphorus (P) in the form of superphosphate fertilizer did not increase seedling P concentration or affect shoot height or weight, although available P in the nursery soil was raised by the addition.

The overall result of increasing seedling size with N fertilization has been to increase the number of plantable seedlings produced from a nursery bed [8, 95].

Table 8. Effects of fertilization at various rates of N on morphology of first- and second-year Douglas-fir seedlings (adapted from [104]).

Morphological measurement, by seedling age	N rates, kg/ha				
First year	0	25	50	75	100
Shoot height, cm	4.5[1]a	5.8b	6.4b	6.2b	6.3b
Shoot dry weight, g	0.11a	0.16b	0.18c	0.18c	0.18c
Root dry weight, g	0.08a	0.11b	0.12b	0.12b	0.12b
Shoot:root ratio (dry wt.)	1.30a	1.39ab	1.47bc	1.49bc	1.54c
Second year	0	50	100	150	200
Shoot height, cm	8.8a	13.9b	16.5c	16.1c	16.8c
Shoot dry weight, g	0.35a	0.99b	1.42c	1.34c	1.45c
Root dry weight, g	0.31a	0.87b	1.01b	0.91b	0.92b
Shoot:root ratio (dry wt.)	1.14a	1.14a	1.39b	1.45bc	1.54c

[1]Means followed by the same letter within a row are not significantly different at the 5% level.

15.6.1.2 Fertilization and hardening

Seedling growth patterns can be altered by withholding—as well as adding—nutrients (see chapter 7). Armson [6] found that fertilized trees grow longer during the growing season than unfertilized trees. Most nurseries in the Northwest stop fertilizing in July or early August in both the first and second growing seasons because they believe that this arrested fertilization, along with restricted watering, helps seedlings harden properly (OSU Nursery Survey).

15.6.1.3 Fall fertilization

Fertilizing seedlings in fall after growth ceases has been shown to have no effect on seedling height and stem diameter at the time of harvesting the following winter [16, 98]. However, bud height, a possible indicator of next year's growth, varied significantly with fall application of N and P [98]; P decreased bud height, whereas N in the absence of P increased it.

15.6.2 Seedling physiology

15.6.2.1 Frost hardiness

The most commonly known effect of fertilization on seedling quality is the reduction of frost hardiness when N is applied during the growing season. N fertilization can prolong seedling growth in the nursery, delaying hardening or the onset of dormancy and later resulting in frost damage in the nursery or damage to inadequately hardened stock during lifting or cold storage [105]; it can also cause earlier budbreak the following spring, resulting in possible frost damage [16]. High levels of P applied to Sitka spruce seedlings extended their active growth period and caused frost damage [58].

Potassium (K) nutrition also may play an important role in the development of frost hardiness [105], although findings have been mixed. Adequate K levels in Douglas-fir slightly increased frost hardiness in winter [48, 49], though K levels in Sitka spruce had no influence on frost damage at two heavily damaged field sites [16]. Timmis' [99] work with Douglas-fir container seedlings showed frost hardiness to be more closely related to the K:N balance than to the level of any single nutrient; a lower K:N ratio (0.6) resulted in hardier seedlings.

Low boron (B) levels have been reported to increase frost damage to tree species (see chapter 7).

Late-season or fall fertilization, which does not usually affect seedling growth and diameter (see 15.6.1.3), has been found to affect frost hardiness. K and N applied as a top dressing decreased December frost damage of Sitka spruce and western hemlock [*Tsuga heterophylla* (Raf.) Sarg.] seedlings [15]. By contrast, in another more recent report [16], N application increased frost damage of Sitka spruce on one of two severely damaged field sites. However, on these same sites, four other species—Norway spruce [*Picea abies* (L.) Karst], western hemlock, grand fir [*Abies grandis* (Dougl. ex D. Don) Lindl.], and lodgepole pine—sustained less severe frost damage, with no increase in injury due to late-season N application. Fall application of K also did not affect frost damage. Applying N in late fall increased frost hardiness of Douglas-fir seedlings, whereas fertilizing with P had no effect [98].

15.6.2.2 Drought resistance

Optimal N levels in seedlings generally can improve their ability to endure and grow during drought in the field. However, N levels that are too low or too high for optimum growth can cause damage during drought and inhibit recovery and growth afterward [73].

Two+0 jack pine (*Pinus banksiana* Lamb.) tested under a drought regime typical of field conditions was significantly more drought resistant when fertilized with intermediate levels of N [91]. Loblolly pine seedlings, when grown at varying N levels in sand culture, were most drought resistant when provided an optimum supply for growth [73]. Longleaf pine (*Pinus palustris* Mill.) had improved drought resistance when grown under a balanced supply of N, P, and K [2].

15.6.3 Growth and survival

15.6.3.1 Growing-season fertilization

Though little is known about the effects of nursery fertilization on seedling performance in the field, where studies have been done, positive effects of fertilization on either height growth or survival have often been reported. van den Driessche's [104] previously mentioned study, in which N fertilization increased Douglas-fir seedling size in the nursery (see 15.6.1.1), also revealed substantially improved Douglas-fir seedling performance in the field. Two years after outplanting, all seedlings fertilized with N in the nursery survived significantly better than unfertilized controls (Fig. 5). The percentage of N in the 2+0 foliage was positively correlated with increased N fertilization levels of 0, 75, 150, 225, and 300 kg/ha. Interestingly, survival dropped off when the percentage of N was greater than 2; often, overfertilization with N results in taller seedlings with high shoot:root ratios which may have poorer survival, especially on dry sites. In van den Driessche's [104] study, surviving seedlings from all treatments had similar growth rates in the field; thus, the fertilized seedlings were still significantly taller at the end of 2 years.

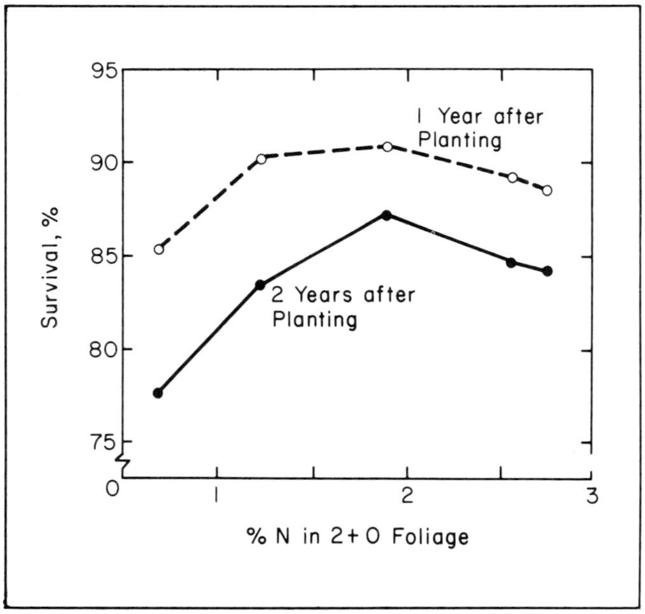

Figure 5. Survival of 2+0 Douglas-fir seedlings 1 and 2 years after outplanting at different foliar N concentrations (adapted from [104]).

In a more recent study, nursery fertilization at 235 kg N/ha increased new shoot growth in the first field season by 51% for coastal Douglas-fir, 36% for interior Douglas-fir, and 58% for Sitka spruce, compared to that at 60 kg N/ha. After three growing seasons in the field, the effect of N on new shoot growth diminished, ranging from 0 to 42% [107]; high N level increased survival of coastal Douglas-fir and Sitka spruce slightly but decreased that of interior Douglas-fir. In another study with Douglas-fir, inorganic fertilizers applied at the nursery increased seedling size, then increased field survival from 70 to 95% and field height after 4 years from 74 to 94 cm [88]; the authors noted in 1966 that improved use of nursery fertilization could increase field growth by 26%.

A series of experiments in the Lake States involving seedlings from four nurseries showed a slight but consistent gain in field survival of jack, red (*Pinus resinosa* Ait.), and white (*Pinus strobus* L.) pine when fertilized with N, P, and K, but no differences in growth were found after 5 to 8 years in the field [9]. In another study, jack, red, and Scotch (*Pinus sylvestris* L.) pine survived the same in the field but had 20 to 30% greater field height growth after fertilization in the nursery [112].

Nursery fertilization improved field height growth of white spruce but did not affect survival [67]; red and white pine also were unaffected [68]. Height of loblolly pine after 3 years in the field was positively correlated with foliar N content increased by nursery fertilization (Fig. 6) [95].

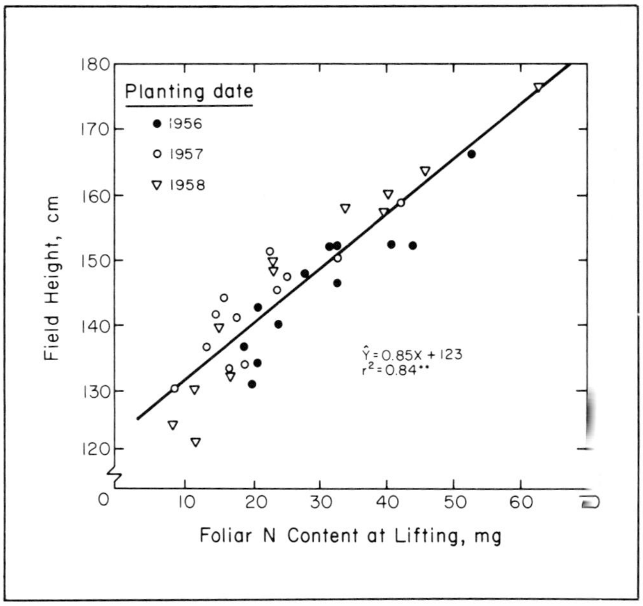

Figure 6. Regression line indicating the relationship between field height at 3 years and foliar N content at lifting of loblolly pine (adapted from [95]).

15.6.3.2 Fall fertilization

The few studies investigating fall fertilization have shown varied effects on field performance. Loblolly and slash pine outplanted after receiving up to 400 lb N/acre in late October survived the same as unfertilized seedlings [38]. Fall fertilization increased N concentration in five tree species [16]; N advanced budbreak of all species except grand fir during the first summer in the field and had no negative effect on survival. Although seedling size in the nursery was unaffected by this fall fertilization, height growth after field planting of Sitka spruce was improved up to 18%, with similar improvement in diameter. In another study, seedlings from five Douglas-fir seed sources, fall fertilized in the nursery with 50 lb N/acre, had improved field survival after 2 years (Fig. 7a) and grew 0.03 to 0.05 m (0.11 to 0.16 ft) taller than unfertilized trees in each of the 5 years after outplanting (Fig. 7b) [4].

15.6.4 Conclusions

The status of nursery soil and crop nutrition should be constantly examined and modified as necessary (for more detail on soil and foliar analysis, see chapter 8, this volume). Recommended fertilizer applications for a 2+0 seedling crop range from 112 to 285 kg of N, 67 to 200 kg of P, and 75 to 150 kg of K per ha (see chapter 7) but should be calibrated for each crop species and nursery site. However, increased fertilization during the growing season generally results in taller and heavier seedlings with larger diameters. Because seedlings grow longer in the growing season when fertilized, most nurseries stop fertilizing in July or early August to harden seedlings. Fall fertilization, usually around October, does not affect seedling height or diameter but can affect terminal bud size and frost hardiness (either positively or negatively). Nurseries fertilizing in fall should apply 30 to 50 kg of N/ha.

Most studies show improved field growth and survival as a result of growing-season fertilization; in addition, fall fertilization may increase field growth of Northwest species. However, effects of seedling nutritional status on field performance require further investigation.

15.7 Root Culturing

Root culturing is the broad term for describing the various nursery practices implemented in the seedbed to alter seedling root growth. Two practices, undercutting and wrenching, involve the mechanical cutting of the root system with a blade drawn horizontally under the seedbed. Two other practices, lateral pruning and box pruning, involve the cutting of the root system with vertical blades. Due to lack of information about undercutting, lateral pruning, and box pruning, most of this section will emphasize results obtained from root-wrenching studies.

15.7.1 Undercutting and wrenching

Undercutting is the drawing of a thin, sharp blade under the seedbed parallel to the surface. The blade severs the taproot and all other roots extending beyond the regulated depth of the undercut. Ninety-five percent of the nurseries in the Northwest undercut, mainly to stimulate root growth in the upper zone of soil so that seedlings gain a more fibrous root system (OSU Nursery Survey).

Nurseries undercut their 2+0 seedlings in fall (of the first growing season), or spring (of the second growing season), or sometimes summer (of the second growing season). Most undercut only once, although some undercut once in spring and then again in early summer. The depth for undercutting ranges from 4 to 12 inches (10 to 30 cm), with most at 5 to 6 inches (13 to 15 cm) (OSU Nursery Survey).

Wrenching, which usually follows undercutting, is done with a thicker, broader blade tilted at an angle (20 to 30°) when drawn under the seedbed. Wrenching cuts off any newly penetrating roots and lifts seedlings, loosening and aerating the soil. Eighty percent of nurseries in the Northwest root-wrench their seedlings to (1) stimulate root growth and enhance fibrous root development, (2) stress and harden seedlings in summer, (3) control shoot height, and (4) aerate and loosen the soil (OSU Nursery Survey). A few nurseries wrench in fall to prevent late flushing and promote root growth.

Wrenching at Northwest nurseries is done in the 2+0 year, usually after undercutting, with the angled blade drawn at a depth of 8 to 10 inches (20 to 25 cm). About ⅓ of the surveyed nurseries wrench only once, usually in June or July; the rest wrench from 2 to 10 times during the summer of the second growing season, usually beginning in June or July and ending in August or September (OSU Nursery Survey). Seedlings may be wrenched once a month, once every 2 or 3 weeks, or even once a week. Multiple wrenching varies considerably in its timing and frequency; the pattern of this variation seems unrelated to species or nursery location. Some nurseries also wrench their transplants (1+1s and 2+1s) approximately 6 weeks after spring transplanting, again a second time for hardening, and perhaps a third time in fall.

One of the most critical factors affecting wrenching is soil moisture. If seedlings are wrenched when the soil is dry and (or) the weather is hot and dry, high plant-moisture stress (PMS) can result. However, seedlings wrenched when the soil is moist or watered immediately after wrenching experience only moderate to low PMS.

Seedlings are undercut and wrenched with a fixed or reciprocating blade attached to a tractor (Fig. 8; see also chapter 3, this volume). Most nurseries in the Northwest use a fixed blade. A specialized root-culturing implement that both undercuts and wrenches with a reciprocating blade reportedly cuts roots without pulling or dragging tree seedlings [57, 110]; its drawbacks have been related to its slow speed, blade breakage, and inability to control the depth of the cut.

15.7.2 Lateral and box pruning

Lateral pruning, also called side pruning or side cutting, is the passing of cutting blades or colters between the drill rows on both sides of the seedlings to sever excessively long lateral

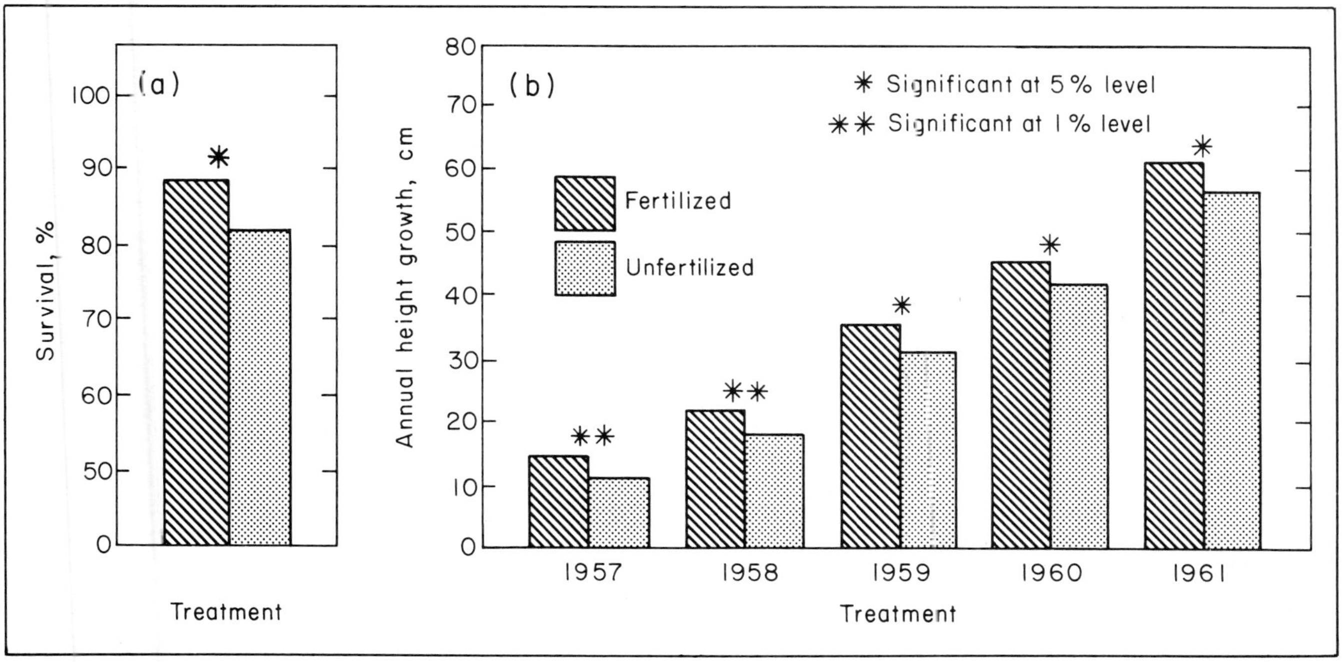

Figure 7. (a) Second-year field survival and (b) 5-year height growth of fall-fertilized and unfertilized Douglas-fir seedlings from five seed sources (adapted from [4]).

Figure 8. Root wrenching 2+0 Douglas-fir seedlings with a fixed blade tilted at a 20 to 30° angle.

roots (Fig. 9). Ninety-five percent of Northwest nurseries lateral-prune their 2+0 seedlings to (1) decrease root tangling among seedlings and facilitate lifting, (2) encourage root growth and fibrosity, and (3) retard shoot height growth (OSU Nursery Survey). Most nurseries (75%) lateral-prune only once, but the others prune 2 or 3 times during the second year. When lateral pruning once, 40% of the nurseries prune in April to May, 40% in June to July, and 20% in September to October. Nurseries that prune more than once usually do so in late spring and then again in late summer or fall. The depth of lateral pruning varies from 4 to 10 inches (10 to 25 cm), with no single depth most common.

Box pruning, a new practice in New Zealand, is the vertical cutting of lateral roots in a four-sided box shape around a seedling. Seedlings to be box-pruned must be in drill rows and must be equally spaced within each row [21]. This alternative root-culturing technique is being investigated because of the concern over root distortion to Monterey pine caused by undercutting, wrenching, and lateral pruning. Although their effects are not yet quantified, undercutting and wrenching may cause poor taproot development after planting [108]. Seedlings need a well-balanced root system: laterals must be evenly distributed and the taproot sturdy if seedlings are to remain stable and not topple in the field [21].

Lateral root pruning is done with tractor-mounted stationary blades or rolling colters drawn between each drill row. The same machine is used to box-prune between drill rows, but the beds are crosscut by hand with a spade [21].

15.7.3 Seedling morphology

15.7.3.1 Shoots

Any of these root-culturing practices, which cuts roots before or during the growing season, arrests or retards seedling height growth. Wrenched Monterey pine seedlings stopped or continued growing at a very slow rate for 3 months after cutting (Fig. 10) [77]. In New Zealand, Monterey pine do not set bud during their first year in the nursery; root wrenching therefore is used to stop growth and to condition or harden nursery stock [108].

Northwest nurseries also wrench to limit shoot growth and harden seedlings. Earlier budset—and shorter seedlings—result in most cases (Fig. 11) [29, 45]. Other studies have shown that root wrenching can control height of southern pines [85, 96] and reduces height of white pine and white spruce [66]. Though very little information is available on how undercutting and lateral root pruning affect shoot morphology, Tanaka et al. [93] note that in their studies undercutting in spring reduced height growth and, therefore, final seedling height. Lateral root pruning at different times from May to September did not reduce height growth of western hemlock, Sitka spruce, or Douglas-fir seedlings in the nursery [34, 35].

Both the timing and frequency of root wrenching can greatly affect final shoot height of seedlings. Generally, shoot height decreases with increased severity or frequency of wrenching [14, 29, 110]. Duryea and Lavender [29] found that single-wrenched trees had slightly reduced shoot height, compared to unwrenched controls, and that multiple-wrenched trees were even shorter (Fig. 12). Benson and Shepherd [14] reported that multiple-wrenched Monterey pine seedlings were ½ the height of unwrenched controls and that the number of culls increased due to wrenching.

Another factor that is important in influencing final seedling height is the timing of the root wrenching. If roots are cut too early in the growing season and height growth is arrested, seedlings may not reach plantable size [110]. Rook [77] found

Figure 9. Lateral root pruning between the drill rows to sever excessively long lateral roots.

that seedlings should be near their desired height and diameter before wrenching because any further growth will mainly be root growth. On the other hand, wrenching late in the growing season may have little effect on height growth in Northwest species because most of their growth is completed by then [96]; this late wrenching may, however, stop lammas growth [1].

The shorter shoot may have an accompanying decrease in diameter caused by wrenching early in the growing season [29, 45], although in another study wrenching caused no decrease in diameter of Douglas-fir [96]. Diameter growth of Monterey pine continued for 1 month and then stopped, resulting in smaller shoot diameters for wrenched seedlings (Fig. 10) [77].

Due to their smaller shoot heights and diameters, wrenched seedlings most often have lighter shoot dry weights [14, 29, 45].

15.7.3.2 Roots

Monterey pine has a carrotlike taproot. When this taproot is severed by undercutting and wrenching, root growth rates on an oven-dry basis are similar for wrenched and unwrenched seedlings, but the final root system is quite different in form. Cutting the taproot causes a loss in apical dominance in the entire root system; lateral root growth increases and many new tertiary roots grow, resulting in a more compact, fibrous root system [14, 77, 108, 110].

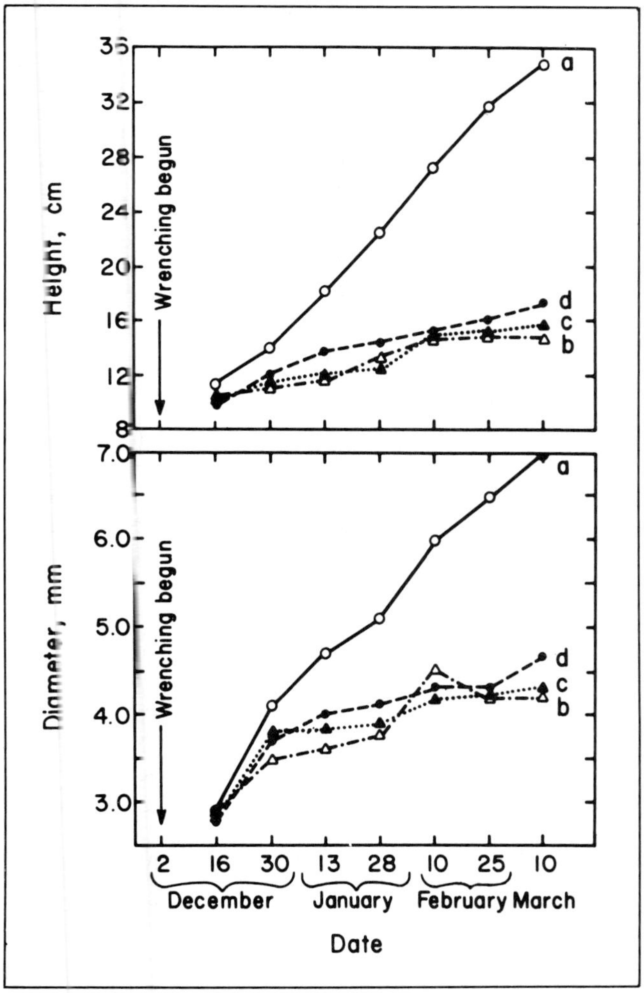

Figure 10. Height and diameter growth of Monterey pine seedlings that were (a) neither undercut nor wrenched; (b) undercut, then wrenched weekly; (c) undercut, then wrenched every 2 weeks; and (d) undercut, then wrenched monthly. Treatment differences were significant at the 1% level (adapted from [77]).

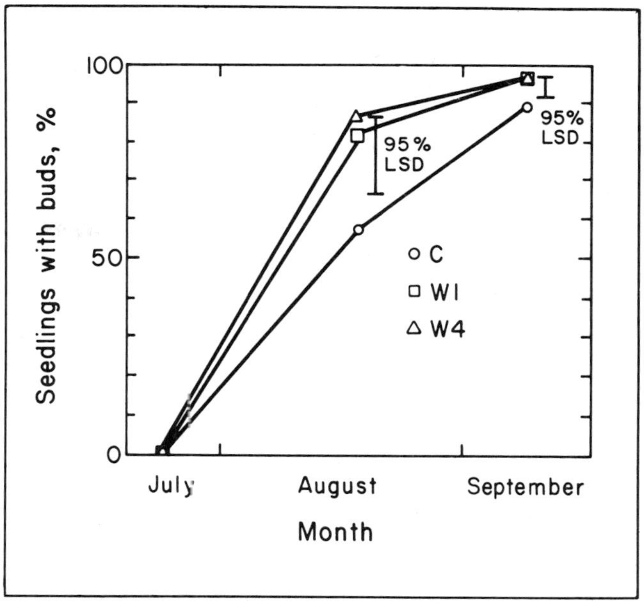

Figure 11. Budset in Douglas-fir nursery beds of unwrenched controls (C), single-wrenched seedlings (W1), and multiple-wrenched seedlings (W4) in the Northwest [unpubl. data, 28].

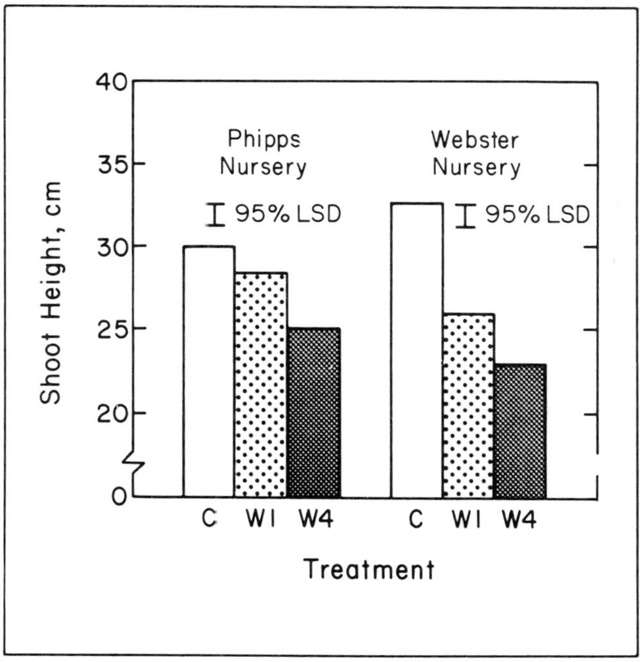

Figure 12. Shoot height at lifting of unwrenched controls (C) and single-wrenched (W1) and multiple-wrenched (W4) Douglas-fir seedlings at two nurseries (adapted from [29]).

Slower growing evergreen species such as firs, spruces, Douglas-fir, and hemlocks tend to have a more fibrous root system than Monterey pine [29, 108]. Results from root cutting of these species have been much more variable. In root-wrenching experiments with different seed sources of Douglas-fir grown at two nurseries in Oregon and Washington, wrenching did not change the total length or dry weight of lateral roots, the total number of root tips, or the total dry weight of the root system [29]. Another study in California also showed that the weights of Douglas-fir root systems did not change with wrenching [45]. Yet these results contradict those of an earlier study [96] in which root dry weights of wrenched seedlings were greater than those of unwrenched seedlings; though root tips were not counted, the authors reported that previous studies (unpublished data) showed repeated wrenchings in late summer to stimulate root growth.

Lateral root pruning has reportedly increased fibrosity in Douglas-fir, Sitka spruce, and western hemlock [34-36], the differences in fibrosity being largely due to the timing of the lateral pruning. Pruning early in the second growing season did not affect lateral root development unless blades were drawn within 6 cm of seedlings. However, Eis and Long [36] found that lateral pruning of Douglas-fir after early August caused formation of callus tissues around wounds where roots had been cut. When these seedlings were lifted in October, new roots had not regenerated nor had secondary roots grown but by the next spring, an abundance of new roots had originated from the callus. This study and Tanaka et al.'s [96] may show that root wrenching or lateral pruning in late summer or early fall can promote root growth.

Recently, there has been some concern about the effect of root form on early tree growth and windthrow in the field [21, 55]. Differences in nursery culturing can impact seedling root systems up to 7 years after field planting [55]. Wrenched and unwrenched seedlings from Tanaka et al.'s [96] study were excavated after 5 years in the field; root-system constriction (Table 9) and L-rooting were significantly greater in the wrenched trees [55]. Long [55] noted that the increase in L-shaped bending may have resulted from the difficulty in planting the larger, more fibrous root system resulting from wrenching.

In New Zealand, there has also been concern over malformities in wrenched root systems—specifically, the absence of a taproot to securely anchor the Monterey pine root system in the field [21]. Studies have shown that box pruning can produce an unbranched taproot with evenly distributed laterals. Root excavations after field planting have shown that box pruning results in less root distortion and better root form than wrenching [21].

Toppling and taproot malformation of lodgepole pine also have been observed in plantations in British Columbia. Lodgepole pine may not regenerate a taproot when undercut deeply, which makes it more prone to toppling after outplanting [30].

15.7.3.3 Shoot:root ratio

Root-culturing practices most often result in reduced shoot:root ratios of seedlings. The lower shoot:root ratio of wrenched Monterey pine is usually due to its lighter shoot [14, 108]. Duryea and Lavender [29] and Koon and O'Dell [45] found that wrenching of Douglas-fir did not change the root dry weight but did reduce the shoot:root ratio due to the lighter shoot. Tanaka et al. [96] found a decrease in the shoot:root ratio of Douglas-fir, attributable to a heavier root system.

15.7.4 Seedling physiology

15.7.4.1 Carbohydrate distribution

Undercutting and wrenching of Monterey pine caused an increase in the proportion of total carbohydrates (^{14}C-photosynthate) translocated to seedling root systems [77]. After roots were cut, carbohydrates were diverted from the foliage to form new root tissue; thus, roots were grown at the expense of shoots [108]. Once again, timing of the wrenching is important: a smaller proportion of current photosynthate went to roots of Monterey pine undercut and wrenched after late summer than earlier in the summer, when height growth was more vigorous [77].

15.7.4.2 Hardening

In New Zealand, Monterey pine seedlings do not normally form a bud and become dormant during their first year [110]. When seedlings are wrenched, growth is arrested, buds are formed, and shoots become more lignified [14]. Field survival of hardened, more dormant, wrenched stock has been shown superior to that of unwrenched stock that is actively growing [110].

Although, in the nursery, wrenched Douglas-fir seedlings set bud earlier than unwrenched seedlings, dormancy of wrenched and unwrenched seedlings has not been measured. However, wrenched and unwrenched Douglas-fir seedlings lifted and stored in fall (November 1) survived and grew equally under moist and dry planting conditions as did those lifted in winter (January), perhaps indicating that wrenching does not affect stage of dormancy [29].

15.7.4.3 Nutrition

Wrenching reduces the foliar concentrations of N, P, and sometimes K. Concentration of N and P in Monterey pine shoot tissue decreased as wrenching severity increased (Table 10) [14]; however, fertilizing seedlings that had been root-wrenched 5 times increased their N and P levels. Menzies [63] also found wrenched Douglas-fir seedlings to have lower foliar concentrations of N, P, and K. Accompanying these lower concentrations can be a yellowing of the needles due to a significant reduction in chlorophylls a and b and in carotenoids [77]. Fertilizer should therefore be applied when wrenching because such nutrient-deficient seedlings have been known to stagnate when outplanted [108].

15.7.4.4 Drought resistance

Wrenching stresses seedlings while in the seedbed, causing them to close their stomata [29]. When planted in the field, however, wrenched Monterey pine have been found to have higher rates of transpiration than unwrenched seedlings, though wrenched and unwrenched Douglas-fir did not differ [29, 76]. During drought, wrenched Monterey pine seedlings maintained higher relative turgidities and more active root growth than unwrenched seedlings [76, 108]. Under an imposed drought in pots simulating drought in the field, wrenching did not affect Douglas-fir seedlings' ability to withstand drought, even though this ability varied among seed sources; however, wrenched seedlings did not set bud as promptly during drought and had fewer active roots than unwrenched seedlings [29]. This lessened ability to regenerate roots and to set bud during drought could lower field survival and growth potential for wrenched Douglas-fir seedlings.

Table 9. Root and shoot characteristics of wrenched and unwrenched 2+0 Douglas-fir 5 years after outplanting (adapted from [55]).

Seedling type	Root: Constriction	Root: Symmetry	Root: Balance	Diameter, mm	Height, cm	Current height growth, cm	Dry weight, g: Root	Dry weight, g: Shoot	Shoot:root ratio
Wrenched	1.28	2.84	1.74	36.9	176.7	48.8	121.3	794.7	0.20
Unwrenched	0.94	2.89	1.62	30.0	159.3	44.0	76.6	510.0	0.16
	*	NS	NS	*	*	*	*	*	NS

NS = not significant
* = significant at the 5% level

Table 10. Foliar nutrient concentrations of wrenched and unwrenched Monterey pine seedlings at harvest (adapted from [14]).

Treatment	N, %	P ppm	K ppm
Unwrenched control	1.92[1]a	1,750a	10,290a
Root-wrenched twice	1.39b	1,320b	8,240a
Root-wrenched 5 times	1.21c	890c	7,540a
Root-wrenched 5 times; fertilizer added twice	1.31bc	1,120bc	7,270a

[1]Means followed by the same letter within a column are not significantly different at the 5% level.

15.7.4.5 Growth regulators

The roots of wrenched Monterey pine had lower levels of inhibitory abscisic acid (ABA) and higher levels of root-growth-promoting indole-3 acetic acid (IAA) than unwrenched controls [94].

15.7.5 Growth and survival

15.7.5.1 Monterey pine

Rook [76] demonstrated that wrenched seedlings survive planting, especially into dry field conditions in New Zealand, better than unwrenched seedlings. van Dorsser and Rook [110] later concluded that Monterey pine must be undercut and wrenched if consistently high survival rates are to be obtained from planting bareroot seedlings.

Wrenched Monterey pine also grow better in the field. In one study, height increment of unwrenched and wrenched seedlings after the first two field seasons was 41.5 and 63.25 cm [109]. Unwrenched seedlings may suffer from severe leader damage (due to their nonlignified shoot) after planting in the field, which retards their growth [14]. In Australia, seedlings wrenched twice during summer had increased stem volume after 5 years, compared to unwrenched controls and to seedlings wrenched 5 times [14]. Box-pruning Monterey pine seedlings to decrease root distortion resulted in superior field growth compared to that for undercut and wrenched stock [21].

15.7.5.2 Other pines

Undercutting increased field survival of longleaf pine, but survival of loblolly and slash pine has varied [85]. In one study, first-year survival of wrenched loblolly pine improved from 70 to 93%, the increase greater at droughty than moist sites [96]; however, height did not differ after one growing season in the field.

In Australia, most planting of Caribbean pine (*Pinus caribaea* Mor.) has been with container seedlings, due to the poor survival of bareroot stock [9]; however, wrenching has been found to increase field survival of bareroot seedlings to an acceptable level [9, 14].

A single wrenching of white pine either did not affect survival or, if done during the growing season of the 3+0 year, decreased it [66]; wrenching during the growing season also decreased white pine growth for 5 years in the field. Tanaka et al. [96] noted that at several locations in southern Oregon, field survival of ponderosa pine was not improved when wrenched biweekly. Preliminary results of another study also indicate no differences in survival for undercut ponderosa pine seedlings and controls planted on a south-facing slope in southern Oregon [unpubl. data, 41].

15.7.5.3 White spruce

When 3+0 white spruce were single-wrenched at four different times in the growing regime, one wrenching during June's flush of growth improved field survival and growth, but fall wrenching in the 2+0 year decreased survival [66].

15.7.5.4 Douglas-fir

In studies at two nurseries with five different seed sources, Duryea and Lavender [29] found that wrenching did not improve survival of Douglas-fir seedlings under either favorable (moist) or unfavorable (dry) planting conditions. Tanaka et al. [96] reported that wrenched Douglas-fir seedlings from a Cascade Mountains source, when planted on a droughty south-facing slope, survived better than unwrenched seedlings (88 vs. 65%), though on the nearby north slope, survival did not differ (58 vs. 57%). Similarly, Koon and O'Dell [45] reported that wrenching Douglas-fir seedlings at 20-cm depth improved their survival from 31 to 56% on a droughty site in California, but wrenching at 15 cm did not. Preliminary results of a recent study indicate that undercutting Douglas-fir seedlings at different times in spring and during the growing season did not improve survival on a south-facing slope in southern Oregon [unpubl. data, 41].

Duryea and Lavender [29] found no case in which root wrenching improved field growth of Douglas-fir seedlings, and in one year's planting of trees from four seed sources, first-year growth was consistently greater for unwrenched than for wrenched seedlings under a number of planting-site conditions. After 3 years in the field, height difference between wrenched and unwrenched seedlings increased (Fig. 13a), indicating that wrenching continued to negatively affect growth in the field. Here it should be pointed out, however, that this wrenching was a multiple one (4 times at 2-week intervals) stressing the seedlings during the growing season. In another study, wrenched trees were superior to unwrenched trees in height [96] (Fig. 13b) and diameter [55] (Table 9) after 5 years in the field; but this wrenching was done later in the growing season to seedlings which had not been stressed but were well watered throughout the summer. It may be that wrenching overstressed the seedlings in Duryea and Lavender's [29] study, whereas those of Tanaka et al.'s [96] were in need of the moderate stress imposed by wrenching to aid hardening.

15.7.6 Conclusions

All species mentioned above, excluding Monterey pine, have an extremely mixed response to root-culturing practices. This response can probably be attributed to variation in: (1) wrenching regime—timing, frequency, and depth; (2) other nursery cultural practices—irrigation, seedbed density, and fertilization; (3) initial size of seedlings to be wrenched; (4) nursery soil and climate; and (5) seed-zone differences. Therefore, calibrating root culturing to the specific nursery site and conditions is extremely important.

However, some general results and recommendations are the following:

- Root wrenching during the growing season results in earlier budset and smaller seedlings.
- Wrenching too soon—that is, before the crop has met its size parameters—may result in a final crop that is too small and therefore has a lower yield.
- Wrenching in dry soil and (or) hot, dry weather without immediate irrigation can greatly stress Douglas-fir seedlings and should be avoided because there may be seedling mortality in the beds or reduced growth later in the field.
- Wrenching to mildly stress seedlings can help induce budset and hardening and may aid field growth and survival.
- Late-summer or fall root wrenching may promote seedling root fibrosity and should be further investigated.

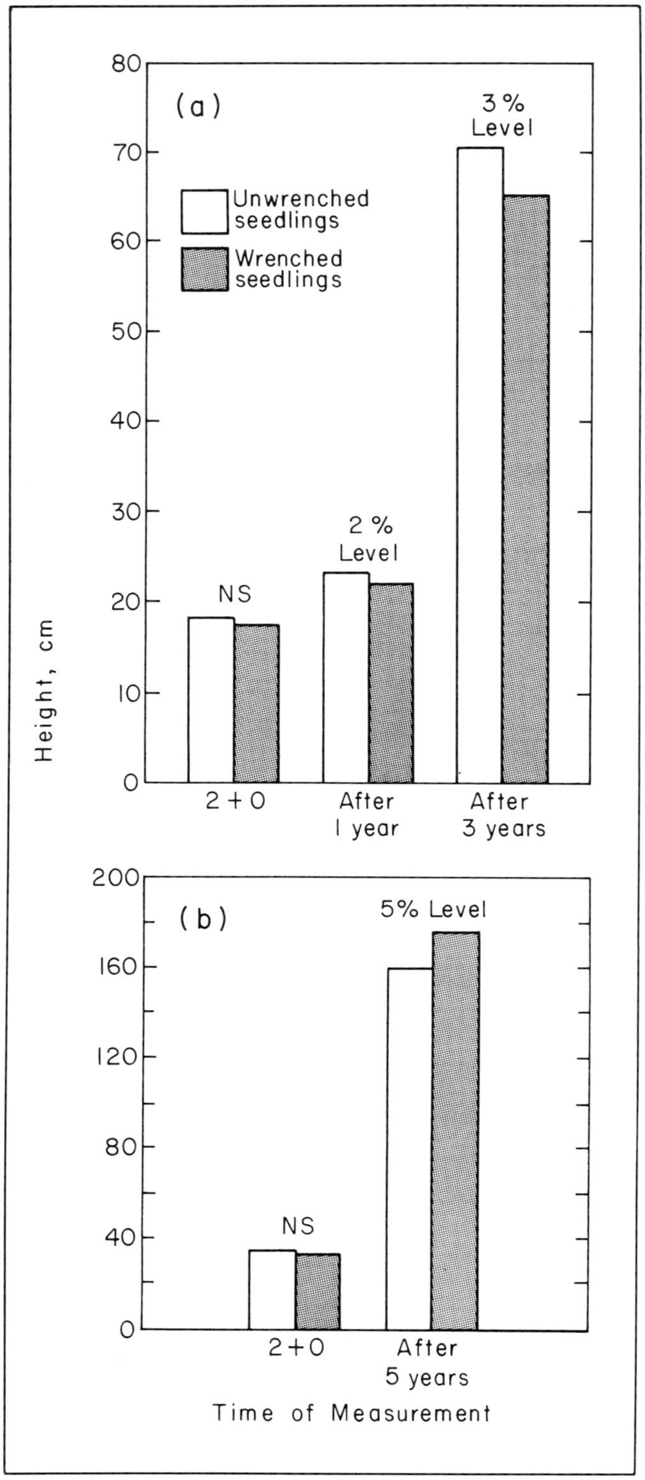

Figure 13. Two contrasting studies showing height of wrenched and unwrenched 2+0 Douglas-fir seedlings at planting and (a) 1 and 3 years (adapted from [29]) and (b) 5 years (adapted from [55, 96]) after planting in the field. In (a), heights were pooled because wrenching affected all seed sources equally. Significance levels are given above paired bars; NS = not significant.

15.8 Top Pruning

Top pruning, also called top mowing or clipping, is the passing of a cutting blade over the seedbed to sever seedling terminal leaders (Fig. 14). Although this nursery practice is common for altering the shoot:root ratio of hardwood species, it has rarely been used for conifers in temperate regions until recently [53].

Today, 92% of the U.S. Douglas-fir nurseries top-prune, though none of the Canadian nurseries surveyed or the U.S. nurseries producing predominantly pine seedlings do so (OSU Nursery Survey). Douglas-fir is the major Northwest species being top-pruned. Most top pruning occurs on 2+0 seedlings to be outplanted or to be transplanted and grown as 2+1s; 3+0 seedlings are pruned in the second year and sometimes again in the third. One nursery top-prunes fall-transplanted plug+1 seedlings, and two nurseries are testing top pruning of 1-year-old seedlings. Occasionally, bareroot transplants (1+1s and 2+1s) also are pruned after transplanting and during the growing season (OSU Nursery Survey).

The main reasons for top pruning in the Northwest are: (1) to control shoot height, (2) to facilitate nursery transplanting of seedlings (1+1s and 2+1s), (3) to achieve crop-size uniformity, and (4) to decrease the shoot:root ratio, especially of seedlings to be planted on harsh sites (OSU Nursery Survey). Nurseries that top-prune mainly for crop-size uniformity do so as many as 3 times in a growing season; usually, 10 to 20% of the tallest seedlings are cut to the desired crop size. Other nurseries prune once or twice during the season. Cutting height ranges from 9 to 16 inches (23 to 41 cm) for 2+0s and 6 to 14 inches (15 to 36 cm) for 3+0s (pruned in their second year); 2+0s for 2+1s are usually cut shorter, 6 to 12 inches (15 to 30 cm).

Figure 14. Top pruning 2+0 Douglas-fir seedlings with a rotary mower.

15.8.1 Timing

Seedlings in the Northwest should be pruned during the flush of growth in early summer. Such timing ensures the proper development of wound calluses and terminal buds. Pruning wounds made during the early-summer flush heal better than those made at other times of the year [42]. Furthermore, stems pruned in late summer after budset may flush again, producing new, succulent, unhardened shoots that are susceptible to winter injury [42]. Even though hardwoods not pruned in the nursery may not be adversely affected when cut at the time of field planting, Douglas-fir seedlings pruned after lifting may suffer [78]. Survival of seedlings pruned to 15 to 20 inches (38 to 51 cm) after lifting was 0%, whereas that of unpruned trees ranged from 41 to 78%, according to root-system size [40]. The most appropriate time to top-prune will differ at each individual nursery location and should be determined by time of budbreak, crop growth rate, and crop-size specifications.

15.8.2 Disease

Because open wounds, in general, are potential sites for disease infection, it is advisable to take certain precautions when top pruning. Clean tools and equipment decrease the chances of carrying spores throughout the nursery. Actively growing seedlings that are top-pruned, adequately watered, and adequately fertilized close wounds promptly, reducing the chance of infection. In a Mississippi nursery, 3,000,000 longleaf pine seedlings became infected with brown-spot needle blight, caused by *Scirrhia acicola* (Dearn) Siggers, after top pruning [44]. Spores from infected seedlings were transported throughout the nursery on the cutting blades; in addition, infected seedling tops left in seedbeds and paths contaminated water, which caused other parts of the nursery to become diseased. Infection and tissue mortality on each seedling occurred near the location where needles had been pruned.

Although top pruning has not been connected with any disease infection in the Northwest, it is important to be aware of this possibility. The most likely problem with disease could be increased risk of gray mold [*Botrytis cinerea* (Fries) Persoon] due to dead, pruned plant material lying in the beds, which creates a favorable microclimate and site for gray mold sporulation. To avoid development of this disease, pruned plant material should be raked and removed from nursery beds immediately after pruning.

15.8.3 Seedling morphology and physiology

Most top pruning work has been done with loblolly pine in the southern U.S. Because the longer growing season there can result in tall, spindly loblolly pine seedlings, top pruning was first investigated as a way to control this kind of top growth and produce a better-balanced seedling with higher survival potential [26]. Because little is known about the effect of top pruning on Northwest species, most of the information reported here is from southern pruning studies.

As in the Northwest, seedlings in the South are pruned to an average crop height. Clipping arrests height growth for 3 weeks, several fascicle buds develop just below the cut, and height growth then resumes. In the meantime, seedlings not pruned because they are shorter than the pruning height selected are said to be "released"; they continue to grow during the 3 weeks when the top growth of the other larger, pruned seedlings is suspended. Diameter growth of the pruned seedlings also is reduced [26, 27]. The overall result is a crop that varies less in both height and diameter.

In the field, top-pruned seedlings may conserve moisture better because their transpiring surface area has been reduced in relation to their root area [42]. This altered shoot:root ratio may aid in reducing transplant shock and in promoting successful plant establishment [42]. Top pruning does not seem to affect the morphology of loblolly pine root systems; seedlings with the same size root collar have similar root systems [26]. Due to the smaller shoot, however, top pruning produces seedlings with decreased shoot:root ratios.

15.8.4 Growth and survival

Top pruning most often improves survival of loblolly pine. Dieurauf [26] found that top pruning of taller seedlings significantly improved field survival (Table 11) but did not influence field height growth of surviving seedlings. Furthermore, forking of top-pruned loblolly pine is not a problem. When seedlings are planted in the field, many leaders are present; but one leader soon dominates, and after two or three seasons, no forks originate from the point of pruning [26, 61].

Table 11. Third-year field survival of top-pruned loblolly pine seedlings planted at two field sites (adapted from [26]).

Treatment	Date	Field survival, %: Pocahontas State Forest	Cumberland State Forest
		Taller seedlings	
Control		51.7[1]a	43.3a
Pruning height, in.			
6	8/12	78.3ab	95.0b
6 and 7	8/12 and 9/9	78.3ab	91.7b
7	9/9	93.3b	93.3b
		Shorter seedlings	
Control		91.7b	96.7b
Pruning height, in.			
4.5	8/12	98.3b	93.3b
4.5 and 5.5	8/12 and 9/9	88.3ab	96.7b
5.5	9/9	96.7b	96.7b

[1]Means followed by the same letter within a column are not significantly different at the 1% level.

For longleaf pine, a difficult species to regenerate, needle clipping (cutting needles to a length of 13 cm just before planting) increased field survival of seedlings from four separate nurseries. This increase was attributed to reduced moisture loss from a decreased needle-surface area [71].

Four hardwood species, top-pruned immediately before planting, differed in their field survival on a droughty site; severe top pruning did not lower survival of sycamore (*Platanus occidentalis* L.) and ash (*Fraxinus* spp.), decreased survival of water oak (*Quercus nigra* L.) only slightly, but substantially lowered survival of cherrybark oak (*Quercus falcata* var. *pagodifolia* Ell.) [114]. Northern red oak (*Quercus rubra* L.) pruned at the time of planting did not differ in survival from unpruned stock; pruned seedlings had faster initial growth in the field, but height of pruned and unpruned stock was equal after two growing seasons [78].

Ponderosa pine top-pruned during active growth in the 3+0 year at a North Dakota nursery had lower survival (34%) than unpruned trees (49%) [93]. In another study, 3+0 ponderosa pine seedlings pruned to 6 inches in the third year and then outplanted showed little effect of top pruning on multiple leaders, disease, or insect damage [47]. As previously mentioned, Hermann's [40] study showed that pruning ponderosa pine seedlings after lifting caused 100% mortality after outplanting.

There are very few studies on effects of top pruning Northwest species. Therefore, we can only speculate about possible pruning effects. Trees intended for droughty sites might be better balanced and survive better if top-pruned. The number of culls in a crop might drop because top pruning releases the smaller trees. But many questions about Northwest species remain unanswered: How are shoot diameter and height, needle surface area, and shoot and root dry weights of individual trees and the entire crop affected when seedlings are top-pruned? How does top pruning affect field growth? Does it affect the number of needle primordia laid down in the new bud and, therefore, growth the next year? How does top pruning affect field survival on moist and dry sites?

The Nursery Technology Cooperative at OSU is currently investigating the effects of top pruning on Douglas-fir seedlings. Studies installed at six nurseries using nine seed sources are designed to answer some of these questions about top pruning.

15.8.5 Conclusions

Top pruning helps the nursery manager control shoot height and achieve crop-size uniformity. Seedlings should be pruned when actively growing in early summer to ensure the proper development of terminal buds. Dead, pruned plant material should be removed from beds immediately after pruning to reduce chances of disease.

So far, little information is available on the effects of top pruning on individual seedling or crop characteristics for Northwest species. Although pruning is an effective alternative when a crop has excessively tall trees, the nursery manager should examine the overall coordination of cultural practices and should be alert to those, such as moderate moisture stress or root wrenching, which might help prevent overly tall crops in future years.

15.9 Transplanting

Transplanting is lifting seedlings from their original seedbed to plant in another location in the nursery (the transplant bed). Seedling density in the transplant bed is lower than that in the original seedbed, allowing more growing space for each seedling. Compared to a seedling of the same age grown at higher densities in a seedbed and not transplanted, transplanted seedlings have larger stem diameters, shoot heights, and root and shoot dry weights. Increased size may give them an advantage on difficult outplanting sites, such as those with heavy brush or severe animal browsing (see chapter 24, this volume). Because of the increased success with these larger seedlings, nurseries report an increasing trend in transplant orders (OSU Nursery Survey); in 1980, 19% of all seedlings produced in the Northwest nurseries surveyed were transplants. Although, in the past, only cull seedlings or those that were too small for field planting were transplanted [33], today seedlings are most often grown specifically for that purpose.

Although very little data are available, transplant seedlings are reported to have (1) more fibrous root systems, (2) larger stem diameters, and (3) lower shoot:root ratios than seedlings of comparable age grown at a standard density (25 to 35 seedlings/ft^2) [8, 43, 93, 103]; their height is usually equivalent or slightly greater than that of comparable, nontransplanted seedlings [8, 43].

15.9.1 General procedure

Transplant steps vary slightly with the time of the year seedlings are transplanted and with nursery location and conditions. Generally, seedlings are (1) watered well, (2) lifted (usually with the same lifting method as for field planting), (3) graded, (4) root-pruned, (5) packed (if storage is necessary), and (6) (sometimes) stored. The transplant bed is prepared (sometimes including fumigation and bed forming); transplant seedlings are planted in prewatered ground and then irrigated.

15.9.1.1 Irrigation and lifting

Care should be taken to minimize stress when transplanting, using guidelines and practices similar to those for lifting for field planting. If moisture stress is high, as it is likely to be in fall or summer transplanting, seedlings should be well irrigated before lifting; because hot, dry weather conditions are likely at the time of lifting, it is important to protect seedlings from overheating and drying. When fall or summer transplanting, many nurseries either lift only as many seedlings as can be processed in the same day or plan to store seedlings only up to 4 days to avoid transplant shock (OSU Nursery Survey).

15.9.1.2 Grading, root pruning, packing, and storing

After grading out culls and trees too large for the transplanter, roots of seedlings to be transplanted are pruned to facilitate the transplanting process and to avoid L-rooting in the transplant bed. Most roots are pruned to 5 to 6 inches (13 to 15 cm); two Canadian nurseries prune all roots to 4 inches (10 cm) (OSU Nursery Survey).

If seedlings are to be transplanted immediately, as in summer and fall, they are returned to field containers and covered with wet burlap (see chapter 22, this volume). Seedlings to be transplanted in spring, which must be stored up to 6 months, are packaged in containers (bags or boxes) similar to those for field planting. Storage, mainly of spring transplants, is most often under the same conditions as for seedlings to be outplanted—that is, in refrigerators at 1 to 2°C with 90% relative humidity or in freezers at −1 to −2°C (OSU Nursery Survey).

15.9.2 Equipment and bed density

Almost all transplanting is done by machine, either with a tractor-drawn mechanical transplanter or a self-propelled transplanter (OSU Nursery Survey); one custom-made machine plants four trenches to a bed instead of the normal six.

Common complaints about transplanters are (1) they are slow, (2) the trench or furrow is not deep enough, resulting in L-rooting of seedlings, and (3) seedling density cannot easily be altered. Transplant-bed densities range from 4 to 12 seedlings/ft^2 (43 to 129 seedlings/m^2), with most transplants grown at 5 to 6 seedlings/ft^2 (53 to 64 seedlings/m^2) (OSU Nursery Survey).

15.9.3 Timing

Seedlings are transplanted at three times of the year—fall, spring, and early summer; in the Northwest, spring is the most common time. A few nurseries transplant part of their 2+1 Douglas-fir crop in fall and the rest in spring. Canadian nurseries transplant their 2+1, 1+1, and 1+2 stock in spring and their 1½+½ and 1½+1½ stock in early summer (for information on plug+1 transplanting, see chapter 16, this volume).

15.9.3.1 Summer transplanting

Summer transplanting, also called hot transplanting, is common in many Canadian nurseries ([8, 103]; OSU Nursery Survey). In British Columbia, 1½+½ and 1½+1½ interior spruce (*Picea glauca* and *engelmannii*), lodgepole pine, and to a lesser extent Douglas-fir are transplanted in June or July, approximately half way through the growing season of their 2+0 year—hence the 1½ (OSU Nursery Survey). The main consideration in deciding when to summer-transplant is that leaders, though stiff and partly lignified, are still flexible enough to avoid breakage when handled [8]. Summer-transplanted seedlings have larger stem diameters and bushier root systems than 2+1, 1+1, or 1+2 transplants [8, pers. commun., 87]. However, it is advisable to thoroughly irrigate seedlings before lifting and after summer transplanting to avoid the consequences of high plant-moisture stress [103].

15.9.3.2 Fall and spring transplanting

Fall-transplanted Douglas-fir 2+1 seedlings are lifted in September and October at the end of the second growing season, stored only if necessary for a maximum of 4 days, and replanted into transplant beds immediately. Seedlings transplanted in spring are lifted sometime beween December and March, stored from 2 weeks to 6 months, and planted between March and late May (OSU Nursery Survey). No data are available on effects of storage followed by late-spring transplanting on transplant-bed growth and survival. To be safe, however, nurseries should lift during their established "lifting windows" and transplant as early as possible to assure early budburst and adequate growth in the transplant bed. Any seedlings lifted in spring should be stored as little as possible before transplanting.

Data from several Weyerhaeuser Company annual nursery reviews show that there is a critical period for fall transplanting.

Transplanting in late October, compared to late September, improved transplant-bed survival from 70 to 91% for trees from six seed sources [pers. commun., 100]; the differences in seedling condition (the later-transplanted seedlings were more hardened) or in weather conditions immediately after transplanting and during winter interacting with seedlings in different physiological states could be responsible for these results.

Only one published report comparing fall and spring transplants was found. In the plains region of Canada, survival of fall-transplanted Scotch pine, white spruce, and Colorado blue spruce (*Picea pungens* Engelm.) varied tremendously in transplant beds according to transplanting date; spring transplanting resulted in less variable survival and was most often better than transplanting in fall [24]. Inconsistent survival of fall transplants could again be due to different weather conditions or different nursery handling procedures of stock that was not completely dormant. On the other hand, two unpublished Northwest reports showed fall transplanting of Douglas-fir to be as favorable as spring [pers. commun., 50, 100]. Seedlings transplanted on October 25, 1976, had the same transplant-bed survival as those transplanted on April 25, 1977, but had earlier budburst (April 22 vs. May 15), greater height, larger stem diameter, and higher shoot:root ratio at lifting (Table 12) [50]. In the other study, spring and fall transplants from two seed sources survived equally well (99%) in the transplant bed and were equal in height, but fall transplants had slightly larger stem diameters and shoot and root dry weights [pers. commun., 100].

In summary, fall transplanting has several major advantages: (1) fall is a time when nursery activities are at a low; fall transplanting therefore lightens the workload in spring, (2) fall-transplanted seedlings may have larger diameters and root masses, and (3) earlier budburst in spring means that fall transplants reach their desired size sooner, leaving more time for hardening in summer. These advantages warrant further investigation.

However most Northwest nurseries do not transplant in the fall because (1) success is too variable—transplant-bed mortality is greater in fall than spring, (2) frost heaving occurs more frequently on fall transplants because the root system has not yet adequately anchored the seedling, (3) ground is not available in fall, (4) height control is sometimes difficult, and (5) nurseries generally do not see an advantage of fall over spring and summer transplanting (OSU Nursery Survey).

15.9.4 Conclusions

Transplant seedlings have more fibrous root systems, larger stem diameters, and lower shoot:root ratios than seedlings of the same age grown at a standard density (25 to 35 seedlings/ft^2). Important steps to ensure successful transplanting are (1) watering thoroughly before lifting, (2) lifting, grading, and root pruning, (3) packing and storing if necessary, (4) preparing the transplant bed, and (5) transplanting into prewatered ground and irrigating after planting.

Summer transplanting is common in Canadian nurseries; spruce and lodgepole pine are transplanted in June and July, resulting in 1½+½ and 1½+1½ seedlings. Spring transplanting is most common in U.S. nurseries, mainly because success of fall transplanting is too variable. Typical transplant stock types are 1–1s and 2+1s.

15.10 Interactions Among Nursery Practices

When two nursery practices interact, the effect of one practice will depend on the particular level of the other. For example, irrigation levels and different seedbed densities interact; that is, trees grown at lower densities will respond to watering levels differently than those grown at higher densities. In general, if no interaction exists, practices are said to be independent of one another (see chapter 28, this volume, for more information on interactions).

Some examples of interactions are:

- Fertilizer applications increased seedling height if seed was sown early, but not if it was sown late [89].
- Changes in water supply affected fertilized and unfertilized seedlings differently [12].
- Fall fertilization increased the chance of frost damage to Sitka spruce but not to Norway spruce, western hemlock, grand fir, and lodgepole pine [16].

It is important for nursery managers to be aware of interaction among practices in their nurseries. If a decision is made to alter or discontinue a current practice or to add a new one, careful attention should be paid to other practices affected by this change. Thinking ahead about the implications of such changes will avert problems that could lower crop quality and yield.

15.11 Conclusions and Recommendations

- In addition to stock-type description, morphological characteristics and physiological condition of seedlings can have a tremendous impact on seedling quality and ultimate field performance.
- Sowing depth can influence both the rate and the total amount of germination and therefore the final number of seedlings in the seedbed. To ensure good growth and crop uniformity, it is important to choose a proper sowing depth, specific to the tree species and soil conditions; then to prepare a level seedbed; and, finally, to make sure that seed depth is consistent throughout the bed.
- The importance of early sowing cannot be overemphasized. Early-sown first-year seedlings benefit from the entire growing season and are large enough for hardening by July and August. The final 2+0 crop is larger the next year and is again ready for hardening by midsummer.
- Lowering seedbed density produces seedlings with larger stem diameters and heavier shoots and roots. Furthermore, seedlings grown at lower densities may have an improved ability to regenerate new roots. Field survival of seedlings varies regardless of seedbed density, but height growth of seedlings grown at lower densities most often is superior for several growing seasons after planting.
- Irrigation schedules that induce a moderate plant moisture-stress level during hardening result in earlier budset, increased cold hardiness, smaller seedlings, and increased field-survival potential. Extremely low or high plant

Table 12. Morphology (at time of lifting), phenology (in the nursery), and survival of fall- and spring-transplanted 2+1 Douglas-fir seedlings [pers. commun., 50].

Transplanting date	Height, cm	Diameter, mm	Fresh weight, g	Shoot:root ratio, g dry wt.	Budburst date	Survival in transplant bed, %
10/25/76	33.9	7.5	49.9	1.52	4/22/77	95
4/25/77	29.0	6.9	36.9	1.30	5/15/77	91

moisture-stress levels can be harmful to seedlings. However, so far, no published investigations define moderate levels or their effect on field performance.

- Fertilization, especially with N, produces taller, heavier seedlings with larger shoot diameters. The timing of N application seems important to seedling frost hardiness. Though little is known about the effects of nursery fertilization on seedling performance in the field, where studies have been done, effects of fertilization on either growth or survival often have been positive.
- Root-culturing practices that cut roots before or during the growing season arrest or retard seedling height growth. The effects of this on Douglas-fir root systems have varied, with no reported measurements of increased root fibrosity, although root wrenching or lateral pruning in late summer or early fall may promote root growth. Both field survival and height growth of root-wrenched seedlings compare inconsistently with the survival and growth of unwrenched seedlings.
- Seedlings are most often top-pruned to control shoot height when shoots elongate in early summer. However, there are no published reports on the effect of this growing-season top pruning on Douglas-fir morphology, physiology, or field performance.
- Transplanted seedlings have larger stem diameters, shoot heights, and root and shoot dry weights than seedlings of the same age grown at higher seedbed densities and not transplanted. This increased size is reported to give transplants an advantage on difficult outplanting sites, such as those with heavy brush or severe animal browsing.
- When two nursery practices interact, the effect of one practice will vary with the particular level of the other. If a nursery manager decides to alter or discontinue a current practice or to add a new one, careful attention should be given to the effect of this change on other cultural practices in the nursery.

Acknowledgments

I am extremely grateful to Dean Cowles, Weyerhaeuser Company; Tom Landis, U.S.D.A. Forest Service; Denis Lavender, Oregon State University; Steve Omi, Oregon State University; Jeff Snyder, Lava Nursery, Inc.; Barbara Thompson, International Paper Company; and Tim White, International Paper Company, for their excellent reviews of this chapter.

References

1. Aldhous, J. R. 1972. Nursery practice. Her Majesty's Stationary Office, London. Forestry Commun. Bull. 43. 184 p.
2. Allen, R. M., and T. E. Maki. 1955. Response of longleaf pine seedlings to soil and fertilizers. Soil Sci. 79:359-362.
3. Anderson, H. W. 1976. Personal communication, State of Washington, Dep. of Resources, Div. of Forest Land Management, Olympia.
4. Anderson, H. W., and S. P. Gessel. 1966. Effects of nursery fertilization on outplanted Douglas-fir. J. Forestry 64:109-112.
5. Anstey, C. 1971. Survival and growth of 1/0 radiata pine seedlings. New Zealand J. Forestry 16(1):77-81.
6. Armson, K. A. 1966. The growth and absorption of nutrients by fertilized and unfertilized white spruce seedlings. Forestry Chronicle 42(2):127-136.
7. Armson, K. A. 1968. The effects of fertilization and seedbed density on the growth and nutrient content of white spruce and red pine seedlings. Univ. of Toronto, Faculty of Forestry, Ontario. Tech. Rep. 10. 16 p.
8. Armson, K. A., and V. Sadreika. 1979. Forest tree nursery soil management and related practices. Ontario Ministry of Natural Resources, Toronto. 179 p.
9. Bacon, C. J., and P. J. Hawkins. 1979. Intensive root wrenching: a prerequisite for the successful establishment of 1-0 *Pinus caribaea* Mor. seedlings. Tree Planters' Notes 30(1):5-7.
10. Balneaves, J. M., and A. R. McCord. 1976. Precision sowing improves quality and performance of 1/0 radiata pine seedlings. New Zealand Forest Serv., Forest Res. Institute, Rangiora. Production Forestry Rep. 90 (unpubl.). 11 p.
11. Baron, F. J., and G. H. Schubert. 1963. Seedbed density and pine seedling grades in California nurseries. U.S.D.A. Forest Serv., Pacific SW Forest and Range Exp. Sta., Berkeley, California. Res. Note PSW-31. 14 p.
12. Bengtson, G. W., and G. K. Voight. 1962. A greenhouse study of relations between nutrient movement and conversion in a sandy soil and the nutrition of slash pine seedlings. Soil Sci. Society of America Proc. 26:609-612.
13. Benson, A. D., and K. R. Shepherd. 1976. Effect of nursery practice on *Pinus radiata* seedling characteristics and field performance: I. Nursery seedbed density. New Zealand J. Forestry Sci. 6(1):19-26.
14. Benson, A. D., and K. R. Shepherd. 1977. Effects of nursery practice on *Pinus radiata* seedling characteristics and field performance: II. Nursery root wrenching. New Zealand J. Forestry Sci. 7:68-76.
15. Benzian, B. 1966. Effects of nitrogen and potassium concentrations in conifer seedlings on frost damage. Pages 101-102 *in* Rep. of Forest Research 1966. Forestry Commission, London. As cited in van den Driessche [105, this list], p. 111.
16. Benzian, B., R. M. Brown, and S. C. R. Freeman. 1974. Effect of late-season top-dressings of N (and K) applied to conifer transplants in the nursery on their survival and growth on British forest sites. Forestry 47:153-184.
17. Blair, R., and F. Cech. 1974. Morphological seedling grades compared after thirteen growing seasons. Tree Planters' Notes 25(1):5-7.
18. Blake, J., J. Zaerr, and S. Hee. 1979. Controlled moisture stress to improve cold hardiness and morphology of Douglas-fir seedlings. Forest Sci. 25:576-582.
19. Brodie, J. D., and P. L. Tedder. 1982. Regeneration delay: economic cost and harvest loss. J. Forestry 80(1):26-28.
20. Brown, J. H. 1970. Seedling growth of three Scotch pine provenances with varying moisture and fertility treatments. Forest Sci. 16:43-45.
21. Brunsden, G. J. 1981. Box pruning and field results. Pages 142-153 *in* Forest nursery and establishment practice in New Zealand. New Zealand Forest Serv., Forest Res. Institute, Rotorua. FRI Symp. 22.
22. Burns, R. M., and R. H. Brendemuehl. 1971. Nursery bed density affects slash pine seedling grade and grade indicates field performance. U.S.D.A. Forest Serv., Southeastern Forest Exp. Sta., Asheville, North Carolina. Res. Pap. SE-77. 7 p.
23. Chavasse, C. G. R. 1977. The significance of planting height as an indication of subsequent seedling growth. New Zealand J. Forestry 22(2):283-296.
24. Cram, W. H., and A. C. Thompson. 1981. Fall and spring transplanting of conifers in the Plains region. Tree Planters' Notes 32(1):16-19.
25. Day, R. J. 1980. Effective nursery irrigation depends on regulation of soil moisture and aeration. Pages 52-71 *in* Proc., North American forest tree nursery soils workshop (L. P. Abrahamson and D. H. Bickelhaupt, eds.). State Univ. New York, Coll. Environ. Sci. and Forestry, Syracuse.
26. Dieurauf, T. A. 1976. Top clipping in the nursery bed. Pages 37-43 *in* Proc., Southeast Area Nurserymen's conf. (C. Lantz, ed.). Eastern Session, Charleston, South Carolina, Aug. 3-5; and Western Session, Mobile, Alabama, Aug. 17-19.
27. Dieurauf, T. A., and J. W. Garner. 1980. Simulated root wrenching. Virginia Div. of Forestry Occasional Rep. 55. 3 p.
28. Duryea, M. L. 1977. Unpublished data, Dep. of Forest Sci., Oregon State Univ., Corvallis.
29. Duryea, M. L., and D. P. Lavender. 1982. Water relations, growth, and survival of root-wrenched Douglas-fir seedlings. Can. J. Forest Res. 12:545-555.
30. Dykstra, G. F. 1974. Undercutting depth may affect root-regeneration of lodgepole pine seedlings. Tree Planters' Notes 25:21-22.
31. Edgren, J. W. 1966. Bed density and size of ponderosa pine seedlings at the Bend Nursery. Pages 21-29 *in* Proc., 10th Biennial Western Forestry Nursery Council meeting.

32. Edgren, J. W. 1975. Douglas-fir 2+0 nursery stock size and first-year field height growth in relation to seed bed density. Pages 72-79 *in* Proc., Servicewide conf. on planting stock production, Coeur d'Alene, Idaho, Sept. 16-18.
33. Edgren, J. W. 1976. Seedbed density, diameter limit culling, and 2+0 Douglas-fir seedling production. *In* Proc., Western Forest Nursery Council and Intermountain Nurserymen's Assoc., Richmond, B.C., Aug. 10-12.
34. Eis, S. 1968. Lateral root pruning—a promising forest nursery practice. Forestry Chronicle 44(5):12-13.
35. Eis, S., and J. R. Long. 1972. Lateral root pruning of Sitka spruce and western hemlock seedlings. Can. J. Forest Res. 2(3):223-227.
36. Eis, S., and J. R. Long. 1973. Root pruning in the nursery. Tree Planters' Notes 24(1):20-22.
37. Garrett, P. W., and R. Zahner. 1973. Fascicle density and needle growth responses of red pine to water supply over two seasons. Ecology 54:1328-1334.
38. Gilmore, A. R., E. S. Lyle, Jr., and J. T. May. 1959. The effects on field survival of late nitrogen fertilization of loblolly pine and slash pine in the nursery seedbed. Tree Planters' Notes 36:22-23.
39. Griffin, A. R. 1974. Geographic variation in juvenile growth characteristics of Douglas-fir (*Pseudotsuga menziesii* (Mirb.) Franco) from coastal ranges of California. Ph.D. thesis, School of Forestry, Oregon State Univ., Corvallis. 153 p.
40. Hermann, R. K. 1964. Importance of top-root ratios for survival of Douglas-fir seedlings. Tree Planters' Notes 64:7-11.
41. Hobbs, S. D. 1982. Unpublished data, FIR Program, Medford, Oregon.
42. Janick, J. 1979. Directing plant growth. *In* Horticultural Science. 3rd ed. W. H. Freeman and Co., San Francisco, California. 608 p.
43. Jones, E. W. 1968. A note on the dimensions of shoots and roots of planting stock. Forestry 41(2):199-206.
44. Kais, A. G. 1978. Pruning of longleaf pine seedlings in nurseries promotes brown-spot needle blight. Tree Planters' Notes 29(1):3-4.
45. Koon, K. B., and T. O'Dell. 1977. Effects of wrenching on drought avoidance of Douglas-fir seedlings. Tree Planters' Notes 28(2):15-16.
46. Kozlowski, T. T. 1971. The challenges of research in forest biology. Forestry Chronicle 47:192-195.
47. Lanquist, K. B. 1966. Top pruning of ponderosa pine. Tree Planters' Notes 79:3-8.
48. Larsen, J. B. 1976. Untersuchungen uber die Frostempfindlichkeit von Douglasienherkünften ünd uber den Einfluss der Nahrstoffversorgung auf die Frostresistenz der Douglasie. Forst- und Holzwirt 31(15):299-300, 302. As cited in van den Driessche [105, this list], p. 112.
49. Larsen, J. B. 1978. Untersuchungen uber die Bedeutung der Kalium und Stickstoffversorgung fur die Austrocknungsresistenz der Douglasie (*Pseudotsuga menziesii*) in Winter. Flora 167:197-207. As cited in van den Driessche [105, this list], p. 112.
50. Lavender, D. P. 1977. Personal communication, Dep. of Forest Sci., Oregon State Univ., Corvallis.
51. Lavender, D. P., K. K. Ching, and R. K. Hermann. 1968. Effect of environment on the development of dormancy and growth of Douglas-fir seedlings. Botanical Gazette 129(1):70-83.
52. Lavender, D. P., and B. D. Cleary. 1974. Coniferous seedling production techniques to improve seedling establishment. Pages 177-180 *in* Proc., North American containerized forest tree seedling symp. (R. W. Tinus, W. I. Stein, and W. E. Balmer, eds.). Great Plains Agric. Res. Publ. 68.
53. Lavender, D. P., and R. K. Hermann. 1976. Role of forest tree physiology in producing planting stock and establishing plantations. Unpubl. paper presented at XVI IUFRO World Congress, Division II. Oslo, Norway. 12 p.
54. Levitt, J. 1972. Responses of plants to environmental stresses. Academic Press, New York. 697 p.
55. Long, J. N. 1978. Root system form and its relationship to growth in young planted conifers. Pages 222-234 *in* Symp. on root form of planted trees, Victoria, B.C., May 16-19.
56. Lopushinsky, W. 1976. Relationship of shoot-root ratio to survival and growth of outplanted Douglas-fir and ponderosa pine seedlings. U.S.D.A. Forest Serv., Pacific NW Forest and Range Exp. Sta., Portland, Oregon. Res. Note PNW-274. 7 p.
57. Lott, J. R., and G. Hallman. 1973. New Zealand root pruner evaluation for U.S. use. Tree Planters' Notes 24(4):18-20.
58. Malcolm, D. C., and B. C. Y. Freezaillah. 1975. Early frost damage on Sitka spruce seedlings and the influence of phosphorus nutrition. Forestry 48:139-145.
59. McClain, K. M., and K. A. Armson. 1975. Growth responses of pine and spruce seedlings to regimes of soil moisture and fertility. Soil Sci. Society of America Proc. 39(1):140-144.
60. McDonald, S. E., and S. W. Running. 1979. Monitoring irrigation in western forest tree nurseries. U.S.D.A. Forest Serv., Rocky Mt. Forest and Range Exp. Sta., Fort Collins, Colorado. Gen. Tech. Rep. RM-61. 8 p.
61. McLemore, B. F. 1982. Comparison of 1-0 and 2-0 loblolly pine seedlings. Tree Planters' Notes 33(2):22-23.
62. McMinn, R. G. 1980. Root development of white spruce and lodgepole pine seedlings following outplanting. Pages 186-190 *in* Can. Forestry Serv., Pacific Forest Res. Centre, Victoria, B.C. As cited in Schmidt-Vogt [80, this list].
63. Menzies, M. I. 1980. Effect of nursery conditioning on the water relations of two-year-old Douglas-fir seedlings after lifting and outplanting. Ph.D. thesis, Univ. of Washington, Seattle. 208 p.
64. Menzies, M. I., J. C. van Dorsser, and B. W. A. Moberly. 1974. Effect of seedling spacing on grade return in the nursery and growth in the field. New Zealand Forest Serv., Forest Res. Institute, Rotorua. Forest Establishment Rep. 55 (unpubl.). 12 p.
65. Morby, F. E. 1982. Irrigation regimes in a bare-root nursery. Pages 55-59 *in* Proc., 1981 Intermountain Nurserymen's Assoc. meeting, Edmonton, Alberta, Aug. 11-13. Can. Forestry Serv., Edmonton. Inf. Rep. NOR-X-241.
66. Mullin, R. E. 1966. Root pruning of nursery stock. Forestry Chronicle 42:256-264.
67. Mullin, R. E., and L. Bowdery. 1977. Effects of seedbed density and nursery fertilization on survival and growth of white spruce. Forestry Chronicle 53(2):83-86.
68. Mullin, R. E., and L. Bowdery. 1978. Effects of nursery seedbed density and top dressing fertilization on survival and growth of 3+0 red pine. Can. J. Forest Res. 8:30-35.
69. Mullin, R. E., and C. Christl. 1981. Morphological grading of white spruce nursery stock. Forestry Chronicle 57(3):126-130.
70. Mullin, R. E., and C. Christl. 1982. Morphological grading of white pine nursery stock. Forestry Chronicle 58(1):40-43.
71. Parker, J. A., and L. W. Haines. 1981. Longleaf pine seedling survival studies. U.S.D.A. Forest Serv., Southern Forest Exp. Sta., New Orleans, Lousiana. Gen. Tech. Rep. SO-34. 4 p.
72. Pawsey, C. K. 1972. Survival and early development of *Pinus radiata* as influenced by size of planting stock. Australian Forest Res. 5(4):13-24.
73. Pharis, R. P., and P. J. Kramer. 1964. The effect of nitrogen and drought on loblolly pine seedlings. Forest Sci. 10:143-150.
74. Pollard, D. F. W., and K. T. Logan. 1977. The effects of light intensity, photoperiod, soil moisture potential, and temperature on bud morphogenesis in *Picea* species. Can. J. Forest Res. 7:415-421.
75. Richter, J. 1971. Das Umsetzen von Douglasien im Kulturstadium. Allgemeine Forst- und Jagdzeitung 142:65-69. As cited in Schmidt-Vogt [79, this list], p. 75.
76. Rook, D. A. 1969. Water relations of wrenched and unwrenched *Pinus radiata* seedlings on being transplanted into conditions of water stress. New Zealand J. Forestry 14:50-58.
77. Rook, D. A. 1971. Effect of undercutting and wrenching on growth of *Pinus radiata* D. Don seedlings. J. Applied Ecology 8:477-490.
78. Russell, T. E. 1973. Survival and growth of bar-slit planted northern red oak studied in Tennessee. Tree Planters' Notes 24(3):6-9.
79. Schmidt-Vogt, H. 1974. Planting material. Pages 70-86 *in* Symp. on stand establishment. IUFRO Joint Meeting, Div. 1 and 3, Wageningen, The Netherlands, Oct. 15-19.
80. Schmidt-Vogt, H. 1981. Morphological and physiological characteristics of planting stock: present state of research and research tasks for the future. Pages 433-446 *in* Proc., IUFRO XVII World Congress. Freiburg i. Br., Federal Republic of Germany.
81. Schomaker, C. E. 1969. Growth and foliar nutrition of white pine seedlings as influenced by simultaneous changes in moisture and nutrient supply. Soil Sci. Society of America Proc. 33:614-618.
82. Schubert, G. H., and R. S. Adams. 1971. Reforestation practices for conifers in California. State of California, Dep. of Conservation, Div. of Forestry, Sacramento, California. 359 p.

83. Shipman, R. D. 1964. Low seedbed densities can improve early height growth of planted slash and loblolly pine seedlings. J. Forestry 62(11):814-817.

84. Shoulders, E. 1961. Effect of nursery bed density on loblolly and slash pine seedlings. J. Forestry 59(8):576-579.

85. Shoulders, E. 1963. Root-pruning southern pines in the nursery. U.S.D.A. Forest Serv., Southern Forest Exp. Sta., New Orleans, Louisiana. Res. Pap. SO-5.

86. Show, S. B. 1930. Forest nursery and planting practice in the California pine region. U.S. Dep. Agric., Washington, D.C. Circular 92.

87. Simpson, D. G. 1982. Personal communication, B.C. Ministry of Forests, Victoria.

88. Smith, J. H. G., A. Kozak, O. Sziklai, and J. Walters. 1966. Relative importance of seedbed fertilization, morphological grade, site, provenance, and parentage to juvenile growth and survival of Douglas-fir. Forestry Chronicle 42:83-86.

89. Sorensen, F. C. 1978. Date of sowing and nursery growth of provenance of *Pseudotsuga menziesii* given two fertilizer regimes. J. Appied Ecology 15:273-279.

90. Stiell, W. M. 1976. White spruce: artificial regeneration in Canada. Can. Forestry Serv., Forest Management Institute, Ottawa. Inf. Rep. FMR-X-85. 275 p.

91. Stoeckeler, J. H., and H. F. Arneman. 1960. Fertilizers in forestry. Advances in Agronomy 12:127-195.

92. Stoeckeler, J. H., and G. W. Jones. 1957. Forest nursery practice in the Lake States. U.S.D.A. Forest Serv., Washington, D.C. Agric. Handb. 110. 124 p.

93. Stoeckeler, J. H., and P. E. Slabaugh. 1965. Conifer nursery practice in the Prairie-Plains. U.S.D.A. Forest Serv., Washington, D.C. Agric Handb. 279. 93 p.

94. Sweet, G. B., and D. A. Rook. 1972. Inhibitor levels associated with growth in seedlings of *Pinus radiata*. New Phytologist 72:1107-1111.

95. Switzer, G. L., and L. E. Nelson. 1963. Effects of nursery fertility and density on seedling characteristics, yield, and field performance of loblolly pine (*Pinus taeda* L.). Soil Sci. Society of America Proc. 27:461-464.

96. Tanaka, Y., J. D. Walstad, and J. E. Borrecco. 1976. The effect of wrenching on morphology and field performance of Douglas-fir and loblolly pine seedlings. Can. J. Forest Res. 6:453-458.

97. TerBush, F. A. 1977. Tree seedling description code. Tree Planters' Notes 28(3, 4):6.

98. Thompson, B. 1983. Why fall fertilize? Pages 85-91 *in* Proc., Western Nurserymen's conf. Western Forestry Nursery Council, Medford, Oregon, Aug. 10-12, 1982. Southern Oregon State Coll., Ashland.

99. Timmis, R. 1974. Effect of nutrient stress on growth, bud set, and hardiness in Douglas-fir seedlings. Pages 187-193 *in* Proc., North American containerized forest tree seedling symp. (R. W. Tinus, W. I. Stein, and W. E. Balmer, eds.). Great Plains Agric. Res. Publ. 68.

100. Triebwasser, M. 1981. Personal communication, Weyerhaeuser Co., Mima Tree Nursery, Olympia, Washington.

101. U.S.D.A. Forest Service. 1974. Seeds of woody plants in the United States. U.S. Dep. Agric., Washington, D.C. Agric. Handb. 450. 883 p.

102. van den Driessche, R. 1963. Nursery experiments with Douglas-fir. Commonwealth Forestry Review 42(3):242-254.

103. van den Driessche, R. 1969. Forest nursery handbook. B.C. Forest Serv., Victoria. Res. Notes 48. 44 p.

104. van den Driessche, R. 1980. Effects of nitrogen and phosphorus fertilization on Douglas-fir nursery growth and survival after outplanting. Can. J. Forest Res. 10:65-70.

105. van den Driessche, R. 1980. Health, vigour and quality of conifer seedlings in relation to nursery soil fertility. Pages 100-1[illegible]0 *in* Proc., North American forest tree nursery soils workshop (L. P. Abrahamson and D. H. Bickelhaupt, eds.). State Univ. New York, Coll. Environ. Sci. and Forestry, Syracuse.

106. van den Driessche, R. 1982. Overwinter sowing of seed pelleted by Hilleshog. B.C. Ministry of Forests, Victoria. Progress Rep., Project 786. 13 p.

107. van den Driessche, R. 1982. Relationship between spacing and nitrogen fertilization of seedlings in the nursery, seedling size, and outplanting performance. Can. J. Forest Res. 12:865-87[illegible]

108. van Dorsser, J. C. 1981. Seedling conditioning. Pages 128-142 *in* Forest nursery and establishment practice in New Zealand. New Zealand Forest Serv., Forest Res. Institute, Rotorua. FRI Symp. 22.

109. van Dorsser, J. C., and B. W. A. Moberly. 1969. Conditioning of autumn sown *Pinus radiata* planting stock. New Zealand Forest Serv., Forest Res. Institute, Rotorua. Establishment Rep. 6 (unpubl.). 20 p. As cited in van Dorsser and Rook [110, this list], p. 71.

110. van Dorsser, J. C., and D. A. Rook. 1972. Conditioning of radiata pine seedlings by undercutting and wrenching: description of methods, equipment, and seedling response. New Zealand J. Forestry 17:61-73.

111. Wakeley, P. C. 1954. Physiological grades of southern pine nursery stock. Pages 311-322 *in* Proc., Society of American Foresters. Washington, D.C.

112. Wilde, S. A., R. Wittenkamp, E. L. Stone, and H. M. Galloway. 1940. Effect of high rate fertilizer treatments of nursery stock upon its survival and growth in the field. J. Forestry 38:806-809.

113. Wilson, B. C., and R. K. Campbell. 1972. Seedbed density influences height, diameter, and dry weight of 3-0 Douglas-fir. Tree Planters' Notes 23(2):1-4.

114. Woessner, R. A. 1972. Four hardwood species differ in tolerance to pruning. Tree Planters' Notes 23(1):28-29.

115. Zaerr, J. B., B. D. Cleary, and J. Jenkinson. 1981. Scheduling irrigation to induce seedling dormancy. Pages 74-78 *in* Proc., Intermountain Nurserymen's Assoc. and Western Forest Nursery Council combined meeting, Boise, Idaho, Aug. 12-14, 1980. U.S.D.A. Forest Serv., Intermountain Forest and Range Exp. Sta., Ogden, Utah. Gen. Tech. Rep. INT-109.

Chapter 16
Plug + 1 Seedling Production

F. F. Hahn

Abstract

The recently developed plug + 1 seedling begins life as a typical plug in a containerized nursery. At the end of its first year, it is transplanted to a bareroot nursery, where it continues to grow for another year before outplanting. This new stock type, with bushy top and moplike, fibrous root mass, has had good survival and height growth on typical Northwest transplant sites. It has also shown, on these sites, a comparable or better total cost:benefit ratio than any other seedling type currently in use.

16.1 Introduction

The plug + 1 seedling, one of the newest among a variety of seedling types used for reforestation, is a hybrid derived from merging recently developed containerized-seedling production methods with age-old bareroot methods. Plug + 1 production utilizes both container and bareroot types of growing facilities and technologies; but it is often difficult to coordinate both technologies because, in most cases, container and bareroot nurseries are operated separately. Furthermore, it is hard to define which phase is more important in such production; credit for a good crop or blame for a bad one is often disputed by the two different growers. In spite of the newness of plug − 1 seedlings and initial difficulties in producing them, they are gaining rapid acceptance, proving themselves a good stock type for certain species in certain areas. However, plug + 1s are not a cure for all reforestation problems.

In this chapter, I describe the evolution of the plug + 1 and its production in both containerized and bareroot phases and assess its overall applicability for reforestation in the Northwest. Most of the discussion relates to Douglas-fir [*Pseudotsuga menziesii* (Mirb.) Franco], the major commercial species in the Northwest. However, the production methods described can be quite successfully applied to other Northwest species.

16.2 A Brief Description and History

The plug + 1 seedling, or p + 1, starts its life in a containerized nursery (Fig. 1a). Once the seedling has a firm root plug and sturdy top and reaches the right physiological stage, it is transplanted to a bareroot nursery (Fig. 1b). The transplanted plug grows as a bareroot seedling for another year—thus the name "plug + 1." Production of p + 1 seedlings always involves these two distinctly different nursery operations, although the scheduling and rearing of p + 1 seedlings in each nursery differ somewhat from those of regular containerized and bareroot seedlings.

It is difficult for the bareroot nursery operator to produce good p + 1 seedlings if the containerized seedlings are of poor quality or are not scheduled properly for transplanting. On the other hand, even the best produced and scheduled containerized seedlings could become poor p + 1 stock if not reared properly in the bareroot nursery. The two types of nurseries greatly depend on each other and must strive to use good growing technology to achieve the desired end product.

Plug + 1 seedlings were first produced in 1971 by the Ray Leach Nursery in Aurora, Oregon. Because the idea showed very little success at that time, further production attempts were abandoned until 1975. During the regular 1975 spring transplanting season, Georgia-Pacific Corporation transplanted 22,000 Douglas-fir seedlings produced in the company's container nursery at Cottage Grove, Oregon, to the Tyee Tree Nursery (bareroot) near Roseburg, Oregon. Cultural practices and production scheduling applied to the plug and bareroot phases of these seedlings did not differ from the normal rearing practices in either facility. At the end of the growing season, the following fall, the entire seedling crop was quite uniform and showed good yield. Seedling tops and roots looked distinctly different from those of regular bareroot transplant stock, however; tops were bushy, and roots resembled a mop, with a large, fibrous root mass (Fig. 2). This clearly was a

In Duryea, Mary L., and Thomas D. Landis (eds.). 1984. *Forest Nursery Manual: Production of Bareroot Seedlings*. Martinus Nijhoff/Dr W. Junk Publishers, The Hague/Boston/Lancaster, for Forest Research Laboratory, Oregon State University, Corvallis. 386 p.

new product, at least in appearance, although its field performance was as yet unknown.

The following winter, field trials were established on various sites to compare the new p+1 seedlings to regularly used seedling types. A year later the first evaluation showed encouraging results in survival and height growth in spite of the 1976 summer drought.

The drought continued into the fall and winter, delaying the regular winter planting season for months. There was real concern that most of the millions of seedlings produced that year would never reach their destination. It is a well-known fact that seedlings held over for an extra year in containers or bareroot nursery beds may reach a size and condition unsuitable for successful field planting; furthermore, containerized seedlings become "pot-bound." Because transplanting seemed to provide the best solution, some of the surplus containerized seedlings were moved to the Tyee Tree Nursery in November. This was possible because the warm, dry weather provided good conditions for seedling bed preparation and transplanting.

Figure 1. (a) Bird's-eye view of a container nursery, where the p+1 begins life, and (b) transplant beds in a bareroot nursery, where the p+1 grows for another year before outplanting.

Transplanted seedlings were not affected by the drought because proper ground moisture was maintained through watering.

After rain finally came in February 1977, full-scale field planting resumed, but there was not enough time left to move all the remaining containerized seedlings to the field. Therefore, another group of containerized seedlings was transplanted during the regular spring transplanting season.

The fall- and spring-transplanted plugs were reared with normal bareroot transplant cultural practices during the next growing season. Neither batch received special treatment. In spite of this, fall and spring transplants differed distinctly in seedling quality; fall transplants had 4 to 5 mm larger caliper, 20 to 25 cm greater top growth, and a better developed root system than spring transplants. We at Georgia-Pacific have seen similar differences due to fall and spring planting for containerized seedlings in previous field-planting operations, but none as pronounced as those at this nursery.

Thus, the disastrous 1976 drought had some side benefits: the nursery industry learned more about large- (or commercial-) scale p+1 seedling development and production. Technology and usage have continued to be refined since.

Figure 2. A 1-year-old greenhouse-grown plug (left) and a 2-year old bareroot originated from a plug, or p+1 (right). Note the p+1's bushy top and moplike, fibrous root system.

16.3 Plug + 1 Status in the Northwest

Plug+1 seedling production in the Northwest was limited to only a few containerized and bareroot nurseries during the late 1970s in spite of its initial success—mainly because of the newness of the p+1 product and the lack of popularity of containerized seedlings. Although the end product is a bareroot seedling, quality of the initial plug still has a major effect on the ultimate crop. Bareroot nursery operators were not always eager to accept container-grown seedlings for transplanting. Some nurseries had strict rules against accepting any crops, bareroot or containerized, from another nursery to avoid possible disease contamination from an outside source.

Tyee Tree Nursery was first to accept containerized seedlings for transplanting and has continued to do so on a large commercial scale ever since. Other small private bareroot nurseries have also adopted this practice. Large industrial and government-operated nurseries, however, were slow in gearing up for p+1 production. The Industrial Forestry Association's nursery at Canby, Oregon, transplanted its first p+1 crop from an outside source in August 1978; soon after, other large nurseries followed suit.

According to the OSU Nursery Survey (see chapter 1, this volume), p+1 production reached about 6.5 million seedlings in the Northwest by 1980—about 2% of total seedling production. Since the 1980 tabulation, p+1s have gained even better and wider acceptance in the Northwest and other areas.

16.4 Containerized Growing Phase

Containerized seedling production is relatively new and, in general, poorly understood. It has developed rapidly over the last 10 to 15 years, during which time many different containerized systems have been tested. Although some systems produced excellent seedlings for direct field planting or p+1 production, not all systems worked as expected; therefore, containerization recently reached a point of stagnation—and reassessment [3].

Specific growing facilities, climate control units, container types, crop scheduling, growing regimes, hardening methods, and shipping preparation for containerized seedling production can now be better defined from experience, as reflected in the following discussion.

16.4.1 Growing facilities

Containerized seedlings can be grown in three types of facilities:

- **Semicontrolled growing facilities** provide minimum protection for the seedling crop because they are seldom equipped with any environmental control units other than a shade screen or plastic-covered roof. For this reason, these seedlings are the least expensive to produce but are suitable for growing only under mild climatic conditions (Fig. 3a).
- **Shelterhouses** are not permanently enclosed and have an environmental control system that can be fully automated and regulated. They are normally equipped with a cooling system that uses thermostatically controlled, removable wall and roof covers for natural ventilation rather than cooling pads and fans (Fig. 3b).
- **Greenhouses** are similar in structure to shelterhouses but have permanent roof and wall covers. For this reason, they rely heavily on artificial environmental control to hold the temperature at a desired level. Therefore, this is the only facility type suitable for growing seedlings year-round regardless of natural climatic conditions (Fig. 3c).

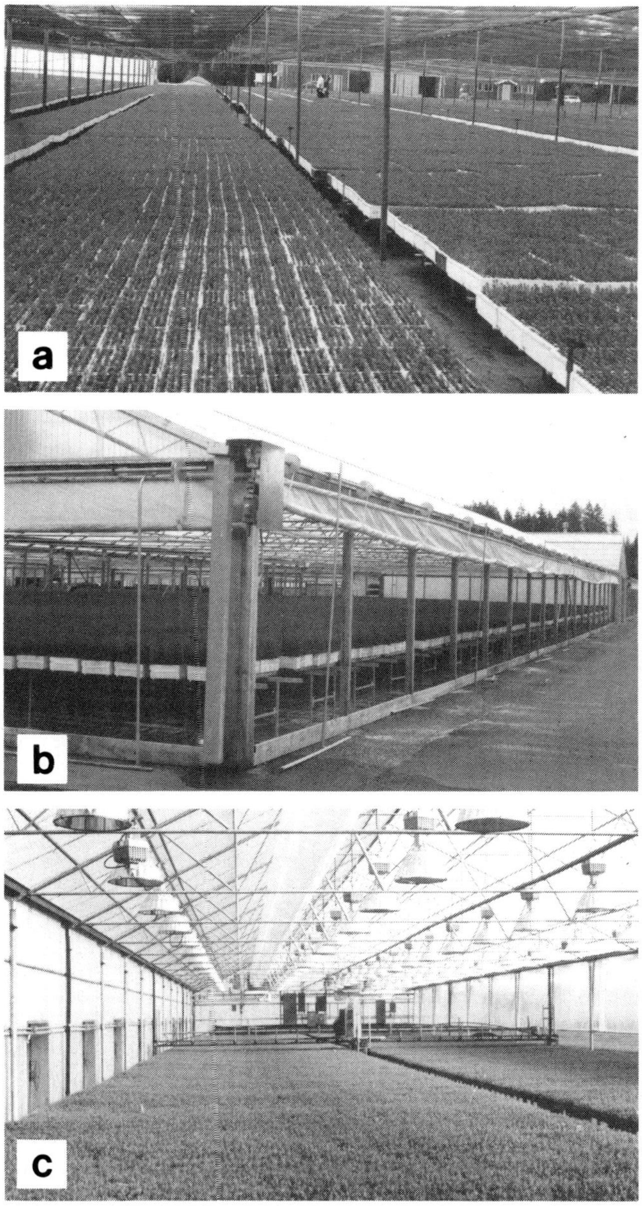

Figure 3. (a) Semicontrolled growing facility; (b) shelterhouse, with thermostatically controlled, removable wall covers; and (c) fully controlled and permanently enclosed greenhouse.

With minor modifications to fit given local conditions, shelterhouses have proved to be the best all-around type of growing facility. They are relatively inexpensive to build and operate and can provide all the needed environmental control for producing good seedlings if a compatible container is used. Such seedlings normally have the required physical and physiological traits to meet the desired transplant quality and date and are performing well when field planted. Because most shelterhouse-grown seedlings are produced on a one-crop-per-year schedule and during the natural growing season, growing is less troublesome and quite low in cost.

16.4.2 Containers

Many different container types and sizes are in use. For successful crop production, the container type must be compatible with the growing facility (see 16.4.1). Normally, almost any type of container is suitable in a greenhouse, in which the environment can be easily regulated to maintain temperatures considered "near optimum." Seedlings raised in shelterhouses or semicontrolled facilities, however, may be exposed to greater climatic extremes. Such conditions promote hardier seedling development within certain limits and, therefore, are more desirable than detrimental, as long as the seedling's most sensitive part—its roots—are protected. For this reason, containers with good insulating capacity, such as styroblocks, are the best choice for producing hardy seedling crops not only in shelterhouses but in all types of growing facilities [3].

Because p+1 seedlings are raised for a total of nearly 2 years in two different nurseries, they generally have plenty of time to reach their desired size if reared properly. Therefore, plugs may be grown in relatively small (30- to 40-cm³) container cavities. The use of small cavities makes the plug phase more economical because fewer resources (soil, water, fertilizer, etc.) are needed, growing space is better utilized, and handling costs during shipping and transplanting are reduced.

16.4.3 Crop growing schedule

Before a p+1 seedling crop is initiated, the plug growing phase has to be timed so that seedlings will reach their optimum size and condition on the desired target date for transplanting. Typical bareroot seedlings reach this optimum in spring, when, traditionally, they have been transplanted. Containerized seedlings, however, differ in this respect because of the way they are grown; raised in semicontrolled or fully controlled facilities, these seedlings have more scheduling flexibility. They can be successfully transplanted to bareroot nurseries almost any time as long as they are in the proper physical and physiological condition and the bareroot nursery is ready to plant them. Past experience shows three frequently used time periods—spring, fall, and late summer—for plug transplanting.

16.4.3.1 Spring transplanting

Spring transplanting is traditional and is well suited to bareroot crops because of the bareroot seedling's favorable physiological condition at that time of year. Nurseries pressed for space also can rotate their transplant beds more favorably in spring than in fall or late summer. But spring transplanting of plugs has had varied success and might be desirable only when:

- Crop rotations or specific nursery practices in a bareroot nursery restrict bedspace availability for late-summer or fall transplanting.
- Winter crops are specifically grown for spring transplanting. This is often done to better utilize expensive greenhouses through multiple-crop production. However, producing a crop during winter requires a lot of costly energy for winter heating and lighting in addition to the ongoing expenses of running a greenhouse. This high energy requirement undoubtedly affects overall p+1 costs.
- Seedling crops initiated in spring and reared during summer are not properly cultured and fail to reach the desired size and condition for late-summer or fall transplanting. Other crops may be deliberately raised during summer for spring transplanting. In either case, seedlings have to be overwintered in greenhouses, which normally results in additional root growth. Producing too many roots for small container-cavity size creates pot-bound plugs; roots of such plugs may have a hard time breaking out of their plug form and developing into a freely growing heavy root mass the following growing season (Fig. 4). Overwintering seedlings in cold storage may avoid the potential problem of pot-bound plugs, but prolonged and improper cold storage may cause deterioration in seedling quality [4].

Because of the stated problems resulting from winter growing, overwintering in greenhouses, and cold storage, spring transplants generally lack the vigor often seen in seedlings transplanted in late summer and fall. Though their height growth is generally acceptable, their roots have a harder time breaking out of the plug form—and little time to do it in—and their tops are often uneven. As a result, some seedlings do not meet minimum standards.

Figure 4. Spring-planted plugs may not have enough time to develop freely growing heavy root mass by lifting time.

16.4.3.2 Fall transplanting

Fall transplanting may work well when plugs are prepared for it and when bareroot nursery conditions are suitable for accepting transplants. Recall that the very first fall transplanting (described in 16.2), in the Douglas-fir region in 19 6, was possible because the unusually dry weather conditions favored bed preparation and transplanting that late in the year. Those fall transplants showed considerably better height, diameter, and root growth than spring transplants a year after transplanting; they had maintained actively growing root systems in their polystyrene containers and were well hardened by the time they were transplanted. Trees with active root tips expand their roots in the nursery bed after transplanting, ensuring rapid development the following growing season.

Normally, however, climatic conditions in fall in the Northwest create considerable problems for transplanting plugs:

- Transplanting may be hindered by high soil moisture due to excessive rains, which limits access to transplant beds and often cuts the transplant season short.
- Early frosts may damage just-transplanted and improperly hardened seedlings.
- Late-transplanted seedlings have very little time to adjust for possibly severe winter conditions.

The stated difficulties were experienced in our operations. To overcome them, we went to late-summer transplanting. We had to readjust our greenhouse growing schedule accordingly, as did the bareroot nursery to which plugs would be transplanted.

16.4.3.3 Late-summer transplanting

Six years of experience in testing and using millions of p+1 seedlings convinced us that late-summer transplanting provides the best growing and transplant logistics and results in good-quality p+1 crops. August-planted plugs can double their caliper and triple their root mass by mid-November (Fig. 5).

For this schedule, the containerized production phase is initiated in late winter or early spring. Most container facilities have some environmental control capabilities to accommodate early sowing, and early germination can be aided by frequent misting and artificial heating. As the weather warms, most seedling growth closely coincides with the natural growing season—which is especially important when plugs are grown in shelterhouses (see 16.4.1). An early sowing date and good cultural practices during summer make it possible for seedlings to reach the desired size, root form, and physiological condition for late-summer transplanting.

Due to inappropriate scheduling or poor cultural practices, container nurseries can have problems readying their p+1 crops for such a transplant date. We succeed in achieving this by following a strict growing schedule and cultural regime (Tables 1 to 4). A similar schedule is used for our regular crop production but a fifth holding, or overwintering, phase is required during the winter planting season (Table 5). Plug+1 seedlings destined for spring transplanting may also be overwintered (see 16.4.3.1). The schedules in Tables 1 to 4 are typical for Douglas-fir in Oregon, though they may vary from year to year, depending on climatic and seedling conditions, and large portions of them also may apply to other species and different growing areas. Therefore, these schedules should be used only as general guidelines for developing growing regimes for specific local conditions.

Seedlings destined for late-summer planting should be sown in early March (Table 1). During the first portion of the early growth phase (Table 2), extra lighting may be necessary to extend the photoperiod to avoid premature budset; seedlings from high-elevation sources and those east of the Cascade Mountains definitely need the extra light. The procedures for seeding, germination, and thinning of p+1 seedlings do not differ from those used for the regular crop; but rearing practices (watering, fertilizing, disease control, etc.) are tailored to p+1 production.

Seedlings for p+1 crops are normally raised in containers with small (30- to 40-cm^3) cavity sizes not only to hold down the production cost but also to encourage rapid seedling growth. Such small containers, densely (900 to 1,000 seedlings/m^2) spaced, promote rapid seedling development during the accelerated growing period (Table 3). Seedlings closely spaced in this stage tend to become taller than those grown in larger cavities at wider spacings. The heavy fertilizer regime outlined in Table 3 also promotes rapid stem and root development. Roots fill the small container cavity rapidly; because of this, height growth normally starts to decline by early July, when hardening can be initiated.

Figure 5. Plugs transplanted in late summer increased their average caliper from 2 to 4 mm and maintained vigorous root growth, tripling their root-mass by mid-November.

During the hardening phase, seedlings are subjected to periodic moisture stress and special fertilizer treatments (Table 4), which promote good budset and stem lignification by early August. By this time, roots have formed a firm plug. Even though not yet completely dormant, seedlings have the needed size and physiological condition to be moved to a bareroot nursery and transplanted, despite the normally hot summer days. (See 16.5.2 for late-summer transplant development and post-transplant care.)

16.4.4 Packaging, handling, and shipping

The crop can be termed ready for transplanting when seedlings are in the proper physiological and morphological stage, when stems are lignified enough to allow extraction of seedlings from containers without injury, and when root plugs are firm enough to hold together during shipping and transplanting.

Seedlings can be packaged and shipped either in a pre-extracted form or in their original containers. Either approach

Table 1. Phase 1, the stratification and germination period, is geared toward proper seed preparation, efficient sowing, germination, and juvenile seedling development.

Growth components	January (Weeks 3–4)	February (Weeks 1–4)	March (Weeks 1–4)	April (Weeks 1–2)
Seed preparation	Soak seed	Stratify at 2°C for 6-8 weeks	Sowing and germination	Juvenile development
Day temperature, °C				
Optimum			25	25
Permissible			18-25	15-30
Night temperature, °C				
Optimum			25	25
Permissible			18-25	15-30
Relative humidity, %				
Optimum			80	40-60
Permissible			50-90	40-70
Light, %				
Natural			60-80 sunlight	60-80 sunlight
Supplemental artificial			None	To avoid budset, extend photoperiod with 400-lux light intensity
Water			Frequent light irrigation	As needed; keep surface dry to avoid disease
Fertilizer			None	Formula A (balanced)
pH			4.0-4.5	4.0-4.5
Fungicide				One preventive application
Recordkeeping	Monitor temperature and humidity in cooler		Keep daily records of all activities	

Table 2. Phase 2, the early growth period, is geared toward seedling establishment, good root and side-branch initiation, and prevention of disease and insect problems.

Growth components	April (Weeks 3–4)	May (Weeks 1–4)
Temperature, °C		
Optimum	20-30	20-30
Permissible	10-33	10-33
Relative humidity, %		
Optimum	40-60	30-60
Permissible	30-70	20-70
Light, %		
Natural	Only sunlight as it penetrates roof covers	
Supplemental	To avoid budset, extend photoperiod with 400-lux light intensity	Continue with extended photoperiod in problem areas
Water	As needed; keep surface dry to avoid disease	Mostly applied with fertilizers as needed; leach each 4-6 weeks
Fertilizer	Chelated iron (15 g/100 liters water), calcium nitrate (60 g/100 liters water)	Apply chelated iron (15 g/100 liters water), calcium nitrate (120 g/100 liters water), and a balanced fertilizer with micronutrients (30-10-10, 30 g fertilizer in 100 liters water), alternately with each watering as needed
pH	4.5-5.0	5.0-5.5
Fungicide	Apply only if needed	One Benlate-Truban preventive fungicide drench (30 g/100 liters water)
Operations	Daily checks on crop	Thin to 1 seedling/cavity
Environmental control	Automatic heating and ventilation as needed	Heat only during cold nights; begin opening roof vents and sidewalls of shelterhouses on warm days for good air circulation and cooling
Recordkeeping	Monitor the crop closely each day; record all activities; note potential problems and their solutions	
Soil and foliar testing	Start routine testing once every 2 weeks; adjust fertilizer regimes to comply with nutrient needs	
Seedling growth	Keep height growth low (4-8 cm); concentrate on side-branch initiation and root development	

Table 3. Phase 3, the accelerated growth period, is geared toward pushing seedling height and diameter growth, root development, and stem lignification. Natural growing conditions are generally optimum in this period.

Growth components	June Weeks 1	2	3	4
Temperature control	Natural air temperature during the day (vents, side walls normally open); vents and walls may be closed on cool nights		Natural air temperature; cool with irrigation water if needed	
Humidity	Same as ambient conditions			
Light				
Natural	Only sunlight as it penetrates roof covers			
Supplemental	None			
Roof covers	Retain on main greenhouse units to keep rain off seedlings; alley roof covers may be removed for added ventilation			
Prewetting	Prewet 5-10 minutes before fertilization to aid water penetration for better fertilizer utilization			
Wetting agent	Use when water penetration becomes poor			
Leaching or nutrient flush	Prevents salt buildup; flush about once a month with about 20 liters water/m^2 of area			
Water	Irrigation water is mostly accompanied by fertilizer			
Fertilizer	Compounds high in nitrogen (3:1:1) and calcium nitrate; 1,000 ppm total fertilizer, 2-3 times a week, and iron chelate (300-500 ppm) once every 2-4 weeks		Compounds high in phosphorus; also those with calcium, magnesium, and potassium (1,200-1,500 ppm total fertilizer solution)	
pH	5.0-5.5			
Height and diameter measurement	Establish sample trees; measure them and plot on graph every 2 weeks			
Soil and foliar testing	Continue testing and adjust nutrient regime as needed			
Disease control	Normally, very little problem in this phase			
Recordkeeping	Same as previous phase			
Seedling growth	The natural growing conditions and fertilizer regime favor fast height growth; push height growth as much as possible to match target curves		Concentrate on rapid height, diameter, and root development; the natural growing conditions and fertilizer regimes favor this	

Table 4. Phase 4, the hardening period, is geared toward initiating and achieving budset, stopping height growth, lignifying stems, boosting diameter and root development, and initiating dormancy.

Growth components	July Weeks 1	2	3	4	August 1	2	3	4
Temperature control	Close to natural air temperature; cool with irrigation water if needed							
Humidity	Similar to phase 3							
Supplemental light	None							
Roof covers	Keep rain off seedlings, eliminating interference with hardening							
Prewetting	Very important after stressing							
Wetting agent	Use after heavy water stress, if needed, to rewet plugs							
Leaching or nutrient flush	Start hardening by flushing nutrients from the soil							
Seedling stress	Dry to wilting point and repeat stressing 2-3 times							
Fertilizer	After first stress, apply 0-52-34 (60 g/100 liters water); after second stress, apply 0-0-62 (30 g/100 liters water); after third stress, apply hardening formulas alternately with 0-52-34 until budset							
Chilling	None						Cool nights may help chill seedling stems	
pH	5.0-5.5							
Soil and foliar testing	Once every 2 weeks or more frequently if needed							
Height and diameter measurement	Similar to phase 3; the two target curves help in adjusting fertilizer regimes							
Disease control	Watch for *Fusarium* and *Botrytis* problems							
Recordkeeping	Similar to the previous phases							
Operations	Spread seedling blocks for better aeration; turn outside blocks to reduce edge effect							
Late-summer transplanting					Package and ship seedlings to transplant nurseries			

has its advantages and disadvantages. Preextracted seedlings are preferable for long-distance shipping and cold storage because they take up less space. Containerized seedlings are preferable for short-distance shipping and direct transplanting because they can be handled economically while remaining well protected.

Seedling extraction and packaging is a routine operation in container nurseries. Most nurseries have an assembly-line set-up (Fig. 6) which makes these processes quick and cost effective. Forty to 50 extracted seedlings generally are placed in small plastic bags to protect the roots (Fig. 7). About 12 to 14 of these bags are packed into cardboard boxes (Fig. 8a), which can then be stacked on pallets and easily moved into enclosed vans or storage areas. Seedlings still in their containers also are placed into cardboard boxes (Fig. 8b), which are stacked on pallets for transport and storage.

Shipping large crops requires large trucks to keep pace with the rapid transplant operation. One transplant machine can plant 70,000 to 80,000 seedlings daily, and some nurseries may operate several transplant machines at the same time, depending on their size and on the urgency of the job. A 40-foot-long van can haul about 380,000 packaged preextracted seedlings or about 160,000 seedlings boxed in their original containers in one trip (Fig. 9). On their return, trucks can bring the empty containers and boxes back to the nursery; well-built boxes can be reused 5 to 6 times. This recycling effects a large savings in packaging and hauling costs.

16.5 Bareroot Growing Phase

As soon as seedlings are moved to transplant nurseries, the second, or bareroot, phase of p+1 production begins. Naturally, by this time, transplant beds and equipment must be prepared and crews organized to start transplanting.

16.5.1 Bed preparation

This phase of p+1 production may not differ much from regular transplant-bed preparation. Cultivating the ground with plows, disks, rototillers, and other soil-loosening devices dur-

Figure 6. Assembly-line operation: plugs are loosened with an extractor, removed from their containers, and then packaged for shipping.

Table 5. Phase 5, the holding (or overwintering) period. During this phase, the crop is held while trees await field planting. Diameter growth continues on a reduced scale; root tips in styroblocks remain active, and transpiration rate is low. The primordial shoot, which will be next season's first flush of growth, is forming inside the bud.

Growth components	September–January (Weeks: September 1 2 3 4; October 1 2 3 4; November 1 2 3 4; December 1 2 3 4; January 1 2)
Temperature	Allow cooling down to −6°C, in case of frost, for short periods; this helps chill seedlings for better hardening and dormancy; avoid freezing of soil plugs
Light	
Natural	Natural daylight as permitted through roof covers
Supplemental	None
Roof covers	All roofs, including alleys, are covered
Watering	As needed; less frequent, because of low transpiration rate
Fertilizer	Resume use of higher nitrogen fertilizer for a balanced fertilizer regime; this promotes more diameter growth and frost hardiness and development of more needles in first growth flush of next season [1]; reduce fertilization frequency
pH	5.0-5.5
Soil and foliar testing	Once a month as needed
Height and diameter measurement	Continue monitoring until growth levels off
Disease control	*Botrytis* generally causes the most problems if not prevented or controlled
Recordkeeping	Same as in all other phases
Operations	Start packaging and shipping seedlings as soon as field planting conditions warrant it; seedlings may be shipped in styroblock containers or in a preextracted form; extracted seedlings also may be cold-stored for later planting

ing dry summer months may require some irrigation to develop the proper soil texture and moisture for transplanting. The ground occasionally has to be fumigated and cultivated before transplanting to minimize disease problems (see chapter 19, this volume).

Often fertilizers are applied before transplanting to develop good nutritional balance in the soil (see chapter 7, this volume). Late-summer-applied fertilizer should contain a very low proportion of nitrogen because too much nitrogen may interfere with the natural hardening process (see chapter 15, this volume).

Field work should be completed before seedlings arrive for transplanting. Normally, there is very little problem accomplishing this during dry summer months, though wet weather may hamper the same operation later in the season or in early spring.

16.5.2 Transplanting

16.5.2.1 Timing

As mentioned earlier, the best time for transplanting plugs in the Northwest is late summer. Experience shows that such timing allows the properly conditioned seedlings to continue to harden under favorable climatic conditions and to enter dormancy. Late-summer transplanting also provides time for active roots to egress from plugs (see Fig. 5). It is important that roots continue their rapid development in nursery soil so that seedlings overwinter safely and begin accelerated development the following growing year.

Figure 7. Extracted seedlings are protected by plastic bags during shipping, storage, and field handling.

Figure 8. Seedlings can be either (a) preextracted, packaged, and boxed or (b) boxed while still in their original containers.

Figure 9. Large, enclosed vans are needed to transport both preextracted and containerized seedlings, to keep pace with today's rapid transplant operations.

A late-summer transplant operation produces not only the best quality p+1 crop but also the best handling logistics for both container and bareroot nurseries. There is generally good weather to prepare transplant beds at a time when both container and bareroot nursery activities are at a low level. Taking advantage of such factors spreads the workload while realizing some savings. Although late-summer planting requires especially close attention to seedling care, well-prepared seedlings will not be damaged by the usually hot weather as long as irrigation systems provide the needed moisture for protection and for initiating seedling development at the new location.

16.5.2.2 Planting and handling

Mechanized transplanters identical to those used for transplanting regular crops are employed for planting p+1s.

Seedlings may be handled on the transplanter while still in their containers or in a preextracted form. Both methods have been widely and successfully used; however, conditions may dictate the method of choice. For instance, if seedlings must be stored for later planting or shipped to distant nurseries, then a preextracted form would be more desirable. For short-distance hauling, handling seedlings still in their containers is preferable, to protect seedling root systems. Either method provides approximately the same planting speed, but handling logistics, handling and shipping costs, and effects on seedling quality may differ.

For planting out of containers, transplant machines are outfitted with racks to hold containers in a convenient position to facilitate direct extraction and transplanting without sacrificing speed (Fig. 10a, b). Though a savings is realized from not preextracting, some additional costs are generated in shipping and from the additional ground support equipment and personnel needed to supply the transplanters with seedlings (Fig. 10a).

Preextracted seedlings are transplanted out of trays (Fig. 11). However, such plugs are handled several times before actually being placed in the ground by the transplanter. Each handling tends to cause root disturbance and exposure. Depending on the severity, such exposure and disturbance may adversely affect seedling performance later on.

During the actual transplanting operation, workers can place seedlings in the clips on the planting wheel rapidly and with minimal effort if the seedlings are well developed and the root plugs hold their form. Clips may have to be adjusted to handle the smaller caliper. Plug-type seedlings eliminate the usual root tangle problem so common with bareroot seedlings. With good moisture content for added weight, plugs enter the ground in a vertical or suspended position. They are firmly anchored by the planter's packing wheels and are free to develop a straight and balanced root system. Because of their widespread roots, bareroot seedlings are difficult to transplant and often develop deformed roots; two of the most common

Figure 10. Transplant machines are (a) equipped with racks to hold containers so that (b) seedlings may be transplanted directly out of their containers.

deformities are "J" and "flat" roots. "J" roots are caused by bending the root mass into a one-sided horizontal position during transplanting; "flat" roots occur when the root mass egresses in a single plane in the transplant trench during seedling development. Plug+1 seedlings very seldom develop "J" roots, and their roots are free to develop in all directions.

16.5.2.3 Post-transplant care

Once seedlings are placed in the transplant bed, a good soaking with irrigation water firms the soil around the plugs and eliminates air pockets. Additional light watering for cooling, especially during warm late-summer days, may also be necessary. Care should be taken not to overwater; too much water, combined with a high nitrogen level, may cause budbreak this time of year if seedlings are not in an advanced stage of hardening.

Past experience shows no problems in rearing seedlings after late-summer transplanting if the seedlings were properly prepared for such an early transplant date. Nevertheless, the nursery operator must pay closer attention to the crop in late summer than after fall or spring planting because of potentially hot, dry weather.

16.5.3 Rearing

16.5.3.1 Seedling development

Containerized seedlings are in a different developmental stage and condition according to whether they are transplanted in spring, fall, or late summer. Such stages and conditions affect how p+1 transplants develop during their bareroot growing phase. Possible effects must be taken into account when cultural practices—slightly different from those used for 2+1 transplants—are designed and applied.

Rearing practices for spring-transplanted p+1s and regular 2-1s probably differ least because these seedling types are often stored the same way and planted at the same time. Their physiological condition and growth initiation closely coincide. Spring-planted p+1s may have some disadvantages in root and stem development during the bareroot growing phase because of possible after-effects of overwintering and less time for root growth from spring to fall. This is evidenced in their top and root quality (see Fig. 4). To counteract this, a more intensive fertilizer regime should be scheduled for spring p-1 transplants to promote faster root and stem development, if needed.

Seedlings successfully transplanted in fall may not differ much in their development stage and rearing practices from late-summer p+1 transplants. However, any development problems due to soil or winter weather conditions should be considered when cultural practices are designed for their rearing during the next growing season.

Late-summer p+1 transplants normally develop into nicely formed trees with numerous strong side branches and a unique moplike root system, compared to other commonly used planting stock (Table 6). Whereas height and caliper were generally smaller for containerized seedlings than for bareroot seedlings, total root mass was greater. The p+3-months seedlings were especially outstanding in this respect. Their root mass tripled

Figure 11. Preextracted seedlings are transplanted from trays.

and their calipers doubled during the 3 months after transplanting, although their height remained the same because they had already set buds in August. Such caliper and root development during fall and winter is typical for late-summer transplants, priming them for outstanding development during the following growing season. Height of p+1s at field planting time was comparable to that of the 2+1s; but their caliper, branch characteristics, and total root mass were considerably better than those of 2+1s or of any other seedling type (Table 6).

The root mass, which supplies seedling tops with water and nutrients, must be well balanced with height and caliper if outplanted seedlings are to perform well. The comparison in Table 6 indicates that 2+0 bareroot seedlings, on the average, had only 91 cm of total root length to each millimeter of caliper; this ratio was 98% higher for 2+1s, 189% higher for styro 2s, 264% higher for styro 5s, 281% higher for styro 8s, 349% higher for p+3-months, and 403% higher for p+1s. Comparing total root mass length to seedling height revealed a similar trend (see Table 6).

One of the most common problems with late-summer p+1 transplants has been their excessively rapid height growth (Fig. 12), which may be due to the following conditions:

- Containerized seedlings grown in greenhouses extend their heights in stages several times without forming buds each time.
- An early sowing date, dense spacing, and small container-cavity size pushes seedling development and promotes early budset in the greenhouse.

Table 6. Top- and root-growth comparisons of nursery stock types.

Stock type	Tops: Average height, cm	Tops: Average caliper, mm	Tops: Number of branches	Tops: Average branch length, cm	Roots: Total length, cm: Per tree	Roots: Total length, cm: Per mm of caliper	Roots: Total length, cm: Per cm of height
Container							
Styro 2	25	2.6	7	2.6	684	263	27
Styro 5	33	3.3	13	2.4	1,024	315	31
Styro 8	39	4.2	17	3.9	1,457	347	37
Bareroot							
2+0	32	5.0	15	4.7	456	91	14
2+1	55	8.0	22	7.8	1,442	180	26
P+3-months[1]	21	4.2	7	3.2	1,717	409	82
Plug+1	51	10.6	32	10.2	4,854	458	95

[1]Plugs planted in August and lifted for evaluation 3 months later.

- Late-summer p+1 transplants continue their diameter growth, bud development, and root growth long after transplanting.

The after-effects of the above promote early flush next spring and more needle and branch development on the rapidly growing new stems. Although vigorous stem, root, and branch development is most desirable, too much height growth could become a handling problem during lifting, packaging, storing, shipping, and field planting. Rapid seedling height development can be controlled through fertilization and other cultural treatments or by early top mowing or root wrenching.

Figure 12. Vigorous height growth of p+1s must be considered when designing fertilizer regimes, to avoid excessive seedling growth.

16.5.3.2 Treatment procedures

Watering.—Routine watering during the growing season will depend greatly on soil composition, local climate, and seedling need. Representative sampling points established throughout the nursery for monitoring soil and seedling moisture will provide guidelines for watering. The "pressure bomb," developed by Waring and Cleary [5], gives good indication of seedling moisture stress (see chapter 12, this volume).

Fertilization.—Transplanted seedlings should be fertilized during the growing season according to soil fertility level (see chapters 7 and 8, this volume) and seedling need. However, seedlings require different combinations of fertilizers during the various seasons and stages of growth [2]. One or two large fertilizer applications during the growing season normally do not satisfy this need and may cause problems in achieving the desired seedling size and quality.

Bareroot nurseries are slowly adopting the fertilizer techniques developed in container nurseries in recent years—and a similar technology is strongly suggested for p+1s. Seedling height and diameter development are plotted on charts and compared to target curves. Foliar and soil nutrition levels are regularly analyzed in the laboratory to ensure that seedlings are being fertilized in the proper amounts and types. Fertilizer should be applied in frequent small doses, rather than in one or two large doses, so that the nursery manager can better control seedling growth to match target curves. It is always possible to add more fertilizer if needed—but impossible to remove it.

As a general rule, seedlings require a fertilizer with higher nitrogen content at the beginning of the growing season. Midseason, more phosphorus is needed for better root and stem development. At the end of the season, a fertilizer low in nitrogen and high in potassium will aid hardening, stem lignification, and bud development [2].

Weed control.—Chemical weed control is becoming a standard practice in bareroot nurseries (see chapter 18, this volume). New chemicals are constantly being developed to provide a large variety of selective treatment methods. Most of these chemicals can be used with p+1s if the directions outlined in the literature and on labels are properly followed.

The nursery operator must remember, however, that transplanted p+1s are younger than the 2+0 seedlings planted for 2+1 stock and may exhibit a slightly different physiological condition. Therefore, some of the standard transplant chemical treatments for weed control should be applied with caution or not applied at all to p+1 transplants.

Manual weed control might still be necessary on a limited scale but could become very expensive because of the high labor costs.

Disease control.—Containerized seedlings generally have few disease problems because of the way they are raised. However, these seedlings still need to be carefully examined before they are shipped to a transplant nursery to avoid transfer of contaminated material. If a disease problem develops, p+1s should be handled like regular bareroot crops (see chapter 19, this volume).

Top mowing.—Many nurseries have had trouble controlling the top growth of p+1 transplants because of their aggressive development. Top mowing is frequently used to hold seedling height at a desired level and to keep the crop more even (Fig. 13); however, the effect of this treatment is still not well understood (see chapter 15, this volume).

A June mowing catches the various Northwest species in their succulent-leader growth stage. These leaders snap off easily if the proper equipment is used without causing much damage to the remaining portion of the leader. Experience has shown that the lateral bud nearest the cut becomes dominant and generally develops into the new leader.

August or September mowing should be avoided because it substantially damages lignified seedling stems. Wounds made at this time do not heal easily and may provide an entry point for diseases. August mowing also often causes a flushing problem; newly flushed tender shoots seldom harden in time to develop cold resistance against early frosts. Late mowing

Figure 13. Plug+1 transplants top-mowed in June achieved a desired height of 19 to 21 inches by lifting time.

also predisposes top buds to excessive forking during the following year.

Root wrenching.—Root wrenching is a new cultural technique in which seedling roots in nursery beds are severed at a given depth by a device equipped with a sharp blade (see chapter 15, this volume). This technique helps control height growth and promotes root development. Properly grown p+1 transplants may not require root wrenching if their heights are controlled with a nutrition regime or top mowing and if seedlings develop their typically strong, fibrous roots and stems.

Hardening.—The treatment for achieving proper cold hardiness—very important in seedling development—is probably no different for p+1s than for other seedling types.

16.6 General Evaluation of the Plug+1

16.6.1 Advantages and disadvantages

No single stock type available can fit all reforestation needs. What may be an advantage in a given condition for one user might be a disadvantage in a different condition for another.

Some general advantages of the p+1 are:

Container growing phase

- Good utilization of genetically improved seed
- Good, even seedling yield
- 1-year lead time before transplanting
- Easy timing and conditioning of seedlings for transplanting in spring, fall, or late summer
- More mechanized handling of seedlings during shipping and transplanting without the need for refrigeration

Bareroot growing phase

- Rapid, good-quality transplanting
- Little problem with root deformities
- Good height, diameter, and root development after transplanting
- Uniform, high seedling yields
- Easy root pruning, packaging, shipping, and field planting because of narrow, fibrous root systems

Field application

- Good survival and growth rate
- Seedling quality and cost:benefit ratio comparable to or better than those of any other seedling type when used on a typical transplant site

Some of the above advantages may become disadvantages if seedlings are not raised, handled, or used properly. Potential problems include:

Container growing phase

- Improper facility and container selection
- Poor crop scheduling and growing practices
- Improper crop conditioning for transplanting
- Seedling-quality deterioration during greenhouse overwintering or prolonged cold storage
- Mishandling during packaging, shipping, or transplanting

Bareroot growing phase

- Improper transplant scheduling
- Heat and frost injury due to poorly prepared and timed transplants
- Rearing problems due to limited experience with p+1 production

Field application

- Improper timing of field planting
- Inappropriate planting-site selection

The above-listed problems do occur—though most are the exception, not the rule—largely because the p+1 growing and application concept is new and very often poorly understood. Though some of these problems are classified as disadvantages, most should disappear with time and experience.

16.6.2 Suggested applications

Plug+1s have been used in routine reforestation under many different conditions and in many locations in the Northwest. As is true for all other stock types, the p+1 will not solve all reforestation problems, and its survival rate and growth performance will vary with conditions.

Comparable survival data for p+1s and other seedling types have been collected by Georgia-Pacific on typical Douglas-fir reforestation sites in western Oregon during routine plantation survival surveys over 7 years. Sites represented a total of 110,000 acres of reforested land and more than 50 million seedlings. Though the data gathered were not from designed and replicated research installations, they nevertheless provide good information for managers and show trends.

Trees planted on coastal, or moist, sites (Fig. 14)—often classified as typical Northwest transplant sites—face strong vegetation competition. This is perhaps why the larger transplants (2+1s and, especially, p+1s) performed so much better than the smaller 2+0 bareroots and container stock types on these moist sites. Trees planted on inland, or drier, sites—for example, the Willamette Valley and southern Oregon—often face a long, dry period during summer. Containerized planting stock performed better than bareroot stock, including p+1s, on these drier sites (Fig. 14).

The containerized seedlings used on the drier sites were raised in styroblocks in shelterhouses, were well hardened, and had a good root mass with active root tips when field planted during routine late-fall and winter planting. Because root tips were active, root growth continued during dormancy; this helped seedlings become established and primed them for good survival and growth the following summer.

Tops of husky bareroots, especially transplant stock, generally are large compared to their root mass. Such large tops often do not get the needed support from their roots for good survival and growth on dry sites after planting. Of all the bareroot stock types, p+1 transplants survived best and came closest to matching the survival of containerized stock—probably due to their root development (see Table 6).

To better quantify and corroborate the above results, well-designed research installations have been established. Of many regeneration experiments, the following replicated test is especially worth mentioning.

In 1979, 2,200 seedlings—p+1s, three containerized stock types (styro 2, 5, and 8), and two bareroot stock types (3+0 and 2+1)—were planted on a north and south slope on typical reforestation-site land near Eugene, Oregon. Concurrently, the same stock types were planted in general reforestation on Georgia-Pacific land. The two Eugene test sites were felt to represent large acreages of Georgia-Pacific land; the north-exposure site was analogous to inland north slopes or to coastal areas (cooler "wet" sites), and the south-exposure site to inland areas (warmer "dry" sites). Each site represents a particular reforestation problem: seedlings must vie with competing vegetation on "wet" sites and suffer moisture stress on "dry" sites. Animal browsing also was monitored in this experiment.

Containerized seedlings showed a remarkable survival rate on both sites (Table 7, Fig. 15), probably due to their superior

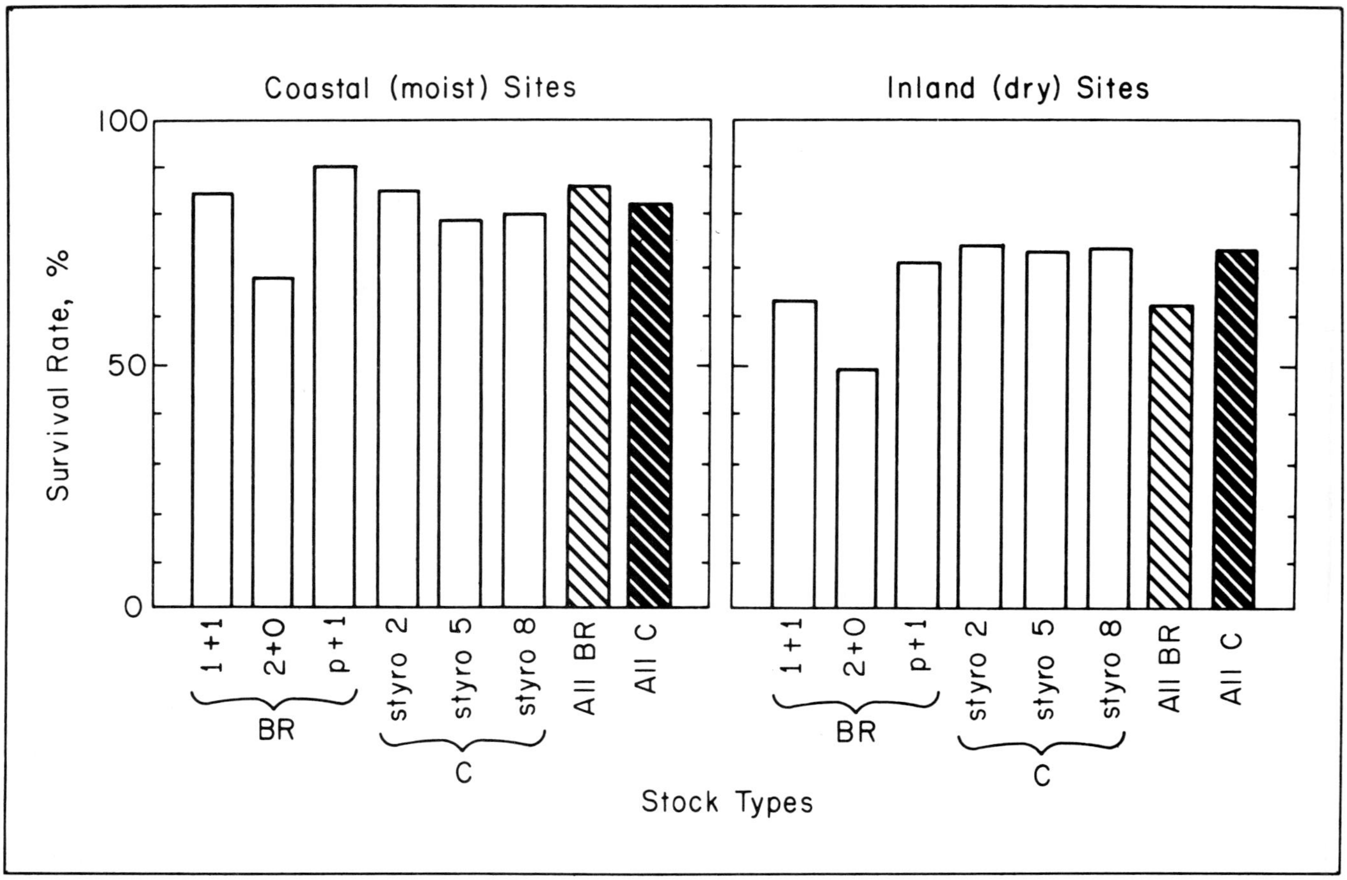

Figure 14. Survival of bareroot (BR) and container (C) stock types on typical Douglas-fir plantations in western Oregon, 1976 to 1982.

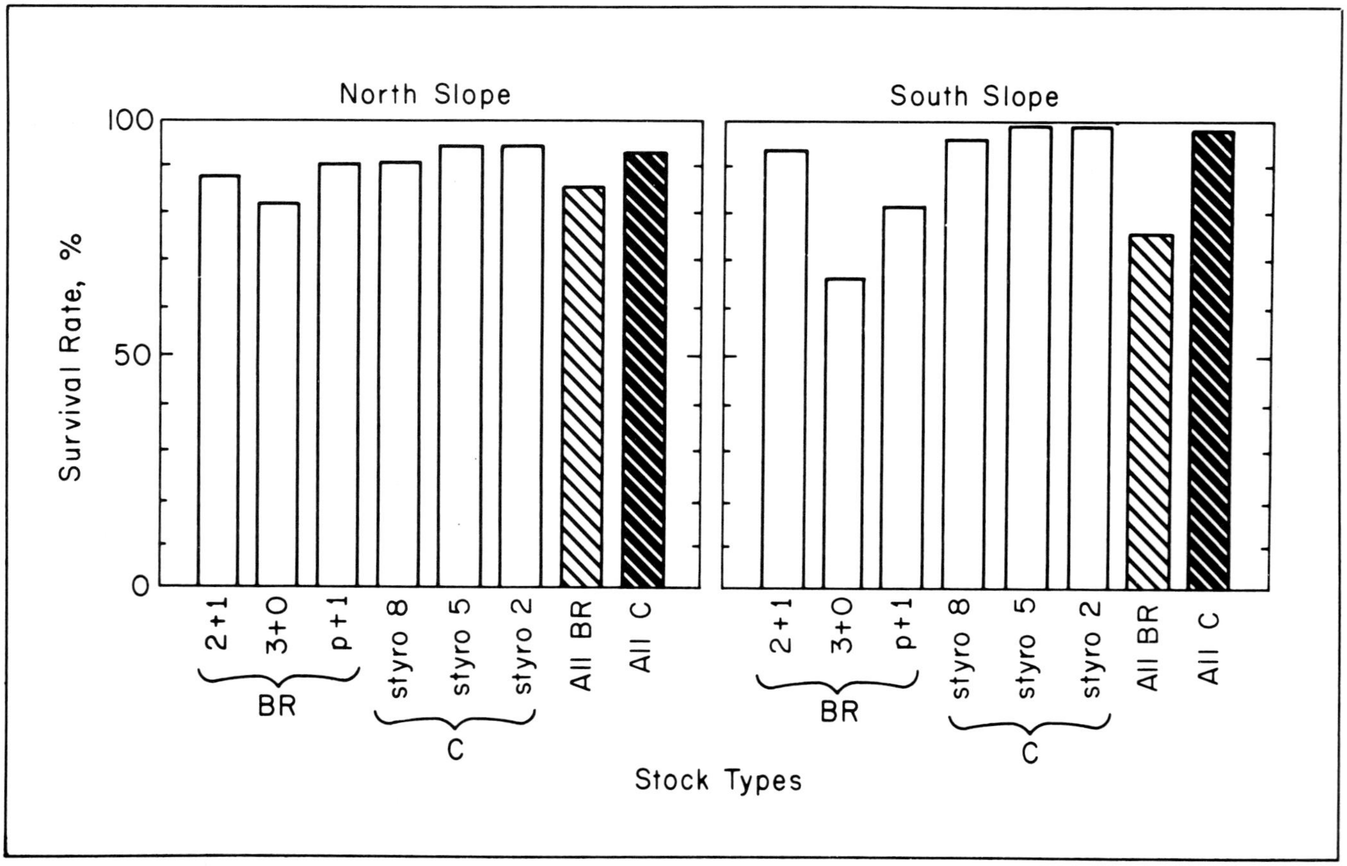

Figure 15. Survival of bareroot (BR) and container (C) stock types in the Georgia-Pacific experiment.

root quality and physiological makeup. In height growth, they had a hard time competing against the initially taller bareroot seedlings on the north slope but made up for their height disadvantage on the south slope (Fig. 16), where they performed extremely well. Containerized seedlings outscored bareroots in growth increment 3:1 on the south slope and 2:1 on the north slope.

The initially taller, "top-heavy" bareroot seedlings generally performed well on the cooler north slope, where they were not exposed to rapid drying conditions after planting and therefore not subjected to typical dry-site planting shock; those seedlings remained above the brush and maintained good height growth. In contrast, bareroot stock did poorly on the south slope (Figs. 15 and 16); their performance trend was exactly the opposite of that on the north slope.

Table 7. Survival and growth of six stock types on north- and south-facing test sites in the Georgia-Pacific experiment.

Stock type, by exposure	Average height/tree, cm		Survival rate, % (12/81)	Growth increment per average tree, %
	Starting (2/79)	Ending (12/81)		
		North slope		
Bareroot				
2+1	46	135	86	193
3+0	59	109	80	85
P+1	36	108	89	200
Total	49	111	84	127
Container				
Styro 2	17	74	89	335
Styro 5	22	82	92	273
Styro 8	33	96	93	191
Total	24	84	91	250
		South slope		
Bareroot				
2+1	48	97	92	102
3+0	61	102	65	67
P+1	38	106	80	179
Total	50	103	74	106
Container				
Styro 2	17	95	94	459
Styro 5	25	100	97	300
Styro 8	33	112	97	239
Total	25	103	96	312

Animal damage on both sites was less than normal for freshly planted areas. Generally, the south slope suffered about twice as much damage as the north slope. Furthermore, bareroot seedlings on both sites were browsed about twice as heavily and twice as often as containerized seedlings. In spite of this, overall height growth was not significantly reduced due to browsing on either site or for any one stock type.

We concluded that the large p+1 transplants, like other large bareroot seedlings, should not be used on hot, droughty sites. Such sites are far better suited for well-developed plug seedlings. Large bareroot seedlings, including p+1s, do rather well on moist slopes and on cool, coastal lands with high site productivity. There, the tall seedlings have the needed ground moisture for good survival and excellent height growth and a better chance of competing with the fast-growing weeds normally present.

16.6.3 Species correlations

Most of the discussion in this chapter directly relates to Douglas-fir seedling production because Douglas-fir is the major species in the Northwest. However, the p+1 production method has also been used quite successfully, on a smaller scale, with other Northwest species, including western hemlock [*Tsuga heterophylla* (Raf.) Sarg.], Sitka spruce [*Picea sitchensis* (Bong.) Carr.], western redcedar (*Thuja plicata* Donn ex D. Don), white fir [*Abies concolor* (Gord. & Glend.) Lindl. ex Hildebr.], grand fir [*Abies grandis* (Dougl. ex D. Don) Lindl.], noble fir (*Abies procera* Rehd.), ponderosa pine (*Pinus ponderosa* Dougl. ex Laws.), and lodgepole pine (*Pinus contorta* Dougl. ex Loud.). So far, there is no real evidence that the stated species and others would not be suitable for p+1 production, though this has yet to be proven.

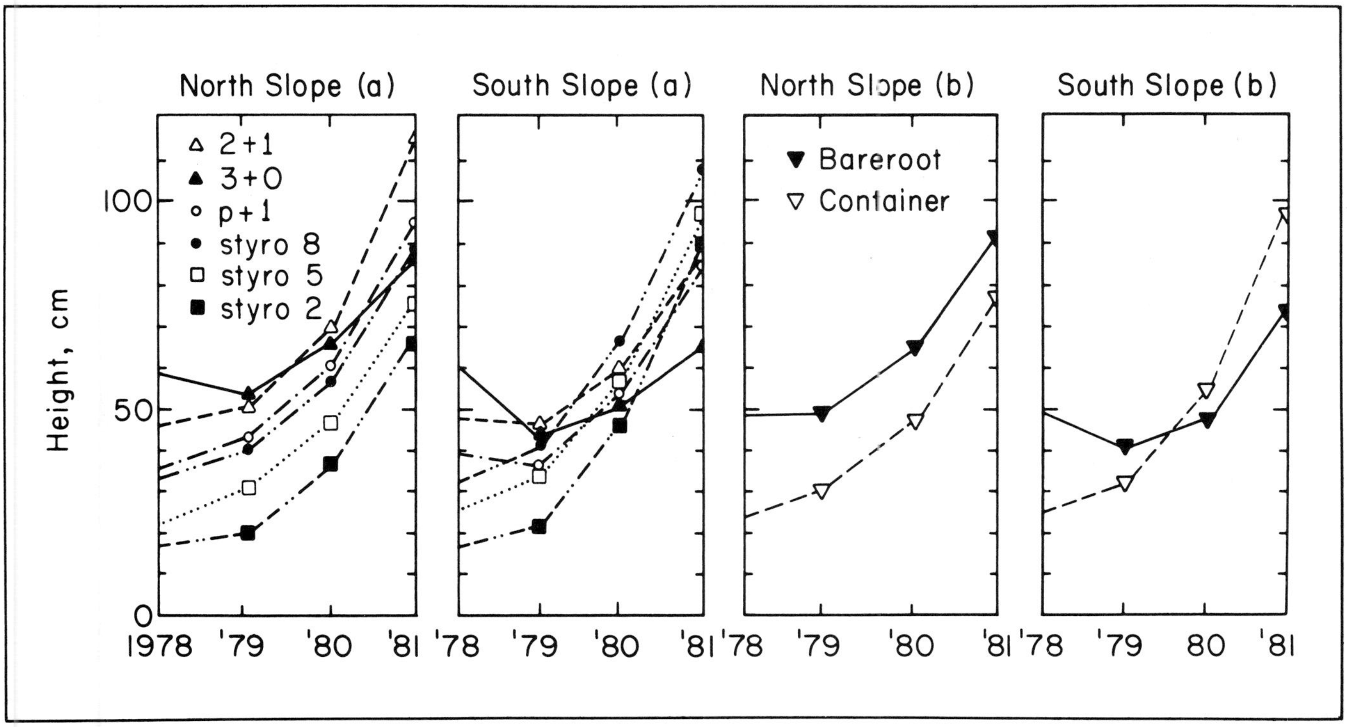

Figure 16. Total (height x survival) performance of individual (a) and combined (b) bareroot and container stock types for the Georgia-Pacific experiment.

16.6.4 Cost comparisons

Containerized seedlings destined for p+1s normally cost slightly more than 1+0 or 2+0 seedlings regularly used for transplanting, although the actual production cost of the transplant phase is normally the same for all transplanted stock. But managers must look at the **total** costs and savings for producing and using stock types—right up through establishment and field performance—to arrive at a realistic cost comparison.

Careful analysis indicates that p+1s have the following advantages:

- Uniform, high-yield stock, in both container and bareroot phases, resulting in low cull factor
- High transplant speed
- More efficient packaging, storing, and shipping
- Easier field handling, resulting in higher planting speed and quality
- Excellent root and top development, supporting good seedling establishment for high survival and growth rates.

Although some of the savings generated at bareroot nurseries are not always passed on to users, other savings resulting from a good-quality product are easily measured in a cost:benefit calculation. **All** costs, from land preparation through seedling establishment, should be included and the total cost then related to seedling performance.

Cost:benefit ratios were computed on the basis of stocking-survey survival results gathered on about 3,000 acres of typical, comparable transplant sites planted with 1.3 million p+1s and 2+1s in equal ratios. The actual total reforestation costs, including site-preparation, seedling, and outplanting costs, were available for both stock types. Total reforestation cost per 1,000 seedlings was 11% higher for the 2+1 transplants than for p+1s right after planting—and 22% higher when survival was considered 3 years later. This illustrates quite well that p+1 stock, seemingly more expensive at the nursery stage, may look quite favorable later because of its good quality and performance.

The Georgia-Pacific experiment with six stock types (see 16.6.2) also was subjected to cost:benefit analysis. Commercial, large-scale reforestation cost figures for each stock type—more useful than the installation cost of the experiment—were available. Cost:benefit ratios were calculated by dividing the establishment (seedling and outplanting) cost for each stock type by its total performance.

The result was a straight-line correlation (±2% variation) for the containerized seedling types. As the establishment cost

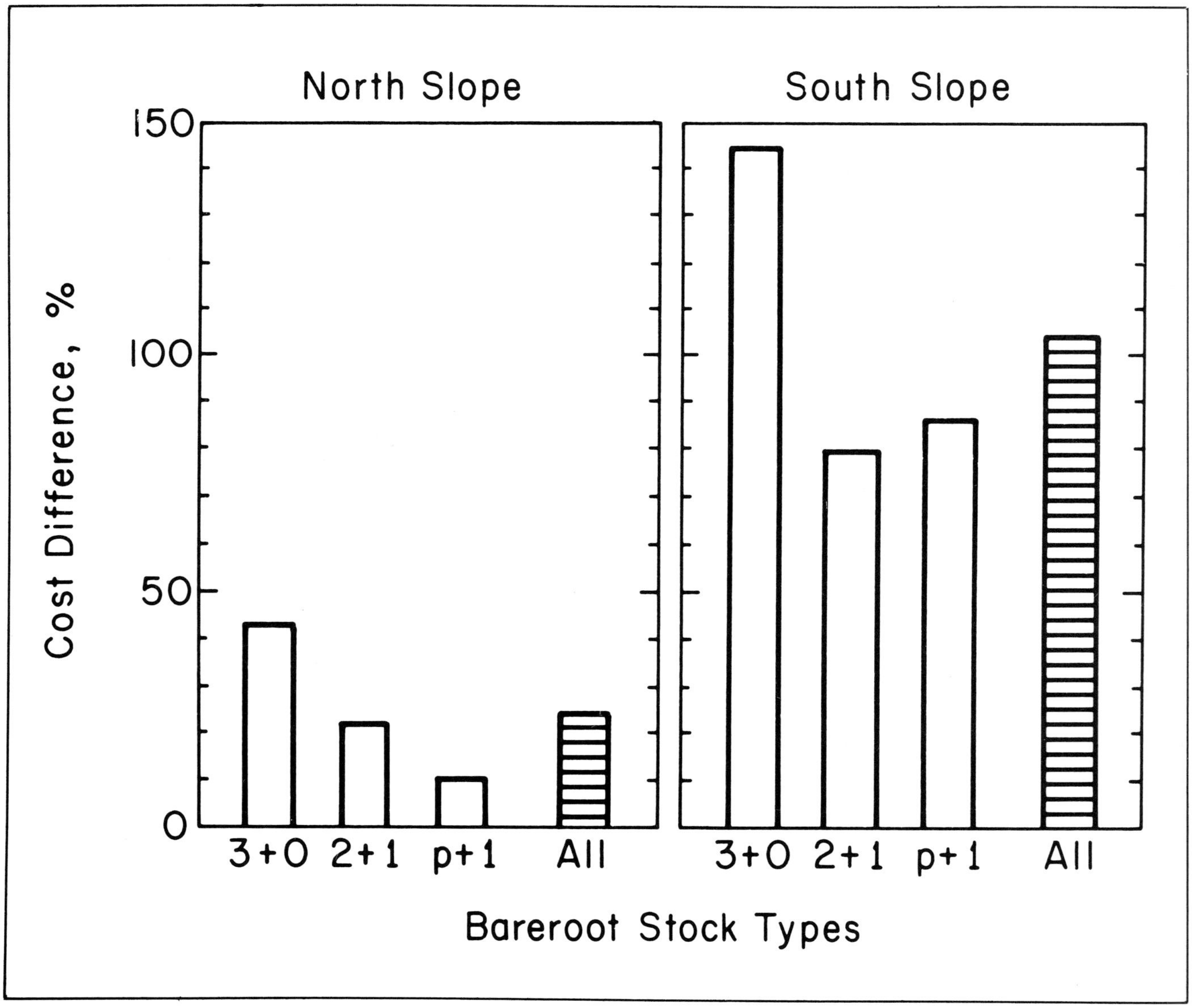

Figure 17. Relative differences in cost:benefit ratio for bareroot stock types when compared to the cost:benefit ratio of containerized seedlings for the Georgia-Pacific experiment.

increased for a given container type, so did its performance; styro 2s had the lowest cost and performance, and styro 8s the highest. However, the cost:benefit ratio varied greatly for the bareroot seedling types. The 3+0s had the lowest initial cost but performed poorly; therefore, their ratio was the least satisfactory. The 2+1s did slightly better than the p+1s on the north slope, but the opposite was true on the south slope.

Because the cost:benefit ratios for the three containerized stock types showed almost no variation, they were used as a base of comparison for the three bareroot stock types. Combined figures for the bareroots indicate that the cost of using bareroot seedlings was about twice that of containerized seedlings on the south slope, but only about 25% greater on the north slope (Fig. 17).

In summary, the experiment showed not only that bareroot transplants survive and grow well on north slopes or cooler sites but also that their cost:benefit ratio is quite low, compared to that of container-grown seedlings. A 12 to 22% higher cost for p+1 and 2+1 stock, respectively (Fig. 17), is justifiable on transplant sites to ensure successful reforestation where vegetation competition is strong.

16.7 Conclusion

Plug+1 seedlings are just one more seedling type available for reforestation—and should always be viewed as such. They have demonstrated special qualities and abilities to perform well if produced and used properly, and they are economical.

This seedling type has good application at present and should perform even better, after improvements, in the future.

References

1. Carlson, W. C., and C. L. Preising. 1981. Effects of controlled bed-release fertilizers on the shoot and root development of Douglas-fir seedlings. Can. J. Forest Res. 11(2):230-242.
2. Hahn, P. F. 1978. Nutrient requirements of containerized nursery stock. Section B. Pages 7-15 *in* Proc., Western Forest Nursery conf., Eureka, California.
3. Hahn, P. F. 1983. Containerized seedling production in a shelterhouse system. Pages 110-128 *in* Proc., Western Forest Nurserymen's conf., Medford, Oregon, Aug. 10-12, 1982. Southern Oregon State Coll., Ashland.
4. Schalla, S. L. 1979. Transplanting the Douglas-fir plug. Tyee Tree Nursery, Roseburg, Oregon.
5. Waring, R. H., and B. D. Cleary. 1967. Plant moisture stress: evaluation by pressure bomb. Science 155(3767):1248-1254.

Chapter 17
Genetic Implications of Nursery Practices

R. K. Campbell and F. C. Sorensen

Abstract

The genetic adaptation of forest trees to plantation sites can be impaired by nursery practices that favor the survival of some seedlings over others, thus producing a seedlot with a genetic makeup different from that of the original seedlot. Seed grading has considerable potential for directly altering the genetic mixture in the seedlot. Stratification period, sowing date, watering regime, lifting date, and other scheduling may have important but less direct influences on adaptation. For most seedlots, the risk of poor adaptation caused by nursery practices is probably no greater than risks caused by several current seed-collection practices. But for seedlots in which only a small percentage of seeds become seedlings that can survive outplanting, the risk may be as large as that in moving seeds between seed zones.

17.1 Introduction

The productivity of a forest plantation depends partly on the adaptation of seedlings to environmental conditions at the plantation site. If some seedlings, for example, break bud before the last spring frost, regenerate roots slowly when competition for moisture is high, or cease elongating when soil moisture is relatively high, their productivity will be low. Adaptation of seedlings to the planting site—or lack of it—reflects choices made by the forester, such as seed source, tree-breeding strategy, and planting-site preparation. But it also reflects seed- and seedling-related nursery practices, which can cause some plants to be culled or lost from the seedlot. In this chapter, we examine genetic implications of such practices in bareroot nurseries.

17.2 Genetic Principles

Although the genetic constitution of an individual plant (its **genotype)** cannot be measured directly, observable characteristics such as size and form (its **phenotype)** can. Phenotype is the result of the genotype's response to a particular environment. The average performance that would result if a seedling could be grown in a variety of environments is called its **genotypic value.** In most species of forest trees, each seedling has its own individual genotype and genotypic value.

Two assumptions are necessary to this discussion. First, the original seedlot provides the optimum genetic makeup for adaptation to the future planting site. The forest population has evolved through centuries of **natural selection** to contain a mixture of kinds of trees that match the climates of a seed zone and probably even microclimates of specific locations within that zone. This mixture is apparently balanced so that individuals can survive not only the stresses of the first few years but also competition from other plants and the rare climatic disasters that can occur during a tree's long life. The least risky procedure in the nursery is to maintain the original genetic mix in the seedlot.

Second, culling and inadvertent favoring of certain plant types in the nursery result in **directional selection.** Selection is directional when one extreme type of seedling is saved, and the opposite extreme dies or is discarded. For example, culling all trees below a certain diameter limit selects for the larger diameter seedlings. Because phenotype depends partly on genotype, directional selection of phenotypes changes the genotypic mixture in a seedlot.

The power of directional selection (whether caused naturally or by humans) to change populations is well illustrated in a study of beech (*Fagus* spp.) in Germany [23]. Seeds from two locations (provenances), Bavaria and Rumania, were sown in a greenhouse and under a natural beech stand in Lower Saxony. After germination in the two habitats, the seedling populations were compared on the basis of genes expressed as enzymes (isozymes), and the two original seedlots were compared. Differences were measured in terms of genic distance, the frequency with which specific genes appear, and of genotypic distance, the frequency with which combinations of genes appear. The seed populations that germinated in the woods were genetically quite different from those that germinated in the greenhouse. The genetic distance between populations caused by natural selection in the two environments was from ¼ to ⅔ the distance that originally existed between seedlots of the two provenances.

Because phenotype depends partly on genotype, directional selection of phenotypes changes the genotypic mixture

In Duryea, Mary L., and Thomas D. Landis (eds.). 1984. *Forest Nursery Manual: Production of Bareroot Seedlings*. Martinus Nijhoff/Dr W. Junk Publishers, The Hague/Boston/Lancaster, for Forest Research Laboratory, Oregon State University, Corvallis. 386 p.

in a seedlot. The genotypic values for the population of genotypes in a seedlot received at a nursery can be illustrated graphically (Fig. 1a). In this idealized figure, mean height ($\bar{x}_1$) of the original population occurs exactly in the center; the standard deviation, which measures variation of genotypic values around the mean, is represented by the distance s_o on the x-axis. Though variation differs from trait to trait, remember that, for our discussion, all traits will be treated as having equal variation in the original seedlots.

Because phenotype describes response of a genotype to a specific environment, phenotypes are not precisely correlated with genotypes. Some seedlings are small because they have genotypes that produce small seedlings in an average environment. If, however, seedlings with "small" genotypes are blessed with more than average space, nutrients, or water, they may grow taller than seedlings with "tall" genotypes growing in an average environment. Culling of phenotypes, therefore, does not result in an exact culling of genotypes. In our example, culling the taller half of the seedlings in a seedlot does not remove all genotypes for tallest seedlings; it does cull most genotypic values from the taller half of the distribution as well

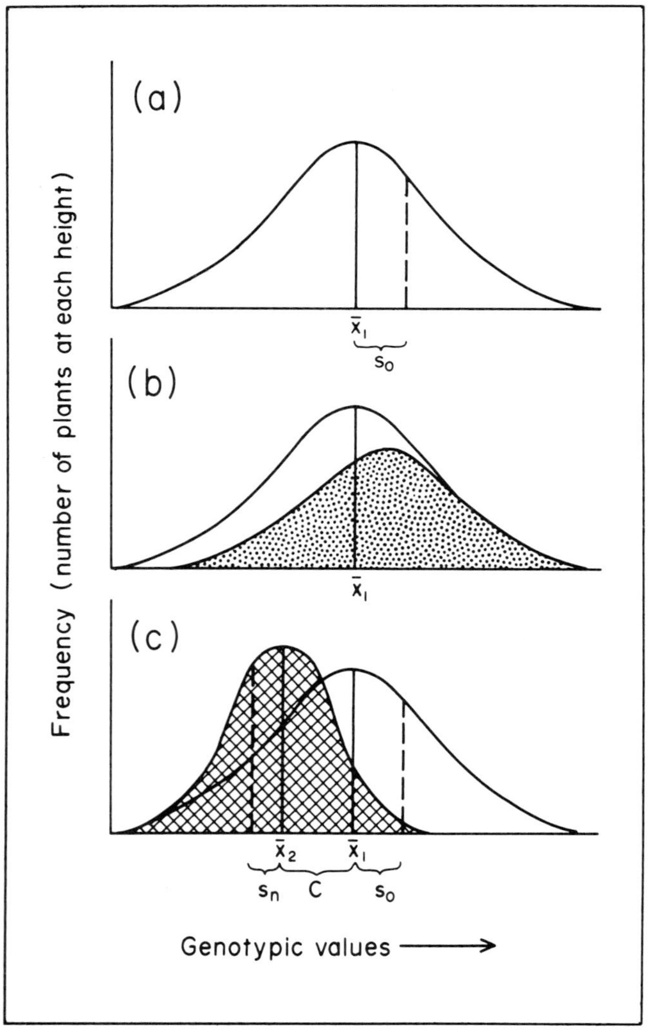

Figure 1. Frequency distributions illustrating changes in the genotypic mixture for height of 2-year-old seedlings caused by culling the taller 50% of seedlings in a seedlot: (a) shows frequency of genotypic values (mean is $\bar{x}_1$ and standard deviation s_o) in the original seedlot; (b) indicates genotypic values removed by culling tallest seedlings (stippled area); (c) shows frequency of genotypic values after culling (cross-hatched area) —$\bar{x}_2$ is the new mean, s_n the new standard deviation, and C the change in mean genotypic value.

as a substantial proportion from the lower half, as illustrated by the stippled area in Figure 1b.

When natural or artificial selection removes seedlings at either the lower or upper end of the distribution of genotypic values, the distribution is changed in two ways. First, the remaining population has a different mean ($\bar{x}_2$); the change in mean value is represented by C in Figure 1c. Second, the variation around the new mean (s_n) is smaller than that around the original mean (s_o); note the narrower curve in Figure 1c. The cross-hatched area represents the shift in genotypic mixture; in this example, relatively more genotypes for short seedlings remain after culling than in the original mixture.

The method for calculating genetic changes is presented later in this chapter (17.5).

17.3 How Nursery Practices Alter Population Structure

Numerous separate practices—seed storage, stratification, sowing, fertilization, watering, weeding, lifting, culling, and packing—can alter nursery populations. Any alteration in population structure associated with these practices can be evaluated by assessing four factors:

- What proportion of the seedlings is lost
- What proportion of the loss is directional
- What correlation exists between phenotypes and genotypic values of the seedlings lost
- How large a correlation exists between the culled trait and some other trait that may cause growth loss or mortality at a later date

Other indirect effects may not appear until several years after outplanting; these include changes in growth-rhythm traits such as budset, which can cause changes in wood quality and disease resistance. Effects may not become evident until the planted stand is ⅓ or ½ rotation age. Types and sizes of such effects depend on correlations of phenotypes with genotypes (Table 1), which vary with traits, on correlations among genotypes (Table 2), which vary with combinations of traits, and on species; those given here are for Douglas-fir [*Pseudotsuga menziesii* (Mirb.) Franco].

Ultimately, seedling survival—and mortality—must be assessed. The most common term used to report nursery mortality is tree percent—the percentage of seeds that germinate and develop into usable seedlings ([5, p. 149]; see also Chapter 5, this volume). Tree percent represents seedling survival to planting age and thus reflects both germinability of seed and the results of all nursery practices. But tree percent differs

Table 1. Correlations between genotypic values and phenotypes for selected representative traits in Douglas-fir, based on nursery measurements of individual seed or seedlings.

Trait	Correlation
Seed weight[1]	0.35
Germination rate[2]	
4-week stratification	0.43
10-week stratification	0.50
Budset date, 1st year[3]	0.57
Budbreak date, 2nd year	0.65
Budset date, 2nd year	0.53
Height, 2nd year	0.44
Diameter, 2nd year	0.48

[1] Upper limit of correlation estimated by intraclass correlation coefficient calculated from data in Silen and Osterhaus [35].
[2] Estimates based on seeds from 200 open-pollinated families, west slope of the Cascade Mountains [unpubl. data, 8].
[3] Estimates of budset, budbreak, height, and diameter from 135 open-pollinated families from southern Oregon [unpubl. data, 8].

greatly among species, among seedlots within species (Table 3) and even among nurseries. Data from the Duncan Seed Centre, British Columbia Ministry of Forests [6], indicates that about 27% of the variation in average tree percent for several species was due to variation among nurseries.

Notwithstanding the voluminous literature on many aspects of nursery technology, we cannot now quantitatively estimate the genetic implications of individual nursery practices and may never be able to separate the effects of various practices. Seedling growth patterns reflect genotypes and their responses to the environment and competitive status with respect to neighboring seedlings. Because nursery conditions and practices—and chance—interact to create a seedling's environment, the effect of various practices on growth and survival is probably impossible to trace. Nevertheless, the genetic principles (see 17.2) are valid, and by keeping them in mind we can make reasonable guesses concerning which practices have the greatest potential for changing genetic structure. Seed-size grading, stratification period, sowing date, and culling are important factors that may foster directional selection. Lifting date also may be important.

Table 2. Genetic correlations between selected, representative traits of Douglas-fir observed in the nursery.[1]

	2nd year				
	Budbreak	Budset	Height	Diameter	Lammas shoot[2]
Budset, 1st year	0.59	0.56	0.96	0.46	0.63
Budbreak, 2nd year	....	0.52	0.65	0.10	−0.07
Budset, 2nd year	....	....	0.67	0.38	0.46
Height, 2nd year	....	....	....	0.79	0.95
Diameter, 2nd year	....	....	....	....	0.56

[1] Based on analysis of 135 open-pollinated families from southern Oregon [unpubl. data, 8].
[2] These correlations occurred only in a warm, moist environment that encouraged lammas-shoot growth in about half the seedlings.

17.3.1 Seed-related practices

Within the nursery, mortality is most likely at or preceding germination and during culling. In species with poor germination capacity, most losses occur before germination, but in species with good germination capacity, losses are associated with culling [pers. commun., 31]. When seeds germinate poorly, seedbed density is low, and most seedlings exceed the culling standard; when seeds germinate well, seedbed density is high, and many seedlings are substandard.

Nursery managers attempt to adjust for differences in germination capacity between seedlots by calculating sowing density (see chapter 5, this volume). This undoubtedly helps but is not completely satisfactory. For some species, results of laboratory germination almost invariably overestimate field germination; the discrepancy is smallest for higher quality seeds that germinate rapidly [7]. Unfortunately, a sowing factor to adjust for such a discrepancy cannot easily be devised, particularly in view of the variability in germination time and percentage from year to year in most nurseries. Therefore, nursery practices which can potentially influence germination success should receive attention.

In Douglas-fir, at least two other pulses of selection come into play. Damping-off in the first few weeks after germination and factors associated with heat stress in late summer may contribute 20 to 30% of the mortality in the first growing season [pers. commun., 41]. Germination timing may be important because temperature in the first month after sowing influences disease incidence [4]. Environment and genetics determine the amount of selection, which varies by seedlot, year, nursery, seedbed, and fungus strain [3].

Table 3. Tree percents and corresponding indexes of selection intensity for seedlots of representative western conifers after mortality and culling in the nursery.

		Selection indexes[1] for change in	
Species	Tree percent[2]	Mean genotypic value (i_1)	Variation (i_2)
White fir [*Abies concolor* (Gord. & Glend.) Lindl. ex Hildebr.]	6-60	1.98-0.64	0.85-0.58
Grand fir [*A. grandis* (Dougl. ex D. Don) Lindl.]	25-70	1.27-0.50	0.76-0.51
California red fir (*A. magnifica* A. Murr.)	4-75	2.15-0.42	0.87-0.47
Noble fir [*A. procera* (Rehd.)]	30-70	1.16-0.50	0.74-0.51
Sitka spruce [*Picea sitchensis* (Bong.) Carr.]	11-36	1.71-1.04	0.82-0.70
Engelmann spruce (*P. engelmannii* Parry ex Engelm.)	30	1.16	0.74
Lodgepole pine (*Pinus contorta* Dougl. ex Loud.)	48-75	0.83-0.42	0.64-0.47
Sugar pine (*P. lambertiana* Dougl.)	21-80	1.37-0.83	0.77-0.42
Western white pine (*P. monticola* Dougl. ex D. Don)	32-90	1.14-0.19	0.72-0.29
Ponderosa pine (*P. ponderosa* Dougl. ex Laws.)	48-80	0.83-0.35	0.64-0.42
Douglas-fir [*Pseudotsuga menziesii* (Mirb.) Franco]	25-77	1.27-0.39	0.76-0.43
Western redcedar (*Thuja plicata* Donn ex D. Don)	25-43	1.27-0.91	0.76-0.67
Western hemlock [*Tsuga heterophylla* (Raf.) Sarg.]	2-75	2.42-0.42	0.89-0.47

[1] Selection index values from Shelbourne [34].
[2] Tree percents from U.S.D.A. Forest Service [42].

Quantitative estimates of seedling losses associated with individual nursery practices have not been reported. The few data published indicate tremendous variation among seedlots. For example, for *Pinus* species, Krugman and Jenkinson [25] reported that average nursery germination has ranged from 20 to 85% of the germination capacities found in laboratory tests; of seed germinated, as little as 19% and as much as 90% (average 55%) produced usable seedlings. For western hemlock [*Tsuga heterophylla* (Raf.) Sarg.], nursery germination ranged from 4 to 79% of viable seeds [7]; of seed germinated, as little as 12% and as much as 96% produced usable seedlings. Similar variation is found from nursery to nursery and seedlot to seedlot. At two Northwest nurseries [pers. commun., 31, 46], the consensus seems to be that patterns of mortality are distinct within seedlots and highly variable among seedlots, even within a single nursery. Whether the pattern for any given seedlot is consistent from year to year is unknown.

17.3.1.1 Seed-size grading

Seed-size grading provides an example of the complexity of predicting the effects of nursery practices. Retaining only part of the seedlot for sowing—usually the heavier seeds—has been used as a way to increase uniformity in seedling size [18, 28] and has received considerable attention in the literature. The practice provides a good example of directional selection. Seed size, however, is also strongly influenced by environment [18, 35] and year of collection [38]. Consequently, the correlation between genotypic value and phenotype is relatively small (0.35) (Table 1). In addition, seed size is affected by seed maturity—and the effects of maturity, in turn, are influenced by other nursery practices, such as duration of seed storage, stratification period, and germination temperature (Fig. 2).

Size grading that removes light seed, for example, also partly selects for genotypes encoding early seed maturity because mature seeds are usually heavier. Size grading thereby selects for other growth or developmental traits genetically correlated with early maturity. As Douglas-fir seeds mature, they gain in weight and germination capacity, right up to seed fall [32]. In some species, however, germination capacity may decrease at full maturity [15]. Maturation also varies from cone to cone within the same tree [10, 17, 30] and from tree to tree and stand to stand [17, 35]. Consequently, variation in weight of seed in bulked lots undoubtedly reflects differences in seed maturity among trees and stands as well as time of collection. Furthermore, grading within a bulked seedlot may eliminate almost all seeds from some trees within seed zones and lots, reducing genetic variation within the lot [35]. Absence of grading, however, may have similar effects. Plants grown from smaller seeds in close mixture with plants grown from larger seeds often have smaller leaf areas in relation to their growing space and suffer disproportionate mortality [1].

17.3.1.2 Stratification period

Seed stratification and sowing date have considerable potential for changing genetic structure because they are correlated with so many aspects of growth and survival (see chapters 4 and 5, this volume).

Length of stratification affects germination energy (Fig. 2) and helps determine when an individual seed will germinate. Early germination usually increases risk of injury by spring frost, but individuals that survive are in a more favorable position to capture environmental resources [20]. These seedlings probably become more vigorous in the seedbed and are less likely to be culled or die by damping-off or heat stress. Therefore, any factor that increases variability of germination time in a seedlot favors early germinating seeds. A short stratification period, particularly when followed by cool germination temperatures, tends to promote such variability.

Germination time varies among seedlots from different geographic sources but also among seeds from trees within a single source and among seeds from a single tree. For example, consider the results from an experiment using seedlots from 185 seed trees from 100 sources in Oregon and Washington [unpubl. data, 8]. Seedlots from individual trees were stratified for 28 and 70 days and germinated at a constant 17°C. In seed stratified 28 days, variation among sources was 4 times greater than that in seed stratified 70 days. To a lesser degree, short stratification also increased variation among seeds from individual trees within sources. For seed stratified 28 days, 50% of the seed from the earliest germinating lots had germinated by 6.2 days, and 50% of the seed from the latest germinating lots by 11.9 days; comparable figures for seed stratified 70 days were 5.5 and 10.2 days. The greatest difference, however, was

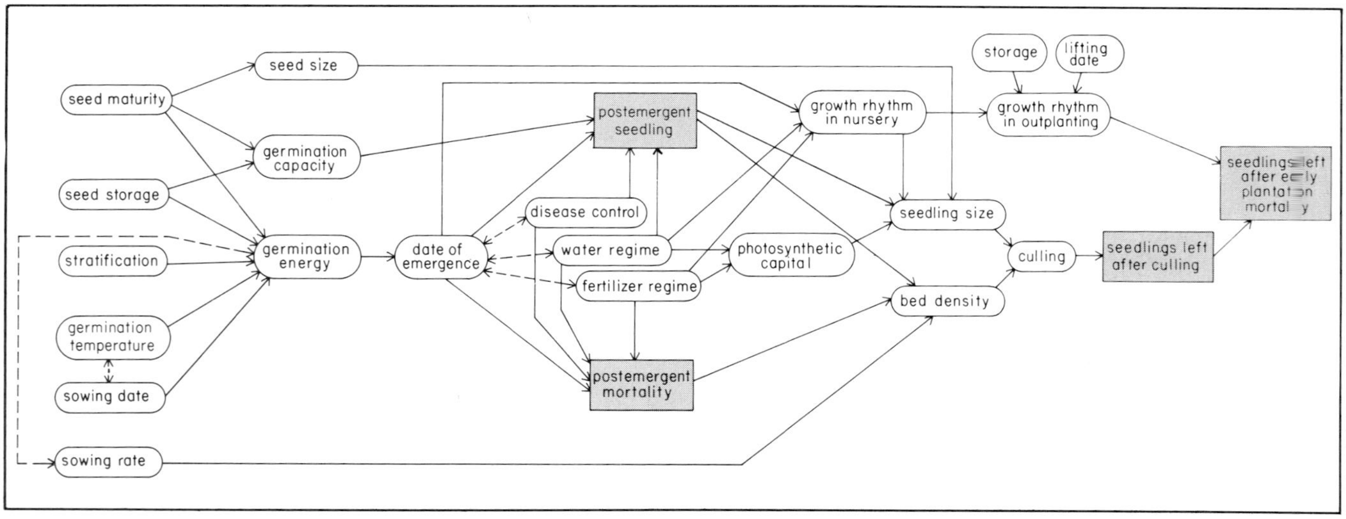

Figure 2. The many paths by which conditions partly under human control may either directly (solid lines) or indirectly (broken lines) influence the number of seedlings remaining after culling or outplanting. For example, seed maturity and storage partly determine germination capacity, which influences the number of seedlings remaining after germination; some of those seedlings may then die, the number depending on several nursery practices. Total postemergent mortality affects seedling size and bed density, which together largely determine culling percentage. In a related path, sowing date affects emergence date. Disease and its control, and water regime, correlate with emergence date and affect mortality before and after germination. Once again, mortality influences seedling size, bed density, and culling percentage. Shaded boxes are nodes at which data are sometimes available for calculating selection intensity.

among individual seeds from single trees. After 28 days of stratification, 99% of the seed had germinated by 34.12 days; after 70 days of stratification, 99% had germinated by 19.74 days.

The practical implication of length of stratification period is that mortality for both preemergent and postemergent seedlings is usually greater in seeds that are sown late or germinate late [2, 27]. Furthermore, seeds that germinate early have an advantage in competition and therefore tend to produce larger seedlings. In Campbell's [8] experiment, if only the seeds germinated in the first 7 days had survived, survival would have been 30% for seed stratified 28 days and 53% for seed stratified 70 days. But the generalization that early sowing produces more and better seedlings can be carried too far. There seems to be an optimum germination time in the nursery—sowing too early or too late results in suboptimal emergence [pers. commun., 40].

Differences in germination date probably would be even greater in the nursery because temperature during the germination period modifies the effects of stratification. Most nurseries sow in the spring throughout a period of rising temperatures; therefore, sowing date partly determines germination temperature. Variations in germination time among seeds, trees, and sources are exaggerated in cool temperatures [9]. Consequently, the shorter the stratification time and the cooler the temperature, the greater the variability in emergence date. We would expect the average temperature during nursery germination to be considerably lower than the 17°C used in Campbell's [8] experiment and the variability in days to germination to be correspondingly greater. Note, however, that stratification is likely to influence nursery performance significantly only in seedlots stratified for less than 60 to 90 days.

17.3.1.3 Sowing date

Sowing date modifies seedling growth in other ways. It is not surprising that seedlings from early sowings are larger [44], but other, more complex responses have been reported [37]. Sorensen [37] planted newly germinated seeds in seedbeds at intervals from April to June and found that the earlier the planting date, the longer the elongation period and the earlier the date of budset. At the end of the second season, this effect on budset persisted; seeds planted early set buds 17 days earlier than seeds planted later. But the early planting also shortened the elongation period in the second year. For germinants planted early, fertilizing the seedbed delayed first-year budset. Consequently, size of first-year seedlings was increased more by early sowing in fertilized plots than in unfertilized ones. The combined effect of fertilizers and early sowing persisted into the second year, increasing diameter disproportionately and thereby decreasing the height:diameter ratio of plants from seeds sown early.

Sorensen concluded that date of emergence influences not only a seedling's size and shape but also its growth rhythm. Such alterations of growth rhythm are not uncommon [21] and carry through at least 2 years and perhaps even beyond outplanting. Furthermore, the effect of sowing date can be modified by fertilization and other practices. But perhaps Sorensen's most significant finding is that genotypes reacted differently to sowing date. For some provenances, the amount and rhythm of seedling growth varied greatly according to sowing date; for others, variation was negligible. Sowing date apparently is an important factor affecting growth because it determines whether the climatic requirements of the seedling are well or poorly met by the nursery environment.

Because fertilizer affects seedling response to sowing date, the question arises as to whether moisture might also. In most nurseries, water is applied on a predetermined schedule. The average date of emergence of seedlings may or may not influence the start of the schedule, but, for the individual seedling, there will be some correlation between its emergence and its exposure to moisture saturation and depletion caused by the schedule. Therefore, emergence date partly determines the pattern of seedling moisture supply. To our knowledge, no reports describe interactions between seedling emergence and soil moisture.

Moisture stress in even the low to moderate range, however, affects budset date in 1-year-old Douglas-fir [19]. The practice of withholding water to induce early budset in nurseries is based on this reaction. Stresses are reflected in loss of potential height growth and thus photosynthetic capital (Fig. 2; also see chapters 12 and 15, this volume). As discussed earlier, treatments that change budset date may affect other aspects of growth rhythm and influence seedling development after outplanting (see 17.3.3). Furthermore, seed sources and, presumably, individual genotypes vary in their response to moisture stress [19, 45].

Length of stratification period and temperature during germination affect the variation in germination rate within seedlots. Consequently, practices or decisions that affect stratification and germination temperature tend to control variability among seedlings and opportunities for directional selection among genotypes. According to the OSU Nursery Survey (see chapter 1, this volume), Northwest nurseries prescribe stratification periods that vary greatly—for example, from 21 to 90 days for Douglas-fir. Shorter stratification periods, in particular, may induce considerable variability in seedbeds sown early or during unusually cool springs. Some seedlots are likely to react more strongly than others. The genetic implications, of course, are greater with poorer germination capacity and lower tree percent.

17.3.2 Seedling-related practices

17.3.2.1 Culling

Almost all Northwest nurseries surveyed (see chapter 1, this volume) reported culling for at least five traits: physical damage, multiple tops, height, stem caliper (diameter), and root mass. But not all losses due to culling cause directional selection.

Culling for physical damage probably results in selection without direction because most damage occurs randomly. Multiple tops may result from damage by natural agents such as frost and insects or from lammas-shoot growth (second flushing). If frost or lammas growth is the cause, selection is probably strongly directional. Frost damage is associated with phenological traits, such as budbreak and budset, which are strongly controlled genetically. Lammas growth tends to occur on the same individuals in successive years [12] and probably correlates well with genotype.

Culling for height and diameter will partly remove inbred plants and others that are weak or aberrant because of chance combining of bad genes from both parents [16, 36, 39]. This culling, although genetic, is not directional as we are using the term. Inbred seedlings tend to be small and more susceptible to disease and produce trees that cannot compete under forest conditions. Culling inbreds, therefore, probably improves the adaptedness and growth capability of the seedlot. Other culling for size will be directional but only partly effective in changing the mean genotypic value of the seedlot; the correlation values (Table 1) indicate only moderate genetic control of height (0.44) and diameter (0.48). Culling for root mass is probably similar to culling for top dimensions unless root diseases are involved, in which case culling may be more directional [11].

Even the mortality caused by natural selection in the nursery before and after seedlings emerge probably is not completely directional and certainly cannot be considered to be selection for any single trait. Because researchers and managers lack knowledge about the complicated relationships among traits and acts of selection, however, the best procedure is to assume that directional selection applies to a single trait. On this assumption, we present indexes (explained in 17.5) of selection intensity (the proportion of seedlings remaining after culling and natural selection) that affect changes in genotypic value and variation among seedlots and species (Table 3). These indexes correspond with tree percents; for example, a tree percent of 6 indicates a high selection intensity of 1.98 for change in mean genotypic value and of 0.85 for change in variability. Because of the above qualifications, however, these intensities must be viewed as indicating the upper limit of directional selection resulting from all natural losses and directed culling from seed sowing to seedling packaging.

Some of the directional selection facilitated by variability among seedlings occurs at culling. Although the amount of selection contributed by culling is significant, 70 to 90% of seedlings normally are saved. Sometimes, however, the seedlings culled as being too small are sold as substandard lots or are transplanted (double-grading). Depending on the degree of culling, these seedlings represent 10 to 30% of the original seedlot. Foresters who want larger seedlings occasionally request heavier than normal culling, saving only 20 to 40% of the original lot. In these special cases, selection intensities are almost double normal intensities, shifting mean genotypic values of a seedlot twice as far as normal culling.

17.3.2.2 Lifting and storage

The final operation in which nursery practices contribute to directional selection is lifting (see chapter 21, this volume). Lifting date, modified by storage length [29] and storage conditions [28], accounts for a substantial part of first-year field mortality. Field survival of western conifers strongly correlates with lifting date [22]. Survival is affected by lifting date through its apparent control of root-growth capacity, which, in turn, depends largely on the chilling seedlings receive before lifting [26] and the effect of photoperiod on response to chilling [29]. Length of storage, independent of lifting date, can affect seedling height growth [33] and date of budbreak [24] for 1 to 3 years after outplanting.

Root-growth capacity seems highly correlated with budbreak timing [26] and with capacity for top growth after outplanting [22]. If root-growth capacity varies genetically within a seedlot, field mortality due to poor root-growth capacity undoubtedly favors some genotypes over others. The seedlings at greatest risk are those whose root-growth capacity has not been sufficiently enhanced by prelift chilling. Although natural selection only indirectly selects the genotypes that survive, phenotypes and genotypic values for developmental traits such as budbreak and budset are usually closely correlated (Table 1). Root-growth capacity probably is also strongly controlled genetically, and selection against incompletely chilled seedlings will change the genotypic mix of a seedlot.

Lifting and storage practices strongly affect natural selection after outplanting. Large changes in genetic structure are not expected from this selection, however, because poorly stocked plantations are often replanted. Failure rates in plantations in the Northwest may approach 30% [13], but survival through the first 2 years usually exceeds 50%, partly because most failed plantations (survival less than 20 to 25%) of public agencies are replanted. Thus, the influence of selection cannot be large unless established seedlings are the survivors of several successive regeneration attempts, each providing only a few seedlings from the much larger number planted. In lots made up of seedlings remaining after heavy nursery mortality, however, the intensity of accumulated selections from nursery and outplanting could be quite large.

17.3.3 Post-outplanting effects

Selection can also be fostered indirectly by any nursery practice that tends to produce seedlings that cannot survive after outplanting. The natural selection that occurs after outplanting can be attributed to nursery practices to the extent that the nursery has engendered a growth rhythm or physiological balance incompatible with that required by the plantation environment. For example, the proportion of nutrients allocated to roots, shoots, and needles may not be appropriate for the season or conditions at outplanting. Or the plant's dormancy cycle may be slightly out of phase with existing environmental conditions (see chapter 14, this volume). Therefore, any nursery practice or environment should be examined for potential causes of growth-rhythm incompatibilities if it produces seedlings with phenotypes characteristic of either very short or very long growing seasons, or if it favors (during culling, for example) seedlings with extra long or extra short vegetative cycles—which may, in turn, favor seedlings with extreme dormancy cycles after outplanting.

The proportion of plantation losses caused by nursery practices cannot be estimated on the basis of available information. Survival after 1 year in the field generally ranges from about 50 to 100% but can be lower. An additional 20 to 30% of surviving seedlings are sometimes lost during the next 2 years. By then, survival in some plantations may be less than 20%. Depending on factors such as site class, cause of loss, and economic constraints, forest managers usually decide to replant plantations with less than 20% survival [pers. commun., 14, 43]. Thus, survival of about 25% after natural selection—which can be considered the lower limit in young plantations—establishes the theoretical upper limit of selection intensity for post-outplanting effects of nursery practices. This limit corresponds to an index of 1.27 for change in mean genotypic value and 0.76 for change in variation (see 17.5). Thus, natural selection, combined with selection due to nursery practices, can cause significant changes in the genotypic mixture—although the specific impact of each remains indefinable.

17.4 Nursery Location

Because nursery environments influence the rhythm of seedling growth, any nursery with an environment greatly different from that of seedlot origin may induce directional selection within the nursery or after outplanting. Nurseries cannot, of course, be sited so as to optimally satisfy the environmental requirement of every seedlot. The choice then becomes one of selecting the nursery most appropriate for a seedlot.

With respect to genetic implications, the choice should depend on the amount of selection that occurs in the nursery and after outplanting. If, for example, a choice is made to sow eastern Oregon seed in nurseries in both eastern and western Oregon, both nursery tree percent and field survival should be closely followed for the first 5 years or so after outplanting. If seedlots planted in western Oregon nurseries have lower than average tree percents, are subject to heavier than average culling at the time of lifting, or routinely suffer heavier than average mortality after outplanting, there is genetic reason for raising the seedlings in "eastside" nurseries only. Observational evidence indicates that seedlots may react specifically to nurseries. Some lots have higher tree percents in some nurseries than would be expected from their performance at other nurseries [7]. This "nursery effect" may account for some of the large variation in tree percent among seedlots within species.

The genetic implications of nursery location may be greater for some species than others. Sitka spruce, western hemlock, and western redcedar survive poorly in all bareroot nurseries (Table 3). Whether this is due to nursery environment or to cultural practices unsuitable for these species is not clear, but poor survival might be considered a genetic basis for restricting a species to certain nurseries.

17.5 Calculating Genetic Changes

Recall Figure 1. The changes in mean genotypic value (C) and in genetic variation (s_n vs. s_o) produced by culling or other types of selection depend on: (1) the correlation between genotypic value and phenotype, h (see Table 1); the indexes of selection intensity, i_1 and i_2 (see Table 3); and (3) the genetic variation in the original seedlot, s_o.

The value of h ranges from 0 to 1. If, for example, h = 1 for seedling height, the nursery manager who discards the taller half of seedlings, as in Figure 1, would also be discarding exactly the upper half of genotypic values for seedling height—that is, everything above $\bar{x}_1$. In Figure 1b, most of the stippled area representing culled genotypic values is above the mean, but some is below; therefore, h is greater than 0 but less than 1. The size of h can be modified by nursery practice; it can be increased somewhat by reducing the environmental variation within the seedbed. For instance, some seedlings may lack mycorrhizae, depending on chance distribution of fungal spores, and this may cause variation in seedling height. Use of proper inoculum could minimize this source of variation and improve the correlation between phenotype and genotype for height. Generally, however, the relative size of environmental variation seems to be characteristic for each trait. For example, the correlation between genotype and phenotype is usually higher for budbreak or budset date than for height or diameter (Table 1); in other words, the average genotypic value would be changed more if seedlings were selected for budbreak or budset date than if they were selected for height or diameter.

Values for selection-intensity indexes i_1 and i_2, derived from the normal curve, are found in prepared tables [34, p. 42 and 43]. The index i_1 helps predict the average change in genotypic value, which depends on the proportion of plants saved. If, for example, all plants are saved, $i_1 = 0$; if 1% are saved, $i_1 = 2.66$. The index i_2 helps predict the average change in genotypic variability in the seedlings left after selection. If all seedlings are saved, $i_2 = 0$; if 1% are saved, $i_2 = 0.90$.

Yet another value, the genetic correlation between traits, r (see Table 2), is necessary because directional selection for one trait may affect the genetic structure for other traits. Values for correlation between genetic traits range from −1 to 1. A negative correlation implies, for example, that smallness in one trait is associated with largeness in another. Whether the correlation is positive or negative is not important here; we will use only positive values (0 to 1). The degree of correlation seems to be characteristic of traits and is not likely to be modified by nursery practice.

Four simple equations using combinations of h, i_1, i_2, s, and r provide approximations useful for indicating the genetic implications of a nursery practice. Two equations are needed to show the direct effect of culling on a trait; two others are needed to show the indirect effect on traits other than the one or several selected for.

The change in mean genotypic value of the primary trait may be represented:

$$C = i_1 \times h \times s_o$$

Remember that all traits are treated as having equal genetic variation in the original seedlot. If, for example, $s_o = 1$ for all traits, it can be eliminated from the equation:

$$C = i_1 \times h$$

If 50% of seedlings are saved ($i_1 = 0.80$), as in Figure 1b, and if h for seedling height is, say, 0.45, then

$$C = 0.80 \times 0.45$$
$$= 0.36 \text{ standard units}$$

Thus, in this example, the mean genotypic value for height of seedlings left would be shifted from s_o by about ⅓ of a standard unit (Fig. 1c).

The variation left in the new population after seedlings have been culled either artificially or naturally may be represented:

$$s_n = s_o \times \sqrt{1 - (h^2 \times i_2)}$$

Then, on the basis of the above illustration:

$$s_n = 1 \times \sqrt{1 - (0.45^2 \times 0.64)}$$
$$= 0.93$$

Therefore, s_n is expected to be about 7% (1.00 − 0.93 = 0.07) smaller than s_o.

To indicate the indirect effects of culling on a related trait, two equations very similar to those just noted can be used. The change in mean genotypic value of the correlated trait (C_2)—in this case, diameter—may be represented:

$$C_2 = i_1 \times h \times r \times s_o$$

where s_o is the genotypic standard deviation for diameter. But if $s_o = 1$ for all traits,

$$C_2 = i_1 \times h \times r$$

Assume r = 0.81 between seedling height and diameter. Then, by selecting 50% of seedlings for height, we change the mean genotypic value for diameter as follows:

$$C_2 = 0.80 \times 0.45 \times 0.81$$
$$= 0.29 \text{ standard units}$$

At the same time, s_n for diameter would be reduced by about 5%:

$$s_n = s_o \times \sqrt{1 - (h^2 \times r \times i_2)}$$
$$= 1 \times \sqrt{1 - (0.45^2 \times 0.81 \times 0.64)}$$
$$= 0.95$$

In this section we removed certain genotypes from the population present in the original seedlot to show how nursery practices can influence genetic structure. Results are the same whether removal is caused by natural or artificial selection. If culling is to change the average genotypic value of a population, the selection must be directional. If selection is directional, the effects of culling on mean genotypic values and variability may be quantified by indexes i_1 and i_2.

We have used equations including i, h, and r primarily to illustrate that the effects of nursery practices operate through all three factors. Because the assumptions are only partly met by these equations, the predictions are only approximations. The calculations encompass only the major, additive components of genetic variation. Nevertheless, the equations point out the type of information needed to identify nursery practices that have potential for altering genetic structure.

If a practice directly or indirectly causes directional loss of a large proportion of seedlings, then i_1 or i_2 is large. If the trait affected by the practice (budset date, for example) also has a large h, then culling (or causing natural culling of) phenotypes for early budset accurately culls genotypes for early budset. We would expect such a practice to cause significant changes in the genotypic mixture—mean budset date would be delayed and variability of budset reduced. If early budset is closely correlated with another trait (frost resistance, for example), we would also expect similar changes in genotypic values for the secondary trait—the mean frost resistance would be decreased and would be less variable than in the original seedlot.

17.6 Genetic Risks of Nursery and Other Forestry Practices

Clearly, many nursery practices can change the genotypic mixture represented in a seedlot. We have hypothesized that this will increase the risk that outplanted seedlings will be poorly adapted to the planting site. But how much does nursery-caused selection increase the risk? Because risk cannot be measured directly without long-term and expensive tests, we will attempt to evaluate it indirectly by assuming that selection is partly adaptational. We will assume that the original seedlot is best adapted to the field planting site, that directional selection in the nursery decreases adaptation to the field site, and that adaptation decreases as nursery losses increase. The loss can be equated to an adaptational risk and compared with adaptational risks associated with other forestry practices. Here, we compare adaptational changes associated with nursery practice with those arising from seed transfers during reforestation.

Let's look at two alternative nursery effects: an average loss which is partially directional and an extreme loss which is completely directional. Average loss might occur in a seedlot with relatively poor germination, lower than average seedbed density (and, consequently, light culling), but good lifting and storage conditions. Extreme loss might occur in a seedlot resulting from double-grading, planted on a severe site. For the average-loss situation, nursery tree percent is 50% and field survival 70%; half of the loss in both nursery and field is directional—correlation between phenotype and genotypic value is moderate (h is 0.45). For the extreme-loss situation, nursery tree percent is 25% and field survival 20%; all loss is directional—correlation between phenotype and genotypic value is fairly good (h is 0.65). By sequentially solving the equations given in 17.5—first for nursery selection, then for field selection—we find that the mean genotypic value of surviving plantings has been changed by about 0.31 standard units in the average-loss situation and 1.58 standard units in the extreme-loss situation.

A difference of approximately 0.3 standard units, as for the average-loss case, is equivalent to the difference in mean genotypic value expected between Douglas-fir stands in southern Oregon separated east to west by about 10 km, or north to south by about 44 km, or in elevation by about 125 m [unpubl. data, 8]. Equivalent distances for the extreme-loss case (difference of 1.58 standard units) are about 45 km east to west, or 170 km north to south, or 500 m in elevation. The increased risk in the average-loss case is probably no greater than would be expected from transferring seeds from collection location to plantation site within some of the standard seed zones in southern Oregon. In the extreme-loss case, however, risks might be as great as those encountered in moving seeds between adjacent or even more distantly separated seed zones. Unfortunately, we lack the long-term studies needed to judge whether the risk caused by moving seeds between zones is either negligible or important. In addition, standard deviations of genotypic values would be reduced by about 8% in the average-loss case and 33% in the extreme-loss case. This change might have additional and unknown effects on adaptation over the length of a rotation, such as providing less opportunity for the chance fitting of genotypes to suitable environments.

Other forestry practices may have genetic implications at least as important as those of nursery practices. One is collecting seed for reforestation in marginal seed years. Seedlots obtained in such years may include seeds from only a small, nonrepresentative sample of parent trees. Unfortunately, it is not possible to make valid comparisons between forest and nursery practices because the effects of forest practices on genotypic mixtures are even less understood than those of nursery practices. However, the nursery phase, which includes many intensive, interacting cultural operations that can alter the genotypic mix in seedlots, is critical to preserving the adaptedness of planted trees. Nursery managers should recognize the importance of their decisions and encourage the accumulation of knowledge about the genetic implications of nursery practices.

17.7 Conclusions and Recommendations

Nursery procedures that maximize tree percent and seedling survival after outplanting will minimize changes in the genotypic mixture of a seedlot. Because most nursery practices are designed to maximize the proportion of seeds that become healthy seedlings, the better the nursery management, the smaller the genetic change in the genotypic mixture.

Predicting seed recovery is a basic problem which, if solved, would answer many questions about the genetic implications of nursery practices. The problem has two parts. First, and most important, germination tests and field germination are poorly correlated. Because managers cannot predict accurately, they usually compensate by oversowing, which often produces too many small seedlings with poor root systems which are then culled after lifting. Not only is seed used inefficiently, but inadvertent genetic selection also occurs. Second, germination in the nursery is usually lower than that in the laboratory. This implies a loss of potentially healthy seedlings before germination and, perhaps, selection against genotypes adapted to the field situation.

Therefore, the first information needed is whether seedling loss caused by our inability to predict nursery germination changes the genotypic mix in seedlots. Isozyme analysis, which can monitor the fate of individual genes from newly germinated seed to outplanted seedling, may be a good approach for generating this information [23]. If overly dense beds are found to cause changes in the genotypic mix, then better prediction of germination will be important. The relationships among stratification period, germination temperature, and germination rate of seeds from many seedlots from known individual trees must be clarified. In particular, answers are needed for species or seedlots commonly producing low tree percents.

Another fruitful research strategy may be to study the seed recovery of specific seedlots planted in several nurseries over a number of years. We may find that some seed sources perform better or more predictably in some nurseries than in others. As long as any species or seedlot survives poorly in the field or nursery, any increase in survival can ameliorate potentially dysgenic effects caused by nursery practices.

References

1. Black, J. N. 1958. Competition between plants of different initial seed sizes in swards of subterranean clover (*Trifolium subterraneum* L.) with particular reference to leaf area and the light microclimate. Australian J. Agric. Res. 9:299-318.
2. Black, J. N., and G. N. Wilkinson. 1963. The role of time of emergence in determining the growth of individual plants in swards of subterranean clover (*Trifolium subterraneum* L.). Australian J. Agric. Res. 14:628-638.
3. Bloomberg, W. J. 1971. Diseases of Douglas-fir seedlings caused by *Fusarium oxysporum*. Phytopathology 61:467-470.
4. Bloomberg, W. J. 1973. Fusarium root-rot of Douglas-fir seedlings. Phytopathology 63:337-341.
5. Bonner, F. T. 1974. Seed testing. Pages 136-152 *in* Seeds of woody plants in the United States. U.S. Dep. Agric., Washington, D.C. Agric. Handb. 450.
6. Bowden-Green, R., and H. Rooke. 1980. Projected recovery of plantable trees per bed foot. Unpubl. rep. Mimeo. 1 p.
7. Buszewicz, G., and G. D. Holmes. 1960. A summary of ten years seed testing experience with western hemlock, *Tsuga heterophylla*. Rep. on Forestry Res., Great Britain J. Forestry Commun.: 110-119.
8. Campbell, R. K. 1982. Unpublished data, U.S.D.A. Forest Serv., Forestry Sciences Lab., Corvallis, Oregon.
9. Campbell, R. K., and S. M. Ritland. 1982. Regulation of seed germination timing by moist chilling in western hemlock. New Phytologist 92:173-182.
10. Ching, T. M. 1960. Seed production from individual cones of grand fir. J. Forestry 58:959-961.

11. Copes D. L. 1982. Varied resistance to Phytophthora infection shown by cuttings of fourteen Douglas-fir clones. Annual meeting, Western Forest Genetics Assoc., Corvallis, Oregon, Aug. 3-5. Abstract.
12. de Champs, J. 1971. Pousses d'aout du Douglas. Pages 173-217 *in* Rapport Annuel L. Association Foretcellulose. Nangis, France.
13. Edgren, J. W. 1980. The reforestation system—a team effort. Pages 12-20 *in* Proc., North American forest tree nursery soils workshop (L. P. Abrahamson and D. H. Bickelhaupt, eds.). State Univ. New York, Coll. Environ. Sci. and Forestry, Syracuse.
14. Edgren, J. 1982. Personal communication, Region 6, U.S.D.A. Forest Serv., Portland, Oregon.
15. Edwards, D. G. W. 1978. Maturity and seed quality: a state of the art review. Pages 233-263 *in* Proc., A symposium on flowering and seed development in trees (F. Bonner, ed.). Mississippi State Univ., Starkville.
16. Eriksson, G., and D. Lindgren. 1975. Några genetiska reflexioner kring plantsortering. (Some genetic aspects on the grading of plants.) Sveriges Skogsvårdsförbunds Tidskrift 73:615-621.
17. Fowells H. A. 1949. An index of ripeness for sugar pine seed. U.S.D.A. Forest Serv., Pacific SW Forest and Range Exp. Sta., Berkeley, California. Res. Note 64. 5 p.
18. Griffin, A. R. 1972. The effects of seed size, germination time and sowing density on seedling development in radiata pine. Australian Forest Res. 5:25-28.
19. Griffin, A. R., and K. K. Ching. 1977. Geographic variation in Douglas-fir from the coastal ranges of California. I. Seed, seedling growth and hardiness characteristics. Silvae Genetica 26:149-157.
20. Harper, J. L. 1965. Establishment, aggression and cohabitation in weedy species. Pages 243-268 *in* The genetics of colonizing species (H. G. Baker and G. L. Stebbins, eds.). Academic Press, New York.
21. Heide, O. M. 1974. Growth and dormancy in Norway spruce ecotypes. II. After-effects of photoperiod and temperature on growth and development in subsequent years. Physiologia Plantarum 31:131-139.
22. Jenkinson, J. L. 1980. Improving plantation establishment by optimizing growth capacity and planting time of western yellow pines. U S.D.A. Forest Serv., Pacific SW Forest and Range Exp. Sta., Berkeley, California. Res. Pap. PSW-154. 21 p.
23. Kim, Z.-S. 1980. Veranderung der genetischen Struktur von Buchenpopulationen durch Viabilitätsselektion im Keimlingsstadium. Göttingen Res. Notes in For. Genetics, Lehrstuhl für Forstgenetik und Forstpflanzenzüchtung der Universität Göttingen. 88 p.
24. Kleinschmit, J., and J. Schmidt. 1974. Ergebnisse von Versuchen zur Kuhlhauslagerung von Forstpflanzen. Forstarchiv 45:165-170.
25. Krugman S. L., and J. L. Jenkinson. 1974. *Pinus* L. Pine. Pages 598-638 *in* Seeds of woody plants in the United States. U.S. Dep. Agric., Washington, D.C. Agric. Handb. 450.
26. Krugman S. L., and E. C. Stone. 1966. The effect of cold nights on the root-regenerating potential of ponderosa pine seedlings. Forest Sci. 12:451-459.
27. Kummel, J. T., C. A. Rindt, and J. T. Munger. 1944. Forest planting in the Douglas-fir region. U.S.D.A. Forest Serv., Portland, Oregon. 154 p.
28. Lavender, D. P. 1978. Seed collection, handling, and damage. Pages 53-62 *in* Regenerating Oregon's forests (B. D. Cleary, R. D. Greaves, and R. K. Hermann, eds.). Oregon State Univ. Ext. Serv., Corvallis.
29. Lavender, D. P., and P. F. Wareing. 1972. Effects of daylength and chilling on the responses of Douglas-fir [*Pseudotsuga menziesii* (Mirb.) Franco] seedlings to root damage and storage. New Phytologist 71:1055-1067.
30. Maki, T. E. 1940. Significance and applicability of seed maturity indexes for ponderosa pine. J. Forestry 33:55-60.
31. Morgan, P. 1982. Personal communication, D. L. Phipps Oregon State Forest Nursery, Elkton.
32. Olson, D. L., and R. R. Silen. 1975. Influence of date of cone collection on Douglas-fir seed processing and germination: a case history. U.S.D.A. Forest Serv., Pacific NW Forest and Range Exp. Sta., Portland, Oregon. Res. Pap. PNW-190. 10 p.
33. Rasanen, P. K. 1970. The effect of lifting date, packing, storing and watering on the field survival and growth of Scots pine seedlings. Acta Forestalia Fennica 112:1-27.
34. Shelbourne, C. J. A. 1969. Tree breeding methods. Forest Res. Institute, New Zealand Forest Serv., Rotorua. Tech. Pap. 55. 43 p.
35. Silen, R., and C. Osterhaus. 1979. Reduction of genetic base by sizing of bulked Douglas-fir seed lots. Tree Planters' Notes 30:24-30.
36. Snyder, E. B. 1972. Five-year performance of self-pollinated slash pines. Forest Sci. 18:246.
37. Sorensen, F. C. 1978. Date of sowing and nursery growth of provenances of *Pseudotsuga menziesii* given two fertilizer regimes. J. Applied Ecology 15:273-279.
38. Sorensen, F. C., and J. F. Franklin. 1977. Influence of year of cone collection on seed weight and cotyledon number in *Abies procera*. Silvae Genetica 26:41-43.
39. Sorensen, F. C., and R. S. Miles. 1974. Self-pollination effects on Douglas-fir and ponderosa pine seeds and seedlings. Silvae Genetica 23:135-138.
40. Tanaka, Y. 1982. Personal communication, Western Forestry Res. Center, Weyerhaeuser Co., Centralia, Washington.
41. Thompson, B. 1982. Personal commmunication, Western Forestry Res. Center, International Paper Co., Lebanon, Oregon.
42. U.S.D.A. Forest Service. 1974. Seeds of woody plants in the United States. Part 2. Dep. Agric., Washington, D.C. Agric. Handb. 450. 889 p.
43. Van Den Berg, D., and I. Holen. 1982. Personal communication, Medford District, Bureau of Land Management, Medford, Oregon.
44. van den Driessche, R. 1969. Forestry nursery handbook. B. C. Forest Serv., Victoria. Res. Note 48. 44 p.
45. White, T. L. 1981. Genecology of Douglas-fir from southwestern Oregon. Ph.D. thesis, Oregon State Univ., Corvallis. 103 p.
46. Wilson, B. 1982. Personal communication, Webster Nursery, Olympia, Washington.

Chapter 18
Weed Management in Forest Nurseries

P. W. Owston and L. P. Abrahamson

Abstract

Weeds often create problems for forest-nursery managers. Left unmanaged, weeds can drastically lower crop quality. Conversely, weed control at any cost can devastate a budget. Proper weed management requires careful development of an integrated program—not mere reaction to problems after they occur. Program components include planning, implementing, documenting, and evaluating results of both prevention and control operations. Factors to consider include crop and weed species, nursery environment, weather, control technology, personnel and equipment, environmental impacts, regulations, and safety of workers and the crop. Principles of program development are discussed; the biology of weeds, as well as physical, biological, and chemical methods of control, are described; and current practices in Northwest forest-tree nurseries are summarized. The need to test prospective techniques is emphasized: (1) small but thorough tests should be conducted before large-scale use of any treatments new to a particular nursery—a myriad of interacting factors make extrapolating results from one nursery to another unwise, and (2) results of tests or operational treatments should be carefully evaluated—subtle but critical damage to crop seedlings may escape notice in cursory examinations.

18.1 Introduction

A weed is any plant growing out of place—especially one that grows faster than crop plants. Weed invasion in nurseries is exacerbated by the common practices of leaving gaps of bare ground and growing single-species crops that do not utilize all of the site resources. Intensive soil management adds to the problem; for example, more intensive irrigation and fertilization almost always require more intensive weed management. Furthermore, most conifer seedlings grow slower than many weed species. Left unmanaged, nursery weeds can virtually destroy entire reforestation programs by greatly reducing crop yield and quality. At the other extreme, some control measures can be biologically effective but economically destructive because of high treatment costs. This chapter emphasizes the need for well-planned weed-management programs in nurseries and provides guidelines for establishing effective, environmentally safe, economical control programs.

18.2 Impact of Weeds on Crops

The primary impact of weeds is reduced crop yield resulting from competition for water, nutrients, light, and space. Weed species vary in their competitive ability, but they characteristically have fast-growing root systems that give them an early advantage in competing for water and nutrients. In addition, use of light and space by weeds can reduce photosynthesis and ultimate crop yield [23]. Weeds can also have an allelopathic effect on crop species—that is, they can harm crops through the production of chemical compounds that escape into the environment.

Other negative impacts of weeds are their potential for harboring insects or disease organisms; for slowing induction of dormancy and cold hardiness by reducing radiation and air movement, subsequently lowering plant moisture stress; and

In Duryea, Mary L., and Thomas D. Landis (eds.). 1984. *Forest Nursery Manual: Production of Bareroot Seedlings*. Martinus Nijhoff/Dr W. Junk Publishers, The Hague/Boston/Lancaster, for Forest Research Laboratory, Oregon State University, Corvallis. 386 p.

for making lifting and sorting procedures more damaging to seedlings and more expensive. Furthermore, weedy nurseries may have an adverse psychological impact on workers and customers, thus potentially reducing productivity and profits.

A national survey of forest-nursery practices conducted in 1974 [1] showed that weed control constituted a major production cost. Of 99 nurseries surveyed, more than ½ reported that the cost of weed control accounted for 10% or more of their costs; ⅓ reported about 25%; and ⅛ reported more than 50%.

18.3 Components of Weed-Management Programs

It is more efficient to anticipate problems than to react to them after they occur. Managing nursery weeds is no different. Weed management should be considered in terms of a complete, designed, integrated pest-management program consisting of four main components: education and planning, implementing, documenting, and evaluating.

18.3.1 Education and planning

Planning long before sowing is critical for developing a balanced attack that is efficient both by itself and when coordinated with other nursery operations. Advance planning permits the nursery manager to have supplies, equipment, and personnel on hand when needed; to have administrative details such as contracts, environmental assessments, and herbicide approvals or registrations[1] completed on time; and to pay adequate attention to safety and coordination.

The need to develop and continually update expertise should be considered in the education and planning phase. Creating and using a library should be part of the effort; many of the sources cited in this chapter—weed science textbooks; handbooks giving technical information on herbicides, pesticide use, and safety [30]; and plant identification guides—would be valuable references. Keeping up with periodical literature is important. *Tree Planters' Notes*, published by the U.S.D.A. Forest Service, and *American Nurseryman*, published by American Nurseryman Publishing Co., are good sources of weed-management information. Consultants can also provide expertise.

Proper identification of weeds by species and in all their stages of growth is an important part of education. Sources such as Hitchcock and Cronquist [16] and Dennis [13] are useful. Learning scientific names avoids the confusion of varied common names. Developing a nursery herbarium also could be beneficial for aiding later identification and training nursery workers. Once a species is identified, information gained from studying its life history and ecology can form the basis for prescribing prevention or control techniques.

During the education and planning phase, nursery managers should be alert to new technological developments, research information, and experiences of others.

18.3.2 Implementing

A sound weed-management program normally includes some aspect of each of the basic techniques: prevention or sanitation and control by physical, chemical, or biological means. Total dependence on a single method will seldom solve all of a nursery's weed problems.

First consideration should be given to preventing weeds from becoming established. Preventive measures tend to be safer and longer lasting than direct control [21]. Effective practices include preventing weeds from going to seed anywhere on nursery grounds; making sure that weed seeds are not carried into seedbeds by clothing, equipment, irrigation water, mulches, or soil amendments or along with transplants from other nurseries; and preventing spreading perennials from entering seedbeds from nonseedbed areas. Vegetative windbreaks can serve as barriers to windblown seed as long as the species used does not create insect or disease problems and does not shed seeds that are easily disseminated by wind.

Prevention by itself is only a partial answer, however. Some type of direct control is necessary in most situations. Interacting factors to consider are: (1) types and species of weeds and crop seedlings, (2) types of control that are feasible at a particular nursery, (3) operations that can serve multiple purposes, (4) costs, and (5) environmental impacts and other secondary effects of weed-control treatments [2]. Remember that the main objective of weed control is to grow more vigorous tree seedlings—not to kill weeds.

Control methods useful in nursery seedbeds may be classified as physical, biological, or chemical. Physical control includes mechanical cultivation, hand weeding, and mulching; biological control includes crop rotation and reliance on natural enemies; and chemical control includes use of inorganic and organic herbicides. Descriptions of the various control methods and how they relate to each other and to other nursery operations are detailed in later sections.

18.3.3 Documenting

Every weed-management program should include provisions for documenting all its pertinent aspects: recording ideas, decisions and their rationale, procedures, descriptions of conditions, and results. Documentation should be done continuously throughout the year—not from memory at the end of the season. Both biological and economic considerations should be included. Results should be measured, not estimated, and should include determinations of effects of treatments on crops as well as on weeds.

18.3.4 Evaluating

Documentation provides the information needed to evaluate decisions and results. Documentation and evaluation should be ongoing during the course of a program; but a final, end-of-season evaluation of all program aspects also should be conducted. Furthermore, a 3- to 5-year evaluation should be made to account for varying weather patterns and to look for long-term trends in such factors as weed population or increasing phytotoxicity to the crop due to prolonged use of a particular herbicide. Conclusions reached should then be considered in the planning phase for the next version of the program.

18.4 Weed Biology

Much of the material in this section is from Crafts [12], Klingman and Ashton [19], and Muzik [23], all of which are good sources of additional information on weed biology.

18.4.1 Types of weeds

Weeds are commonly classified as annual (winter or summer), biennial, or perennial. Annuals (those living less than 1 year) are generally the easiest but often the most expensive to control because of their abundance and rapid growth. Winter annuals germinate in the fall or early winter and produce seed early the next summer; summer annuals germinate in the spring and seed in the fall. Biennials (those living 1 to 2 years) consist of only a few species and are generally treated the same as annuals. Perennials may live indefinitely, and many reproduce by vegetative means as well as by seed; these are

[1]This chapter discusses pesticides. It does not contain recommendations for their use, nor does it imply that the uses discussed have been registered. All uses of pesticides must be registered by appropriate state and (or) federal agencies before pesticides are approved for application.

the hardest to control. In some cases, the same species may be in different categories in different parts of its range or under different growing conditions. Weed species found in Northwest nurseries are listed in Table 1 (OSU Nursery Survey; see chapter 1, this volume).

18.4.2 Seeds

Weeds are notorious producers of large quantities of seed. Single plants of some species produce thousands of seeds annually—some produce a million or more. A few species, for example, dandelion (*Taraxacum officinale* Weber), may set seed without having been pollinated.

Seeds are disseminated by wind, water, humans and other animals, and machinery and as impurities in straw or crop seed. Seeds of many species have special adaptations for wind dissemination; parachutelike structures, cottonlike coverings, and thin wings are common examples. Light seeds may drift in the air for miles. Most seeds will float, and some remain viable in water for 3 to 5 years; some have air-filled envelopes or corky structures as particular floating adaptations. Other species have hooks or other clinging structures on their seeds that aid their dissemination by humans and other animals. Many seeds remain viable even after passing through animal digestive tracts or being regurgitated by birds.

Seed longevity varies by species—from a few weeks to 1,000 years. Enough seeds remain viable when buried in soil that complete germination and destruction of residual weed seeds in crop fields may take several years of cultivation.

18.4.3 Vegetative reproduction

Most perennial weeds and a few annuals spread vegetatively as well as by seeds. These weeds, which include many grasses and broadleaves, cause some of the most serious competition problems and are often the most difficult to control.

Types of underground reproductive structures include rhizomes (underground stems), tubers, roots, corms, bulbs, and bulblets. Stolons (stems that grow on top of the soil) are another type. Most plants spread relatively slowly—less than 10 feet/year in many cases—if left alone; however, cultivation spreads plant pieces, and some vegetative parts root quickly in moist soil.

Many perennials root very deeply, especially in cultivated fields with deep soils lacking hardpans. Depth from which roots may regenerate is the important factor. For example, quackgrass [*Agropyron repens* (L.) Beauv.] does not regenerate from depths of more than 1 foot, but field bindweed (*Convolvulus arvensis* L.) can do so from a depth of 4 feet.

18.5 Methods of Control

18.5.1 Physical

18.5.1.1 Mechanical cultivation

Drill-sown seedbeds can be cultivated by tractor-drawn equipment when crop seedlings and weeds are small (see chapter 3, this volume). Cultivation should be shallow and careful, to avoid physical damage to the seedlings, which reduces growth and provides avenues for entrance of pathogenic fungi. In addition, splash erosion of cultivated soil can suffocate small seedlings or promote damping-off. The threat of injury to seedlings as they grow larger limits between-row cultivation to an early, partial component of weed-control programs unless spacing between rows is wider than the normal 6 inches in conifer seedbeds.

Table 1. Some common weeds found in Northwest forest nurseries[1] (OSU Nursery Survey).

Family	Species	Common name(s)	Life cycle
Equisetaceae	*Equisetum* spp.	Horsetails	Perennial
Gramineae	Many (not specified)	Grasses	Summer or winter annual or perennial
Cyperaceae	*Cyperus* spp.	Flatsedges, nutsedges	Perennial
Salicaceae	*Populus trichocarpa*	Black cottonwood	Perennial
	Salix spp.	Willows	Perennial
Polygonaceae	*Polygonum convolvulus*	Wild buckwheat	Summer annual
	P. persicaria	Black bindweed, ladysthumb, smartweed	Summer annual
Amaranthaceae	*Amaranthus* spp.	Pigweeds	Summer or winter annual
Portulacaceae	*Portulaea oleraceae*	Common purslane	Summer annual
Caryophyllaceae	*Spergula arvensis*	Corn spurry	Summer annual or biennial
	Spergularia rubra	Red sandspurry	Annual or perennial (rarely)
	Stellaria media	Common chickweed	Annual, biennial, or perennial
Cruciferae	*Brassica campestris*	Wild mustard	Winter annual or biennial
	Capsella bursa-pastoris	Shepherd's purse	Summer annual or biennial (rarely)
Leguminoseae	*Lupinus* spp.	Lupines	Annual or perennial
	Trifolium spp.	Clovers	Annual or perennial
	Vicia spp.	Vetches	Annual, biennial, or perennial
Geraniaceae	*Erodium cicutarium*	Common storksbill, redstem filaree	Winter annual or biennial
Onagraceae	*Epilobium angustifolium*	Fireweed	Perennial
Hippuridaceae	*Hippuris* spp.	Mare's tails	Perennial
Asclepiadaceae	*Asclepias* spp.	Milkweeds	Perennial
Convolvulaceae	*Cuscuta* spp.	Dodders	Summer annual
Solanaceae	*Solanum* spp.	Nightshades	Summer annual or perennial
Compositae	*Solidago occidentalis*	Western goldenrod	Perennial
	Senecio jacobaea	Tansy ragwort	Biennial or perennial
	S. vulgaris	Common groundsel	Annual or biennial
	Hypochoeris radicata	False dandelion, catsear	Perennial
	Taraxacum officinale	Common dandelion	Perennial
	Sonchus spp.	Sowthistles	Annual, biennial, or perennial

[1]Scientific and common names are from Hitchcock and Cronquist [16].

Approximately ½ the large forest-tree nurseries in the Northwest cultivate for weed control in pathways between seedbeds, and only about ¼ use between-row cultivation, most on a limited basis, of 1+0 seedbeds (OSU Nursery Survey). Cultivation within seedbeds is more common in nurseries growing hardwood seedlings at wider spacings throughout the United States.

18.5.1.2 Hand weeding

Hand weeding has been the mainstay of forest-nursery weed-control programs. Done properly, it can be safe and effective. Its main drawback is the high labor cost. Currently, hand weeding is often used to supplement chemical methods—to remove weeds that were missed or resistant to herbicides or those in seedbeds of crop species that are particularly sensitive to registered chemicals. The technique is most useful for weeds that propagate by seed; those that spread vegetatively usually need repeated weeding because it is difficult to pull all of a plant's roots.

To be most effective, weeds should be removed before they go to seed, spread vegetatively, or become so large or numerous that they interfere with tree growth or damage trees when the weeds are pulled. Soil should be moist enough so that weeds pull readily without breaking underground. Some types of hand cultivators are helpful if weeds are too small or numerous to readily grasp by hand or when pulling by hand causes roots to break underground. Weeds should be carried off nursery beds and thrown away or composted.

All nurseries surveyed report some use of hand weeding (OSU Nursery Survey). Amounts reported vary from 1 to 80 person-hours/acre over an entire season, the variation resulting from differences in weed populations, management philosophies, and other practices. However, all nursery managers would like to reduce the need for hand weeding because of the high costs involved.

18.5.1.3 Mulching

Mulches have a variety of purposes: they protect soil from erosion, crusting, and puddling; reduce splash erosion and frost heaving; help retain soil moisture; minimize soil temperature fluctuations; and suppress weed growth (see chapter 9, this volume). Mulches control weeds by preventing light penetration to underlying weeds and (or) by imposing a thick, dry layer through which germinating weeds cannot grow. Hand weeding has been reduced by as much as 60 to 90% because of mulches [5], though a much smaller effect is more common.

Mulches are generally not used for weed control in Northwest nurseries, and we do not recommend their use for that purpose. They are not cost effective relative to other types of control. If mulches are used for other objectives at a nursery, however, then gains in weed control are a bonus, as long as mulches are free of weed seeds so that they do not add to weed-control problems.

18.5.2 Biological

18.5.2.1 Crop rotation

Periodically leaving ground fallow or using green manure or cover crops to improve soil conditions (see chapter 10, this volume) can be effective ways to reduce the populations of weed seeds. With either technique, residual seeds germinate and can be tilled under before the next crop is sown. Fallow areas can be cultivated as often as necessary to prevent germinated weeds from going to seed and to expose additional residual seeds to germination. Dense cover crops discourage invasion by weeds, but if weed control is a major objective, it is better to combine the fallow technique with irrigation to stimulate germination between tillings. Furthermore, weeds in cover crops may become a serious problem if ignored. A few nurseries in the Northwest gain some weed control using these techniques (OSU Nursery Survey).

18.5.2.2 Natural enemies

Insects have generally been the most successful biological agents used in weed control [18]. They are usually host specific and slow acting, however—characteristics not suited to nurseries.

One biological agent—Chinese weeder geese—has shown promise at the U.S.D.A. Forest Service Wind River Nursery [14]. These geese, especially developed for use in rice paddies, are used in mint and cotton fields and in organic gardens in the United States. Dutton [14] reported that the geese at Wind River eat mainly seeds but also young plants of grasses and broadleaves such as sandspurry (*Spergularia* spp.) and dandelion. They seldom injure tree seedlings and can be used, if carefully watched, in 1+0 seedbeds after conifers are about 1 month old. Young geese are best. They are fenced in with 1-foot-high chicken wire in areas of 5 acres or less and are allowed to wander. They must be protected from predators, however, and are easier to replace each year than to keep over winter. Nursery personnel are pleased with the results of their trial and plan to increase the program.

18.5.3 Chemical

Herbicides—chemicals that suppress or kill plants—have been applied in forest-tree nurseries for many years. Materials used from the 1930s into the 1950s included inorganic compounds such as sulfuric acid, zinc sulfate, carbon disulfide, and sodium chlorate, as well as organic chemicals such as allyl alcohol, parachlorophenyl dimethylurea, methyl bromide, chloropicrin, and mineral spirits [33]. Except for the last three, the above have dropped from use as safer, more cost-effective, modern organic herbicides have been developed.

Because of the reliance being placed on herbicides, most of the remainder of this chapter is devoted to herbicide technology (see 18.6). The basics of the technology and use of herbicides in Northwest nurseries are described. In addition, safety to crops and effectiveness of the chemicals in controlling weeds are discussed.

18.6 Herbicides

18.6.1 Registration and use

Like other pesticides, herbicides are controlled by law. From the customer's standpoint, the product label is an important legal document. The label describes registered uses (those uses approved by the U.S. Environmental Protection Agency or a state agency); active ingredients and their concentrations and formulations; instructions for mixing and applying; guidelines for handling and storing the herbicide and for protecting the environment; and information on safety for humans and other animals.

A herbicide must be applied in one of several ways: (1) for the use pattern and site specified on the label, according to the directions and precautions stated; (2) for a proposed use pattern, on registered sites, after prudent interpretation of the label; or (3) under experimental permits issued by a state or the U.S. Environmental Protection Agency. In the last instance, the permit is usually issued to a manufacturer's representative, who gives general experimental guidance. The use of some pesticides, restricted because of the potential hazard to human health or environmental contamination, must be directly supervised by a certified applicator. Bohmont [9] presents a good summary of pesticide regulations.

Registrations of herbicides for forest-tree nurseries have historically been limited because of the small quantities applied and the chemical companies' potential for high liabilities in case of phytotoxicity to crop trees. Recognizing the potential benefits of herbicides, however, the U.S.D.A. Forest Service began programs to obtain experimental results to support federal and state registrations and to demonstrate the safety and effectiveness of promising herbicides to nursery managers. The first program started in 1970 when the Forest Service and Auburn University began the Cooperative Forest Nursery Weed Control Project for the 13-state area of the southeastern United States [15]. Between 1976 and 1980, a Western Nursery Herbicide Project was conducted with cooperators from state, federal, and private nurseries; the Forest Service; and the State University of New York, Syracuse [28]. Twenty-eight nurseries in 12 states were involved. In 1979, the Forest Service started an Eastern Nursery Herbicide Project in five states in cooperation with Purdue University and the State University of New York [17]; in 1981, this project was expanded to eight nurseries in three Great Lakes states.

All of these programs have similar objectives and methodologies, and information from one region often helps support that from others. More than 25 herbicide registrations for forest-tree nurseries have been obtained as a result of these studies [2, 22], and the number grows yearly. Although further improvements are needed, nursery managers now have a reasonable number of herbicides approved for production of conifer seedlings and a few for hardwood seedlings.

18.6.2 Characteristics

The material in this section is primarily from the textbooks of Klingman and Ashton [19] and Ashton and Crafts [8].

18.6.2.1 Action of herbicides on plants

Effects are determined by interactions among the herbicide, environmental conditions, and morphological and physiological characteristics of the plant. First, herbicides have to be absorbed through leaves, roots, stems, or seeds, depending on the characteristics of the particular chemical and how it is applied. Environmental conditions at the time of application affect the rate and amount of absorption. Instructions on the product label describe conditions under which the chemical can be effectively applied.

Some herbicides act on contact: the tissues that absorb the material are killed, but none of the herbicide is translocated to other parts of the plant. This type of chemical is useful for controlling small annual weeds, usually with no residual effect or danger of herbicide being absorbed by nontreated crop plants through the soil.

Noncontact herbicides translocate within the plant in much the same way as other solutes. Translocation is particularly important in controlling plants with underground reproductive structures. To apply an overdose of some herbicides can actually reduce herbicidal effect by damaging sprayed parts so much that disruption of tissues prevents translocation.

The phytotoxicity of most modern organic herbicides is caused by their disruption of plant metabolism. Biochemical reactions that may be affected are photosynthesis, respiration, carbohydrate metabolism, lipid metabolism, protein synthesis, and nucleic acid metabolism. The reaction disrupts plant growth and structure and is expressed as injury or death, depending upon the intensity of effect.

18.6.2.2 Selectivity

Selectivity refers to the differential effect of a particular herbicide on different crops. For example, a very selective chemical will retard or kill only a small group of plants at a particular stage of growth. A nonselective chemical is phytotoxic to all species. The product label of a given herbicide lists the plant species affected.

Selectivity involves interactions among the herbicide, plant, and environment. Herbicide factors include chemical structure, concentration and formulation used, and method of application (for example broadcast vs. directed sprays). Plant factors include age, growth rate, morphology, physiology, and genetics of both weed and crop species. Main environmental factors are soil texture, amount of organic matter, rainfall or irrigation, and temperature. For example, water is necessary to activate soil-applied herbicides; high humidity usually makes foliage-applied herbicides more effective; high organic matter reduces effectiveness of most soil-applied chemicals; and some materials work better when air and soil temperatures are cool [34], whereas others kill weeds only at high temperatures.

One type of selectivity involves the interaction between leaching characteristics of a herbicide and rooting depth of a plant. For example, a deep-rooted plant is not affected by a chemical that stays near the soil surface, whereas shallow-rooted plants are killed.

18.6.2.3 Persistence in the soil

Herbicides vary in the length of time they remain active in the soil. Their persistence is important to the duration of weed control and to possible crop phytotoxicity, which might result from multiple applications of persistent chemicals. Potential environmental pollution is also a concern with persistent herbicides. Herbicides generally used in forest nurseries vary from those with little or no soil persistence (for example, glyphosate) to those providing full-season weed control (for example, napropamide).

Factors that affect persistence are microbial, chemical, and physical decomposition; adsorption on soil colloids; leaching; volatility; photodecomposition; and removal by plants when harvested [19]. It is important to know the general characteristics of persistence for each chemical used; the manufacturer and the *Herbicide Handbook* [38] are good sources of this information. More detailed information for a particular nursery requires conducting chemical analyses or biological assays (bioassays) with sensitive plant species. Anderson [7] and William [39] describe bioassay techniques that nursery managers can employ themselves.

18.6.2.4 Classification

Herbicides may be classified in a variety of ways—for example, by chemical type, mode of action, or time of application. Classification by chemical type is of minor interest to nursery managers—those wanting such information should consult previously mentioned textbooks. Classification by general type of action [9] may be more useful (Table 2). Classification by time of application in relation to the growing cycle of both weeds and crop may be exemplified by the following general scheme:

Preplant: Herbicides to be applied anytime *before* sowing seeds of crop species or transplanting crop seedlings.

Preemergence: Herbicides to be applied *after* sowing but *before* emergence of crop or weed seedlings.

Postemergence: Herbicides to be applied *after* emergence of crop or weed seedlings.

Because the terms "preemergence" and "postemergence" can relate to either crops or weeds [38], an alternate scheme may be less confusing:

Preseeding: Herbicides to be applied *before* sowing or transplanting crops.

Table 2. Classification of herbicides by type of action (adapted from [9]).

Where applied	Type of action	Selectivity
Foliage	Contact	Nonselective
	Contact	Selective
	Translocated	Nonselective
	Translocated	Selective
Soil	Short residual	Nonselective
	Short residual	Selective
	Long residual	Nonselective
	Long residual	Selective

Incorporation: Herbicides to be physically incorporated into soil *before* crop seeding.

Postseeding: Herbicides to be applied *after* sowing but *before* germination of crop seedlings.

Postgermination: Herbicides to be applied *after* germination of crop seedlings.

18.6.3 Types of treatments and their use

18.6.3.1 Fumigation

Fumigants are chemicals that volatilize, penetrating as gases into air spaces and films of water around soil particles. They are generally nonselective, making them useful in controlling pathogenic fungi, soil-inhabiting insects, and nematodes as well as weed seeds [27]. Their nonselectivity, however, makes fumigants detrimental to the beneficial mycorrhizal fungi and nitrogen-fixing and symbiotic bacteria in the soil [2, 36]. Fumigants are also very expensive. Considering the ready availability of effective postseeding and postgermination herbicides, the nonselectivity and high cost of fumigants make it difficult to justify their use primarily to eliminate weeds. If fumigation is necessary for other pests, then weed control early in the first season can be a bonus [10]; however, fumigants provide no residual control.

Two Northwest nurseries fumigate primarily to remove weeds; three fumigate both for weeds and for other pests (OSU Nursery Survey). The other nurseries in the northwestern United States use annual fumigation primarily for controlling pathogens. Nurseries in British Columbia normally fumigate former agricultural land during its establishment. (For further information about fumigation and its use in Northwest nurseries, see chapter 19, this volume.)

18.6.3.2 Mineral spirits

Sold under a variety of trade names, mineral spirits has been used for weed control in conifer nurseries since the 1940s. This herbicide, derived from naphthenic petroleum, contains 10 to 20% aromatic hydrocarbons. The following information on mineral spirits, unless specified otherwise, comes from Stoeckeler and Jones [33] and Wakeley [37].

Generally effective on a broad spectrum of weeds, mineral spirits is used most successfully after weeds germinate and when they are no more than 2 inches in height or breadth. Preemergence control of weeds is sometimes attained in late spring. However, earlier applications are probably ineffective because dormant weed seeds are resistant to the chemical.

Most hardwood species and larches (*Larix* spp.) are sensitive to mineral spirits at all stages of growth. Pines (*Pinus* spp.) are least sensitive, and Douglas-fir [*Pseudotsuga menziesii* (Mirb.) Franco], true firs (*Abies* spp.), spruces (*Picea* spp.), and junipers (*Juniperus* spp.) are usually resistant. As a general rule, mineral spirits applied to conifers before seedlings are 4 to 6 weeks old is likely to damage them, though applications prior to conifer germination do not appear harmful.

Mineral spirits has been nearly abandoned as a weed-control treatment in large forest nurseries in the United States for several reasons: (1) large, repeated doses, 25 to 100 gallons/acre 3 to 8 times/season, are necessary; (2) environmental regulations require a reduced percentage of aromatic hydrocarbons, the active ingredients for weed control; and (3) more effective, less costly herbicides are now available.

Nurseries in British Columbia report extensive use of mineral spirits (OSU Nursery Survey). These nurseries can still obtain material with high levels of aromatic hydrocarbons, and Canadian regulations prevent application of the newest herbicides. One nursery uses mineral spirits before germination of spruce and Douglas-fir and for 2+0 and 2+1 spruce beds. Three others use it as a postgermination herbicide, and one of the latter also applies it between rows of transplants to kill annual weeds. Application rates in all cases are approximately 50 gallons/acre.

18.6.3.3 Modern herbicides

Selective organic herbicides applied as low-pressure, liquid sprays are the most common and effective chemicals for nursery weed control. They are not a panacea, however: (1) one application at the recommended rate generally does not provide season-long control; (2) no single herbicide safe for crop seedlings will control all weed species; and (3) if one species or type of weed is controlled, another will likely take its place [21]. Given these herbicide characteristics, the best attack is usually to use combinations (though not necessarily mixtures) or to alternate them during a season.

Nomenclature.—Each product has three names: a chemical name that describes its makeup to an organic chemist [for example, 2-chloro-1-(3-ethoxy-4-nitrophenoxy)-4-(trifluoromethyl)benzene]; a common name accepted by one or more societies or standard-setting organizations that is usually a shortened, more easily remembered version of the chemical name (for example, oxyfluorfen); and a trade or product name given by the manufacturer (for example, Goal®).[2]

Formulations.—Herbicides are marketed in various physical forms to make application easier and (or) to make the chemical more effective. The following discussion is adapted from Newton and Knight [24], who give more detailed information about herbicide formulations.

Water-soluble liquids and powders need to be mixed when preparing a spray solution but do not settle out with time. They tend to bead on waxy foliage, so require a surfactant for efficient absorption. Wettable powders—typically, fine dusts mixed with inert materials and applied in water—are used for herbicides that have low water solubility. Flowable concentrates—very finely ground wettable powders in a liquid matrix—are easier to handle than powders. Wettable powders and flowable concentrates should be mechanically agitated during application. Oil-soluble liquids and emulsifiable concentrates are useful for penetrating waxy foliage. They may be mixed with water if emulsifiers are included in the formulation to disperse oil droplets in the water. Granular and pelleted herbicides are suitable for dry application to soil. Gaseous fumigants, another type of formulation, were mentioned earlier (see 18.6.3.1).

Application techniques.—Most nursery herbicides are applied by spraying small amounts of chemical diluted in water carrier at 20 to 60 lb pressure. Spraying, done properly, provides much more uniform application than does mechanically

[2]Mention of trade name or product does not imply endorsement by the U.S. Department of Agriculture.

spreading and incorporating granular or pelleted materials. Spray volumes of 20 to 50 gallons/acre are normally used, although 10 to 20 gallons/acre usually suffice for translocated materials. Instructions for specific compounds are found on product labels.

The nozzle is the key component of any spray system; its type and condition affect the uniformity and rate of application and the amount of drift. Nozzles are made in a variety of spray patterns. The most common type used for broadcast applications is a flat-fan spray with tapered edges and 30 to 50% overlap from adjacent nozzles. A cone-shaped spray is best for spot treatments, a flat-fan spray with even edges for band or strip spraying.

Nozzles are manufactured from a variety of materials, and choice depends largely on the type of chemical being sprayed. Some chemical formulations are corrosive; others, such as wettable powders, are abrasive. The *Herbicide Handbook* [38] lists use precautions for each specific herbicide. In any case, proper operation requires clean nozzles. Water and bristle brushes or wooden pegs should be used to clean nozzles instead of wire or knives, which might damage carefully milled edges. Filters or strainers, located on the intake of the spray tank, in the line, or as part of the nozzle, help keep nozzles clean. Coarse filters (50 mesh or larger) are usually needed for wettable powders.

The pressure of the spray affects nozzle output and spray pattern, so a pressure regulator is necessary. It is best to use the pressure recommended for a given nozzle. If application rates must be changed, use different size nozzles rather than changing the operating pressure.

Other main components of a spray system are a spray tank or reservoir, a pump, and plumbing designed for the pressures and materials to be used. Sprayers can be mounted on either a tractor or trailer (see chapter 3, this volume).

Most of the large forest-tree nurseries in the Northwest apply herbicides with tractor-mounted sprayers: of the nurseries surveyed (OSU Nursery Survey), 11 use tractors only (100- to 200-gallon capacity), four use trailer-mounted sprayers (150- to 500-gallon capacity), and five use both types. Several nurseries also reported using hand sprayers for spot treatments.

Mixing several herbicides together can be an efficient way to control different types of weeds simultaneously. For example, so-called tank mixes can do "double duty" when both grasses and broadleaf weeds are a problem, or when preemergence and postemergence weed control is needed at the same time. Although such mixes are legal unless specifically restricted on the label of one of the chemicals, it is best to use only herbicide combinations that are recommended on the labels to assure that the chemicals are compatible. Mixes should always be tested for any new use—even if a nursery manager has already tried the same chemicals separately. The combination could possibly change the phytotoxic characteristics of the separate chemicals, such as when oxyfluorfen or bifenox is combined with glyphosate [3]. Rates of application may be reduced if two chemicals are used together [21]. Experimenting with new mixes of separately registered herbicides is critical—there may be a chemical incompatibility that would clog equipment and reduce or eliminate weed control.

Granular materials can be spread like dry fertilizers. Some work better if mixed into the soil, to reduce volatility and to place the herbicide close to weed seeds. Instructions on the label concerning proper depth of incorporation and equipment should be followed closely for proper results. Also, if the soil is too wet or equipment is not operating at recommended speed, poor mixing may occur. Granular herbicides are more difficult and expensive to apply and are not any safer or more effective than preseeding broadcast sprays [31].

Bohmont [9] and Klingman and Ashton [19] give more information on application equipment and describe calibration techniques and formulas for calculating doses. Proper and careful calibration, mixing, and application play an important part in preventing adverse environmental impacts from herbicide treatments.

18.6.4 Herbicide use in Northwest nurseries

Each nursery surveyed (OSU Nursery Survey) relies on herbicides to some extent. Several managers stated they use herbicides with reluctance and only as a "last resort." Others have intensive programs that utilize three to five different chemicals for different weed problems or different species or age classes of crop seedlings. Including fumigants and mineral spirits, use of 14 different chemicals was reported (Table 3; OSU Nursery Survey). All of these nurseries also do some hand weeding.

Table 3. Herbicides used in Northwest forest-tree nurseries (OSU Nursery Survey).[1]

Common name[2]	Trade name[3]	No. of nurseries reporting use
Methyl bromide/chloropicrin	MBC-33®	16
Oxyfluorfen	Goal®	9
Bifenox	Modown®	8
Mineral spirits	Various	5
Napropamide	Devrinol®	4
Paraquat	Gramoxone® or Ortho Paraquat®	4
Diphenamid	Enide®	3
Hexazinone	Velpar®	3
Glyphosate	Roundup®	2
DCPA	Dacthal®	2
Propazine	Milogard®	2
Simazine	Princep®	2
Amitrole	Various	1
Atrazine	Various	1

[1]This listing does not imply any specific registration. Uses may have been experimental or operational in one or more of four states in the U.S. or in British Columbia.
[2]Listed in order of decreasing use.
[3]The use of a product name is for identification only and does not imply product endorsement.

Two of the most commonly applied chemicals, oxyfluorfen and bifenox, can control a wide spectrum of broadleaf weeds and grasses [38]. Oxyfluorfen, a contact herbicide effective both during preemergence and postemergence periods, has a very low translocation rate, impacting tops more than roots. It resists removal by rain, is strongly adsorbed on soil, has negligible leaching through soil, and is nonpersistent in the environment. Bifenox, an effective preemergence herbicide that also can be used postemergence when weeds are no more than 2 to 3 inches tall, has a low translocation rate, is rapidly absorbed by foliage, and is herbicidally active for 6 to 8 weeks. Not easily removed by rain, it is less affected by weather and by clay and organic matter in soil than most preemergence herbicides. Bifenox has negligible leaching through soil and is nonpersistent in the environment.

18.6.5 Herbicide effectiveness

Evidence is plentiful that herbicides can effectively reduce weed populations in nurseries and thereby lower costs of hand weeding. For example, in the Western Nursery Herbicide Project mentioned earlier (see 18.6.1), time spent hand weeding was reduced 39 to 98% in first-year seedbeds treated with one of three different herbicides, compared with nontreated seedbeds (Table 4). Abrahamson [2-4] reported average gross savings of $4,000 to $7,000/acre of seedbed over hand weeding alone.

These data also illustrate how the effectiveness of the same chemical at different nurseries or different chemicals at the

Table 4. Reductions in time spent hand weeding in first-year seedbeds treated with three herbicides (adapted from [26]).

Nursery, by state	Bifenox	DCPA	Napropamide
	Percent reduction[1]		
Oregon			
Aurora	47	40	88
Klamath Falls	98	93	47
Lava	86	43	57
Phipps	27	20	10
Stone	73	25	24
Average	66	44	45
Washington			
Greeley	58	43	63
Toledo	93	39	80
Webster	78	48	67
Wind River	93	98	91
Average	80	57	75

[1]Reduction in time spent hand weeding compared with time required for nontreated seedbeds.

same nursery can vary widely. Differences in weed populations, soil, weather, and procedures are important factors in this variation. Steward [31] and Abrahamson [3] have shown that postseeding treatments are usually more effective than post-germination applications for total-season weed control. Early-season weed control is important in forest nurseries because winter annuals have had several months to become established in seedbeds, and prolific summer annuals have their main flush of germination in the spring.

Herbicides, even if used properly and cost effectively, do not provide 100% control. Some hand weeding or spot treatments will probably always be necessary for areas inadvertently skipped in application and for resistant weeds that should not be allowed to spread or go to seed. Actually, there is a secondary advantage to having a crew periodically go through nursery beds—they can spot problems with insects, diseases, and fertility, as well as weeds, that might otherwise go unnoticed. Furthermore, applying herbicides in doses heavy enough to eliminate all weeds increases the risk of crop damage [25].

18.6.6 Phytotoxicity to crop trees

It is senseless to use a herbicide in such a way that crop seedlings are damaged. Because herbicides are designed to be toxic to plants, the potential for crop damage is often high. Problems can result from improperly applying the chemical, applying too much chemical, or treating too frequently [21].

Phytotoxic effects can take many forms. Possible symptoms in crop seedlings include germination failure; needle chlorosis or burn; stem swelling or lesions; stunted or distorted growth of needles, shoots, roots, or the whole seedling; and mortality. Sometimes the damage is obvious (for example, heavy mortality or severe stunting); at other times, the effects are small losses in growth that can only be detected by careful analyses. In other cases, close observation is needed in the field. An example of the latter is the stem swelling on Douglas-fir and several other species west of the Cascade Mountains caused by DCPA [11]. Casual observers had attributed the swelling to heat damage, but careful workers—who compared the occurrence with seedlings in untreated seedbeds—found that the herbicide was the cause. On the other hand, nursery managers must also be careful not to mistakenly identify disease or nutrition problems as herbicide phytotoxicity just because a herbicide was used. Kozlowski and Sasaki [20] have a good discussion about the difficulties of assessing subtle phytotoxic effects.

One systematic approach to assessing damage to crop seedlings was developed by Anderson [6], who used the following rating scheme:

Rating	Description
10	No seedlings damaged.
7-9	Slight damage; seedlings will recover and achieve near-normal growth.
4-6	Moderate damage; few seedlings have died, but some show chemical effects and reduced growth.
1-3	Severe damage; many seedlings have died, and others are discolored and stunted.
0	All seedlings dead.

For consistency, the same individual should do all of the rating at a nursery. In addition, the person should briefly describe and record specific factors used to determine the ratings as an aid to later analyses. Table 5 is an example of damage ratings from a screening test in Northwest nurseries.

Systematic germination counts and seedling measurements such as height, stem caliper, and root weight, all supported by statistical analyses, are further steps that may reveal subtle phytotoxic effects.

The OSU Nursery Survey provided information on phytotoxicity, but there were too many unknown variables to draw conclusions about specific treatments. Some general observations are possible, however: (1) all the chemicals used at more than one nursery (Table 3), except methyl bromide/chloropicrin, were identified as phytotoxic in one or more instances; (2) hexazinone has caused the most severe, widespread problems; (3) oxyfluorfen and bifenox were often reported to cause needle burn or curl and sometimes were implicated in reduced germination or low-level seedling mortality; and (4) such symptoms as slight mortality, growth reduction or deformities needle burn, chlorosis, and stem swelling (from DCPA) were reported for the other herbicides. Most of the burning seems to occur on new, active growth. Another subtle type of problem that has been reported in crops is that of herbicides predisposing Douglas-fir seedlings to diseases such as *Fusarium* top blight (see chapter 19, this volume).

Conifer species that seem most sensitive to herbicides include coast redwood [*Sequoia sempervirens* (D. Don) Endl.], lodgepole pine (*Pinus contorta* Dougl. ex Loud.), western hemlock [*Tsuga heterophylla* (Raf.) Sarg.], and western larch (*Larix occidentalis* Nutt.) (OSU Nursery Survey; [31]).

The effect on mycorrhizae is another type of phytotoxicity damage that must be considered. The importance of mycorrhizal development on most conifer nursery stock is becoming increasingly recognized (see chapter 20, this volume), and any practices that impede mycorrhizal formation should be discouraged. In a study at six western nurseries, neither bifenox, DCPA, nor napropamide significantly reduced the proportion of feeder roots colonized by mycorrhizal fungi or the number

Table 5. Effect of postseeding application of herbicides on conifer seedlings (adapted from [32]).

Herbicide	Dosage, lb active ingredient/acre	Average damage rating[1]		
		Douglas-fir	Lodgepole pine	Western hemlock
Nontreated	0	8.8	10	9.7
Bifenox	3	8.8	10	10
DCPA	10.5	8.2	10	3.7
Diphenamid	4	8.2	9.7	10
Napropamide	3	8.5	10	5.4
Hexazinone	0.5	8.3	1.7	4.7

[1]Based on scale developed by Anderson [6]: 0 means complete kill, 10 means no damage; see text.

of mycorrhizal types, compared to controls of 1+0 Douglas-fir and ponderosa pine seedlings [35]. In contrast, several of the herbicide-species combinations had greater numbers or types of mycorrhizae, compared with the controls. Trappe [35] showed that: (1) different herbicides or application rates can have different effects on mycorrhizae—thus nursery managers and scientists should assess this factor when monitoring herbicide effects; (2) more variation occurred between nurseries than between herbicide treatments; and (3) it may develop that some herbicide treatments will have mycorrhizal benefits that make them useful beyond just weed control.

18.6.7 Conducting screening tests

The safety and effectiveness of any herbicide should be tested at each nursery before operational use. Testing is urged because there is a strong possibility of differential results from varied interactions of different mixtures of tree and weed species, soil and climatic factors, and cultural practices at different nurseries. Furthermore, herbicides used at tree nurseries represent such a small market that nursery managers must often collect or arrange for collection of their own efficacy and phytotoxicity data for registration rather than depend on chemical companies.

Because research agencies spend relatively little time studying weed control in Northwest forest nurseries, much of the work falls to nursery managers. To assist in this process, a detailed plan has recently been published [29] describing the layout of a study, procedures for applying herbicides, and requirements for gathering and analyzing data on weed control and phytotoxicity; sample data forms are included. In addition, the basic plan may be used to conduct administrative studies of other types of weed-control practices.[3]

Several years of testing are advisable because of variations in effects caused by different weather. Tests should include "double doses" to evaluate the safety limits on crop seedlings and incorporate sound design procedures, such as leaving an untreated control and randomizing and replicating treatments to reduce bias (see chapter 28, this volume). Finally, close scrutiny for possible phytotoxicity is essential; subtle effects may go unnoticed without a combination of visual observations and measurements.

18.7 Recommendations

Weed management is an important component of a nursery manager's job. To be successful, it must be done with thoroughness and professionalism. In summary, we recommend that nurseries

- Plan and implement an integrated weed-management program
- Gain expertise in weed science and be alert for new developments
- Document and evaluate conditions, treatments, and results on a thorough and continuous basis
- Test new prevention and control treatments and analyze them carefully for safety and cost effectiveness

References

1. Abbott, H. G., and S. D. Fitch. 1977. Forest nursery practices in the United States. J. Forestry 75:141-145.
2. Abrahamson, L. P. 1980. Pesticides in forest tree nurseries—primary and secondary effects. Pages 191-204 *in* Proc., North American forest tree nursery soils workshop (L. P. Abrahamson and D. H. Bickelhaupt, eds.). State Univ. New York, Coll. Environ. Sci. and Forestry, Syracuse.

[3]Guidelines and forms for such tests are also available from the Horticulture Extension Office, Oregon State University, Corvallis.

3. Abrahamson, L. P. 1981. Herbicide trials for weed control in Great Plains forest tree nurseries. Pages 61-102 *in* Proc., 33rd Annual meeting of the Forestry Committee. Great Plains Agric. Council, Lubbock, Texas. Great Plains Agric. Council Publ. 102.
4. Abrahamson, L. P. 1983. Herbicides, an important component of the weed control program at Oklahoma State (Norman) Nursery. Pages 171-191 *in* Proc., 1982 Southern nurserymen's confs. Southeastern Area, State and Private Forestry, U.S.D.A. Forest Serv., Atlanta, Georgia.
5. Aldhous, J. R. 1972. Nursery practice. Her Majesty's Stationery Office, London. Forestry Commun. Bull. 43. 184 p.
6. Anderson, H. W. 1963. A system for evaluating effective weed control in forest nurseries. Tree Planters' Notes 61:19-23.
7. Anderson, H. W. 1970. A bioassay technique for measuring herbicide residuals in forest nursery soils. Tree Planters' Notes 21(2):1-5.
8. Ashton, F. M., and A. S. Crafts. 1973. Mode of action of herbicides. John Wiley and Sons, New York. 504 p.
9. Bohmont, B. L. 1981. The new pesticide user's guide. B & K Enterprises, Inc., Fort Collins, Colorado. Approx. 309 p.
10. Boyd, R. J. 1971. Effects of soil fumigation on production of conifer nursery stock at two Rocky Mountain nurseries. U.S.D.A. Forest Serv., Intermountain Forest and Range Exp. Sta., Ogden, Utah. Res. Pap. INT-91. 19 p.
11. Callan, N. W. 1980. Dacthal injury on Douglas-fir and true firs at the Medford Forest Nursery. Pages 23-24 *in* Proc., Intermountain Nurserymen's Association meeting, Aspen, Colorado, Aug. 13-16, 1979. U.S.D.A. Forest Serv., Mount Sopris Nursery, Carbondale, Colorado.
12. Crafts, A. S. 1975. Modern weed control. Univ. of California Press, Berkeley and Los Angeles. 440 p.
13. Dennis, L. J. 1980. Gilkey's weeds of the Pacific Northwest. Oregon State Univ. Press, Corvallis. 382 p.
14. Dutton, D. W. 1983. Chinese weeder geese, do they or do they not weed in the nursery? Pages 44-48 *in* Proc., 1982 Western Nurserymen's conf. Western Forestry Nursery Council, Southern Oregon Regional Services Institute, Southern Oregon State Coll., Ashland.
15. Gjerstad, D. H., D. B. South, C. W. Brewer, and W. D. Kelley. 1980. 1979-80 Annual report—Auburn University Forest Nursery Cooperative. Dep. of Forestry, Auburn Univ., Auburn, Alabama. 109 p.
16. Hitchcock, C. L. and A. Cronquist. 1973. Flora of the Pacific Northwest. Univ. of Washington Press, Seattle. 730 p.
17. Holt, H. A., and L. P. Abrahamson. 1980. Developing weed control programs for forest nurseries in central U.S. Page 52 *in* Abstracts—1980 meeting. Weed Sci. Society of America, Toronto, Canada Feb. 5-7.
18. Huffaker, C. B. [1978?]. An overview of biological control, with particular commentary on biological weed control. Pages 3-12 *in* 1976 Proc., IV International symp. on biological control of weeds (T. E. Freeman, ed.). Univ. of Florida, Gainesville. 299 p.
19. Klingman, G. C., and F. M. Ashton. 1975. Weed science: principles and practice. John Wiley and Sons, New York. 431 p.
20. Kozlowski, T. T., and S. Sasaki. 1968. Effects of direct contact of pine seeds or young seedlings with commercial formulations, active ingredients, or inert ingredients of triazine herbicides. Can. J. Plant Sci. 48:1-7.
21. Kuhns, L. J. 1982. How to control weeds—the basics and the herbicides. American Nurseryman 155(5):29-49.
22. Moon, B. L. 1982. Alliance for progress. Dow Industrial Vegetation Management 14(1):6-8.
23. Muzik, T. J. 1970. Weed biology and control. McGraw-Hill Book Co., New York. 273 p.
24. Newton, M., and F. B. Knight. 1981. Handbook of weed and insect control chemicals for forest resource managers. Timber Press, Beaverton, Oregon. 213 p.
25. Newton, M., J. Lemhouse, and R. K. Hermann. 1977. Chemical weed control in western conifer nursery beds—research and program development. *In* Proc., Joint meeting, Western Forest Nursery Council and Intermountain Forest Nurserymen's Assoc., Richmond, B.C., Aug. 10-12, 1976. B. C. Forest Serv., Victoria. Not paginated.
26. Owston, P. W., and H. G. Weatherly. 1979. Semi-operational trials of three herbicides in Pacific Coast forest nurseries. U.S.D.A. Forest Serv., Forestry Sciences Lab., Corvallis, Oregon. Unpublished rep.

27. Portman, R. W., and G. Santo. 1978. Soil fumigation manual. Washington State Univ. Ext. Serv., Pullman. Ext. Manual 4387. 12 p.
28. Ryker, R. A. 1980. Western nursery herbicide study—1979 update. Pages 16-22 *in* Proc., Intermountain Nurserymen's Association meeting, Aspen, Colorado, Aug. 13-16, 1979. U.S.D.A. Forest Serv., Mount Sopris Nursery, Carbondale, Colorado.
29. Sandquist, R. E., P. W. Owston, and S. E. McDonald. 1981. How to test herbicides at forest tree nurseries. U.S.D.A. Forest Serv., Pacific NW Forest and Range Exp. Sta., Portland, Oregon. Gen. Tech. Rep. PNW-127. 24 p.
30. Singer, J. 1980. Pesticide safety, guidelines for personnel protection. U.S. Forest Serv., Forest Insect and Disease Management, Methods Application Group, Davis, California. 45 p.
31. Steward [Stewart], R. E. 1977. Herbicides for weed control in western forest nurseries. Pages 78-89 *in* Proc., Western Society of Weed Sci., vol. 30.
32. Stewart, R. E., P. W. Owston, and H. G. Weatherly. 1979. Evaluation of six herbicides for weed control in Pacific Coast forest nurseries. Pages B-127 to B-133 *in* Conf. and workshop proc., Western Forest Nursery Council and Intermountain Nurserymen's Assoc. (R. W. Gustafson, ed.). Eureka, California, Aug. 7-11, 1978. U.S.D.A. Forest Serv., Pacific SW Region, San Francisco, California. 288 p.
33. Stoeckeler, J. H., and G. W. Jones. 1957. Forest nursery practice in the Lake States. U.S.D.A. Forest Serv., Washington, D.C. Agric. Handb. 110. 124 p.
34. Ticknor, R. L. 1981. Weed control in nurseries and ornamental plantings. Ornamentals Northwest 5(4):5-8.
35. Trappe, J. M. 1983. Effects of the herbicides bifenox, DCPA, and napropamide on mycorrhiza development of ponderosa pine and Douglas-fir seedlings in six western nurseries. Forest Sci. 29:464-468.
36. Trappe, J. M., and R. F. Strand. 1969. Mycorrhizal deficiency in a Douglas-fir Region nursery. Forest Sci. 15(4):381-389.
37. Wakeley, P. C. 1954. Planting the southern pines. U.S.D.A. Forest Serv., Washington, D.C. Agric. Monograph 18. 233 p.
38. Weed Science Society of America. 1983. Herbicide handbook. 5th ed. Weed Sci. Society of America, Champaign, Illinois. 51 p.
39. William, R. D. (compiler). 1983. Oregon weed control handbook. Oregon State Univ. Ext. Serv., Corvallis. 184 p.

Chapter 19

Pest Management in Northwest Bareroot Nurseries

J. R. Sutherland

Abstract

A survey of Northwest bareroot nurseries revealed that root rots (*Macrophomina*, *Fusarium*, and *Phytophthora*) and *Sirococcus* blight are the major diseases, that the cranberry girdler (*Chrysoteuchia topiaria*) and various aphids are the most important insect pests, and that numerous rodents and seed-eating birds are frequent animal pests. Many site features affect both pest occurrence and successful management. However, the types of pest-management practices used depend upon factors including the availability of advice from pest-management specialists, the use of surveys, and numerous cultural practices. Nontarget effects of pesticides are also of concern. More information is needed on many aspects of pest management so that nursery managers can grow better seedlings.

19.1 Introduction

Most of the disease, insect, and animal problems affecting bareroot seedlings in the Northwest also occur in other nurseries throughout North America. However, several pests are specific to the Northwest or to hosts grown there.

Regardless of the pest's nature, the types and severity of damage and the pest-management practices used in the Northwest are often unique. This is undoubtedly due to factors such as seedling species, cultural practices, soils, and—of utmost importance—overall effect of the area's maritime climate. Although many local problems have received cursory attention in publications on forest-nursery management [2, 3, 30, 31], forest pathology [10, 20], entomology [17], and soils [42] or in-depth coverage in bulletins devoted solely to seedling diseases [27] or diseases and insects [36], there is a continuous need to update the status of pests and pest-management practices, especially in the Northwest, where seedling production has increased dramatically in recent years.

Consequently, during the fall and winter of 1981, 21 Northwest nurseries were surveyed to determine current disease, insect, and animal problems and the techniques being employed to manage them (OSU Nursery Survey; see chapter 1, this volume). In this chapter, these survey results are summarized and trends noted so that Northwest bareroot nurseries can update their current pest-management practices.

19.2 Status of Nursery Pests in the Northwest

19.2.1 Diseases

Thirteen diseases or disease groups are present in Northwest nurseries (Table 1). Their causes, symptoms, and other pertinent information are given by Peterson and Smith [27] and Sutherland and Van Eerden [36]. Charcoal root disease [*Macrophomina phaseoli* (Maubl.) Ashby], cortical rot (*Fusarium roseum* Link), *Fusarium* (*F. oxysporum* Schlect.) and *Phytophthora* (P. spp.) root rots, and *Sirococcus* blight (*Sirococcus strobilinus* Preuss) appeared to be increasing at certain nurseries. All but the last of these are soil-borne root diseases.

Douglas-fir [*Pseudotsuga menziesii* (Mirb.) Franco], spruces (*Picea* spp.), true firs (*Abies* spp.), and the two- and three-needle ("hard") pines (*Pinus* spp.) were the major seedling species affected by diseases. However, severity estimates and host data (Table 1) may be biased somewhat because Douglas-fir is the most common seedling species grown in the Northwest and because several of the diseases, such as the gall and foliage rusts and *Sirococcus* blight, mainly affect pines and spruces. Therefore, severity ratings may reflect host production figures as much as seriousness of a particular disease. Only four, or

In Duryea, Mary L., and Thomas D. Landis (eds.). 1984. *Forest Nursery Manual: Production of Bareroot Seedlings*. Martinus Nijhoff/Dr W. Junk Publishers, The Hague/Boston/Lancaster, for Forest Research Laboratory, Oregon State University, Corvallis. 386 p.

	% nurseries rating problems as						Controls	
	Severe	Moderate	Slight	Negligible	Hosts	Damage trends	Chemical	Cultural
Diseases								
Charcoal root disease	0	0	15	85	Douglas-fir, pines	Static or increasing	Frequently, methyl bromide (MBr) soil fumigation	None specified
Cortical rot	0	10	20	70	Douglas-fir, pines	Mainly increasing	MBr soil fumigation; also benomyl drenches and sprays	Sanitation, site selection
Damping-off	0	9	81	9	Many species	Static or decreasing	Mostly MBr soil fumigation; captan or benomyl seed treatments, drenches, or sprays.	Improve drainage
Fusarium root rot	5	11	42	42	Douglas-fir, pines, spruces, true firs	Static or increasing	Mostly MBr soil fumigation; captan, benomyl, and ethridiazola drenches; captan and benomyl sprays	Regulate seedling density and nitrogen fertilization, dry fallow
Fusarium top blight	5	19	29	48	Douglas-fir, spruces, true firs	Static or decreasing	MBr soil fumigation; captan and benomyl drenches and sprays	Regulate nitrogen fertilization; irrigate to cool soil
Gray mold	0	30	45	25	Many, especially Douglas-fir	Static or decreasing	Mostly captan, benomyl, and chlorothalonil sprays	Regulate seedling density
Melampsora foliage rusts	0	0	5	95	Douglas-fir, pines	Decreasing	Unspecified fungicide sprays	None specified
Nematodes	0	15	15	70	Many, especially Douglas-fir	Decreasing	Mostly MBr soil fumigation	Bare fallow infested soils
Phytophthora root rots	5	5	26	63	Douglas-fir, true firs	Mostly increasing	Unspecified spray and MBr soil fumigation	Improve drainage, sanitation
Seed fungi	0	5	45	50	Many, including spruces, true firs, pines	Mostly decreasing	Chlorothalonil seed treatments	Running water soak (prestratification?)
Sirococcus blight	5	5	29	62	Pines, sometimes spruces	Static or increasing	Chlorothalonil, captafol, and benomyl sprays	Presowing identification of infested seedlots; collect and destroy diseased seedlings
Storage molds	0	10	57	33	Many; least on spruces	Static or increasing	Captan and benomyl sprays	Store stock at −1 to −2°C; store dry
Western gall rust	0	0	15	85	"Hard" pines	Static or increasing	Unspecified sprays	Cull affected stock
Insects								
Aphids	0	19	52	29	Many, especially Douglas-fir	Generally increasing	Diazinon, endosulfan, carbaryl, insecticidal soap, or acephate sprays	Predaceous beetles
Cranberry girdler	0	20	45	35	Many, including Douglas-fir, true firs	Mostly increasing	MBr soil fumigation; chlorphrifos, diazinon, or fenvolerate sprays	Encourage beneficial birds around nursery
Cutworms	0	5	43	52	Many	Static or decreasing	Mainly carbaryl sprays and insecticidal baits	None specified
European pine shoot moth	0	0	15	85	Pines	Decreasing	Carbaryl or diazinon sprays	None specified
Marsh crane fly	0	0	10	90	Many	Generally decreasing	Diazinon sprays	None specified
Root and vine weevils	0	6	17	78	Many	Generally decreasing	Acephate sprays	Bare fallow infested soils
Springtails	0	5	25	70	Many, especially spruces	Static	Diazinon and mineral oil sprays	None specified
Animals								
Birds	10	38	19	33	Many, especially pine seeds	Static or decreasing	None specified	Screening, shooting, noise; early seed sowing
Deer	0	0	6	94	No preference	None	None specified	Fencing
Miscellaneous rodents	5	5	35	55	Many	Static or increasing	Unspecified baits	Shoot
Moles	0	0	14	86	Many	Decreasing	None specified	None specified
Rabbits	0	5	0	95	Many	Static	None specified	None specified

about 30%, of the 13 diseases reported were considered severe. At most nurseries, these and the nine remaining diseases were rated moderate to nonexistent.

Northwest nursery managers use numerous chemicals and cultural procedures for seedling disease control (Table 1). Methyl bromide soil fumigation is the most extensively used control for soil-borne diseases of seeds, germinants, and roots of older seedlings. Captan, benomyl, and chlorothalonil are the major fungicides sprayed for shoot diseases; sometimes, they are applied as drenches for soil-borne pathogens. Of the numerous cultural practices employed against diseases, some (e.g., improving drainage) make the environment less conducive for pathogen survival or dissemination, whereas others (e.g., withholding nitrogen fertilizers during the early part or all of the first growing season to reduce *Fusarium* root rot) increase host resistance.

19.2.2 Insects

Seven insects or insect groups were reported by Northwest nurseries (Table 1). Pertinent information on their biology, life histories, hosts, and damage are given by Furniss and Carolin [17], Sutherland and Van Eerden [36], and Triebwasser and Overhulser [38]. Aphids, especially Cooley spruce gall aphid [*Adelges cooleyi* (Gillette)] and giant conifer aphids (*Cinara* spp.) as well as the cranberry girdler (*Chrysoteuchia topiaria* Zeller) caused the most damage. The latter, a relatively new (or recently recognized) problem in forest nurseries [38], is one of the few insects whose importance nursery managers feel is increasing. All insects, except the European pine shoot moth [*Rhyacionia buoliana* (Schiffermuller)] which is host specific on "hard" pines, affected many seedling species. Obviously, nursery insects have much wider host ranges than diseases do; indeed, most of these insects also damage many ornamental and agricultural crops.

Various insecticides, including carbaryl and diazinon, are sprayed to control insects on seedling shoots. Insecticidal soaps are sometimes used against aphids and probably should be tried for controlling other insects, especially springtails (order Collembola). Some nurseries use predaceous lady beetles (order Coleoptera) as a biological control for aphids whereas others encourage insectivorous birds. Bare fallowing is mentioned for root and vine weevil (order Coleoptera) control.

19.2.3 Animals

Animals, particularly rodents and seed-eating birds, are troublesome in many nurseries (Table 1). Animal damage tends to be a sporadic problem, which likely accounts for the preponderance of mechanical and disruptive control devices (e.g., fencing and noise) rather than chemicals.

19.3 Site Factors Affecting Pests

Nursery soil characteristics, climate, water quality, and general locality are the main site factors determining what disease, insect, and to some extent, animal problems will plague a nursery. Any one or a combination of these factors will influence the kind and amount of damage caused by pests and the types, timing, and effectiveness of controls [32].

19.3.1 Soil

The ideal conifer nursery soil should have a light sandy loam texture, free drainage, and a pH of 4.5 to 6.0 and should contain 3 to 5% organic matter [41]. Northwest nursery managers rank compaction, organic matter maintenance, poor drainage, and unfavorable texture as major soil problems (Table 2). Obviously, much effort goes toward amending these conditions (Table 2). An often overlooked aspect of such soil problems is their relationship to the incidence, damage, and successful control of many pests, particularly soil-borne diseases. For example, soil compaction or heavy texture, which may be co-related, can increase preemergence damping-off losses by delaying germinant emergence, thereby increasing the period of susceptibility to the disease. Nonuniform texture within or among fields can decrease the efficacy of soil fumigants which must be applied on the basis of average soil moisture, temperature, and pest numbers. Poor drainage is one of the main factors contributing to *Phytophthora* root rot [19].

19.3.2 Climate

Although most Northwest nurseries are located in the milder coastal zone, seedlings nonetheless are often damaged by inclement weather, particularly in winter (Table 2). This occurs because many nurseries grow several seedling species and ecotypes—e.g., seedlots whose origins are coastal ("westside") or interior ("eastside"), or high or low elevation—whose tolerances vary to harsh weather. For example, in coastal British Columbia nurseries, shoot damage from winter desiccation is normally less for interior than for coastal Douglas-fir. The important point about climate-caused damage is that it often predisposes seedlings to pathogens, including storage (numerous fungi) and gray [*Botrytis cinerea* (Fries) Persoon] molds. Although unreported by nursery managers in the OSU Survey, the cool, wet weather of spring and early summer often favors damping-off and shoot blights such as *Sirococcus*. Aphid populations frequently increase during late summer and early fall, when it is dry. Though animal damage and climate are seldom related, severe weather in interior British Columbia has caused birds to migrate to the coast, where they consume large numbers of recently sown seeds.

19.3.3 Water

Water quality was not considered a problem in Northwest nurseries, but high water tables were reported by several nurseries. Impermeable layers beneath the plow zone, either occurring naturally or created by plowing, cultivation, or unsuitable soil texture, certainly contribute to this problem. High water tables favor certain root rots [19] and the marsh crane fly (*Tipula paludosa* Meigen). Surprisingly, no nurseries reported water-borne pathogens such as *Phytophthora*, which may be common in irrigation ponds.

19.3.4 Locality

None of the nursery managers surveyed related nursery locality and pests. However, in retrospect, many would probably consider their nursery "off site" as far as pests are concerned because potential pest problems are seldom considered when selecting a nursery locality or even when expanding an existing nursery. Normally, indigenous pests are not severe enough to disqualify a proposed site, but they should be identified and controlled before seedling production begins.

Both the broad ecological, climatic, or elevational zone and the specific locality within a zone can dictate what the pest problems might be. The nematode *Xiphinema bakeri* Williams, cause of corky root disease [6], is a pest that occurs only in coastal nurseries from British Columbia [33] to California [11]. *Sirococcus* blight and the marsh crane fly also predominate in coastal Northwest nurseries. Other pests are more prevalent in specific locations adjacent to nurseries; damage from the cranberry girdler, for example, is most severe in nurseries surrounded by grassland [38]. Sometimes, forests adjacent to nurseries contain trees affected with western gall rust [*Endocronartium harknessii* (J. P. Moore) Y. Hiratsuka] or certain foliage rusts (*Melampsora* spp.), which produce inocula for nursery seedlings [36].

Table 2. Site problems, amending practices, and related pest-control difficulties in Northwest bareroot nurseries.

Problem[1]	Amending practice	Pest difficulties related to site problem
Soil		
Compaction	Subsoiling, improve drainage; add organic matter; limit vehicle access.	Preemergence damping-off increased; germinant emergence inhibited; pathogen survival enhanced; pesticide effectiveness possibly affected.
Organic matter maintenance	Add organic matter via sawdust, peat, etc.; use cover crops.	Many, including damping-off and root-rot pathogens.
Poor drainage	Subsoiling, ditching, tiling, etc.	Root rots and damping-off, especially *Pythium*- and *Phytophthora*-caused diseases; marsh crane fly; pesticide timing and effectiveness possibly affected.
Unfavorable texture (too heavy or too light)	Improve drainage, e.g., by subsoiling; add organic matter or sand; withdraw from production or change usage; use fertilizers and pesticides judiciously; compact for seeding or transplanting.	Preemergence damping-off losses increased as germinant emergence hindered; better survival of many soil-borne pathogens, e.g., *Phytophthora* and *Fusarium*; more favorable to certain nematodes, e.g., *Xiphinema*; insect larvae more prevalent.
Nonuniform quality	Carry out cultural and pest-control practices for the "average" soil; replace clay "pockets" with lighter soil; transplant into poorer areas.	Pockets or areas with nonuniform control; variation in germination and growth may affect disease incidence.
Climate		
Wind abrasion	Plant windbreaks or use snowfence windbreaks.	Winter desiccation of shoots increases storage molding.
Frost pockets	Irrigate for frost protection; improve air drainage.	Frost damage of shoots increases gray and storage mold losses.
Frozen soil	Mulch, delay lifting, etc.	Possible winter desiccation of shoots can increase storage mold losses.
Water		
High water table	Improve drainage by tiling, subsoiling, or ditching.	Damping-off; many root-rot pathogens and certain insects, e.g., marsh crane fly.

[1]Major problems ranked by severity within categories (see chapter 1, this volume).

Some pests, especially soil-borne diseases and insects, are indigenous to the nursery site itself. In the mid-1960s, indigenous white grubs (order Coleoptera) damaged numerous seedlings in a new (cleared forest-grassland) nursery near Prince George, British Columbia. Nurseries established on formerly agricultural land often inherit fungus and nematode problems.

19.4 Disease and Insect Management

Nurseries should use an integrated pest-management approach comprising alternative strategies such as cultural controls and sensible use of pesticides. Which of these strategies predominates or how closely they are integrated depends upon (1) economics (not discussed here), (2) the overall philosophy of the nursery manager, (3) sources of advice and on-site expertise available to the nursery manager, (4) use of pest-detection surveys, and (5) cultural practices.

19.4.1 Philosophy

The philosophy of the nursery manager and staff towards various pest-management practices is of utmost importance. Sometimes, pesticides are used without considering integrated approaches or alternative strategies such as pest prevention. For example, none of the nursery managers surveyed gave any of the following as alternatives to using pesticides for preventing or controlling the following three important pest problems in Northwest nurseries (Table 1):

Damping-off: Delay seed sowing until soil temperature is high enough to promote rapid germination; sow stratified seeds for quicker germination; cover seeds with noncompacting material; maintain soil pH at 4.5 to 5.5; use areas prone to damping-off for transplants or resistant seedling species.

Storage molds: Prevent abiotic damage (e.g., frost and fertilizer burns) that predispose seedlings to storage molds; minimize the storage period; survey stored stock frequently to detect incipient molding; when possible, freeze-store stock.

Aphids: Where applicable, remove alternate or reservoir hosts from the nursery area (e.g., eliminate spruce from around Douglas-fir nurseries affected by Cooley spruce gall aphid).

Most nursery managers would insist that they do use some or most of these alternative strategies for these three and other pests, but how often does the manager relate these practices to some nonpest aspect of nursery management?

19.4.2 Sources of advice

Oftentimes, the source of advice on nursery pests will dictate the kind of pest-control procedure used. Invariably, agricultural pathologists and entomologists recommend pesticides, particularly when the tree nursery is in a vegetable- or fruit-growing area. Pesticide sales or technical representatives may be biased in their recommendations. The advice of some experts may be too academic—they may want to do a long-term research project—when in fact the nursery manager needs a quick, "best possible" answer.

To evaluate the quality of advice from outside sources and to effectively deal with pest problems, each nursery should have a pest manager on staff or on contract. This specialist should know the indigenous and potential pest problems of the nursery and the on-site factors (e.g., soil textures and drainage) that affect pest occurrence, damage, and control and should understand pest biology, pesticides (including application equipment and safety procedures), and nursery practices.The specialist should obtain up-to-date information from scientific and trade journals and by attending nursery meetings, pesticide-certification courses, and so forth. Besides the specialist, other nursery staff should attend training courses that emphasize pesticide-application equipment, toxicity, and safety.

19.4.3 Surveys

The success of any pest-management program primarily depends upon detecting problems before damage occurs—that is, usually when the pest is in an innocuous or incipient stage. This is best done by routinely surveying nursery soil, soil amendments, water, seeds, and seedlings (both growing and stored) for pathogens and insects. Pheromone or light traps, for example, can be used to detect insects. Transplants and equipment from other nurseries also should be checked for pests. Surveys pinpoint the presence, locality, developmental stage, and abundance of pathogenic fungi and insects so that preventive measures can be applied. Though field workers usually first notice impending problems, all nursery staff should be alerted to watch for initial indications of pests.

The nature and complexity of the pest problem usually determine how survey samples should be collected. Sometimes, a sampling expert must be consulted; generally, however, any system that produces a representative sample of the entire unit (e.g., field or seedlot) to be assayed will work. Soil samples may be collected according to a grid pattern, which helps nursery staff in relocating problem areas. Soil and seedling samples frequently are collected from affected and adjacent disease-free areas and then compared. Seedling samples should contain several individual seedlings, preferably in separate containers, whose symptoms vary from severe to nonexistent. Normally, samples should be kept at 10 to 15°C, protected from desiccation, and assayed soon after collection. A written account of seedlot numbers and past cultural histories should accompany samples, especially seedlings, to the diagnostic laboratory. Samples that can be used for more than one purpose are best; for example, in British Columbia, each soil sample from fallow fields is halved: one portion is assayed for soil nutrients, the remainder for pathogens and insects.

19.4.4 Cultural practices

Nearly any cultural practice may favor or deter pests. The nursery manager should avoid or modify practices that promote pests but integrate useful practices into the overall nursery-management program.

19.4.4.1 Seedling and windbreak species

The seedling species produced at a nursery profoundly affects the kinds and severity of pest-caused losses because many pests tend to be host specific. For example, a nursery growing ponderosa pine (*Pinus ponderosa* Dougl. ex Laws.) would probably have negligible losses from charcoal root disease because ponderosa pine is not severely affected by the disease [28]. Conversely, the same nursery could suffer significant losses from *Sirococcus* blight, which affects ponderosa pine.

Most nurseries contain areas that harbor or are somehow prone to certain pests. Sometimes the problem can be overcome by cropping the area with a nonhost seedling species; thus, an area known for *Fusarium* root rot, primarily a Douglas-fir disease [36], might be planted to less susceptible spruce. If possible, it is best to avoid certain pests altogether; in British Columbia, for instance, *Abies* seedlings are grown only in nurseries outside the zone infested by the balsam woolly aphid [*Adelges piceae* (Ratzeburg)].

Establishing windbreaks or ornamental plantings with species that promote pests is a common mistake. Nursery managers seem to particularly like *Populus* spp., which are alternate hosts for various conifer foliage rusts [36]. Needle distortion and chlorosis of Douglas-fir, caused by the Cooley spruce gall aphid, are most prevalent in nurseries with nearby spruces, the alternate host; a little detective work frequently reveals that the spruce were purposely planted.

19.4.4.2 Cover crops

Cover crops add organic matter to soil and retain nutrients ([15]; see chapters 9 and 10, this volume). However, the overall relationships of cover crops and pests in the Northwest are largely unknown. Davey and Krause [15] cite several examples, from elsewhere in North America, of cover crops that affect seedling diseases; for example, *Fusarium* root rot of conifers increased after a buckwheat cover crop whereas damage from charcoal root disease decreased after cover crops of rye or millet, both nonhosts.

The OSU Nursery Survey revealed that many of the U.S. Northwest nurseries sow lupines and other leguminous cover crops. The terminal crook fungus (*Colletotrichum acutatum* Simmonds f. sp. *pinea* Dingley & Gilmour) on Monterey pine (*Pinus radiata* D. Don) in New Zealand supposedly originated from lupine cover crops [16]. In the Northwest, gramineous cover crops might lead to buildup of and seedling damage by the cranberry girdler.

19.4.4.3 Irrigation

Besides satisfying the seedlings' water needs, irrigation can alter the nursery environment for pest control (see chapters 11 and 12, this volume). Judicious use of irrigation water can reduce preemergence damping-off by speeding seed germination and germinant emergence. Irrigation can also be employed to cool seedbeds and thereby diminish certain fusaria-caused root rots, though withholding water helps check gray mold outbreaks. Irrigation can prevent high moisture stress, which predisposes Douglas-fir seedlings to *Phomopsis* canker (*Diaporthe lokoyae* Funk) [29]. Using irrigation water to prevent frost damage controls disease indirectly because frost-damaged seedlings are prone to storage molds [21].

19.4.4.4 Crop rotation

Many nongovernment and (particularly) transplant nurseries rotate various agricultural or ornamental crops with forest-nursery seedlings. However, pest populations can build up on these other crops and subsequently damage tree seedlings. Seedling damage from root-feeding larvae of vine weevils, following a crop of strawberries, would not be unusual. Constant monoculture of tree seedlings also can lead to pest buildup. Conditions permitting, seedling species should be rotated in such nurseries.

19.4.4.5 Mulches, seed coverings, and soil amendments

Pests can be brought into nurseries on mulches and seed covers. *Meria laricis* Vuill. can be introduced on larch needles, *S. strobilinus* on pine needles, and the seed pathogen *Caloscypha fulgens* (Persoon) Boudier in forest duff. Seed coverings such as sand, grit, or peat mulches can contain pathogenic *Pythium* and *Fusarium*. New sources of these materials should be assayed for such fungi.

Mulches can also improve pathogen survival. Bloomberg [7] showed decomposing sawdust (from mulch) readily yielded the *Fusarium* root-rot pathogen of Douglas-fir seedlings. Cutworms (species of *Euxoa* and *Peridroma*) often hide in sawdust mulches. Besides pest problems, decomposing sawdust or other organic materials may produce phytotoxic compounds or tie up nutrients, resulting in seedling nutrient deficiencies [15].

Pests also enter nurseries in soil amendments. In British Columbia, *Xiphinema bakeri* nematodes were brought into the

Duncan Nursery in infested sand used to lighten a clay soil. Peat is another good source of pest fungi and insects. All soil amendments, especially those from new sources or suppliers, should be assayed for pests before being used.

19.4.4.6 Equipment and transplants

Pests can be introduced both within and among nurseries via equipment or transplants. To prevent the spread of pathogens and insects among nurseries, all used equipment should be steam cleaned or otherwise sanitized before moving it to another site. Preventing within-nursery spread of pests on equipment is nearly impossible.

Pests are commonly introduced in nurseries by transplants. A classic example in the Northwest is *Phytophthora* root rot [19]. Transplants should not be moved among nurseries, and pest-infested trees should not be transplanted to "clean" areas within a nursery.

19.4.5 Pesticides

All Northwest nurseries use pesticides, mainly (in order of importance) herbicides, fungicides, and insecticides. Use at a particular nursery depends upon the economic importance of pests and the general philosophy of the manager and staff towards pesticides. Pesticides are applied as seed dressings, sprays, and drenches or as soil fumigants. The major pesticides used in the Northwest, application methods and timing, rates, safety, storage, and other pertinent information are contained in publications by Berg [4], Capizzi et al. [12], Hamel [18], and MacSwan and Koepsell [23] for U.S. nurseries and by Miller and Craig [25] for those in British Columbia. The last-mentioned publication has good sections on safety, handling, formulations, and application equipment as well as a pesticide-terms glossary.

19.4.5.1 Seed treatments

Fungicides may be dusted or stuck onto seeds with methyl cellulose, for example, to protect them from damping-off both before and after germinants emerge. Captan, thiram, and benomyl are the fungicides most often used on seeds in the Northwest. The main drawback to fungicide seed treatment is that fungicide phytotoxicity often outweighs the beneficial effects of disease control, particularly when damping-off losses are static [22]. The difficulty of finding the ideal fungicide—one high in fungi toxicity but low in phytotoxicity—is illustrated in studies by Carlson [13] and Vaartaja [39], who tested 326 fungicides in total and found only six new materials suitable for use on tree seeds. Interestingly, four of these contained the two old standbys, captan and thiram.

Fungicide seed treatments have other disadvantages: (1) they often cause sticking or clumping of seeds and impede sowing; (2) because their fungicidal spectrum is too narrow, they are ineffective when several pathogens are present (a common situation); (3) they are too short-lived to protect seeds and germinants throughout the entire preemergence-through-postemergence susceptibility period; and (4) they may lose their effectiveness by killing susceptible strains of the pathogen, leaving only resistant ones. This is evident when the effectiveness of a newly used fungicide gradually declines over a few years. Such a problem might be overcome by changing fungicides every few years or by using a mixture of materials each year.

Past research on fungicide treatment has overlooked the fact that conifer seeds, unlike many agricultural seeds, vary greatly in both germination capacity and speed, even within seedlots. Consequently, some seedlots, which germinate well and rapidly, may need little or no fungicide or may simply require a different fungicide than a seedlot that germinates poorly and slowly. Fungicide prescriptions are needed for seedlots, based on seedlot characteristics such as seedcoat damage, germination speed, and capacity [9, 35]. The problem here is that because slower germinating seeds require longer protection, they are also subjected to prolonged fungicide phytotoxicity.

In sum, to date, no one has been able to find the magic chemical to protect conifer seeds from damping-off. Moreover, "bad" seeds cannot be made better by treating them with some chemical. At present, the best solution for most Northwest nurseries seems to be to collect high-quality seeds, which probably do not need a presowing fungicide treatment.

19.4.5.2 Sprays

In most Northwest nurseries, fungicides and insecticides are applied as sprays. Generally, fungicide sprays are for shoot diseases such as gray mold on Douglas-fir, and insecticide sprays are for aboveground feeding insects. Sprays usually give satisfactory results provided the prerequisites (e.g., formulation, timing, application method) are correct; otherwise, pests may not be controlled, phytotoxicity may result, or both.

Phytotoxicity commonly arises when materials are applied to seedling species or age classes for which phytotoxicity data are lacking. Changing formulations can cause problems too, especially for emulsifiable formulations where the emulsifier, rather than the active ingredient, may be phytotoxic. When phytotoxicity risk is unknown, apply the material to a few seedlings for one to several days before operational spraying. Compared to other application methods, sprays are the most likely to contaminate nearby crops or nursery workers. Precautions should be taken to prevent pesticide drift from sprays.

19.4.5.3 Drenches

Fungicides, insecticides, and nematicides are often applied as drenches to the soil. Regardless of the pest, drenches are generally used only in desperation and are frequently unsatisfactory for large-scale applications.

In the Northwest, fungicide drenches have varied from being totally ineffective—for controlling spruce damping-off [34]—to being somewhat promising—at very high dosages for *Fusarium* root rot of Douglas-fir [8]. The inconsistencies are probably due to the short soil life of drenches resulting from chemical or microbial decomposition and dilution or from chemical binding. Inherently, all pesticide drenches pose the risk of phytotoxicity, which is especially serious for postplanting applications but also can occur with preplanting applications unless soil conditions are suitable and sufficient time is allowed for drenches to disappear before sowing or planting.

19.4.5.4 Soil fumigation

Soil fumigation with 67% methyl bromide/33% chloropicrin is the most widely practiced pest-control procedure in U.S. Northwest nurseries (Table 1). In contrast, soil fumigation is seldom used in British Columbia nurseries, except to initially eradicate pests (including weeds) in nurseries being established on formerly agricultural land. This difference may occur (1) because the methyl bromide formulation registered in Canada—only 2% chloropicrin—provides inadequate control, (2) because conventional herbicide practices may be better established, thereby eliminating the need to fumigate soil, or (3) because fumigation is expensive. Soil fumigation costs $1,000 per acre ($2,500 per ha) and controls (usually) weeds, diseases, and insects. Weed control alone (with herbicides) costs $200 per acre, leaving a theoretical $800 per acre for controlling diseases and insects. Sites requiring this level of maintenance might best be abandoned or paved for a container nursery! Less drastic courses include using the $800 per acre "savings" to buy or clear an additional acre of land that is pest free—the enlarged nursery might better lend itself to implementation of alternative pest-management strategies—or to correct conditions such as poor drainage that favor pests.

Proponents of fumigation argue, on the other hand, that it is a valuable insurance measure (an extremely difficult position to refute), that the high value of seedlings warrants (equals or exceeds) the fumigation costs, and that the nature of certain pests (e.g., sclerotia-forming pathogens) makes fumigation the only practical control. The cost competitiveness of soil fumigation might be improved with fumigants other than methyl bromide [5, 40] that require less expensive application procedures. Solar pasteurization [14, 26] also could be useful for controlling soil-borne pests in Northwest nurseries.

19.4.5.5 Nontarget effects

Though primarily used to control pests, pesticides also affect other organisms, including seedlings [1].

Methyl bromide is the most commonly used biocide in U.S. Northwest nurseries (Table 1), but this and other soil fumigants are seldom used in British Columbia. Soil fumigation has the undesirable effect of eliminating (at least temporarily) ectotrophic mycorrhizal fungi; this can lead to stunted, phosphorus-deficient seedlings. Cool, wet soils can retain fumigants, which can then damage seedlings. Research is needed to determine how repeated long-term fumigation of soil affects soil microbe populations, including soil-borne pathogens.

Pesticide phytotoxicity to seedlings is not uncommon and may vary depending upon many factors including seedling development stage [37]. Under certain circumstances, such as low incidence of disease, pesticide phytotoxicity may outweigh the beneficial effects of the pesticide. For example, in British Columbia, seeds are not treated with fungicides because the chemicals kill as many seeds and germinants as they protect from damping-off [22]. Pesticide damage, even though not visible, may predispose seedlings to pathogens—for example, the apparently increased susceptibility of Douglas-fir to *Fusarium* top blight following herbicide applications [36]. Not all secondary effects are undesirable, however; herbicides that can reduce Collembola populations eliminate the need for insecticide spraying [24].

19.5 Animal Management

Management of animals is largely by exclusion or eradication [18, 25, 30]. Large animals such as deer can be excluded by animal-proof fencing around the nursery. Though trying to keep small animals and birds out of the nursery is impractical, some physical methods, such as covering seedbeds with window screening, may be used to protect recently sown seeds and germinants from birds, mice, and squirrels. The habitat of windbreaks and areas around the nursery should not favor rodents, but encouraging predatory birds might be worthwhile. In extreme situations, traps, shooting, or poisoning might be used to eliminate such pests. However, many poisons may no longer be considered environmentally acceptable or safe.

19.6 Recommendations for Future Research

To improve pest management in Northwest bareroot nurseries, we now need to:

- Determine effects of various cover crops on pest enhancement
- Assess long-term effects of soil fumigation on beneficial and harmful soil microorganisms
- Investigate possible use of solar pasteurization of soil for pest control
- Treat seeds with fungicides for disease control on the basis of prescriptions for specific seedlots
- Use more biological control agents
- Reduce pesticide usage with better application techniques
- Determine effects of diseases such as damping-off on seedling distribution and, in turn, on seedling quality
- Use auditory and olfactory repellents for protecting seeds and seedlings from animals
- Design better survey and predictive techniques
- Coordinate efforts to identify and manage pests and disseminate information about them to nurseries

References

1. Abrahamson, L. P. 1980. Pesticides in forest tree nurseries—primary and secondary effects. Pages 191-204 *in* Proc., North American forest tree nursery soils workshop (L. P. Abrahamson and D.H. Bickelhaupt, eds.). State Univ. New York, Coll. Environ. Sci. and Forestry, Syracuse.
2. Aldhous, J. R. 1972. Nursery practice. Her Majesty's Stationery Office, London. Forestry Commun. Bull. 43. 184 p.
3. Armson, K. A., and V. Sadreika. 1979. Forest tree nursery soil management and related practices. Ontario Ministry of Natural Resources, Toronto. 179 p.
4. Berg, G. L. (ed.). 1982. Farm chemicals handbook. Meister Publ. Co., Willoughby, Ohio. (Published annually.)
5. Bloomberg, W. J. 1965. The effect of chemical sterilization on the fungus population of soil in relation to root disease of Douglas-fir seedlings. Forestry Chronicle 41:182-187.
6. Bloomberg, W. J. 1968. Corky root of Douglas-fir seedlings. Can. Dep. of Forestry and Rural Dev. Bi-Monthly Res. Notes 24(1):8.
7. Bloomberg, W. J. 1976. Distribution and pathogenicity of *Fusarium oxysporum* in a forest nursery soil. Phytopathology 66:1090-1092.
8. Bloomberg, W. J., and W. Lock. 1974. Importance of treatment timing in the control of *Fusarium* root rot of Douglas-fir seedlings. Phytopathology 64:1153-1154.
9. Bloomberg, W. J., and J. Trelawny. 1970. Effect of thiram on germination of Douglas-fir seed. Phytopathology 60:1111-1116.
10. Boyce, J. S. 1948. Forest pathology. McGraw-Hill, New York. 550 p.
11. California Forest Pest Control Action Council. 1971. Forest pest conditions in California—1970. California Dep. of Forestry, Sacramento. 25 p.
12. Capizzi, J., G. Fisher, H. Homan, C. Baird, A. Retan, and A. Antonelli. 1982. Pacific Northwest insect control handbook. Ext. Services of Oregon State Univ., Corvallis; Washington State Univ., Pullman; and Univ. of Idaho, Moscow. (Published annually.)
13. Carlson, L. W. 1972. Forest tree nursery disease control. Pages 111-114 *in* Proc., Joint meeting, Western Forest Nursery Council and Intermountain Forest Nurserymen's Assoc., Olympia, Washington.
14. Cooley, S. 1981. Progress report: solar pasteurization of nursery soil. Forest Pest Management, U.S.D.A. Forest Serv., Portland, Oregon. 17 p.
15. Davey, C. B., and H. H. Krause. 1980. Functions and maintenance of organic matter in forest nursery soils. Pages 130-165 *in* Proc., North American forest tree nursery soils workshop (L. P. Abrahamson and D. H. Bickelhaupt, eds.). State Univ. New York, Coll. Environ. Sci. and Forestry, Syracuse.
16. Dingley, J. M., and J. W. Gilmour. 1972. *Colletotrichum acutatum* Simmds. f. sp. *pinea* associated with terminal crook disease of *Pinus* spp. New Zealand J. Forestry Sci. 2:192-201.
17. Furniss, R. L., and V. M. Carolin. 1977. Western forest insects. U.S.D.A. Forest Serv., Washington, D.C. Misc. Publ. 1339. 654 p.
18. Hamel, D. R. 1981. Forest management chemicals. U.S.D.A. Forest Serv., Washington, D.C. Agric. Handb. 585. 512 p.
19. Hansen, E. M., P. B. Hamm, A. J. Julis, and L. F. Roth. 1979. Isolation, incidence and management of *Phytophthora* in forest tree nurseries in the Pacific Northwest. Plant Disease Reporter 63:607-611.

20. Hepting, G. H. 1971. Diseases of forest and shade trees of the United States. U.S.D.A. Forest Serv., Washington, D.C. Agric. Handb. 306. 658 p.
21. Hopkins, J. C. 1975. A review of moulding of forest nursery seedlings in cold storage. Can. Forestry Serv., Pacific Forest Res. Centre, Victoria, B.C. Rep. BC-X-128. 16 p.
22. Lock, W., J. R. Sutherland, and L. J. Sluggett. 1975. Fungicide treatment of seeds for damping-off control in British Columbia forest nurseries. Tree Planters' Notes 26(3):16-18, 28.
23. MacSwan, I. C., and P. A. Koepsell. 1982. Pacific Northwest plant disease handbook. Ext. Services of Oregon State Univ., Corvallis; Washington State Univ., Pullman; and Univ. of Idaho, Moscow. (Published annually.)
24. Marshall, V. G., and S. Ilnytzky. 1976. Evaluation of chemically controlling the collembolan *Bourletiella hortensis* on germinating Sitka spruce and western hemlock in the nursery. Can. J. Forest Res. 6:467-474.
25. Miller, A. V., and S. M. Craig. 1979. Handbook for pesticide applicators and pesticide dispensers. Province of British Columbia, Ministry of Environ., Victoria. 233 p.
26. Old, K. M. 1981. Solar heating of soil for the control of nursery pathogens of *Pinus radiata*. Australian Forest Res. 11:141-147.
27. Peterson, G. W., and R. S. Smith, Jr. 1975. Forest nursery diseases in the United States. U.S.D.A. Forest Serv., Washington, D.C. Agric. Handb. 470. 125 p.
28. Smith, R. S., Jr. 1975. Charcoal root disease. Pages 11-13 *in* Forest nursery diseases in the United States (G. W. Peterson and R. S. Smith, Jr., tech. coordinators). U.S.D.A. Forest Serv., Washington, D.C. Agric. Handb. 470.
29. Smith, R. S., Jr. 1975. *Phomopsis* canker of Douglas-fir. Pages 42-44 *in* Forest nursery diseases in the United States (G. W. Peterson and R. S. Smith, Jr., tech. coordinators). U.S.D.A. Forest Serv., Washington, D.C. Agric. Handb. 470.
30. Stoeckeler, J. H., and G. W. Jones. 1957. Forest nursery practice in the Lake States. U.S.D.A. Forest Serv., Washington, D.C. Agric. Handb. 110. 124 p.
31. Stoeckeler, J. H., and P. E. Slabaugh. 1965. Conifer nursery practice in the Prairie-Plains. U.S.D.A. Forest Serv., Washington, D.C. Agric. Handb. 279. 93 p.
32. Sutherland, J. R., and R. L. Anderson. 1980. Seedling disease and insect problems related to nursery soil conditions in North America. Pages 182-190 *in* Proc., North American forest tree nursery soils workshop (L. P. Abrahamson and D. H. Bickelhaupt, eds.). State Univ. New York, Coll. Environ. Sci. and Forestry, Syracuse.
33. Sutherland, J. R., and T. G. Dunn. 1970. Nematodes in coastal British Columbia forest nurseries and association of *Xiphinema bakeri* with a root disease of Douglas-fir seedlings. Plant Disease Reporter 54:165-168.
34. Sutherland, J. R., and W. Lock. 1977. Fungicide-drenches ineffective against damping-off of Sitka and white spruces. Can. Forestry Serv., Dep. of Environ. Bi-Monthly Res. Notes 33(1):6-7.
35. Sutherland, J. R., W. Lock, and L. J. Sluggett. 1975. Damping-off in British Columbia forest nurseries: control trials with fungicides applied to different quality seeds. Can. Forestry Serv., Pacific Forest Res. Centre, Victoria, B.C. Rep. BC-X-125. 20 p.
36. Sutherland, J. R., and E. Van Eerden. 1980. Diseases and insect pests in British Columbia forest nurseries. B.C. Ministry of Forests/Can. Forestry Serv., Victoria. Joint Rep. 12. 55 p.
37. Sutherland, J. R., T. A. D. Woods, and S. Ilnytzky. 1977. Pesticide phytotoxicity studies with seeds, germinants and seedlings of British Columbia conifers. Can. Forestry Serv., Pacific Forest Res. Centre, Victoria, B.C. Rep. BC-X-165. 7 p.
38. Triebwasser, M.E., and D. C. Overhulser. 1981. The cranberry girdler in conifer nurseries of western Washington and Oregon. Pages 80-83 *in* Proc., Intermountain Forest Nurserymen's Assoc. and Western Forest Nursery Assoc. combined meeting, Boise, Idaho, Aug. 12-14, 1980. U.S.D.A. Forest Serv., Intermountain Forest and Range Exp. Sta., Ogden, Utah. Gen. Tech. Rep. INT-109.
39. Vaartaja, O. 1956. Screening fungicides for controlling damping-off of tree seedlings. Phytopathology 46:387-390.
40. van den Driessche, R. 1963. Partial sterilization of Douglas-fir seedbeds with formalin and chloropicrin. Forest Sci. 9:330-334.
41. van den Driessche, R. 1980. Health, vigour and quality of conifer seedlings in relation to nursery soil fertility. Pages 100-120 *in* Proc., North American forest tree nursery soils workshop (L. P. Abrahamson and D. H. Bickelhaupt, eds.). State Univ. New York, Coll. Environ. Sci. and Forestry, Syracuse.
42. Wilde, S. A. 1958. Forest soils. The Ronald Press, New York. 537 p.

Chapter 20
Mycorrhiza Management in Bareroot Nurseries

R. Molina and J. M. Trappe

Abstract

Mycorrhizae, or "fungus-roots," involve the intimate association of plant roots with specialized soil fungi. Forest-tree seedlings depend upon their mycorrhizae for adequate nutrient uptake; those lacking mycorrhizae can be severely stunted and their growth in newly sown beds uneven. Nursery managers should avoid practices that cause mycorrhiza deficiency. For example, because soil fumigation destroys mycorrhizal fungus populations, alternative pest-control measures should be substituted whenever possible. Careful seedling manipulations and handling also will reduce damage to mycorrhizae. Soil disturbance should only be necessary to meet management goals so as to minimize disruption of delicate fungus-soil networks. Fertilization can both foster and inhibit mycorrhiza development; appropriate levels are best determined by experience. The integrated use of mycorrhiza-management tools with other cultural practices and the potential use of selected beneficial fungi for mycorrhizal inoculation of seedlings will help ensure the successful production of vigorous planting stock.

20.1 Introduction

Nursery managers have long recognized the importance of well-developed root systems for producing resilient planting stock. We now realize that adequate development of mycorrhizae on seedling roots is equally important—indeed, essential—for healthy seedling growth in the nursery and desired performance after outplanting.

In this chapter, we focus on the major benefits seedlings derive from mycorrhizae, how environmental factors and management practices affect mycorrhizal fungus populations and subsequent development of mycorrhizae, and methods to foster mycorrhiza development in bareroot nurseries. We also provide an update on prospects for artificially inoculating seedlings with selected, highly beneficial mycorrhizal fungi.

20.2 Mycorrhizae Defined

Mycorrhiza is a Greek word meaning "fungus-root." Nearly all land plants form some type of mycorrhiza with specialized soil fungi. Mycorrhizal associations are classic examples of mutualistic symbioses because the fungus and host plant depend on each other for survival in natural ecosystems.

The mycorrhizal fungus is best considered as a far-reaching extension of the root system. A fine network of fungus threads (hyphae) explores and extracts nutrients from a volume of soil far beyond the bounds of the roots' capabilities. Many of these nutrients are translocated through the hyphal network to the mycorrhizae, where they are released to the roots for host utilization. In exchange, the host serves as primary energy source for the fungus, providing simple sugars and possibly other compounds derived from host photosynthates.

Several different types of mycorrhizae are known, but ectomycorrhizae and vesicular-arbuscular (VA) mycorrhizae are the most common and most relevant to trees. Ectomycorrhizae are the most important to western bareroot nurseries because all members of the Pinaceae—true fir (*Abies*), larch (*Larix*), spruce (*Picea*), pine (*Pinus*), Douglas-fir (*Pseudotsuga*), and hemlock (*Tsuga*) spp.—form ectomycorrhizae. Members of the Fagaceae [e.g., beech (*Fagus*) and oak (*Quercus*) spp.] and Betulaceae [e.g., birch (*Betula*) and alder (*Alnus*) spp.] as well as madrone (*Arbutus*) and basswood (*Tilia*) spp. also form ectomycorrhizae. Most other land plants form VA mycorrhizae. The Cupressaceae (cedars) and Taxodiaceae (including redwoods) are the most important in this regard in western forest nurseries; hardwoods such as sweetgum (*Liquidambar*), maple (*Acer*), and yellow-poplar (*Liriodendron*) spp. are important VA mycorrhizal hosts in eastern nurseries. Alder, eucalyptus

In Duryea, Mary L., and Thomas D. Landis (eds.). 1984. *Forest Nursery Manual: Production of Bareroot Seedlings*. Martinus Nijhoff/Dr W. Junk Publishers, The Hague/Boston/Lancaster, for Forest Research Laboratory, Oregon State University, Corvallis. 386 p.

(*Eucalyptus*), willow (*Salix*), and poplar (*Populus*) are among the genera that readily form both ectomycorrhizae and VA mycorrhizae.

20.2.1 Ectomycorrhizae

Several different forms of ectomycorrhizae are shown in Figures 1 to 6. The fungi colonize the surfaces of the short feeder roots, often forming a thick mantle around them. Ectomycorrhizae can frequently be seen with the unaided eye or a hand lens because many are white or brightly colored. Similarly, if ectomycorrhizae are abundant, a dense moldlike fungal growth is visible in the soil when seedlings are lifted. When examined microscopically in cross section (Figs. 9 and 10), the fungus is seen to enter the root, penetrating between the cortical cells to form an interconnecting network called the Hartig net. It is within this extensive hypha-root cell contact zone that nutrient exchange occurs.

The fungi also produce plant hormones that stimulate root branching and elongation, thereby increasing the root's absorptive surface. Branching patterns of ectomycorrhizae are often host determined and are therefore characteristic of the host-seedling species. For example, pine ectomycorrhizae are typically forked or dichotomously branched (Figs. 2, 4, and 21), whereas other hosts may predominantly form structures that are pinnate (Figs. 5 and 6), coralloid (Fig. 3), tuberculate, or variably branching (Fig. 1). It is important to realize that thousands of species of mushroom, puffball, and truffle fungi (higher Basidiomycetes and Ascomycetes) (Figs. 13, 14, 15, 17, 19, and 20) can form mycorrhizae on a large array of host species. Thus, the physical and physiological diversity of ectomycorrhizal forms is enormous.

20.2.2 Ectendomycorrhizae

A subtype of ectomycorrhizae is the ectendomycorrhiza. Because the mantle it forms is thin and translucent, feeder roots display the brown color of underlying epidermal cells. Ectendomycorrhizae branch like ectomycorrhizae but lack root hairs; in addition to forming a Hartig net, the fungi also penetrate scattered cortical cells (Figs. 11 and 12). Small Discomycetes (cup fungi), which form ectendomycorrhizae [3], are often common and beneficial in temperate bareroot nurseries [12].

20.2.3 Vesicular-arbuscular mycorrhizae

Vesicular-arbuscular mycorrhizae do not differentiate morphologically from nonmycorrhizal roots and are therefore not discernible by the unaided eye. Roots must be selectively stained [39] to highlight the fungus within and then examined microscopically to determine its presence and structure. In roots thus prepared (Figs. 7 and 8), hyphae of the VA mycorrhizal fungus can be seen to ramify throughout the roots and often to form the characteristic vesicles and arbuscules for which the mycorrhiza is named. Vesicles (Fig. 7) are storage organs containing carbohydrates and also serve as reproductive structures. Arbuscules (Fig. 8) are very finely branched, short-lived, intracellular structures which partake in nutrient exchange. Although these fungi are often said to "infect" the roots, they cause no apparent harm.

As with ectomycorrhizal fungi, the main portion of the VA fungus lies outside the root, exploring the surrounding soil for nutrients and translocating them to the roots. Unlike ectomycorrhizal fungi, however, VA fungi do not produce large mushroom-like reproductive structures. Instead, they produce large, mostly soil-borne, globose spores (Fig. 16). Furthermore, VA fungus spores cannot be dispersed for long distances by air movement, as can mushroom spores; spore dispersal is limited primarily to mechanical movement of soil. The consequences of this important feature on VA mycorrhizal development in nurseries will be considered later (see 20.3.4, 20.4.2).

20.2.4 Benefits of mycorrhizae

In addition to greatly enhanced uptake of nutrients, especially phosphorus, mycorrhizae confer other benefits to their hosts. They can take up water [5] and increase drought resistance of young seedlings [38]. Some mycorrhizal fungi can also detoxify certain soil toxins [53] or enable seedlings to withstand high soil temperatures [22] or extreme acidity [18]. Of practical importance to nursery management, some mycorrhizal fungi can protect roots against certain pathogens [17]. For example, the mycorrhizal fungus *Laccaria laccata* Scop. ex Fr. has been shown to protect feeder roots from *Fusarium* infection [44].

Historically, the absolute dependence of forest trees on their mycorrhizae was repeatedly demonstrated when ectomycorrhizal Pinaceae were introduced to the Southern Hemisphere. Only when accompanied by their associated ectomycorrhizal fungal symbionts could these exotics survive and thrive (see [32, 33]). A classic example was Monterey pine (*Pinus radiata* D. Don). Initial attempts to establish pine seeding nurseries in Australia and New Zealand failed. When mycorrhizal fungi native to pine stands were unknowingly introduced into seedling beds, however, seedlings grew vigorously and survived outplanting. Today, the pine plantations of Australia and New Zealand are among the world's more productive forests.

Similar examples of mycorrhizal dependency were evident in afforestation attempts in the treeless grasslands of the United States [29] and on the steppes of Russia [9]. More recently, Schramm [42] and Marx [18, 21] have shown the need for mycorrhizal planting stock inoculated with specifically adapted fungi for tree establishment on strip-mined and other severely disturbed sites. Thus, tree seedlings must be accompanied by their mycorrhizal fungi when planted in areas lacking suitable mycorrhizal fungi.

20.3 Mycorrhiza Management

Because each nursery is unique, each must develop its own specific mycorrhiza-management strategies through experimentation and good recordkeeping. Examples given here are general cases. In no case has research been intensive enough to provide more than fragmentary understanding of what is taking place in the soil.

20.3.1 Mycorrhiza development and occurrence in bareroot nurseries

Mycorrhiza development—or the lack of it—in bareroot nurseries is affected by several biologic and environmental factors, many of which we cannot control.

Nurseries established in forest zones or surrounded by ectomycorrhizal hosts usually produce seedlings with abundant and diverse ectomycorrhizae. If nursery beds are not fumigated, all seedlings will develop mycorrhizae early in the first growing season. Even if beds are fumigated, regular and prolific fruiting of sporocarps in neighboring forests provides abundant spore inoculum, as does fruiting of sporocarps in established nursery beds. Under these conditions, wherever spores enter the soil, ectomycorrhizae can begin developing with the first production of feeder roots 6 to 10 weeks after germination and continue developing through the growing season as fungi extend into surrounding soil and colonize roots of adjacent seedlings. By the end of the first year after fumigation, most seedlings will usually be mycorrhizal. During the second growing season, nearly all seedlings will be abundantly mycorrhizal. Thus, a rich supply of natural fungus inoculum will promote the early development of ectomycorrhizae needed to ensure uniformly healthy planting stock.

In contrast, nurseries developed away from native forests or on new ground with no history of ectomycorrhizal hosts can experience mycorrhiza deficiency, resulting in serious financial and reforestation setbacks. Such seedlings are stunted, chlorotic, and severely nutrient deficient. Mycorrhiza deficiency symptoms may persist well into the second year; even after recovery, seedling size may vary considerably within the seedbed.

Trappe and Strand [52] report a striking example of this situation in Oregon's Willamette Valley (Fig. 18). In a new nursery established on fumigated, formerly agricultural land, the first crop of Douglas-fir [*Pseudotsuga menziesii* (Mirb.) Franco] seedlings exhibited an unexpected, severe, phosphorus deficiency not detected by soil analysis. Only in the second year, after natural inoculation by wind-blown spores, did many seedlings recover and begin to grow; others remained stunted through the second growing season. Although other environmental and biologic factors can cause stunting or nutrient deficiency symptoms, such symptoms are characteristic of mycorrhiza deficiency. When they appear, nursery managers should carefully examine seedling roots or have them evaluated by an expert.

Different tree species vary in susceptibility to mycorrhiza deficiency. Douglas-fir and true firs appear especially mycorrhiza dependent and show symptoms of mycorrhiza deficiency more quickly than pines.

Although effects of management practices on mycorrhiza development will be discussed later in 20.3, soil fumigation as a cause of mycorrhiza deficiency deserves emphasis here. Properly applied fumigation with methyl bromide/chloropicrin gases usually eliminates mycorrhizal fungus populations along with targeted pests. Even in nurseries with large native fungus populations, fumigation can cause a lengthy delay in mycorrhiza development, resulting in substantial growth loss the first growing season. Availability of fungal spores for natural reinoculation of fumigated beds can be reduced during droughty years when mushroom production is low or when prolonged heavy rains wash spores from the air during the mushroom fruiting season [49].

Soil fumigation is particularly devastating to VA mycorrhizal fungi; we have observed that VA mycorrhizal redwoods and cedars especially suffer the consequences. Because VA fungus spores are not dispersed by air, once the population is eliminated, such spores are returned to fumigated beds only erratically through movement of spore-containing soil by machines and on shoes.

In most bareroot nurseries, root systems are dominated by a few nursery-adapted mycorrhizal fungi—in stark contrast to the hundreds of fungi common to even small areas of forest. By far the most common ectomycorrhizal fungi in bareroot nurseries are species of the genus *Thelephora*. *Thelephora terrestris* (Ehrh.) Fr. is especially common worldwide and fruits conspicuously at the base of seedlings (Fig. 20); its ectomycorrhizae are very smooth, usually a pale cream-brown (Fig. 21). The ectomycorrhizal fungi *Laccaria laccata* (Fig. 14) and *Inocybe lacera* (Fr.) Kummer (Fig. 19) and ectendomycorrhizal fungi also are common in Northwest nurseries. We have occasionally observed truffle fungi of the genus *Rhizopogon* (Fig. 15) and boletes of the genus *Suillus* (Fig. 17) to be common in a few nurseries. Nurseries surrounded by dense forest stands, such as the U.S.D.A. Forest Service Wind River Nursery in Washington, often harbor diverse ectomycorrhizae. Even in those nurseries, however, *Thelephora*, *Laccaria*, and ectendomycorrhizal species tend to predominate.

Results from the OSU Nursery Survey (see chapter 1, this volume) reaffirm many of the phenomena described above. Of the responding nurseries, about 75% report good to abundant ectomycorrhiza development on lifted seedlings. *Thelephora* and *Laccaria* species fruit most commonly, but a scattering of other species was observed. In several instances, respondents ascribed a recurring stunting problem for some tree species, particularly during the first season, to a lack of mycorrhizae. Thus, nursery managers must continue to be alert to the possibilities of mycorrhiza deficiency.

20.3.2 The nursery soil system

In general, nursery soils that are good for tree seedling growth are also good for mycorrhiza development on those seedlings. Good organic matter content, good tilth, good drainage, and adequate but not excessive nutrient levels are all associated with good mycorrhiza formation [33]. Much has been written about effects of fertilization on mycorrhiza formation, but only one conclusion can be drawn at present: because each soil is unique, levels of fertilizer which might promote mycorrhiza formation—or that might inhibit it—must be determined through experience.

It is when soil-management problems arise and when steps are taken to alleviate those problems that mycorrhizal populations are most often disrupted. This is because we often treat symptoms rather than causes of problems for lack of information on what is occurring in the soil. For example, spots of root rot may develop in a nursery because of localized poor drainage. If fungicides are applied to control the root rot, mycorrhizal fungi also may be decimated. The resulting mycorrhiza deficiency is then reflected by nutrient deficiency. If that symptom is treated by extra fertilization, mycorrhiza formation may be even further depressed. Once the nutrient-starved seedlings stop growing, their root systems are open to attack by yet other pathogens for lack of protection by mycorrhizal fungi. But the cause of the problem—poor drainage—remains uncorrected.

To minimize the chances for these kinds of deleterious chain reactions, the soil must be regarded as a system of interacting biological and physical components (see chapters 6, 7, and 9, this volume). Disrupt one component and all others are affected. Planned disruptions can be used to advantage in furthering management goals, but consistent success requires experience and care. The goals must be carefully defined because different goals may require different approaches. Managing mycorrhizae for an ultimate goal of good, uniform seedling growth in the nursery may entail different long-range plans and procedures than managing mycorrhizae for an ultimate goal of optimum survival of stock outplanted on stressful sites.

20.3.3 Uses and abuses of soil fumigation and other pesticides

As noted earlier, properly applied soil fumigation decimates beneficial organisms along with target pests. Loss of beneficial bacteria may be as serious as loss of mycorrhizal fungi. Seedlings will not grow satisfactorily until the mycorrhizal fungi, and possibly associated microorganisms, are replaced. Replacement of the VA mycorrhizal fungi required by cedars and redwoods can be slow and erratic; poor and nonuniform growth of these species on fumigated soil is common in western nurseries. Ectomycorrhizal fungi may be replaced more rapidly through aerially dispersed spores, but replacement depends both on weather favorable for spore production and on timing of the fumigation. As long as cold or dry weather does not inhibit spore production, beds fumigated in late summer or early autumn will be exposed to natural spore dispersal of mycorrhizal fungi in autumn; however, microbes antagonistic toward mycorrhizal fungi can establish concurrently in the fumigated beds. In contrast, beds fumigated in spring, just before sowing, will contain only mycorrhizal propagules that the fumigant missed.

The goal of optimum seedling growth in the nursery with minimal mycorrhiza management thus calls for minimizing fumigation. Pests should be controlled by alternative methods whenever possible (see chapter 19, this volume). When fumigation is deemed necessary, late summer is better timing than the spring in which seeds will be sown.

A more sophisticated goal than the passive approach outlined above is inoculation of planting stock with fungi selected to promote the best survival and growth in plantations. Successful inoculation can be expected to result in good, uniform growth of seedlings in the nursery as well. In this approach, soil could be fumigated to minimize populations of wild mycorrhizal fungi and microbial antagonists, preferably just before inoculation with a selected fungus; spring fumigation is preferable where weather permits. If late-summer or autumn fumigation is unavoidable, aggressive native mycorrhizal and antagonistic organisms may reinvade the soil over winter. In that case, only antagonist-resistant and highly competitive mycorrhizal fungi can be successfully inoculated. Evidence is also mounting that "helper" bacteria can be important in promoting inoculation success and that these bacteria can be cultured and inoculated along with the desired fungi [pers. commun., 14].

Selective biocides can be used instead of or in conjunction with soil fumigation. Herbicides do not generally appear to depress mycorrhiza formation and in some cases even seem to increase it, possibly by increasing exudation of sugars from roots [43, 51]. Weed control thus seems compatible with mycorrhiza management (see chapter 18, this volume). Insecticides and nematicides at field-application levels generally appear not to harm mycorrhizae or depress mycorrhiza formation. Some fungicides, on the other hand, are inhibitory, although those that inhibit ectomycorrhizal fungi do not necessarily affect VA fungi, and vice versa, at least not at concentrations occurring in soil after field applications [8, 30].

No matter how much is revealed by research about effects of pesticides on mycorrhizal fungi and mycorrhiza formation, it is important to realize that most pesticides used in nurseries are synthetic compounds that organisms have never before encountered. Moreover, a given chemical will not necessarily produce the same responses in all species of fungi or hosts or in all nurseries. Hence, first use of a chemical in a nursery should always be in trials of limited scope that include evaluation of its effects on mycorrhiza development.

Inoculating beds with mycorrhizal fungi selected for their strong protection of roots against pathogens is a potential alternative to routine fumigation or use of fungicides in some cases. A highly promising example is *Laccaria laccata*, a fungus with excellent potential for inoculation in western nurseries [34, 36]; this fungus strongly suppresses *Fusarium oxysporum* Schlecht. in nursery conditions [44].

20.3.4 Hazards of crop rotation

Switching rotations from ectomycorrhizal to VA mycorrhizal trees can produce mycorrhiza deficiency because the fungi of the two mycorrhizal types are totally different. For example, western redcedar (*Thuja plicata* D. Don), incense-cedar [*Calocedrus decurrens* (Torr.) Florin], or coastal redwood [*Sequoia sempervirens* (D. Don) Endl.], which are VA mycorrhizal, will encounter few or no propagules of VA mycorrhizal fungi in beds with a preceding crop of ectomycorrhizal Douglas-fir or pine. The deficiency will be further compounded if beds are fumigated before VA hosts are sown. Because spores of VA mycorrhizal fungi do not disperse by air, recolonization of beds can be slow and the tree crop accordingly poor. If crop rotation is deemed necessary for some reason, steps to inoculate beds with VA fungi are in order. Cover crops of VA mycorrhizal hosts may be useful in building up VA inoculum in a bed, provided that the fungi are initially present and that the cover crop is grown long enough for mycorrhizae to form on it.

The problem can occur in reverse when ectomycorrhizal hosts are sown in beds with a previous history of VA hosts (including most cover crops). If recolonization by aerially dispersed ectomycorrhizal fungi is rapid, adverse effects on seedling growth may be minimal; however, such rapid recolonization cannot be counted on. Again, inoculation of the beds with appropriate mycorrhizal fungi may prevent growth loss and unacceptable variation in seedling size within the bed.

20.3.5 Seedling manipulations

Procedures such as wrenching, undercutting, and mowing are not known to inhibit mycorrhiza formation, but they cost seedling energy. Such procedures are used either out of necessity or because their benefits to seedlings are believed to outweigh their costs. In the case of mycorrhizae, practices such as wrenching break up much of the nutrient-absorbing network of fragile hyphae that grow from the mycorrhizae into surrounding soil. These hyphae will regrow but at the cost of seedling-produced energy that would otherwise have been available to increase seedling size.

Procedures such as lifting, sorting, packing, storing, and transporting seedlings should be performed with care to minimize damage to the fine-root system. Mycorrhizae destroyed by rough handling, desiccation, or heating will have to be replaced at the planting site at a cost of seedling energy and nutrients.

20.3.6 Managing VA mycorrhizal hosts

Although most management considerations discussed for ectomycorrhizae also apply to VA mycorrhizal hosts, a somewhat different strategy is needed to foster VA mycorrhizae. For example, if certain nursery beds have been known to raise vigorous crops of cedars or if surveying indicates that the soil harbors good populations of VA fungi, nursery managers may want to use those beds exclusively for VA hosts and forego intermittent fumigation unless pathogens become a serious problem. If those beds are not to be used for a season or two to grow trees, they should be planted with a cover crop which will maintain the VA fungus populations as well as add good organic matter when plowed under. In fact, some VA mycorrhizal cover crops have been purposely planted to increase the populations of VA fungi, thus ensuring good mycorrhiza development on the next tree crop [11]. If pot-cultured VA fungus inoculum (see 20.4.2.2) is used in nursery beds to eliminate mycorrhiza deficiency or introduce more efficient fungus strains, subsequent fumigation should be avoided and intermittent cover crops planted to maintain the populations of the introduced fungi.

20.4 Mycorrhizal Inoculations in Bareroot Nurseries

Nursery managers may choose to artificially inoculate beds with mycorrhizal fungi either to eliminate potential or current mycorrhiza deficiencies or to improve outplanting performance of seedlings. Several procedures are available for introducing either ectomycorrhizal or VA mycorrhizal fungi. In this section we discuss several general methods and strategies for inoculating, first, ectomycorrhizal hosts, then VA mycorrhizal hosts. Refer to Mikola [33], Trappe [49], Marx [20], and Schenck [41] for detailed discussions of past and current technological advances in this field.

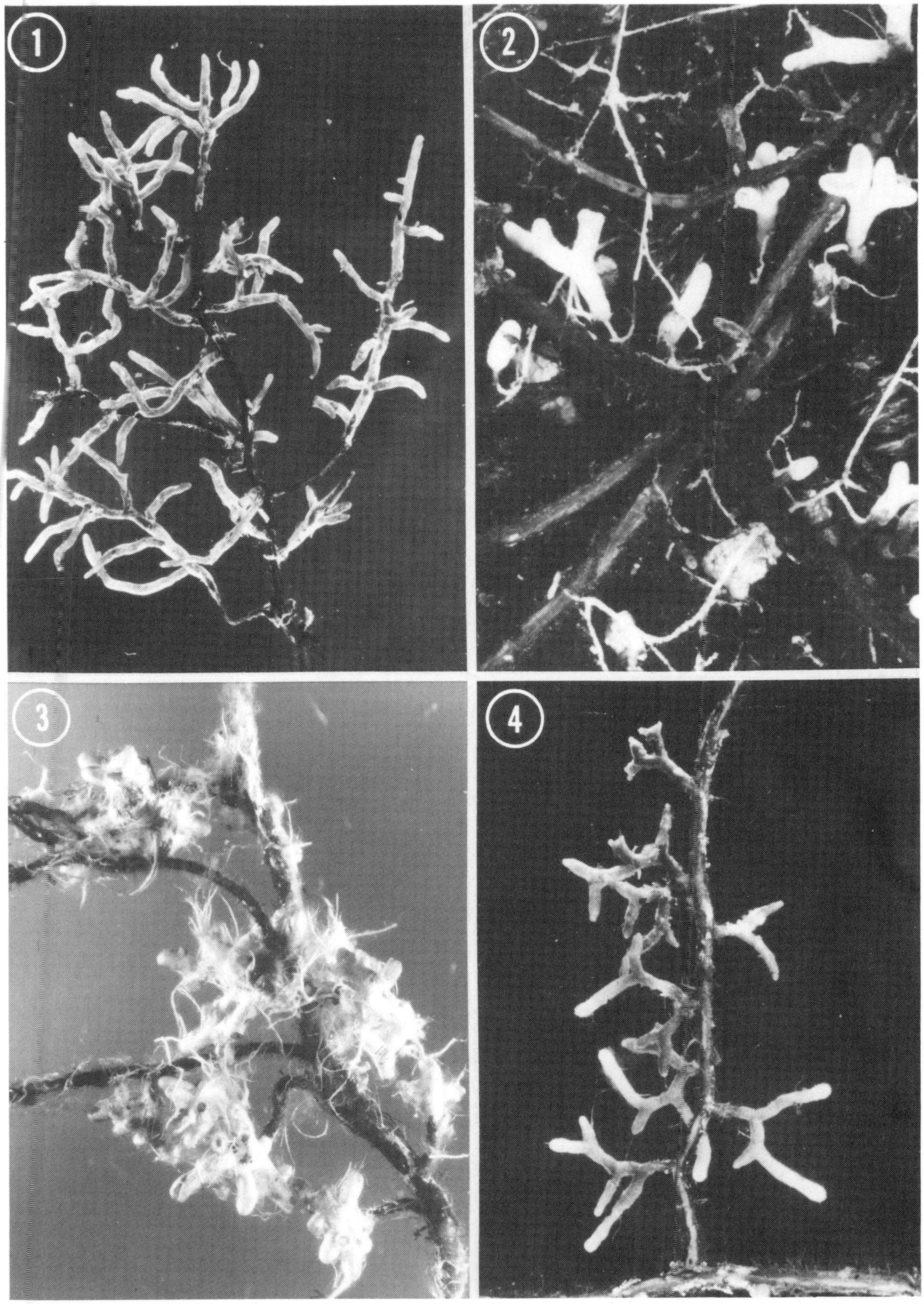

Figures 1-4. Ectomycorrhizal forms.

(1) Variably branched ectomycorrhizae formed between Sitka spruce [*Picea sitchensis* (Bong.) Carr.] and the fungus *Amanita muscaria*; 3.1x.

(2) Ectomycorrhizae formed with pine (*in vivo*); 3.3x. Note the characteristic forklike dichotomous branching and colonization of the soil by fungus strands called rhizomorphs.

(3) Compact, coralloid ectomycorrhizae formed between lodgepole pine (*Pinus contorta* Dougl. ex Loud.) and *Scleroderma laeve*; 4.2x. Note the strands of mycelia attached to the ectomycorrhizae (photo by B. Zak).

(4) Dichotomously branched ectomycorrhizae formed between western white pine (*Pinus monticola* Dougl. ex D. Don) and the fungus *Gastroboletus subalpinus*; 3.8x.

Figures 5-8. Douglas-fir ectomycorrhizae and fescue VA mycorrhizae.

(5, 6) Pinnately branched ectomycorrhizae formed between Douglas-fir and unknown fungi; 3.4x.

(7) Vesicles (V) within a selectively stained fescue root; 150x.

(8) Arbuscules (A) within a selectively stained fescue root; 600x.

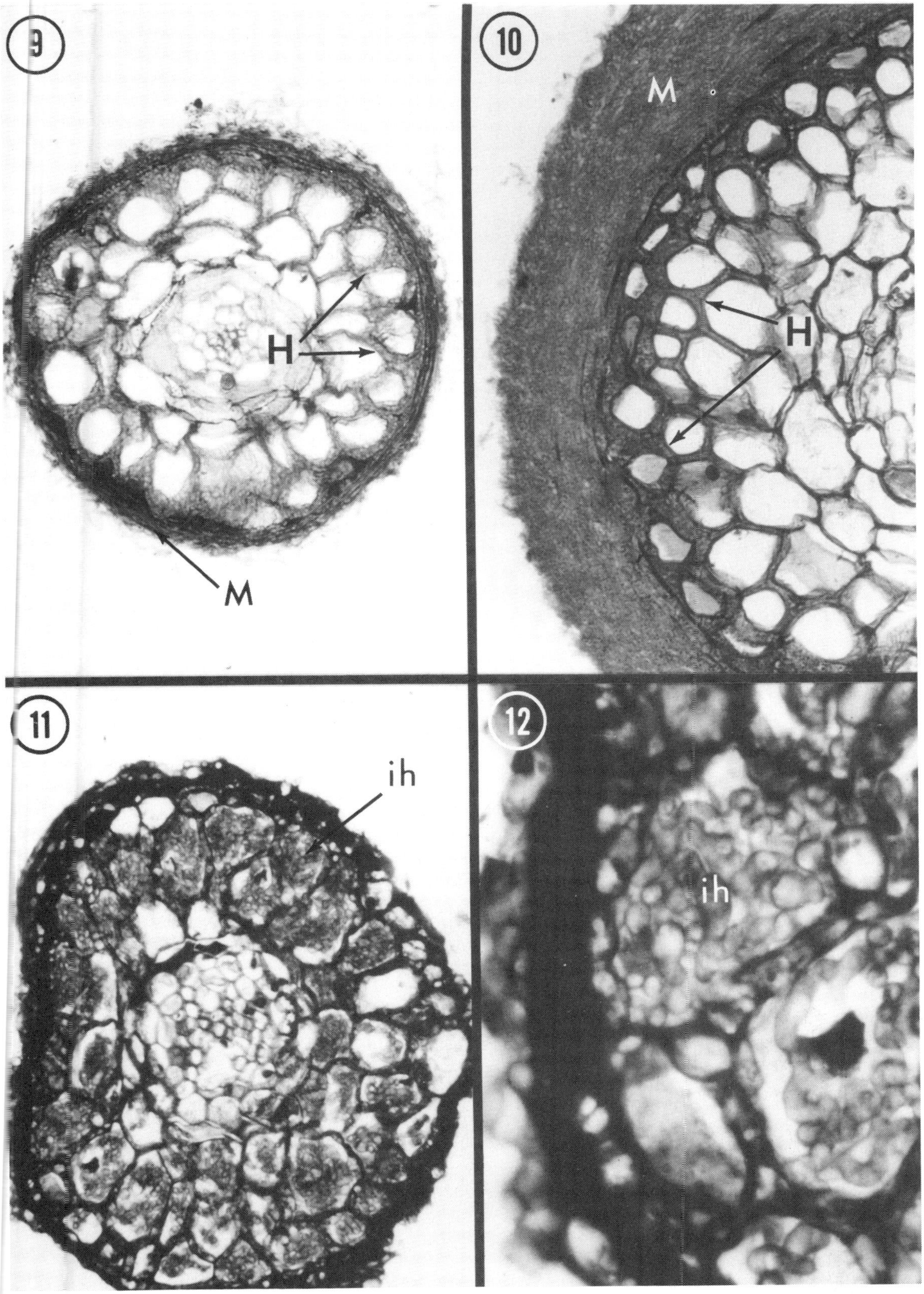

Figures 9-12. Cross sections of ectomycorrhizae and ectendomycorrhizae (H = Hartig net, M = mantle, ih = intracellular hyphae).

(9) Ectomycorrhizae formed between lodgepole pine and the fungus *Rhizopogon fuscorubens*; 50x.

(10) Ectomycorrhizae formed between western larch (*Larix occidentalis* Nutt.) and the fungus *Amanita muscaria*; 50x.

(11, 12) Pine ectendomycorrhizae; 160x and 630x, respectively. Note the abundant intracellular hyphae filling many cortical cells.

Figures 13-17. Ectomycorrhizal fungus fruiting bodies and VA fungus spores.

(13) ***Amanita muscaria*** **mushrooms, common throughout conifer woodlands.**

(14) ***Laccaria laccata*** **mushrooms fruiting in a Douglas-fir bed.**

(15) ***Rhizopogon vulgaris*** **truffles found fruiting in pine beds at the U.S.D.A. Forest Service Pine Nursery at Bend, Oregon.**

(16) Spores of the VA mycorrhizal fungus ***Glomus epigeaum*****; spores range from 75 to 140 μm in diameter.**

(17) Bolete mushrooms of the genus ***Suillus*** **found fruiting among pine seedlings; these mushrooms are recognized by the presence of pores, rather than gills, on their underside.**

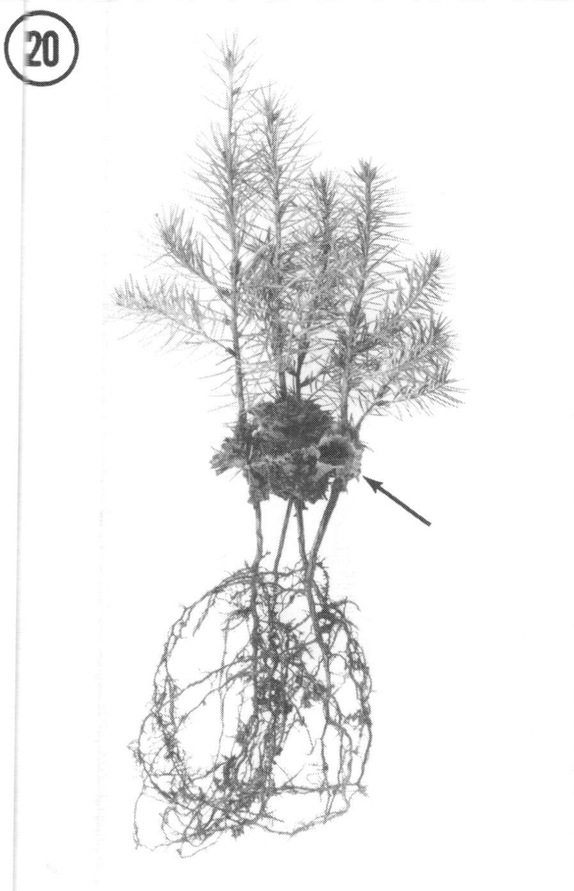

Figures 18-21. Mycorrhiza deficiency in Douglas-fir beds and two common ectomycorrhizal fungi in Northwest nurseries.

(18) Apparent mycorrhiza deficiency in 2+0 Douglas-fir beds. Scattered clumps of tall seedlings are mycorrhizal, whereas nonmycorrhizal seedlings remain severely stunted (see [52] for greater detail).

(19) Sporocarps of the ectomycorrhizal fungus *Inocybe lacera* found fruiting in the clumps of mycorrhizal Douglas-fir seedlings in Figure 18.

(20) Crustlike sporocarp of the very common ectomycorrhizal fungus *Thelephora terrestris* on the stems of Douglas-fir seedlings.

(21) Ectomycorrhizae formed between pine and *Thelephora*; 3.2x.

20.4.1 Ectomycorrhizal inoculation

Four primary sources of ectomycorrhizal inoculum are available: soil inoculum, mycorrhizal "nurse" seedlings interplanted in beds, spores and sporocarps, and pure fungus cultures. Each has advantages and disadvantages, so nursery managers should carefully weigh each option before selecting which approach best suits their needs.

20.4.1.1 Soil inoculum

The most commonly used and probably the most reliable inoculum is forest soil taken from beneath ectomycorrhizal hosts. About 10% by volume of soil inoculum is incorporated into approximately the top 10 cm of nursery-bed soil before sowing or transplanting [33]. Inoculation of new or fumigated beds by soil taken from established beds (beds previously supporting seedlings with good mycorrhiza development) is also feasible. The major drawback is the logistics of collecting and transporting the large quantities of soil needed. Unfortunately, weed seeds, rhizomes, and potential pathogens may be introduced along with the beneficial fungi. Nonetheless, soil inoculation continues to be regularly and successfully used in many areas of the world to promote healthy mycorrhiza development [33].

20.4.1.2 "Nurse" seedlings

Planting mycorrhizal "nurse" seedlings (mycorrhizal seedlings from which the fungus can spread and colonize new seedlings) or incorporating chopped roots of ectomycorrhizal hosts into nursery beds can provide a source of ectomycorrhizal fungus inoculum. However, mycorrhizal colonization may spread slowly and unevenly, the large "nurse" seedlings can interfere with cultural practices, and the risk of introducing unwanted pests remains.

20.4.1.3 Spores and sporocarps

Spores and chopped sporocarps (mushrooms, puffballs, and truffles) of some ectomycorrhizal fungi provide an excellent source of natural inoculum. The Gasteromycetes (puffballs and related fungi) with abundant spore masses offer better sources of large numbers of spores than the gilled fungi. Several recent studies have shown spores of the puffball fungus *Pisolithus tinctorius* Coker and Couch to be effective inoculum for southern pines [19, 20, 28]. Large quantities of spores are easily collected, and a variety of application methods, including dusting, spraying, coating seeds, and applying in a hydromulch, have been effective. Marx [20] reports acceptable levels of mycorrhiza formation, improved seedling growth in the nursery, and improved outplanting success following inoculation with *P. tinctorius* spores. *P. tinctorius* also fruits abundantly in many areas of the Northwest; however, our experimental nursery inoculations with its spores have produced erratic results [1]. Further research is needed before this plentiful source of natural inoculum can be recommended.

Good inoculation success has also been noted when seeds coated with dried *Rhizopogon* spores [45, 46, 47] or pulverized *Rhizopogon* sporocarps [4] have been introduced into nursery beds. In recent experiments, Castellano and Trappe [unpubl. data, 2] found fresh and dried spore suspensions of *Rhizopogon vinicolor* Smith and *R. colossus* Smith to be effective in inoculating bareroot and container-grown Douglas-fir. Success with *R. vinicolor* is particularly promising; Pilz [40] and Parke et al. [38] have shown that fungus to be an important mycorrhizal symbiont of newly outplanted Douglas-fir seedlings. *R. vinicolor* also improves drought resistance of inoculated seedlings [38], an important consideration for hot, dry sites.

Unfortunately, it can be difficult to collect large enough quantities of spores of most fungi for large-scale nursery inoculations. Application methods and rates for effective inoculation as well as methods of spore storage need further research before spore inoculation can be operational.

20.4.1.4 Pure fungus cultures

The final inoculum source is pure cultures of specially selected, beneficial ectomycorrhizal fungi; intense research is currently in progress worldwide for developing this promising source. A pure culture of a specific fungus is first isolated, usually from a sporocarp or, occasionally, directly from its ectomycorrhiza (see [37]). The nutritional and growth requirements of such a fungus and its ability to form ectomycorrhizae, stimulate growth, or offer other benefits, such as disease protection or drought resistance, to its hosts can then be evaluated. This background information is vital for selecting the best isolates for attaining specific nursery goals.

20.4.1.5 Selection criteria

The thousands of ectomycorrhizal fungi are characterized by tremendous physiological diversity, including ease of isolation, growth in pure culture, effectiveness as mycelial inoculum, and benefit to the host. Consequently, criteria have been developed for selecting the most promising fungi for small- and large-scale testing so that, ultimately, nursery goals can be met. The major selection criteria are summarized by Molina ([35]; see [20] and [49] for greater detail):

- **Good growth in culture:** Most ectomycorrhizal fungi grow slowly; relatively fast-growing isolates are preferred.
- **Effectiveness in forming mycorrhizae:** Many fungi can easily be grown in culture for inoculum production, but only some of these consistently perform well as vegetative inoculum.
- **Special ecological adaptations:** For example, the common ectomycorrhizal fungus *Cenococcum geophilum* Fr. is well known for its drought resistance and is also an important symbiont of trees growing at timberline. Similarly, some fungi are more effective than others in producing enzymes important for nutrient absorption.
- **Competitive ability:** Marx [20] emphasizes that the introduced fungus must compete well against the resident mycorrhizal fungi and dominate the root systems of inoculated stock. Our preliminary studies also point to a need for the introduced fungus to resist antagonistic soil microorganisms that can build up over winter after autumn soil fumigation. The isolate should also protect roots against pathogens such as *Phytophthora* or *Fusarium* spp.
- **Host range:** Many fungi can form mycorrhizae with most ectomycorrhizal hosts, whereas others will form ectomycorrhizae only with specific hosts such as Douglas-fir or pines. Because modern nurseries often raise many tree species, it is important that ectomycorrhizal hosts and fungi be compatible.
- **Improved seedling performance in plantations:** This is the ultimate criterion to be met before an isolate can be recommended for wide-scale nursery inoculation.

Marx and Kenny [27] review past and recent research developments on production of ectomycorrhizal fungus inoculum. Basically, Marx and Bryan [23] refined a system to grow pure cultures of specific fungi in a vermiculite substrate moistened with nutrient solution. After about 3 months' incubation, the vegetative inoculum is washed, dried, and refrigerated until used. Just before sowing, the inoculum is worked into the rooting zone of nursery beds where it remains quiescent until planted seeds germinate and seedlings produce feeder roots, a period of about 6 to 8 weeks. The fungus is sheltered within the vermiculite particles during this period.

Limited success has been achieved with *Pisolithus tinctorius* by Donald Marx and coworkers at the U.S.D.A. Forest Service Institute for Mycorrhiza Research and Development (Athens, Georgia). Inoculation of nurseries in the southern United States has yielded excellent establishment of *P. tinctorius* on seedling root systems. As a result, seedling growth in the nursery has significantly increased, at times doubling that of noninoculated controls [24]. More importantly, *P. tinctorius* inoculation has significantly increased survival and growth of outplanted inoculated seedlings on extremely disturbed sites such as mine spoils [18, 21], as well as on routine regeneration sites [25]. Experimentation is continuing, to render this technology operational.

Such results prompted efforts to produce *P. tinctorius* inoculum for large-scale nursery inoculations. From 1977 through 1980, Marx et al. [26] conducted complex nationwide tests of *P. tinctorius* vegetative inoculum (Mycorrhiz®) produced by Abbott Laboratories (Chicago, Illinois) in 30 conventional bareroot nurseries located in 25 states. Final results indicated that one isolate of *P. tinctorius* could be produced in large industrial fermentors for use in bareroot nurseries. A broadcast rate of approximately 1 liter inoculum per square meter of soil surface gave the best results. Large tractor-drawn seeders have been modified to rapidly incorporate such inoculum into the rooting zone when seed is sown [16]; unfortunately for western nurseries, inoculation was satisfactory only on pine species grown in southern and southeastern nurseries, the region from which the single *P. tinctorius* strain originated. That this strain did poorly in northwestern nurseries reinforces the premise that fungus strains adapted to particular regions and habitats should primarily be selected for use in those regions.

Encouraged by the commitment of industrial representatives and interest of nursery managers and foresters, several groups of mycorrhiza researchers are now collecting, selecting, and testing promising species and strains of ectomycorrhizal fungi for nursery inoculations. In the Northwest, we have had encouraging results in ongoing studies with the ectomycorrhizal fungus *Laccaria laccata*. It has performed well on container-grown seedlings [34, 36] and in bareroot nurseries [unpubl. data, 7]. The inoculum was produced for experimental use by Sylvan Spawn Laboratory of Butler County Mushroom Farms (Worthington, Pennsylvania); this firm can produce small to large amounts of vegetative inoculum of diverse ectomycorrhizal fungi. At the Pacific Northwest Forest and Range Experiment Station, long-range research plans include continued work to select and test new and promising fungus strains with the hope of finding dependable strains to meet both nursery production and reforestation goals. Nursery managers are thus encouraged to remain alert to future developments in this field.

20.4.2 VA mycorrhizal inoculation

Unlike ectomycorrhizal fungi, VA mycorrhizal fungi have not yet been isolated and grown in pure culture because they must be attached symbiotically to their hosts to grow and reproduce. This presents a major obstacle to aseptic mass production of VA mycorrhizal fungi for large-scale nursery inoculations. Methods are available, however, to circumvent these difficulties and ensure VA mycorrhizal colonization of nursery stock.

20.4.2.1 Soil and root inoculum

As with ectomycorrhizal inoculation, the easiest method is to incorporate soil (plus root fragments) taken from under VA mycorrhizal hosts. Fortunately, VA mycorrhizal fungi show little or no host specificity; those associated with grasses, legumes, and several herbs and shrubs can form VA mycorrhizae with cedars, redwoods, sweetgums, and maples. Thus, locating soil with VA mycorrhizal fungi is relatively easy. The same drawbacks noted for soil inoculation of ectomycorrhizal hosts (see 20.4.1.1) apply here: the risk of introducing pests is ever present, and the need to move large quantities of soil can be impractical.

20.4.2.2 Pot-cultured inoculum

Refined techniques to multiply and introduce selected VA fungi are becoming available through intense research efforts in pot culturing [5]. In this technique, soil-borne spores which are very large are first sieved from the soil, examined microscopically, and identified to species (see [50]). Spores are then surface sterilized and mixed with sterilized soil in which a host plant such as sorghum is greenhouse grown. As host roots penetrate the inoculated substrate, the spores germinate and colonize the roots to form mycorrhizae. After about 4 to 6 months, the fungus has established its hyphae-soil network and has produced more spores. Once such pot cultures are established, the soil containing spores, mycelium, and colonized root fragments can be used to inoculate nursery or field crops or start new pot cultures, thus multiplying available inoculum for future use.

Pot culturing also affords the opportunity to select species, strains, or mixtures thereof that offer the greatest benefit to the targeted host species. As with ectomycorrhizal fungi, research is underway to produce commercial quantities of dependable VA fungus inoculum for large-scale nursery and field inoculations. Fortunately for forest-tree nurseries, the gains made on research directed towards VA inoculation of agricultural crops provide information directly applicable to forest-tree seedling inoculations.

20.4.2.3 Application of VA inoculum

Given the availability of the above inoculum source, Menge and Timmer [31] list several field-inoculation procedures. VA fungus inoculum can be broadcast and rototilled into seedbeds, a method that has worked well with citrus seedlings [48]; however, a major disadvantage is that large amounts of inoculum are needed to obtain rapid root colonization. VA fungus inoculum can also be banded or side dressed next to seeds or seedlings. This is particularly effective when inoculum quantities are limited [31]; for best results, bands should be placed in an area of root proliferation, usually about 5 to 15 cm from seedlings or seeds. Placing inoculum in layers or pads directly beneath seeds where developing roots will penetrate the inoculum is the most effective. Layering of inoculum has been successful for peach [13] and citrus [10]. If enough inoculum is available, it can be applied with commercial tractor-drawn seeders or fertilizer banders [31].

Seed has been pelleted with VA fungus inoculum, but success of the technique has been erratic so far [31]. Optimum placement of inoculum for rapid root colonization is a problem yet to be solved.

As with ectomycorrhizal inoculations, two major questions must be addressed before large-scale VA inoculations are feasible. First, what specific fungus species or mixture of species is best for particular hosts grown under various nursery conditions? Second, how much inoculum is needed to provide adequate mycorrhiza development and ensure healthy seedling growth? The second question is crucial for establishing the cost effectiveness of mycorrhizal inoculation. Fortunately, much of the current practical application of VA mycorrhiza research is focusing on these questions as well as on methods of producing mass inoculum.

20.5 Conclusions and Recommendations

Tree seedlings have evolved a beneficial, mutual dependency upon mycorrhizal fungi for normal root functions. Recognition, utilization, and management of mycorrhizae are part of the skillful production of resilient planting stock. In developing mycorrhiza-management tools, nursery managers and staff must learn to recognize the presence—and absence—of various mycorrhizal types and understand how mycorrhizal fungus populations are affected by nursery operations.

We recommend books on mycorrhizae by Marks and Kozlowski [15] and Schenck [41] as excellent references for nursery staff. Nursery managers are urged to keep abreast of current mycorrhiza research aimed toward practical use in nursery production and reforestation. Mycorrhiza research has truly "mushroomed" over the last decade, and knowledgeable mycorrhiza specialists are available nationwide to assist. The continuing interest and research support we have received from several nurseries convince us that the time is right for garnering the full benefits of mycorrhiza management.

Specific recommendations

- Include mycorrhiza management into the entire nursery management scheme.
- Become familiar with the various types of mycorrhizae and groups of fungi involved in mycorrhizal associations.
- Regularly examine seedling roots to monitor and record mycorrhiza development throughout the nursery.
- Observe and record the effects of new or experimental management practices on mycorrhiza development as well as on other seedling characteristics.
- Be alert to and avoid practices that cause mycorrhiza deficiency.
- Recognize that fumigation destroys mycorrhizal fungus populations in addition to pathogens and weed seed. Consider alternative, selective biocides to eliminate specific pests.
- If mycorrhiza deficiency becomes a problem with newly planted seedlings or if newly cultivated or fumigated ground presents a high risk for developing mycorrhiza deficiency, consider one of the mycorrhizal inoculation options discussed in this chapter.
- Remain alert to research developments on mycorrhizal inoculation of nursery seedlings with pure cultures of fungi proven effective and beneficial.
- Obtain the assistance of a mycorrhiza specialist to help optimize mycorrhiza-management practices.

References

1. Alvarez, I. F., and J. M. Trappe. 1983. Effects of application rate and cold soaking pretreatment of *Pisolithus* basidiospores on effectiveness as nursery inoculum on western conifers. Can. J. Forest Res. 13(3):533-537.
2. Castellano, M., and J. M. Trappe. 1982. Unpublished data, U.S.D.A. Forest Serv., Forestry Sciences Lab., Corvallis, Oregon.
3. Danielson, R. M. 1982.Taxonomic affinities and criteria for identification of the common ectendomycorrhizal symbiont of pines. Can. J. Botany 60:7-18.
4. Donald, D. G. M. 1975. Mycorrhizal inoculation for pines. South African Forestry J. 92:27-29.
5. Duddridge, J. A., A. Malibari, and D. J. Read. 1980. Structure and function of mycorrhizal rhizomorphs with special reference to their role in water transport. Nature 287:834-836.
6. Ferguson, J. J., and S. H. Woodhead. 1982. Production of endomycorrhizal inoculum. A. Increase and maintenance of vesicular-arbuscular mycorrhizal fungi. Pages 47-54 *in* Methods and principles of mycorrhizal research (N. C. Schenck, ed.). American Phytopathological Society, St. Paul, Minnesota.
7. Hung, L., and R. Molina. 1982. Unpublished data, U.S.D.A. Forestry Serv., Forestry Sciences Lab., Corvallis, Oregon.
8. Iloba, C. 1979. The effect of fungicide application on the development of ectomycorrhizae in seedlings of *Pinus sylvestris* L. Flora 168:352-357.
9. Imshenetskii, A. A. (ed.). 1955. Mycotrophy in plants. U.S. Dep. Commerce, Washington, D. C. Translations TT 67-51290. 362 p.
10. Kleinschmidt, G. D., and J. W. Gerdemann. 1972. Stunting of citrus seedlings in fumigated nursery soils related to the absence of endomycorrhizae. Phytopathology 62:1447-1453.
11. Kormanik, P. P., W. C. Bryan, and R. C. Schultz. 1980. Increasing endomycorrhizal fungus inoculum in forest nursery soil with cover crops. Southern J. Applied Forestry 4:151-153.
12. Laiho, O. 1965. Further studies on the ectendotrophic mycorrhiza. Acta Forestalia Fennica 79:1-35.
13. LaRue, J. H., W. D. McClellan, and W. L. Peacock. 1975. Mycorrhizal fungi and peach nursery nutrition. California Agric. 29:6-7.
14. Linderman, R. 1982. Personal communication, U.S.D.A. Horticultural Res. Lab., Corvallis, Oregon.
15. Marks, G. C., and T. T. Kozlowski (eds.). 1973. Ectomycorrhizae—their ecology and physiology. Academic Press, New York. 444 p.
16. Maronek, D. M., and J. W. Riffle. 1982. Ectomycorrhizal inoculation procedures for greenhouse and nursery studies. Pages 147-155 *in* Methods and principles of mycorrhizal research (N. C. Schenck, ed.). American Phytopathological Society, St. Paul, Minnesota.
17. Marx, D. H. 1973. Mycorrhizae and feeder root diseases. Pages 351-382 *in* Ectomycorrhizae—their ecology and physiology (G. C. Marks and T. T. Kozlowski, eds.). Academic Press, New York.
18. Marx, D. H. 1976. Use of specific mycorrhizal fungi on tree roots for forestation on disturbed surface areas. Pages 47-65 *in* Conf. on forestation of disturbed areas (K. A. Utz, ed.). Birmingham, Alabama, April 14-15. U.S.D.A. Forest Serv., Atlanta, Georgia.
19. Marx, D. H. 1976. Synthesis of ectomycorrhizae on loblolly pine seedlings with basidiospores of *Pisolithus tinctorius*. Forest Sci. 22:13-20.
20. Marx, D. H. 1980. Ectomycorrhiza fungus inoculation: a tool for improving forestation practices. Pages 13-71 *in* Tropical mycorrhiza research (P. Mikola, ed.). Oxford University Press, Oxford.
21. Marx, D. H. 1980. Role of mycorrhizae in forestation of surface mines. Pages 109-116 *in* Symp. on trees for reclamation in the eastern United States. U.S.D.A. Forest Serv., Lexington Kentucky.
22. Marx, D. H., and W. C. Bryan. 1971. Influence of ectomycorrhizae on survival and growth of aseptic seedlings of loblolly pine at high temperatures. Forest Sci. 17:37-41.
23. Marx, D. H., and W. C. Bryan. 1975. Growth and ectomycorrhizal development of loblolly pine seedlings in fumigated soil infested with the fungal symbiont *Pisolithus tinctorius*. Forest Sci. 21:245-254.
24. Marx, D. H., W. C. Bryan, and C. E. Cordell. 1976. Growth and ectomycorrhizal development of pine seedlings in nursery beds infested with the fungal symbiont *Pisolithus tinctorius*. Forest Sci. 22:91-100.
25. Marx, D. H., W. C. Bryan, and C. E. Cordell. 1977. Survival and growth of pine seedlings with *Pisolithus* ectomycorrhizae after two years on reforestation sites in North Carolina and Florida. Forest Sci. 23:363-373.
26. Marx, D. H., C. E. Cordell, D. S. Kenny, J. G. Mexal, J. D. Artman, J. W. Riffle, and R. J. Molina. 1984. Commercial vegetative inoculum of *Pisolithus tinctorius* and inoculation techniques for development of ectomycorrhizae on bare-root tree seedlings. Forest Sci. (in press).
27. Marx, D. H., and D. S. Kenny. 1982. Production of ectomycorrhizal fungus inoculum. Pages 131-146 *in* Methods and principles of mycorrhizal research (N. C. Schenck, ed.). American Phytopathological Society, St. Paul, Minnesota.
28. Marx, D. H., J. G. Mexal, and W. G. Morris. 1979. Inoculation of nursery seedbeds with *Pisolithus tinctorius* spores mixed with hydromulch increases ectomycorrhizae and growth of loblolly pines. Southern J. Applied Forestry 3:175-178.
29. McComb, A. L. 1938. The relation between mycorrhizae and the development and nutrient absorption of pine seedlings in a prairie nursery. J. Forestry 36:1148-1153.
30. Menge, J. A. 1982. Effect of soil fumigants and fungicides on vesicular-arbuscular fungi. Phytopathology 72:1125-1132.

31. Menge, J. A., and L. W. Timmer. 1982. Procedures for inoculation of plants with vesicular-arbuscular mycorrhizae in the laboratory, greenhouse and field. Pages 59-68 *in* Methods and principles of mycorrhizal research (N. C. Schenck, ed.). American Phytopathological Society, St. Paul, Minnesota.
32. Mikola, P. 1970. Mycorrhizal inoculation in afforestation. International Review Forestry Res. 3:123-196.
33. Mikola, P. 1973. Application of mycorrhizal symbiosis in forestry practices. Pages 383-411 *in* Ectomycorrhizae—their ecology and physiology (G. C. Marks and T. T. Kozlowski, eds.). Academic Press, New York.
34. Molina, R. 1982. Use of the ectomycorrhizal fungus *Laccaria laccata* in forestry. I. Consistency between isolates in effective colonization of containerized conifer seedlings. Can. J. Forest Res. 12:469-473.
35. Molina, R. 1982. Mycorrhizal inoculation and its potential impact on seedling survival and growth in southwest Oregon. Pages 86-91 *in* Proc., Reforestation of skeletal soil (S. D. Hobbs and O. T. Helgerson, eds.). Medford, Oregon, Nov. 17-19, 1981. Forest Research Lab., Oregon State Univ., Corvallis.
36. Molina, R., and J. Chamard. 1983. Use of the ectomycorrhizal fungus *Laccaria laccata* in forestry. II. Effects of fertilizer forms and levels on ectomycorrhizal development and growth of container-grown Douglas-fir and ponderosa pine seedlings. Can. J. Forest Res. 13:89-95.
37. Molina, R., and J. G. Palmer. 1982. Isolation, maintenance, and pure culture manipulation of ectomycorrhizal fungi. Pages 115-129 *in* Methods and principles of mycorrhizal research (N. C. Schenck, ed.). American Phytopathological Society, St. Paul, Minnesota.
38. Parke, J. L., R. G. Linderman, and C. H. Black. 1983. The role of ectomycorrhiza in drought tolerance of Douglas-fir seedlings. New Phytologist 95:83-95.
39. Phillips J. M., and D. S. Hayman. 1970. Improved procedures for clearing and staining parasitic and vesicular-arbuscular mycorrhizal fungi for rapid assessment of infection. Trans. of the British Mycological Society 55:158-161.
40. Pilz, D. P. 1982. Management impacts on the ectomycorrhizal associations of *Pseudotsuga menziesii* var. *menziesii* seedlings: field and greenhouse bioassays. M.S. thesis, School of Forestry, Oregon State Univ., Corvallis. 58 p.
41. Schenck, N. C. (ed.). 1982. Methods and principles of mycorrhizal research. American Phytopathological Society, St. Paul, Minnesota. 244 p.
42. Schramm, J. R. 1966. Plant colonization studies on black wastes from anthracite mining in Pennsylvania. American Philosophical Society Trans. 56:1-194.
43. Schwab, S. M., E. L. V. Johnson, and J. A. Menge. 1982. Influence of simazine on formation of vesicular-arbuscular mycorrhizae in *Chenopodium quinona* Willd. Plant and Soil 64:283-287.
44. Sinclair, W. A., D. M. Sylvia, and A. O. Larsen. 1982. Disease suppression and growth promotion in Douglas-fir seedlings by the ectomycorrhizal fungus *Laccaria laccata*. Forest Sci. 28:191-201.
45. Theodorou, C. 1971. Introduction of mycorrhizal fungi into soil by spore inoculation of seed. Australian Forestry 35:17-22.
46. Theodorou, C., and G. D. Bowen. 1970. Mycorrhizal responses of radiata pine in experiments with different fungi. Australian Forestry 34:183-191
47. Theodorou, C., and G. D. Bowen. 1973. Inoculation of seeds and soil with basidiospores of mycorrhizal fungi. Soil Biology and Biochemistry 5:765-771.
48. Timmer, L. W., and R. F. Leyden. 1978. Stunting of citrus seedlings in fumigated soils in Texas and its correction by phosphorus fertilization and inoculation with mycorrhizal fungi. J. American Society for Horticultural Sci. 103:533-537.
49. Trappe, J. M. 1977. Selection of fungi for ectomycorrhizal inoculation in nurseries. Annual Review of Phytopathology 15:203-222.
50. Trappe, J. M. 1982. Synoptic keys to the genera and species of zygomycetous mycorrhizal fungi. Phytopathology 72:1102-1108.
51. Trappe, J. M. 1983. Effects of the herbicides bifenox, DCPA, and napropamide on mycorrhiza development of ponderosa pine and Douglas-fir seedlings in six western nurseries. Forest Sci. 29(3):464-468.
52. Trappe, J. M., and R. F. Strand. 1969. Mycorrhizal deficiency in a Douglas-fir region nursery. Forest Sci. 15:381-389.
53. Zak, B. 1971. Detoxification of autoclaved soil by a mycorrhizal fungus. U.S.D.A. Forest Serv., Pacific NW Forest and Range Exp. Sta., Portland, Oregon. Res. Note PNW-159. 4 p.

Harvesting and Outplanting the Bareroot Seedling

Chapter 21
Lifting, Grading, Packaging, and Storing

A. N. Burdett and D. G. Simpson

Abstract

While being lifted, tree seedlings are subject to mechanical damage. At the same time, their foliage may be inoculated with soil-borne spores of storage-mold fungi. During subsequent handling and storage, stock condition may change rapidly as a result of desiccation, molding, metabolic activity, or developmental processes. Damage to or deterioration of stock during and after harvest may be minimized through: choice of the method and date of lifting; prelift root pruning, fungicide application, and physiological conditioning; and post-lift control of stock water potential and temperature. Grading can improve stock quality either by identifying inferior seedlings or batches of stock for culling or by revealing deficiencies in quality that can be avoided in the future through changes in stock-production techniques. To be useful, grading must be in accordance with standards which reflect the stock characteristics necessary for satisfactory field performance. To improve stock quality through changes in cultural and handling practices, nursery managers must investigate the effects of alternative practices on the field-performance potential of the stock they produce.

21.1 Introduction

The quality of bareroot nursery stock depends greatly on the way it is lifted, graded, packaged, and stored. During lifting, delicate absorbing roots are easily lost, while larger roots may be stripped of bark or fractured. Once out of the ground, trees can become desiccated rapidly if not protected. Grading can improve stock quality by eliminating inferior seedlings or batches of stock; but it can be detrimental if seedlings are desiccated or physically damaged in the process or if culling standards are inappropriate. During storage, stock condition can change greatly due to metabolic (e.g., respiration) and developmental (e.g., loss of bud dormancy) processes or to the action of pathogenic or saprophytic fungi. The rate of change is strongly influenced by the storage environment (e.g., temperature and relative humidity).

How, and to what extent, the quality of bareroot stock is affected by methods of handling and storage depends on the condition of trees at the time they are lifted. This, in turn, depends on the cultural regime under which stock has been raised and the time of year at which lifting occurs. Thus, to develop the best procedures for harvesting and storing bareroot stock, the following questions must be answered:

(1) What cultural regime should be used to prepare or condition stock for lifting?

(2) What is the optimum lifting date?

(3) How can mechanical damage to stock during lifting and subsequent handling be minimized?

(4) By what means and within what ranges should the temperature and water potential of lifted stock be maintained?

(5) What criteria should be used for grading stock?

(6) How can molding of stored stock be prevented?

This chapter discusses each of these questions [see chapter 15 and chapters 23 and 24 for detailed treatment of questions (1) and (5), respectively]. Variation in treatment effect on stock quality due to differences in species, seed source, and nursery environment is too great, however, to allow specific recommendations on many points. Here, the emphasis will be on general principles and the kind of nursery trials needed to develop optimum procedures for particular circumstances.

In Duryea, Mary L., and Thomas D. Landis (eds.). 1984. *Forest Nursery Manual: Production of Bareroot Seedlings*. Martinus Nijhoff/Dr W. Junk Publishers, The Hague/Boston/Lancaster, for Forest Research Laboratory, Oregon State University, Corvallis. 386 p.

21.2 Preparing Stock for Lifting

21.2.1 Physiological conditioning

The ability of stock to withstand transplanting and storage varies seasonally, and this has led to the belief that stock should be dormant when lifted. Aside from the vagueness of the term **dormant** in this context (see 21.3.3.1), this is a misleading proposition. Bareroot stock can be transplanted successfully at almost any time of year, provided that conditions at the planting site are favorable. It can also be stored, at least briefly, at almost any time of year. Thus, it is by no means essential that stock be "dormant" when lifted.

Nevertheless, it is true that stock is best able to withstand transplanting and storage when it is adapted to winter conditions. Furthermore, cultural treatments applied during the summer or fall which promote the adaptation of stock to winter conditions tend to increase the success with which stock can be transplanted and stored. For example, moisture- or nutrient-stress treatments which increase frost hardiness or induce bud dormancy have been observed to improve the ability of stock to withstand transplanting and storage [2, 10].

It is far from clear, however, why winter-adapted stock is best able to withstand transplanting and storage. Many seedling characteristics change both seasonally and in response to cultural treatments which condition stock for lifting. Care must be taken, therefore, to avoid unjustified assumptions about the causal connection between readiness for lifting and particular seedling characteristics such as frost hardiness or bud dormancy.

Until more is known about the physiology underlying the ability of stock to withstand transplanting and storage, the approach to conditioning stock for lifting must remain largely empirical. Knowing the way in which cultural treatments affect seedling physiology may indicate their value as conditioning treatments. But the only conclusive evidence of the value of a conditioning regime is a demonstration that, for a particular species raised at a particular nursery, it does improve stock quality. For a detailed discussion of conditioning treatments, see chapter 15, this volume.

21.2.2 Undercutting and lateral root pruning

Tree seedlings must have a compact root system if they are to be planted properly. This means that the root system of field-grown seedlings should be pruned. Though roots can be pruned after stock has been lifted (table pruning; see 21.8), it is advantageous to prune before lifting; this may not only modify seedling morphology and physiology in a beneficial way (see chapter 15, this volume) but also reduce the amount of stretching and stripping of roots that occurs during lifting.

Root pruning in the nursery bed may be restricted to undercutting. Drill-sown stock is often root-pruned laterally, however, by passing tractor-mounted knives (colters) through the soil between the drill rows. By preventing intermeshing of roots of trees in adjacent rows, this type of pruning reduces the extent of root stripping that occurs when lifted seedlings are separated from one another.

Two-way root pruning of grid-spaced seedlings entirely prevents intermeshing of roots of adjacent trees [8]. This technique, sometimes referred to as box pruning, is still only experimental, however.

21.2.3 Fungicide treatment

Stock that is to be stored for any length of time is prone to damage by mold. The extent of molding depends on a number of factors including the length and duration of storage, the physiological condition of the stock, and the degree to which the stock has been inoculated with spores of mold fungi such as *Botrytis cinerea* and *Rhizoctonia solani*. It may be possible to reduce mold damage by treating stock with a fungicide immediately before lifting (see chapter 19, this volume). Benomyl is reported to be effective against many of the soil-borne pathogens responsible for molding on bareroot stock [18].

21.3 Choosing a Lifting Date

If cold-storage facilities are not available, the nursery manager has little discretion in choosing a lifting date. Trees must be lifted when conditions at the planting site favor plantation establishment. Stock must be adapted to site conditions, but this should be ensured by adjusting the cultural regime (i.e., conditioning; see chapter 15, this volume) rather than by waiting for seasonal changes in stock condition. Otherwise, the optimum planting date will be missed.

If, however, cold-storage facilities are available, then lifting date is an important independent variable affecting stock quality because cold storage modifies the normal pattern of seasonal changes in stock physiology. For example, if stock is placed in cold storage in late winter or early spring, budbreak and the loss of cold hardiness are delayed [29]. The seasonal decline in root-growth capacity may also be delayed [25]. Depending on circumstances, such effects can greatly enhance the ability of stock to survive and grow when planted. Deteriorative changes in stock condition may also occur during cold storage (see 21.5), however, and the nature and extent of these are influenced by lifting date. Often, therefore, the optimum date for lifting stock is not readily apparent. Alternative approaches to estimating optimum lifting date are considered in the following paragraphs.

21.3.1 Past experience

One way of estimating optimum lifting date is the purely empirical. Stock is lifted on different dates and stored until the normal planting season. Its condition is then assessed by field testing or other means of predicting field performance (e.g., root-growth capacity testing).

Studies by Stone and Jenkinson [25] and Jenkinson and Nelson [19] with Douglas-fir [*Pseudotsuga menziesii* (Mirb.) Franco] raised at nurseries in California illustrate both the benefits and limitations of this approach. They show that, in the case of stock for late winter or spring planting, root-growth capacity is highest (at the time of planting) if lifting occurs considerably before the planting date. They also indicate, however, that the optimum time of year for lifting stock (i.e., lifting window) varies with provenance and nursery and from one year to another. It also can be expected to vary according to the cultural regime used to condition the stock for lifting.

Thus, if the past is to be of use as a guide to when stock should be lifted, experience must be gained over several years for each combination of seed source, cultural regime, and nursery. This is both costly and time consuming. Moreover, because optimum lifting date varies from year to year, it can never be predicted by the calendar with complete certainty. Nevertheless, nursery managers should not abandon past experience as a basis for choosing lifting dates until the superiority of an alternative guide has been demonstrated under the particular circumstances of their own nursery.

21.3.2 Prelift chilling

Seasonal changes in the ability of stock to withstand transplanting and storage seem due, at least in part, to seasonal changes in temperature [22]. Thus, it may be possible to obtain a good indication of the optimum date for lifting stock by monitoring temperature in the nursery.

Support for this idea was gained by Stone and Jenkinson [25] in a study with Douglas-fir in California. They found that stock for late winter or early spring planting had the highest root-growth capacity (at the time of planting) if lifted after 1,500 to 1,800 hours of exposure to temperatures below 10°C (after August 1). They also found that, in one nursery, the date by which 1,500 hours of chilling occurred varied by 2 months during a 3-year period. This seems to demonstrate the superiority of prelift chilling over the calendar as a guide to the optimum date for lifting stock that is to be cold stored.

However, the relationship between prelift chilling and prestorage stock quality cannot be assumed the same for all species or even for all provenances of a single species. It can be expected to vary with cultural regime and may vary from one nursery to another.

Furthermore, the value of prelift chilling as an estimator of optimum lifting date will depend on how it is measured. As suggested by our observations in nurseries in interior British Columbia, the best index of chilling may not be the same for all species. We found that, in Douglas-fir that was fall lifted and cold stored until spring, root-growth capacity was more closely related to prelift chilling below a threshold of 10°C than of 5°C. In lodgepole pine (*Pinus contorta* Dougl. ex Loud.) and white spruce [*Picea glauca* (Moench) Voss], however, prelift chilling below 5°C seemed a better estimator of post-storage root-growth capacity than chilling below 10°C (Table 1).

Table 1. Relationship between the root-growth capacity and prelift chilling of fall-lifted, cold-stored stock in three species of 2 + 0 stock. The trees were lifted on five dates from September 12 to November 6 and stored at −2°C until May 15. The coefficient of determination (r^2) is for the best-fit regression line. Because different formulas were used to fit curves to the different sets of data, r^2 values followed by different letters are not strictly comparable.

Hours after August 1 below:	r^2		
	Douglas-fir	Lodgepole pine	White spruce
10°C	0.86a	0.52b	0.58a
5°C	0.77a	0.74b	0.71a

Methods of measuring prelift chilling which might prove superior to summation of hours when nursery air temperature is below a threshold include summation of hours when: air temperature is within a certain range (e.g., 0 to 10°C); when soil temperature is either below a threshold or within a certain range; and when both air and soil temperatures fall within specified bounds (used successfully in eastern Canada to estimate readiness for lifting in several conifers [24]).

Thus, before prelift chilling can be adopted as a guide to the optimum date for lifting stock, calibration data must be obtained for each nursery, species, cultural regime, and, even perhaps, seed source. At the same time, different methods of measuring prelift chilling must be evaluated.

But even with adequate calibration data, prelift chilling may not always provide a reliable indication of readiness for lifting—it has been shown that a warm period can negate the effect of earlier chilling [26]. Possibly, the effects of alternating warm and cool weather can be integrated in some way. But this remains to be demonstrated.

21.3.3 Seedling physiology

The effect of lifting date on post-storage stock quality must reflect seasonal changes in seedling physiology (see chapters 14 and 23 for more information on seedling physiology and quality assessment). The nature of the relevant changes is unknown. Nevertheless, if the causal variable or one of its close correlates were identified, it would be possible to measure directly whether stock is in a condition to be lifted and stored. A number of the physiological attributes of tree seedlings are known to vary seasonally, and a relationship between certain of these changes and readiness for lifting has been demonstrated or assumed.

21.3.3.1 Dormancy

It is commonly asserted that the ability of tree seedlings to withstand transplanting and storage depends on their dormancy [e.g., 11, 17]. The meaning of this is obscure, however, because the meaning of dormancy in this context is quite unclear. In the broadest sense, dormancy is a state of growth inactivity in the absence of environmental constraints to growth. Dormancy, therefore, pertains only to meristematic tissue and not to the whole plant. Tree seedlings have three meristematic zones: the shoot apex, the root apex, and the vascular cambium. Growth potential in these tissues is not normally correlated. The maximum degree of bud domancy occurs during fall, when root dormancy (if defined as the inverse of root-growth capacity) may be at a minimum [25]. In the vascular cambium, rest, or dormancy, may not occur at all [33].

Whether a close relationship exists between some phase of bud dormancy and readiness for lifting is a possibility worth examining. As yet, however, no evidence of such a relationship has been reported in the literature. Moreover, if it does exist, its practical significance is questionable because a quick method of measuring bud dormancy remains to be developed. The possibility of a relationship between root dormancy and readiness for lifting is discussed in the following section.

21.3.3.2 Root-growth capacity

In a study with Douglas-fir in California, Stone and Jenkinson [25] found that the root-growth capacity of fresh-lifted stock increased during fall, reached a peak in late fall or early winter, and then declined to a low level by late winter. Paralleling these changes were changes in the ability of stock to maintain its root-growth capacity during 3 months of storage. Stock lifted early in the fall underwent a sharp decline in root-growth capacity during storage, whereas stock lifted later on—just before the root-growth capacity of the fresh-lifted stock reached its peak—maintained or increased its root-growth capacity during storage. As lifting date was delayed further, the ability of stock to maintain its root-growth capacity during storage declined.

We did not find a similar relationship between storability and root-growth capacity at lifting in Douglas-fir, white spruce, or lodgepole pine raised at nurseries in interior British Columbia. In all cases, the root-growth capacity of fresh-lifted stock was relatively constant throughout the fall, although storability, as measured by the ability of stock to maintain root-growth capacity during 6 months of cold storage, increased sharply from early September until freeze-up in early November (Fig. 1). Evidently, the value of root-growth capacity as a guide to the storability of tree seedlings is limited.

21.3.3.3 Frost hardiness

In an experiment with white spruce and lodgepole pine, we observed a close relationship between frost hardiness at lifting and the ability of 2 + 0 seedlings to maintain their root-growth capacity during storage (Fig. 2). The method we used for measuring frost hardiness took too long (4 weeks) to provide a signal as to when stock should be lifted. There are, however, a number of quick tests for estimating frost hardiness (see chapter 23, this volume) which may be suitable for this purpose. For example, measuring stem electrical impedance provides a rapid means of estimating relative frost hardiness; the closest correlation between stem impedance and seedling frost hardiness is observed when impedance is measured after trees have been subjected to overnight freezing at a standard temperature [30].

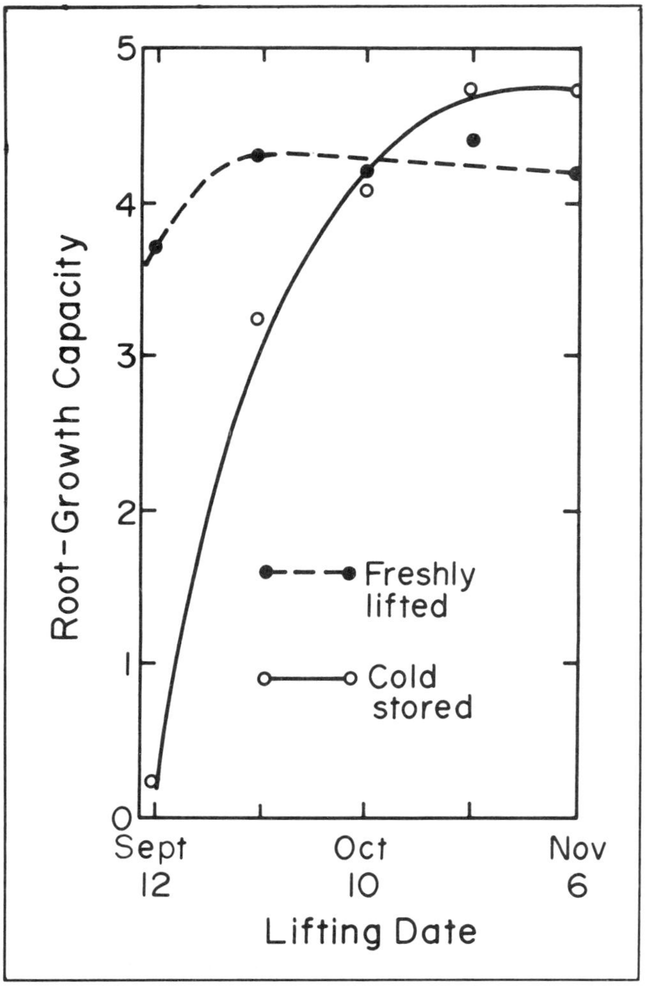

Figure 1. Effect of lifting date on the root-growth capacity of 2+0 bareroot lodgepole pine before and after 6 months of storage at −2°C. Root-growth capacity was measured with a 1-week test, described by Burdett [5], in which the scale for measuring root growth was: 0 = no new roots; 1 = no new roots > 1 cm long; 2 = 1 to 3 new roots > 1 cm; 3 = 4 to 10 new roots > 1 cm; 4 = 11 to 30 new roots > 1 cm; 5 = > 30 new roots > 1 cm.

A high correlation between the stem impedance of freshly lifted stock measured after overnight freezing at −10°C and post-storage survival has been observed in Douglas-fir [32]. Working with 2+0 Douglas-fir, lodgepole pine, and white spruce, we observed a close correlation between stem impedance of freshly lifted seedlings measured after overnight freezing and the ability of the stock to maintain its root-growth capacity during 6 months of cold storage. Apparently, therefore, frost hardiness and its correlates (e.g., stem impedance) have promise as estimators of readiness for lifting. However, these estimators should not be implemented until calibration data have been gathered. Ideally, the data should indicate the consistency of the relationship between frost hardiness at lifting and post-storage stock quality from year to year, from nursery to nursery, and with different species, provenances, and cultural regimes.

21.3.3.4 Electrical wave-form modification

An electrical square-wave signal applied to the stem of a tree seedling is modified in various ways depending on the seedling's condition. It has been suggested that this wave-form modification (oscilloscope technique) may indicate whether stock is ready to lift. This hypothesis seems to have been refuted, however, by Askren and Hermann [1], who found no consistent relationship beween wave-form modification of freshly lifted Douglas-fir seedlings and their ability to survive when planted after cold storage.

21.3.4 Weather and soil conditions

Weather and soil conditions must also be taken into account when choosing a lifting date. If stock is not watered after lifting, its ability to survive and grow when planted may be related to its water potential at the time of lifting [13]. The best time to lift stock, therefore, is when plant moisture stress is low. This occurs early in the morning, or when the weather is cool and humid. These conditions also favor lifting because they minimize the potentially harmful rise in plant moisture stress during field handling.

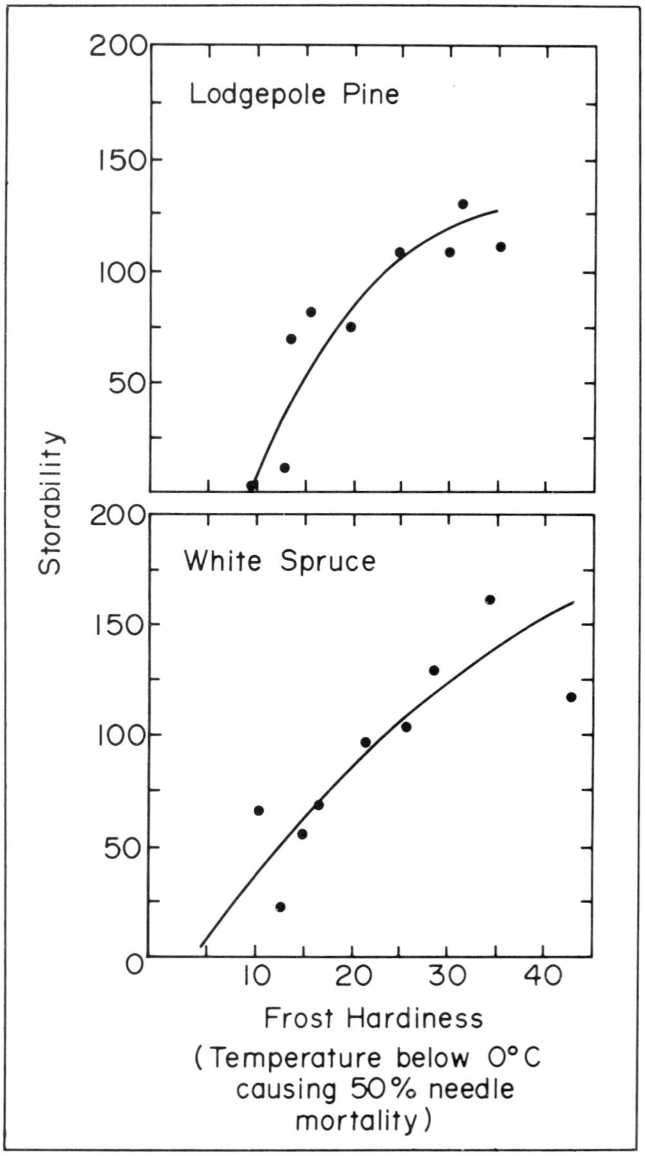

Figure 2. Relationship between storability (estimated by root-growth capacity after 6 months of storage at −2°C as a percentage of root-growth capacity at lifting) and frost hardiness in 2+0 bareroot lodgepole pine and white spruce. Test for measuring root-growth capacity and scale for scoring root growth are described in Figure 1 caption. Frost hardiness of trees was estimated by subjecting samples of stock to one of a range of subfreezing temperatures and observing the temperature causing 50% needle mortality after 4 weeks in a warm greenhouse.

To reduce mechanical damage, lifting should be restricted to occasions when the ground is not frozen. If soil is even lightly crusted with frost, severe root damage is likely during lifting. Lifting when the soil is wet, and therefore relatively heavy, should also be avoided. Another reason for not lifting when the soil is wet is to keep foliage clean and, hence, uncontaminated with soil-borne fungal spores.

21.4 Lifting

21.4.1 Minimizing mechanical damage

To minimize mechanical damage during lifting, roots can be pruned before lifting (see 21.2.2). Damage can also be controlled through the choice of lifting method. To select the best method for the prevailing conditions, trials are necessary. The degree of damage resulting from each method can be assessed visually (e.g., according to the number of stripped or broken roots). Physical symptoms of injury are not necessarily well correlated with functional damage, however. Thus, it is essential to assess effects on the functional integrity of the stock as indicated by, for example, its root-growth capacity or survival and growth after planting.

The importance of lifting method as a determinant of stock quality was demonstrated in experiments with 2 + 0 lodgepole pine at a government nursery in British Columbia. In eight trials, the root-growth capacity of stock lifted with an Egdal machine was consistently higher than that of stock lifted with a Grayco machine (Table 2) [unpubl. data, 4]. Survival tests on 1 + 0 Douglas-fir lifted with four types of machines at another British Columbia government nursery also indicated appreciable impacts of lifting machines on seedling quality [pers. commun., 3]. (Table 3). Methods of operating lifting machines also affect stock quality. Testing is, therefore, a prerequisite to developing satisfactory lifting procedures. Even if stock is hand lifted, care is essential if serious root damage is to be avoided.

Table 2. Effect of lifting method on the root-growth capacity of freshly lifted 2 + 0 bareroot lodgepole pine. Root-growth capacity was estimated with a semiquantitative 1-week test described by Burdett [5]. The difference between treatment means was statistically significant at the 1% level [4].

Date of trial	Root-growth capacity as % hand-lifted stock[1]	
	Grayco	Egdal
September 14	61	78
September 20	65	134
October 5	68	99
October 5	78	92
October 26	72	93
October 26	86	103
November 7	79	100
November 7	85	92
Mean	74 (± 9)	99 (± 16)

[1]To eliminate bias due to between-plot differences in stock quality, the root-growth capacity of the machine-lifted stock is shown as a percentage of that of a sample of trees carefully lifted from the same plot by hand immediately before machine lifting took place.

Table 3. Survival of 1 + 0 bareroot Douglas-fir seedlings lifted with four machines at Surrey Nursery, in British Columbia, and planted in 50-tree plots established in the nursery at the time stock was shipped [pers. commun., 3].

Lifting machine	Number of seedlots lifted	Mean 1st-year mortality, %
Löve	3	0.7
Egdal	147	1.1
Fobro	32	6.3
Grayco	5	11.2

21.4.2 Minimizing desiccation

Desiccation of stock during lifting can be reduced in several ways. These include: protecting lifted stock from direct exposure to the sun (e.g., with a canopy over the lifting machine); reducing to a minimum the time required to transfer lifted stock into field containers; misting stock in the field immediately after lifting; and limiting lifting operations to times when the evapotranspiration rate is low (see 21.3.4).

21.5 Presort Storage

Once stock has been lifted, its condition can change rapidly as a result of desiccation, molding, metabolic activity (e.g., respiration), or developmental processes (e.g., bud flushing) (see also chapter 22, this volume). Of these, the last three are highly temperature dependent. Although there are certain exceptions (e.g., loss of bud dormancy in fall-lifted stock stored for spring planting [31]), most changes in stock condition resulting from these processes are deleterious. Success in minimizing the deterioration of stored stock depends primarily, therefore, on controlling plant water potential and temperature.

21.5.1 Controlling water potential

Tree seedling roots, especially the fine roots [12], can become desiccated rapidly when exposed to air. Lifted stock should, therefore, be placed in field containers as soon as possible. Containers must be covered to prevent moisture loss. Damp burlap, canvas, or felt is often used for this purpose. An advantage of this method of covering is that it results in some evaporative cooling. A disadvantage is that the covering material may dry quickly and thus require constant rewetting. For this reason, it may be better to use a field container with a tight-fitting lid (e.g., as commonly used with waxed cardboard or plastic tote boxes) or a polyethylene liner. Some moisture may be lost even from containers with lids, in which case trees should also be covered with damp burlap or some other moisture-holding material.

Sometimes it is necessary to lift trees which have relatively low water potential (i.e., below − 10 bars). Daniels [13] found that this reduced survival and growth in Douglas-fir planted after 55 days of cold storage. He found, however, that adverse effects of low water potential at lifting were eliminated by spraying the trees with water immediately after lifting. This is consistent with, though not necessarily explained by, Hopkins' report [18] that molding in stored bareroot Douglas-fir is reduced by dipping the stock in water before placing it in storage.

Adding water to stock that is to be stored may not always be beneficial, however. According to Eliason [15], excess moisture promotes molding of cold-stored white spruce. Moreover, several studies have shown no relationship between moisture potential and field performance of cold-stored Douglas-fir [16, 32]. Clearly, more information about the relationship between planting-stock water potential and field performance is essential.

In practice, many nurseries in the Northwest do water lifted stock. Sometimes, plants are watered according to need, as judged from plant water potential measured with the "pressure bomb." This appears to be a sound approach, but plant moisture-stress levels do not remain constant in lifted stock even if the stock is protected from desiccation. Experiments with Douglas-fir have shown that the water potential of stock in field containers can rise, without the addition of water, from − 10 to − 1 bars within 3 days of lifting, the rate of increase depending on storage conditions [pers. commun., 14]. Evidently, if water potential is to be monitored, the time of measurement in relation to both time of lifting and post-lift storage conditions must be standardized.

The distribution of moisture within the containers of stored stock also must be considered. Unless stock is frozen, metabolic heat production will create a temperature gradient from the middle of the container to the container wall. Convection in the container will take warm, moisture-laden air from within the mass of trees to the container wall, where the moisture will condense [6]. Through this process, stock at the center of the container may become dehydrated. To minimize this effect, the temperature gradient across a container of stock must be kept small. Means of doing this include refrigeration (to reduce metabolic heat production) and use of small containers.

21.5.2 Controlling temperature

In sunny weather, solar heating of tree seedlings in field containers can be rapid. Field containers should, therefore, be placed in the shade as soon as they have been filled. The length of time during which stock can withstand storage at temperatures much above 0°C varies with stock condition. As a general rule, however, trees should not be stored without refrigeration for more than a few hours.

If an exception to this rule is contemplated, the consequences should be evaluated experimentally. There is no adequate theoretical basis for predicting how a particular species, raised at a particular nursery and lifted at a particular time of year, will be affected by a period of warm storage. Any attempt to provide general guidelines would simply be misleading.

Even when stock is held without refrigeration only briefly (up to 24 hours), care should be taken to keep it cool. In large containers (e.g., field bins), metabolic heating can occur rapidly. This may be controlled through evaporative cooling of the ambient air (e.g., with a mist system) or, even more effectively, by watering the trees thoroughly (i.e., hydrocooling).

Once stock is placed under refrigeration, it should cool quickly to storage temperature. Temperature within containers should be monitored and appropriate action taken when necessary to hasten cooling.

A number of factors influence the rate at which stock cools. One is rate of air circulation around containers. For good air circulation, the containers must be adequately spaced. If gravity convection is relied upon for air movement, wider spacing is required than for forced air movement.

Container size has a major influence on the rate at which stock cools. The upper limit to the size of a cubical container in which stock will cool rapidly from 20°C or less to an ambient temperature of 0°C (i.e., to within 1 or 2°C of ambient in 48 hours) is approximately 1 m^3 [6]. The limit will vary considerably, however, depending on factors such as thermal conductivity of the container wall, metabolic activity of the stock, and density with which the stock is packed.

The density of packing affects not only the amount of heat to be removed, but also the freedom of air movement within the container. Convection is important in transferring heat from the contents of the container to the container wall. Its effectiveness is enhanced by the movement of moisture, with the circulating air, from the warm trees to the cool container wall (see 21.5.1). This results in evaporative cooling of trees without moisture loss from the container [6]. Condensation of moisture on the container wall transfers to the container wall the heat lost by the trees.

Stock is usually stored for only a short time before sorting. To freeze stock and thaw it again, without exposing it to extremes of temperature, can take many days. Consequently, stock is not normally frozen during presort storage. The risk of stock deterioration during storage increases with temperature, however. The optimum temperature for holding stock before sorting is, therefore, no higher than is necessary to prevent freezing. With a well-designed and properly loaded cooler, stock can be held below 1°C without risk of freezing it.

21.6 Grading

Grading is intended to improve stock quality either by identifying inferior stock for culling or by revealing deficiencies in quality to be avoided in the future through changes in stock-production techniques. Tree seedlings can be graded individually, according to certain morphological standards, or batch graded in accordance with tests and measurements on only a sample of each batch of stock. Batch grading permits the use of both destructive and relatively expensive (compared with visual grading) evaluation techniques, including tests for physiological as well as morphological characteristics (see chapter 23, this volume).

21.6.1 Grading standards

Ideally, a grading (stock quality) standard should specify the type of stock able to perform satisfactorily under conditions of normal use. In reality, stock standards are often somewhat arbitrary. They may be based on little more than the knowledge of what a nursery has produced in the past, or they may originate in very questionable assumptions about the relationship between stock characteristics and plantation performance.

Great benefit can be gained from applying recently acquired knowledge about the physiology of plantation establishment to the definition of stock-quality standards. Physiological characteristics to which particular attention should be paid include root-growth capacity, drought and frost hardiness, and bud dormancy [27]. Important morphological and anatomical characteristics include stem unit number in resting buds [28], height:diameter ratio [9], foliage anatomy (e.g., whether sun- or shade-adapted), and root form [7].

21.6.2 Single-tree grading

Traditionally, stock is graded by hand according to a visual assessment of characteristics such as height, stem diameter, root and shoot form, root and bud damage, frost, or desiccation injury (i.e., foliage discoloration). Trees may be graded in the field as soon as they have been lifted. Usually, however, unsorted stock is moved in containers to a sorting building, where it is distributed to graders either in field containers or from field containers by way of a moving belt. Most often, a moving belt is used to take the graded stock from the sorters to a packing station. The period during which trees are exposed while being graded is usually brief (less than 2 minutes). Sorting sheds are kept cool, and a high humidity is sometimes maintained by watering the floor.

The cost of single-tree grading is high in terms both of labor and of trees discarded on the basis of morphological characteristics that may have only a tenuous relationship with field performance. In the future, the cost of single-tree grading may be reduced by using labor-saving grading machinery [23]. More desirable, however, is the development of cultural techniques which make possible the production of stock of such uniform quality that the need for single-tree grading is eliminated [27].

21.6.3 Batch grading

Batch grading serves two purposes. One is to identify inferior batches of stock for culling. The other is to obtain information about the seedlings currently produced as a basis for directing research and development to improve the quality of future nursery stock.

At present, batch culling is not widely practiced. Its potential is illustrated, however, by a program in British Columbia government nurseries to grade stock according to its root-growth capacity. On one occasion, many batches of bareroot lodgepole pine, totaling several million trees, were discarded because of their low root-growth capacity. A small quantity of the culled stock was trial planted. First-season survival was only 2% [pers. commun., 20].

21.7 Bundling

Graded stock is often tied in counted bundles. This makes it easier to inventory the stock and to measure planter productivity. Various materials are used for tying bundles, including jute, masking tape, and elastic bands. Plastic stretch film is probably best, however, because it is least likely to cause stem abrasion.

Because bundling costs money and may damage seedlings, other methods of quantity control may be preferable. For example, seedling numbers can be estimated by weight, the number of trees per unit weight being determined by sampling. Alternatively, seedlings can be counted directly into shipping cartons. If the shipping carton is designed to be carried by the planter, planter productivity is easily monitored.

21.8 Table Root Pruning

Lateral root pruning of drill-sown stock only severs roots growing more or less perpendicular to the drills. Roots growing parallel to the drills cannot be pruned until stock has been lifted (table pruning). Stock is table-pruned at most Northwest nurseries to make trees easier to plant. Roots are trimmed with a variety of saws and other cutting tools, usually to a length of 20 to 25 cm, after trees have been bundled. The customer may specify a different length, however. A longer root system may be needed for auger planting on dry sites and a shorter one needed for planting in shallow soils.

21.9 Packaging and Post-Sort Storage

The principles that apply to the storage of unsorted stock (see 21.5) apply equally to the storage of sorted stock. Trees must be protected from desiccation, and they must be kept cool.

21.9.1 Packaging and moisture retention

The simplest way of protecting stock from desiccation is to package it in bags or cartons with a vapor barrier. One such container is the multiwall kraft/polyethylene bag. Another is a cardboard carton with a polyethylene liner. These are the most commonly used containers for storing and shipping tree seedlings. Such containers limit gas exchange by the stock (although polyethylene is permeable to some gases, including carbon dioxide); however, there is no evidence that this has harmful consequences (see review by Hocking and Nyland [17]).

If, for some reason, stock must be packaged in unsealed cartons, bales, or crates, it can be protected from moisture loss by the maintenance of a high ambient humidity (> 98%) or packaged with a water-saturated material such as peat moss or wood shavings. Cedar shavings should be avoided, however, because they release compounds that are toxic to seedlings [21].

21.9.2 Packaging and temperature control

The largest container in which seedlings will cool rapidly to ambient air temperature is around 1 m^3 (see 21.5.2). Most storage and shipping containers are an order of magnitude smaller than this and so allow rapid equilibration between seedling temperature and that of the ambient air. Controlling stock temperature may be difficult, however, if storage containers are placed tightly together. Containers are usually stored on pallets with provision for good air circulation around the pallets. Nevertheless, warm spots may occur within the stack of containers on a pallet unless there is room for air movement (i.e., 2 to 5 cm) between the containers.

21.9.3 Storage duration and temperature

Storage duration—not an independent variable—is determined by the interval between the optimum lifting date and the optimum planting date (see 21.3).

The most favorable storage temperature appears to be either just above or just below 0°C [17]. Frozen storage prevents molding [17]. But if molding does not occur, the performance of stock held just above 0°C may be as good as or superior to that of frozen stock [31]; this may be because, at temperatures below 0°C, respiration is too slow for adequate cellular maintenance.

While stock is frozen or thawed, a temperature gradient must be maintained from the edge to the center of the storage container. The gradient should be no more than 1 or 2°C, otherwise seedlings near the edge of the container will be subjected to unacceptably low (during freezing) or high (during thawing) temperatures. Thus, while stock is thawed or frozen, the ambient air temperature should be no more than 1 or 2°C above or below freezing, respectively. Under these conditions, freezing or thawing will take days or even weeks. Care must be taken, therefore, to allow adequate time to thaw frozen stock before it is needed for planting.

21.10 Conclusions

The opportunities for damage to stock while it is being lifted, graded, packaged, and stored are numerous. Many causes of stock damage can be guarded against by applying elementary physical and biological principles. Much remains to be learned, however, about the relationship between nursery cultural regimes and stock handling practices, on the one hand, and plantation performance, on the other. Nursery managers must strive, therefore, to determine how alternative handling and storage treatments affect the field-performance potential of the stock they produce. This will often require nursery-specific trials because, in its combination of soil, climate, species and provenances grown, cultural regimes, and handling practices, every bareroot nursery is unique.

References

1. Askren, C. A., and R. K. Hermann. 1979. Is the oscilloscope technique suitable for predicting survival of planting stock? Tree Planters' Notes 72:7-11.
2. Blake, J., J. Zaerr, and S. Hee. 1979. Controlled moisture stress to improve cold hardiness and morphology of Douglas-fir seedlings. Forest Sci. 25:576-582.
3. Brazier, A. L. 1982. Personal communication, B. C. Ministry of Forests, Victoria.
4. Burdett, A. N. 1978. Unpublished data, B. C. Ministry of Forests, Victoria.
5. Burdett, A. N 1979. New methods for measuring root growth capacity: their value in assessing lodgepole pine stock quality. Can. J. Forest Res. 9:63-67.
6. Burdett, A. N. 1981. Bin storage of unsorted bare root stock. Page 18 *in* Research review 1980-81. B. C. Ministry of Forests, Victoria.
7. Burdett, A. N., D. G. Simpson, and C. F. Thompson. 1983. Root development and plantation establishment success. Plant and Soil 71:103-110.
8. Chavasse, C. G. R. 1978. The root form and stability of planted trees, with special reference to nursery and establishment practice. Pages 54-62 *in* Proc., Symp. on root form of planted trees (E. Van Eerden and J M. Kinghorn, eds.). B. C. Ministry of Forests/Can. Forestry Serv , Victoria. Joint Rep. 8.
9. Chavasse, C. G. R. 1981. Planting stock quality: a review of factors affecting performance. New Zealand J. Forestry 25:144-171.

10. Cheung, K. 1973. Induction of dormancy in container-grown western hemlock [*Tsuga heterophylla* (Raf.) Sarg.]. B. C. Forest Serv., Victoria. Res. Note 59. 5 p.
11. Cleary, B. D. 1978. Seedlings. Pages 63-97 *in* Regenerating Oregon's forests (B. D. Cleary, R. D. Greaves, and R. K. Hermann, eds.). Oregon State Univ. Ext. Serv., Corvallis.
12. Coutts, M. P. 1981. Effects of root or shoot exposure before planting on the water relations, growth, and survival of Sitka spruce. Can. J. Forest Res. 11:703-709.
13. Daniels, T. G. 1978. The effects of winter plant moisture stress on survival and growth of 2 + 0 Douglas-fir seedlings. M.S. thesis, Oregon State Univ., Corvallis. 86 p.
14. Darbyshire, R. 1982. Personal communication, School of Forestry, Oregon State Univ., Corvallis.
15. Eliason, E. J. 1962. Damage in overwinter storage checked by reduced moisture. Tree Planters' Notes 55:5-7.
16. Hermann, R. K., D. P. Lavender, and J. B. Zaerr. 1972. Lifting and storing western conifer seedlings. Oregon State Univ., Forest Res. Lab., Corvallis. Unpubl. rep. 8 p.
17. Hocking, D., and R. D. Nyland. 1971. Cold storage of coniferous seedlings. Applied Forestry Res. Institute, New York State Univ., Coll. Environ. Sci. and Forestry, Syracuse. Rep. 6. 70 p.
18. Hopkins, J. C. 1980. Storage molds. Page 58 *in* Proc., Western forestry conf. Western Forestry and Conservation Assoc., Portland, Oregon.
19. Jenkinson, J. L., and J. A. Nelson. 1978. Seed source lifting windows for Douglas-fir in the Humboldt nursery. Pages B-75-B-77 *in* Proc., Western Forest Nursery Council and Intermountain Nurserymen's Assoc., Eureka, California, Aug. 7-11.
20. Kooistra, C. M. 1978. Personal communication, B. C. Ministry of Forests, Victoria.
21. Kreuger, K. W. 1963. Compounds leached from western red cedar shingle tow found toxic to Douglas-fir seedlings. U.S.D.A. Forest Serv., Pacific NW Forest and Range Exp. Sta., Portland, Oregon. Res. Note PNW-7.
22. Lavender, D. P., and P. F. Wareing. 1972. Effects of daylength and chilling on the responses of Douglas-fir [*Pseudotsuga menziesii* (Mirb.) Franco] seedlings to root damage and storage. New Phytologist 71:1055-1067.
23. Lawyer, J. N. 1981. Mechanization of nursery production of bare root deciduous planting stock. Pages 30-37 *in* Proc., Symp on engineering systems for forest regeneration. American Society of Agric. Engineers, St. Joseph, Michigan.
24. Mullin, R. E., and R. E. Hutchinson. 1978. Fall lifting dates, overwinter storage, and white pine seedling performance. Forestry Chronicle 54:261-264.
25. Stone, E. C., and J. L. Jenkinson. 1971. Physiological grading of ponderosa pine nursery stock. J. Forestry 679:31-33.
26. Stone, E. C., and E. A. Norberg. 1979. Root growth capacity: one key to bare-root seedling survival. California Agric. (May) :14-15.
27. Sutton, R. F. 1979. Planting stock quality and grading. Forest Ecology and Management 2:123-132.
28. Thompson, S. 1981. Shoot morphology and shoot growth potential in 1-year-old Scots pine seedlings. Can. J. Forest Res. 11:789-795.
29. van den Driessche, R. 1968. Measurement of frost hardiness in two-year-old Douglas-fir seedlings. Can. J. Plant Sci. 49:159-172.
30. van den Driessche, R. 1973. Prediction of frost hardiness in Douglas-fir seedlings by measuring electrical impedance in stems at different frequencies. Can. J. Forest Res. 3:256-264.
31. van den Driessche, R. 1977. Survival of coastal and interior Douglas-fir seedlings after storage at different temperatures and effectiveness of cold storage in satisfying chilling requirements. Can. J. Forest Res. 7:125-131.
32. van den Driessche, R., and K. Cheung. 1979. Relationship of stem electrical impedance and water potential of Douglas-fir seedlings to survival after cold storage. Forest Sci. 25: 507-517.
33. Worrall, J. 1971. Absence of "rest" in the cambium of Douglas-fir. Can. J. Forest Res. 1:84-89.

Chapter 22
Nursery Storage to Planting Hole: A Seedling's Hazardous Journey

J. W. Edgren

Abstract

The nursery environment can be heavily manipulated by nursery personnel, but the field environment into which seedlings are outplanted is less controllable and more diverse. Therefore, nursery manager and customer must work hand in hand to promote careful seedling processing and handling to ensure plantation success. Once lifted, the extremely vulnerable seedlings must be protected from temperatures above 2°C, freezing, relative humidities below 90%, plant moisture stress above 5 bars, and rough handling. Any damage incurred has a cumulative effect on seedling vigor. Seedlings must be shipped in containers that maintain proper environmental conditions and protect them from physical abuse; kraft/polyethylene bags are the most satisfactory container currently in use. Refrigerated vans with racks that allow air to circulate and heat to dissipate are the most dependable, trouble-free shipping vehicle. Once in the customer's hands, seedlings should be stored in refrigerated facilities for best results, although snow caches can provide excellent storage; mechanical refrigeration units can be either commercial or customer owned, portable or permanent. Planting-site storage requires coolers or tarps to protect seedlings from wind and sun. Planters must handle seedlings with extreme care to avoid injury and potential plantation failure. Nursery managers and customers must maintain close communications at all times to assure that substandard seedling performance is accurately diagnosed so that proper corrective measures can be taken.

22.1 Introduction

The nursery manager's responsibility in the reforestation of cutover timber lands does not end with the production and on-site storage of planting stock. Nursery environment, cultural practices, and seedling processing influence stock performance after planting just as surely as shipping, handling, and planting practices of nursery customers do.

This chapter presents a customer's view of the nursery manager's share of responsibility for plantation success and suggests ways that nursery managers and their customers can work together to promote mutual trust and understanding. It reinforces points already made in other chapters of this *Manual* concerning the microenvironments suitable for young dormant trees and advises nursery managers on points to monitor when visiting customer plantations. The points emphasized apply whether nursery managers and customers are federal, private, or a combination of these.

22.2 The Nursery Environment

The previous chapters in this *Manual* deal almost exclusively with the nursery environment—from nursery-site selection before seed is ever sown (see chapter 2, this volume) to cold storage of trees at the very end of the production process (see chapter 21). Soil structure and fertility can be altered considerably with organic matter and chemical fertilizers (see chapters 6 through 10). Climate can be manipulated by shading devices and watering strategies to protect seedlings from

In Duryea, Mary L., and Thomas D. Landis (eds.). 1984. *Forest Nursery Manual: Production of Bareroot Seedlings.* Martinus Nijhoff/Dr. W. Junk Publishers, The Hague/Boston/Lancaster, for Forest Research Laboratory, Oregon State University, Corvallis. 386 p.

both heat and cold (see chapter 12). Although absolute control of the nursery environment is not possible, a high degree of uniformity and regulation is essential for producing high-quality seedlings. The American Heritage Dictionary includes among its definitions of "nurse" the phrase ". . . to take special care of; foster. . . ." That is what should happen in a forest-tree nursery.

22.3 The Field Environment

By contrast, the physical environment into which seedlings must be planted is less controllable and more diverse than that of the nursery. Seedlings must be able to cope with drought, heat, cold, vegetative competition, and animals—for the most part without assistance from irrigation systems and other artificial means of protection. Seedlings grown in an average-size federal nursery may be destined for 10 National Forests. Whereas one individual, the nursery manager, has been solely responsible for them in the nursery, the seedlings may now be distributed to as many as 40 or 50 National Forest Districts with an equal number of individuals assuming responsibility for their care. The customer range of most private nurseries is even larger. The sheer size of the program assures many opportunities for problems.

22.4 The Seedling's Priorities

The seedling requires rigid control of temperature, moisture, and physical handling at all times during shipping, field storage, and planting. These needs have evolved over literally millions of years and cannot be changed. While separated from the soil, the seedling is in a hostile environment—like a man in space or a fish out of water—and is must vulnerable at this time because adequate protection is difficult to provide.

Compliance from all persons involved with strict handling standards cannot be overemphasized. Other priorities must give way when they begin to infringe upon the seedling's needs. For instance, if truck loads of trees arrive from the nursery late in the day, they must be placed into local storage immediately rather than the next day. At no time should trees be left on trucks over a weekend.

From lifting to outplanting, seedlings must go through 18 to 20 steps during which failure to control temperature and moisture or physical abuse can occur. Each instance of substandard treatment at any point in the production and handling sequence—exposure to high temperature, low humidity, or rough handling—accumulates to the detriment of the seedling [17]. Seedlings in transit or storage have no opportunity to recover from one instance of substandard handling before another occurs.

Although one or two minor violations of seedling priorities may not be critical, several together will almost surely cause some degree of irreversible damage. For instance, seedlings may tolerate an extra hour or two in the packing shed, a short trip in an open truck, or brief root exposure by careless planters, but all of these occurrences combined may result in physiological deterioration. Though outplanting survival has customarily been the measure of seedling quality and performance, growth reduction due to poor handling is a more serious consequence [1, 19]. It is therefore extremely important that all seedling handlers be aware of seedling priorities and of the need for their rigid observance.

22.5 Shipping Seedlings from the Nursery

The following discussion assumes that seedlings have been lifted while dormant during very late fall or winter for spring planting. Seedlings lifted for early fall planting will probably not be hardened off or fully dormant, so special shipping and handling procedures must be followed. Field foresters must plan regeneration so that trees can be lifted for spring planting while dormant (see chapters 14, 21, and 23, this volume).

During shipping and subsequent field storage, the seedling's immediate environment (inside packages) must be maintained at high relative humidity and low temperature. The standards here are the same as those which must be maintained in nursery storage. Humidity must be in the 90 to 95% range to assure that plants do not become desiccated [11]. However, free moisture must be avoided in seedling packages or the probability of storage-mold growth will be greatly increased [17]. Temperature must be in the 1 to 2°C (33 to 35°F) range to assure a low level of physiological activity and maintenance of dormancy [14]. Seedlings must not be frozen accidentally because cellular damage may result.

Fifty-two percent of seedling producers responding to the OSU Nursery Survey (see chapter 1, this volume) maintain seedling temperatures in transit as follows:

Temperature, °C (°F)	Respondents, %
1-2 (33-35)	14
2-3 (35-37)	24
3-4 (37-40)	14

Although physiological activity may not actually begin until temperatures reach 5°C (41°F) [14], it is extremely risky to permit seedling temperatures to rise that high. If seedlings begin growth in storage, they produce more heat, which induces more growth, which produces more heat, and so on. Once begun, this progression is difficult to reverse. Both shipping containers and vehicles are critical to maintaining the proper environment within packages.

22.5.1 Containers

Shipping packages must protect seedlings from desiccation by maintaining high relative humidity and from physical damage by shielding them from crushing pressure and hard blows. A moisture-holding medium such as sphagnum moss may be included in the shipping container, depending on either the policy of the nursery or the wishes of the customer. Shipping packages must serve both the nursery manager and field forester adequately or they are not acceptable. Close coordination between seedling producers and customers will assure a container that is acceptable to each and that adequately protects the trees.

The OSU Nursery Survey revealed the following proportionate use of packing containers and moisture-retaining media:

Container type	Respondents, %	Medium type	Respondents, %
"Poly" bags	90	None	53
Waxed boxes	43	Sphagnum moss	33
Bales	5	Shingle toe	9
		Peat moss	5

22.5.1.1 Bales

Open-ended bales are the least desirable containers for shipping and long-term storage because seedling tops are exposed to the atmosphere, permitting moisture loss and physical damage (stem breakage and loss of terminal buds). Their one advantage over closed boxes or bags is that seedlings can easily be rewetted if they become dry. Nursery customers who cannot control planting schedules or field-storage duration because of unpredictable planting-site weather should not specify open-ended bales.

22.5.1.2 Waxed boxes

Rigid, waxed cardboard boxes protect seedlings from physical damage and can be well sealed against moisture loss if a plastic liner is used. They are difficult to seal, however, without a plastic liner. A moisture-holding medium is not effective over long storage periods if boxes are not sealed, and rewetting is difficult and time consuming because each box must be opened to do the job. In their favor, boxes are easy to stack and use storage space efficiently. However, closely stacked boxes inhibit heat transfer and can lead to heat buildup within boxes. Special care must be taken in stacking if boxes are used (see also chapter 21, this volume).

22.5.1.3 "Poly" bags

Kraft/polyethylene, or "poly," bags appear to be the most popular and satisfactory storage and shipping container currently in use. Properly sealed bags retain moisture without a moisture-holding medium around roots and protect seedlings from physical damage if handled with normal care; they can keep seedlings moist for storage up to 3 months [4]. Bags are commonly sewn shut or banded by a banding machine with plastic strips (Fig. 1). However, care must be exercised to avoid packing bags so tightly with trees or so close together on stacking racks that heat generated by live seedlings cannot be properly dissipated [14, 17].

Figure 1. Nursery worker closes a "poly" storage bag, leaving room within for air circulation and heat transfer.

22.5.2 Vehicles

Seedlings are quite vulnerable during shipping. Shipping vehicles which contain seedling racks and which can maintain storage packages at temperatures near freezing [11] should always be used.

Temperature and moisture stress within storage packages must not be permitted to increase during shipping. Rising temperature stimulates physiological activity (respiration) in seedlings, which reduces carbohydrate reserves and produces more heat. Seedlings which become active during shipping will be most likely to mold during subsequent storage and continue to lose stored food reserves [17].

The OSU Nursery Survey revealed the following proportionate use of refrigerated vans and transit times:

Refrigerated vans			
Production shipped, %	Respondents, %	Transit time, hr	Respondents, %
0-49	33	2-4	66
50-79	14	5-8	5
80-100	53	9-12	24
		48 or more	5

Figure 2. Seedlings may be hauled (a) to planting sites in insulated trailers or (b) from nurseries to local long-term storage in refrigerated vans.

22.5.2.1 Open trucks

Open trucks or stock trailers are not adequate shipping vehicles. Cloth or other flexible covers alone, including space blankets, cannot protect seedlings adequately from sun and wind during the long trips frequently required to reach customer-owned storage. Seedlings also are subject to freeze injury in such vehicles. If open vehicles must be used, shipping containers should be securely covered with both a damp tarp and a radiation shield such as a space blanket.

22.5.2.2 Insulated vans

Insulated truck or trailer boxes with racks (Fig. 2a), though not the preferred vehicle, are adequate for short trips if the canopy interior is cold at the start and if temperature inside storage packages does not increase. But insulated vans may not be acceptable for long trips (6 to 8 hours or longer) on warm days or nights. If unavoidable and unpredictable delays occur in geographically isolated areas, the probability of seedling damage is great. Again, seedlings can easily freeze in such vehicles, though it would take longer than in rigs with less protection. Insulation of the truck bed is extremely important because the muffler is a primary heat source; an insulated canopy will trap muffler heat if it is not excluded by an insulated bed.

22.5.2.3 Refrigerated vans

A refrigerated van with racks (Fig. 2b) is the most dependable, trouble-free method of shipping seedlings from the nursery [6]. A refrigeration unit with both electric and liquid fuel capacity is probably best because it can operate at any time and place. Therefore, delays enroute usually are not critical.

Nurseries or customers need not own refrigerated vans because reliable commercial units are available on contract. Such haulers handle meat and seafood and can maintain a desired temperature indefinitely. During 1983, commercial refrigeration vans could be hired for $2.10/loaded mile—or about 1 cent/1,000 seedlings/mile; 200,000 seedlings (approximately one 40-foot van load) could be shipped 100 miles for $210 [pers. commun., 16].

22.6 Monitoring Storage Conditions off the Nursery Site

22.6.1 Ambient temperature and relative humidity

As mentioned earlier, temperature should be held between 1 and 2°C (33 and 35°F) and humidity between 90 and 95% within seedling packages so that seedlings do not break dormancy or become desiccated.

Probe thermometers are best for monitoring temperature within storage containers because they can be used without opening boxes or bags (Fig. 3). However, any holes created in containers by probe thermometers must be taped shut to avoid moisture loss. Thermometers must be carefully calibrated so their accuracy is known; then temperatures should be taken and recorded when seedlings leave nursery storage and again when they enter customer storage.

22.6.2 Within-package monitoring

The critical area for seedling environment is within storage packages, not in the ambient spaces of tree coolers (see 22.7.2). Coolers with ambient temperatures of − 1 to 0°C (30 to 32°F)

Figure 3. Monitoring within-package temperature with a probe thermometer.

can be expected to maintain seedling packages at 1 to 2°C (33 to 35°F). Even though cooler spaces may be − 1°C (30 °F), within-package temperatures as high as 4°C (39°F) have been measured when inadequate racking systems or closely stacked boxes have prevented air circulation. Root-mass temperatures 8°C (14°F) above ambient have been reported under conditions of poor air circulation [21].

The OSU Nursery Survey revealed that 95% of the nurseries questioned do some kind of within-package monitoring:

Condition monitored	Respondents, %
Temperature	90
Mold	52
Plant moisture stress	38
Root condition	24

Customers should also monitor conditions within packages to reduce the number of unknowns when attempting to analyze plantation failure.

22.6.3 Alarm systems

Alarm systems are essential on tree coolers to warn of refrigeration breakdowns. Flashing lights, sirens, or bells are usually installed but may not be adequate if coolers are in remote areas, as many U.S.D.A. Forest Service coolers are. A dialer programmed to ring a responsible person's telephone should also be installed where no one will be close to hear bells or sirens or see flashing lights on cooler boxes. Damaging temperatures can occur in full coolers in a matter of hours if breakdowns are not reported immediately.

For example, a single cooler containing 5,000 ft^3 of storage space can frequently hold up to 300,000 or more seedlings worth, let's say, $30,000. At 500 trees/acre, these seedlings may be used to plant 600 acres. If trees are damaged by inadequate refrigeration, the entire 600 acres may require replanting at a cost of $100/acre—or $60,000 for the planting contract alone. Overhead and contract administration could add another $100/acre, bringing total losses to $150,000. This is a worst-case example on an acre basis (all 600 acres failed), but probably not on a cost basis because these cost estimates are conservative and will almost surely inflate in the future.

22.6.4 Plant moisture stress

The pressure chamber technique [20] is best for monitoring plant moisture stress, or PMS ([5]; see also chapters 12 and 23, this volume). If a relatively dry atmosphere exists, creating a transpirational demand in the absence of a water supply to the roots, atmospheric pull creates tension in the seedling water columns. The chamber measures the pressure necessary to reduce this tension; a high instrument reading means high moisture stress in the seedling. The pressure chamber can be used to determine moisture stress in established as well as stored seedlings.

Stresses above 12 atmospheres (atm)[1] at lifting time are detrimental to seedling survival after outplanting [8]. This, therefore, can be interpreted as a maximum level. A 5-atm PMS has been recommended as a maximum during processing [2] and for seedlings in storage bags; this PMS level is not difficult to maintain and gives an acceptable margin of safety. Stresses higher than 5 atm indicate a problem requiring immediate correction. Humidities of 90 to 95% in shipping containers should be adequate to maintain storage PMS below 5 atm.

The J-14 hydraulic press is not considered a satisfactory tool for monitoring PMS at this time [5, 13]. Whereas the pressure chamber uses intact dead plant cells to reveal PMS,

[1] 1 atm = 14.7 psi = 1 bar.

the J-14 press crushes both live and dead cells. Pressure chamber use is based on proven physical theory [20]; use of the J-14 press is not. To date, tests have revealed little correlation between pressure chamber and J-14 press readings of PMS (see also chapter 23, this volume).

Nursery managers should take an active part in educating seedling customers on temperature and moisture stress and on the best equipment for monitoring PMS.

22.7 Customer Facilities

If nursery customers pick up or receive seedlings more than 1 or 2 days before planting dates, they must have storage capable of maintaining the temperature and humidity conditions previously identified. If the customer does not own such storage, commercial facilities are available in most cities, or refrigerated vans used to haul frozen foods over the highway can be rented and positioned near planting sites. Even if the customer intends to plant trees immediately on delivery, arrangements should be made to store seedlings under controlled conditions. For instance, if planting begins immediately but requires a week for completion, those trees planted after the second day probably will suffer without controlled storage. Furthermore, planting schedules frequently do not proceed as planned.

22.7.1 Nonrefrigerated storage

Most types of nonrefrigerated storage in which environmental conditions are not controlled (such as unheated buildings or root cellars) do not provide adequate protection for dormant seedlings. Snow caches are an exception to this rule.

22.7.1.1 Unheated buildings

In the past, seedlings were planted as they came from the nursery or were stored in shady areas or well-ventilated, unheated buildings. During the winter, these methods may have been adequate for periods up to a week. Over longer periods, however, normal daily temperature fluctuations above or below freezing can cause irreversible damage.

22.7.1.2 Root cellars

Root cellars, also used as temporary storage, will protect seedlings from freezing but are not cold enough to maintain dormancy or retard mold growth. Although root-cellar temperatures do not fluctuate widely, dampness at temperatures in the 10 to 13°C (50 to 55°F) range creates an excellent environment for mold growth, which presents a serious threat to the quality of stored seedlings [12].

22.7.1.3 Snow caches

Properly designed and constructed snow caches (Fig. 4) can provide excellent storage conditions for tree seedlings [7]. Temperature fluctuations are almost absent; the level will stay near 0°C (32°F). Relative humidity will be near 100%. Seedlings remain dormant, and mold growth is negligible.

To be practical, snow caches should be planned only where snow can naturally be expected. However, some U.S.D.A. Forest Service units have built them using snow machines like those used by ski-resort concessionaires.

22.7.2 Refrigerated storage

Mechanical refrigeration is probably the best choice for large planting programs. Storage need not be customer owned, and either portable or permanent units can be used.

22.7.2.1 Portable units

Portable units may be inadequate for removing heat due to inferior air circulation and unsatisfactory for maintaining proper

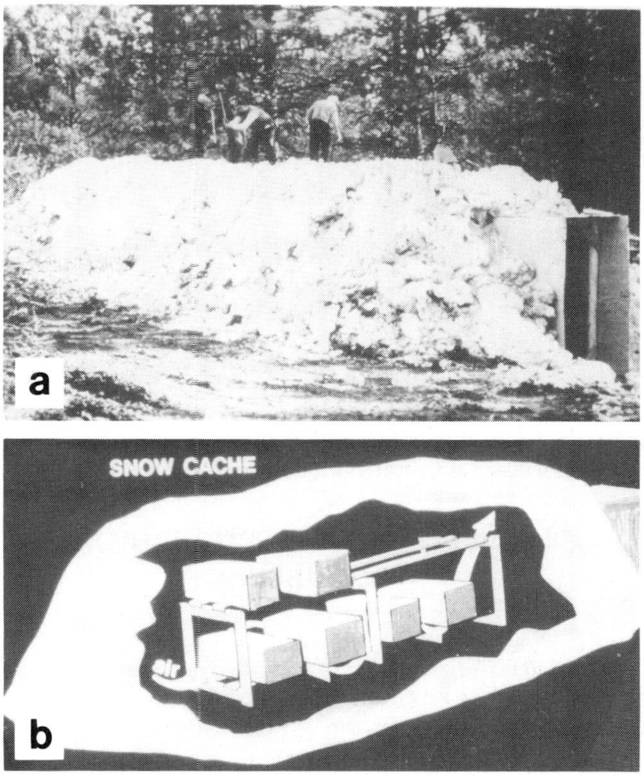

Figure 4. (a) A snow cache used for long-term field storage; (b) racks in snow caches (as in shipping vehicles) permit the circulation of air necessary for heat transfer.

relative humidity. Though controlling humidity is less critical for short-term storage of sealed packages, low humidity can still be damaging, and the potential for such damage increases with increased storage time. Frequently, because storage time cannot be accurately predicted, methods of humidity control become critical.

If portable units must be used, wet burlap on the floor or draped over storage racks is an acceptable way to maintain adequate humidity. Care must be exercised to keep seedling storage bags dry, however, because bags are more vulnerable to handling damage when wet. Free moisture in contact with storage containers also must be avoided because it is an invitation to mold problems.

22.7.2.2 Permanent units

Permanent storage units (Fig. 5) are best for carefully controlling temperature and humidity during long storage periods. Seedlings which are lifted during late winter but which cannot be planted until spring need the best storage that can be provided. Frequently, nurseries do not have enough cooler space in which to store their entire crop until snow leaves planting sites. If this is the case, other long-term facilities must be sought. For example, excellent equipment normally used to store fruit in Oregon and Washington may be rented to store seedlings in the off-season for fruit.

22.7.3 Frozen storage

For the time span most seedling customers need storage (1 to 3 months), frozen storage should not be necessary, though it may be beneficial for longer times (3 to 5 months). But as storage length increases, the probability of equipment failure increases. While this is also true of conventional nonfreezing storage, the consequences of equipment failure in conventional storage are less detrimental and easier to handle.

Figure 5. A permanent U.S.D.A. Forest Service tree cooler for storing seedlings at field units.

22.8 Seedling Protection during Outplanting

Seedlings remain in their original shipping package, where they are relatively easy to protect, until planting time. They must then be removed from the package and exposed to the atmosphere. Just a few seconds' exposure to a desiccating atmosphere can cause irreversible root damage because seedling root tips have no protective cuticle covering them as needles do. At this point, it is not as important to keep seedlings cool as it is to keep them wet. However, because not all seedlings taken to the field on a given day may be planted that day, they must also be kept cool.

Transportation from customer storage to the planting site should be either in refrigerated vans, which can be parked at the site, or in insulated truck canopies.

22.8.1 Planting-site storage

Planting-site storage is usually for 1 day or less. If 2 days are required, the seedlings should be returned to refrigerated storage overnight, mostly for protection from freezing, unless they are kept in a portable insulated or refrigerated unit at the planting site. Packages must always be kept in the shade regardless of storage method.

22.8.1.1 Desert cooler

The desert cooler technique (heat loss through evaporation) is satisfactory for keeping seedlings cool if evaporation surfaces can be kept moist. Wet burlap is a good moisture-holding medium, and wet storage bags at this point in the planting process are not a problem. Bags direct from refrigeration can be kept well below 4°C (39°F) in shade all day even when ambient temperatures approach 16 to 18°C (61 to 64°F).

22.8.1.2 Tarps

Reflective covers such as space blankets or the Seedling Heat Shield® do an excellent job of maintaining a cool atmosphere and shielding seedling packages from drying winds and radiation (Fig. 6). Such tarps reflect sunlight and heat outward and also trap the residual cold of refrigeration. Canvas tarps are not acceptable seedling covers because they tend to transmit heat inward as well as permit cold to escape [9].

22.8.1.3 Acclimatization

Acclimatization was developed in conjunction with "jelly rolling" (see 22.8.2.3) to permit seedlings to become gradually accustomed to planting-site climate. Seedlings are stored overnight at the planting site, usually in a tent, to reduce the shock of going directly from storage temperature to warmer planting-site temperatures. Though this sounds like a good idea and though planting systems using the acclimatization principle in conjunction with "jelly rolling" are quite successful, no studies are known which demonstrate the physiological desirability of acclimatization [pers. commun., 15]. Dormant seedlings appear not to be damaged by fairly wide fluctuations in normal temperature, though actively growing seedlings would be damaged by freezing [4]. If benefits truly do accrue from acclimatization, they are probably associated with the moisture-rich atmosphere in which seedlings are held. "Jelly rolling" may be advantageous where even a slight increase in moisture within the tree could benefit planting situations.

22.8.2 Planter handling

Careless handling and planting can undermine the most flawless stock production, shipping, and storage practices [3, 10]. Of those involved, planters probably have the smallest stake in long-term performance of planted stock. Therefore, those charged with the reforestation responsibility must help planters learn proper care and handling procedures for seedlings before and during planting.

22.8.2.1 "Bagging up"

The highest probability of damaging root exposure occurs when planters fill planting bags with trees. This operation must be done quickly. Planters counting trees must do so with trees in the packing bag, their roots immersed in a bucket of water, or in a tent shielded from the wind—not on the tailgate of a pickup. If storage containers are not completely emptied of seedlings, care must be taken to close them after bagging up.

22.8.2.2 Root dips

Dipping roots in aqueous solutions of various moisture-retaining products or in plain water immediately before trees are bagged for planting helps protect roots from desiccation during the short time they are exposed during planting [18]. Ground peat moss, horticultural-grade vermiculite [6], or a mixture of the two is frequently used. Other moisture-holding products such as Terra Sorb® also have been tried, though experience with them is limited.

22.8.2.3 "Jelly rolling"

The term "jelly roll" has been used to describe a seedling protection method developed by the U.S.D.A. Forest Service in Utah. Seedlings are rolled into a water-saturated burlap

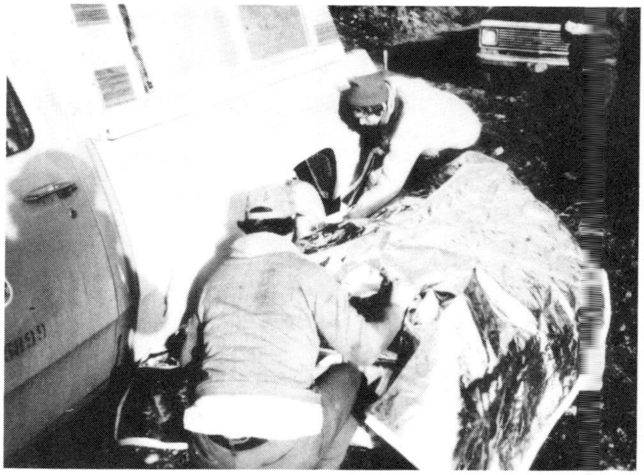

Figure 6. Forest workers store seedlings temporarily under a reflective tarp.

sheet (36 by 18 inches), with tops exposed and roots enclosed (Fig. 7). A moisture-holding medium such as sphagnum or peat moss is then packed around roots for additional protection. This "jelly roll" is slipped into the planter's bag, and trees are extracted one at a time for planting [6].

"Jelly rolling" increases costs (about $6 to $8/1,000 trees, 1981 prices) and probably is not necessary in locations or during periods where atmospheric stress is low. However, such protection is worth the cost when planting is anticipated during periods of relatively high climatic stress [6].

22.8.2.4 Insulated planting bags

Planting bags insulated with polyurethane foam (Fig. 8) or similar materials can be used to protect seedling roots when climatic conditions become warm and dry. Waterproof bags with foam linings are particularly effective in keeping trees moist and cool during the hours required to plant the number of trees they contain. Uninsulated canvas bags may permit root drying during this time.

22.8.2.5 The planting operation

During planting, seedlings may be beyond the direct control of planting supervisors and at the mercy of the planters. To avoid damaging root exposure, planters must remove seedlings from planting bags one at a time. Carrying several seedlings in the hand from planting spot to planting spot will permit roots to dry out.

Planters must not constrict roots into a tight mass before planting or strip laterals from the root system. These actions

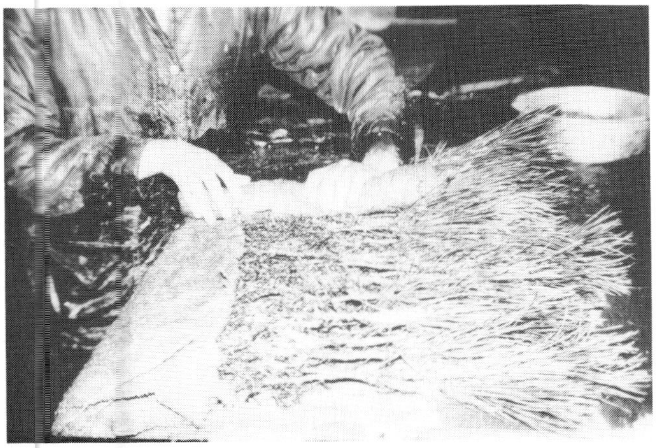

Figure 7. Seedlings being "jelly rolled" to protect roots from desiccation during planting.

Figure 8. A waterproof planting bag with insulating insert, used for protecting seedlings during planting.

damage roots and reduce the ability of the tree to cope with problems associated with all planting sites.

The importance of careful seedling handling was recently demonstrated with bareroot Monterey pine (*Pinus radiata* D. Don) seedlings [1]. Control seedlings were carefully hand lifted from nursery beds, while all others were removed by a belt lifter. All seedlings were packed directly into rigid boxes. Root abrasion caused by normal lifting and soil removal was noted, and survival and growth data were recorded for several classes of root damage. The seedlings were then transported to planting sites and planted from the boxes, greatly reducing the number of handling steps. Growth and survival were compared after 6 months for batches of seedlings stored for 13 and 20 days. Storage conditions were not described. Control (hand lifted) seedlings in both storage categories averaged 21.6 cm tall and 4.5 mm in diameter; survival was 100%. Machine-lifted seedlings stored 13 days suffered the most damage; they averaged 11.1 cm tall and 2.2 mm in diameter, with 93% survival; machine-lifted trees stored 20 days averaged 8.1 cm tall and 2.2 mm in diameter, with 37% survival. These findings clearly indicate that seedlings handled carefully can be stored for at least 3 weeks with no adverse effects and that trees handled roughly deteriorate even after very short storage periods.

22.9 Conclusions and Recommendations

- Once bareroot seedlings are lifted, they are extremely vulnerable to adverse environmental conditions and difficult to protect in the field. The seedling's need for tender loving care must be given highest priority.
- Seedling damage is cumulative. Each instance of substandard treatment decreases the seedling's ability to perform up to the capability of the planting site.
- Adverse environmental conditions during processing, storage, and transportation include temperatures above 2 to 3°C (35 to 37°F); relative humidities below 90% or PMS above 5 bars; and rough handling resulting in physical damage.
- Nursery managers and customers must maintain close communications to assure that substandard seedling performance is accurately diagnosed so that proper corrective measures can be taken.

Responsibility for plantation success or failure must be shared. Planters and silviculturists cannot perform miracles with stock that has been carelessly produced any more than nursery managers can guarantee stock performance in spite of sloppy field handling. The understandable tendency of nursery managers and customers alike to hold the other responsible for plantation failure must be tempered by mutual understanding of each other's problems and mutual trust in each other's professional abilities. The practice of field foresters visiting nurseries in Oregon and Washington to observe operations and discuss stock characteristics is quite well established; nursery managers must put forth an equal effort to visit customer plantations.

Production of forest-tree seedlings must be viewed *not* as an end in itself, but as a means to an end: seedlings are produced to replace trees harvested from the forest. Everyone involved in production and planting must look beyond the green, healthy-looking seedling to the ultimately more meaningful measure of success—the established plantation.

References

1. Anonymous. 1978. Pine seedlings—handle with care! What's new in forest research. Forest Res. Institute, Rotorua, New Zealand. No. 67. 4 p.
2. Cleary, B. D. 1978. Low internal water stress during nursery processing increases survival and growth of bare root seedlings. Forestry Update 4(4):3.
3. Cleary, B. D., and D. R. DeYoe. 1982. Seedling care and handling. Oregon State Univ. Ext. Serv., Corvallis. Ext. Circular 1094. 4 p.
4. Cleary, B. D., R. D. Greaves, and P. W. Owston. 1978. Seedlings. Pages 63-97 *in* Regenerating Oregon's forests (B. D. Cleary, R. D. Greaves, and R. K. Hermann, eds.). Oregon State Univ. Ext. Serv., Corvallis.
5. Cleary, B. D., and J. B. Zaerr. 1980. Pressure chamber techniques for monitoring and evaluating seedling water status. New Zealand J. Forestry Sci. 10(1):133-141.
6. Dahlgreen, A. K. 1976. Care of forest tree seedlings from nursery to planting hole. Pages 205-238 *in* Proc., Tree planting in the inland Northwest (D. M. Baumgartner and R. J. Boyd, eds.). Washington State Univ. Cooperative Ext. Serv., Pullman.
7. Dahlgreen, A. K., R. A. Ryker, and D. L. Johnson. 1974. Snow cache seedling storage: successful systems. U.S.D.A. Forest Serv., Intermountain Forest and Range Exp. Sta., Ogden, Utah. Gen. Tech. Rep. INT-17. 12 p.
8. Daniels, T. G. 1978. The effects of winter plant moisture stress on survival and growth of 2+0 Douglas-fir seedlings. M.S. thesis, School of Forestry, Oregon State Univ., Corvallis.
9. DeYoe, D. 1981. Seedling bags get hot. Forestry Update 6(4):2.
10. Hobbs, S. 1979. Handling seedlings on the planting site. FIR Rep. 1(4):9.
11. Hocking, D., and R. D. Nyland. 1971. Cold storage of coniferous seedlings, a review. Applied Forestry Res. Institute, State Univ. New York, Coll. Forestry, Syracuse. Res. Rep. 6. 70 p.
12. Hopkins, J. C. 1975. A review of moulding of forest nursery seedlings in cold storage. Can. Forestry Serv., Pacific Forest Res. Centre, Victoria, B.C. Rep. BC-X-128. 16 p.
13. Jaramillo, A. 1981. Review of techniques used to evaluate seedling quality. Pages 84-95 *in* Proc., Joint Meeting, Western Forest Nursery Council and Intermountain Nurserymen's Assoc., Boise, Idaho, Aug. 12-14, 1980. U.S.D.A. Forest Serv., Intermountain Forest and Range Exp. Sta., Ogden, Utah. Gen. Tech. Rep. INT-109.
14. Jenkinson, J. L. 1975. Cold storage of bare-root planting stock. Service-wide Conf. on Planting Stock Production, Coeur d'Alene, Idaho. 8 p.
15. Lavender, D. P. 1980. Personal communication, Dep. of Forest Sci., Oregon State Univ., Corvallis.
16. Morby, F. E. 1981. Personal communication, U.S.D.A. Forest Serv., J. Herbert Stone Nursery, Central Point, Oregon.
17. Navratil, S. 1973. Pathological and physiological deterioration of planting stock in cold storage (silviculture review). Timber Seminar, Thunder Bay District, Thunder Bay, Ontario. Unpubl. rep. 27 p.
18. Owston, P. W., and W. I. Stein. 1972. Coating materials protect Douglas-fir and noble fir seedlings against drying conditions. Tree Planters' Notes 23(3):21-23.
19. Pierpoint, G., J. M. Patterson, J. G. Boufford, and C. Glerum. 1977. Irregular growth of outplanted red pine. II. The influence of handling and planting on 1-year performance. Ministry of Natural Resources, Ontario. Forest Res. Note 6. 4 p.
20. Scholander, P. F., H. T. Hammel, E. D. Bradstreet, and E. A. Hemmingsen. 1965. Sap pressure in vascular plants. Science 148:339-346.
21. Shearer, R. C. 1970. Problems associated with western larch planting stock. Pages 27-29 *in* Proc., Joint Meeting, Western Forest Nursery Council and Intermountain Forest Nurserymen's Assoc., Coeur d'Alene, Idaho.

Chapter 23
Assessing Seedling Quality

G. A. Ritchie

Abstract

Characteristics of planting stock which reflect quality (defined here as performance potential) are categorized as either "performance" attributes or "material" attributes. Performance attributes, such as root-growth potential, cold hardiness, and stress resistance, are assessed by subjecting whole seedlings to certain environmental regimes and evaluating their response. Because performance attributes are integrators of all or many seedling subsystems, they often correlate well with seedling performance potential; however, they tend to require laborious and time-consuming procedures. Material attributes, such as dormancy status, water relations, nutrition, and morphology, are assessed by measuring the attribute in question by any number of direct or indirect methods. Although material attributes are often more easily and rapidly measured than performance attributes, the former generally yield little definitive information on seedling quality unless values fall well outside of some established range. Of the Northwest nurseries responding to the OSU Nursery Survey, many reported using various methods to assess seedling conditions. However, most methods were used to indicate the desirability of carrying out certain cultural operations, such as irrigation or lifting, rather than to measure seedling quality itself.

23.1 Introduction

The final test of a forest-tree seedling is its performance after outplanting. Every observer of plantation establishment is aware that survival and adequate early growth of planted seedlings cannot be taken for granted. Some seedlings survive and prosper even on difficult sites, whereas others die soon after planting or remain in check for several years. These differences in performance reflect differences in factors which collectively make up what is known as "seedling quality." As defined at the New Zealand IUFRO workshop, "Techniques for Evaluating Planting Stock Quality" (August 1979), the quality of planting stock is the degree to which it realizes the objectives of management—"Quality is fitness for purpose." If the purpose of planting stock is to become established and grow successfully in a plantation, then fitness is a function of survival and growth potential. Seedling quality, then, is defined in these terms in this chapter.

Seedling quality is prerequisite to intensive forest practice because upon it depends the initial architecture of the forest. Hence, it has been the subject of much research and several recent reviews. Bunting [7] discussed morphological and physiological aspects of seedling quality. Jaramillo [53] evaluated several electrical and chemical indicators of planting-stock condition. Chavasse [15] reviewed cultural techniques for maintaining seedling quality with emphasis on New Zealand production systems and species. Schmidt-Vogt [97] reviewed much of the European work, and Cleary et al. [18] gave a brief overview pertinent to Northwest nurseries. A special issue of the *New Zealand Journal of Forestry Science* (vol. 10, no. 1) is dedicated entirely to the subject of planting-stock quality. Finally, Sutton

In Duryea, Mary L., and Thomas D. Landis (eds.). 1984. *Forest Nursery Manual: Production of Bareroot Seedlings.* Martinus Nijhoff/Dr W. Junk Publishers, The Hague/Boston/Lancaster, for Forest Research Laboratory, Oregon State University, Corvallis. 386 p.

[114] presented an especially thoughtful, yet concise, synthesis of the subject. All of the above make excellent reading.

Seedling quality reflects the integration of a multitude of physiological and morphological characteristics of the seedling—much as human health reflects a vast array of human physiological and morphological properties—and an instructive analogy can be drawn here. When examined by a physician, the patient is subjected to a battery of measurements—some simple and others highly sophisticated. It is from the collective results of these tests, not just one test alone, that the physician is able to characterize the patient's general health. As there is no one index of human health, there is no one yardstick of seedling quality. Furthermore, the likelihood of finding one is low. Like the physician, we have at our disposal an array of procedures which can be applied to develop information on certain aspects of seedling quality. From these tests and the informed interpretation of their results, it is possible to predict, with some reliability, the survival and growth potential of any seedling on any site.

For this review, attributes of seedling quality are grouped into two categories. **Performance attributes** are measured by subjecting whole seedlings to some test condition and measuring their performance; examples are root-growth potential and stress resistance. These attributes integrate the combined functioning of many physiological and morphological subsystems within the seedling. **Material attributes** include certain of these subsystems; examples are root starch concentration, leaf osmotic potential, and shoot:root ratio. These attributes, taken in mass, ultimately determine seedling performance but, considered individually, have relatively low predictive value unless they fall far outside some normal range. The relationship among material and performance attributes, and their influence on seedling quality, are illustrated in Figure 1.

In this chapter, I review in detail techniques proposed for assessing seedling quality towards defining the state-of-the-art of this technology, contrast current practices in Northwest nurseries with the state-of-the-art, and present practical information for forest-nursery and regeneration personnel.

Unfortunately, providing balanced coverage of the various seedling attributes discussed is not always possible. For example, a detailed section on frost-hardiness testing is followed by a brief page on stress testing. This apparent lack of balance does not necessarily indicate the relative importance of the former and unimportance of the latter, but rather reflects the simple fact that the scientific literature on frost hardiness is vast whereas that on stress testing is limited.

Finally, much of the quantitative information presented here was developed in research on Douglas-fir [*Pseudotsuga menziesii* (Mirb.) Franco]. This, again, reflects the nature of the available literature. A solid data base for this very important species is highly desirable. Such a data base is also needed, however, for many other important species, particularly the interior pines (*Pinus* spp.), which have not been the subject of such intensive investigation. However, the biological similarities among conifers native to the Northwest are generally strong enough to render this review relevant to most, if not all, commercially important species.

23.2 Performance Attributes

23.2.1 Root-growth potential

A key to seedling survival and establishment is rapid resumption of water and mineral uptake after outplanting. Resumption depends on the rate at which seedlings renew intimate soil-root contact by initiating and elongating roots into the soil matrix. Stone [108] first reported that tree seedlings vary widely in their ability to regenerate new roots after planting into an optimum environment—which depends upon their physiological status. This ability, called root-growth potential (RGP) [85], is a key seedling-quality attribute for the above reason; it is also a good general indicator that all systems in the seedling are functioning properly. High RGP is often correlated with high field survival [e.g., 85; also 71].

A seedling develops RGP while it is growing in the nursery. If seedlings are not to be stored, RGP should be measured immediately after lifting. However, because RGP can change dramatically during storage [47, 69, 84, 140], it should be measured after storage as well as before. Expression of RGP is mediated by conditions on the planting site, especially soil moisture and temperature. This sequence, recently reviewed by Ritchie and Dunlap [85], is summarized in Figure 2.

23.2.1.1 Standard measurement method

The standard method of measuring RGP is similar to that first described by Stone et al. [112, 113]. After all white root tips are removed, seedlings are potted in a light soil or potting mix (peat:vermiculite forestry mix is recommended) and held for a specific period, usually 28 days, under conditions favorable for root growth. Though these conditions vary somewhat for different species, 20°C air and soil temperature and 16-hour photoperiods are often used. Seedlings are then carefully washed out of the pots and new roots measured, counted, or both. Three pots of five seedlings each per treatment are normally sufficient to give valid statistical comparisons.

Test conditions can be tailored to species (e.g., boreal conifers may have lower optimum soil temperatures), but it is particularly important that conditions be consistent among tests. Most critical are soil temperature and moisture, air temperature, humidity, and photoperiod [85, 115], each of which can affect test results.

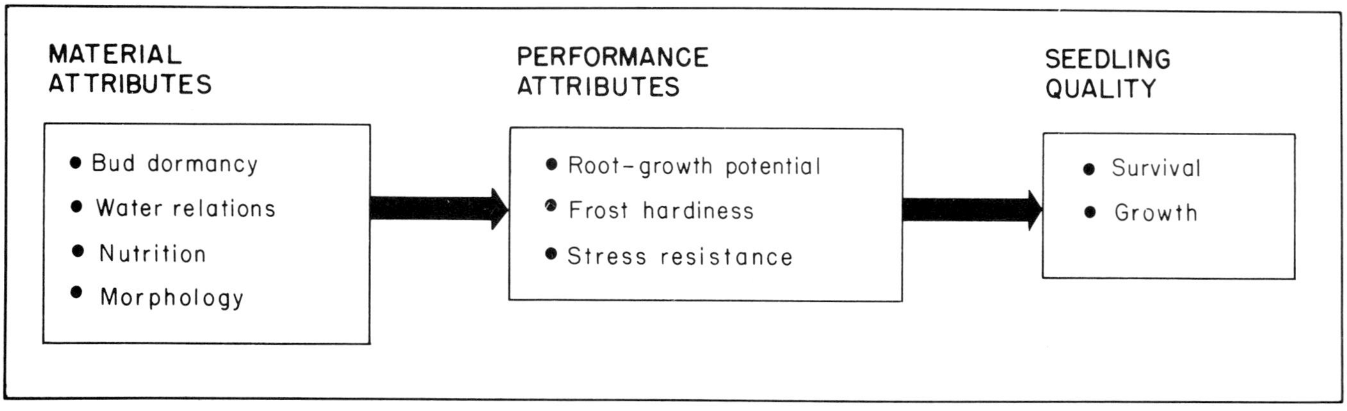

Figure 1. Seedling quality can be assessed in terms of measurable performance attributes which, in turn, reflect the sum of innumerable material attributes. Performance attributes are normally better predictors of seedling survival and growth than material attributes.

23.2.1.2 Short cuts

The standard method just discussed (23.2.1.1) has three major disadvantages: (1) it requires substantial quantities of potting soil and considerable greenhouse space, (2) root measuring and counting are laborious and time consuming, and (3) results are not available for 1 month. Several approaches that circumvent these problems follow:

Hydroponic growing.—RGP tests need not necessarily be carried out in pots of soil mix. We have had good results with aerated water baths made from 38-liter (10-gal.) fish aquariums, painted black, and covered with plywood lids into which 5.5-cm (2.2-in.) holes had been drilled. Seedlings were suspended into the tanks through #12 rubber stoppers drilled and slit radially and placed in the holes. Baths were filled with tap water, which was continuously aerated with a small aquarium pump and bubble stone. No nutrients were added, but a copper penny was placed in each tank to impede algae and mold growth. When held in a greenhouse next to seedlings in standard root-growth trials, seedlings in the baths produced nearly the same length and number of new roots as those in the pot trials in 11 separate tests.

Some advantages of hydroponic growing are: (1) less space is required, (2) there is no need for pots or potting mix, (3) root temperature and moisture conditions are readily controlled and remain nearly constant, (4) roots are neither broken nor lost during extraction, (5) roots are clean and very easily measured, and (6) root growth can be observed during the test.

Shortening testing time.—Several workers have experimented with reducing testing time of the standard method from 1 month to only 1 or 2 weeks. According to Burdett [pers. commun., 9], 1- and 2-week results are well correlated with 4-week results in some species, hence greatly reducing the time needed for testing. Burdett's test conditions, which accelerate root growth, are:

Day temperature	30 ± 0.5°C
Night temperature	25 ± 0.5°C
Daily photoperiod	16 hours
Light intensity	11,000 ± 1,000 lux
Relative humidity	75 ± 5%

It has been our experience with coastal Douglas-fir (var. *menziesii*) that new roots do not appear until near the end of the second week at 20°C air and soil temperature. It may be possible to accelerate this process with forcing conditions such as Burdett describes. Stone [unpubl. data, 110] has tried accelerated conditions with white fir [*Abies concolor* (Gord. & Glend.) Lindl. ex Hildebr.] with only limited success.

Streamlining measurement procedures.—Typically, number and total length of new roots per seedling are measured to estimate RGP. Number gives an estimate of initiation rate, and length an estimate of elongation rate. Both are normally needed for detailed physiological studies but may not be necessary for gross estimates of RGP.

Some short cuts are available: (1) counting the number of roots which exceed some critical length (e.g., 1 cm); (2) measur-

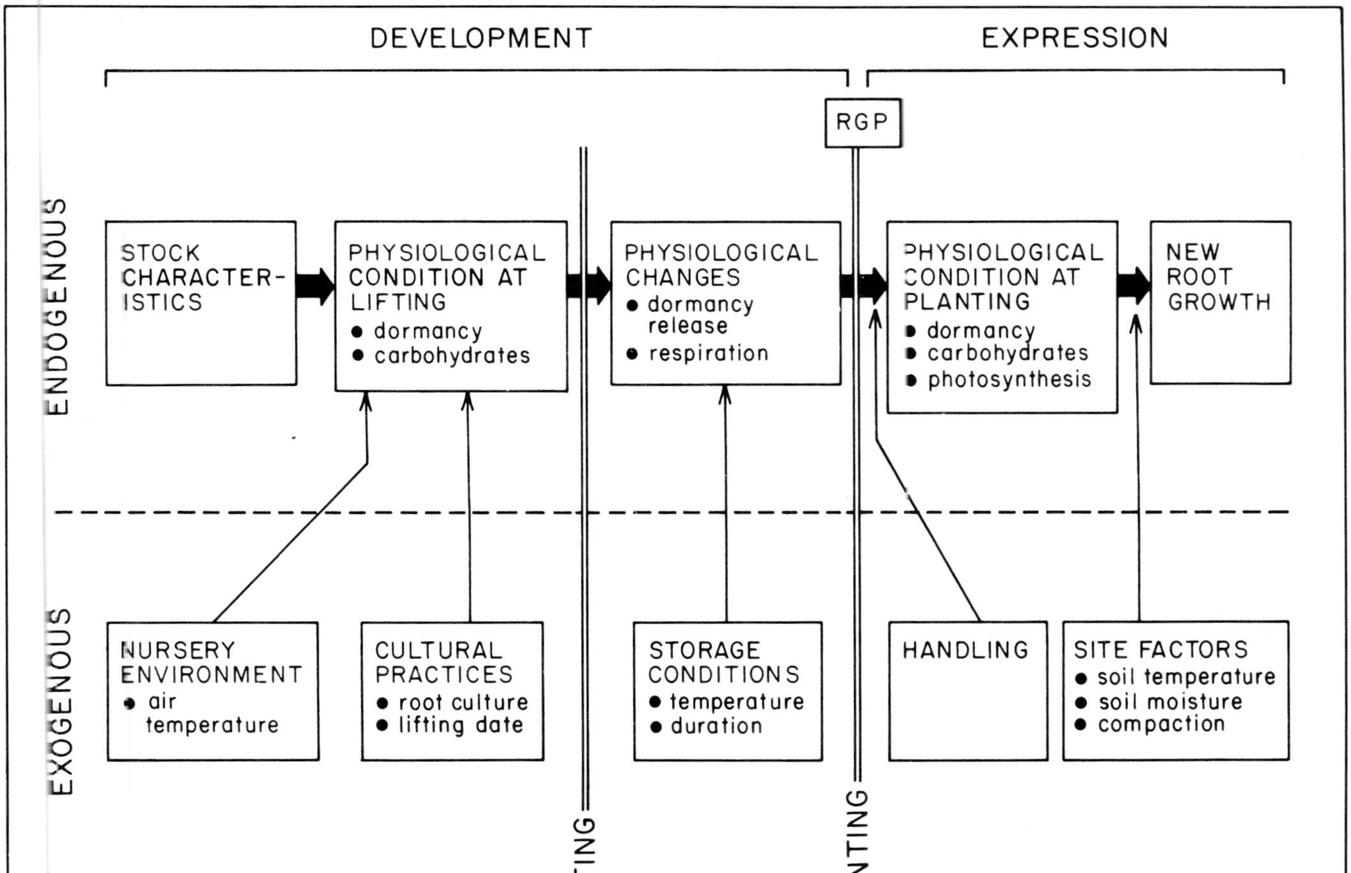

Figure 2. Development and expression of root-growth potential (RGP). Development is affected by endogenous (internal) seedling properties which reflect exogenous (external) forces; these forces act upon the seedling during nursery growth and storage. After planting, expression is limited by factors at the planting site. The most appropriate point at which to measure RGP is immediately before planting (adapted from [85]; reproduced with permission from the *New Zealand Journal of Forestry Science*).

ing the length of only the three longest roots; (3) clipping, drying, and weighing the new roots; (4) developing a scoring index based upon numbers of roots exceeding a certain length; (5) developing a set of "reference" photographs of root systems of known lengths for visual comparisons; and (6) measuring root volume before and after the 30-day test [8]. Burdett [pers. commun., 9] recommends a scoring system based on the following scale:

Class	Description
0	No new root growth
1	Some new roots, but none over 2 cm long
2	1 to 3 new roots over 1 cm long
3	4 to 10 new roots over 1 cm long
4	11 to 30 new roots over 1 cm long
5	More than 35 new roots over 1 cm long

Each of these methods is useful, but information content usually falls with measurement cost. It is important to design the measurement strategy with objectives and resource constraints clearly in view.

23.2.2 Frost hardiness

Frost hardiness may be defined as the minimum temperature at which a certain percentage of a random seedling population will survive or will sustain a given level of damage [102, 121, 128]. The term LT_{50} (lethal temperature for 50% of a population) is commonly used to define the hardiness level.

During the growing season, tree seedlings are normally killed by temperatures near freezing. During fall, hardiness increases rapidly in response to changing photoperiod, low temperatures, and other factors [136] and reaches a seasonal minimum in midwinter. For coastal Douglas-fir, this minimum is around −25°C; in many timberline species, it is near −40°C [3]; and in boreal species such as spruces (*Picea* spp.) and firs (*Abies* spp.), it may be −70°C or lower [93]. With a return to springlike conditions, hardiness is rapidly lost.

If tree seedlings are subjected to temperatures below their hardiness limit after planting, mortality will be substantial. Hence, frost hardiness can be a major factor affecting survival and establishment [134] and must be regarded as a key seedling-performance attribute.

The mechanisms of frost hardiness are very complex and involve many interacting factors, including (1) the ability of plant tissues to (a) avoid or tolerate freeze desiccation, (b) prevent lethal intracellular ice-crystal formation, and (c) withstand nonlethal extracellular ice-crystal formation, and (2) the propensity of cell water to reach subfreezing temperatures without freezing, i.e., supercooling [67, 136]. Elaboration of these mechanisms is beyond the scope of this chapter, but for the interested reader, Mazur [68] and Levitt [65] offer thorough analyses, Weiser [136] gives a concise review pertinent to woody plants, and Glerum [34] and Brown [6] review aspects of frost hardiness in forest trees.

Assessing frost hardiness has two steps: (1) subjecting plant material to subfreezing temperatures and (2) evaluating the effect of this treatment. Frost-hardiness determination can then usefully be applied in the nursery (1) as a guide to providing frost protection during autumn and spring and (2) as an indicator of stock hardiness at planting time. Because the effects of cold storage on hardiness are poorly understood, hardiness rating of seedlings when lifted may not be valid after storage.

23.2.2.1 Freezing treatments

The classical procedure for freeze testing is to (1) randomly select a sample of seedlings from the population of interest, (2) place them into a freeze chamber of some type, (3) lower the temperature at a given rate until the test temperature is reached, (4) hold the test temperature for a given time period, (5) then return at a given rate to the starting temperature. This is repeated across a range of temperatures believed to bracket the hardiness of the seedlings.

Several aspects of this procedure warrant attention. First, sample size should be carefully determined because seedlings (and transplants) vary genetically with respect to hardiness development and phenology. Generally, between 20 and 40 plants are used depending upon species and experience of the evaluator. Second, the rate of temperature decrease should be monitored. Timmis [unpubl. data, 119] recommends that a 5°C/hour temperature decrease not be exceeded because higher rates may compound injury induced by the minimum temperature. Note, however, that the rate of temperature increase can exceed that of temperature decrease, e.g., 20°C/hour vs. 5°. Third, duration of the minimum temperature also is important because longer exposures normally increase damage. Two hours at minimum temperature is common. Most crucial is that, for results to be comparable, all tests must be carried out in precisely the same manner [65]. Repeated freezing can result in increased damage, especially when the minimum temperature is low enough to cause injury [36]. It is also important, when using whole seedlings, to insulate the roots because they are likely to be far less hardy than the shoots [39, 80].

Numerous types of freezing chambers are available, ranging from simple units which can be taken to the field and placed over seedlings [e.g., 35] to sophisticated laboratory chambers with precise programmable temperature controllers [e.g., 101]. Such chambers include radiation [2] and advective [89] frost chambers and freezing bars [92] which provide temperature gradients. Advantages and disadvantages of the various types of units are discussed in a comprehensive review by Warrington and Rook [134].

23.2.2.2 Evaluating frost damage

The only procedure for unequivocally evaluating damage after freezing tests is to hold the seedlings in a greenhouse or growth chamber for several weeks and then visually to inspect them, including roots, for damage. It is also critical to understand which tissues are likely to be least hardy, which varies seasonally. Menzies and Holden [73] recommend the following index to evaluate freeze damage in Monterey (*Pinus radiata* D. Don) and bishop (*Pinus muricata* D. Don) pines and Douglas-fir seedlings:

Index value	Damage
0	None
1	Buds undamaged, needles reddening
2	Buds may be damaged, 10 to 30% of needles killed
3	40 to 60% of needles killed
4	70 to 90% of needles killed
5	All needles killed, stem dead

The obvious and formidable disadvantage to this approach is the often excessive time required for damage to become apparent, during which seedlings must be cared for and observed.

Several methods have been proposed for avoiding this waiting period by indirectly assessing frost damage immediately after the freezing test. Most are based upon measuring the degree of inactivation of enzymatic or metabolic functions or measuring changes in membrane properties. Timmis [117] has critically evaluated the applicability of five such techniques to tree seedlings: (1) direct measurement of photosynthesis, (2) leaf-segment flotation on phosphate buffer solution as an estimate of photosynthetic rate [122], (3) dehydrogenase enzyme activity assessed with the tetrazolium chloride test [106], (4) changes in membrane ion permeability detected by electrical impedance [5, 33, 125, 126], and (5) plant water potential measured with a pressure chamber [4]. Timmis found that each method was useful to some degree in detecting freezing damage. However, accuracy of the determination depended upon the stage of hardiness of the tissue when freeze tested. On balance, the electrical impedance method gave the most reliable results across all levels of hardiness, confirming findings of van den Driessche [126]. Differences in electrical impedance ratio in the upper stem predicted survival after freezing with 87% accuracy and enabled LT_{50} values to be predicted within 2°C at all phases of hardening and dehardening.

Impedance ratio measurement.—The following method is recommended for coastal Douglas-fir nurseries [unpubl. data, 120]. The meter used, designed by W. D. Perry, Weyerhaeuser Co., is enclosed in a small hand-held plastic box.[1] Impedance ratios (IR) obtained on freeze-treated seedlings are interpreted differently according to the stage of hardening or dehardening when seedlings are frozen. During early stages of hardening (until late November), LT_{50} values can be estimated within 1°C if an IR of 3 is used to discriminate between live and dead seedlings. That is, seedlings with ratios lower than 3 will be dead and those with higher than 3 will survive. After November the discriminating ratio increases gradually to about 5 in late January. IR values are less reliable in midwinter because low temperatures tend to kill buds before stems and because bud injury is not detected by stem impedance ratios. Therefore, to estimate freeze damage during this period, seedlings should be held in a warm greenhouse for 3 days and the buds then cut open and examined for obvious browning in relation to (1) uninjured buds and (2) buds definitely killed by deep freezing (lower than −30°C). The extent of bud mortality in the test seedlings is then judged and classified.

In April, or in prematurely dehardened seedlings, the meter again gives good estimates of LT_{50} values if a discriminating ratio of 2.5 is used.

Diffusate conductivity method.—This widely used method—possibly more accurate, but also more laborious, than measuring IR—is based upon the principle that freeze-injured cells contain damaged membranes which allow cell fluid to escape into the xylem. Cell fluid contains dissolved materials and therefore has higher electrical conductivity than xylem water, which is relatively pure. Comparing the conductivity of xylem diffusate from among uninjured, injured, and dead seedlings provides an estimate of the amount of injury, if any, that occurred. The method, pioneered by Dexter et al. [21, 22], has been used successfully on a number of woody plant species [e.g., 107, 125, 128, 139].

In the following procedure (after [36]), stem segments 2.5 cm long are collected from freeze-treated seedlings immediately below the apical bud. These are placed into capped glass vials containing 15 ml of distilled water and held in a water bath at 25°C for 24 hours. They are then shaken, and the conductivity of the water (and xylem diffusate) is measured with a suitable device. The stem segments are then killed (frozen at −15°C for 24 hours), replaced into the 25°C water bath for 24 hours, and remeasured. Relative conductivity, R_t, is calculated as

$$R_t = L_t/L_k \qquad (1)$$

where L_t is the specific conductivity of the diffusate from the sample subjected to temperature (t), and L_k is the specific conductivity of the diffusate from the sample frozen at temperature and then killed. The R_t of frozen seedlings can be confounded, however, by changes in the R_t of unfrozen seedlings. To eliminate this source of error [31], an injury index, I_t, must be calculated:

$$I_t = 100(R_t - R_o)/(1 - R_o) \qquad (2)$$

where R_o is the relative conductivity of the control (unfrosted) seedling given by L_o/L_d, L_o is the conductance of diffusate from controls, and L_d is the conductance of diffusate from controls killed as indicated above.

Green and Warrington [36] reported excellent results with this method on Monterey pine. R_t values determined 3 days after freezing treatments accurately predicted freezing damage as assessed visually 1 month later. This correlation was improved to $r^2 = 0.92$ with the I_t value. Green and Warrington determined that an R_t value of 0.5 or greater indicated seedling death was imminent. van den Driessche [128] applied the diffusate conductivity method to Douglas-fir seedlings with some success but was not able to identify a critical index of injury, as were Green and Warrington. Nevertheless, the method predicted well ($r^2 = 0.77$) the lethal temperature of whole plants subjected to freezing temperatures.

23.2.3 Stress resistance

A simple technique for assessing a seedling's overall "physiological soundness" has been pioneered at Oregon State University [46][2] and is currently offered by the university as a service. Sixty seedlings are randomly selected from a lot and divided into two equal groups. The first group (controls) is planted directly into 25- x 25-cm (10- x 10-in.) fiber pots, 10 per pot, placed in a greenhouse or growth room, and watered. The second group is washed, blotted to remove excess water from the roots, and then suspended in a growth cabinet for 15 minutes at 30% relative humidity and 32°C (90°F). Following this stressing treatment, seedlings are removed and their roots soaked in water for 5 minutes. They are then potted and placed alongside the controls, where both groups are watered regularly and maintained under fairly constant 20°C (68°F) temperature and a 15-hour photoperiod.

Seedlings are evaluated after 2 weeks, 1 month, and 2 months. Mortality is noted when it occurs. After 2 months, seedlings are classified as follows:

Mortality among stressed stock, %	Classification
0-10	Excellent
11-20	Good
21-30	Fair
31-100	Poor

If there is mortality in the control group or if abnormal budbreak is noted, these classifications can be modified. The length of time required for stress damage to become apparent varies. Some seedlings will show no damage for 4 weeks, then begin to turn brown and die; others will begin to show damage after 10 days. Generally, lots in the poorest condition will show damage symptoms earliest.

The goal of the testing procedure is ultimately to predict field survival, hence the test is designed so that mortality of the stressed trees should correspond roughly with expected field mortality. Under normal conditions, "poor" lots should not be planted at all, and "fair" lots should be planted only in areas where severely stressful conditions will not be encountered.

Tests have been administered to over 1,000 seedling lots representing virtually all important Northwest conifers during the past 4 years. Unfortunately, it has not been possible to quantitatively assess the accuracy of all test predictions.

[1] Circuit design and operating procedure are available from the author on request.

[2] For more information, contact Douglas McCreary, Department of Forest Science, Oregon State University, Corvallis, Oregon 97331.

However, during 3 years of testing in cooperation with the Bureau of Land Management, field performance correlated well with lots displaying either very high or very low test survival. Correlations were not as strong in intermediate lots [pers. commun., 70].

There have apparently been no published attempts to relate performance in the stress test with other performance attributes. Possibly, peak periods of stress resistance may not coincide with those of other properties such as RGP.

In the future, those using this test procedure may be asked to furnish information on the history of each lot submitted for testing (e.g., lifting date, storage time and temperatures, etc.). In time, and with this information, developing valuable correlations among these variables and stress resistance may be possible.

Finally, the physiological mechanisms underlying stress resistance are not well understood. They may be related to the seedling's ability to grow roots, to control water loss, to increase water uptake, to endure internal water deficits, or to other mechanisms (see analysis of [118]). This may be a profitable area for future research.

23.3 Material Attributes

23.3.1 Bud dormancy

Perennial plants which have evolved in regions with strongly seasonal climates can adapt to a wide range of environmental temperature and moisture regimes with changing seasons. Plants "anticipate" these changes by keying in on reliable environmental cues such as photoperiod and soil temperature. As seasons change, plants cycle through various physiological states, each adaptively tuned to ambient conditions; this is referred to as the dormancy cycle and has been a major area of inquiry in plant-biology research ([e.g., 81, 90, 95, 131]; see also chapter 14, this volume).

In conifer seedling crops, the dormancy cycle comprises several "stages" [18]. Dormancy is **induced** from midsummer to late summer (dates given by Cleary [18] are specific to western Oregon) as overwintering buds are formed. These may break and form lammas shoots if seedlings are fertilized, given long photoperiods or heavy irrigation, or experience heavy late summer or early autumn rain after a droughty period. Dormancy **deepens** in late summer and early fall. During this period, buds will not flush if exposed to favorable conditions, but seedlings are not yet resistant to frost or lifting damage and cannot be successfully cold stored [47, 64]. Dormancy **peaks** (true dormancy) in early winter, when it is characterized by (1) an almost total absence of growth anywhere on the seedling and (2) a requirement for several hundred hours of low temperatures (0 to 10°C) before buds can break in response to higher temperatures [81]. This **chilling requirement** [133] is an adaptive mechanism which ensures against buds breaking during a midwinter warm spell and being subsequently killed by a return of cold weather.

The length of the chilling requirement has been determined experimentally for coastal Douglas-fir by Wommack [141], Lavender and Hermann [63], and van den Driessche [127]; for interior Douglas-fir (var. *glauca*) by Wells [137] and van den Driessche [127]; for western hemlock [*Tsuga heterophylla* (Raf.) Sarg.] by Nelson and Lavender [74]; and for several spruces by Nienstadt [75, 76]. The chilling requirement of these species is generally fulfilled by exposure to temperatures at or below 5°C for 2,000 hours and may also be fulfilled by cold storage. After this requirement has been satisfied, buds will break rapidly once exposed to springlike conditions; in this state, seedlings are called **postdormant.** The interactions among chilling, flushing temperature, photoperiod, and time required for budbreak have been elegantly demonstrated for Douglas-fir by Campbell [10] and Campbell and Sugano [12].

Of most interest to the forest-nursery manager is the stage of **true dormancy.** It is generally felt that seedlings lifted before or after the period of true dormancy are high risk and prone to suffer serious damage from cold storage [47]. If so, it is important to know when true dormancy begins and ends. Beyond this, Hermann [42, 43, 44] demonstrated, in a series of important experiments, that Douglas-fir seedlings vary greatly with respect to their ability to withstand environmental stresses as they pass through the stage of true dormancy itself. Hence, it is not only important to know when seedlings enter true dormancy, but also to know the intensity of dormancy at any point in time. Because seedlings do not change visibly from late summer to early spring, determining their exact dormancy status (or intensity) has been troublesome and the subject of much experimentation. Some suggested techniques follow. Four of these—dry-weight fraction, mitotic index, hormone analysis, and electrical resistance—if developed and verified, would offer rapid, inexpensive methods of assessing dormancy status and clearly deserve further study.

23.3.1.1 Budbreak tests

The most reliable measure of the intensity of dormancy is the time required for terminal buds to break in a forcing environment [50]. In practice, this is determined by bringing a sample of seedlings indoors, potting them in a suitable medium, and holding them in a standard test environment simulating springlike conditions (e.g., 12- to 14-hour days, 20°C air temperature). Seedlings are checked daily; when terminal bud scales part to expose new needles, the date is recorded. After terminal buds have broken on all seedlings, the average number of days to terminal budbreak (DBB) is calculated. Because dormancy intensity weakens as winter chilling accumulates, buds will break faster the later in winter seedlings are forced (Fig. 3).

23.3.1.2 Dormancy release index

The relationship between DBB and chilling sum (see 23.3.1.3) can be fitted with a reciprocal function (i.e., 1/DBB) (Fig. 3). Campbell and Sugano [11] employed this relationship to quantify dormancy intensity in Douglas-fir seedlings. They devel-

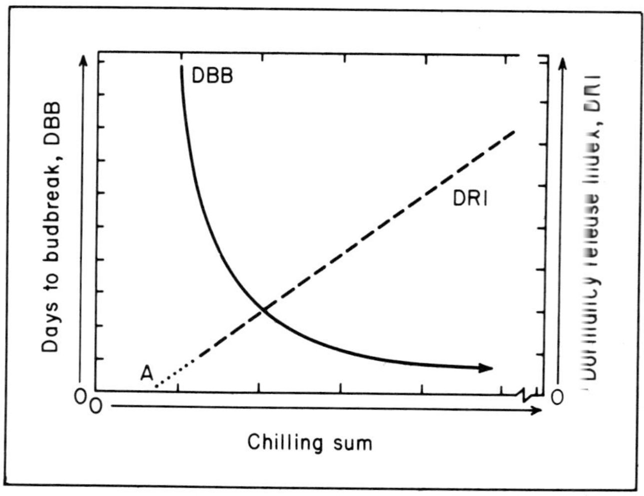

Figure 3. Chilling sum at time of lifting is related to days to terminal budbreak, or DBB (solid curve), of coastal Douglas-fir seedlings under 16-hour photoperiod and 20°C day and night temperature. Resulting values can be expressed as a dormancy release index, or DRI (straight dashed line); for Douglas-fir, DRI = 10/DBB. Extrapolation of DRI to point A on the x-axis may provide an estimate of the earliest date to begin lifting for storage.

oped the term DARD (daily average rate of development), calculated as:

$$DARD = 100/DBB \quad (3)$$

which gives an estimate of the developmental rate of the seedling at any time during dormancy release.

This concept has been extended [85] into what is called a dormancy release index (DRI):

$$DRI = DBB_r/DBB \quad (4)$$

where DBB_r is the number of days required for budbreak in a fully chilled seedling.[3] For coastal Douglas-fir, DBB_r is 10; hence, for this species, DRI = 10/DBB. The value of DBB_r must be determined experimentally for each species but probably does not vary among species by more than a few days. Dormancy release can be compared among species with DRI because it always varies from 0 (in a seedling just entering true dormancy) to 1 (in a seedling fully released from dormancy). DRI values for stored and unstored Douglas-fir seedlings have been shown to be good indicators of physiological condition in our own unpublished experiments.

23.3.1.3 Chilling sums

The disadvantage of DRI as a nursery manager's tool is the excessive time required to get results. Seedlings lifted early in winter may not break bud in the test environment for 100 days or more. However, we have found with Douglas-fir that the relationship between DRI and chilling sum (number of hours a seedling spends at ≤5°C) does not vary appreciably among seedlots at a given nursery from year to year. Therefore, once this relationship has been empirically established for a given species and nursery, dormancy status during winter can be accurately predicted from monitoring chilling sums.

The chilling sum is determined simply by monitoring air temperature at about 1 m above the ground and summing the hours during which the temperature is within some range known to be effective at releasing dormancy in the species of interest. Our experience has been that the range 0 through 5°C is useful for northern conifers. In California nurseries, however, the range 0 through 10°C may be more appropriate. Once hourly temperature data have been collected, chilling sums may be tallied within any temperature range desired. Data collection should begin around October 1 in coastal nurseries and in early to mid-September in more northern or interior regions.

This approach has been highly refined for predicting time of budbreak in fruit crops, especially in regions where late frosts are common. By taking into account the relative efficiencies of different chilling temperatures and the effects of warm interruptions (which can negate chilling) and other factors, chilling equations are available which predict budbreak time with an error of only ±2 days for some crops [27, 83]. Whether this level of accuracy is warranted in forest-seedling crops, however, is yet to be established.

By calculating chilling sums, it may also be possible to estimate the earliest date at which lifting for storage can begin. In Figure 3, extrapolation of the DRI curve to the x-axis (point A) indicates that the first several hundred hours of autumn chilling do not actually contribute to physiological dormancy release. Point A generally coincides with the last week in November in western Washington, which is viewed by some nursery personnel as the earliest date on which successful lifting for storage can occur. However, this relationship needs to be developed for other species and regions.

[3] Determined experimentally with seedlings lifted from the nursery in late winter, stored at −1°C for 6 months, then tested for budbreak at 20°C under a 16-hour photoperiod.

23.3.1.4 Oscilloscope technique

Zaerr [143] reported the interesting observation that square-wave electrical signals are propagated differently through living plant tissue than through dead tissue. The form of this propagation can be determined with an oscilloscope. Following up on this work, Ferguson et al. [29] tested a wide range of species, including some conifers, at different times of year and found that the types of oscilloscope waves observed seemed to be related to periods of plant activity and inactivity. This finding led to speculation that the oscilloscope technique may be the long-awaited "dormancy meter," and a number of investigators set out to verify Ferguson's results.

Disappointingly, this work has not been very successful [1, 50, 53, 79] due to lack of reproducibility, interspecific variability, and artifacts produced by touching or moving sample branches. These problems probably reflect the unknown complexity of plant-tissue circuitry, and it is likely that a dormancy-related change in a particular capacitive or resistive component will have only a very small effect on the overall response. Furthermore, changes in properties unrelated to dormancy will also influence tissue electrical properties [116]. Though this technique may hold promise with further development, its present operational usefulness for assessing seedling dormancy status is limited [50, 53].

23.3.1.5 Dry-weight fraction

The dry-weight fraction (DWF) of seedling shoots may be a simple, rapid method of assessing dormancy. Dry-weight fraction is calculated as

$$DWF = DW/TW \quad (5)$$

where DW is the oven-dry weight of the seedling shoot, and TW is its turgid weight.

Dry-weight fraction changes annually in a predictable manner in many woody plant species. In Douglas-fir seedlings, DWF increases gradually during fall and early winter, peaks in January, then falls rapidly during spring [88]. If this pattern reflects seedling physiological condition and is relatively independent of weather, then it might be used as an indirect measure of seedling dormancy status during winter. DWF is now used routinely in some Swedish seedling nurseries to determine when to begin lifting [pers. commun., 91].

23.3.1.6 Mitotic index

During autumn, mitotic activity in conifer buds declines rapidly as dormancy deepens [77, 78]. This phenomenon has been exploited by Carlson et al. [13] as a tool for determining when Douglas-fir seedlings have become dormant. Using a squash and stain technique, they microscopically ascertain the percentage of cells in the terminal meristem which show mitotic figures. This "mitotic index" (MI) declines steadily throughout autumn, reaching zero apparently at about the time seedlings enter dormancy. Hence, it might serve as an indicator of the onset of dormancy. But because MI remains near zero until mid-March, it would not be useful in assessing the progress of dormancy release.

23.3.1.7 Hormone analysis

Dormancy induction and release are hormone-mediated processes. In principle, then, it should be possible to assess the status of dormancy by measuring concentrations, or ratios of concentrations, of various dormancy-regulating hormones. Good correlations have been observed, for example, between free abscisic acid concentration and apparent dormancy intensity throughout winter in buds of European beech (*Fagus sylvatica* L.) trees [142]. Hatch and Walker [38] were able to assess the dormancy intensity of peach and apricot buds on the basis of the concentration of gibberellic acid required to make them break. Zaerr and Lavender [144] have evaluated the potential for using hormone tests as a litmus for dormancy status in seedlings and concluded that rapid advances in analytical techniques might make this a real possibility in the future—but not now.

23.3.1.8 Electrical resistance

Cyclic seasonal changes in the electrical resistance of the inner bark of maple (*Acer* spp.), oak (*Quercus* spp.), and pine trees have been reported [20]. In some cases, these changes were dramatic, with resistance decreasing from spring to summer and increasing from summer through autumn. Unfortunately, the period of greatest interest to the nursery manager—December through March—was not sampled due to frozen stems.

23.3.2 Water relations

Most aspects of seedling physiology influence, and are influenced by, seedling water status (see also chapter 12, this volume). Its effects on plant growth and function and the technology available for measuring it are subjects of a voluminous and complex literature [e.g., 56-61, 66, 103, 104, 123]. Here, a few central concepts are briefly reviewed, and then several measurement methods that may be of practical value to the nursery grower are summarized.

23.3.2.1 Water potential

The status of water (W) in a seedling reflects the imbalance between the rate at which water is absorbed by its roots (A) and the rate at which it is transpired (T) through the leaves:

$$W \simeq (A - T + S) \qquad (6)$$

where S, a relatively small term, represents the storage of water within the seedling itself. During the day, and sometimes at night, T exceeds A so that the water in the seedling comes under tension, or "stress." When stress is sufficiently great or prolonged, growth and photosynthesis cease, metabolic systems break down, and mortality follows.

As used above, W represents the water content of the seedling. But to be physiologically precise, water status should be quantified in terms of its free energy, or "water potential," Ψ_w. Water potential is defined thermodynamically as the ability of water to do work in comparison to free, pure water at standard pressure and temperature, whose water potential is zero. Units of water potential are dimensionally equivalent to pressure units; therefore, Ψ_w can be expressed in pounds per square inch, atmospheres, or bars. In the metric system, the appropriate units are joules per kilogram or Pascals. Here I will use the unit megaPascal, MPa, which is recommended for plant research [52].[4]

23.3.2.2 Components of water potential

The water potential of a seedling has several component potentials. Here, we are interested primarily in two, osmotic potential (Ψ_π) and turgor potential (Ψ_p), which are related to Ψ_w as follows:

$$\Psi_w = \Psi_p + \Psi_\pi \qquad (7)$$

Turgor potential is a positive force exerted inward on the cell contents by the rigid cell wall, much as the skin of a balloon exerts a force on the air inside the balloon. As the cell loses water, the force weakens. Osmotic potential is a negative force resulting from the effect of dissolved solutes (e.g., sugars, salts) and other materials on the free energy of water. As solute concentration increases, osmotic potential decreases; in pure water, it is zero.

[4] 1 MPa = 10 bars ~ 10 atm ~ 150 psi.

Turgor potential is a **very important** property. Virtually every physiological process in the seedling is sensitive to turgor such that a turgor drop below a given level, if sustained, can result in death [51].

These concepts are integrated into a diagram (Fig. 4), originally conceived by the German scientist Karl Höfler [49], illustrating the manner in which the components of water potential change as the seedling gains or loses water. When a seedling is fully hydrated (100% water content), Ψ_w is zero, by definition, and the value of Ψ_p is equal and opposite in sign to the value of Ψ_π (due to Equation 7). As a net loss of water is experienced, solutes are trapped in the cells by the cell membrane while water escapes into the cell walls and xylem. Thus, cell solute concentration increases and osmotic potential decreases. Turgor also falls because the cells lose volume. Therefore, water potential falls and stress increases. In the example shown in Figure 4, with a loss of, say, 10% water content, Ψ_w is -0.5 MPa, Ψ_π is -1.7 MPa, and Ψ_p is 1.2 MPa. When water loss is 30%, Ψ_p becomes zero and Ψ_w is -2.5 MPa. This value of water potential at zero turgor (Ψ_z) is a **critical point** because it presumably indicates at what stress level death is imminent.

These concepts underline the central point that seedling water status cannot be adequately described by measuring water potential alone. Complete assessment must also include estimates of its components. Having said that, I now turn to a survey of available techniques, only two of which, psychrometry and the pressure chamber, are capable of determining the value of component potentials.

23.3.2.3 Measurement techniques

Gravimetric methods.—Gravimetric methods of measuring seedling water status yield information on water content only and not on water potential. However, if other alternatives

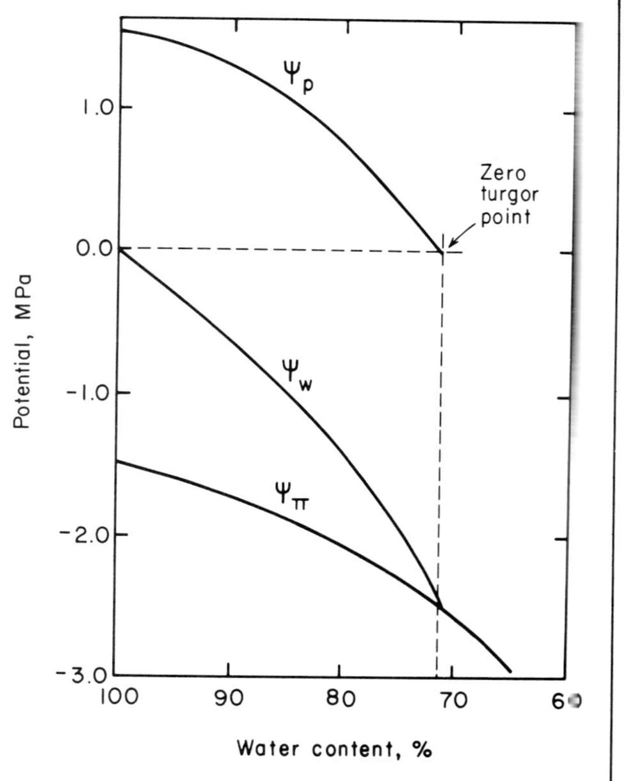

Figure 4. A Höfler diagram showing the manner in which water potential, Ψ_w, and its component potentials (osmotic, Ψ_π, and turgor, Ψ_p), change with respect to a change in cell water content.

are not available, or if calibrations between water content and potential have been established, then measuring seedling water content may be useful.

A widely used measure is Weatherley's [135] Relative Water Content (RWC). It is determined by weighing a sample (normally a leaf) immediately after it is collected and again after it has been brought to full turgidity by floating on water in the dark until it ceases to gain weight. The sample is then oven dried for 48 hours and weighed again. RWC is calculated as:

$$\text{RWC} = \frac{\text{fresh wt.} - \text{dry wt.}}{\text{turgid wt.} - \text{dry wt.}} \times 100 \qquad (8)$$

In a fully turgid sample, RWC is 100%. A corollary measure is the "water deficit," in which the same steps are performed, and water deficit (WD) is calculated:

$$\text{WD} = \frac{\text{turgid wt.} - \text{fresh wt.}}{\text{turgid wt.} - \text{dry wt.}} \qquad (9)$$

A particular disadvantage of using these methods on conifer seedlings is the difficulty sometimes experienced in bringing sample material to full turgor.

Psychrometric methods.—Psychrometric techniques of estimating water potential are based on the principle that if a tissue sample is placed in a small chamber, the humidity of the air in the chamber will come to equilibrium with tissue Ψ_w. Hence, with appropriate calibration, a measure of humidity gives an estimate of water potential [26, 105, 138]. Though this principle has been understood for many years, only recently have affordable devices been developed for accurate, reproducible measurement.[5] They generally consist of a small chamber containing a thermocouple psychrometer (to measure humidity) and the associated electronics to generate, read, and transmit an electrical signal.

A significant advantage of psychrometry is that it permits separate estimates of the osmotic and turgor components of water potential. To do this, Ψ_w is measured in the normal way; then the tissue sample is frozen and remeasured. Freezing destroys the cell membranes, eliminating turgor potential. Therefore, the value obtained on the frozen sample is equal to osmotic potential. Turgor can then be calculated as the difference between water and osmotic potentials.

Success of this method with conifers has been mixed. There are several sources of difficulty: (1) conifer needles come to equilibrium very slowly in the sample chamber due to their waxy surface and propensity to tightly close stomata, (2) resins exuded from cut needles tend to gum up the chamber and thermocouples, and (3) cutting needle tissue releases extracellular water which dilutes the sample and yields water-potential values that are erroneously high. Although psychrometry is the technique of choice for determining water potential in the laboratory [104], it has not yet found use as an operational nursery tool.

Density method.—The density method (also called the dye method) was first described in Russian by Shardakov [100] and later in English by Knipling [55]. A series of graded water solutions having a range of osmotic potentials is prepared. Each of the solutions is then divided in two parts and a dye (e.g., methylene blue) introduced into one part. Next, the solutions are placed into a series of test tubes, each pair containing a solution of known osmotic potential, one clear and one dyed. A small sample of plant tissue is placed into each of the clear solutions and held there for several minutes. Samples with lower osmotic potentials than the solution will take up water; therefore, the density of the solution will increase due to its increased solute concentration. Samples with higher osmotic potentials than the solution will lose water, thus diluting the solution and decreasing its density. The samples are then removed, and a drop of the dyed solution is introduced into the middle of each test tube containing the clear solutions. The solution having the same osmotic potential as the sample will not have changed density; thus, the drop of dye will remain in the middle of the tube.

This method is economical, portable, and rapid. It requires neither electricity nor gas pressure and can be made with very simple parts. But it does not provide estimates of water-potential components, and there is little, if any, information on its applicability to conifers.

Freezing-point depression method.—The freezing point of a solution is a function of its osmotic potential, a measure of the former providing an estimate of the latter. Cary and Fisher [14] and Fisher [30] describe an inexpensive, portable device capable of accurately measuring the freezing point of plant sap, provided that appropriate temperature corrections are made.

The major limitation to this method is in obtaining the sample of plant sap for analysis. Squeezing tissue yields a sap sample which is nearly pure because it contains extracellular and filtered cellular water. Grinding or blending plant material to obtain sap contaminates cell water with extracellular water and raises the osmotic potential. Freezing-point tests on sap collected by these different methods from the same tissue sample have yielded results which differ by as much as 50% [94]. Because of this problem, and because data on conifers are very limited, the technique is not recommended for nursery use.

J-14 hydraulic press.—A relatively recent innovation in rapid water-potential determination is the hydraulic press.[6] In principle, the device uses hydraulic pressure to press a sample of plant material against a clear plexiglass screen. As pressure increases, water is exuded from the cut tissue or leaf edges, or the tissue changes color. Childs [17] evaluated the press against the pressure chamber technique (see below), reporting good correlations (r^2 ranged from 0.66 to 0.90) in calibrations with bareroot and container stock and field-grown seedlings. However, similar comparisons by Cleary and Zaerr [19] gave poor results. More work is needed before this technique can be recommended for nursery use.

Pressure chamber.—Reintroduction of the Dixon pressure chamber by Scholander et al. [99] has provided an ingenious and invaluable tool for measuring plant water potential. Since then, considerable experience has been gained with the pressure chamber, much of it summarized by Ritchie and Hinckley [87]. The apparatus and procedures required to use this technique with forest-tree seedlings have been recently detailed elsewhere [18, 19] and will not be repeated here. Rather, I will focus on aspects of pressure-chamber use not addressed by the above papers—specifically, on measuring roots and individual needles and on generating "pressure-volume" curves.

Although not given much attention in the literature, the root system is an integral part of a seedling's anatomy and is generally far more susceptible to cold and desiccation than the shoot. Hence, the physiological condition of the root system should be assessed as part of overall seedling quality. The pressure chamber can be used to develop such information.

We have successfully measured root water potential using apparatus and procedures identical to those described for shoots [19]. Normally, the seedling is severed at the root

[5] One such unit can be purchased from Wescor, Inc., Logan, Utah 84321.

[6] Available from Campbell Scientific, Inc., Logan, Utah 84321.

collar; then the entire root system, with soil removed, is placed in the chamber for measurement. With larger seedlings, such as 2+1s, it may be necessary to remove a major lateral root for measurement; its value will be nearly identical to that of the entire root system. Values of root water potential are normally much higher than those of shoots and exhibit far less seasonal and diurnal fluctuation [88].

Measuring water potential of leaves (needles) rather than that of branches enables repeated determinations on a seedling and greatly reduces compressed gas consumption. To measure needles directly, the rubber gland (#6 rubber laboratory stopper) holding the sample in the chamber lid must be modified (Fig. 5). Note that the stopper has been slit to the radius on one side so that a needle can be placed into the central hole without pushing it through the stopper. Note also that a portion of the underside of the stopper has been hollowed out with a cork borer so that the needle is not crushed during pressurization.

To prepare pine samples for measurement, collect a fascicle of needles and strip off the fascicle sheath. Then sever the base of the fascicle crosswise with a razor blade; this will cut the xylem traces and permit sap to escape during measurement. For conifers other than pines, the needle should be cut crosswise just above its point of attachment to the branch. The measurement is then performed as on a small branch, except that viewing should be with a 15 or 20x magnifier and a light.

Needle water potential was nearly identical to that measured on the branch from which needles were taken in several pine species by Johnson and Nielson [54] and in ponderosa (*Pinus ponderosa* Dougl. ex Laws.) and Jeffrey (*Pinus Jeffreyi* Grev.& Balf.) pines by Ritchie and Hinckley [86]; however, in Douglas-fir, Pacific silver fir (*Abies amabilis* Dougl. ex Forbes), and noble fir (*Abies procera* Rehd.), needle values were higher than equivalent branch values. Calibrations for leaf and stem water potential of these species are [86]:

Species	Equation[1]	r^2
Pacific silver fir	$\Psi_s = -0.59 + 1.48\,\Psi_l$	0.91
Noble fir	$\Psi_s = -0.47 + 1.34\,\Psi_l$	0.82
Douglas-fir	$\Psi_s = -0.77 + 1.28\,\Psi_l$	0.92

[1] Ψ_s = stem water potential (bars).
Ψ_l = leaf water potential (bars).

A highly valuable feature of the pressure chamber is that it enables osmotic and turgor potentials (Equation 7) to be measured through generation of what is called a "pressure-volume" (P-V) curve [16, 37, 40, 124]. A simplified procedure for generating a P-V curve is given in Appendix 1, this chapter. A P-V curve represents the relationship of reciprocal water potential ($1/\Psi_w$) with water content. The curve has two distinctly different regions, one that is curvilinear and one that is linear (Fig. 6). The linear region can be extrapolated to the y-axis (to point A) with a straight line to give the osmotic potential when the seedling is at full turgor. The point where the linear and curvilinear regions meet is the water potential at which turgor is lost, or the "zero turgor point." Its value can be determined by extrapolating horizontally to point B.

An important finding has been that both these properties change dramatically from month to month in Douglas-fir seedlings [88]. Water potential at zero turgor in shoots and roots was lowest (seedlings tend to be more drought tolerant) in midwinter and late summer and highest in spring over the course of a year (Fig. 7). This may partly explain why seedlings are so sensitive to handling and planting when lifted in March and April.

This technique also has considerable potential for detecting certain types of hidden seedling damage. For example, frost-damaged seedlings typically have membrane lesions which result in solute leakage from cells (see 23.2.2.2). This disrupts

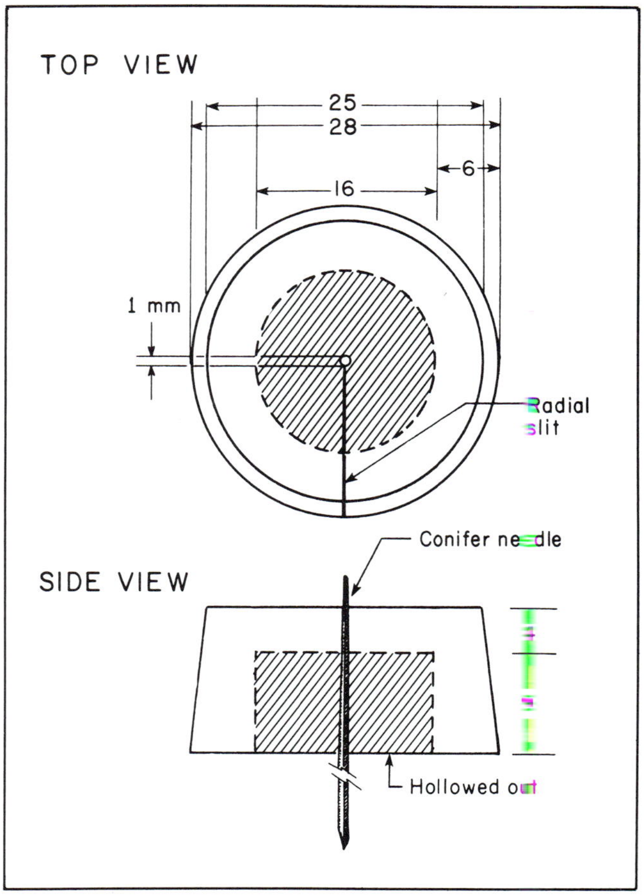

Figure 5. Diagram of a #6 rubber laboratory stopper modified to accept a conifer needle for pressure-chamber measurement. All dimensions are in millimeters.

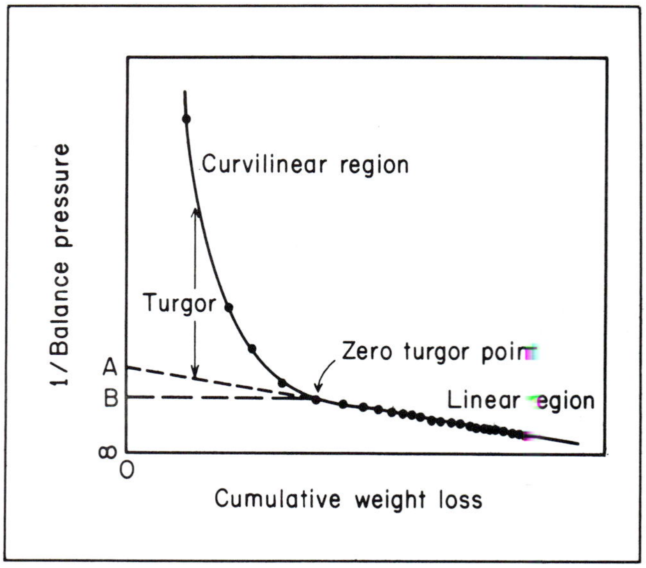

Figure 6. A pressure-volume (P-V) curve. The linear region represents the relationship between cell volume and osmotic potential when turgor equals zero. Extrapolation to point A gives an estimate of the osmotic potential at full turgor, to point B an estimate of the osmotic potential at zero turgor (see Appendix 1, this chapter).

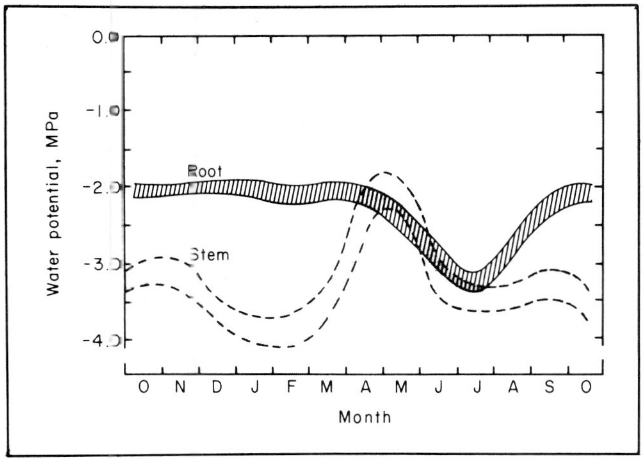

Figure 7. Seasonal changes in the "critical water potential" for a Douglas-fir seedling. This value is approximately equal to the water potential at which turgor becomes zero (zero turgor point; see Fig. 6).

tissue osmotic properties. A P-V curve from such a seedling would not show a well-defined linear region. Or seedlings with severely depleted carbohydrate reserves resulting, say, from long-term storage would have abnormally high osmotic potentials at full turgor.

Cleary and Zaerr [19] have suggested some general guidelines for interpreting pressure-chamber readings on tree seedlings. For bareroot stock from a nursery bed, water potential should not fall below −1.0 MPa and, ideally, should be above −0.5 MPa; if it falls below −2.0 MPa, the seedling may suffer severe physiological damage. Figure 7, however, indicates that the above values are not fixed. Seedlings in midwinter are apparently far more tolerant of low water potentials than they are in spring.

23.3.3 Nutrition

The literature on plant nutrition in general is voluminous. Our interest, however, is plant nutrient analysis as a direct indicator of seedling quality. The literature on this topic, unfortunately, is weak. Two aspects of nutrition are considered here—mineral nutrients and food reserves.

23.3.3.1 Mineral nutrients

All physiological processes, as well as morphological ones, are influenced by mineral nutrition [114]. There has been hope, then, that some simple measure of seedling nutrient status might be developed as an index of seedling quality [e.g., 96].

In reviewing nutrient status of Northwest conifer seedlings, van den Driessche [129] found that potassium (K), phosphorus (P), and nitrogen (N) affect various species differently with respect to frost hardiness. Increased K generally improves hardiness whereas excess P has been shown to decrease hardiness in some species. N can improve hardiness if applied late in summer after height growth has ceased; if applied earlier, N can prolong shoot growth, retard dormancy development, and delay the onset of hardiness (see chapter 7, this volume).

Stress resistance also may be affected by nutritional status. For example, N and K can reduce transpiration rate, whereas P tends to increase it. N and K may also improve tissue water relations by enhancing turgor maintenance through osmotic adjustment. As to the effects of mineral nutrition on RGP, the data are too limited to warrant discussion [85].

van den Driessche [130] was able to show an improvement in survival of Douglas-fir seedlings following N fertilization in the nursery. It was not clear, however, whether this effect was direct or due to a general increase in seedling size (see 23.3.4) brought about by the extra N.

Menzies [72] analyzed foliar nutrient content of Douglas-fir seedlings grown at a large Northwest coastal nursery. December-lifted seedlings had adequate to low N, low to very low P, and adequate to low K, according to van den Driessche's [125] classification. By March, all nutrients had fallen to low or very low concentrations. Yet these seedlings had 98% survival and excellent growth 2 years after outplanting. It may be that, except in cases of severe deficiency, growth and performance reflect the intricate interplay between seedling nutrition and other factors governing physiology, morphology, and site conditions. Because the effects of mineral nutrition on seedling physiology are complex and interacting, no consistent relationship has yet been demonstrated between any aspect of seedling nutrient content and seedling quality, except in cases of severe deficiency.

23.3.3.2 Food reserves

Seedlings store food reserves in the form of sugars, starch, hemicelluloses, proteins, fats, oils, and other compounds (for discussion, see [62] and also chapter 14, this volume). The sugars and, especially, starch are key forms. Many workers have stressed the importance of adequate food reserves to seedling performance [see reviews by 48, 132], and some have suggested that a measure of starch content might be used as an indicator of seedling vigor [28, 32].

Hellmers [41] noted a correlation between a decline in root starch during storage (determined by iodine staining) and reduced survival in ponderosa pine seedlings after outplanting. Winjum [140] suggested that concentration of reducing sugar might serve as an index of seedling quality in Douglas-fir and noble fir. Puttonen [82] mentioned the possibility of using the carbohydrate pool as a measure of seedling physiological condition. Others [e.g., 109, 111] have proposed a cause-effect relationship between carbohydrate reserves and RGP, although more recent evidence [84] does not support this view.

Unfortunately, this relationship does not seem to have been examined by systematic rigorous experimentation. This is disappointing because some carbohydrate components (e.g., starch) are easily determined and because the idea that "food reserves" are critical to seedling quality seems sound. However, carbohydrate chemistry is exceedingly complex. Interconversions among carbohydrates occur continuously, and the various metabolites function in different ways at different times. Therefore, although carbohydrate assessment would seem to hold promise as a future tool for indicating seedling quality, such technology is not now available.

23.3.4 Seedling morphology

In a strict biological sense, morphology means form and structure. In practice, however, any seedling characteristic that can be readily observed or measured is normally construed as morphological. The most commonly cited morphological properties are those that are most easily measured: shoot height and weight, root-system weight or volume, root fibrosity (often subjectively assessed), stem diameter at the root collar, bud "set," foliage color, and various ratios such as shoot:root weight or top height:stem diameter (sturdiness ratio). Each of the above characteristics can be manipulated to some extent in the nursery by controlling seedbed density, undercutting and wrenching, transplanting, top mowing, irrigation, and nutrient management (see chapter 15, this volume). Because they are relatively easy to control and measure, morphological characteristics have been used extensively over the years to define seedling quality [7, 114]. Indeed, some European nations have adopted legislation establishing morphological grading standards for tree seedlings [98].

More recently, however, considerable research attention has been focused on "physiological grading" of planting stock (as previously discussed). Results of this work indicate that: (1) seedling physiological condition exerts a strong influence on seedling survival and growth potential; (2) components of physiological condition are numerous, change rapidly over time, and can change independently of one another; and (3) physiological condition cannot be visually determined.

It follows from this that comparisons of seedling performance based upon **morphological** traits are valid only when seedlings are in the same **physiological** condition when tested. This simple deduction probably invalidates much of the published research on the effects of morphology on seedling performance and accounts for the inconsistency and variability which pervade this literature (see, e.g., [45]). I know of no published work on seedling morphology and performance in which the condition of physiological homogeneity has been quantitatively satisfied. Therefore, the following comments on the relation between morphology and performance are based upon generalizations and personal observation and, as a result, must be viewed as qualitative and biased.

Operational experience tends to indicate that, other factors equal, seedlings with large stem calipers tend to outperform those with smaller calipers [15, 18, 25, 98, 114]. Furthermore, stem caliper tends to be well correlated with other seedling size characteristics, as illustrated by unpublished data from four Douglas-fir stock types (Table 1). Note, however, that stem caliper was not well correlated with shoot:root ratio in these seedlings (see also chapter 24, this volume).

Dobbs [24] examined the relationship between mass (fresh weight) and field performance of white spruce [*Picea glauca* (Moench) Voss] and lodgepole pine (*Pinus contorta* var. *latifolia* Engelm.) seedlings and transplants in interior British Columbia. Large individuals tended clearly to outperform small ones regardless of species or stock type. Furthermore, the size advantage was amplified by site characteristics: differences in performance were more pronounced on unscarified sites. Although not quantitatively documented, his seedlings seem to have been physiologically similar.

Table 1. Linear relationships between stem caliper (diameter, mm) and five morphological characteristics of Douglas-fir seedlings of four stock types from a single Twin Harbors, Washington, seedlot.

Characteristics	Stock type[1]	r^2
Height (cm) and stem caliper	2+0S	0.41
	2+0L	0.31
	1+1	0.26
	2+1	0.45
Root dry weight (g) and stem caliper	2+0S	0.80
	2+0L	0.78
	1+1	0.69
	2+1	0.82
Shoot dry weight (g) and stem caliper	2+0S	0.89
	2+0L	0.81
	1+1	0.71
	2+1	0.85
Total dry weight (g) and stem caliper	2+0S	0.88
	2+0L	0.83
	1+1	0.76
	2+1	0.87
Shoot:root ratio (dry) and stem caliper	2+0S	0.00
	2+0L	0.00
	1+1	0.03
	2+1	0.01

[1] Standard bed density 2+0 seedlings (2+0S) were grown at ~209/m²; and low density 2+0 seedlings (2+0L) at ~143/m²; transplants were grown at ~209/m², transplanted at ~55/m².

Cleary et al. [18] also have reported that on sites where animal browsing, brush competition, and snow press were severe, larger seedlings had an advantage, presumably because they can sustain browsing and can grow above weeds more effectively than smaller seedlings. However, largeness can be disadvantageous under some conditions. For example, on high-elevation sites where brush competition is not excessive but desiccating winds are prevalent, a large foliar surface would tend to place greater transpirational demand on the root system (see 23.3.2.1); in that case, smaller seedlings may be preferred.

But size alone is not meaningful if the seedling is out of balance [18, 72]. A large top requires a large root system to supply water and nutrients; hence, some measure of shoot:root ratio balance is indicated. The pitfall here is that root weight or volume is not a very good indicator of the root system's ability to provide water and minerals. Total surface area of the root system or some measure of root-system fibrosity or absorption capacity is needed, but, unfortunately, such quantities are difficult and costly to determine. Furthermore, the root surface must effectively contact the soil after planting.

Dickson et al. [23] attempted to develop an integrated approach to quantifying morphological quality by formulating an index which included several morphological features. Their quality index (QI) was calculated as:

$$\mathrm{QI} = \text{seedling dry wt. (g)} \Big/ \left[\frac{\text{height (cm)}}{\text{diameter (mm)}} + \frac{\text{top wt. (g)}}{\text{root wt. (g)}} \right] \qquad (9)$$

where the higher the index, the better the seedling. When I applied this index to Douglas-fir seedlings of four stock types (plug, 2+0, 1+1, and 2+1), it gave values of 0.99, 1.79, 1.88, and 2.30, respectively. This ranking corresponds quite closely (and perhaps coincidentally) with observed performance rankings of these stock types in many of our field trials.

In conclusion, morphological characteristics probably exert the ultimate influence on seedling performance only when physiological characteristics do not differ significantly among seedlings.

23.4 Current Practice

Of the nurseries surveyed in the OSU Nursery Survey (see chapter 1, this volume), all reported making some routine measurements of seedling quality. Each was contacted by letter or telephone to obtain detailed information on the nature, application, and interpretation of the tests used. A synthesis of this information follows.

23.4.1 Root-growth potential

Four nurseries reported using RGP to assess seedling quality. One uses the standard measurement method developed by Stone, described in 23.2.1.1, and three use the more rapid scoring system developed by Burdett, described in 23.2.1.2. One nursery indicated that RGP was measured only for troubleshooting and not routinely. Another indicated that RGP was used initially to establish the optimum lifting window and later only as an annual spot check.

23.4.2 Frost hardiness

Nine nurseries reported assessing frost hardiness on a more or less routine basis. All use the test to determine when frost protection is needed in fall and spring, and all but one reported testing 1+0s only. The most commonly used method is to place potted seedlings in an on-site freezer chest and reduce the temperature to some level, then remove the seedlings and assess damage after a few days. Temperatures used generally bracket the expected LT_{50} value. Methods of assessing damage ranged from odor, to general visual appearance, to cutting buds, to scraping bark to detect dead cambium. One nursery reported using frost hardiness as an indicator of when to begin fall lifting, but none reported using it as an indicator of seedling quality before shipping stock to customers.

23.4.3 Stress resistance

Only three nurseries measure stress resistance. They use the services of Oregon State University and the test methods described in 23.2.3. One nursery reported that results of stress tests did not agree well with results of RGP tests and that RGP correlated better with seedling survival in the field. Most stress tests are conducted for reforestation personnel rather than for nurseries.

23.4.4 Dormancy

Seven nurseries assess seedling dormancy. Three use the oscilloscope, as described by Ferguson (see 23.3.1.4). One uses visual appearance of buds and foliage color; another performs a budbreak test. Two coastal nurseries monitor chilling sums and begin lifting when cumulative hours of air temperature below 6°C approach 600. One reported experimenting with chilling sums. All of the above assess dormancy only as a guide to determine when to begin lifting, and none use it as a measure of seedling quality itself.

23.4.5 Water relations

Water status was the most common measure of seedling quality. Thirteen nurseries routinely measure water status with a pressure chamber. The most common application is monitoring stress buildup in seedlings during lifting, grading, packing, and storage.

Each nursery has guidelines regarding acceptable levels of stress. Generally, nurseries do not lift when stress exceeds 1.0 or 1.5 MPa and do not permit stress to exceed 0.5 MPa when grading and packing. Some nurseries use predawn pressure-chamber measurements to indicate the need for irrigation and to manage development of late-summer stress for dormancy induction. But none reported measuring stress in root systems or on individual needles or using P-V curves or other more advanced techniques, although one Oregon nursery is beginning some preliminary work in this area.

23.4.6 Nutrition

Eight nurseries reported monitoring seedling foliar nutrient content. In all but one, this is restricted to the 1+0 crop. In all cases, samples are sent to a regional laboratory (either the Ministry of Forests Laboratory in Victoria or Oregon State University in Corvallis) for testing and interpretation. Most samples are taken in late summer or fall and the results used to fine-tune fertilizer prescriptions for the following year. Many foliar analyses are used in conjunction with soil nutrient analyses (see chapter 8, this volume). Again, in no case was nutrient content used as an index of seedling quality itself. No mention was made of carbohydrate analysis.

23.4.7 Morphology

Virtually all nurseries grade seedlings based upon their morphological characteristics. In almost every case, stem caliper at the root collar and shoot height (from the root collar to the terminal bud) are the characteristics measured. Also, seedlings showing any visible signs of damage such as torn roots, scarred bark, or broken tops are normally culled. Cull standards vary with nursery and species and are often determined by the buyer of the stock.

23.5 Summary

Attributes of seedling quality are categorized as either performance attributes (RGP, frost hardiness, stress resistance) or material attributes (bud dormancy, water relations, nutrition, morphology). Performance attributes are assessed by placing samples of seedlings into specified controlled environments and evaluating their responses. Although some effective short-cut procedures are being developed, performance tests tend to be time consuming; however, they produce results on whole-plant responses which are often closely correlated with field performance. Material attributes, on the other hand, reflect only individual aspects of seedling makeup and are often poorly correlated with performance.

Bud dormancy status seems to be correlated, at least phenologically, with the three performance attributes. Unfortunately, no rapid method of measuring dormancy intensity is yet available, although several are promising. Nursery chilling sums seem to offer a good method of indirectly estimating dormancy in some species.

Seedling water status also is related to all three performance attributes, but in complex and interacting ways. Although several methods are available for measuring water status, the pressure chamber is the method of choice because it (1) is rapid, accurate, and simple to use, (2) measures water status in energy terms, and (3) permits estimation of the turgor and osmotic components of water potential.

Seedling nutrition affects all aspects of seedling performance. However, measurements of nutritional status (usually made on foliage) are poor indicators of seedling quality.

Seedling morphology is a widely used grading criterion. More often than not, larger seedlings tend to outperform smaller seedlings on many sites. However, physiological factors generally override size effects.

Of 21 Northwest nurseries surveyed, all reported using at least one of the above measurement techniques. However, with the exception of morphology, these techniques were generally not used to assess seedling quality itself, but rather to monitor the effects of some cultural operation (e.g., lifting, grading, wrenching) on the seedling crop or to determine the optimum manner in which to perform such operations.

23.6 Recommendations

23.6.1 Operations

It is not realistic to recommend that nurseries routinely monitor the physiological condition of their planting stock, given the complex, time-consuming nature of the available methods. With respect to performance attributes, which give the most useful predictions, test results would be available too late to be of much practical use. Nearly all nurseries do monitor seedling morphology, and this practice is certainly worthwhile. Emphasizing root quality as well as traditional height and caliper standards also might be desirable.

For troubleshooting, when seedling damage is suspected, a good approach seems to be the accelerated RGP test (see 23.2.1.2). Any serious damage would generally be expected to show up under the forcing conditions described, although there are never any guarantees. Of course, this assumes that nurseries have access to controlled-environment chambers. Measurements of water potential, by pressure chamber or any other method, seem to be of limited use because a dead seedling can have either high or low water potential. Cold damage to the stem can sometimes be detected by sectioning buds, and cold damage to the roots by scraping bark to detect dead cambium. A more laborious but more definitive method seems to be the P-V curve, where freeze-caused cell lesions are evidenced by abnormally high osmotic potentials or the lack of a linear portion of the curve. However, the relationship between degree of damage and impact on performance has yet to be established.

A broader and perhaps ultimately more useful approach to assessing seedling quality has two parts:

(1) Nurseries should systematically collect air temperature data beginning in early autumn so that, over 3 or more years, a typical chilling curve for that nursery could be developed. From this information, plus a record of the lifting date and time in storage for each stock order, nursery managers could infer the degree of stock dormancy. Such data should accompany each stock order shipped, along with any other information that might bear on the performance potential of that stock (e.g., cold-storage temperature, climatic abnormalities during lifting, etc.).

(2) In planning the performance tracking for each year's plant, regeneration personnel should select for tracking stock that spans a range of lift-store combinations. Performance of this stock—whether successes or failures—should be systematically reported back to the nursery each year. In addition, woods personnel should note site weather conditions and any abnormalities which might have affected stock performance (e.g., inadvertent overheating of stock, poor performance of planting crew, etc.). With nursery and woods personnel cooperating in such a manner, it should be possible over time to build a data base to assist nursery managers not only in fine-tuning their lift-store operations but also in accurately predicting, rather than directly assessing, stock quality.

23.6.2 Research

In my judgment, past research on assessing seedling quality has overemphasized developing a "black box" which could be used to give an immediate, categorical evaluation of a given seedling based upon some measurable property. Had this work been successful, this chapter could have been written on one page. Considering the complexity of the seedling, the planting site, and seedling-site interactions, it is doubtful that such a black box will ever be developed.

A seemingly more intelligent approach would be to establish empirical relationships between seedling quality (assessed as cold hardiness, RGP, and some measure of drought resistance) and seedling history. It is already well known that these properties change seasonally in predictable ways, all three tending to be low in fall, high in winter, and low again in spring. Hence, winter-lifted seedlings tend always to be of the best physiological condition. Although it has not yet been rigorously demonstrated, these properties are probably related to the bud-dormancy intensity of the seedling as it weakens through winter in response to chilling. Exploring the relationships between chilling history and the above properties seems a potentially valuable avenue for research.

One complicating factor is the effect of cold storage on these performance attributes. It is known, for example, that cold storage affects RGP. Depending upon the lifting date, RGP can increase, decrease, or remain constant in storage. Why does this happen? Are there predictable patterns? Could RGP be predicted from chilling history? The same questions apply to cold hardiness and drought resistance. With these empirical relationships established for given regions and species, it would be possible operationally to make educated predictions of seedling quality without ever examining the seedlings themselves.

Acknowledgments

I am indebted to Drs. Mary Duryea, Oregon State University, and Thomas D. Landis, U.S.D.A. Forest Service, for reviews of an early version of the manuscript. A special note of thanks to Dr. Roy F. Sutton of the Canadian Forestry Service for an especially thorough critical reading of a later draft, and to Dr. Roger Timmis, Weyerhaeuser Company, for providing information on frost-hardiness testing.

This work and its publication were generously supported by Weyerhaeuser Company, Tacoma, Washington.

References

1. Askren, C. A., and R. K. Hermann. 1979. Is the oscilloscope technique suitable for predicting survival of planting stock? Tree Planters' Notes 30:7-12.
2. Aston, M. J., and D. M. Paton. 1973. Frost room design for radiation frost studies in Eucalyptus. Australian J. Botany 21:193-199.
3. Becwar, M. R., C. Rajashekar, K. J. Hansen Bristow, and M. J. Burke. 1981. Deep undercooling of tissue water and winter hardiness limitations in timberline flora. Plant Physiology 68:111-114.
4. Bixby, J. A., and G. N. Brown. 1974. Rapid determination of cold hardiness in black locust seedlings using a pressure chamber. 3rd North American Forest Biology workshop, Boulder, Colorado. Abstr. No. 12.
5. Blazich, F. A., D. R. Evert, and D. E. Bee. 1974. Comparison of three methods of measuring winter hardiness of internodal stem sections of *Forsythia intermedia* "Lynwood." J. American Society for Horticultural Sci. 9:211-214.
6. Brown, G. N. 1980. Control of frost hardiness in tree shoots. Pages 297-304 *in* IUFRO Proc., Joint workshop on xylem and shoot growth physiology (C. H. A. Little, ed.). Fredericton, New Brunswick, Canada.
7. Bunting, W. R. 1980. Seedling quality: growth and development—soil relationships, seedling growth and development, density control relationships. Pages 21-42 *in* Proc., North American forest tree nursery soils workshop (L. P. Abrahamson and D. H. Bickelhaupt, eds.). State Univ. New York, Coll. Environ. Sci. and Forestry, Syracuse.
8. Burdett, A. N. 1979. New methods for measuring root growth capacity: their value in assessing lodgepole pine stock quality. Can. J. Forest Res. 9:63-67.
9. Burdett, A. N. 1981. Personal communication. B. C. Ministry of Forests, Victoria.
10. Campbell, R. K. 1978. Regulation of bud-burst timing by temperature and photoregime during dormancy. Pages 19-33 *in* Proc., V. North American forest biology workshop (C. A. Hollis and A. E. Squillace, eds.). Univ. of Florida, Gainesville.
11. Campbell, R. K., and A. I. Sugano. 1975. Phenology of bud-burst in Douglas-fir related to provenance, photoperiod, chilling, and flushing temperature. Botanical Gazette 136:290-298.
12. Campbell, R. K., and A. I. Sugano. 1979. Genecology of bud-burst phenology in Douglas-fir: response to flushing temperature and chilling. Botanical Gazette 140:223-231.
13. Carlson, W. C., W. D. Binder, C. O. Feenan, and C. L. Preisig. 1980. Changes in mitotic index during onset of dormancy in Douglas-fir seedlings. Can. J. Forest Res. 10:371-378.
14. Cary, J. W., and H. D. Fisher. 1969. Plant moisture stress: a portable freezing point meter compared with the psychrometer. Agronomy J. 61:302-305.
15. Chavasse, C. G. R. 1980. Planting stock quality: a review of factors affecting performance. New Zealand J. Forestry 25:144-171.
16. Cheung, Y. N. S., M. T. Tyree, and J. Dainty. 1975. Water relations parameters on single leaves obtained in a pressure bomb and some ecological interpretations. Can. J. Botany 53:1342-1346.
17. Childs, S. (Date unknown.) Evaluation of a rapid technique to estimate plant water potential for applications in forestry. Unpublished manuscript, Washington State Univ., Pullman. 7 p.
18. Cleary, B. D., R. D. Greaves, and P. W. Owston. 1978. Seedlings. Pages 63-97 *in* Regenerating Oregon's forests (B. D. Cleary, R. D. Greaves, and R. K. Hermann, eds.). Oregon State Univ. Ext. Serv., Corvallis.

19. Cleary, B. D., and J. B. Zaerr. 1980. Pressure chamber techniques for monitoring and evaluating seedling water status. New Zealand J. Forestry Sci. 10:133-141.
20. Davis, W., A. Shigo, and R. Weyrick. 1979. Seasonal changes in electrical resistance of inner bark in red oak, red maple, and eastern white pine. Forest Sci. 25:282-286.
21. Dexter, S. T., W. E. Tottingham, and L. F. Graber. 1930. Preliminary results in measuring the hardiness of plants. Plant Physiology 5:215-223.
22. Dexter, S. T., W. E. Tottingham, and L. F. Graber. 1932. Investigations of the hardiness of plants by measurement of electrical conductivity. Plant Physiology 7:63-78.
23. Dickson, A., A. L. Leaf, and J. F. Hosner. 1960. Quality appraisal of white spruce and white pine seedling stock in nurseries. Forestry Chronicle 36:10-13.
24. Dobbs, R. C. 1976. Effect of initial mass of white spruce and lodgepole pine planting stock on field performance in the British Columbia interior. Can. Forestry Serv., Dep. Environ. Rep. BC-X-149. 14 p.
25. Edgren, J. W. 1980. The reforestation system—a team effort. Pages 12-20 *in* Proc., North American forest tree nursery soils workshop (L. P. Abrahamson and D. H. Bickelhaupt, eds.). State Univ. New York, Coll. Environ. Sci. and Forestry, Syracuse.
26. Ehlig, C. F. 1962. Measurement of energy status of water in plants with a thermocouple psychrometer. Plant Physiology 37:288-290.
27. Erez, A., G. A. Couvillon, and C. H. Hendershott. 1979. The effect of cycle length on chilling negation by high temperatures in dormant peach leaf buds. J. American Society for Horticultural Sci. 104:573-576.
28. Farmer, R. E., Jr. 1978. Seasonal carbohydrate levels in roots of Appalachian hardwood planting stock. Tree Planters' Notes 29:22-24.
29. Ferguson, R. B., R. A. Ryker, and E. D. Ballard. 1975. Portable oscilloscope technique for detecting dormancy in nursery stock. U.S.D.A. Forest Serv., Intermountain Forest and Range Exp. Sta., Ogden, Utah. Gen. Tech. Rep. INT-26. 16 p.
30. Fisher, H. D. 1972. An inexpensive method of determining plant moisture stress using freezing-point depression. Soil Sci. 113:383-385.
31. Flint, H. L., B. R. Boyce, and D. J. Beattie. 1967. Index of injury—a useful expression of freezing injury to plant tissues as determined by the electrolytic method. Can. J. Plant Sci. 47:229-230.
32. Gilmore, A. R. 1962. Root growth of transplanted loblolly pine (*Pinus taeda* L.) seedlings in relation to chemical root reserves. Illinois State Academy of Sci. Trans. 55:38-41.
33. Glerum, C. 1970. Vitality determinations of tree tissue with kilocycle and megacycle electrical impedance. Forestry Chronicle 46:63-64.
34. Glerum, C. 1976. Frost hardiness of forest trees. Pages 403-420 *in* Tree physiology and yield improvement (M. G. R. Cannell and F. T. Last, eds.). Academic Press, New York.
35. Glerum, C., J. L. Farrar, and R. L. McLure. 1966. A frost hardiness study of six coniferous species. Forestry Chronicle 42:69-75.
36. Green, L. M., and I. J. Warrington. 1978. Assessment of frost damage in radiata pine seedlings using the diffusate conductivity technique. New Zealand J. Forestry Sci. 8:344-350.
37. Gross, K. 1980. The pressure-volume technique as an aid to the characterization of freshness of forest planting stock. Pages 137-151 *in* Proc., Characterization of plant material (H. Schmidt-Vogt, ed.). IUFRO meeting, Div. 1, Freiburg i. Br., Federal Republic of Germany.
38. Hatch, A. H., and D. R. Walker. 1969. Rest intensity of dormant peach and apricot leaf buds as influenced by temperature, cold hardiness and respiration. J. American Society for Horticultural Sci. 94:304-307.
39. Havis, J. R. 1976. Root hardiness of woody ornamentals. HortScience 11:385-386.
40. Hellkvist, J., G. P. Richards, and P. G. Jarvis. 1974. Vertical gradients of water potential and tissue water relations in Sitka spruce trees measured with the pressure chamber. J. Applied Ecology 11:637-667.
41. Hellmers, H. 1962. Physiological changes in stored pine seedlings. Tree Planters' Notes 53:9-10.
42. Hermann, R. K. 1962. The effect of short-term exposure of roots on survival of 2-0 Douglas-fir stock. Tree Planters' Notes 52:28-30.
43. Hermann, R. K. 1964. Effects of prolonged exposure of roots on survival of 2-0 Douglas-fir seedlings. J. Forestry 62:401-403.
44. Hermann, R. K. 1967. Seasonal variation in sensitivity of Douglas-fir seedlings to exposure of roots. Forest Sci. 13:140-149.
45. Hermann, R. K., and D. P. Lavender. 1976. Large stock: is it really superior? IUFRO, 14th World Congress. Contributed paper, Working Group 1.05.04. Mimeo. 8 p.
46. Hermann, R. K., and D. P. Lavender. 1979. Testing the vigor of coniferous planting stock. Oregon State Univ., Forest Res. Lab., Corvallis. Res. Note 63. 3 p.
47. Hermann, R. K., D. P. Lavender, and J. B. Zaerr. 1972. Lifting and storing western conifer seedlings. Oregon State Univ., Forest Res. Lab., Corvallis. Res. Pap. 17. 8 p.
48. Hocking, D., and R. D. Nyland. 1971. Cold storage of coniferous seedlings: a review. State Univ. New York, Coll. Forestry, Syracuse. Applied Forestry Res. Institute Rep. 6. 70 p.
49. Höfler, K. 1920. Ein Schema für die osmotische Leistung der Pflanzenzelle. Deutsche Botanische Gesellschaft, Berlin 38:288-298.
50. Holbo, H. R., C. A. Askren, and R. K. Hermann. 1981. Does the oscilloscope technique reveal the dormancy status of Douglas-fir and ponderosa pine? Forest Sci. 27:405-412.
51. Hsaio, T. C., E. Acevedo, E. Fereres, and D. W. Henderson. 1976. Stress metabolism: water stress, growth and osmotic adjustment. Philosophical Trans. of the Royal Society of London 273:479-500.
52. Incoll, L. D., S. P. Long, and M. R. Ashmore. 1977. SI units in publications in plant science. Current Advances in Plant Sci. 28:331-343.
53. Jaramillo, A. 1980. Review of techniques used to evaluate seedling quality. Pages 84-95 *in* Proc., Intermountain Nurseryman's Assoc. and Western Forest Nursery Association. U.S.D.A. Forest Serv., Intermountain Forest and Range Exp. Sta., Ogden, Utah. Gen. Tech. Rep. INT-109.
54. Johnson, N. E., and D. G. Nielson. 1969. Pressure chamber measurements of water stress in individual pine fascicles. Forest Sci. 15:452-453.
55. Knipling, E. B. 1967. Measurement of leaf water potential by the dye method. Ecology 48:1038-1040.
56. Kozlowski, T. T. (ed.). 1968. Water deficits and plant growth. Vol. I. Development, control and measurement. Academic Press, New York. 390 p.
57. Kozlowski, T. T. (ed.). 1968. Water deficits and plant growth. Vol. II. Plant water consumption and response. Academic Press, New York. 333 p.
58. Kozlowski, T. T. (ed.). 1972. Water deficits and plant growth. Vol. III. Plant responses and control of water balance. Academic Press, New York. 368 p.
59. Kozlowski, T. T. (ed.). 1976. Water deficits and plant growth. Vol. IV. Soil water measurement, plant responses, and breeding for drought resistance. Academic Press, New York. 383 p.
60. Kozlowski, T. T. (ed.). 1978. Water deficits and plant growth. Vol. V. Water and plant disease. Academic Press, New York. 323 p.
61. Kozlowski, T. T. (ed.). 1981. Water deficits and plant growth. Vol. VI. Woody plant communities. Academic Press, New York. 582 p.
62. Kramer, P. J., and T. T. Kozlowski. 1979. Physiology of woody plants. Academic Press, New York. 811 p.
63. Lavender, D. P., and R. K. Hermann. 1970. Regulation of the growth potential of Douglas-fir seedlings during dormancy. New Phytologist 69:675-694.
64. Lavender, D. P., and P. F. Wareing. 1972. Effects of daylength and chilling on the responses of Douglas-fir (*Pseudotsuga menziesii* (Mirb.) Franco) seedlings to root damage and storage. New Phytologist 71:1055-1067.
65. Levitt, J. 1980. Responses of plants to environmental stresses. Vol. I. Chilling, freezing and high temperature stresses. Academic Press, New York. 497 p.
66. Levitt, J. 1980. Responses of plants to environmental stresses. Vol. II. Water, radiation, salt and other stresses. Academic Press, New York. 607 p.
67. Li, P. H. 1978. Plant cold hardiness research. HortScience 13:222-224.
68. Mazur, P. 1969. Freezing injury in plants. Annual Review of Plant Physiology 20:419-448.
69. McCracken, I. 1978. Carbon dioxide uptake of pine seedlings after cool storage. Forest Sci. 24:17-25.

70. McCreary, D. 1982. Personal communication, Department of Forest Science, Oregon State Univ., Corvallis.
71. McMinn, R. G. 1980. Root growth capacity and field performance of various types and sizes of white spruce stock following outplanting in the central interior of British Columbia. Pages 37-41 *in* Proc., Characterization of plant material (H. Schmidt-Vogt, ed.). IUFRO meeting, Div. 1, Freiburg i. Br., Federal Republic of Germany.
72. Menzies, M. I. 1980. Effect of nursery conditioning on the water relations of two year old Douglas-fir seedlings after lifting and outplanting. Ph.D. thesis, Univ. of Washington, Seattle. 208 p.
73. Menzies, M. I., and D. G. Holden. 1981. Seasonal frost tolerance of *Pinus radiata*, *Pinus muricata* and *Pseudotsuga menziesii*. New Zealand J. Forestry Sci. 11:92-99.
74. Nelson, E. A., and D. P. Lavender. 1979. The chilling requirement of western hemlock seedlings. Forest Sci. 25:485-490.
75. Nienstaedt, H. 1966. Dormancy and dormancy release in white spruce. Forest Sci. 12:374-384.
76. Nienstaedt, H. 1967. Chilling requirements in seven *Picea* species. Silvae Genetica 16:65-68.
77. Owens, J. N.1968. Initiation and development of leaves in Douglas-fir. Can. J. Botany 46:271-278.
78. Owens, J. N., and M. Molder. 1973. A study of DNA and mitotic activity in the vegetative apex of Douglas-fir during the annual growth cycle. Can. J. Botany 51:1395-1409.
79. Parmelee, K. 1979. The relationship of electrical impedance to vegetative maturity and dormancy in red-osier dogwood. M.S. thesis, Oregon State Univ., Corvallis. 87 p.
80. Pellett, N. E., and D. B.White. 1969. Soil-air temperature relationships and cold acclimation of container-grown *Juniperus chinensis* "Hetzi." J. American Society for Horticultural Sci. 94:453-456.
81. Perry, T. O. 1971. Dormancy of trees in winter. Science 171:29-36.
82. Puttonen, P. 1980. Effect of temporary storage temperature on carbohydrate levels in Scots pine seedlings and planting success. Pages 169-179 *in* Proc., Characterization of plant material (H. Schmidt-Vogt, ed.). IUFRO meeting, Div. 1, Freiburg i. Br., Federal Republic of Germany.
83. Richardson, E. A., S. D. Seeley, and D. R. Walker. 1974. A model for estimating the completion of rest for "Redhaven" and "Elberta" peach trees. HortScience 9:331-332.
84. Ritchie, G. A. 1982. Carbohydrate reserves and root growth potential in Douglas-fir seedlings before and after cold storage. Can. J. Forest Res. 12:905-912.
85. Ritchie, G. A., and J. R. Dunlap. 1980. Root growth potential: its development and expression in forest tree seedlings. New Zealand J. Forestry Sci. 10:218-248.
86. Ritchie, G. A., and T. M. Hinckley. 1971. Evidence for error in pressure-bomb estimates of stem xylem potentials. Ecology 52:534-536.
87. Ritchie, G. A., and T. M. Hinckley. 1975. The pressure chamber as an instrument for ecological research. Advances in Ecological Res. 9:165-254.
88. Ritchie, G. A., and R. G. Shula. 1983. Seasonal changes of tissue water relations in shoots and root systems of Douglas-fir seedlings. Forest Sci. (in press).
89. Robotham, R. W., J. Lloyd, and I. J. Warrington. 1978. A controlled environment room for producing advective white or black frost conditions. J. Agric. Engineering Res. 23:301-311.
90. Romberger, J. A. 1963. Meristems, growth and development in woody plants. U.S.D.A. Forest Serv., Washington, D.C. Tech. Bull. 1293. 214 p.
91. Rosvall-Ahnebrink, G. 1982. Personal communication, Institution för Skogsforyngring Sveriges Lantbruksuniversitet, Garpenberg, Sweden.
92. Rowley, J. A., C. G. Tunnicliffe, and A. O. Taylor. 1975. Freezing sensitivity of leaf tissue of C4 grasses. Australian J. Plant Physiology 2:447-451.
93. Sakai, A. 1978. Low temperature exotherms of winter buds of hardy conifers. Plant and Cell Physiology 19:1439-1446.
94. Salisbury, F. B., and C. W. Ross. 1978. Plant physiology. Wadsworth Publishing Co., Inc., Belmont, California. 438 p.
95. Samish, R. M. 1954. Dormancy in woody plants. Annual Review of Plant Physiology 5:183-204.
96. Sandvik, M. 1980. Environmental control of winter stress tolerance and growth potential in seedlings of *Picea abies* (L.) Karst. New Zealand J. Forestry Sci. 10:97-104.
97. Schmidt-Vogt, H. (ed.). 1980. Characterization of plant material. Proc., IUFRO meeting, Div. 1. June 23-26. Freiburg i. Br., Federal Republic of Germany. 193 p.
98. Schmidt-Vogt, H. 1981. Morphological and physiological characteristics of planting stock: present state of research and research tasks for the future. Pages 433-446 *in* Proc., IUFRO XVII World Congress. Kyoto, Japan.
99. Scholander, P. F., H. T. Hammel, E. D. Bradstreet, and E. A. Hemmingsen. 1965. Sap pressure in vascular plants. Science 148:339-346.
100. Shardkov, V. W. 1948. New field method for the determination of the suction pressure of plants (in Russian). Doklady Akademii Nauk. SSSR 60:169-172.
101. Shepherd, K. R. 1980. The use of controlled environments in forestry research. New Zealand J. Forestry Sci. 10:105-115.
102. Siminovitch, D. 1963. Evidence from increase in ribonucleic acid and protein synthesis in autumn for increase in protoplasm during the first frost hardening of black locust bark cells. Can. J. Botany 41:1301-1308.
103. Slatyer, R. O. 1967. Plant-water relationships. Academic Press, New York. 366 p.
104. Slavik, B. 1974. Methods of studying plant water relations. Springer-Verlag, New York. 449 p.
105. Spanner, D. C. 1951. The Peltier effect and its use in the measurement of suction pressure. J. Experimental Botany 1 :145-168.
106. Steponkus, P. L., and F. O. Lanphear. 1967. Refinement of the triphenyl tetrazolium chloride method of determining cold injury. Plant Physiology 42:1423-1426.
107. Stergios, B. G., and G. S. Howell, Jr. 1973. Evaluation of viability tests for cold stressed plants. J. American Society for Horticultural Sci. 98:325-330.
108. Stone, E. C. 1955. Poor survival and the physiological condition of planting stock. Forest Sci. 1:90-94.
109. Stone, E. C. 1970. Variation in the root-growth capacity of ponderosa pine transplants. Pages 40-46 *in* Regeneration of ponderosa pine (R. K. Hermann, ed.). Oregon State Univ., Corvallis.
110. Stone, E. C. 1980. Unpublished data, Univ. of California, Berkeley.
111. Stone, E. C., and J. L. Jenkinson. 1970. Influence of soil water on root growth capacity of ponderosa pine transplants. Forest Sci. 16:230-239.
112. Stone, E. C., J. L. Jenkinson, and S. L. Krugman 1962. Root regenerating potential of Douglas-fir seedlings lifted at different times of the year. Forest Sci. 8:288-297.
113. Stone, E. C., G. H. Schubert, R. W. Benseler, F. J. Baron, and S. L. Krugman. 1963. Variation in the root regenerating potentials of ponderosa pine from four California nurseries. Forest Sci. 9:217-225.
114. Sutton, R. F. 1979. Planting stock quality and grading. Forest Ecology and Management 2:123-132.
115. Thompson, B. E., and R. Timmis. 1978. Root regeneration potential in Douglas-fir seedlings: effect of photoperiod and air temperature on its evaluation and control. Pages 86-109 *in* Proc., IUFRO symp. on root physiology and symbiosis (A. Riedacker and J. Gagnaire-Michard, eds.). Nancy, France.
116. Timmis, K. A., L. H. Fuchigami, and R. Timmis. 198 . Measuring dormancy: the rise and fall of square waves. HortScience 16:200-202.
117. Timmis, R. 1976. Methods of screening tree seedlings for frost hardiness. Pages 421-435 *in* Tree physiology and yield improvement (M. G. R. Cannell and F. T. Last, eds.). Academic Press, New York.
118. Timmis, R. 1980. Stress resistance and quality criteria for tree seedlings: analysis, measurement and use. New Zealand J. Forestry Sci. 10:21-53.
119. Timmis, R. 1980. Unpublished data, Weyerhaeuser Technology Center, Tacoma, Washington.
120. Timmis, R. 1981. Unpublished data, Weyerhaeuser Technology Center, Tacoma, Washington.
121. Timmis, R., and J. G. Worrall. 1974. Translocation of dehardening and bud-break promoters in climatically 'split' Douglas-fir. Can. J. Forest Res. 4:229-237.
122. Truelove, B., D. E. Davis, and L. R. Jones. 1974. A new method for detecting photosynthesis inhibitors. Weed Sci. 22:15-17.
123. Turner, N. C., and P. J. Kramer (eds.). 1980. Adaptation of plants to water and high temperature stress. John Wiley and Sons, New York. 482 p.

124. Tyree, M. T., and H. T. Hammel. 1972. The measurement of the turgor pressure and the water relations of plants by the pressure-bomb technique. J. Experimental Botany 23:267-282.
125. van den Driessche, R. 1969. Influence of moisture supply, temperature and light on frost-hardiness changes in Douglas-fir seedlings. Can. J. Botany 47:1765-1772.
126. van den Driessche, R. 1973. Prediction of frost hardiness in Douglas-fir seedlings by measuring electrical impedance in stems at different frequencies. Can. J. Forest Res. 3:256-264.
127. van den Driessche, R. 1975. Flushing response of Douglas-fir buds to chilling and to different air temperatures after chilling. B.C. Forest Serv., Victoria. Res. Note 71. 22 p.
128. van den Driessche, R. 1976. Prediction of cold hardiness in Douglas-fir seedlings by index of injury and conductivity methods. Can. J. Forest Res. 6:511-515.
129. van den Driessche, R. 1980. Health, vigor and quality of conifer seedlings in relation to nursery soil fertility. Pages 100-120 *in* Proc., North American forest tree nursery soils workshop (L. P. Abrahamson and D. H. Bickelhaupt, eds.). State Univ. New York, Coll. Environ. Sci. and Forestry, Syracuse.
130. van den Driessche, R. 1980. Effects of nitrogen and phosphorus fertilization on Douglas-fir nursery growth and survival after out-planting. Can. J. Forest Res. 10:65-70.
131. Vegis, A. 1964 Dormancy in higher plants. Annual Review of Plant Physiology 15:185-224.
132. Venn, K. 1980. Winter vigor in *Picea abies* (L.) Karst. VII. Development of injury to seedlings during overwinter cold storage. A literature review. Reports of the Norwegian Forest Res. Institute 35.9:485-529.
133. Wareing, P. F. 1969. The control of bud dormancy in seed plants. Symposia of the Society of Experimental Biology 23:241-262.
134. Warrington, I. J., and D. A. Rook. 1980. Evaluation of techniques used in determining frost tolerance of forest planting stock: a review. New Zealand J. Forestry Sci. 10:116-132.
135. Weatherley, P. E. 1950 Studies in the water relations of the cotton plant. I. The field measurement of water deficits in leaves. New Phytologist 49:81-97.
136. Weiser, C. J. 1970. Cold resistance and injury in woody plants. Science 169:1269-1278.
137. Wells, S. P. 1979. Chilling requirements for optimal growth of Rocky Mountain Douglas-fir seedlings. U.S.D.A. Forest Serv., Intermountain Forest and Range Exp. Sta., Ogden, Utah. Res. Note INT-254.9 p.
138. Wiebe, H. H., G. S. Campbell, W. R. Gardner, S. L. Rawlins, J. W. Cary, and R. W. Brown 1971. Measurement of plant and soil water status. Utah Agric. Exp. Sta. Bull. 484. 71 p.
139. Wilner, J. 1960. Relative and absolute electrolytic conductance tests for frost hardiness of apple varieties. Can. J. Plant Sci. 40:630-637.
140. Winjum, J. K. 1963. Effects of lifting date and storage on 2+0 Douglas-fir and noble fir. J. Forestry 61:648-654.
141. Wommack, D. E. 1964. Temperature effects on the growth of Douglas-fir seedlings. Ph.D. thesis, Oregon State Univ., Corvallis. 176 p.
142. Wright, S. T. C. 1975. Seasonal changes in the levels of free and bound abscisic acid in blackcurrant (*Ribes nigrum*) buds and beech (*Fagus sylvatica*) buds. J. Experimental Botany 26:161-174.
143. Zaerr, J. B. 1972. Early detection of dead plant tissue. Can. J. Forest Res. 2:105-110.
144. Zaerr, J. B., and D. P. Lavender. 1980. Analysis of plant growth substances in relation to seedling and plant growth. New Zealand J. Forestry Sci. 10:186-195.

Appendix 1: Simplified Procedure for Constructing Pressure-Volume Curve

(1) Prepare in advance about 40 sections of 3- or 5-mm inside-diameter plastic tubing by cutting it into 5-cm-long sections and filling each section with dry tissue paper.

(2) In late afternoon or evening, select the seedling to be tested and sever it at the root collar. The shoot should be small enough that it can be placed into a pressure chamber. With a large seedling, use only the terminal portion.

(3) Partially submerge the shoot in room-temperature water overnight so that it becomes saturated (reaches full turgor).

(4) Early the following morning, remove the shoot and surface-dry it with a soft towel.

(5) Remove the bark from the basal 1 cm of the shoot and enclose the foliage in a plastic bag. The bag should be perforated and tied to the stem near the base.

(6) Place the bagged shoot into the pressure chamber and measure the balance pressure (P*); balance pressure is the same as plant moisture stress or shoot water potential. Record this value in space "A_1" on the data sheet (Fig. A1-1). If the pressure is greater than 0.1 MPa (1 bar), it indicates that the shoot is not at full turgor; it must be discarded and another sample selected.

(7) Weigh a piece of tissue-filled tubing to the nearest 0.001 g and record this value in space "B_1" (Fig. A1-1). Place the tubing over the end of the shoot which is protruding from the pressure chamber so that the dry tissue is in contact with the xylem surface.

(8) Increase the chamber pressure 0.5 MPa (5 bars) and hold it constant for 10 minutes. Because the time period is important, it is desirable to use a laboratory timer.

(9) After 10 minutes, remove the tube and record its weight in space "C_1." The weight gain in grams is due to the weight of the sap absorbed by the tissue and equals the incremental volume of sap lost in cubic centimeters at that pressure.

(10) Slightly reduce the chamber pressure to draw any sap away from the cut surface, then slowly increase the pressure, determine a new balance pressure, and record it in space "A_2."

(11) Weigh another piece of plastic tube, record its weight in space "B_2," and place it atop the cut stem as in step (7).

(12) Repeat steps (7) through (11) about 25 times.

The data sheet will then contain a series of P* values, along with pairs of corresponding initial and final tube weights. Calculate the reciprocal of each pressure (1/P*) and the tube-weight difference at each pressure increment. Then calculate the cumulative tube-weight differences beginning at the first pressure and at each successive pressure to the end.

The pressure-volume (P-V) curve is constructed by plotting the values of 1/P* against the corresponding cumulative weight-loss value and should resemble the curve shown in Figure 6 of the text.

Note: The same procedure may be used on root systems if 0.3 MPa pressure increments are substituted for 0.5 MPa increments in step (8).

Seedling number ____________ Date ____________

Root or shoot ____________ Name ____________

P*	1/P*	Tubing weight, g Initial	Final	Difference	Cumulative wt. loss, g
A_1		B_1	C_1		
A_2		B_2	C_2		
...		...	...		
...		...	...		
...		...	...		
A_{25}		B_{25}	C_{25}		

Figure A1-1. Sample pressure-volume data sheet.

Chapter 24
Planting-Stock Selection: Meeting Biological Needs and Operational Realities

R. D. Iverson

Abstract

The tremendous variability in Northwest planting sites requires a variety of planting stock. Genetic, physiological, and morphological seedling characteristics must be matched to site and organizational objectives. Vegetation classification schemes help foresters select species, and seed zones help them determine areas where seedlings of any species can safely be moved from their source. Local species and seed sources should always be preferred unless documented research proves otherwise. Large seedlings, if properly conditioned, will grow faster on favorable sites. Seedlings with tall shoots are better suited to brushy areas and where animal damage may be a problem; large-caliper seedlings will perform better where heat, insects, or physical bending are problems. Droughty conditions require seedlings with well-developed roots. Container-grown seedlings can be used to extend the planting season, but spraying soil-active herbicides over them immediately after planting is risky.

24.1 Introduction

Careful selection of planting stock is critical in any reforestation prescription. A good choice of stock may even compensate for inadequate site preparation. But what may be considered high-quality stock for adequately prepared areas might not prove suitable for those that are inadequately prepared.

Foresters and nursery managers are jointly responsible for producing high-quality nursery stock. "Quality" here is defined as the ability of stock to realize management objectives at planting sites [61]. The forester knows what morphological and physiological characteristics of seedlings can maximize performance at planting sites. The nursery manager is charged with producing seedlings that meet those specifications economically. Use of ideal seedlings will result in plantations that have the lowest cost per surviving tree or, better yet, the highest estimated present net value [58]. Using present net value as a criterion bases comparisons on growth as well as survival.

A variety of species and stock types is grown to fit the highly variable topography, soil, and climate of the Northwest. Over 20 species and seven different stock types were produced in 1980 (Table 1) (OSU Nursery Survey; see chapter 1, this volume). Plans for nursery production through 1985 continue to be tailored to meet customer needs. For example, the OSU Survey indicated a trend toward growing larger trees; as a result, more transplants and 2 + 0 seedlings grown at low densities will be produced.

Over the years, foresters have selected stock on the basis of their experience with its performance and research results. Even though stock performance is at times contradictory, most

Table 1. Estimated 1980 seedling production at major Northwest nurseries by species group and stock type (OSU Nursery Survey).

	Production, in 1,000s of seedlings
Species group	
Douglas-fir	168,047
Western hemlock	1,123
Spruce	37,657
True firs	17,441
Ponderosa pine	26,795
Lodgepole pine	16,323
Other pine	1,899
Western larch	2,832
Miscellaneous species	5,385
Total	277,502
Stock type	
1 + 0	1,184
1 + 1	7,356
1 + 2	1,400
2 + 0	219,892
2 + 1	38,479
3 + 0	1,754
Plug + 1	6,575
Miscellaneous types	862
Total	277,502

In Duryea, Mary L., and Thomas D. Landis (eds.). 1984. *Forest Nursery Manual: Production of Bareroot Seedlings*. Martinus Nijhoff/Dr W. Junk Publishers, The Hague/Boston/Lancaster, for Forest Research Laboratory, Oregon State University, Corvallis. 386 p.

anomalies can be explained and general guidelines for stock selection offered. In this chapter, I review factors influencing stock selection and discuss relationships between seedling characteristics and performance on specific site types.

24.2 Considerations in Selecting Stock

Organizational objectives, planting-stock availability, and environment at the planting site are factors influencing stock selection. Although environment at the planting site is very important biologically for selecting the right stock, there are situations in which organizational objectives and stockavailability may be overriding factors.

24.2.1 Organizational objectives

Organizational objectives influence stock choice by dictating the reforestation system or species. Some organizations, for example, may adopt a container-grown seedling system to mechanize reforestation, extend the planting season, shorten production time, or facilitate production of species difficult to grow as bareroot stock [57]. Planting Douglas-fir [*Pseudotsuga menziesii* (Mirb.) Franco] in lieu of western hemlock [*Tsuga heterophylla* (Raf.) Sarg.] on sites suited for either is an example of managing for a preferred species. Even though hemlock may be suited ecologically, it is generally considered less desirable than Douglas-fir for both lumber and pulpwood. The potential of hemlock, however, gradually is being recognized [51], and the planting of Douglas-fir on hemlock sites is becoming less common. Other resource values also may influence species choice. Wildlife considerations, for example, may justify using a species not normally planted for fiber production [50].

24.2.2 Planting-stock availability

A shortage of the preferred species or stock type often causes foresters to use less desirable planting stock. Shortages can and do occur because of nursery pests, damaging weather conditions, inadequate seed supplies, or poor germination.

24.2.3 Planting-site environment

Classifying the planting-site environment is important, in the long term, for selecting species and seed source and, in the short term, for determining morphological and physiological seedling characteristics. Using the correct species and seed source will ensure that seedlings are adapted to infrequent climatic extremes or diseases which could affect plantation performance in the future but go unnoticed during establishment. Seedlings with the correct morphological and physiological characteristics are better adapted to meet initial threats to survival and optimal growth, such as animals, falling debris, or heavy brush competition.

24.2.3.1 Species

Each species will be best adapted to a given range of environmental conditions. Proper species selection may require more than just surveying native tree species and their relative frequency in the previous stand. Some species may have higher yield potential than others [23] or may be better adapted to the environments created by harvesting or other site disturbances.

Vegetation classifications provide useful aids to selecting species. Habitat types delineate sites with equivalent environments where plant succession leads to the same climax species. If the habitat type is known, identifying the seral or pioneer species on a given site is possible. A generally accepted rule is that the seral or pioneer species will survive better and grow faster on clearcuts or burned areas than the climax species [47].

Environments can be classified on the basis of soil characteristics or other site features as well as vegetation. For example, on the Vancouver Forest District, the British Columbia Forest Service considers vegetation types and soil nutrient and moisture characteristics to guide species selection and intensity of prescribed burning [35].

Vegetation descriptions and keys are available for much of the Northwest. Major vegetational units in Oregon and Washington have been described by Franklin and Dyrness [18] and numerous regional plant communities and habitat types by Bailey [4], Daubenmire and Daubenmire [14], Reed [48], Pfister [46], Hall [22], Dyrness et al. [17], Cooper [12], and Wirsing and Alexander [66].

Other site factors also are important in selecting species. Using certain species is risky because of their vulnerability to insects and diseases. Sugar pine (*Pinus lambertiana* Dougl.), for example, is particularly well suited to many sites in southwestern Oregon but is susceptible to blister rust. Therefore, it is not recommended for drainages where rust spores remain viable and can travel long distances in the humid night air [26]. Douglas-fir and ponderosa pine (*Pinus ponderosa* Dougl. ex Laws.) are preferred to sugar pine in such situations.

24.2.3.2 Seed source

Merely selecting the correct species will not guarantee that the stock suits the site. Adaptive differences also occur within a species. Coastal Douglas-fir (var. *menziesii*), for example, seems adapted to specific environments even though, as a species, it ranges from British Columbia to northern California. Campbell [5] found differing genetic potentials among different seed sources of Douglas-fir from the same watershed, and Hermann and Lavender [28] noted differences among seed sources from north and south aspects of the same mountain. Using only adapted seedlings will lessen the risk of widespread mortality because of climatic extremes, disease, or insects and will reduce the probability of growth loss.

The best adapted seeds originate from stands close to the area to be reforested—assuming that trees in the immediate vicinity developed naturally and are not plantations from an off-site seed source. How far seed can be moved from its original source depends on how closely the planting-site environment matches that where the seed originated. Limits to seed transfer could be defined as geographic, altitudinal, ecologic, or physiographic intervals across which adaptive differences among populations can be detected [49].

Actual experience is the best way to determine what distance seed can be moved from its source without losing general adaptation. Nevertheless, zones of similar environment have been delineated on maps for the Northwest (available from the Western Forest Tree Seed Council, Portland, Oregon) to guide seed-transfer limits. Such zones are designated by a 3-digit code in which the first digit identifies the physiographic and climatic region within a state, the second identifies the physiographic and climatic subregion within a region, and the third identifies the zone within a subregion. The local seed zone should be selected as a first choice. Seed from adjacent zones will substitute only if the environment there is similar to the local one.

As a general rule, seedlings should also originate from within 150 m (500 ft) in elevation of the planting site. In some areas, however, adhering to such subzones may be less critical than adhering to other gradients. With Douglas-fir, for example, elevational and north-south seed transfers are less dangerous (have lower risk of maladaptation) than east-west transfers [6]. Different species may also have varying elevational intervals. Suggested intervals for some species in the northern Rocky Mountains are 150 m (500 ft) for lodgepole pine (*Pinus contorta* var. *latifolia* Engelm.) and Douglas-fir (northern Idaho), 300 m

(1,000 ft) for Douglas-fir (western Montana), 460 m (1,500 ft) for ponderosa pine, 610 m (2,000 ft) for western larch (*Larix occidentalis* Nutt.), and infinite for western white pine (*Pinus monticola* Dougl. ex D. Don) [49]. Seed collected from a particular stand should be transferred (either upward or downward in elevation) about ½ the interval indicated.

Use of nonadapted species and seed sources can result in total, immediate mortality—or satisfactory, initial plantation establishment but reduced growth or failure later because of disease, insects, or periodic climatic extremes [8]. Numerous examples demonstrate the need for using adaptive species and seedlings from local sources [13, 25, 33, 37].

24.2.3.3 Seedling morphology and physiology

Unfortunately, planting stock is often prescribed only by stock type This practice makes it impossible to correlate seedling characteristics with field performance and does not sufficently describe seedling morphology so that foresters can order what they need. Within a stock type, seedling size can vary between nurseries or within the same nursery in different years. Describing stock only by type in research trials may be one reason that inconsistent performance has been reported for the same and different types (see also chapter 15, this volume).

Some nurseries use a seedling description system [45] that includes stock type, shoot height, root-collar diameter, shoot:root ratio, and sowing date, all based on characteristics for 75% of the seedlings in a lot. This system allows foresters to specify seedling needs in more detail and nursery managers to describe more precisely the characteristics of seedlings. Though the need to grow a variety of seedlings to match planting-site conditions has long been recognized [9, 10, 34], it is now possible to better manipulate seedling morphology and physiology to meet specified needs. Table 2 shows the median values of morphological targets for species and stock types produced at major Northwest nurseries.

Results of past studies do not clearly indicate what seedling characteristics ensure optimum performance. Studies relating performance to seedling size, for example, support one of three conclusions: (1) large seedlings are best; (2) seedling size at time of planting is not of primary significance; and (3) small seedlings are best. Such contradictions are attributable to wide variations in nursery conditions, stock treatment and handling, and site conditions among studies [67]. Yet despite some inconsistent results, past studies show that matching seedling physiological and morphological characteristics to the planting site improves results, perhaps by ensuring fuller utilization of nutrients, water, and light. Though greater flexibility in seedling characteristics is permissible on favorable sites, greater vigor is particularly important on marginal or difficult ones.

Table 2. Median morphological targets for representative species and stock types grown at major Northwest nurseries (OSU Nursery Survey).[1]

Species	Stock type	Shoot height, cm (in.)	Caliper, mm	Shoot:root ratio
Douglas-fir	1 + 0	11.5 (4.5)	3	1.8
	1 + 1	38.0 (15)	8	Not specified
	1 + 2	76.0 (30)	10	1.5
	2 + 0	30.5 (12)	5	2.0
	2 + 1	46.0 (18)	7	1.8
	3 + 0	61.0 (24)	8	Not specified
	Plug + 1	46.0 (18)	9	1.9
True firs	2 + 0	15.0 (6)	4.5	1.5
	2 + 1	23.0 (9)	6	2.0
Lodgepole pine	2 + 0	13.0 (5)	4	1.3
Ponderosa pine	2 + 0	13.0 (5)	4	1.8
Spruce	2 + 0	18.0 (7)	4	1.5
	1 + 2	76.0 (30)	10	1.5
	2 + 1	24.0 (9)	5.5	1.5
Larch	2 + 0	20.0 (8)	4	Not specified

[1]Targets for some species and stock types may be represented by only one nursery, and some nurseries did not specify targets.

24.3 Matching Stock and Site

24.3.1 Favorable sites

Favorable sites include those that have a long growing season, sparse residual vegetation, low probability of animal or insect damage, gentle slopes, and sufficient moisture so that seedlings are not severely stressed. Though seedlings do not need special characteristics to ensure their survival on favorable sites, fast initial growth rates are desirable.

Under favorable conditions, large seedlings, regardless of size standard used, demonstrate more growth than small seedlings. For example, large white spruce [*Picea glauca* (Moench) Voss] and lodgepole pine seedlings outgrew smaller ones of the same stock types [15]. Similarly, Douglas-fir, western hemlock, and Sitka spruce [*Picea sitchensis* (Bong.) Carr.] grown in 125-cm^3 (8-in.^{3}) styroblocks outperformed seedlings grown in 40-cm^3 (2-in.^{3}) styroblocks after 5 years in the field [2]. Shoot and root dry weights of seedlings produced in the larger container were substantially greater than those of seedlings from the smaller container.

Seedling stem caliper, also related to initial growth and other seedling parameters, is often considered the best morphological index of planting-stock quality [7, 38]. Ponderosa pine seedlings 3.6 mm or more in stem caliper grew more after 2 years on favorable sites in northern California than seedlings 2.5 mm or less in caliper [31]. Similar results were obtained for Douglas-fir and Sitka spruce in Washington [64]. Large-caliper Douglas-fir seedlings (defined as ≥ 12 mm in diameter for 2 + 1 transplants, ≥ 5 mm for 2 + 0 seedlings, and ≥ 2 mm for plug seedlings) grew more than smaller caliper ones within the same stock types. Wierman [64] concluded that 2 + 0 Douglas-fir seedlings less than 3 mm should not be used. Smith [55] found that, on the basis of potential returns, the optimum Douglas-fir seedling is at least 5 mm in diameter at the root collar and 38 cm (15 in.) tall. Because the most cost-effective seedling size is a function of tree performance and production costs, the increased survival and growth of larger seedlings must more than compensate for the additional cost of producing them. Consequently, optimum size will change as either cost or performance changes.

Large seedlings can grow more than smaller ones because of their greater photosynthetic area. However, faster growth will occur only if seedling development is properly synchronized. Nursery practices that induce dormancy at the correct time and fulfill chilling requirements will provide for early, rapid shoot growth and high root-growth capacities [39]. That is, a seedling must be physiologically prepared for planting (see chapters 14 and 15, this volume).

24.3.2 Difficult sites

Difficult sites are defined as those requiring careful effort to reforest successfully by planting. High-elevation sites, for example, have a short growing season and a short period of favorable planting conditions in the spring. Similarly, sites with heavy cover of residual woody and herbaceous weeds, areas populated with animals or insects that feed on seedlings, droughty sites, steep slopes prone to soil and debris movement, and frost pockets all require specialized planting stock.

24.3.2.1 High-elevation sites

On high-elevation sites, cold soils or spring snow may make it necessary to extend normal planting periods. Soil temperatures at 900 m (2,953 ft) in the Northwest can remain below 10°C (50°F) at 22 cm (8.7 in.) depth until June, whereas soil temperatures reach 10°C in April at 200 m (656 ft) [16]. At some high elevations, favorable conditions for both soil moisture and temperature are short lived because once soils warm, moisture is rapidly depleted.

Container-grown seedlings are often used when the planting season must be extended. They may adjust better in less favorable planting conditions because their roots are not stressed by pruning or handling, and they have immediate access to some moisture and fertilizer in the enclosing medium [57]. Bareroot seedlings, on the other hand, must reestablish all root-soil contacts after planting. Containerized stock has been used successfully in eastern Oregon by Weyerhaeuser Company to extend the planting season [65]: container-grown seedlings reportedly achieved 85% survival, and growth rates equaled those of bareroot stock. It is not certain, however, whether the present net value of the container-grown stock exceeded that of the bareroot stock because initial costs were higher.

24.3.2.2 Sites with competing woody vegetation

Where woody vegetation threatens to overtop seedlings, large seedlings and those with long shoots are particularly desirable. In heavy brush, initial seedling height may be more important than initial growth rate [56]; if seedlings are quickly overtopped, there is little chance they will outgrow competing vegetation in a reasonable time without release [30, 53]. Initial height of Douglas-fir seedlings was shown to be especially important to seedling establishment on Oregon coastal sites with overtopping woody vegetation [32]. Arnott [1] also found that large bareroot Douglas-fir seedlings grew more than smaller containerized stock in areas where vegetative competition was severe. In both studies, the large seedlings were 1 + 2 transplants averaging 43 cm (17 in.) tall.

The tallest seedlings, however, may not always grow more. In a test that compared 2 + 1, 3 + 0, plug + 1, and several sizes of containerized seedlings, Hahn and Smith [21] found that 3 + 0 seedlings grew slower after 3 years in the field than bareroot transplants even though the 3 + 0 seedlings were taller initially. However, the 2 + 1 and plug + 1 seedlings, which were initially taller than the containerized seedlings, did grow faster after 3 years than the containerized stock on a north slope with vegetative competition. The poor performance of the 3 + 0 seedlings may be attributable to differences in physiology or shoot:root ratio of the 3 + 0 and other stock types. I recommend using Douglas-fir seedlings that are at least 43 cm (17 in.) tall and have at least a 7-mm root-collar diameter in areas where woody vegetation is a problem and water is not limiting.

In brushy environments, initial shoot height is important for other species as well as Douglas-fir. Newton [42] studied the performance of western hemlock wildlings and concluded that success decreased rapidly below a height of 61 cm (24 in.). Similarly, large white spruce and lodgepole pine seedlings planted on nonscarified plots outperformed small seedlings on scarified plots [15]; shoots of large pines averaged 15.2 cm (6.0 in.) long for 2 + 0 stock and 21.1 cm (8.3 in.) for 2 + 1 stock and those of large spruce 20.4 cm (8.0 in.) for 2 + 0 stock and 19.8 cm (7.8 in.) for 2 + 1 stock.

Studies of containerized stock have shown that large seedlings are needed where vegetative competition is a problem. White spruce seedlings grown in styroplug 8 (8-in.^{3}) containers were compared with seedlings grown in styroplug 2 (2-in.^{3}) and styroplug 4 (4-in.^{3}) containers on prepared areas and those with dense competing vegetation [41]. The small seedlings (initially 16 cm, or 6.3 in., tall) grew less than the larger ones (initially 22 cm, or 8.7 in., tall) on sites where competition was stiff. Large stock growing on untreated sites or small stock growing where soil had been tilled instead of just scarified gained 50 to 100% in total seedling mass by the end of five growing seasons, compared to small stock planted on untreated sites. In this example, the better performance of large stock on unprepared sites could more than compensate for its higher initial cost.

24.3.2.3 Sites with downslope movement and falling debris

Large-caliper seedlings are more suitable than small-caliper ones in areas prone to downslope soil movement or falling debris. Soil deposition or debris lodged on seedlings can create static bending stress that will reduce height growth [52] or even bury seedlings [19]. Large-caliper 2 + 0 Douglas-fir seedlings grew 82% more after 3 years than smaller caliper ones on unstable granitic soils where 153 of 200 trees had varying degrees of soil deposited around them [60]. Stems of larger seedlings averaged 8 mm, those of smaller ones 6 mm.

Large stem caliper provides other advantages in addition to bending resistance. The thicker bark on larger stems may allow heat to dissipate along and away from the stem, making large-stemmed seedlings more heat tolerant on sites where high temperature is a problem. The succulent thin stems of newly germinated Douglas-fir and hemlock seedlings generally die at temperatures between 51°C (123°F) and 60°C (140°F) [54].

24.3.2.4 Sites with insects and animals

Insects are not a common threat to plantation establishment. Nevertheless, *Stremnius carinatus* (Boheman), a weevil native to Pacific Coast forests, caused 2 to 11% mortality in plug plantations sampled in a 1975 survey [20]. Thick bark is believed to discourage attack by weevils. Therefore, large stock is recommended where sizable weevil populations are suspected. A total catch of 30 weevils from 10 traps collected over a 2-week period indicates a potential hazard.

Large planting stock also is needed in areas where animals may damage trees. Large shoots lessen the frequency and consequences of clipping by hares [24, 44]. After exposure to a large number of hares for 4 months in a 1-acre enclosure [24], Douglas-fir seedlings 74 cm (29 in.) tall were not reduced in size, but shorter seedlings were. In another test in the same study, seedlings 48 cm (19 in.) or more in height withstood hare damage when an effective repellent (e.g., TMTD) was applied before planting; the initial protection enabled the terminal shoot to rapidly grow above the reach of hares. Large seedlings also are more likely to recover from clipping because of their greater photosynthetic surface. Further, numerous branches on seedlings may be helpful because they provide a choice of browse; if the terminal is browsed, a lateral branch is available to replace it.

24.3.2.5 Droughty sites

Seedlings with well-developed root systems are needed on moisture-limited sites so that the absorptive capacity of roots can counteract transpirational losses from shoots [3, 27, 29]. Douglas-fir seedlings with large roots (shoot:root ratio 1.25, oven-dry basis) had 22 to 26% higher survival than those with small roots (shoot:root ratio 0.71) on dry sites in north-central Washington [40]. Similarly, wrenched Douglas-fir seedlings developed higher shoot:root ratios and survived well on a dry south slope in the Cascade Mountains near Springfield, Oregon [62] and on a dry Coast Range site in northern California [36], as compared to seedlings not wrenched. Even though shoot:root ratio may not always indicate root size or absorptive capacity, it is a convenient index of seedling balance.

Root characteristics such as surface area and root form also could be important on droughty sites. Comparing nursery and natural stock, Stein [58] noted that lateral roots of nursery stock are usually trimmed to the same length as tap roots or are shaped in containers to parallel the tap root, whereas lateral roots of natural seedlings develop a short distance from the soil surface and branch extensively to tap both surface and lower soil layers. Differences in the form and balance of nursery stock and seedlings that develop on site should be evaluated further to determine which of these differences may be critical to a tree's normal top and root development.

Foresters have long recognized that controlling competing grasses and other herbaceous vegetation with herbicides will reduce moisture stress in seedlings. The use of container-grown seedlings on sites treated with soil-active herbicides could be risky, however. Potting medium apparently does not adsorb the chemical, which deposits near fine seedling roots, causing severe damage [43]. Bareroot seedlings or plug transplants would less likely be damaged on sites where soil-active herbicides are applied the year of planting.

24.3.2.6 Sites with frost pockets

The physiological condition of seedlings is crucial on sites where frost is likely. Nursery cultural operations should be synchronized with seedling dormancy to ensure dormancy requirements are fulfilled (see chapters 14 and 15, this volume). Failure of seedlings to complete the requirements of each dormancy phase will result in decreased seedling vigor and increased vulnerability to environmental stress, including frost. Frost hardiness of seedlings might also be improved by manipulating fertilizer regimes (see chapter 7, this volume). Evidence suggests that the ratio of nitrogen to potassium influences cold hardiness [63].

24.4 Conclusions

The genetic, morphological, and physiological characteristics of planting stock should be designated for individual planting sites. Biologically, the goal is to plant seedlings that will most fully utilize site resources and will be least constrained by animals, vegetation, debris, or other factors. Operationally, the goal is to maximize the present net value of the plantation. Using the correct planting stock will ensure that returns on other reforestation investments such as site preparation and maintenance also are maximized.

Planting-stock prescriptions can only be based on results from past field performance. Prescriptions will change as seedling production techniques improve and as more is learned about the interactions of seedlings with their environment. As foresters become more confident of their needs, they must alert nursery managers so that the best seedling may be produced at the lowest cost.

References

1. Arnott, J. T. 1974. Performance in British Columbia. Pages 283-290 *in* Proc., North American containerized forest tree seedling symp. (R. W. Tinus, W. I. Stein, and W. E. Balmer, eds.). Great Plains Agric. Res. Publ. 68.
2. Arnott, J. T., and D. Beddows. 1982. Influence of styroblock container size on field performance of Douglas-fir, western hemlock, and Sitka spruce. Tree Planters' Notes 33(3):31-34.
3. Baer, N., F. Ronco, and C. W. Barney. 1977. Effects of watering, shading, and size of stock on survival of planted lodgepole pine. U.S.D.A. Forest Serv., Rocky Mt. Forest and Range Exp. Sta., Fort Collins, Colorado. Res. Note RM-347. 4 p.
4. Bailey, A. W. 1966. Forest associations and secondary plant succession in the southern Oregon Coast Range. Ph.D. thesis, Oregon State Univ., Corvallis. 164 p.
5. Campbell, R. K. 1979. Genecology of Douglas-fir in a watershed in the Oregon Cascades. Ecology 60:1036-1050.
6. Campbell, R. K., and F. C. Sorensen. 1978. Effect of test environment on expression of clines and delimination of seed zones in Douglas-fir. Theoretical and Applied Genetics 51:233-246.
7. Chavasse, C. G. R. 1977. The significance of planting height as an indicator of subsequent seedling growth. New Zealand J. Forestry 22(2):283-296.
8. Ching, K. K., and D. P. Lavender. 1978. Seeds. Pages 47-62 *in* Regenerating Oregon's forests (B. D. Cleary, R. D. Greaves, and R. K. Hermann, eds.). Oregon State Univ. Ext. Serv., Corvallis.
9. Cleary, B., and R. Greaves. 1976. Determining planting stock needs. Pages 60-81 *in* Tree planting in the inland Northwest (D. M. Baumgartner and R. J. Boyd, eds.). Washington State Univ. Ext. Serv., Pullman.
10. Cleary, B., and R. Tinus. 1980. Preservation of nursery stock quality through packaging, storage, transport and planting. New Zealand J. Forestry Sci. 10(1):295-296.
11. Colby, M. K., and G. D. Lewis. 1973. Economics of containerized conifer seedlings. U.S.D.A. Forest Serv., Rocky Mt. Forest and Range Exp. Sta., Bottineau, North Dakota. Res. Pap. RM-108. 7 p.
12. Cooper, S. V. 1975. Forest habitat types of northwestern Wyoming and contiguous portions of Montana and Idaho. Ph.D. thesis, Washington State Univ., Pullman. 190 p.
13. Dahl, H. A., and L. A. Nicholson. 1970. Management decisions relating to off-site plantations. Pages 26-27 *in* Regeneration of ponderosa pine (R. K. Hermann, ed.). Oregon State Univ., Forest Res. Lab., Corvallis.
14. Daubenmire, R., and J. B Daubenmire. 1968. Forest vegetation of eastern Washington and northern Idaho. Washington Agric. Exp. Sta. Bull. 60. 104 p.
15. Dobbs, R. C. 1976. Effect of initial mass of white spruce and lodgepole pine planting stock on field performance in the British Columbia interior. Can. Forestry Serv., Pacific Forest Res. Centre. Rep. BC-X-149. 14 p.
16. Dougherty, P. M. 1980. The influence of site and climate factors on regeneration at high elevations in the Pacific Northwest. Page 68 *in* Executive summaries, Proc., 1980 Western forestry conf. Western Forestry and Conservation Assoc., Portland, Oregon.
17. Dyrness, C. T., J. F. Franklin, and W. H. Moir. 1974. A preliminary classification of forest communities in the central portion of the Western Cascades in Oregon. U.S./International Biological Program, Coniferous Forest Biome. Bull. 4. 123 p.
18. Franklin, J. F., and C. T. Dyrness. 1973. Natural vegetation of Oregon and Washington. U.S.D.A. Forest Serv., Pacific NW Forest and Range Exp. Sta., Portland, Oregon. Gen. Tech. Rep. PNW-8. 417 p.
19. Franklin, J. F., and J. S. Rothacher. 1962. Are your seedlings being buried? Tree Planters' Notes 51:7-9.
20. Gutzwiler, J. R. 1976. Plantation damage by the weevil *Stremnius carinatus* (Boheman). Page 85 *in* Executive summaries, Proc., 1976 Western forestry conf. Western Forestry and Conservation Assoc., Portland, Oregon.
21. Hahn, P. F., and A. J. Smith. 1983. Douglas-fir planting stock performance comparison after the third growing season. Tree Planters' Notes 34(1):33-39.
22. Hall, F. C. 1973. Plant communities of the Blue Mountains in eastern Oregon and southwestern Washington. U.S.D.A. Forest Serv., Pacific NW Region Area Guide. Guide 3-1. 62 p.
23. Handley, D. L. 1976. The yield potential of western hemlock. Pages 221-227 *in* Western hemlock management (W. A. Atkinson and R. J. Zasoski, eds.). Coll. of Forest Resources, Univ. of Washington, Seattle.
24. Hartwell, H. D. 1969. Control of damage by snowshoe hares on forest plantations. Pages 80-83 *in* Wildlife and reforestation in the Pacific Northwest (H. C. Black, ed.). Oregon State Univ., Corvallis.
25. Harvey, G. M. 1974. Epidemiology of *Lophodermella morbida* in plantations of ponderosa pine in western Oregon. Ph.D. thesis, Oregon State Univ., Corvallis. 103 p.
26. Hayes, G. L., and W. E. Hallin. 1962. Tree species to grow in the south Umpqua drainage. U.S.D.A. Forest Serv., Pacific NW Forest and Range Exp. Sta., Portland, Oregon. Res. Note 221. 7 p.
27. Hermann, R. K. 1964. Importance of top-root ratios for survival of Douglas-fir seedlings. Tree Planters' Notes 64:7-11.
28. Hermann, R. K., and D. P. Lavender. 1968. Early growth of Douglas-fir from various altitudes and aspects in southern Oregon. Silvae Genetica 17:143-151.

29. Hobbs, S. D., R. H. Byars, D. C. Henneman, and C. R. Frost. 1980. First year performance of 1-0 containerized Douglas-fir seedlings on droughty sites in southwestern Oregon. Oregon State Univ., Forest Res. Lab., Corvallis. Res. Pap. 42. 15 p.
30. Howard, K. M. 1978. The influence of seral coast range vegetation on the growth habit of juvenile Douglas-fir. M.S. thesis, Oregon State Univ., Corvallis. 208 p.
31. Iverson, R. D. 1981. Effects of site preparation and seedling caliper on initial performance of ponderosa pine. International Paper Co., Western Forest Res. Center, Lebanon, Oregon. Res. Letter 48. 7 p.
32. Iverson, R. D., and M. Newton. 1980. Large Douglas-fir seedlings perform best on Oregon coastal sites. International Paper Co., Western Forest Res. Center, Lebanon, Oregon. Tech. Note 55. 9 p.
33. James, G. A. 1958. Port Orford cedar plantations in the Pacific Northwest. U.S.D.A. Forest Serv., Pacific NW Forest and Range Exp. Sta., Portland, Oregon. Res. Note 164. 2 p.
34. Jaworsky, J. M. 1958. Some problems concerning Douglas-fir nursery stock reflected in experimental planting at Sutton Creek, spring 1954 and 1955. Assoc. of B.C. Foresters thesis, Faculty of Forestry, Univ. of British Columbia, Vancouver. 118 p.
35. Klinka, K. 1977. Guide for the tree species selection and prescribed burning in the Vancouver forest district. B.C. Forest Serv., Res. Div., Vancouver. 42 p.
36. Koon, K. B., and T. O'Dell. 1977. Effects of wrenching on drought avoidance of Douglas-fir seedlings. Tree Planters' Notes 28(2):15-16.
37. Krygier, J. T. 1958. Survival and growth of thirteen tree species in coastal Oregon. U.S.D.A. Forest Serv., Pacific NW Forest and Range Exp. Sta., Portland, Oregon. Res. Pap. 26. 20 p.
38. Kummel, J. F., C. A. Rindt, and T. T. Munger. 1944. Forest planting in the Douglas-fir region. U.S.D.A. Forest Serv., Portland, Oregon. 154 p.
39. Lavender, D. P., and B. D. Cleary. 1974. Coniferous seedling production techniques to improve seedling establishment. Page 177 *in* Proc., North American containerized forest tree seedling symp. (R. W. Tinus, W. I. Stein, and W. E. Balmer, eds.). Great Plains Agric. Res. Publ. 68.
40. Lopushinsky, W. 1976. Relationship of shoot-root ratio to survival and growth of outplanted Douglas-fir and ponderosa pine seedlings. U.S.D.A. Forest Serv., Pacific NW Forest and Range Exp. Sta., Portland, Oregon. Res. Note PNW-274. 7 p.
41. McMinn, R. G. 1982. Size of container-grown seedlings should be matched to site conditions. Pages 307-312 *in* Proc., Canadian containerized tree seedling symp. (J. B. Scarrett, C. Glerum, and C. A. Plexman, eds.). Sault Ste. Marie, Ontario.
42. Newton, M. 1978. Test of western hemlock wildlings in brushfield reclamation. Oregon State Univ., Forest Res. Lab., Corvallis. Res. Pap. 39. 22 p.
43. Newton, M. 1982. Herbaceous weeds—new information. Unpubl. paper presented at OSU workshop on the use of herbicides in forest vegetation management, Oregon State Univ., Corvallis, March 2-4.
44. Newton, M., and H. C. Black. 1965. Large planting stock of Douglas-fir helps evade damage by animals and sprouting brush on favorable sites. Western Weed Control Conf. res. progress reps. 3 p.
45. Nicholson, L. A. 1974. Seedling size description. Pages 74-86 *in* Proc., Western Forest Nursery Council meeting, Portland, Oregon.
46. Pfister, R. D. 1972. Vegetation and soil in the subalpine forest of Utah. Ph.D. thesis, Washington State Univ., Pullman. 98 p.
47. Pfister, R. D. 1976. Choosing tree species for planting. Pages 193-204 *in* Tree planting in the inland Northwest (D. M. Baumgartner and R. J. Boyd, eds.). Washington State Univ. Ext. Serv., Pullman.
48. Reed, R. M. 1969. A study of vegetation in the Wind River Mountains, Wyoming. Ph.D. thesis, Washington State Univ., Pullman. 77 p.
49. Rehfeldt, J. 1981. Seed transfer guidelines for the northern Rocky Mountains. Paper presented at Western Reforestation Coordinating Committee meeting, Western Forestry conf., Dec. 1-3.
50. Revel, J. 1974. Planting stock-planting site interactions. B. C. Forest Serv. Reforestation Notes 1(1):7-8.
51. Richen, C. W. 1976. Attitudes of management toward western hemlock. Pages 7-9 *in* Western hemlock management (W. A. Atkinson and R. J. Zasoski, eds.). Coll. of Forest Resources, Univ. of Washington, Seattle.
52. Riech, F. P. 1966. Influence of static bending stress on growth and wood characteristics of nine-year-old Douglas-fir from two geographic sources. M.S. thesis, Oregon State Univ., Corvallis. 58 p.
53. Ruth, R. H. 1956. Plantation survival and growth in two brush-threat areas in coastal Oregon. U.S.D.A. Forest Serv., Pacific NW Forest and Range Exp. Sta., Portland, Oregon. Res. Pap. 17. 14 p.
54. Silen, R. R. 1956. Use of temperature pellets in regeneration research. J. Forestry 54:311-312.
55. Smith, J. H. G. 1975. Big stock vs. small stock. Paper presented at Western Reforestation Coordinating Committee meeting, Western Forestry conf., Dec. 3.
56. Smith, J. H. G., and G. S. Allen. 1962. Improvement of Douglas-fir planting stock. Faculty of Forestry, Univ. of British Columbia, Vancouver. Res. Pap. 55. 16 p.
57. Stein, W. I. 1976. Prospects for container-grown nursery stock. Pages 89-103 *in* Tree planting in the inland Northwest (D. M. Baumgartner and R. J. Boyd, eds.). Washington State Univ. Ext. Serv., Pullman.
58. Stein, W. I. 1978. Naturally developed seedling roots of five western conifers. Pages 28-35 *in* Proc., Root form of planted trees symp. (E. Van Eerden and J. M. Kinghorn, eds.). B.C. Ministry of Forests/Can. Forestry Serv., Victoria. Joint Rep. 8.
59. Streeby, L. L. 1978. Economic aspects of reforestation decisions. Pages 39-45 *in* Regenerating Oregon's forests (B. D. Cleary, R. D. Greaves, and R. K. Hermann, eds.). Oregon State Univ. Ext. Serv., Corvallis.
60. Strothman, R. O. 1980. Large stock and fertilizer improve growth of Douglas-fir planted on unstable granitic soil in northern California. U.S.D.A. Forest Serv., Pacific SW Forest and Range Exp. Sta., Berkeley, California. Res. Note PSW-345. 7 p.
61. Sutton, R. F. 1979. Planting stock quality and grading. Forest Ecology and Management 2(2):123-132.
62. Tanaka, Y., A. J. Long, and J. K. Winjum. 1974. Effects of undercutting and wrenching on 2-0 Douglas-fir seedlings. Pages 60-71 *in* Proc., Western Forest Nursery Council meeting, Portland, Oregon.
63. Timmis, R. 1974. Effect of nutrient status on growth, bud set, and hardiness in Douglas-fir seedlings. Pages 187-193 *in* Proc., North American containerized forest tree seedling symp. (R. W. Tinus, W. I. Stein, and W. E. Balmer, eds.). Great Plains Agric. Res. Publ. 68.
64. Wierman, C. A. (Date unknown.) Early survival and growth of planted seedlings graded by root collar diameter. Univ. of Washington, Seattle. Unpubl. pap.
65. Williams, W. T. 1976. Containerized seedlings in eastern Oregon. Page 82 *in* Executive summaries, Proc., 1976 Western Forestry conf. Western Forestry and Conservation Assoc., Portland, Oregon.
66. Wirsing, J. M., and R. R. Alexander. 1975. Forest habitat types on the Medicine Bow National Forest, southeastern Wyoming. U.S.D.A. Forest Serv., Rocky Mt. Forest and Range Exp. Sta., Fort Collins, Colorado. Gen. Tech. Rep. RM-12. 11 p.
67. Zaerr, J. B., and D. P. Lavender. 1976. Size and survival of 2-0 Douglas-fir seedlings. Oregon State Univ., Forest Res. Lab., Corvallis. Res. Pap. 32. 6 p.

Improving Nursery-Management Style: Selected Topics

Chapter 25
Sales and Customer Relations

W. B. Ellington

Abstract

Sales and customer relations are of increasing importance in bareroot nursery management. A nursery should determine product demand through independent market analysis, establish its own nursery concept (or self-image), and then promote its products and services by personally contacting potential customers and by advertising. Once a sale has been made, good customer relations between nursery management and customers help assure repeat sales and expose problems and opportunities to serve customers better. Research and testing programs at the nursery and in field plantations enhance customer relations and demonstrate a mutual willingness to solve problems.

25.1 Introduction

Sales and customer relations are gaining importance in bareroot nursery management as the number of independent nurseries increases. Before the mid-1970s, very little bareroot reforestation stock was grown by independent producers. Public agencies grew most of the seedlings they needed in their own nurseries or under cooperative agreements with other agencies. Private industry's needs were satisfied by their own nurseries, industrial association nurseries, or state nurseries under contract.

However, the emphasis on reforestation during the 1970s resulting from economic factors and the influence of forest-practice laws created a demand for bareroot seedlings that considerably exceeded the combined capacities of established Northwest nurseries. Some of this demand was satisfied by independent container nurseries, but the problems associated with container-grown stock caused agencies and private-forest land managers to look for new sources and techniques. As a result, several independent bareroot nurseries producing mostly reforestation seedlings were established around 1975. If these nurseries were to successfully compete in a market already dominated by public agencies and private industry, then sales and customer relations clearly merited high management priority.

This chapter addresses factors influencing sales and customer relations in bareroot forest-tree nurseries so that managers can better evaluate the effectiveness of these activities in their own operations.

25.2 Sales and Customer Relations Defined

A distinction should be drawn between **sales** and **customer relations,** even though the two tend to overlap and mutually support each other. Sales activities relate mostly to promoting nursery products and services through personal contact and advertising. Customer relations activities relate to the ongoing dialogue between nursery managers and seedling customers about stock quality and performance once a sale has been made. The regularity and success of customer relations will significantly affect subsequent sales.

Sales and customer relations are not only of concern to independent, private nurseries. Public and industrial nurseries have customers with the same or similar needs as those of independent nurseries and must pay equal attention to customer satisfaction. Active contact between customers and nursery managers is essential before, during, and after the nursery production period, regardless of nursery ownership type, if good customer relations are to be established and future sales assured.

25.3 Sales

25.3.1 Market analysis

Market analysis is essential for determining the demand for bareroot seedlings as well as their quality. Sites for new nurseries should be selected, at least in part, according to their suitability for producing the stock types in demand. Already existing nurseries should frequently assess their markets to evaluate whether their balance of stock types and range of services are properly tuned to current customer needs. Economic changes, legal requirements, and special situations such as forest disasters caused by fire, wind, volcanoes, or pathogens all bear on market conditions for seedlings.

Price levels for species and classes of stock must be periodically assessed for competitiveness. Most stock prices are published either in state information bulletins or in special advertisements from individual private nurseries. The most authoritative compilation of market prices is prepared by the states of California and Oregon, which regularly publish lists of

In Duryea, Mary L., and Thomas D. Landis (eds.). 1984. *Forest Nursery Manual: Production of Bareroot Seedlings.* Martinus Nijhoff/Dr W. Junk Publishers, The Hague/Boston/Lancaster, for Forest Research Laboratory, Oregon State University, Corvallis. 386 p.

available stock (generally, by asking price per specific lot) from all nurseries wishing to contribute that information. Surplus-stock lists and prices of federal and state nurseries are based on compiled costs of certain classes of stock and are not considered realistic expressions of "market" values.

Market analysis must be tuned to the distinction between contract growing, in which the customer asks the nursery to grow a certain number of seedlings, and speculative growing, in which the nursery manager estimates market demand and grows seedlings on the basis of that estimate. Speculative seedling production must be carefully planned to avoid over-growing, and astute market analysis is the only basis for doing so. Seed-supply levels and seedling stocks growing in other nurseries can sometimes be keys to marketing decisions when substantial shortages develop in specific types or age classes in a given species. Records of annual sales of Christmas trees as well as popularity of certain species can be good pointers in a market analysis.

Independent market analysis is preferable to internal analysis because it likely will be less biased. Nursery owners and managers may weigh future market choices in light of past decisions to avoid embarrassing themselves and thereby compounding their earlier mistakes.

25.3.2 Nursery concept

Concept—an important consideration for every nursery whether public or private—is formulated from the specific characteristics of each individual nursery. Sales activities should be related to nursery concept; for example, at Lava Nursery we concentrate on high-elevation and arid-zone stock. Most nurseries are located in climates that closely match those of certain planting areas and that favor the requested stock types; this factor should be emphasized by sales personnel to help assure customers that their seedlings will be exposed to the least climatic risks and will probably be available for lifting at a time compatible with customers' planting schedules. Proximity of a nursery to customers' plantations also is an asset and should be emphasized in sales. Customers generally are more comfortable when seedlings are being grown close to their plantations because they can visit the nursery frequently, communicate directly with the nursery manager, and have less anxiety about transportation costs and unknown factors in an unfamiliar area.

Elemental to any nursery concept is the goal of modern forest management, which focuses on rapid and effective regeneration of areas recently harvested or of underproductive forest land.

25.3.3 Selling approach

Smaller nurseries rely on their owners or managers to do most of their selling, whereas larger corporations utilize sales specialists as well as their nursery managers. Although motivations of private and public nurseries may differ, selling methods in all classes of bareroot nurseries are much alike. Generally, sales personnel with a thorough grasp of cultural techniques affecting seedling morphology and physiology—and the ability to articulate it—enhance sales. Conversely, sales personnel who promise seedling specifications and performance that are clearly impossible or who guarantee uniformly good results from seedlings grown off site damage nursery credibility—and sales.

25.3.3.1 Communication

Personal contact is the best sales approach; it is the most time consuming but the most rewarding. In-person contact assures good communication and the greatest opportunity for "give and take." It is the best way to instill confidence in a customer and the best means of learning about customer needs. This face-to-face contact is most effective in the actual nursery environment or customer's plantation; office visits are less satisfactory because they often are interrupted by other business and lack the immediate presence of soil, plants, equipment, and employees.

Telephone communication allows each party more flexibility in timing, avoids the considerable commitment and expense of travel, and is essential for arranging in-person sales and for follow-up work. Ideally, personal contact should precede phone contact. Although the remote nature of telephoning is a disadvantage, the carry-over of confidence from an in-person contact can enhance the value of many subsequent phone calls.

Written communication suffers from the delays inherent in composing, typing, mailing, and reading but has the advantage of providing exact, retrievable records for both parties. In the case of complicated technical information or business procedures, written contact is absolutely essential. The growing sophistication of bareroot nursery operations, the maintenance of seed-source integrity, and the increasing range of seedling specifications demand clear understanding of what is being bought and sold. Good written records on consistent formats using consistent terms provide both the customer and nursery manager with the basis for mutual understanding of a seedling production order. Personal rather than form letters are preferable, though they are less time efficient. Form letters containing general information such as seedling availability, special services, and prices are useful but should be followed up by personal contact.

25.3.3.2 Advertising

Advertising can take the form of special publications, pamphlets, form letters, periodical ads, and convention booths and programs. Electronic media advertising is not economically feasible for individual nurseries (though it may be more so through nursery-association sponsorship) because the size of the audience—bareroot seedling users—is not all that large. Local radio or TV advertisements may be warranted in heavily populated areas where many small woodland owners or Christmas tree growers could be expected.

Printed advertising should be directed at the most likely outlet for seedlings. Nurseries producing stock for Christmas trees should concentrate on grower-association publications and compile a mailing list of members who advertise in them. Associations of small woodland owners generally have local and statewide newsletters. State forestry publications have been willing to mention availability of private and public nursery stock. Professional publications such as the *Journal of Forestry*, *Western Forester*, and *American Forests* are excellent places to advertise, and trade publications such as *Forest Industries* provide a broad range of potential buyers. Regardless of specific format, however, all printed advertising should contain two basic features. First, the nursery and its concept should be briefly introduced. Second, the special nature of the product prompting the ad should be clearly and concisely defined, accompanied by prices, terms, and ordering information.

Attending meetings and field trips can combine advertising with personal contact. Booths at conventions where printed material, photos, videocassettes, and samples are available and where personal representatives are accessible can be very effective. However, giving away sample trees probably should be avoided; although a lasting reminder of their donor, these trees are likely to be overstressed and die shortly before or after planting. Forestry- and Christmas-tree-oriented meetings as well as forest-industry and horticultural exhibitions all can yield business. Nursery owners or managers can increase their exposure considerably by presenting a special subject at one of these gatherings; care should be taken to make the subject fit the convention orientation or theme.

For good results, advertising should be specific. A general ad format through which special messages can be promoted is probably best; customers will readily recognize the nursery through the ad's format and easily identify the current "special." Even the general format can reflect a particular emphasis—for example, a unique location or product, special services, or unusual capabilities.

25.3.3.3 Pricing

Pricing practices are the stickiest aspect of sales. Basically, costs must be recovered and incentive provided by revenues from seedling production. If pricing were related only to costs plus a reasonable return on investment, it would be fairly easy—although inflation will always be problematic on long-term growing orders. But other factors must be considered—and the more competitive the seedling market, the more acute these factors become.

Competition is very important in pricing policy. Nursery managers must meet what their competitors are charging for the same stock or outproduce them in quality. The market analysis suggested earlier (see 25.3.1) should weigh heavily in determining prices and assessing competition in the marketplace. However, competitive prices are meaningless without adequate margins to cover costs, reward risk-taking, and provide operational cash flow. Sales can be enhanced by giving discounts for volume, organizational affiliation, or long-term repeat orders. But discounts must be cost effective, at least recovering a saving in unit cost of production or unit cost reduction in overhead to justify the reduced unit revenue.

Erratic pricing policies can create credibility problems with customers. For instance, selling growing services at a certain rate to regular customers and then bidding a much lower rate on a government contract would be a good reason for regular customers to feel they were being gouged. If nursery managers keep good cost records and have a clear goal for profit and risk (including inflation), they will not compromise prices for the sake of trading dollars.

25.3.4 Recordkeeping

Sales records are not only essential for financial management of the nursery—they are the basis for sales projections. Records are valuable for determining profit margins for specific customers, for species, for age classes, for sections of the nursery, and for different cultural practices as well as for pinpointing seedling performance problems. Shipping records should indicate all peculiarities of shipments to customers.

Records should be as simple and accurate as possible yet tell the story. A recordkeeping system **must** be adopted to assure that there are no gaps between what customers want and what they receive. A small nursery can do well with a handwritten system coordinated among sales, operations, and administration as long as all personnel understand and consistently follow it. Larger nurseries—those producing more than 6 or 7 million seedlings per year, depending upon the number and diversity of growing orders and speculative sales—should consider computerization (see chapter 27, this volume). A properly selected and programmed computer system can provide all elements of nursery management and operation with a greater range of data in rapid fashion. Answering customers' questions regarding the status of their accounts and production requests can be greatly facilitated.

One of the most serious problems in nursery recordkeeping is the failure to make modifications to reflect shifting customer needs; for example, changes in harvest programs frequently call for changes in seedling production, processing, and shipment. Any changes should be entered on nursery records immediately, and staff should be trained to check all records before proceeding.

25.3.5 Employee morale

Sales of nursery stock and services are not only the responsibility of sales personnel, managers, or owners. Every employee who cares about his or her job and its future should be sales conscious. The team effort engendered by high employee morale is the best sales image that can be presented to a customer. In fact, employees who work well together and care about the results of their work cannot help but produce high-quality seedlings from reasonably good nursery facilities. Management's task is to instill this "team spirit" and keep it alive and well by setting understandable goals and objectives, providing adequate training to assure quality production, rewarding dedication and high standards, and, most of all, listening to employees and taking seriously their comments and concerns (see chapter 26, this volume). Employees who take pride in their work and their employer keep their premises neat and clean and their equipment in top condition; they display an air of confidence and satisfaction in reaching job objectives. Extra dollars spent on developing this "team spirit" are just as effective as those spent on advertising and other sales activities.

25.4 Customer Relations

Regardless of ownership type, continuing contact with customers to whom a nursery has sold trees or provided growing services is essential and a strong complement to any sales program. This contact not only assures repeat sales but exposes problems and opportunities. Follow-up contact, preferably in person at the customer's plantation, is the best way to assess seedling performance and to analyze and determine the most effective means of supplying well-matched stock. Similarly, reasonably frequent customer visits to the nursery are valuable; mutual understanding of nursery specialities and limitations is fostered in this way. Subsequent phone conversations and letters are so much more effective once the nursery manager and customer have exchanged visits and developed rapport.

Aside from personal contact, customers should regularly be provided with written reports on the status of their seedlings. Where problems occur, color photos, preferably in time sequences, should be used to identify these problems. For example, Lava Nursery has a form that reports the status of each seedling lot for each customer. On the back of this form is a questionnaire requesting items such as desired packing dates and material, size specifications for sorting, and disposition of surplus trees. This combination form enhances nursery recordkeeping and provides customers with timely information, prompting them to make decisions vital to the nursery program.

Research and testing programs at the nursery and in the field are good customer-relations tools. These mutual efforts at solving problems defuse the old animosity that developed when nurseries just grew trees and customers just planted them. Common recognition of problems and joint efforts at solution not only are more effective, they are the basis for developing confidence and respect between the two parties—and, consequently, increasing sales.

25.5 Conclusions and Recommendations

Nurseries growing bareroot seedlings are faced with the task of attracting and keeping customers. Sales are best effected through direct personal contact, by phone, or by personal letter. Indirect approaches via advertising and convention displays reinforce any personal contact previously established. To sell its products and services most effectively, each nursery should present a strong concept emphasizing its unique features. High employee mòrale is a key component to sales success; in general, customers will move their growing orders to nurseries where employees take a professional and caring approach.

Overselling both quantity and quality should be avoided. Potential customers should be made aware of hazards as well as benefits to secure trust and confidence between seller and purchaser.

Customer relations are an essential adjunct to a sales program. The rapport developed between nursery manager and customer aids communication, assuring better seedling quality and performance and more sales.

Chapter 26
Improving Productivity in Forest Nurseries

S. M. Hee

Abstract

In recent years, planting-stock costs have nearly doubled and in some cases tripled, largely due to increased labor costs. Improvements in productivity are a means of offsetting these cost increases. Though poor productivity is frequently treated as a labor problem, it is really management's responsibility. Establishing specific productivity goals, formulating practical plans and strategies, executing plans in a well-organized manner, and effectively tracking and evaluating results are components of a long-term program to improve productivity. Important elements underlying the process include input from first-line supervisors and workers, establishment of effective training and motivation programs, and frequent feedback to workers on productivity performance.

26.1 Introduction

The cost of planting stock in recent years has nearly doubled and in some instances tripled. Today, that cost represents ⅓ to ½ the total cost of regeneration. Extremely tight cash flows due to the prevailing depression in the U.S. forest products industry make financing regeneration difficult. Furthermore, regeneration requires a sizable capital investment from which there are no cash returns for extended periods, perhaps up to 60 years. These conditions—which will probably persist for the foreseeable future—dictate that forest nurseries operating in the Northwest become especially proficient at producing high-quality, low-cost planting stock.

The typical Northwest forest nursery spends about 80% of its total operating budget on labor. Over the years, increased costs of wages and benefits have driven seedling costs close to planting costs—with little or no gain in productivity. Therefore, substantial improvements in productivity will be required if Northwest nursery managers are to keep regeneration affordable.

Our experience at Weyerhaeuser Company bears this out. Over the past 5 years, we have seen significant improvements in lifting, packing, and transplanting (Fig. 1), attained in spite of increasingly stringent culling criteria. As productivity has increased, seedling costs have been relatively stable or in some cases reduced during a period when inflation and wage increases were at or near double-digit levels. This chapter addresses problems underlying poor or marginal productivity in forest nurseries and offers measures for improving it.

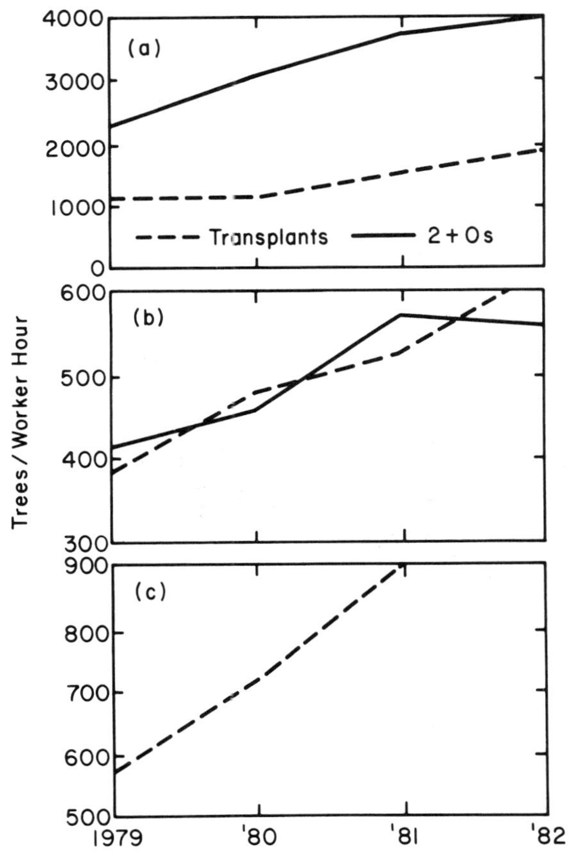

Figure 1. Productivity trends in (a) machine lifting, (b) packing, and (c) transplanting at Weyerhaeuser Company nurseries since 1979.

26.2 Defining Productivity

Nursery productivity can be defined as units of output obtained per hour of worker labor:

Productivity = units of output/hour of labor

In Duryea, Mary L., and Thomas D. Landis (eds.). 1984. *Forest Nursery Manual: Production of Bareroot Seedlings.* Martinus Nijhoff/Dr W. Junk Publishers, The Hague/Boston/Lancaster, for Forest Research Laboratory, Oregon State University, Corvallis. 386 p.

In the nursery packing operation, for example, units of output would be 1,000s of seedlings packed per hour of labor. If a crew of 10 workers can pack 80,000 seedlings per 8-hour workday, then their productivity would be:

$$\frac{80{,}000 \text{ seedlings}}{10 \text{ workers} \times 8 \text{ hours}} = \frac{80{,}000 \text{ seedlings}}{80 \text{ hours}} = 1{,}000 \text{ seedlings/hr}$$

Seedling output in the equation refers to seedlings that meet or surpass predetermined quality specifications set or agreed to by the customer. These specifications usually include both quantitative aspects, such as height or caliper, and qualitative aspects, such as seedling health and physiology. Defective seedlings are never included in the output. High productivity (output) rates are impossible unless crop quality and yields are kept at an acceptable level. Similarly, productivity cannot be increased without improving quality and yield.

26.3 Four Steps to Increasing Productivity

Poor productivity—often considered a labor problem—is usually the result of poor management or lack of concern by management. Most nursery managers know the specific costs of various nursery activities but often do not know their specific productivity rates.

Labor costs are a function of wage rates and the number of worker hours needed to do a job. Although managers frequently have little or no direct control over wage rates, which often are determined at a higher level by the nursery owner, corporate policy, or bargaining-unit agreement, they do have control over how worker hours are allocated. Unlike cost ratios, productivity ratios (such as number of seedlings lifted or packed per worker hour) are a direct measurement of the work process. Improvements in this area lead to dollar savings that are immediately manifested as increased cash flow and reduced unit costs.

Increased productivity does not just happen—the nursery manager makes it happen. Improvements require a four-step process: (1) establishing goals, (2) formulating plans, (3) executing plans in a well-organized manner, and (4) tracking and evaluating results.

26.3.1 Establishing goals

Establishing specific goals to improve productivity requires in-depth knowledge of current nursery processes and productivity levels. This knowledge resides in the local nursery staff. This group, therefore, must play a leading role in developing productivity objectives and related strategies. If outside consultants are used, they should be deployed in a manner which supplements but does not overshadow the efforts of nursery staff. Ownership in these goals and strategies is extremely important and should not be underrated.

At Weyerhaeuser in 1976, we shifted the focus to becoming more cost effective in our nurseries. At one of our locations, an outside management consulting firm was brought in to help reduce costs and increase productivity. A typical industrial engineering approach, including time and motion studies, was taken, and numerous reports were made to the head office in Tacoma. In spite of the flurry of activity, real progress was virtually nonexistent because the on-site nursery staff had been largely excluded from active participation in the process. They had no ownership in the program and viewed the chosen goals as those of upper management and their well-paid consultants—not of the people actually doing the work at the nursery.

Specific productivity goals and targets are best developed by the nursery staff. It is then the manager's responsibility to make certain that these goals are challenging, yet realistic and achievable. Goals must be expressed in quantitative terms and communicated effectively to workers at all levels to secure their commitment.

26.3.2 Formulating plans

Specific plans and strategies on how to achieve productivity goals are absolutely essential. To be effective, planning should be focused on a few key areas rather than broadly aimed across all nursery activities. For example, in our initial efforts to improve productivity at Weyerhaeuser nurseries, we concentrated on lifting, packing, and transplanting (Fig. 1). These activities are highly labor consumptive, utilizing 80 to 90% of the total worker hours required to actually produce a crop. We felt that even modest increases in productivity could generate significant dollar savings.

Planning for productivity improvement means change and should involve workers from various levels at the nursery. The concept here is that the people who actually do a job know best how that job should be done. Proposed changes that are reviewed and critiqued by first-line supervisors and key people on crews will generally have a high probability for success. In many instances, however, managers fail to fully utilize these key staff people; consequently, change is often met by stiff resistance in the work areas. A higher degree of worker involvement in the planning stage is effective in diminishing this resistance because both supervisor and crew feel some ownership in the plan through involvement in the decision-making process.

Managers and staff alike must remember, however, that production systems in nurseries are integrated and require balance throughout. For example, packing depends on lifting, which precedes it, and on storage and shipping, which follow it. Improvements in one area must not be bottle-necked by inefficiencies in another. Furthermore, because nursery operations are highly dependent on weather conditions, developing contingency plans is essential.

In the end, meaningful plans must be practical and realistic. Crews should be able to execute plans using the available system and the equipment on hand or that which can readily be obtained. Because each nursery is unique, comparing one nursery to another is relatively pointless. The real important issue is what is appropriate for each individual nursery, given its physical, operational, and financial conditions.

26.3.3 Executing plans

Setting goals and planning are academic issues if plans are not executed properly. Getting the job done is primarily the responsibility of first-line supervisors and crew members. How well they function ultimately determines productivity level, hence cost. Therefore, training and motivation of employees are especially important contributing factors here.

26.3.3.1 Training

Much of the work in nurseries is seasonal, which creates a certain turnover rate among employees. To achieve and maintain high levels of productivity, training and retraining must be an ongoing process. Proper job instruction and orientation are a crucial requirement for new employees, and refresher training is often helpful for regular employees returning after an extended layoff. Lack of training seriously handicaps workers in achieving maximum productivity.

26.3.3.2 Motivation

Good productivity heavily depends not only on how well the work is organized but also on the extent to which the work group is motivated. Effective motivators are specific productivity performance goals, some level of participation in decisions affecting a worker's job, and both positive and negative feedback—which is probably the most important. Employees who are simply ignored do not know whether their work is satisfactory and therefore cannot be expected to operate anywhere near their maximum productivity levels.

Incentive programs can be highly effective motivators in cases where baseline productivity is well defined and the results are at an acceptable level. For example, at one of Weyerhaeuser's nurseries where incentive programs are aimed toward production teams or work groups rather than toward individuals, we have found that cash awards are not the only method of providing incentive. Alternatives such as gift certificates, leaving work earlier on Fridays, or a company-paid catered lunch are often more effective than their equivalent in cash. Employers must understand what will motivate the work team and should ask team members to find out. Before an incentive program is implemented, however, managers must be sure that the production system and its associated hardware are not limiting factors and that the proper steps have been taken in work organization and training. If not, incentives often result in added costs rather than incremental benefits.

Everyone in the work group should be included in the incentive program. Excluding support personnel, for example, will be perceived as unfair. In addition, communication of incentive-program rules and details must be absolutely clear and well understood by all involved; otherwise, misunderstandings may lead employees to believe that management is "cheating," and the incentive program will immediately lose its effectiveness.

The nature of incentives is to increase output—but not at the cost of product quality. Therefore, quality control, always important in the nursery, becomes even more so when productivity incentives are operating. An effective way to place equal emphasis on both quality and quantity is to tie incentives to certain quality criteria. In the event that these criteria are not met, incentive payoffs would be automatically terminated.

26.3.4 Tracking and evaluating results

Keeping track of progress toward productivity goals throughout the season is vitally important. This information must be available on at least a daily basis to enable first-line supervisors to correct problems as they arise and provide quantitative performance feedback to crews. Accumulated from year to year, productivity records allow for the development of more intelligent production planning based on observed trends. Budgeting can then be based primarily on work measurements rather than on inflation.

26.4 Conclusions and Recommendations

Although poor productivity is often treated as a labor problem, it is really management's responsibility. Managers should:

- Define productivity in specific terms for each nursery operation, then focus on those operations for which changes will create the most significant overall improvements. Operations that are most labor intensive are likely candidates.
- Use a step-by-step process in which goals are established, plans formulated and executed, and results evaluated. Consult first-line supervisors and key personnel on crews; remember that those most directly affected by changes better accept them if they have participated in the decision-making process.
- Ensure that employees receive adequate instruction and orientation and that they remain highly motivated through continued participation, feedback, or incentive programs; remember that trained and motivated workers do the best job. Tie quality control to productivity incentives.
- Bear in mind that results do not occur overnight but require long-term commitment and perseverance.

Chapter 27
Nursery Record Systems and Computers

C. B. Royce

Abstract

Nineteen Northwest bareroot nurseries were polled concerning the crop, site, and administrative records they keep and their methods of recordkeeping. Seventy-nine percent of those nurseries currently use computers in some of their recordkeeping, and 95% expect to be using computers within the next 5 years. Sample printouts from the D. L. Phipps Oregon State Forest Nursery highlight features of its computerized crop recordkeeping system. Though computers have numerous applications for bareroot nurseries, they are not a panacea for good recordkeeping. Each nursery must assess its own needs to determine whether computerization is appropriate.

27.1 Introduction

In the nursery business, producing a 2- or 3-year-old seedling involves a vast array of interactive variables. When the proper combinations generate the ideal end product, nursery managers want to be able to reproduce them. One of the best ways to ensure this is through good recordkeeping.

Each nursery is unique in its physical components, objectives, administrative techniques, financial capabilities, customers, and management style. Therefore, its particular recordkeeping styles and needs probably also differ. In this chapter, I describe record types and recordkeeping methods for bareroot nurseries in the Northwest and point up advantages—and some limitations—of computerized recordkeeping.

27.2 Nursery Record Survey

I developed a separate questionnaire and circulated it to nursery managers representing the 21 nurseries participating in the OSU Nursery Survey (see chapter 1, this volume). The following 19 nurseries responded:

Canada

Surrey Nursery
Skimikin Nursery
Red Rock Nursery
Chilliwack River Nursery

United States

Private

Weyerhaeuser: Mima Nursery, Aurora Nursery, Bonanza Nursery
Industrial Forestry Association: Canby Nursery, Toledo Nursery, Greeley Nursery
Tyee Tree Nursery
Lava Nursery

Federal (U.S.D.A. Forest Service)

Coeur d'Alene Nursery
Humboldt Nursery
Lucky Peak Nursery
J. Herbert Stone Nursery
Wind River Nursery

State

Webster Forest Nursery, Washington State Department of Natural Resources
D. L. Phipps Oregon State Forest Nursery, Oregon State Department of Forestry

Most of the questionnaire dealt with manual or computerized recordkeeping of three broad types: (1) crop, (2) site, and (3) administrative. Crop records (Table 1) include all operations for a particular crop from seed procurement through delivery of the seedling. Site records (Table 2) include any operation or physical alteration impacting the nursery site which may or may not be crop specific or have a long-range impact on the site itself. Administrative records (Table 3) include all other necessary nursery records that are not crop or site specific. Within each of these three record types are categories and subcategories listed in descending order—that is, from "most kept" to "least kept." Nurseries not keeping certain records are not necessarily disinterested or remiss. In many cases, the record is not applicable.

Summing the number of responses, by recordkeeping method, for each record type (from Tables 1, 2, and 3) and expressing each sum as a percentage of the total responses per record type give the following averages:

Record type	None kept	Kept manually	Kept by computer	Kept manually and by computer
	%			
Crop	31	45	13	11
Site	33	65.5	1	0.5
Administrative	14	68	8	10

In Duryea, Mary L. and Thomas D. Landis (eds.). 1984. *Forest Nursery Manual: Production of Bareroot Seedlings*. Martinus Nijhoff/Dr W. Junk Publishers, The Hague/Boston/Lancaster, for Forest Research Laboratory, Oregon State University, Corvallis. 386 p.

These percentages show that: (1) Most of the records listed on the questionnaire are being kept by most of the nurseries; (2) the most popular recordkeeping method is manual; and (3) the most popular records for computerization seem to be crop. Thirty-five percent of the crop records are being computerized, as are 21% of the administrative records and 2% of the site records.

Fifteen of the 19 nurseries use computers in some of their recordkeeping operations. Of those 15, three have only indirect access to a computer via central processing.[1] Twelve have direct access to a computer from the nursery: five have "stand-alone" computers,[2] and seven have nursery terminals.[3]

Direct computer access from the nursery, whether by stand-alone computer or nursery terminal, is fairly new. Although three of the nurseries have been using terminals for the past 6 years, nine have been using computers for only 18 months or less. Of the nurseries represented by this questionnaire, 95% expect to be using computers within the next 5 years, and 90% expect to have direct computer access from the nursery.

[1]All information to be computerized is mailed to another (i.e., central) location for data processing; final printouts are then mailed back to the nursery.

[2]Computer or "intelligent" terminal at the nursery.

[3]Computer terminal at the nursery linked to a central computer; this allows for direct data entry and retrieval at the nursery.

Table 1. Crop records from 19 bareroot nurseries in the Northwest.

Record type	None kept	Kept manually	Kept by computer	Kept manually and by computer
	Number of nurseries			
Seed Data				
Seedlot number	0	8	5	6
Owner	0	8	5	6
Species	0	8	5	6
Elevation	0	8	5	6
Seed zone	1	8	5	5
Germ. test lab	2	8	7	2
Date seed received	2	6	6	5
Inventory number	4	7	4	4
Seed crop year	4	3	7	5
Seed data information source (lab or estimate)	4	5	7	3
Certification class	5	4	6	4
% seed moisture	7	2	6	4
Germ. test date	7	1	8	3
Collection site	7	4	7	1
X-ray	11	2	4	2
Vigor class	16	0	3	0
Sowing Formula Data				
Kilograms to sow	1	9	5	4
Amount ordered	1	9	5	4
Planned bed feet	2	8	5	4
Harvest density	2	8	5	4
Theoretical germ.	3	7	5	4
Thousand seed weight	3	7	5	4
Germ. - chill	4	6	5	4
Theoretical falldown	4	7	5	3
Purity	6	4	5	4
% block loss	6	7	4	2
No chill germ.	11	3	2	3
Stratification Data				
Seedlot number	2	8	5	4
Date soaked	2	13	2	2
Kilograms expected from customer	5	11	1	2
Kilograms received from customer	5	10	1	3
Date chilled	6	10	1	2
Days chilled	7	9	1	2
Hours soaked	8	9	1	1
Seed treatment	9	8	1	1
Date dried	9	9	1	1
Problems	9	9	1	0
% moisture end stratification	14	4	1	0
% moisture mid-stratification	15	3	1	0
% moisture end soak	15	3	1	0
Calibration				
Total weight to sow	6	10	2	1
Number of bags	8	10	1	0
Test bag weight	9	9	1	0
Øyjörd Calibration Information				
Gear number	6	9	2	2
Grams/revolution	7	11	1	0
Bed feet/revolution	7	11	1	0
Zero max turns	8	7	2	2
Problems	8	10	1	0
Funnel gap	9	8	2	0

Table 1. **Crop records from 19 bareroot nurseries in the Northwest.—(Continued)**

Record type	None kept	Kept manually	Kept by computer	Kept manually and by computer
	Number of nurseries			
Actual Sowing				
Location	1	12	2	4
Actual bed feet sown	1	12	2	4
Lot number	2	11	2	4
Drill adjustments	2	15	1	1
Time or date	6	9	2	2
Problems	6	11	1	1
Transplanting				
Lot number	4	9	1	5
Location	4	9	1	5
Actual bed feet	4	9	3	3
Time or date	5	12	1	1
Problems	6	8	1	4
Weeding				
Chemical	0	15	1	3
Hand	10	8	0	1
Mechanical	10	8	0	1
Thinning				
Prethinning density	10	6	0	3
Post-thinning density	11	5	0	3
Mulching				
Weed control	12	5	0	2
Frost heaving	13	5	0	1
Moisture conservation	14	3	0	2
Seed protection	15	3	0	1
Soil splash	16	2	0	1
Soil stabilization (hydromulch)	17	1	0	1
Fertilizing				
Soil	0	17	1	1
Foliage	11	7	0	1
Pruning				
Root, horizontal	5	11	0	3
Root, vertical	5	10	1	3
Top	7	9	1	2
Irrigation Hours by Location				
Growth	2	17	0	0
Water stress	3	16	0	0
Wash off fertilizer	7	7	2	3
Wash off chemical	9	10	0	0
Wrenching	3	12	2	2
Protection				
Pesticide applied	1	13	2	3
Shade screening	10	8	0	1
Biological control	15	4	0	0
Plant Moisture Stress Records				
Location	4	15	0	0
Time	5	14	0	0
Tensiometer Records				
Location	8	10	1	0
Time	8	10	1	0
Seedling Inventory				
Lot number	0	10	5	4
Location	0	10	5	4
Net	0	10	5	4
Plot interval	1	12	4	2
Sample size	3	9	4	3
Gross	3	7	5	4
Cull	5	7	4	3
Dead	8	6	4	1

Table 1. Crop records from 19 bareroot nurseries in the Northwest.—(Continued)

Record type	None kept	Kept manually	Kept by computer	Kept manually and by computer
	Number of nurseries			
Seedling Lifting				
Lot number	0	11	4	4
% lot	0	13	4	2
Time	1	12	4	2
Location	1	10	4	4
Plant moisture stress	7	7	3	2
Packing				
Date	0	12	4	3
Minimum caliper	0	12	4	3
Minimum height	0	12	4	3
Root pruning length	0	12	4	3
Lift date	1	12	3	3
Quality-control remarks	1	13	3	2
Volume	3	9	4	3
Plant moisture stress	7	5	4	3
Special Service				
Double grade	3	11	3	2
Packing material	5	9	3	2
Top prune	10	5	3	1
Storage				
Prepack				
Temperature	12	6	0	1
Humidity	12	7	0	0
Duration	12	7	0	0
Post-Pack				
Temperature	1	15	0	3
Humidity	2	15	0	2
Duration	2	15	1	1
Quality-Control Remarks	4	15	0	0
Shipping				
Picked up	1	11	4	3
Delivered	3	9	4	3
Quality-control remarks	4	14	0	1
Type vehicle	8	11	0	0
Temperature	12	6	0	0
Humidity	12	6	0	1
Harvest Analysis				
Acceptable seedlings				
Height	0	12	5	2
Caliper	0	12	5	2
Shoot:root ratio	6	6	5	2
Unacceptable seedlings				
Primary cull factor	4	9	4	2

Table 2. Site records from 19 bareroot nurseries in the Northwest.

Record type	None kept	Kept manually	Kept by computer	Kept manually and by computer
	Number of nurseries			
Soil Analysis				
Nutrient	0	19	0	0
pH	1	18	0	0
Organic matter	1	18	0	0
Soil series	4	15	0	0
Texture	3	16	0	0
Ground Water				
Drain tile	6	13	0	0
Drainage problems	9	10	0	0
Ground water table	11	8	0	0

Table 2. Site records from 19 bareroot nurseries in the Northwest.—(Continued)

Record type	None kept	Kept manually	Kept by computer	Kept manually and by computer
	Number of nurseries			
Soil Amendments				
Cover crop	1	16	1	1
Organic	1	16	1	1
Fertilization (presowing)	1	15	1	2
Fumigation				
Cost/acre	4	15	0	0
Soil temperature	6	13	0	0
Plastic seal	7	11	1	0
Rates	7	11	0	1
Problems	10	9	0	0
Reshoots	10	9	0	0
Ground Preparation				
Subsoiling	6	12	1	0
Disk	7	12	0	0
Harrow	7	12	0	0
Chisel plow	8	11	0	0
Bed forming	8	11	0	0
Land plane	9	10	0	0
Float	10	9	0	0
Moldboard plow	10	9	0	0
Roller harrow	10	9	0	0
Rototiller	12	7	0	0
Roterra	13	6	0	0
Debris removal	14	5	0	0
Weather				
Date	0	18	1	0
Temperature	0	18	1	0
Rainfall	0	18	1	0
Wind speed	10	8	1	0
Solar radiation	12	6	1	0
Insects and Disease				
Problems identified	2	17	0	0
Surveys	3	16	0	0
Underground Irrigation				
Mainlines	1	18	0	0
Road and Fence Lines	7	12	0	0

Table 3. Administrative records from 19 bareroot nurseries in the Northwest.

Record type	None kept	Kept manually	Kept by computer	Kept manually and by computer
	Number of nurseries			
Nonexpendable Property Inventory (buildings, vehicles, equipment, etc.)	0	15	1	3
Personnel Records	0	18	1	0
Payroll	0	9	2	8
Purchasing	0	14	0	5
Fiscal Records (budgets, tree prices, billings)	0	11	3	5
Historical Crop Records (transplant, sowing requests, actual transplant, sowing and production, etc.)	0	18	1	0
Studies	1	17	1	0
Laws, Directives, and Policy	2	15	2	0
Customer Lists	3	12	4	0
Public Relations (tours, gifts, news releases, etc.)	4	15	0	0
Expendable Property Inventory (office supplies, seed, etc.)	6	11	1	1
Motor Pool	9	5	1	4
Cone Handling and Seed Extraction	9	7	3	0

27.3 Computerized Crop Records: An Example

At the D. L. Phipps Oregon State Forest Nursery, we have indirect access to a large IBM model 370 computer that services four agencies. Until 4 years ago, we had not used the computer in any of our crop recordkeeping. Since that time, we have developed six major data groupings, or files, for our crop records: a seed data file, stratification data file, sowing data file, sowing location file, 1 + 0 inventory file, and 2 + 0 inventory file. Data for each of these files are entered into the computer by keypunching, edited for accuracy, and stored on hard disks for each crop. Producing printouts like those in this chapter does not require the sophistication of the IBM model 370; the data can easily be handled by several of the minicomputers available today.

Figure 1 represents a format displaying most of the components of each of our six crop files, by seedlot. (This format is currently not available on the computer but is being programmed.) Data can be manipulated and displayed by any of the components shown in Figure 1. For example, by combining our inventory and sowing location files and portions of our seed data files, we can generate a printout (Fig. 2) detailing inventory by seedlot. Informative to the customer, this printout also can be grouped with others to form a larger picture of nursery crop data.

Aggregated crop inventories are available in several formats. The one shown in Figure 3 summarizes all 2 + 0 inventory by density and species. We base seedling charges on a planned production-per-bed-foot basis. Thus, this run enables us to quickly compare actual with planned production per sowing density and to establish final seedling prices. With only minor program modifications, the computer can calculate those prices.

Our bed-analysis format (Fig. 4) shows the net number of seedlings in each of our increasingly larger aggregations, i.e., seedlings per inventory plot, per seedbed, per block, and per seedlot. We use this printout in preparing lifting orders. Most of our customers prefer to pick up portions of their seedlots at several different times during the winter. This format allows us to lift as close as possible to a customer's requested seedlings per seedlot.

We use an analysis of the sowing and inventory files to compare planned and actual yields for germination, 1 + 0 and 2 + 0 inventories, and harvest. Such data are useful in fine-tuning the sowing formula and gauging production trends. A similar analysis (Fig. 5) compares planned and actual bed footages sown. Predetermined levels of difference are highlighted with asterisks (note "Percent Difference" column), which has proved useful in flagging lots to be checked for thinning. By manipulating our sowing location file, we can produce Figure 6, a chronological listing by block, pipe, and bed of each seedlot in the nursery to identify ownership, lot number, or any other known component from a bed location.

Other kinds of formats are available. A frequency distribution in table form (Fig. 7) lists the number of seedlots by seed-collection year. A frequency distribution in bar-graph format (Fig. 8) represents the number of seedlots by percent purity. This format has good visual impact as does that of Figure 9, a

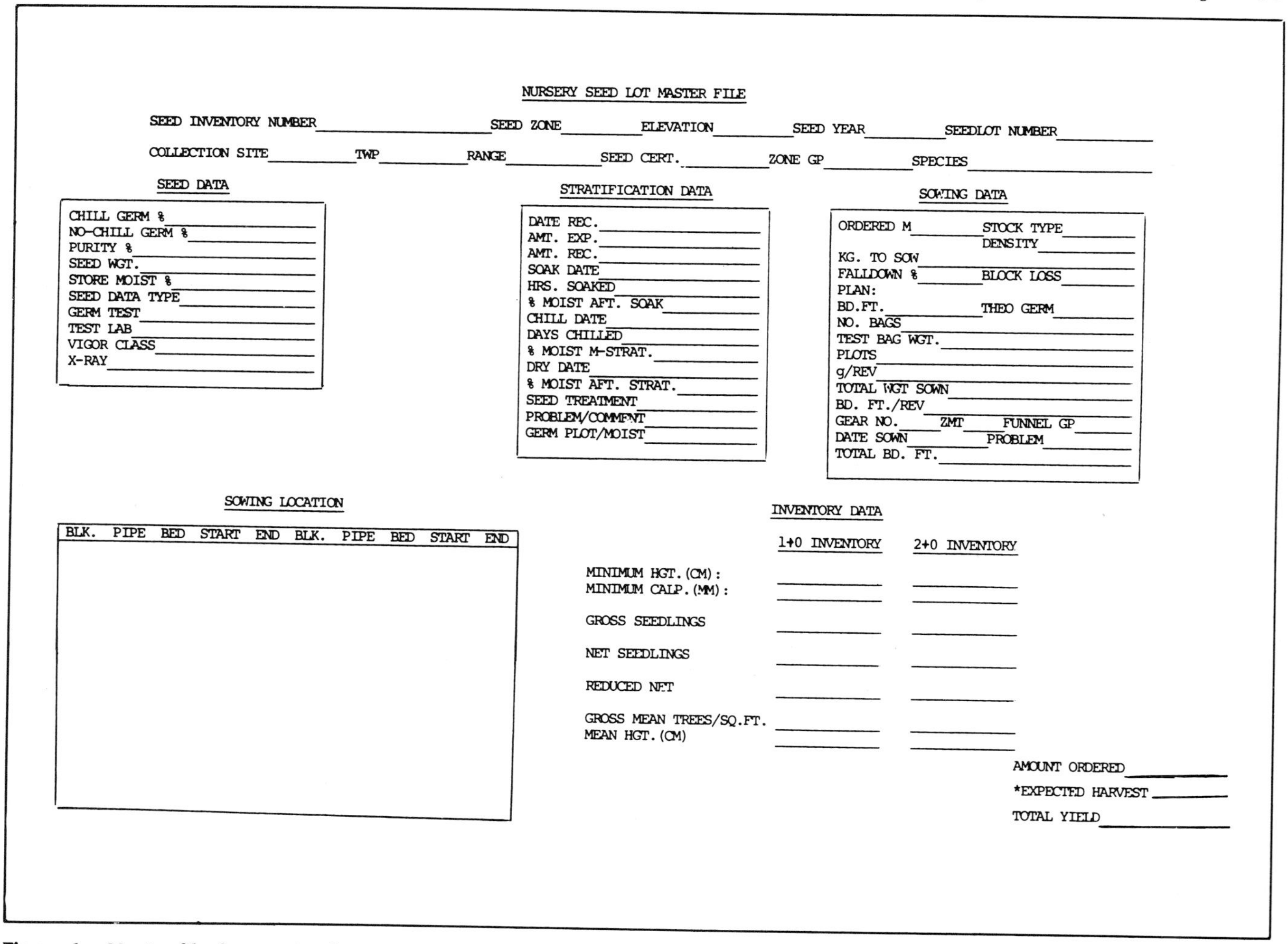
NURSERY SEED LOT MASTER FILE

SEED INVENTORY NUMBER ____ SEED ZONE ____ ELEVATION ____ SEED YEAR ____ SEEDLOT NUMBER ____

COLLECTION SITE ____ TWP ____ RANGE ____ SEED CERT. ____ ZONE GP ____ SPECIES ____

SEED DATA

CHILL GERM %
NO-CHILL GERM %
PURITY %
SEED WGT.
STORE MOIST %
SEED DATA TYPE
GERM TEST
TEST LAB
VIGOR CLASS
X-RAY

STRATIFICATION DATA

DATE REC.
AMT. EXP.
AMT. REC.
SOAK DATE
HRS. SOAKED
% MOIST AFT. SOAK
CHILL DATE
DAYS CHILLED
% MOIST M-STRAT.
DRY DATE
% MOIST AFT. STRAT.
SEED TREATMENT
PROBLEM/COMMENT
GERM PLOT/MOIST

SOWING DATA

ORDERED M ____ STOCK TYPE ____
DENSITY ____
KG. TO SOW
FALLDOWN % ____ BLOCK LOSS ____
PLAN:
BD.FT. ____ THEO GERM ____
NO. BAGS
TEST BAG WGT.
PLOTS
g/REV
TOTAL WGT SOWN
BD. FT./REV
GEAR NO. ____ ZMT ____ FUNNEL GP ____
DATE SOWN ____ PROBLEM ____
TOTAL BD. FT.

SOWING LOCATION

BLK.	PIPE	BED	START	END	BLK.	PIPE	BED	START	END

INVENTORY DATA

	1+0 INVENTORY	2+0 INVENTORY
MINIMUM HGT. (CM):		
MINIMUM CALP. (MM):		
GROSS SEEDLINGS		
NET SEEDLINGS		
REDUCED NET		
GROSS MEAN TREES/SQ.FT.		
MEAN HGT. (CM)		

AMOUNT ORDERED ____
*EXPECTED HARVEST ____
TOTAL YIELD ____

Figure 1. Master-file format (in the process of being computerized) displaying most of the components of the six crop files used by the Phipps Nursery.

scattergram displaying harvest analysis data of caliper and height for the entire crop. It quickly provides an impression of seedling morphology and its distribution throughout the crop.

The dollar savings in computations alone have been substantial with computerization of our crop records. Although costs will vary with individual systems, ours might be useful indicators. Data entry by keypunching all six of our crop files, averaging 250 seedlots costs approximately $500 for the entire crop, or $2 per seedlot. Inventory summaries such as that shown in Figure 2 cost approximately $3.60 for the entire crop, or about $0.014 per seedlot. Crop summaries like the one shown in Figure 3 cost less than $0.10 per page. However, these costs only reflect input and output of data; they do not include prorated costs for the overall computer-system purchase or rental, computer programming, or data storage.

In addition to direct financial savings, our nursery has benefited from the increased accuracy, speed, and versatility of the computerized system, from its long-term data storage and retrieval capabilities, and from its ability to quickly summarize and compare current data with those from previous seasons.

```
D.L.PHIPPS FOREST NURSERY                    INVENTORY BY SEED LOT                                    RUN DATE 10/04/82

SEED LOT NUMBER: A100081                     |INVENTORY NUMBER: 74-012-01-061-1.0-76   |SEED LOT NUMBER: A100081
SEED ZONE:       061                         |COLLECTION SITE:                         |AMOUNT ORDERED:      75,000
ELEVATION:       1.0                         |HARVEST DENSITY:  25.0                   |STOCK TYPE:          01
SPECIES:         0010 DOUGLAS FIR            |TOWNSHIP:
ZONE GROUP:      0 2 CENTRAL COAST           |RANGE:
OWNERSHIP:       02010 NON-CONTRACT          |SEED CERT. CLASS: 0
PLOT SIZE: 1.0   BED WIDTH: 4.0   STANDARD HGT: 15.0 CM   STANDARD CAL: 3.0 MM   NO. PLOTS: 78   BED FEET IN LOT: 946

                                  |**************|**************|           | - - - - - - - -SOWING LOCATION- - - - - - -
                                  | 1+0 INVENTORY| 2+0 INVENTORY|           |BLK PIPE BED START END  BLK PIPE BED START END
                                  |**************|**************|           |
GROSS INVENTORY:                  |     132,316  |     135,497  |           | 19  34   6     9  330
                                  |              |              |           | 19  35   1   330    1
DEAD:                             |       2,285  |       9,135  |           | 19  35   2     1  295
PERCENT:                          |         1.7  |         6.7  |           |
                                  |              |              |           |
CULL:                             |              |      28,186  |           |
PERCENT:                          |              |        20.8  |           |
- - - - - - - - - - - - - - - - - | - - - - - - -|- - - - - - - | - - - - - |
NET TOTAL INVENTORY:              |     130,031  |      98,176  |           |
                                  |              |              | * AMOUNT *|
TREES ORDERED:                    |      75,000  |      75,000  | * ORDERED*|
                                  |              |              |           |
DIFFERENCE (ORDERED - NET):       |      55,031  |      23,176  |           |
PERCENT:                          |        73.3  |        30.9  |           |
- - - - - - - - - - - - - - - - - | - - - - - - -|- - - - - - - | - - - - - |
REDUCED TOTAL NET:                |     119,363  |      90,125  |           |
PERCENT:                          |       159.1  |       120.1  |*EXPECTED *|
                                  |              |              |*HARVEST  *|
- - - - - - - - - - - - - - - - - | - - - - - - -|- - - - - - - | - - - - - |
GROSS MEAN TREES PER FOOT OF BED: |       139.8  |       143.2  |           |
GROSS MEAN TREES PER SQUARE FOOT: |        34.9  |        35.8  |           |
GROSS MEAN HEIGHT PER LOT (CM):   |              |              |           |
- - - - - - - - - - - - - - - - - | - - - - - - -|- - - - - - - |           |
STANDARD DEVIATION:               |        6.06  |        7.39  |           |
STANDARD ERROR:                   |         .67  |         .82  |           |
STANDARD ERROR PERCENT:           |        3.81  |        4.53  |           |
COEFFICIENT OF VARIATION:         |       17.35  |       20.66  |           |
- - - - - - - - - - - - - - - - - | - - - - - - -|- - - - - - - |           |
NO. PLOTS 5% SAMPLE ERROR:        |        45.7  |        64.5  |           |
- - - - - - - - - - - - - - - - - - - - - - - - - - - - - - - - - - - - - - - - - - - - - - - - - - - - - -
```

Figure 2. Printout detailing inventory by seedlot.

D.L.PHIPPS FOREST NURSERY — 2+0 INVENTORY - 1981 — RUN DATE 9-14-82

DENS.	SPP. CODE	TREES ORDERED	BED FEET	GROSS	DEAD	CULL	NET	DIFF. ORD-NET	REDUCED NET	MEAN TREES /B.F.
12.5	0010	20000.	493.	32314.	3538.	6350.	22426.	-2426.	20587.	45.5
17.5	0010	4295200.	72156.	5638491.	639465.	1631098.	3367928.	927272.	3091758.	46.7
17.5	0210	11200.	198.	13993.	49.	4062.	9882.	1318.	9072.	49.9
25.0	0010	12223200.	151817.	15818186.	4338764.	2835110.	8644312.	3578888.	7935478.	56.9
25.0	0210	45000.	602.	51801.	1657.	15930.	34214.	10786.	31408.	56.8
25.0	0220	10000.	120.	5362.	0.	926.	4436.	5564.	4072.	37.0
25.0	0240	9700.	140.	3425.	16.	1515.	1894.	7806.	1739.	13.5
25.0	0260	5000.	62.	3148.	0.	1021.	2127.	2873.	1953.	34.3
25.0	0320	37000.	559.	23350.	655.	3202.	19493.	17507.	17895.	34.9
25.0	0410	50000.	598.	46582.	3804.	10212.	32566.	17434.	29896.	54.5
25.0	0800	20000.	261.	27217.	153.	2886.	24258.	-4258.	22269.	92.9
30.0	0010	1963600.	21811.	2816088.	425250.	425284.	1965554.	-1954.	1804379.	90.1
30.0	0310	22000.	222.	13908.	369.	1991.	11548.	10452.	10601.	52.0
30.0	0320	10100.	139.	12075.	243.	1529.	10303.	-203.	9458.	74.1
NUR. TOTAL		18722000.	249178.	24505940.	5413963.	4941036.	14150941.	4571059.	12990564.	56.8

Figure 3. Printout summarizing all 2+0 inventory by density and species.

D.L.PHIPPS FOREST NURSERY   2+0 INVENTORY BED ANALYSIS (MEAN TREES-NET INVENTORY)   RUN DATE 8-24-81   PAGE 49

SEED LOT NUMBER-A710080   OWNER-02000   SPECIES-0010   ZONE GROUP-073   SEED ZONE-491   STOCK TYPE-01   DENSITY-25.0   ELEVATION-1.5

BLK	PIP	BED	PLT INT	BED LEN	TOTAL TREES IN BED	MEAN TREES/ BD.FT.	5	25	50	75	100	125	150	175	200	225	250	275	300	325	350	375	400	425	450	475
							-MEAN TREES PER BED FOOT BY PLOT-																			
08	19	3	25	431	42187	97		118	130	92	122	110	96	82	84	92	100	72	76	88	84	72	128	118		
08	19	4	25	450	51550	114	116	116	120	102	144	86	112	122	122	136	130	118	138	76	96	112	100	116		
08	19	5	25	450	45750	101	122	130	102	108	104	88	96	106	112	116	78	102	102	108	76	94	92	94		
08	19	6	25	450	52150	115	130	98	110	88	132	110	106	134	134	122	124	116	108	116	122	106	128	102		
BLOCK TOTAL				1781	191637	107																				
08	20	1	25	450	46250	102	76	90	118	84	120	106	106	120	94	110	82	96	112	108	102	106	118	102		
08	20	2	25	450	38150	84	94	82	96	100	86	90	116	108	118	44	46	56	86	78	80	82	86	78		
08	20	3	25	167	7395	44	44	56	38	48	54	36	34													
BLOCK TOTAL				1067	91795	86																				
LOT TOTAL				2848	283432	99																				

Figure 4. Printout indicating bed analysis for all 2+0 inventory, used in preparing lifting orders.

D.L.PHIPPS FOREST NURSERY   UNDERSOW AND OVERSOW PERCENT BY SEED LOT

SEED LOT NUMBER	SPP. CODE	OWNER CODE	ZONE GROUP	ELEV.	PLANNED BED FEET	ACTUAL BED FEET	PERCENT DIFFERENCE	
A-40-10-80	0010	02000	020	1.0	721.1	752.0	4.29	
A-40-21-80	0010	02000	020	1.0	889.0	1262.0	41.96	
A-41-00-80	0010	02000	011	1.5	1092.5	1108.0	1.42	
A-42-00-80	0010	02000	011	1.5	1150.0	1190.0	3.48	
A-43-10-80	0010	02000	011	0.5	3017.6	3035.0	0.58	
A-43-21-80	0010	02000	011	0.5	834.9	843.0	0.97	
A-44-00-80	0010	02000	011	0.5	130.9	136.0	3.90	
A-45-00-80	0010	02000	011	1.0	552.0	566.0	2.54	
A-46-10-80	0010	02000	012	0.5	348.5	360.0	3.30	
A-46-21-80	0010	02000	012	0.5	318.6	333.0	4.52	
A-47-10-80	0010	02000	012	1.0	1173.0	1204.0	2.64	
A-47-21-80	0010	02000	012	1.0	883.2	900.0	1.90	
A-48-10-80	0010	02000	012	1.0	1467.4	1478.0	0.72	
A-48-21-80	0010	02000	012	1.0	188.6	197.0	4.45	
A-49-00-80	0010	02000	012	1.0	467.2	486.0	4.02	
A-50-10-80	0010	02000	012	1.0	273.7	277.0	1.21	
A-50-21-80	0010	02000	012	1.0	1405.3	1427.0	1.54	
A-51-00-80	0010	02000	012	1.0	143.8	99.0	-31.15	****
A-52-00-80	0010	02000	013	1.0	3473.0	3466.0	-0.20	
A-53-00-80	0010	02000	013	1.0	143.8	149.0	3.62	
A-54-00-80	0010	02000	013	1.5	448.5	433.0	-3.46	
A-55-00-80	0010	02000	013	1.5	2200.0	2154.0	-2.09	
A-56-00-80	0010	02000	030	0.5	461.2	476.0	3.21	
A-57-10-80	0010	02000	030	1.0	593.4	592.0	-0.24	
A-57-21-80	0010	02000	030	1.0	4733.4	4728.0	-0.11	
A-58-10-80	0010	02000	030	1.0	6586.1	6663.0	1.17	
A-58-21-80	0010	02000	030	1.0	1892.9	1922.0	1.54	
A-59-10-80	0010	02000	030	1.5	282.9	322.0	13.82	
A-59-20-80	0010	02000	030	1.5	710.7	735.0	3.42	
A-59-31-80	0010	02000	030	1.5	328.9	364.0	10.67	
A-60-00-80	0010	02000	040	0.5	255.3	260.0	1.84	
A-61-10-80	0010	02000	040	1.0	940.7	941.0	0.03	
A-61-20-80	0010	02000	040	1.0	5429.2	5449.0	0.36	
A-61-31-80	0010	02000	040	1.0	159.9	162.0	1.31	
A-62-00-80	0010	02000	040	1.0	359.4	353.0	-1.78	
A-63-10-80	0010	02000	040	1.5	3290.2	3198.0	-2.80	
A-63-21-80	0010	02000	040	1.5	1060.3	1051.0	-0.88	
A-64-00-80	0010	02000	040	0.5	614.1	617.0	0.47	
A-65-10-80	0010	02000	040	1.0	774.0	774.0	0.0	
A-65-20-80	0010	02000	040	1.0	201.3	204.0	1.34	
A-65-30-80	0010	02000	040	1.0	1105.2	1103.0	-0.20	
A-65-41-80	0010	02000	040	1.0	928.1	953.0	2.68	
A-66-10-80	0010	02000	040	1.5	412.9	415.0	0.51	
A-66-21-80	0010	02000	040	1.5	265.7	277.0	4.25	
A-67-00-80	0010	02000	030	1.0	862.5	859.0	-0.41	
A-68-00-80	0010	02000	030	1.5	862.5	894.0	3.65	
A-69-00-80	0010	02000	060	1.0	816.5	783.0	-4.10	
A-70-00-80	0010	02000	060	1.5	575.0	566.0	-1.57	
A-70-01-80	0010	02000	060	1.5	1920.5	1899.0	-1.12	
A-71-00-80	0010	02000	073	1.5	2875.0	2848.0	-0.94	
A-72-00-80	0010	02000	073	2.5	1035.0	960.0	-7.25	
A-77-00-80	0010	02000	030	1.5	12014.1	12031.0	0.14	
B-01-00-80	0010	01110	030	2.0	575.0	611.0	6.26	
B-02-00-80	0010	01110	011	1.5	575.0	565.0	-1.74	

Figure 5. Printout allowing comparison of planned and actual bed footages sown.

D.L. PHIPPS FOREST NURSERY — SEED LOT DATA BY SOWING LOCATION

BLOCK	PIPE	BED	START	END	SEED LOT NUMBER	STOCK TYPE	DENS	ZONE GROUP	SEED ZONE	ELEV	SPP. CODE	OWNER CODE	DATE SOWN
20	01	1	001	090	J640082	09	25.0	011	053	2.0	0010	03250	03-19-82
20	01	2	002	080	J640082	09	25.0	011	053	2.0	0010	03250	03-19-82
20	01	3	002	090	N620082	09	25.0	050	511	3.0	0010	03240	03-19-82
20	01	4	001	088	N520082	09	25.0	050	502	2.0	0010	03240	03-19-82
20	01	5	007	015	N520082	09	25.0	050	502	2.0	0010	03240	03-19-82
20	01	5	027	090	A600082	09	25.0	013	072	0.5	0010	02000	03-19-82
20	01	6	006	090	A600082	09	25.0	013	072	0.5	0010	02000	03-19-82
20	02	1	022	150	J610082	03	17.5	011	052	1.5	0010	03250	04-26-82
20	02	2	001	150	J610082	03	17.5	011	052	1.5	0010	03250	04-26-82
20	02	3	001	150	J610082	03	17.5	011	052	1.5	0010	03250	04-26-82
20	02	4	001	150	J610082	03	17.5	011	052	1.5	0010	03250	04-26-82
20	02	5	001	150	J610082	03	17.5	011	052	1.5	0010	03250	04-26-82
20	02	6	013	150	J610082	03	17.5	011	052	1.5	0010	03250	04-26-82
20	03	1	001	240	B950082	03	17.5	011	052	1.5	0010	01130	04-26-82
20	03	2	001	237	B950082	03	17.5	011	052	1.5	0010	01130	04-26-82
20	03	3	001	240	B950082	03	17.5	011	052	1.5	0010	01130	04-26-82
20	03	4	001	240	B950082	03	17.5	011	052	1.5	0010	01130	04-26-82
20	03	5	001	240	B950082	03	17.5	011	052	1.5	0010	01130	04-26-82
20	03	6	001	240	B950082	03	17.5	011	052	1.5	0010	01130	04-26-82
20	04	1	001	300	B950082	03	17.5	011	052	1.5	0010	01130	04-26-82
20	04	2	001	300	B950082	03	17.5	011	052	1.5	0010	01130	04-26-82
20	04	3	001	227	B950082	03	17.5	011	052	1.5	0010	01130	04-26-82
20	04	3	240	300	B940082	03	17.5	011	052	1.5	0010	01130	04-23-82
20	04	4	001	300	B940082	03	17.5	011	052	1.5	0010	01130	04-23-82
20	04	5	001	300	B940082	03	17.5	011	052	1.5	0010	01130	04-23-82
20	04	6	001	300	B940082	03	17.5	011	052	1.5	0010	01130	04-23-82
20	05	1	001	375	B940082	03	17.5	011	052	1.5	0010	01130	04-23-82
20	05	2	001	375	B940082	03	17.5	011	052	1.5	0010	01130	04-23-82

Figure 6. **Printout providing chronological listing of seedlot data by bed location.**

27.4 Assessing Computer Needs

It has been said that the personal computer will totally revolutionize our private lives and the small business world within the next 5 years. Potential computer applications for bareroot nurseries are numerous, including word processing, payroll, personnel, purchasing, production records, inventories, billing, ordering, budget projections, sowing calculations, and literally dozens of other daily nursery tasks. Computer systems in use today number in the hundreds. How, then, can nursery managers find out what is available to them and appropriate for their particular needs?

First: Are detailed records important to your nursery? If the answer is no, a computer probably will not benefit you. Computers can store, retrieve, and manipulate large volumes of data rapidly and accurately, but cannot turn unmotivated or disinterested recordkeepers into good recordkeepers or generate meaningful data from poor records.

Second: Can your nursery afford to computerize? The major cost associated with a small computer during its first 5 years of use is not the computer itself or its programs, but the personnel costs of collecting, keypunching, interpreting, and using the data. The tasks and functions to be computerized and the financial efficiencies expected must be thoroughly evaluated and a computerized system then compared with present recordkeeping methods.

Third: What does the market offer? If you are not familiar with computer terminology and technology, finding out what systems are available and best suited to your needs can be

PAGE 1

NUMBER OF SEED-LOTS
BY SEED-LOT YEAR AND CROP-YEAR

CROP YEAR	SEED,LOT,YEAR 74	75	76	77	78	79	80	81	82	TOTAL
	134	205	148	180	0	0	61	30	0	758
00	0	0	0	0	81	39	0	0	35	155
62	0	0	0	0	1	3	2	0	0	6
64	0	0	0	0	0	1	1	0	1	3
65	0	0	0	0	2	4	3	1	1	11
66	0	0	0	0	8	12	3	7	2	32
67	0	0	0	0	1	0	0	0	0	1
68	0	0	0	0	9	11	7	4	5	36
70	0	0	0	0	1	3	3	0	1	8
71	0	0	0	0	21	17	13	10	14	75
73	0	0	0	0	0	0	0	4	1	5
74	0	0	0	0	1	0	3	1	1	6
75	0	0	0	0	0	1	0	3	0	4
76	0	0	0	0	35	44	32	28	11	150
77	0	0	0	0	7	13	10	6	3	39
78	0	0	0	0	0	115	102	80	111	408
79	0	0	0	0	0	0	0	0	1	1
80	0	0	0	0	0	0	0	87	73	160
81	0	0	0	0	0	0	0	0	4	4
TOTAL	134	205	148	180	167	263	240	261	264	1862

Figure 7. Printout of frequency distribution of the number of seedlots by seed-collection and crop years.

NUMBER OF SEED-LOTS BY PERCENT PURITY
ALL SEED-LOTS FROM 1974 THRY 1982
NUMBER_OF_SEED-LOTS

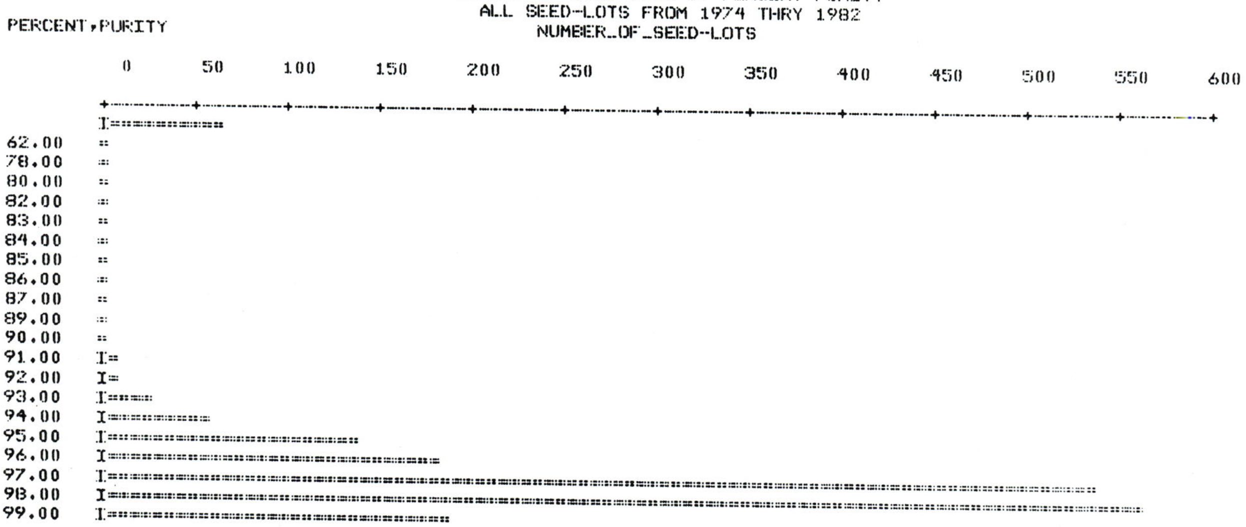

Figure 8. Printout of frequency distribution showing number of seedlots by percent purity.

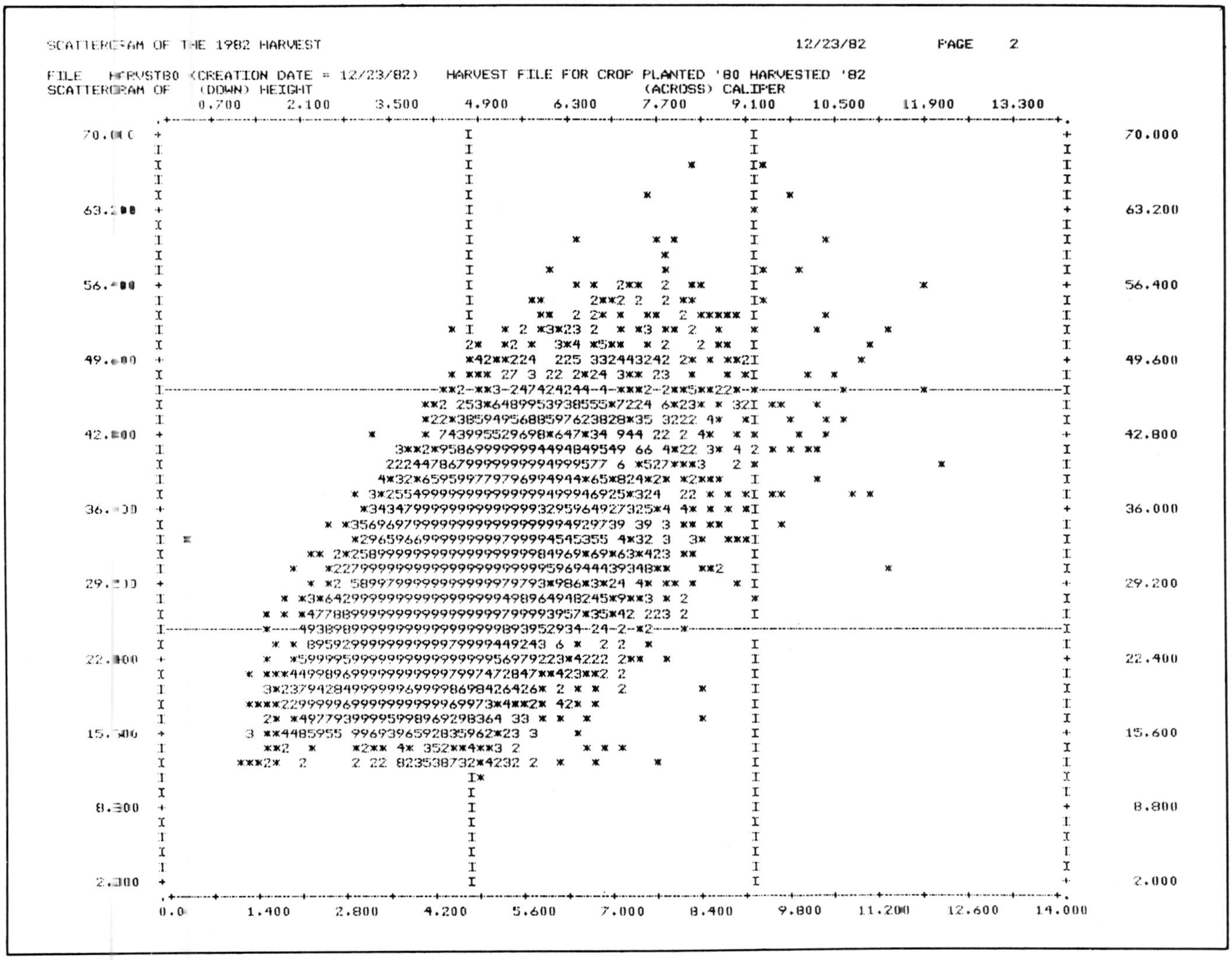

Figure 9. Printout of scattergram displaying harvest analysis data by caliper and height.

extremely frustrating. My advice is to employ a computer consultant. A consultant can give you unbiased advice about all aspects of computer hardware (machines) and software (programs) including prices, capabilities, and dependability. Other available sources of information include other nurseries already using computers, hardware and software sales personnel, local and state colleges, libraries, and a large assortment of regularly published computer magazines.

27.5 Conclusions

- The bareroot nursery business offers diverse and complex recordkeeping opportunities.
- Of the 19 nurseries responding to my questionnaire, 79% currently use computers in some of their recordkeeping operations, and 95% expect to be using computers within the next 5 years.
- Computer terminals and stand-alone computers at the nursery are relatively new. Of the 15 nurseries using computers, 12 have either a nursery terminal or a stand-alone computer. Nine of these have been in use less than 18 months.
- The rapid and diverse developments in the field of personal computers are revolutionizing data processing, providing nurseries with more options for recordkeeping.
- Computer consultants can serve a valuable function by assisting nurseries in evaluating their recordkeeping needs and determining what computer system might best meet those needs.
- Computers are not a panacea for good recordkeeping. They can store, retrieve, and manipulate data accurately and efficiently but cannot make good recordkeepers out of poor ones or produce meaningful analyses from inadequate or incorrect data.

Upgrading Nursery Practices

Chapter 28
Designing Nursery Experiments

T. L. White

Abstract

Nursery managers face a wide variety of problems that lend themselves to research methods. This chapter (1) describes fundamental statistical concepts—inference, replication, randomization; (2) discusses methods for controlling experimental error—pairing, blocking, split-plot design, covariance—to increase the sensitivity of experiments; and (3) traces the research process applied to forest-tree nurseries—defining problems, designing and conducting experiments, and implementing solutions based on sound interpretation of data. Combining the statistical concepts presented here with personal experience and biological intuition strengthens the nursery researcher's ability to meet the major goal—and challenge—of nursery research: to develop new methods for producing high-quality seedlings at low cost.

28.1 Introduction

Nursery managers and growers are researchers both by need and inherent nature. Keenly observant and inquisitive, they continually seek to improve seedling quality and cost-effectiveness of their nursery practices. They face a wide variety of problems that lend themselves to research methods—for example, whether to use a new piece of equipment or a new herbicide, how dense and when to sow seed for various stock types, or how to determine optimum fertilizer and irrigation regimes. Necessarily, however, recommendations to alter nursery practices are nearly always based on incomplete information, which successful nursery managers evaluate in light of their experiences and instincts to make sound, effective decisions. The science of **statistics** deals with drawing conclusions from incomplete information, whereas **biometrics** is the application of these statistical techniques to biological problems. Statistically designed experiments often provide important information upon which nursery managers can base their decisions and calculate the degree of uncertainty associated with their conclusions.

The design and execution of experiments are often team efforts involving biometricians and researchers. Information in this chapter should help make the nursery researcher a stronger team member, better able to balance statistical considerations with practical and biological aspects of nursery problems. Specifically, the chapter objectives are to (1) trace the research process as applied to forest-nursery problems (see 28.3), (2) contrast operational trials with statistically designed experiments (see 28.4), (3) describe, intuitively, the statistical concepts of designed experiments (see 28.5), and (4) delineate the processes involved in designing, executing, analyzing, and interpreting those experiments (see 28.6).

In Duryea, Mary L., and Thomas D. Landis (eds.). 1984. *Forest Nursery Manual: Production of Bareroot Seedlings*. Martinus Nijhoff/Dr W. Junk Publishers, The Hague/Boston/Lancaster, for Forest Research Laboratory, Oregon State University, Corvallis. 386 p.

28.2 How to Use This Chapter

A few words are in order regarding use of this chapter.

The chapter is intended as a complete introductory treatise on experimental design for nursery workers; no statistical training is assumed. It is not a formula-oriented discussion; excellent texts by Freese [8] and Little and Hills [13] provide formulas for analyzing basic experimental designs. Nor does it specifically address the plantation-testing phase of nursery research, although the statistical concepts described may be widely applied to agricultural and forestry experimentation.

The chapter may be useful both for a first-time reading, to gain an understanding of statistical concepts of experiments, and for future reference. Some concepts are treated more fully than needed for a first reading and should probably just be skimmed. In particular, a quick reading of definitions (Table 1) and sections 28.3 to 28.5 is good preparation for section 28.6, the main focal point, describing how to design and execute experiments and interpret their results. For future reference, section 28.6 can be used as a checklist of items to consider when actually planning nursery experiments.

28.3 The Research Process: An Overview

An **experiment** is a planned inquiry to obtain new information or to confirm or deny previous results for the purposes of making recommendations [18]. The process of experimentation is profitably applied to many problems encountered in forest-tree nurseries (Fig. 1). Cause-effect relationships [19, p. 86] are established by observing how certain **response variables** (say, caliper) are influenced by specified levels of one or more **factors** (say, fertilizer). Although nursery managers may not necessarily think of it as experimentation, they commonly (1) encounter a problem, (2) seek out existing information, (3) refine the problem and set hypotheses or objectives, (4) plan and conduct experiments to obtain new data pertinent to their nursery conditions, (5) draw conclusions based on their interpretation of the new data in light of existing information and their instincts, and (6) implement a change in nursery procedure. Often, because conclusions lead to new problems or questions, several stages of experimentation may be required. (See also chapter 29, this volume, for more details on problem-solving techniques.)

Table 1. Definitions of common terms used in the design of experiments.

Term	Definition	Examples/Comments
Experiment	Planned inquiry to obtain new information or to confirm or deny results from previous investigations for the purposes of making recommendations [18, p. 88].	Nursery experiments gain information on new field techniques, equipment, packing-shed alignments, storage facilities, etc.
Operational trial	Preliminary experiment in which each treatment is applied to only one plot (nonreplicated).	Useful when treatment effects will be large relative to background "noise" (uncontrolled variation).
Designed experiment	Detailed, critical investigation in which precise, unbiased conclusions and measures of uncertainty associated with those conclusions are required.	Treatments are *nearly always* replicated more than once on separate plots and allocated to plots at random.
Inductive reasoning	Drawing conclusions or making predictions about a wide sphere of interest (a population) from particular cases or observations (samples).	The sun has risen every day for millennia (a large sample); therefore, it will rise tomorrow.
Deductive reasoning	Drawing conclusions or making predictions based on well-defined principles from which those conclusions or predictions logically follow.	That the sun will rise tomorrow logically follows from the principles of astronomy.
Factor	An item, element, or process under investigation in an experiment. Effects of a given factor are examined by testing each factor at more than one level (factor level).	Sowing date, irrigation, seed source, bed density, etc. are factors; three rates of nitrogen (N1, N2, N3) and two sowing dates (D1, D2) are factor levels.
Treatment	All factors and their levels applied to an experimental plot.	From above, N1/D2 is a treatment plot sown on the second date receiving the lowest nitrogen level.
Experimental plot	Smallest physical unit to which a treatment is allocated independent of all other treatments.	A specific length of nursery bed.
Observational unit	Observed or measured items within an experimental plot [11, p. 9].	Tree seedlings within a nursery experimental plot.
Measurement plot	Portion of the experimental plot actually measured; unmeasured portion serves as buffer or border.	The center of a nursery plot; seedlings on either side of center are *not* measured.
Response variable	Variable (characteristic) measured on each experimental plot to assess influence of treatments.	Number of plantable seedlings, percent germination, height, caliper, shoot:root ratio.
Precision	Relative dispersion or clustering of measurements or estimates.	A precise measurement is one of low dispersion; if remeasured, it will be nearly the same.
Accuracy	Absolute correctness of measurements or estimates.	A scale that *always* weighs 2 g too heavy is inaccurate even if precise (consistent).
Bias	Directional (up or down) measure of inaccuracy.	The above scale is biased upward 2 g.
Confounding	Condition in which the effects of two of more factors on a given response variable are confused and *cannot* be separated.	A nursery researcher finds larger seedlings from the field with both higher N and P levels; the effects of N and P *cannot* be separated.
Experimental error	A measure of the variation among experimental plots receiving identical treatments [18, p. 90].	Experimental error will be high if field plots are inherently variable or if experimental technique is sloppy.

For example, perhaps during an initial experiment, a new rotating root-pruning table is found to save money on the packing line, but seems to damage too many roots. How can the table be redesigned to hold the seedlings in place better so fewer roots are damaged? Existing information may be sought from engineers and manufacturers, and objectives formulated for designing one or two modifications to test in a second experiment. Having then tried the new modifications for a period of time, the manager decides that one of them "causes" the desired "effects" (less root damage) and operationally implements the new root-pruning procedure on the packing line.

Consider, as a further example, the process of experimenting with a new herbicide to control weed species in forest nurseries [16]. Experiments are set up to compare various application rates of a new chemical to the standard weed-control method to determine relative phytotoxicity and effectiveness. Note that *no* amount of experimentation can totally prove beyond all doubt that this chemical will be suitable for all nurseries under all conditions. This is common of problems requiring **inductive reasoning**—in which inferences are made about a larger sphere of interest from a smaller data base. The broader and more intensive our **sample**—that is, the more years and nurseries in which the chemical is used and tested—the more comfortable (certain) we feel about applying the results to the **population of interest**, perhaps all nurseries of similar soil type. We further use **deductive reasoning**, based on underlying biological, chemical, and physical principles, to extend and rationalize the inductive inferences drawn from experimentation. For example, the new chemical will probably not be particularly suitable for nurseries suffering severe grass competition if it is chemically ineffective against grasses.

Because experimental processes do not absolutely prove the hypotheses being investigated, the amount of data (sample size) required to make a decision becomes a personal choice. How certain must the conclusions be? Some problems require

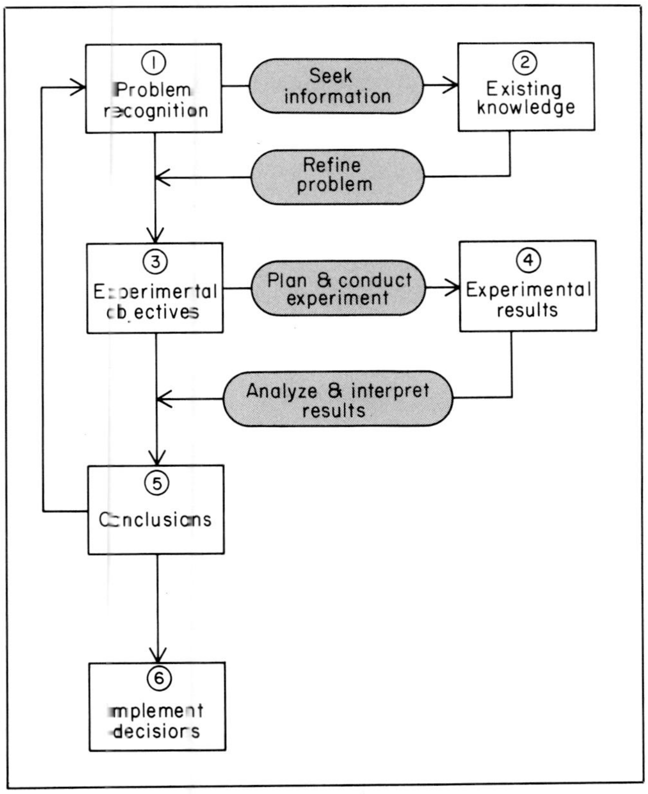

Figure 1. The research process, as it might be applied in forest-tree nurseries (adapted from [19, p. ix]).

stronger evidence than others, perhaps because the decision would be more costly to implement. Thus, the type and amount of experimentation and the formality with which the research is applied to a problem vary with both the investigator and the problem.

28.4 Operational Trials versus Designed Experiments

Some problems are appropriately addressed with **operational trials** (preliminary investigations in which each treatment is applied to only one plot). Others must be examined through **designed experiments** (detailed, critical investigations in which precise, unbiased conclusions and associated measures of uncertainty are required). The nature of yet other problems precludes *any* type of experimentation. For example, when a manager discovers a widespread insect or disease problem in the nursery, immediate control by the "recommended" method is more important than experimenting to find the optimum rate or chemical. The manager may decide to experimentally treat some small areas at a lower rate or with an untried chemical, but even this may be unacceptable because of the large inoculum source remaining if the untested **treatment** were unsuccessful.

28.4.1 Operational trials

In general, operational trials have two main uses: preliminary investigations and final-phase, large-scale testing. Operational trials are particularly suitable for preliminary experiments when background variation among **experimental plots** is small relative to expected treatment effects. For example, the effects of a new herbicide formulation on weed mortality will be large relative to the background effects due to other causes. If the preliminary objectives are to see whether the chemical has any promise and warrants further testing, an operational trial is suitable for visually assessing effectiveness and phytotoxicity. Many preliminary (screening) tests of, for example, new chemicals, fertilizer regimes, wrenching blades, and packing-line arrangements are appropriately conducted as unreplicated operational trials.

For final-phase or large-scale testing, the use of small experimental plots needed to obtain sufficient replication may generate experimental artifacts. That is, the new treatment *cannot* be "operationally" applied to the small experimental plots often needed for designed, well-replicated experiments. Operational trials are suitable in these situations. Larger plots, more representative of operational application of the new treatment, are used, while statistical replication is sacrificed.

When an operational trial is chosen as the appropriate method of experimentation, the various treatments should be (1) applied to similar areas to minimize systematic effects due to uncontrolled variation and (2) compared in more than one nursery bed so that the inferences and conclusions drawn will have broader validity.

28.4.2 Designed experiments

Given that an experiment is warranted, a designed experiment is more appropriate than an operational trial for cases benefiting from one or more of the special attributes of designed experiments (Table 2). Designed experiments are particularly useful for detailed investigations—for example, establishing optimum rates of a chemical or procedure, investigating interactions among multiple factors, or revealing the biological principles of a phenomenon under investigation. In these cases, the attributes of designed experiments are well worth the extra effort. Uncontrolled, background variation affecting growth rate within and between nursery beds is large due to differences in fertility, soil type, water drainage, and irrigation.

Well-designed experiments, randomized and replicated over this background variation, are often required to achieve unbiased estimates of treatment response with the appropriate level of precision and range of validity. Most of the remainder of this chapter considers the concepts behind and execution of designed experiments in nursery research.

28.5 Statistical Concepts of Designed Experiments

Statistical-inference concepts, discussed first (see 28.5.1), center around developing and quantifying the uncertainty of the conclusions drawn from designed experiments. Replication and randomization, discussed next (see 28.5.2), are the two core concepts of experimental design and are considered in some depth; many of the benefits of designed experiments derive because they are, by definition, randomized and replicated.[1] Finally, methods for controlling experimental error (see 28.5.3) are presented which can result in more precise estimates of treatment means and more certainty about the conclusions drawn from experiments.

28.5.1 Statistical inference

28.5.1.1 Point estimates, interval estimates, and hypothesis tests

Statistical inference is the process of using sample data to generalize about a population or wider sphere of interest. For designed experiments, this usually implies calculating the level of uncertainty associated with these generalizations. This is one of the main rationales for using designed experiments.

An example should clarify the three main elements of statistical inference (Table 3) and the importance of these in nursery experimentation. If five plots receive a specific fertilizer regime, the mean number of plantable seedlings for that regime would be calculated by summing the total number of plantable seedlings for all five plots and dividing the sum by 5. This treatment mean is subject to experimental error and is only a **point estimate** of the true population mean (the response achieved if that regime were applied to an infinite number of nursery plots). Though further experiments might show this point estimate to be in error, for now, it is a single number that estimates the parameter of interest.

[1]Nonreplicated and fractionally replicated designs are of limited usefulness in nursery research and are *not* considered here; see Cox [7] and Kempthorne [11].

Confidence intervals quantify the uncertainty and state the error associated with point estimates. Note that the span of a confidence interval is closely related to the experimental error. If the just-mentioned fertilization experiment yields highly reproducible results (i.e., precise, low experimental error), then the experimenter can state with a high degree of confidence (say, 95%) that the true mean fertilizer response lies within a narrow range surrounding the estimated treatment mean.

Hypothesis testing allows researchers to quantify the uncertainty with which they accept or reject hypotheses formulated before an experiment. A statistical hypothesis is testable by experimentation in the sense that experimental results will either tend to support or refute it; however, because it can never be totally proven or disproven, researchers calculate the level of confidence placed on the decision to accept or reject.

Statistical hypotheses are formulated as **null hypotheses**—that is, the effects under investigation are assumed to have *no* effect on the response variable. Examples are (1) all fertilizer regimes yield the same number of plantable seedlings, (2) bed density has no effect on stem caliper, (3) the effect of nitrogen (N) fertilization on height growth is the same regardless of the level of phosphorus (P) fertilizer (no interaction). This "innocent-until-proven-guilty" approach has both statistical and scientific underpinnings. From a statistical standpoint, for example, a researcher calculates the probability of the observed differences between two treatment means occurring by chance if, in fact, there are no differences between the true treatment effects. If there is only a small chance (say, 5%) of obtaining the observed differences if the null hypothesis is true, the researcher concludes that treatments differ—and rejects the null hypothesis. From a scientific standpoint, null hypotheses state a skepticism and wariness of the consequences of being wrong. For instance, a nursery researcher may not want to implement a new, more costly fertilizer regime until the evidence points overwhelmingly in its favor; that skepticism is maintained by hypothesizing no effect.

28.5.1.2 Incorrect conclusions from experiments

Because statistical hypotheses can never be proven or disproven, it is inevitable that incorrect conclusions are drawn from experiments. Two types of incorrect decisions (errors) are possible (Fig. 2):

- **Type 1 error (α):** The null hypothesis is rejected when it is actually true. That is, differences among treatments are declared statistically significant when the true treatment effects are, in fact, identical.

Table 2. Attributes of designed experiments (adapted from [7, p. 5]).

Attribute	Explanation	Comments/Examples
Absence of systematic effects	Treatment comparisons are not confounded or biased due to uncontrolled (background) variation.	Comparison of two fertilizer regimes would be biased by consistently applying one regime to plots in a more fertile part of the nursery.
Proper degree of precision	Poor design and large, uncontrolled variation result in large experimental error and imprecise estimates of treatment effects; "overdesign" results in overexpenditure of effort for the necessary data.	High precision occurs when (1) experimental plots have similar background characteristics, (2) experimental procedures are conducted with care and accuracy, (3) a large number of replications are used, and (4) the experimental design is efficient [7, p. 154].
Wide range of validity	Inferences and conclusions will apply to the entire population of interest; replication over time and space broadens the range of validity.	Testing a herbicide for a few years in several nurseries results in broadly applicable conclusions.
Quantification of degree of uncertainty	The "reasonable shadow of a doubt" accompanying experimental conclusions is quantifiable.	There is a 99% chance that the new fertilizer regime results in 2+0 height increase of 3.1 to 5.2 cm above the standard regime.

Table 3. Three main elements of statistical inference.

Term	Explanation	Comments/Examples
Point estimate	A single number that estimates a certain quantity in the population of interest.	Treatment mean, standard deviation, minimum, maximum, and range are all point estimates.
Confidence interval	For a given level of confidence, the specified range within which the quantity of interest lies.	A confidence interval on a treatment mean states with, say, 95% confidence that the true mean response lies between two estimated values.
Hypothesis testing	A statistical technique to accept or reject a hypothesis formulated before the experiment in light of the empirical results.	Designed experiments allow quantification of the level of uncertainty associated with acceptance or rejection.

- **Type 2 error (β):** The null hypothesis is accepted when it is really false. That is, statistically significant differences among the treatments are *not* declared even though they actually exist.

Examples help clarify these two types of error. Consider the United States judicial system; the null hypothesis is "innocent until proven guilty beyond a reasonable shadow of a doubt" [10, p. 167]. The null hypothesis is stated this way purposely because our society is wary of the consequences of convicting innocent people; we therefore view "guiltiness" with some skepticism. Two correct decisions are possible (Fig. 2): (1) exonerating innocent defendants (accepting the null hypothesis when it is true), and (2) convicting guilty defendants (rejecting the null hypothesis when it is false). Convicting an innocent defendant is a Type 1 error, whereas exonerating a guilty one is a Type 2 error. This emphasizes the interrelatedness of the two error types. Because the evidence must be strong for juries to declare a defendant guilty, the Type 2 error rate is large. Similarly, in nursery experiments, if a researcher requires overwhelming evidence to reject the null hypothesis (i.e., declare differences among treatments), the Type 2 error rate will be high (i.e., some important treatment differences will be missed).

Only by increasing experimental precision can the rates of making both types of errors be reduced simultaneously. Increased precision is related to the notion of the **power,** or sensitivity, of the experiment. If Type 2 errors are infrequent (β is small), a researcher can be more confident of declaring treatment differences that actually exist, and the experiment is said to be powerful (sensitive to treatment differences).

In most investigations, the level of α is set by the experimenter; thus, the rate of Type 1 errors is known. Testing hypotheses at the $\alpha = 0.05$ level states explicitly that there is a 5% chance of declaring differences among treatments when they do *not* exist. On the other hand, β is often undetermined and in many cases extremely high. That is, biologically important differences among treatments are often missed (not declared significantly different) because the experiment is not powerful enough to detect them at $\alpha = 0.05$. The nursery researcher should always examine the magnitude of treatment differences and ask the biometrician for an approximation of the power of the experiment.

(a)

Null Hypothesis (H_0): Treatments do not differ; they have the same effect on the response variable.

Decision regarding H_0	State of nature: H_0 is true	State of nature: H_0 is false
Accept H_0	Correct decision	Type 2 error (β)
Reject H_0	Type 1 error (α)	Correct decision

(b)

Null Hypothesis (H_0): Defendant is innocent.

Decision of jury	State of nature: Defendant innocent	State of nature: Defendant guilty
Declared innocent	Correct decision	Guilty person goes free (β)
Declared guilty	Innocent person declared guilty (α)	Correct decision

Figure 2. Correct and incorrect decisions are often made on the basis of incomplete information: (a) hypotheses tested by designed experiments and (b) a defendant judged by a jury (adapted from [10, p. 173]).

28.5.2 Replication and randomization

28.5.2.1 Replication

Replication is the repetition of treatments on more than one experimental plot [13, p. 5]. True replication means that a given treatment is applied independently to the multiple plots receiving that treatment. Because this last point causes considerable confusion in forestry experiments, it is useful to distinguish between *subsampling* and *replication*. For example, a nursery researcher applies two different fertilizer regimes to the entire length of each of two adjacent nursery beds. The researcher then scatters six 2-foot-long plots throughout each bed and counts the number of plantable seedlings lifted from each plot. In this instance, there are *not* six replicates of each treatment because the treatments are not applied independently to the six plots; in fact, all six plots received the same treatment application and are in the same bed. Rather, there is one replicate of each regime (a nursery bed) which is subsampled via subplots. To obtain true replication in the above example, 12 plots would first have to be chosen and the treatments then allocated to them at random (see 28.5.2.2).

The three functions of replication are to (1) increase the precision of estimated treatment effects, (2) provide a measure of experimental error, and (3) broaden the range of validity to which the experimental conclusions apply. The first and third functions were recognized in agricultural experimentation as early as the 1700s, as farmers noticed that uncontrollable variation in yields from field to field and year to year made it impossible to recommend the use of one crop variety over another without comparing the two in a number of fields and years [5].

To illustrate these functions, suppose that the effects of two fertilizer regimes on 2+0 stem caliper are compared in adjacent plots in a number of different places (replicates) in the nursery. Plots receiving the same fertilizer treatment will vary in caliper due to uncontrolled variation in, say, soil fertility, soil texture, date sown, proximity to irrigation lines, and so on. When the fertilizers are compared over a large number of replicates, effects of these uncontrolled factors "average out," and the estimated difference between fertilizer regimes more precisely measures the true difference due to fertilizer. The uncontrolled variation among plots—the experimental error—can be measured by comparing replicates of the same treatment. If, for example, in replicate after replicate, regime 1 consistently results in larger caliper than regime 2, experimental error is small relative to treatment differences, and the researcher is likely to reject the null hypothesis that the two fertilizers affect caliper equally. Ideally, the experiment must be replicated under varying conditions of time and space. Repeating this same experiment over several years and nurseries extends the inference space (population of interest) in these two dimensions, broadening the range of validity of experimental results.

28.5.2.2 Randomization

"**Randomization** is the assignment of treatments to experimental plots so that all plots have an equal chance of receiving a treatment. It functions to assure unbiased estimates of treatment means and experimental error" [13, p. 5]. Put another way, randomization serves to equalize background (uncontrolled) characteristics of experimental plots receiving different treatments and provides a basis for statistical inference [1, p. 32]. Plots receiving one treatment should differ in no systematic way from plots receiving another treatment; this is accomplished in practice by drawing numbers out of a hat or by using random-number tables or random-number generators to match the treatments by chance to their assigned field plots.

Consider the following example to describe the reasons for randomization. Four replications of two fertilizer regimes are compared in the same nursery bed (Fig. 3). Two (of many) alternative field layouts include a systematic layout (Fig. 3a), in which regime 1 precedes regime 2 in each replicate, and a "random" layout (Fig. 3b). Treatment means (say, for caliper) are obtained by averaging the four replicates of each regime, and experimental error is estimated from the variation among plots receiving the same regime. Further suppose that a water gradient in this bed causes drainage to become consistently poorer from left to right. Because regime 1 always occurs before 2 in the systematic layout, it always experiences slightly more favorable drainage; this **bias** does *not* average out. Even if no true differences exist between the fertilizer regimes, mean seedling caliper (the response variable) may always be larger for regime 1 because it consistently experiences better drainage. Therefore, the estimated effect of regime 1 in the systematic layout is biased upwards from the confounding effects of water drainage.

Only the extreme nature of the water gradient allowed recognition of the bias created by this particular systematic design. However, other systematic designs may suffer similar bias associated with gradients that researchers fail to recognize. Thus, random designs are essential as insurance against the possible bias generated by systematic variation in the uncontrolled characteristics of experimental plots.

Problems arise in experiments with few replications and treatments because "extreme" layouts—outcomes of randomization that appear systematic or unfavorable for some reason—occur fairly often, even at random. For example, if the experiment in Figure 3 were treated as a paired experiment with four pairs (see 28.5.3.1), the alternating scheme (12121212 or 21212121) would occur ⅛ of the time at random. Instances of such extreme layouts do *not* vitiate the need for randomization, but rather indicate the need for carefully examining the random layout before its field implementation. Cox [7, p. 86] presents an excellent discussion of methods for dealing with extreme outcomes.

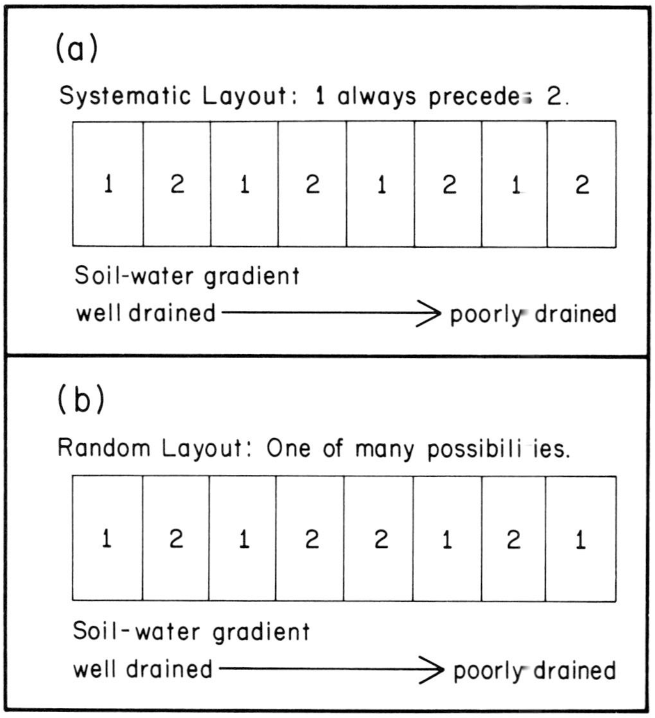

Figure 3. Two possible field layouts—(a) systematic and (b) random—of an experiment comprising four replications of two fertilizer regimes (1, 2) in a nursery bed.

28.5.3 Controlling experimental error

Reducing **experimental error** can greatly increase the sensitivity (power) of experiments to treatment differences. This section describes statistical methods of controlling error by choice of experimental design (see 28.5.3.1 to 28.5.3.3) and by covariance (see 28.5.3.4).

To this point, the discussion of experimental design suggests that treatments are always assigned to experimental plots totally at random. For example, in a nursery experiment with three replications of each of 10 treatments, randomization would ensure that each of the 30 nursery plots had an equal chance of receiving any replicate of any treatment. Such designs, called **completely randomized designs (CRDs)**, are the simplest to lay out and analyze; however, experimental precision can often be increased by employing slightly more complex designs. Thus, imposing certain restrictions on the random assignment of treatments to experimental plots can control experimental error.

28.5.3.1 Pairing

The most intuitively appealing restriction of randomization occurs when an experiment testing the effects of two treatments is designed such that treatment assignments are made in pairs. Two similar experimental plots are identified and called a pair; each plot within the pair randomly receives one of the two treatments. That is, randomization is still employed but is restricted to within-pair allocation of treatments to plots. The number of replicates equals the number of pairs. Because interest centers on comparing the relative, rather than absolute, effects of the two treatments, it is natural to use the *difference* between the paired treatment plots as the measure of response. Because experimental plots within pairs have similar background characteristics, uncontrolled variation (experimental error) is reduced by comparing differences on like plots.

An example should make clear these conceptual advantages of pairing. A nursery researcher, interested in comparing a new fertilizer regime to the standard regime for 2+0 Douglas-fir [*Pseudotsuga menziesii* (Mirb.) Franco], decides to test eight replications of the two treatments. To ensure broad validity of the results to the entire range of nursery conditions, the researcher first chooses eight nursery beds at random from all the beds containing Douglas-fir in their 2+0 season. Each of the 16 experimental plots will be 10 feet long and will receive one of the two regimes during the growing season; the number of plantable seedlings in the inner 4 feet of each plot (leaving a 3-foot border on each end) will be assessed at time of lifting. Because the two plots in each bed occupy only 20 feet, their position along the bed is also randomly located.

Two alternative designs for this experiment are shown in Figure 4. For the CRD (Fig. 4a), each of the 16 plots had an equal chance of receiving one of the two treatment regimes; one (of many) possible schemes is shown. For the paired-plot arrangement, one regime was randomly assigned to either of the two plots per bed by a coin flip, the other to the remaining member of the pair.

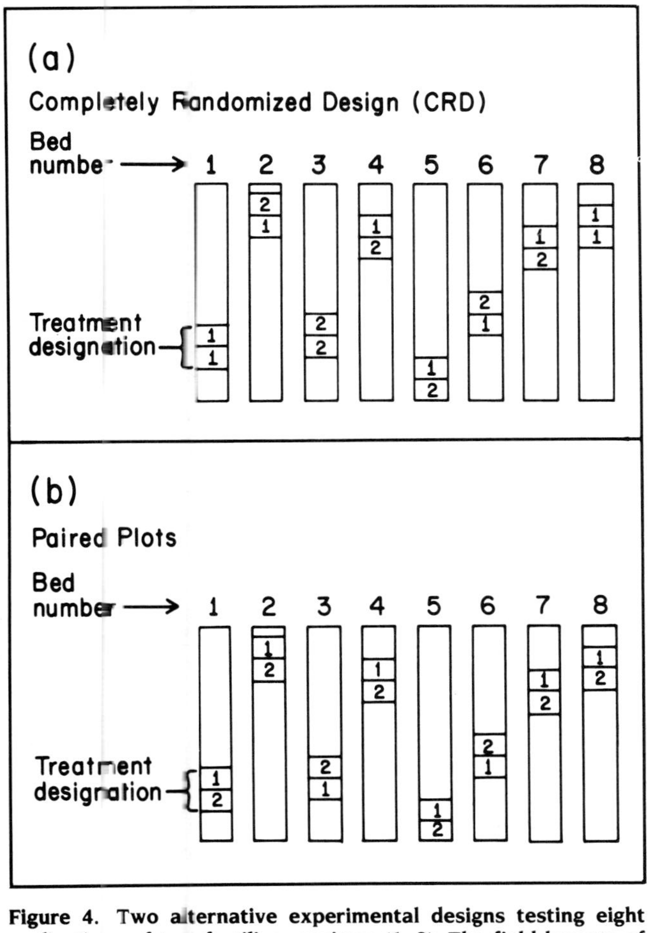

Figure 4. Two alternative experimental designs testing eight replications of two fertilizer regimes (1, 2). The field layouts of nursery beds and the plot locations within beds are chosen at random. Fertilizer regimes are assigned (a) at random to the 16 plots and (b) to the 16 plots as eight pairs, each member of the pair receiving one of the two fertilizer regimes.

Imagine that these 16 plots sample a wide range of drainage conditions, fertility, soil texture, and proximity to irrigation lines. Adjacent plots in the same bed will be more similar with respect to these conditions than will plots in different beds. Pairing plots exploits this similarity by basing the analysis of treatment effects on the difference between treatments occurring in the same bed. Two plots lying in a poorly drained area of a bed necessarily have reduced yield of plantable seedlings. In the CRD, one fertilizer regime may be assigned to both plots of the poorly drained bed location, reducing its average over the eight replicates and increasing experimental error. In the paired-plot arrangement, however, the two regimes will be negatively affected in the same way on a poorly drained location; the differential effect of regime 1 over regime 2 may remain relatively stable, except for other sources of background variation associated with differences between plots within pairs.

28.5.3.2 Blocking

The concept of pairing logically extends to that of blocking for experiments with more than two treatments. In fact, paired plots are the simplest case of blocking. Suppose, for an experiment testing a new herbicide at two application rates (L = low, H = high) against both a control (C = no herbicide) and the standard herbicide (S = standard), that each of the four treatments is replicated 5 times (20 plots). If a CRD is used (Fig. 5a), treatments are assigned totally at random. If a **randomized complete block (RCB)** is used (Fig. 5b), each nursery bed is assigned one complete replicate of the experiment; randomization is restricted to the allocation of treatments *within a block*. If nursery beds are scattered, representing a wide variety of conditions, plots within a bed should be more similar to one another than to those from different beds.

Consider what this does to comparisons of treatment effects for the above-mentioned herbicide experiment. At the end of the first growing season, height of 100 seedlings within each plot is measured to determine the possible phytotoxic effects of the new chemical. Suppose that bed 5 was not well prepared by the bed former and that this, combined with the inherent soil attributes of that bed, reduces height growth in bed 5 regardless of treatment. Bed 4, on the other hand, is located in a part of the nursery recently mulched, which accelerates seedling growth. These "bed effects" are average effects on 1+0 height common to all plots in a given bed. In addition to bed and treatment effects, 1+0 height also is influenced by "plot effects"—uncontrolled variation due to background characteristics of the specific plots within beds. For any of the 20 plots, then, 1+0 height may be expressed as the sum of these three effects:

Height = (treatment effect) + (bed effect) + (plot effect)

The mean for each of the four treatments is estimated as its average over five replicates. In the case of the RCB, bed effects influence each treatment mean equally because each treatment occurs once in each bed. In the CRD, differential bed effects influence treatment averages; specifically, the negative effect of bed 5 reduces the average levels for L and H of the new chemical, whereas the positive effect of bed 4 increases the average height in two plots for S. Thus, the RCB design increases the precision with which treatment effects are estimated by allowing bed effects to be estimated separately and removed from the comparisons of treatments.

This increase in precision is reflected in reduced experimental error. Overall error is estimated by the variability among the five plots receiving the same treatment. In the CRD, the five plots for any one treatment vary both by bed and plot effects, and both contribute to experimental error. In the RCB, bed effects can be directly estimated and can be eliminated from the experimental error.

It is critical to realize conceptually that blocking works any time experimental plots of an entire replication (one plot of each treatment) can be grouped such that they are more similar to one another than to plots of other blocks. Then, differences among the groups are termed block effects (bed effects, in the herbicide experiment). This may mean, for example, that blocks are sown on different days, measured by different observers, or located in different parts of the nursery (see 28.6 for practical implications of blocking).

Finally, the concept of blocking can be extended to more than one dimension through Latin Squares and similar designs. Though sometimes useful in nursery research, these designs suffer from sensitivity to missing data (e.g., plots accidentally destroyed by faulty irrigation) and from restrictions on the number of treatments and replications. Neter and Wasserman [15] present an excellent discussion on design and analysis and on overcoming problems of these designs.

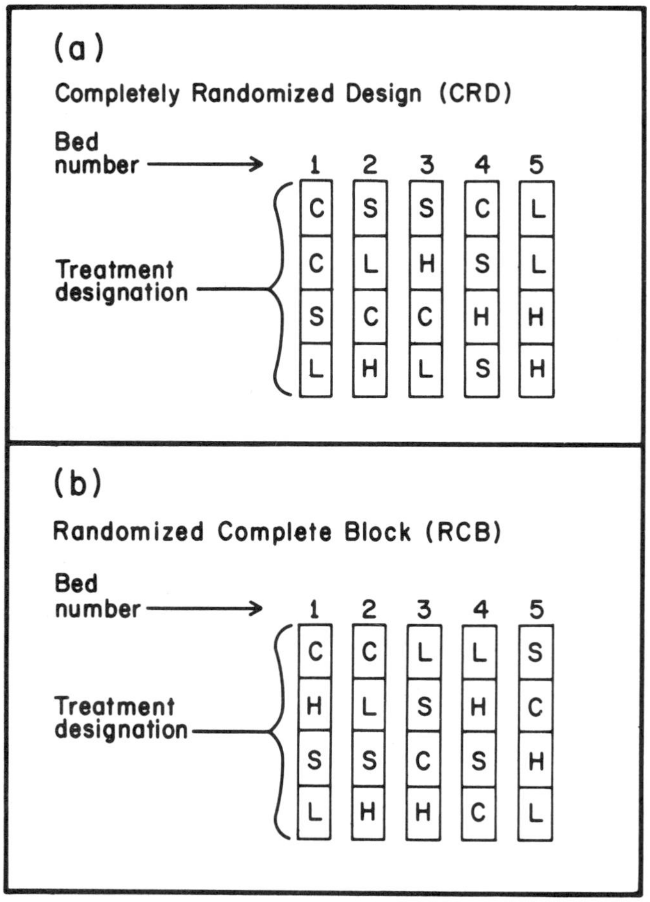

Figure 5. Two alternative experimental designs testing four herbicide treatments (C = control, S = standard herbicide, L = low level of new herbicide, H = high level of new herbicide). The benefit of the RCB (b) over the CRD (a) is that each bed receives one replication of all four treatments.

28.5.3.3 Split-plot principle

When, for practical reasons, some factors of an experiment require larger plot sizes than others, the split-plot principle is often applied. For instance, the minimum size of plots for irrigation treatments and sowing dates is necessarily larger than that for different sowing densities and seed sources. A split-plot design for two factors calls for assigning treatments of one factor to larger plots (called main plots or whole plots) in a CRD, RCB, or other design and then splitting each whole plot into enough subplots to accommodate one replicate of each treatment level of the second factor. Because each whole plot contains a complete replication of the treatments of the second factor, it is a "block" of the second factor. Randomization occurs in two stages: first, in assigning treatments of factor 1 to whole plots, then, in assigning treatments of factor 2 to subplots within each whole plot. Precision is often sacrificed for estimating effects of the whole-plot factor, but increased for subplot treatment comparisons. Split-plot designs can be extended to multiple factors at both the whole-plot and subplot level and even to splitting the subplot (split-split plots). Cox [7, p. 142] and Little and Hills [13, chapters 8, 9] give excellent accounts of the concepts and algebra of split-plot designs; Cochran and Cox [6, chapter 7] present a more advanced treatment.

For the purposes of describing the concepts behind split-plot designs, consider a two-factor nursery experiment investigating the effects of three fertilization regimes (F1, F2, F3) at each of two irrigation levels (H = high, L = low) on stem caliper of 2+1 Douglas-fir (Fig. 6). The irrigation system in the nursery may require that several beds on either side of an irrigation line receive the same irrigation treatment. It may be that the nursery researcher can only devote six lines to the entire experiment (say, two lines in each of three different sections of the nursery). Therefore, a possible RCB design for irrigation (exclusive of fertilization) may be obtained by randomly assigning one of the two irrigation treatments to one of the two lines within each section (block) (Fig. 6a); this is an RCB with two treatments and three blocks. All beds watered by each line receive the same irrigation treatment. The fertilizer treatment may then be added by "splitting" each bed into three lengths (subplots) to which one of the three fertilizer regimes is randomly assigned (Fig. 6b). Thus, the randomization of fertilizer treatments is restricted to allocation within an irrigation whole-plot.

Regardless of the response variable, two types of experimental error are associated with this experiment. The subplot error, resulting from residual variation among subplots, estimates microsite and other background differences influencing the response of subplots within a whole plot. The whole-plot error, resulting from the uncontrolled background variation among whole plots within a block, is usually larger than the subplot error.

28.5.3.4 Covariance

In a previous section (28.5.3.2), blocking of experimental plots into groups of similar soil types, etc. was presented as a method of reducing experimental error. Even after plots are grouped, however, background characteristics of plots within a block may still vary. Knowing that this variation exists may be used to reduce experimental error by the statistical process of covariance [7, chapter 4; 18, chapter 15]. Covariance requires making additional measurements of these other characteristics (called **concomitant variables**) on each experimental plot.

For an experiment testing the effects of different types of root wrenching on seedling caliper, suppose that even after blocking, substantial variation in soil N level exists among plots within blocks. N level may have an average influence on caliper; higher inherent N means larger average caliper. Knowing this relationship may help the researcher adjust treatment means to a common starting value of soil N.

The analysis of covariance may be used for any type of statistical design (e.g., CRD, RCB, Latin Square) as an additional method of increasing precision. Foresters routinely use the simplest form of covariance by analyzing height "growth" instead of total height as a measure of treatment response. The rationale is to remove some of the initial variation in

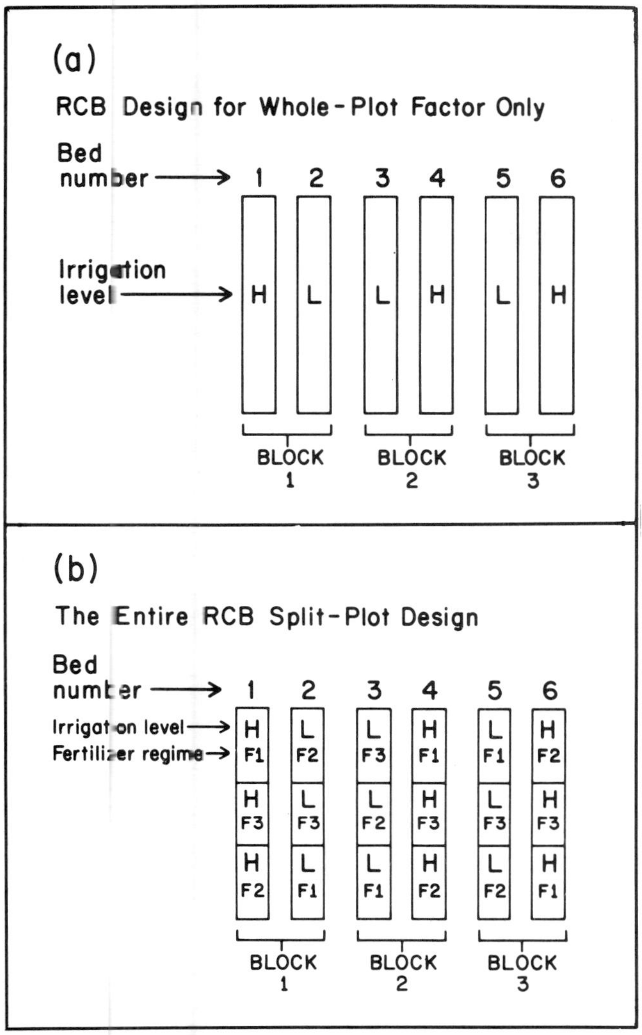

Figure 6. RCB split-plot design testing two irrigation levels (H = high, L = low) and three fertilizer regimes (F1, F2, F3): (a) irrigation, requiring larger plots, is the whole-plot factor; (b) fertilization, allowing smaller plots, is the subplot factor.

height by measuring height before and after treatment application and analyzing just the portion of height (the increment) added after treatment.

Some words of caution are in order, however. The interpretation of covariance analysis can become very complicated, making it difficult to separate the effects of the concomitant variable from those of the treatments. Statistically, this happens for complex experimental designs and when some plots are missing. Biologically, this happens when the concomitant variable is affected by the treatment [7, p. 48]. This last point is extremely important. For example, if soil P level is used as a concomitant variable in an experiment assessing effects of phosphate fertilization on caliper and if it is measured *after* fertilizer application, then the treatment will drastically influence the concomitant variable. Adjusting for the average effect of soil P on caliper might eliminate all treatment (fertilizer) effects. To avoid this type of problem, make sure that concomitant variables are (1) observed before treatments are applied or (2) unaffected by the treatments.

28.6 The Research Process for Designed Nursery Experiments

This section covers in some depth the research process (28.3) applied to designed nursery experiments. Emphasis is placed on the importance of statistical concepts (28.5) in developing, designing, conducting, analyzing, and interpreting these experiments.

The following discussion provides a checklist for managers as they encounter and attack problems in their nurseries. Though presented in chronological order, the steps of the research process are not independent of each other. Often, understanding the rationale of subsequent steps helps accomplish the current one.

28.6.1 Defining the problem

28.6.1.1 Recognizing a problem

Most nursery experiments stem from problems requiring practical solutions (see chapter 29, this volume). Even nursery research investigating the fundamental principles underlying a problem—such as the physiological basis of increased outplanting vigor after fall fertilization of 2+0s—has immediate practical application. Practical problems may be arbitrarily classified as either today's limitations or tomorrow's opportunities. Today's limitations include existing insect, pathogen, or drainage problems, or optimizing packing-line arrangement with existing equipment; tomorrow's opportunities involve incorporating new technology—whether new chemicals, machinery, or cultural practices—to improve seedling quality and cost-effectiveness. Nursery researchers must be adept at recognizing and addressing both problem types.

28.6.1.2 Refining the problem

Once a problem area has been identified, the nursery researcher relies on personal experience and secures existing information from literature, chemical labels, other nursery personnel, and specialists (e.g., extension agent) to answer the following questions:

Does the problem warrant research? The answer may be no if (1) immediate action is required, (2) results from other studies are conclusive and broadly applicable, (3) cost or effort involved in doing the research is high relative to potential benefits, or (4) cost of implementing the results is too high.

If the problem is researchable, what specific questions remain to be answered? Perhaps some parts of the problem seem solved, but others need further investigation. For example, a new herbicide may have been demonstrated safe and effective in several nurseries, but optimal application rates and timing remain to be determined for your nursery conditions.

Is a designed experiment needed? Perhaps the attributes of designed experiments (see 28.4.2) are *not* required and an operational trial (see 28.4.1) will answer the questions for less cost and effort.

To what population should the results apply? The inference space [2, p. 84] or range of validity of the results must be defined. For instance, should the results and conclusions apply to (1) 1+0 Douglas-fir in one particular nursery block, (2) 1+0 Douglas-fir in poorly drained parts of the nursery, (3) 1+0 Douglas-fir in all bareroot nurseries west of the Cascade Mountains, or (4) 1+0 and 2+0 conifers in all bareroot nurseries west of the Cascades?

28.6.2 Setting objectives

Delineating the experimental objectives serves to clearly state the problems to be addressed and sets the framework for specifying experimental methodology [3, p. 38]. Objectives can take several forms: (1) to determine the effects of a certain factor, for example, of fall fertilization with N; (2) to investigate interactions, for example, of irrigation and fertilization; (3) to find optimum application rates, for example, of herbicides or fertilizers; and (4) to develop prediction equations, for example, of the average shoot:root ratio of a batch of seedlings, given their average height, caliper, and seed source. Although statistical null hypotheses are often implied when research objectives are stated, their explicit delineation is left until the experiment has been designed (28.6.3.5).

Objectives should be ranked to ensure that the experiment is designed to answer the most important questions first. Often, it is possible (for example, by split-plot design) to increase the precision with which certain treatment means are estimated or hypotheses tested by sacrificing precision on others.

28.6.3 Planning the experiment

The nursery researcher frequently directs the early efforts of planning an experiment (see 28.6.3.1 to 28.6.3.3), whereas the biometrician may direct the later ones (see 28.6.3.4 to 28.6.3.7). Though the leader may change, a true team effort is required throughout to ensure that the experiment meets its stated objectives. For instance, the nursery researcher may list factors to investigate and the biometrician then help determine the levels at which to test each factor. Conversely, the biometrician may set a minimum number of replications needed and the nursery researcher put an upper practical limit on the number feasible. A constant balance is needed between what might appear statistically favorable and what is practically suited to experimental conditions in the nursery.

28.6.3.1 Choosing treatments

Controls and standards.—Nursery experiments most often require a treatment that serves as the basis against which the effectiveness of other treatments is judged. Controls and standards both serve this purpose.

A **control treatment** is the zero-level application of a factor; no wrenching, 0 pounds of N fertilizer, no irrigation, and no herbicide are examples of controls that might be used to judge whether particular treatment levels of wrenching, N fertilization, irrigation, and herbicide spray, respectively, were effective. Control treatments are particularly useful when an unproven or new factor is being tested but are less so when the experimental objective is to find an optimum level of a "known-effective" factor.

Standard treatments are the "standard operating procedures." A new treatment must often prove itself against the standard to justify altering current practices. For instance, a new alignment of personnel in the packing shed must be proven superior to the current one to justify switching.

Single-factor versus multifactor experiments.—Single-factor experiments, in which only one condition (factor) is varied among the treatments, are commonly used in nurseries at either end of the research process: operational trials or final-stage experiments. In operational trials, the researcher might test three new herbicides for control of grasses or compare two seeders for evenness of sowing; in both cases, only one factor, herbicide or equipment type, is varied. In the final stages of experimentation, the researcher often knows the proper levels at which to control nontreatment conditions and varies only the critical factor of interest (say, herbicide application rate) to find the optimum level [6, p. 152].

Most nursery experiments lie between these two extremes, exploring the effects of one factor (say, bed density) over various levels of other factors (sowing dates and species). Such multifactor experiments often test, for example, whether the most effective bed density is the same for all sowing dates and species.

Factorial experiments.—Factorials are by far the most common arrangement of treatments in multifactor experiments. Factorial experiments test each level of each factor at all levels of the other factors. In a three-factor experiment with two bed densities, three N levels, and two seed zones of Douglas-fir, there are 2 x 3 x 2 = 12 treatments. Each treatment consists of a specified level of each of the three factors—for example, treatment 1 might be low bed density, no N (the control), and seed zone 062; 12 separate treatments are required to test each factor across all levels of the other factors. These 12 treatments can be applied to experimental plots in a variety of experimental designs (CRD, RCB, or Latin Square); "factorial" just defines the number and structure of the treatments, not the field design.

The nature of factorials and the reasons for their importance are discussed fully in Cox [7, p. 94]. Briefly, factorials allow explicit investigation of the interaction among factors. If interactions are *not* significant, then factorial experiments extend the range of validity and increase the precision of estimating factor effects, relative to separate experiments of the individual factors. For example, the effects of N and P may be investigated either in separate experiments or in one factorial experiment. If experiments are done separately, N is held at a constant (standard) level while multiple P levels are investigated; conversely, P is held at a constant level while multiple N levels are investigated. The factorial allows elucidation of interactions because rates of N and P are varied together so that all combinations of both factors are applied; for example, N may increase caliper only in the presence of high P levels. In the absence of interactions (that is, if the effects of N do *not* depend on the level of P, and vice versa), the range of validity is extended because the researcher knows that each nutrient is effective over several levels of the other, *not* just the standard. The precision of estimating effects is also increased [7, p. 94].

Factorials are not without their drawbacks [6, p. 152]. But for most nursery experiments, these are more a matter of the complexity of the problem than the factorial itself. Factorials can often become large (for example, a 3 x 3 x 5 factorial has 45 treatments), making them difficult to implement and, sometimes, interpret; however, the efficiency of factorial arrangements, compared to that of separate experiments, increases for large, multifactor experiments.

Choosing factors.—While mainly directed to factorials, the discussion here applies rather broadly to choosing factors in multifactor experiments [7, p. 134]. For the most part, multifactor experiments examine only one or two factors of primary interest; these **primary factors** are the reason for the experiment. **Supplementary** or **subsidiary** factors [6, p. 151] may be added to (1) shed light on the mode of action, (2) extend the range of validity, and (or) (3) determine interactions with the primary factor(s).

In an experiment testing the effects of fall N fertilization on the frost hardiness of 2+0 Douglas-fir, the primary factor is N; supplementary factors might include seed zone and irrigation. The N levels are tried (1) at various irrigation levels, both to examine interactions (perhaps standard irrigation is *not* best with fall fertilization) and to elucidate N's mode of action in increasing frost hardiness (perhaps moister conditions promote the physiological actions of N relating to frost hardiness), and (2) at different seed zones, to provide a wider range of validity if N acts consistently across all the zones tested. The nursery researcher should choose primary factors that meet the experimental objectives and supplementary factors that ensure general conclusions can be drawn about the primary factors over the intended population.

Choosing factor levels.—In choosing levels at which to test each factor, the nursery researcher must again consider the experimental objectives and intended population. For quantitative factors whose levels represent points on a continuum (e.g., pounds of fertilizer), the range of levels should bracket the range expected to be operationally feasible [7, p. 141]. For example, the lowest level of fertilization or chemical application is often zero—the control; the upper level is chosen on the basis of experience, cost, or other information as the upper practical extreme. How many levels and where to position the levels depend on the nature of the response curve (linear, quadratic, asymptotic) and the purpose of the experiment (see [7, p. 141] for a full discussion). For many nursery experiments, three or four well-spaced levels (including a control, if appropriate) are sufficient.

For supplementary factors included to extend the range of validity or to detect interactions, a few extreme levels often suffice. For example, if N fertilizer causes similar responses in extreme seed sources of coastal Douglas-fir—say, west Cascade, valley, and coastal—it may be safe to extend the experimental conclusions to all coastal Douglas-fir. And if N shows no interaction with P over a wide range of levels, then inferences over the entire range are valid. The only caution here is to use factor levels of interest. If Cascades Douglas-fir is not grown at the nursery, then why include it unless more fundamental questions about the Douglas-fir species are of interest. Again, keep the population of interest in mind when choosing both factors and their levels.

Choosing levels for nontreatment conditions.—Once the treatments have been determined, it is critical that the conditions or factors *not* varied are held constant at meaningful levels. For example, in an experiment testing the relative effectiveness of two root-wrenching depths, only depth is varied. Other factors—such as seed zone, stratification period, sowing date, and fertilization and irrigation regimes—are held constant. In this instance, the rate of irrigation may have a dramatic impact on treatment effectiveness; thus, its level (though not varied) is critical to interpreting the results. Often, but not always, the nontreatment factors are held constant at their normal or standard levels to test the effectiveness of the treatments if everything else is done as usual.

28.6.3.2 Choosing variables to measure

The nursery researcher is faced with a wide array of variables that could be affected by or could affect treatment responses. Which variables to measure depends on how much time and effort are available and how likely it is that a particular variable may be of practical value. Variables measured fall into three broad categories; keeping these categories in mind can often help researchers decide which variables are pertinent.

Response variables.—Response variables are those that the treatments were meant to test. They are usually measured on all **observational units** (items to be experimentally measured or observed; for example, seedlings) in a **measurement plot** (portion of the experimental plot actually measured), then aggregated to obtain a plot mean or sum. The most critical of these are usually delineated directly in the experimental objectives. Height, caliper, shoot:root ratio, number of plantable seedlings, foliar N levels, frost hardiness, and outplanting growth and survival are examples of often-measured nursery response variables.

Response variables may be either quantitative (numeric) or categorical (falling into discrete classes). The numeric are routinely analyzed with a technique termed analysis of variance, the categorical by other techniques [17] or by assigning numbers to the classes (categories) to make them pseudonumeric. For example, six vigor classes might arbitrarily be assigned the numbers 1 (low) to 6 (high). Caution must be used here because this implies that class 6 is 1 unit better than class 5, 2 units better than class 4, and so on. The biological basis of such assignments should be weighed. For categorical data, more classes mean more discrimination among treatments unless responses *cannot* be accurately assigned. Four and six classes are often useful numbers of classes for assigning biological responses; an even number is recommended because of the psychological tendency of observers to overassign responses to a middle category (such as to the third class, if five classes were available).

Concomitant variables.—Concomitant variables are those measured on each experimental plot or observational unit (seedling), for the purposes of using covariance analysis (see 28.5.3.4). The precision of the experiment can be increased by adjusting response variables to a common, average level of the concomitant variable. Concomitant variables—for example, pretreatment soil N levels, initial bed density, or soil textures—must be measured on each experimental plot, be independent of treatment effects, and be numeric.

Explanatory variables.—Explanatory variables are often measured to shed light on underlying principles of action or to document experimental conditions. These variables can be measured at any level. On the experimental-plot level, the nursery researcher might test to see whether fertilizer or chemicals were applied properly by assaying each plot shortly after application. On the block level (if beds are blocks), the researcher might monitor plant water potential at various points in nursery beds situated varying distances from the irrigation lines; large differences in seedling growth from block to block may then be related to water status. Finally, on the nursery level, the researcher might monitor climatic conditions relative to sowing date; early sowing may pay handsome dividends in some years, whereas its effects may be disastrous during other years with different spring weather.

In general, explanatory variables are measured for biological or physical, *not* statistical, reasons. They are often used in the deductive process to extrapolate or "explain" experimental results.

28.6.3.3 Determining plot size

Determining the appropriate size and shape of experimental plots requires both statistical and practical considerations. For a specified amount of land devoted to an experiment, the number of replications necessarily decreases as plot size increases. As a rule, once a *minimum* plot size is reached, precision is increased more effectively by adding replications than by enlarging plots [9, p. 3]. Practical considerations, subsequently described, often loom large in determining this minimum plot size.

Remember that the experimental plot (say, a length of nursery bed) is the smallest physical unit to which a treatment is applied independent of all other treatments. All observational units (seedlings) within a plot do *not* have to be measured. It is largely the responsibility of the nursery manager to ensure that the total size of the experimental plot satisfies practical constraints and of the researcher and biometrician together to determine the size of the measurement plot within each experimental plot.

Plot shape, usually constrained by bed shape in forest-tree nurseries, will not be addressed here; references include Le Clerg et al. [12] and Gomez [9].

Experimental plots.—Practical considerations influencing size of the experimental plot fall into three overlapping categories: operational constraints, representation, and independence.

- **Operational constraints:** These refer to physical limitations in the ability to apply treatments independently to small plots. Such limits often lead to split-plot designs (see 28.5.3.3), in which one factor (the whole-plot factor) is applied to a much larger plot than other factors (subplot factors) in the experiment. Irrigation is a good case in point; the physical nature of the system often requires that several beds on either side of a line *must* receive the same irrigation level. Thus, a group of beds is the smallest physical unit to which the researcher can independently (randomly) assign different irrigation levels. Other treatments (the subplot treatments) in the split-plot design, such as fertilizer levels or different wrenching techniques, can be randomly assigned to smaller plots within this group of beds.

  For multifactor experiments, the researcher should construct a brief list detailing the operationally feasible plot sizes for the factors under investigation. This list also is useful for determining experimental designs and conducting the experiment.

- **Representation:** In most nursery experiments, it is critical that the conditions imposed by the experimental treatments be representative of those same treatments applied operationally [7, p. 194]. For instance, the artificial nature of irrigating by hand may be intolerable even though it allows smaller plot sizes. Many treatments applied in nursery experiments have start-stop problems in the sense that representative treatments are *not* attained at the beginning and ending of each plot; for example, many seeders disperse seed unevenly for the first and last foot or two. Thus, the total plot size must be large enough to leave a representative measurement plot in the "middle," the "ends" serving as borders.

  Representation, as used here, relates to bias and inaccuracy that can result from nonrepresentative plots. Careful, precise experimental technique *cannot* overcome bias resulting in this manner. For example, in time and motion studies such as might be conducted to investigate alternative packing-line arrangements, the experimental time allotted must be long enough to accurately represent the operational situation. Some arrangements, faster in the short run, may cause workers to take more breaks or to suffer more illness or boredom when imposed under normal, operating conditions.

  Representation is less important in fundamental studies investigating the basic biological principles underlying treatment response. There, statistical precision and choice of treatments to illuminate reasons for response are most important; hand application of fertilizer or irrigation and manual sowing (or thinning to desired bed densities) may be entirely suitable.

- **Independence:** Treatment application and (or) response on a particular plot should *not* influence response on adjacent plots [7, p. 196]. For example, spread of sprays, fertilizer, and water can unknowingly affect growth on near-by plots. In time and motion studies, two packing-line arrangements tried on successive dates might allow a carry-over effect from the first day's arrangement to the second. Researchers should make every effort to ensure that experimental plots are large enough for measurement plots to respond independently.

Measurement plots.—In most field experiments, the actual measurement plot is a subplot nestled within the experimental plot. The unmeasured observational units (seedlings) surrounding the measurement plot buffer that plot from edge effects and from effects caused by treatments on other experimental plots.

In addition to independence and representation, the number of trees in a measurement plot and its orientation and location are the major concerns.

- **Edge effects:** These occur both on the ends and sides of nursery-bed plots. As a very general rule, end effects are bad (artifacts of the experiment), and side effects are good (representative of the nursery).

  End effects usually occur as a result of stop-start problems associated with treatment application to small experimental plots (see **Representation,** just discussed). They should be avoided by placing measurement plots away from the ends of experimental plots; these ends should be left to border the measurement plot, buffering it from the external influences of adjacent treatments and making it more nearly like a randomly chosen location in the middle of an operational bed.

  Side effects occur because the outermost drill row along each side of the nursery bed tends to grow and respond to treatment differently than the inner rows. However, this type of edge effect would occur to the side rows if the treatment were applied on an operational scale and in practical nursery experiments. Because the inferences drawn should pertain to the population of *all* seedlings, these outermost rows should be included in the measurement plot to make it as representative as possible of the population of interest. Thus, a measurement plot for practical experiments should be a swath that stretches across the entire seedbed; each row contributes equally to the plot mean just as each row contributes equally to the harvestable crop.

- **Response variable:** Depending upon the trait being measured, fewer or more trees need be included in the measurement plot to obtain a precise plot mean for the treatment. The amount of effort and expense required to obtain each measurement also influences the number included. As a general rule, for traits like height, caliper, and number of plantable seedlings, a 1-foot section of bed provides a more than adequate number of seedlings (~ 100 at 25/ft^2) and is easy to lay out.

- **Subsampling:** Multiple measurement plots are often placed within one experimental plot either to allow for multiple destructive sampling throughout the growing season (e.g., when shoot:root ratio is assessed at multiple times during the growing season) or to provide an estimate of within-plot variability. In the case of such subsampling, the following considerations apply to each measurement plot: (1) handling of one measurement plot should *not* influence response on adjacent ones, often necessitating that buffer areas be left between measurement plots; (2) each measurement plot should represent the population of interest; and (3) enough seedlings should be included in each measurement plot to provide a precise plot mean.

28.6.3.4 Choosing an experimental design

By this point, tentative decisions have been made (mostly by the nursery researcher) regarding factors and factor levels to investigate, variables to measure, and practical limits on experimental- and measurement-plot sizes. The biometrician and researcher now employ the concepts of randomization, replication, and error control (see 28.5) to develop a statistically and operationally appropriate experimental design.

Applying these concepts to nursery field experiments produces frequent use of only a few common designs. CRDs (see Figs. 4a and 5a), in which the random assignment of treatments to plots is unrestricted, are *not* common in nursery experimentation. A logical field grouping of experimental plots almost always exists such that plots within a group are more similar to each other with respect to water drainage, proximity to irrigation, and (or) soil characteristics than are plots from different groups. As a result, the RCB design (see Figs. 5b and 6) is the most commonly used in nurseries. In addition, RCBs are relatively easy to lay out and analyze and are relatively insensitive to the accidental loss or destruction of a plot or two. Latin Squares are used, though less frequently, when bidirectional field gradients exist; however, they suffer from the drawbacks discussed in section 28.5.3.2.

For multifactor experiments, especially those investigating irrigation, sowing date, or wrenching, split-plot designs are common (see Fig. 6). These most often have the whole-plot factor (such as irrigation) arranged in randomized complete blocks, with the treatments of one or two subplot factors (such as three levels of fertilization and two bed densities) arranged totally at random within each whole plot. Although these designs are common, more complex ones are warranted for large experiments that "mushroom" and would occupy too much space and require too much effort if many replications of each treatment were applied. The biometrician and researcher must confer and be innovative to arrive at appropriate designs (such as fractional factorials) for these more complex experimental situations.

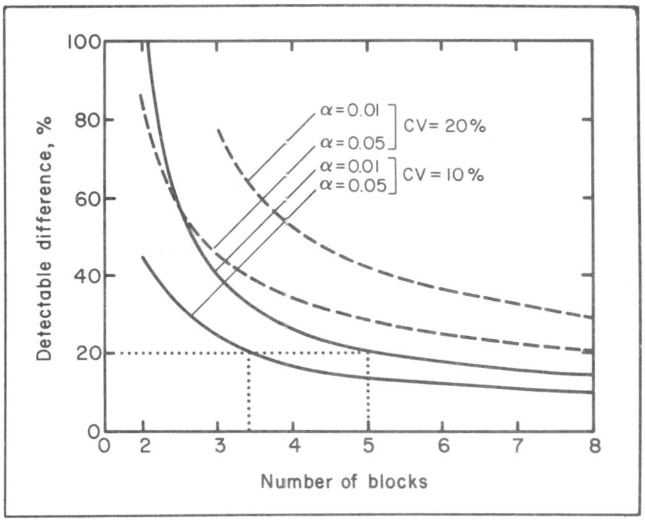

Figure 7. Effects of the number of blocks on the level of detectable difference (D) among treatments for two levels of inherent variability (coefficient of variation, CV, = 10 and 20%) and two levels of Type 1 error rate (α = 0.05 and 0.01) for an RCB with three treatments. Dotted line indicates the number of blocks required at two α levels when D = 20%.

28.6.3.5 Determining the number of replications

Theoretical considerations.—Many factors (practical and statistical) impinge on the number of replications required for a particular experiment. The effects of these factors are described here, both mathematically and intuitively, but understanding their effects does not depend on the mathematical relationships; it is added only for completeness. The practical nursery implications for each factor are delineated.

For RCB designs, the factors influencing the number of replications (i.e., blocks) interact through the formula

$$n = \frac{4t_\alpha^2(CV)^2}{D^2} \qquad (1)$$

where n = number of blocks

D = detectable level of difference (%) between two treatments

t_α = tabular value of t for a specified Type 1 error rate (α) and number of degrees of freedom

CV = coefficient of variation (%) obtained as (mean square error)$^{1/2}$/experimental mean.

The **detectable level of difference** (D) is that difference between two treatment means (expressed as a percentage) which the experiment is able to declare significant. For example, if height of 2+0 seedlings increases from 15 to 18 inches as a result of fertilization, D = 20% (a 20% increase). In general, smaller differences are more difficult to detect (declare statistically significant) and require more replications. The researcher should decide in advance, roughly, the differences among treatments that represent biologically or practically important responses.

Recall that the **Type 1 error rate** (α) is the probability of declaring treatments significantly different when, in fact, they are not. The nursery researcher will necessarily set a low Type 1 error rate if a high degree of confidence is required before drawing conclusions from an experiment. The higher level of confidence required necessitates more replications to declare results significant. For example, for a given level of detection (say, D = 20%), more replications are required to declare results significant at α = 0.01 (99% confidence level) than at α = 0.05 (95% confidence level) (Fig. 7).

In principle, the **coefficient of variation** (CV) measures the background variation among plots receiving the same treatment as a percentage of the treatment mean. Thus, for a given level of detection, more replications are required for traits with higher CVs because the higher variability means larger experimental error (Fig. 7). For nursery experiments, CVs are influenced by (1) the response variable (e.g., root-growth capacity is extremely variable), (2) the experimental material (1+0 heights are more variable on a percentage basis than 2+0 heights), and (3) the variability among field plots. CVs between 10 and 20% are common in nursery field experiments.

In general, experiments with more treatments require fewer replications. In rough terms, each treatment provides an estimate of experimental error via the variation among the experimental plots receiving that treatment; these are pooled (combined) to give an average "experiment-wide" estimate of error. More treatments result in more of these individual estimates and thus a more precise estimate of the experimental error. When both the replications and the number of treatments are

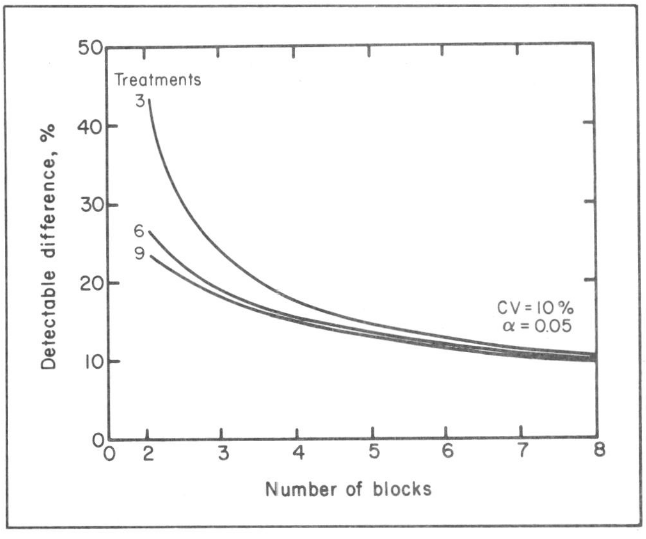

Figure 8. Effects of the number of blocks on the level of detectable difference between two treatment means for an RCB with different numbers of treatments (3, 6, 9).

small, an increase in either is especially effective in increasing the sensitivity of the experiment to detecting differences (Fig. 8).

Appropriate experimental designs can effectively reduce the number of replications needed for a given desired level of detection by eliminating extraneous sources of experimental error from treatment comparisons. This increases precision by reducing the experimental error and thus the CV. Furthermore, subsampling within an experimental plot can sometimes reduce the number of replications needed by providing an additional estimate of experimental error. The use of subsampling depends on many factors [4] but is most beneficial when the number of replications is limited for practical reasons.

Practical recommendations.—In addition to the previously stated statistical considerations, the number of replications required is influenced greatly by several practical factors. If the chance of incurring missing plots is high, more replications are needed. In addition, the accuracy of the experimental work is extremely critical in reducing the CV and thus lowering the number of replications required.

In general, three replications are a minimum, and more are required if (1) the number of treatments is small, (2) the response variable has a high CV, or (3) small differences should be detectable with high confidence. When estimates of the factors in Equation 1 are available, the number of blocks (for an RCB) can be calculated. When data are *not* available, a rule of thumb [14] is to choose the number of blocks, n, such that $(n - 1)(t - 1) \geq 12$, where t is the number of treatments; this ensures 12 degrees of freedom for estimating experimental error.

28.6.3.6 Delineating the field layout

An appropriate field layout matches the experimental design to the field gradients existing in the nursery. For RCBs, plots within a block should be as similar as possible; this necessitates close attention to water and soil gradients. It is often effective to block on nursery beds because plots within a bed are similar. Sometimes, however, more similarity can be achieved by blocking across beds, combining plots in several adjacent beds into the same block. It is frequently necessary to skip over certain local areas in the nursery that are dissimilar to other plots being included in a block. For example, relatively narrow, low-lying areas might be dissimilar to any of the other plots within a block and, if so, should be excluded from the experiment; these serve as buffers and are ignored for the purposes of experimentation.

Often, different blocks are put in different nursery fields. This has the advantage of broadening the range of validity to the entire nursery and makes plots within a block (field) more similar, thereby reducing experimental error. Attention should also be paid to possible carry-over effects resulting from previous nursery treatments. For example, suppose one part of the nursery had been hydromulched and another part mulched with sawdust. Because these two treatments could have lasting carry-over effects, plots within a block should come from areas receiving only one of the prior treatments.

28.6.3.7 Outlining the analysis

At this point, the biometrician should outline the form of the analysis, usually by delineating the sources of variation and degrees of freedom in the analysis of variance. Are the underlying hypotheses and probable precisions associated with the F tests in line with experimental objectives? What are the biological and practical implications of finding either significant or nonsignificant results for each test? If the chances are high that hypotheses will not be tested precisely enough, an alternate design is warranted. If the new design requires more effort or is not feasible, perhaps the experiment should be delayed or cancelled.

28.6.3.8 Documenting the plan

The experimental plan is often documented in the form of a research proposal or study plan by outlining the problem objectives and proposed experimental design as already described (see 28.6.1 to 28.6.3). This allows peer review, aids analysis, and documents the experiment in case of personnel turnover. Methods of writing study plans vary widely, depending upon the level of formality required.

In addition to the written description, the experimental design itself and the field layout are best documented by an analysis of variance table and a schematic diagram of the field plot arrangement. The table describes the form of initial analysis and succinctly states hypotheses under investigation. The schematic diagram, essentially a map, shows the layout of the treatments as they have been randomized and assigned to nursery plots; often, the positions of measurement plots within experimental plots are shown, as are any local areas omitted from the experiment. These diagrams can be simple or detailed; cryptic schematics are shown in Figures 3 to 6. The schematic (1) reinforces the written description of the experimental design by explicitly depicting the assignments of treatments to plots from which the analysis of variance can easily be constructed, (2) is useful during the experiment for applying treatments and collecting data, and (3) allows plot means to be charted as they occur in the field, often revealing spurious local trends.

The importance of documenting the experimental plan cannot be overemphasized; yet, too often, the effort involved hinders executing the research. Each researcher must find the most suitable compromise. Handwritten notes on the objectives and the experimental plan, including a list of factors and variables to be investigated, and a schematic map of the field plot layout are a *minimum*.

28.6.4 Conducting the experiment

28.6.4.1 Employing proper technique

Employing proper technique means conducting the experiment in a manner maximizing both precision and accuracy. High accuracy is achieved by using properly calibrated machinery and experienced, observant workers with proven good judgment and by closely following the experimental plan. Precision is increased by uniform application of treatments, meticulous measurement technique, and, in general, care and common sense.

28.6.4.2 Using the experimental design

Return, for a moment, to the concepts of randomization and blocking (see 28.5). When possible, seedlings should be treated, lifted, and measured according to the randomization scheme documented in the schematic diagram. As an extreme breech of this, consider the bias potentially introduced by first lifting and measuring all replicates of treatment 1, then treatment 2, and so on. As the experiment progresses, lifting conditions may change and measurement techniques become more refined. These effects can be randomized over treatments, thereby minimizing bias among treatment comparisons, by adhering to the original randomization scheme in all phases of the experiment.

To maximize the benefits from blocking, treat and measure all plots within a block before moving to other blocks. If seedlings in an entire experiment cannot be sown, lifted, or measured on the same day, do different blocks on different days. Then, any day-to-day differences in conditions tend to average out, influencing all plots (treatments) within a block similarly. When possible, one team of observers should lift and measure all seedlings in plots within a block. If one worker lifts carelessly and measures trees in inches instead of centimeters, experimental accuracy and precision will be affected; however, any errors introduced will *not* bias treatment comparisons if that worker lifts and measures all treatments within a block because all treatments will have received the same poor technique.

28.6.5 Interpreting experimental results

After editing the data to eliminate data-collection errors, descriptive statistics (point estimates) must be calculated from the data set and inferences made about the population of interest (see 28.5.1). The analysis itself is beyond the scope of this chapter. But if the experiment is properly designed and conducted, the descriptive statistics should be unbiased, precise estimates of population parameters, and the level of uncertainty associated with those estimates and with tests of hypotheses should be low. Nevertheless, interpreting the practical implications of these statistical tests and inferences raises some problems and is nearly always the province of the researcher, not the biometrician. Three problems of interpretation are considered in this section.

28.6.5.1 Statistical significance

The nursery researcher must always interpret the practical significance of experimental results in light of their statistical significance. However, three different situations may arise in which the researcher relies on deductive reasoning based on knowledge and personal experience to question or even ignore the statistical inferences obtained from data analysis.

First, statistically significant differences among treatments may be too small to be practically important. This implies that the experiment was more sensitive (powerful) than required for that particular hypothesis or treatment comparison. For instance, if doubling the amount of N fertilization results in a statistically significant (say, at the 99% confidence level) increase in average 2+0 seedling height of 0.5 ± 0.2 cm, then the experiment was extremely sensitive. Though confident that this increase was *not* due to chance, the researcher may still decide that the small increase does not warrant the extra cost of the additional fertilizer.

Second—the reverse of the first case—a treatment comparison or hypothesis may not be statistically significant; yet the magnitude of the differences involved may be biologically or practically significant. Suppose that fall N fertilization of 2+0 seedlings results in a 20% increase in root-growth capacity and a 40% increase in early-winter frost hardiness, compared to the controls. If neither of these differences approaches statistical significance, one of two alternatives exists. Either the variable nature of the traits has resulted in large differences occurring by chance or a Type 2 error (β) is being made. Recall that Type 2 errors result when the experiment is not powerful enough to declare differences even though they, in fact, exist. When differences of practical importance are not statistically significant (say, at $\alpha = 0.05$), the experimenter can calculate the magnitude of differences required to approach statistical significance. If values of frost hardiness must differ by 100%, the researcher would question the power of the experiment and perhaps plan a better one.

Third, a statistically significant result may contradict biological principles or past experiences. In this situation, (1) results may be spurious (on the average, 1 out of 20 tests at $\alpha = 0.05$ will be incorrectly declared significant), (2) treatments may have been mislabeled, or (3) the biological reasoning may be flawed. The experimenter must be open to all eventualities and reexamine both the planning and conducting of the experiment and the biological theory underlying it to see where the fault lies. Sometimes, the statistically significant difference has a low range of validity—for example, when two treatments declared statistically different were tested in only one part of the nursery for a single growing season. This experimental design may lead to spurious results, especially if plot location or growing season were atypical. Such tests of significance require scrutiny.

28.6.5.2 Correlation versus causation

Experiments are most often conducted to establish cause-effect relationships of practical significance; that is, the presence of a certain level of a factor under investigation causes an identifiable response in a measured variable. The experiment is set up to determine these cause-effect relationships by specifically controlling the factor levels. However, correlations among variables not being controlled may also be found during experiments, and while useful, these must be interpreted with extreme caution.

For example, an experiment testing different levels of N fertilization may indicate that fertilized seedlings are significantly taller than controls. If soil P levels are also measured (but not controlled) on each plot, there may be a strong, statistically significant correlation indicating that high soil P levels are associated with faster growth. However, it is *not* correct to conclude that higher P levels *cause* faster growth. P may not be a limiting element for growth at all but may simply be indicative of (a proxy variable for) the level of organic matter on a plot. If more organic matter causes faster growth and produces more soil P, then the correlation between P and growth will be high even though no causality exists between the two variables. The pitfall of inappropriately assigning causality to such correlations cannot be overstressed.

28.6.5.3 Interactions

Statistically significant interactions among factors are common in the biological sciences. When more than two factors are involved, the practical interpretation of the interaction may be difficult; however, proper interpretation of two-factor interactions is essential to drawing correct conclusions from nursery experiments. A range in magnitude of these interactions can exist, and three different types are considered here.

For a two-factor experiment investigating the effects of three levels of fall N fertilization and two levels of irrigation on early-winter frost hardiness of 2+0 Douglas-fir seedlings, three hypothetical outcomes (Fig. 9) are considered. Suppose that low water levels have the consistent effect of "shutting seedlings down," causing them to enter dormancy early, and therefore increasing early-winter frost hardiness. If N aids this metabolic transition independent of water level, then no interaction between N and water exists (Fig. 9a). Now suppose that increasing N levels increases early-winter frost hardiness regardless of water level, but the rate of increase is faster when water levels are high. That is, N is more effective in the presence of high water levels (perhaps because the additional water is needed for N to better aid the physiological transition). This type of interaction—in which two factors affect each other, but trends within each factor are similar when plotted over a second factor—is called a scale effect (Fig. 9b). Note that the practical interpretation in both cases a and b is very similar: high N and low water levels result in superior early-winter frost hardiness.

But when one factor acts differently in the presence than in the absence of a second factor, levels of the first factor change their ranking, depending upon the level of the second factor. For example, high N levels may be more effective in the presence of high water levels, but low N levels may be more effective in the presence of low water levels (Fig. 9c). Perhaps too much N "burns" the seedlings and retards frost-hardiness development if seedlings are not well watered. This type of interaction, called rank change, makes it impossible to describe the effects of N without considering the particular water level applied and greatly alters the conclusions drawn from the experiment.

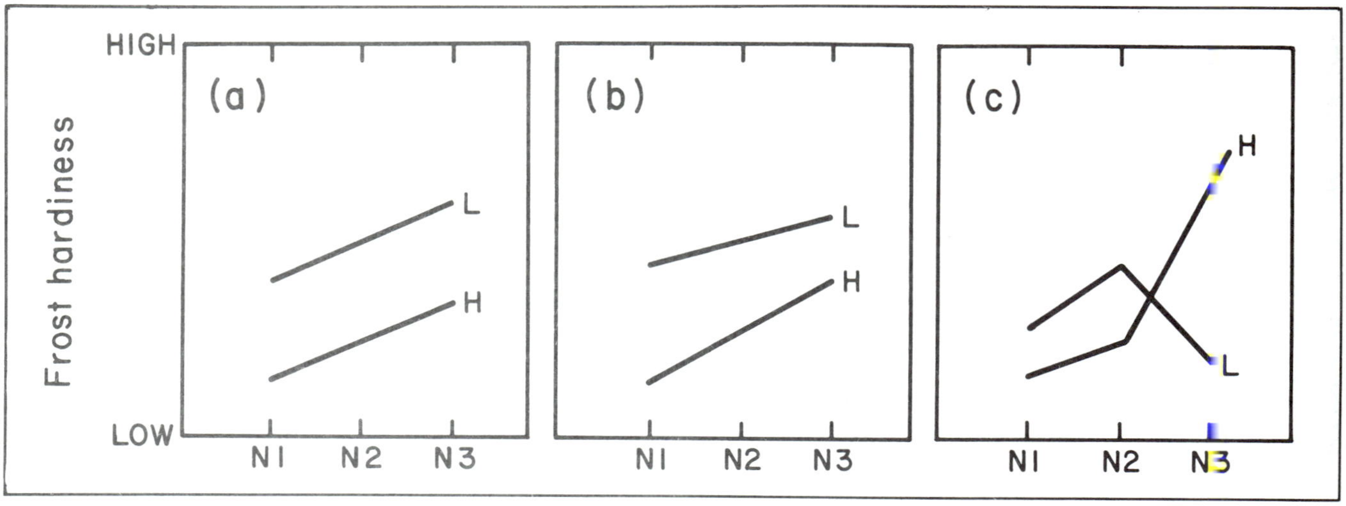

Figure 9. Hypothetical effects of two factors, irrigation level (H = high, L = low) and fall nitrogen fertilization (N1, N2, N3), on early-winter frost hardiness of 2+0 Douglas-fir for (a) no two-factor interaction, (b) scale-effect interaction, and (c) rank-change interaction.

28.7 Conclusions and Recommendations

The main goal of nursery research is to develop new techniques that produce high-quality seedlings in a cost-effective manner. After the initial steps of identifying a problem area and setting experimental objectives, the nursery researcher plans an experiment by choosing treatments to test and variables to measure, determining plot sizes, selecting an appropriate experimental design, determining the number of replications, and delineating the field layout (assigning treatments to plots). The experiment is then conducted in a manner to maximize precision and accuracy of experimental results. Finally, the results are analyzed and interpreted in light of the researcher's intuition and personal experiences, and recommendations are made to implement the conclusions.

Applying a very few statistical concepts (mainly randomization, replication, and blocking) in a common-sense manner can aid researchers at each step of the nursery research process. While implementing the design and interpreting the results, the researcher must always balance statistical with biological and practical considerations to achieve an experiment that meets its objectives with an appropriate expenditure of effort.

Acknowledgments

I sincerely thank the following people for their help with and technical reviews of this manuscript: Mary Duryea, Oregon State University; Jim Fischer, U.S.D.A. Forest Service; Henry Laik, International Paper Company; Tom Landis, U.S.D.A. Forest Service; Paul Morgan, State of Oregon; Susan Stafford, Oregon State University; and Barbara Thompson, International Paper Company.

References

1. Anderson, S., A. Auquier, W. Hauck, D. Oakes, W. Vandaele, and H. Weisberg. 1980. Statistical methods for comparative studies. John Wiley and Sons, New York. 289 p.
2. Anderson, V. L., and R. A. McLean. 1974. Design of experiments: a realistic approach. Marcel Dekker, Inc., New York. 418 p.
3. Andrew, C. O., and P. E. Hildebrand. 1977. Planning and conducting applied research. MSS Information, New York. 116 p.
4. Bancroft, T. A. 1964. Analysis and inference for incompletely specified models involving the use of preliminary test(s) of significance. Biometrics 20:427-442.
5. Cochran, W. G. 1976. Early development of techniques in comparative experimentation. Pages 3-25 *in* On the history of statistics and probability (D. B. Owen, ed.). Marcel Dekker, Inc., New York.
6. Cochran, W. G., and G. M. Cox. 1957. Experimental designs. John Wiley and Sons, New York. 611 p.
7. Cox, D. R. 1958. Planning of experiments. John Wiley and Sons, New York. 308 p.
8. Freese, F. 1967. Elementary statistical methods for foresters. U.S.D.A. Forest Serv., Washington, D.C. Agric. Handb. 317. 87 p.
9. Gomez, K. A. 1972. Techniques for field experiments with rice. International Rice Res. Institute, Los Banos, Philippines. 46 p.
10. Hamburg, M. 1974. Basic statistics. Harcourt Brace Jovanovich, Inc., New York. 451 p.
11. Kempthorne, O. 1973. Design and analysis of experiments. Robert E. Krieger, New York. 631 p.
12. LeClerg, E. L., W. H. Leonard, and A. G. Clark. 1962. Field plot technique. Burgess Publishing Co., Minneapolis, Minnesota. 371 p.
13. Little, T. M., and F. J. Hills. 1972. Statistical methods in agricultural research. Agric. Extension, Univ. of California, Berkeley. 242 p.
14. Mason, E. G. 1981. Improvement of nursery and establishment techniques. Pages 186-198 *in* Forest nursery and establishment practice in New Zealand (C. G. R. Chavasse, ed.). New Zealand Forest Serv., Rotorua. FRI Symp. 22.
15. Neter, J., and W. Wasserman. 1974. Applied linear statistical models. Richard D. Irwin, Inc., Homewood, Illinois. 842 p.
16. Sandquist, R. E., P. W. Owston, and S. E. McDonald. 1981. How to test herbicides at forest tree nurseries. U.S.D.A. Forest Serv., Pacific NW Forest and Range Exp. Sta., Portland, Oregon. Gen. Tech. Rep. PNW-127. 24 p.
17. Snedecor, G. W., and W. G. Cochran. 1967. Statistical methods. Iowa State Univ. Press, Ames. 593 p.
18. Steel, R. G. D., and J. H. Torrie. 1960. Principles and procedures of statistics. McGraw-Hill Book Co., New York. 481 p.
19. Westmeyer, P. 1981. A guide for use in planning and conducting research projects. C. C. Thomas, Springfield, Illinois. 116 p.

Chapter 29
Problem Solving in Forest-Tree Nurseries with Emphasis on Site Problems

T. D. Landis

Abstract

Problems are defined as the difference between "what is" and "what should be," and their definition is based on value judgments. Many production problems in Northwest bareroot nurseries are due to site; 68% of site problems are soil related, 25% climate related, and 7% water related. Many people rely solely on their own training and experience to solve problems, but integrating creative techniques with a systematic approach can reveal new solutions. A creative problem-solving system consists of five steps: the problem is identified and then analyzed, ideas are generated, hypotheses are developed and tested, and the resulting solution is implemented. Nursery managers can become better problem solvers by increasing their knowledge about nursery science and gaining direct and indirect experience in the field.

29.1 Introduction

We all experience problems in our daily lives. Many people develop their own techniques to solve problems reasonably effectively through personal experience and training. Because every problem is different, however, someone's "pet" problem-solving technique may not be the best or the most efficient for all problems. Methods based on trial and error, for example, may be useless in a crisis situation when time is at a premium.

Problems are nothing new to nursery managers. Administrative constraints, site deficiencies, and equipment breakdowns are just a few of the problems that occur daily in tree nurseries. Good managers realize that problems are a natural part of any operation and must be dealt with directly, quickly, and effectively.

This chapter discusses the nature of problems and looks in particular at site problems in Northwest nurseries and some of their solutions. The basic elements of creative problem solving are presented in the hope that practicing forest-nursery managers can use these techniques to improve their problem-solving abilities.

29.2 What Constitutes a Problem?

"Well, I tell you there's no problems, only solutions . . ."
—John Lennon

My favorite definition of a problem is **any situation in which there is a difference between "what is" and "what what should be"** [6]. This definition emphasizes the relative nature of all problems. Defining problems always involves value judgments—what is a problem to one person may not be to another. The values or objectives of an organization will at least partially define the nature of its problems.

A complicating factor is that the differences between "what is" and "what should be" are frequently dynamic. In the case of seedling quality, the "what is" aspect changes with the physiological and developmental status of the seedling during the growing season or with short-term changes in weather. The "what should be" aspect of seedling specifications changes from year to year and from customer to customer. Any time that these "differences" reach significant levels, a problem may arise.

29.3 Site Problems in Northwest Nurseries

Many production problems in bareroot forest nurseries are related to site. As biological systems, nurseries are susceptible to many problems because of the numerous uncontrollable variables that affect seedling production. This high level of variability is a principal source of differences between "what is" and "what should be."

In Duryea, Mary L., and Thomas D. Landis (eds.). 1984. *Forest Nursery Manual: Production of Bareroot Seedlings*. Martinus Nijhoff/Dr W. Junk Publishers, The Hague/Boston/Lancaster, for Forest Research Laboratory, Oregon State University, Corvallis. 386 p.

Although all nursery managers can describe the perfect nursery site, most realize that their particular site has certain deficiencies. Very often, nursery sites are selected for nonbiological reasons such as land cost or availability; these suboptimal locations can lead to site-related problems. Other problems can be attributed to the "off-site" nature of most forest nurseries; that is, many forest-tree seedlings are produced for high-elevation plantings, whereas most nurseries are situated at lower elevations to take advantage of the more level topography and longer growing season. In the mountainous terrain of the Northwest, many nurseries are located in river valleys where the variable alluvial soils can cause problems.

The intensive cultural practices of nurseries today also can lead to site problems. Heavy machinery used during the lifting season when soils are wet can damage soil structure and result in undesirable soil compaction. Frequent irrigation and heavy nitrogen fertilization can hasten organic matter decomposition and thus decrease soil productivity.

On the basis of the response frequency of Northwest nursery managers (OSU Nursery Survey; see chapter 1, this volume), soil-related problems account for 68% of the site problems in Northwest nurseries, climate-related problems for about 25%, and water-related problems for the remaining 7% (Table 1).

29.3.1 Soil-related problems

The large percentage (68%) of site problems related to soil (Table 1) reflects the critical importance of the soil component in forest-nursery operations.

29.3.1.1 Organic matter maintenance

The #1 rating of organic matter maintenance reflects both the perceived importance of organic matter in nursery management and the concern over a projected decrease in available and affordable sources (see chapters 7, 9, and 10, this volume).

Nursery managers are addressing this problem through the traditional techniques—adding amendments and growing a green manure or cover crop during the normal rotation (Table 2). The most commonly used organic matter amendments are raw materials such as sawdust or peat moss, although some alternative sources such as mint sludge are now being applied. Several nursery managers expressed skepticism about the ability of green manure or cover crops to actually increase the soil organic matter level. Several Northwest nurseries are also experimenting with composting organic materials such as sewage sludge before incorporation into nursery soil.

29.3.1.2 Poor internal drainage and soil compaction

Although they can be caused by different factors, these two soil conditions are considered together because the corrective treatments are similar. Poor internal drainage can be caused by soil compaction but also can result from naturally formed impermeable layers that often develop in fine-textured soils.

Subsoiling (deep ripping) is the most common treatment for soil compaction and also improves soil infiltration and percolation rates (Table 2). Subsoiling physically fractures the restrictive layers in the soil with tractor-drawn ripping teeth, usually during the fallow year. Some nurseries even rip the tractor paths between seedbeds during the rotation (see chapters 6 and 13, this volume).

Subsoil drainage systems can relieve drainage problems (Table 2), and surface ditches can control water runoff in nurseries with low infiltration rates and heavy rainfall.

Several cultural practices can help reduce soil compaction and increase internal drainage (Table 2). Incorporating organic matter into the soil profile will improve soil structure and retard the formation of hardpan layers. Grading and leveling nursery blocks and raising seedbeds can help drain surface soils.

Because nursery equipment is a major cause of soil compaction and resultant drainage problems, several nurseries mentioned corrective treatments involving equipment use (Table 2). Limiting the number of times that tractors enter a field and avoiding tractor entry during wet periods can reduce compaction in the tractor paths between seedbeds. Crawler tractors cause less compaction than wheel tractors. Wheel tractors can be equipped with special tracks that more evenly distribute tire pressure; dual wheels could be used when the field is not in seedbeds. Tilling equipment such as rototillers destroy soils structure and should not be used on fine-textured or poorly structured soils.

Table 1. Site problems in Northwest bareroot nurseries as rated by nursery managers (OSU Nursery Survey).

	Response frequency, %	x	Problem priority[1]	=	Priority rating	Top-5 ranking
Soil						
Acidity	4.1		2.9		12.0	
Alkalinity	1.0		2.0		2.0	
Salinity	0.0		0.0		0.0	
Too heavy	6.2		2.2		13.8	
Too light	2.1		3.3		7.0	
Too variable	6.2		2.4		15.0	
Soil compaction	13.4		1.9		25.0	3
Poor internal drainage	9.3		3.3		3[illegible].0	2
Rocks	4.1		2.9		12.0	
Organic matter maintenance	13.4		3.0		40.3	1
Soil splash	3.1		2.2		5.9	
Tilth	0.0		0.0		0.0	
Uneven topography	5.2		3.1		15.0	
% of Total	68.1					
Water						
Poor quality	1.0		3.0		[illegible].0	
High water table	2.0		1.0		[illegible].0	
Availability	4.1		2.7		1[illegible].2	
% of Total	7.1					
Climate						
Intense rainfall	2.1		2.8		[illegible].0	
Frost damage	6.2		2.9		1[illegible].0	5
Frozen soil	5.2		1.7		[illegible].0	
High temperatures	2.1		2.8		[illegible].0	
Wind damage	7.2		3.0		2[illegible].7	4
Late snowfall	1.0		3.0		[illegible].0	
Erosion	1.0		5.0		[illegible].0	
% of Total	24.8					
	100.0%					

[1]Based on 1 (negligible) to 5 (severe) rating.

29.3.2 Climate-related problems

Northwest nursery managers found 25% of their site problems related to climate (Table 1). Wind damage and frost damage were the most common, ranking fourth and fifth in overall importance.

29.3.2.1 Wind damage

This type of injury includes both abrasion from blowing soil particles and winter drying. Windbreaks, either vegetative or mechanical, were the most commonly listed treatment for wind protection (Table 2). Standard woody-plant windbreaks

Table 2. The five most important site-related problems in Northwest nurseries and their current remedies (OSU Nursery Survey).[1]

Problem	Remedy
Organic matter maintenance	Raw organic amendments (sawdust, peat moss, mint pulp, sludge) Green manure or cover crops Composed organic amendments (sludge and commercial mixes)
Poor internal drainage soil compaction[2]	Subsoiling (deep ripping) Subsoil drainage systems Added organic matter to improve soil structure Surface ditches to control runoff Limited tractor use, especially on wet soils Use of wide tires or tracks on tractors Subsoil tractor paths during rotation Raised seedbeds Avoid machinery (e.g., rototiller) that destroys soil structure Land leveling Avoid growing tree crops in problem areas
Wind damage	Windbreaks Snowfence along irrigation lines Corn barriers in fields
Frost damage	Irrigation to protect succulent seedlings Sawdust mulch for frost heaving Perimeter vegetation removed to promote air movement

[1]Problems and remedies ranked according to relative importance; see also Table 1.
[2]Considered together because corrective treatments are similar.

are effective but occupy a considerable amount of growing space and may serve as sources of disease inoculum (see chapter 19, this volume). Some nurseries string snowfencing along the irrigation line between the seedbeds to reduce wind exposure; however, snowfences can interfere with the distribution pattern of the irrigation system (see chapter 11, this volume). One nursery used rows of corn to reduce wind exposure in seedbeds but reported only minimal effectiveness.

29.3.2.2 Frost damage

Both frost heaving and freezing injury are included in this category. Irrigation for frost protection was the most commonly listed solution for frost injury to succulent seedlings (Table 2), but its perceived effectiveness varied considerably, however, probably due to differences in technique (see chapter 12, this volume).

Frost heaving is most common with small seedlings or recent transplants in fine-textured soils. Winter mulches of materials such as sawdust provide protective insulation at the soil surface and reduce the damaging sequence of alternating periods of freezing and thawing. Removing woody vegetation surrounding the nursery promotes cool air drainage and helps eliminate frost pockets.

29.3.3 Water-related problems

Water-related problems were least troublesome to Northwest nurseries (Table 1). Poor water quality or a high water table were not common problems, but water availability was of some concern. One nursery was connected to a domestic water source which increased water cost and sometimes restricted availability. Slow recharge of irrigation wells, another problem, was remedied by drilling additional wells. Water availability also can be limited during winter, which is a problem when it is required for frost protection.

29.4 Problem-Solving Techniques

The soil-, climate-, and water-related problems (see 29.3) faced by Northwest nursery managers and staff are varied and complex. Simple solutions rarely satisfy because problem conditions are often interrelated; eliminating one problem may heighten or even create another. Managers must rely on their own experience and analytical skills, and many have developed their own personal problem-solving techniques. However, those who rely on the more conventional approaches (see 29.4.1) probably will have less success than those who try to solve problems creatively (see 29.4.2).

29.4.1 Conventional problem-solving approaches

As already mentioned, many managers have developed their own personal problem-solving techniques. Before we discuss more scientific problem-solving methods, let's look at some of the more popular approaches.

Those who use the **ostrich approach** ignore problems in the hope that they will go away. Some problems do seem to solve themselves, or, if ignored long enough, may be solved for us. More often, though, problems that are ignored become even more serious or spawn a second generation of problems.

The **panacea approach** is the universal application of a "tried and true" solution without regard to its suitability for different problem situations [6]. This approach is a favorite of experienced managers who have achieved positive results in the past but who overlook the variable nature of most problems and the advent of new technology.

People who use the **shotgun approach** do not take the time to approach problem solving systematically but believe that if enough solutions are tried, one of them should surely work [6]. Many people, when confronted with a problem, feel that it is best to "do something" as quickly as possible; the danger is that some of these haphazard solutions may actually make the problem worse.

29.4.2 Creative problem solving

Creative problem solving can be defined as the incorporation of creative processes into a systematic approach for solving problems.

29.4.2.1 The creative process

"Genius is the capacity for seeing relationships where lesser men see none."
—William James

Some people think of creativity as an artistic attribute and do not associate it with science or technology. Actually, truly revolutionary scientific theories result from creative thinking. In developing his theory of relativity, Einstein used the abstract concept of imagining himself riding on a beam of light [4].

Even though everyone is familiar with the concept of creativity, it still has no generally accepted definition [4]. Creative people are often at a loss to explain their special talent. Yet, in spite of their inability to explain or define it, most people recognize creativity when they see it. For our purposes here, creativity can be thought of as the ability to develop fresh insights about situations and formulate innovative ways of dealing with them.

Campbell [2] views the creative process as a series of separate but sometimes overlapping mental phases:

- The gradual, long-term process of accumulating and updating knowledge from both formal training and personal experience forms the basis for creativity. The more information we have about a particular problem, the better we will be able to solve it. This **preparation** phase never ends because new knowledge that may be relevant to future problems is constantly being generated.

- Once enough information has been gathered, all aspects of the problem are carefully analyzed. The length of this period of **concentration** will depend on the complexity of the problem and the amount of information available.
- Next, the problem should be temporarily abandoned for an unspecified period of time to allow the unconscious mind to mull over the details. The custom of "sleeping on it" before making a decision exemplifies this phase. Unfortunately, this **incubation** process is frequently neglected because many people become obsessive when dealing with a problem and think that they can solve it only by intense and continuous concentration or immediate action.
- Flashes of insight—often symbolized as a brightly shining light bulb in the comics—are the most familiar phase in the creative process. Most inventors and other innovative people have experienced these sudden insights, which often reveal previously unknown relationships. Although this **illumination** phase is the most exciting aspect of creativity, it is virtually impossible without proper preparation.
- Finally, the newly conceived ideas are tested to determine if they really solve the problem. During this **verification** process, many initially attractive ideas are found to be faulty upon closer inspection.

Although many techniques have been developed to stimulate creativity in problem solving, the underlying principle of each is to temporarily suspend critical judgment while developing the widest range of ideas [4].

29.4.2.2 Roadblocks to creativity

"Everyone is a prisoner of his own experiences. No one can eliminate prejudices— just recognize them."

—Edward R. Murrow

Some people are naturally creative, but most of us have to work at it. Unfortunately, the human mind has several inherent processes that inhibit creative thinking;

- Most people develop a certain fixed way of thinking based on their previous knowledge and training [1]. Once a thought process is formed, it is usually very difficult to overcome. Most professional groups are guilty of such **conditioned thinking**—and nursery managers are no exception. Realizing this common pitfall is the first step in dealing with it.
- Having committed an error once, we often have an unconscious tendency to repeat the error again and again [1]. Apparently, the human mind is unable to detect these **persistent errors.** Often, a fresh perspective—or someone else double-checking our work—is needed.
- **Functional fixedness** is the tendency to see only one use for an object. Campbell [2] calls this the "inability to consider uncommon uses for common objects." Unfortunately, the more highly specialized a person's field is, the more likely that person is to fall victim to this trait. People who are good with hammers see every problem as a nail [2].

29.4.2.3 Overcoming roadblocks to creativity

"Facts do not cease to exist because they are ignored."

—Aldous Huxley

Realizing that roadblocks to creative problem solving exist, we can take measures to counteract them by rethinking the problem, discussing it with other people, or abandoning it temporarily. Rethinking the problem requires starting at the very beginning and developing a new perspective. This is often very difficult to do because most people are accustomed to looking at a situation from only one angle. Writing a review of the problem is sometimes helpful because the physical process of translating concepts into written words can provide new insights. Discussing the problem with other people—particularly people not directly involved—can also be a good way of obtaining new perspectives. Temporarily abandoning a problem forces complete detachment from it for a few days. This mental break may allow new ideas to surface by permitting the unconscious mind to consider other alternatives or to put all aspects of the problem into their proper perspective.

29.5 Five Steps to Creative Problem Solving

Creative problem solving incorporates creativity into a basic problem-solving system (Fig. 1) comprising five steps: (1) identifying the problem, (2) analyzing the problem, (3) generating ideas, (4) developing and testing hypotheses, and (5) implementing a solution. Adopting a standard system is essential to preclude the testing of possible solutions before the real problem has been identified.

This five-step system can best be illustrated by following an actual nursery problem through all the steps. Our sample problem is a nutritional disease, characterized by irregular patches of stunted and chlorotic seedlings, which is encountered in bareroot nurseries containing areas of alkaline or calcareous soil.

Before actual problem solving begins, however, make sure that a real problem exists—some **apparent** problems can be resolved by merely taking a closer look at the situation.

29.5.1 Step 1—Identifying the problem

"Trouble that is easily recognized is half-cured."

—St. Francis de Sales

A problem has to be identified before it can be solved. Problem identification requires knowledge and experience because a manager must know what is right before being able to recognize what is wrong: nursery managers must know what a healthy seedling looks like before they can identify a sick one.

Managers must be observant and open minded. They must become sensitized to the differences between "what is" and "what should be"; because problems often develop gradually, these individual differences may go unnoticed until the situation reaches a critical level. Problem identification is also subject to changes in the state of knowledge about an operation. An increased understanding of a certain procedure can expose problems where they either did not exist before or lay unseen.

In our sample illustration, it was clear that a severe problem existed. We had no difficulty determining that a sizable portion of the seedlings was so stunted that those seedlings would not reach merchantable size by the end of the rotation.

29.5.2 Step 2—Analyzing the problem

"Thinking a problem through is hard for the untrained mind."

—Anonymous

Problem analysis begins with the development of a clear statement about the problem. Once identified, the problem should be described as accurately as possible; the terms *what,*

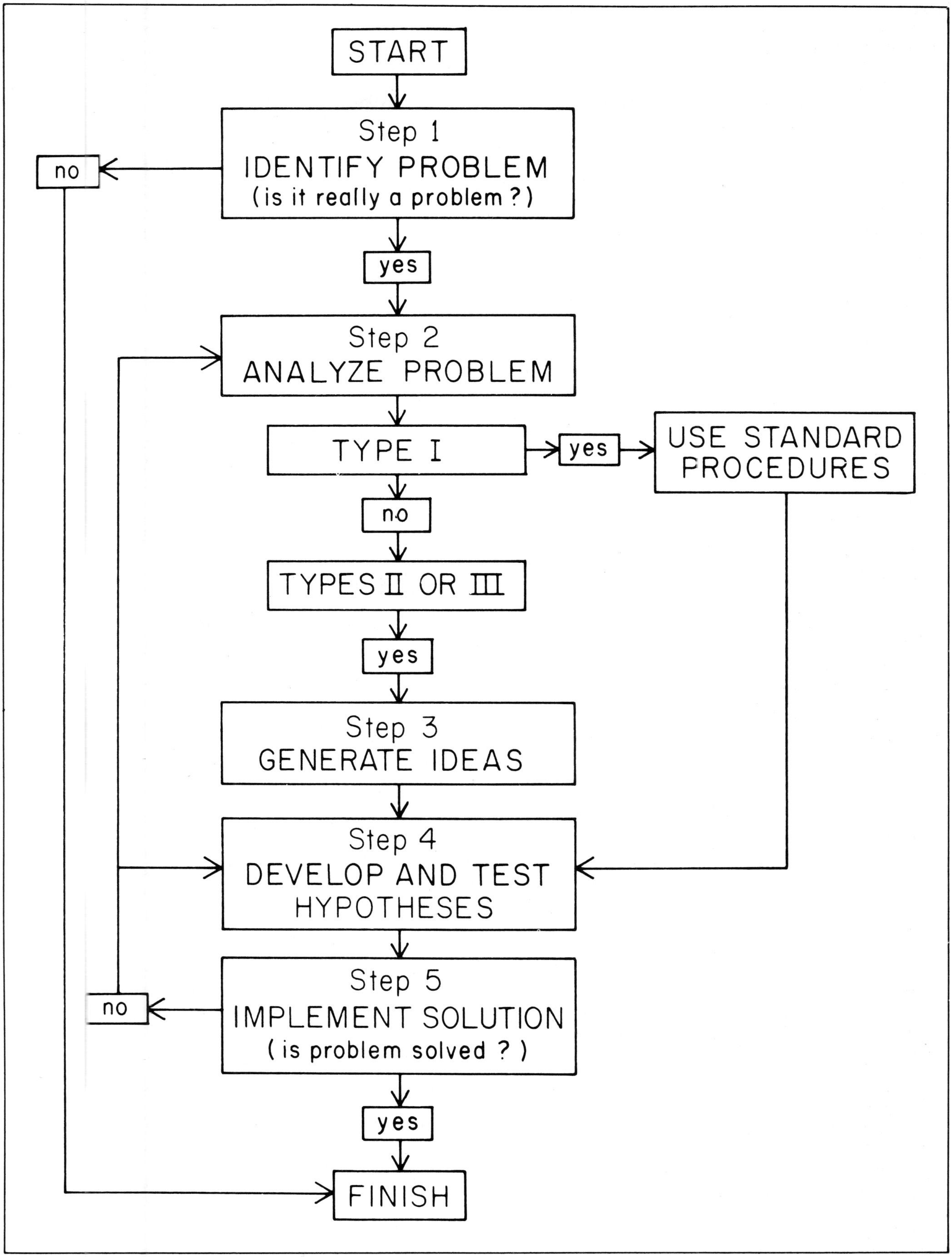

Figure 1. Flow chart illustrating a basic problem-solving system.

when, where, and *how much* are often helpful. Make a list of knowns and unknowns to order your observational data in some way. Carefully delineate the boundaries of the problem before attempting to solve it.

Try to observe with an impartial, open mind and not to confuse symptoms with causes. Because it is impossible to observe everything closely, be discriminating; try to identify the significant characteristics. Often, the exceptional phenomenon is the critical element and can lead to the explanation of the usual [1]. Double-check to be sure that the **stated** problem is the **real** problem; too often it is not. Furthermore, the real problem can be easier to solve than the stated one because it is almost impossible to solve a poorly diagnosed problem [3].

Once the significant information has been gathered and organized, the problem should be ranked in terms of importance, urgency, and change [5]:

- The **importance** of the problem will dictate whether it is worth solving, assuming it is solvable at all [3]. Specify the available resources (money, personnel, time) that can be expended on the solution; some problems cannot be solved economically.
- The **urgency** of the situation will determine whether it must be dealt with immediately or can be postponed. Consider the amount of time that can be allotted to the given problem.
- The **change** in nature, if any, of the problem also must be evaluated. Determine whether the problem is getting better or worse or remaining the same; a situation that is deteriorating will be more threatening than one that is improving.

The problem should next be classified into some sort of order. Van Gundy [6] uses a functional approach, separating problems into three types based on the amount of information available.

- Type I problems are **well structured.** These are the routine problems that occur daily. Their main characteristic is that all the information needed to solve them is already available. Problems in this category have probably occurred before and can usually be solved by standard procedures. Expertise for solving these problems can normally be found at the nursery, so outside help is not required.
- Type II problems are **semistructured.** This is an intermediate category—some information about the problem is available, but some degree of uncertainty also exists. These problems may have occurred before but something about them makes them different. Existing techniques must be adapted to solve this type of problem, and some expert help may be needed. The final solution is probably a combination of standard and newly developed methods.
- Type III problems are **poorly structured.** Their distinguishing characteristic is that little or no information is available about them. These are the problems never encountered before; therefore, expert help should be sought and the information needed to solve these problems generated through the problem-solving process. Solutions to poorly structured problems usually have to be custom made and require creative problem-solving techniques.

The effect of each of these three problem types is illustrated in Figure 1. Type I problems are usually solved with standard operating procedures, whereas Types II and III require more creative steps.

We diagnosed our sample problem as a Type III because it was a new situation we knew little about. The stunted, chlorotic seedlings were restricted to specific areas in the nursery; some nursery managers reported that they had observed the same condition in those same areas in previous crops. Everyone agreed that the problem was serious and should be dealt with immediately. The disease did not appear to be getting worse but would most likely reappear in future crops in susceptible areas of the nursery. Because the type of problem was new, we planned to consult nursery experts.

29.5.3 Step 3—Generating ideas

"In every work of genius, we recognize our own rejected thoughts."

—Ralph Waldo Emerson

29.5.3.1 Information gathering

A good knowledge base—the primary prerequisite for the creative process—can usually be obtained from nursery literature, staff discussions, and experts in the forest-nursery field.

Nursery literature includes manuals, technical books, and research publications. Publications in the fields of agronomy and horticulture or other related sciences can be valuable sources of new ideas; many of the cultural practices now used in tree nurseries were originally developed for other crops and later converted for use in tree-seedling nurseries. Older nursery publications should not be ignored because many "outdated" ideas may be able to be modified for solving the problems of today.

The nursery staff is a valuable source of information. Many of these people have accumulated a considerable amount of experience over the years. By presenting a problem at a staff meeting, nursery managers can benefit from a variety of different experiences and gain valuable new perspectives of the situation.

It is important to realize that one single source of information may not provide the solution to a problem. More often, information from a number of separate sources must be synthesized to generate new ideas. As with medical problems, it is often wise to solicit a second opinion. The amount of time and effort that can be dedicated to information gathering depends, of course, on the importance and urgency of the problem (see 29.5.2).

29.5.3.2 Creative techniques

Ideas can be generated by either single individuals or groups. Group sessions have the benefit of a variety of people with different perspectives, and the interaction of experts and untrained individuals can sometimes result in innovative ideas [4]. Groups that contain individuals of different status in an organization, however, can actually stifle creative expression because lower ranked employees may feel intimidated.

In contrast, certain problems are better suited for solution by individuals because some highly trained people may feel restricted or encumbered by group approaches to problem solving. Creativity may actually be stimulated in isolation due, perhaps, to a sort of sensory deprivation phenomenon [4]; most people would agree that creativity is inhibited in an atmosphere filled with distractions.

Brainstorming is a creative technique that can be used by either individuals or groups and deliberately encourages irrational thinking to produce a wide range of ideas. This process, however, can be quite difficult for highly trained individuals who have been "programmed" into a set pattern of logical thinking and find it hard not to judge ideas prematurely.

Individuals or groups can sometimes use analogies to generate new ideas. This creative technique forces us to look at the parallels or similarities between objects or situations to generate new perceptions about them. Very often, analogies prove effective in circumventing conditioned thinking (see 29.5.2).

Keep in mind that being creative does not always require discovering new facts—it often relies only on seeing new relationships between already known facts. A new use for a piece of already existing equipment is such an example.

We gathered information about our sample problem by first carefully observing the symptoms of the stunted, chlorotic seedlings. The newly developed secondary needles were generally more chlorotic than the older needles. We collected and analyzed soil and foliage samples from both normal and symptomatic seedlings for all nutrient elements. Consulting both the standard nursery literature and that on general plant nutrition added to our information base. The nursery experts provided diagnoses based on their assessments of the situation. Armed with a good perception about the seedling disorder, we were prepared to weave our ideas into a hypothesis.

29.5.4 Step 4—Developing and testing hypotheses

"One of the great tragedies of science is the slaying of beautiful hypotheses by ugly facts."

—T. H. Huxley

Once generated, a list of ideas must be evaluated and decisions made so that ideas can be converted into hypotheses. In most cases, information will not point clearly to one hypothetical solution, and managers will have to make decisions based on incomplete evidence (see chapter 28, this volume). However, most decisions are made under some degree of uncertainty because all the facts are never known [3].

A manager must keep an open mind during the evaluation process and take time to consider all aspects of the situation. The most obvious solutions are not always the best, and once an opinion has been formed, it is more difficult to think of alternatives. Beveridge [1] cautions us to beware of ideas that seem obvious and are accepted without question. In evaluating various ideas, it is important to consider all possible consequences so that the solution to one problem does not generate a new problem.

Managers must accept the fact that some ideas simply are not practical operationally. However, ideas that initially seem impractical may be able to be modified to a more practical form. Ideas should be judged in the light of all their attributes until testing is completed.

Hypotheses are only possible explanations and need to be empirically tested under actual conditions. Beware of the natural tendency to adopt an attractive hypothesis regardless of any data to the contrary. Most people are inclined to judge in the light of their own experience, knowledge, and prejudices rather than on the actual evidence [1]. Initially, hypotheses may not provide a complete solution and may need to be modified. Beveridge [1] warns about adopting an inflexible "all or nothing" attitude whereby a hypothesis that does not provide a complete solution the first time is discarded. Remember that hypothesis testing takes time. If a problem requires an immediate solution, implementing an interim procedure may be wise until adequate testing can be completed.

In our sample problem, we hypothesized that the problem was a micronutrient deficiency. The chlorosis of the younger needles indicated that the deficient element was immobile in the seedling. Soil tests showed that the soil in the affected areas was alkaline or that high levels of calcium were present. Foliar analysis revealed that whereas some micronutrient concentrations were lower in the chlorotic seedlings, iron concentration was usually high compared to that in normal seedlings. The soil-fertility literature stated that many micronutrients are unavailable to plants in soils with a high pH and that a nutritional disorder called iron chlorosis was common when conifers were planted on alkaline or calcareous soils. Even though some of the evidence was contradictory, we hypothesized that our seedlings had an iron deficiency caused by high soil pH or excessive soil calcium levels.

To test this hypothesis, we decided to apply a specially formulated iron-chelate fertilizer to the diseased seedlings. The liquid fertilizer was applied over the seedbeds as a spray that could be absorbed either directly into the foliage or through the roots. Although seedling response was variable, we had generally favorable results. The foliar chlorosis was alleviated, which supported our hypothesis that the seedlings were iron deficient.

29.5.5 Step 5—Implementing a solution

"A man's legs must be long enough to reach the ground."

—Abraham Lincoln

The last step in the problem-solving system is testing the hypothesis operationally. Some hypotheses may seem adequate on an experimental basis but may fail under operational conditions.

Once a hypothesis has been tested and implemented, a decision must be made as to whether the problem is completely solved. If the hypothesis provides an acceptable solution to the problem, then problem solving is complete. If not, then it is necessary to return to Step 4 to develop an alternative hypothesis or to Step 2 to reanalyze the problem (Fig. 1). Several different hypotheses may need to be tested before an acceptable solution is found.

In our sample problem, the fact that the chlorotic seedlings did respond favorably to iron fertilization was not proof that the overall problem was solved. The special chelate fertilizer was very expensive and, even though the foliar chlorosis was cured, the affected seedlings were still too small to make shippable grade. We needed to develop a permanent, practical, and economical solution to the iron-chlorosis problem.

After returning to the nursery literature and discussing the situation with our technical experts, we designed a long-term soil-amendment program to help prevent iron deficiency. We planned to (1) add sulfur and sawdust during the fallow year of the crop rotation to lower soil pH and help reduce the adverse effects of high soil calcium and (2) incorporate or band the iron-chelate fertilizer into the seedbed before sowing to make the fertilizer available to young seedlings and prevent stunting and chlorosis.

29.6 Becoming Better Problem Solvers

"Nature never overlooks a mistake, or makes the smallest allowance for ignorance."

—T. H. Huxley

The basic role of management is to achieve certain specified objectives. The objective in a tree-seedling nursery is obvious: to produce a specific number of healthy seedlings on a given date and at a reasonable cost. Most problems in tree nurseries arise when this objective is not met, either directly or indirectly.

Because problems are a predictable consequence of any operation, managers should attempt to become more adept in the art of problem solving. Nursery managers can become their own problem-solving experts by gaining the knowledge and experience necessary for making those intuitive associations that provide shortcuts to solutions.

Nursery managers must realize that learning is a continuous process but that they can never learn enough about the technical aspects of their operation. New information is constantly being generated, and managers must attempt to stay abreast of new developments.

Pure knowledge about nursery science is not enough, however; it must be tempered by actual field experience. Experts can take shortcuts in problem solving because their knowledge is functionally organized; such organization derives from a great deal of practical experience [4]. Experience can be acquired directly through time on the job, or indirectly, through visits to other nurseries and discussions with other nursery managers. Experience can either help or hinder the creative process, however; for example, it can lead to functional fixedness, which can stifle creativity (see 29.4.2.2).

An excellent (non-nursery) example of the benefit of actual experience in problem solving was the recent success of the Gossamer Condor, a light-weight human-powered aircraft, in navigating a difficult figure-8 pattern course and winning the Kremer prize. The Kremer prize of $86,000 had been unattainable for almost two decades until a Californian named Paul MacCready decided to try a novel approach. Instead of working from "high-tech" engineering design and aeronautical theories, he built a craft out of cheap, available materials. Because the Condor could be repaired quickly, MacCready was able to launch nearly 500 test flights using 12 significantly different models of the aircraft in a little over 1 year. He had found a way to gain experience quickly and inexpensively "instead of merely applying the logical consequences of theory" [7].

Planning is one of the most effective techniques in good problem management. Although problems cannot always be avoided, their effect can be minimized if they are planned for. Yet nursery managers must be flexible and realize that plans may need to be modified during the season. Good plans include contingencies that outline alternative management strategies for circumventing problems as they are encountered. Many problems lose much of their initial impact if a manager deals with them quickly and efficiently.

Realizing that problems are to be expected—and that they can be handled—can make a manager's job much more enjoyable. Problem solving, though frustrating, can also be very rewarding.

29.7 Conclusions and Recommendations

- All problems are relative because their definition depends upon value judgments. Problems in forest-tree nurseries are dependent upon the objectives and expectations of the nursery manager.
- Site-related problems are some of the most common and damaging problems in Northwest nurseries. Over ⅔ of site problems are soil related and about ¼ climate related; water-related problems are of minor consequence.
- Creative problem solving is defined as the incorporation of creative processes into a systematic approach to problem solving. Creative techniques can provide new insights into the nature of a problem and lead to novel solutions. Many specialized people, including nursery managers, have a fixed way of thinking based on previous training and must be aware of such possible roadblocks to creativity.
- Successful problem solving requires a systematic approach: a problem must be identified and analyzed, ideas for solving it generated, hypotheses developed and tested, and the proposed solution implemented operationally. It is important to approach a problem methodically and take each step in proper sequence.
- Nursery managers should strive to become better problem solvers by increasing their knowledge base and gaining as much direct or indirect experience as possible. Because knowledge about tree-nursery management is continually increasing, managers must try to keep abreast of new developments by attending nursery workshops and technical meetings and by reviewing the literature. Although direct on-the-job experience is invaluable, experience can also be gained indirectly by visiting other nurseries and discussing technical matters with other nursery workers.

References

1. Beveridge, W. I. B. 1957. The art of scientific investigation. Vintage Books, New York. 239 p.
2. Campbell, D. 1977. Take the road to creativity and get off your dead end. Argus Communications, Niles, Illinois. 136 p.
3. Furuta, T. 1978. Environmental plant production and marketing. Cox Publ. Co., Arcadia, California. 232 p.
4. Hunt, M. 1982. The universe within. Simon and Schuster, New York. 415 p.
5. Rice, C. A. 1981. Rational problem solving—a training course manual. Programs in Communication, Inc., Boulder, Colorado. 51 p.
6. Van Gundy, A. B. 1981. Techniques of structured problem solving. Van Nostrand Reinhold, New York. 301 p.
7. WGBH-Boston. 1983. Children of Icarus. Pages 196-201 *in* Nova—Adventures in Science. Addison-Wesley Publ. Co., Reading, Massachusetts.

Chapter 30
Nurseries in the Northwest: A Unique Opportunity for Improving Forest Yield

J. C. Gordon

Abstract

Nursery management represents the greatest concentration of technology and investment in the forest growth cycle and rivals wood processing in capital and labor intensity. Northwest nursery managers are thereby in a unique position to increase forest yield by (1) using environmental control techniques unavailable elsewhere in forestry operations to tailor their product, (2) by taking advantage of diverse and productive Northwest conifer species, and (3) by producing stock types adapted to specific site conditions to improve regeneration success. Current nursery research and information transfer have been inadequate to fully realize potential yield increases because of (1) poor communication between researchers and nursery managers and (2) lack of stimulating and well-articulated goals. Future challenges to nursery managers will result from (1) the use of genetically improved seed, (2) the need to produce a wide array of species, and (3) the introduction of new systems and concepts for growing bareroot seedlings.

30.1 Introduction

Nursery management represents the greatest concentration of technology and investment in the forest growth cycle and rivals wood processing in capital and labor intensity. On an area basis, any effect on subsequent seedling or tree growth wrought by nursery decisions and investment is multiplied many fold (Table 1). This extraordinary leverage given to nursery managers' decisions not only makes those decisions the focus of land managers' attention but also provides an unusual (in forestry) opportunity for using research to improve productivity. Furthermore, because the nursery production cycle is short relative to many other forestry operations, the impacts of managers' decisions usually can be quickly seen, and the effects of new research information applied to nursery operations can be rapidly evaluated. Precisely the same factors that make nurseries logical and potentially profitable places to improve yield make them good places to concentrate research investment.

Table 1. Small acreages in the nursery produce many thousands of seedlings to be outplanted on many-times-larger acreages. Data from OSU Nursery Survey (see chapter 1) and U.S.D.A. Forest Service [16].

	Nursery area, acres (ha)	Seedling production, 1,000s	Field area planted, acres (ha)
Oregon	1,745 (727)	92,554	220,987 (92,078)
Washington	1,209 (504)	103,939	147,504 (61,460)
Total	2,954 (1,231)	196,493	368,491 (153,538)

30.2 Management Opportunities

Northwest nursery managers can manipulate a small area—the nursery—to significantly increase productivity over the much larger area to which seedlings are outplanted (1) by taking advantage of environmental control techniques within the nursery and of the productive conifer species native to the Northwest and (2) by selecting for stock types adapted to specific site conditions.

Environmental control is routinely greater in the nursery portion of the production cycle than elsewhere (Table 2). Although control of weeds and animal damage is typical for both nurseries and field sites, the considerable biological, chemical, and physical manipulation of soil and careful monitoring of water levels through irrigation and land drainage are commonplace only in nurseries.

Table 2. Application of intensive practices in Northwest nurseries (OSU Nursery Survey) and on even-aged, Northwest field sites planted to Douglas-fir.

Practice	Nursery	Field site
Mechanical cultivation	C[1]	U
Chemical weed control	C	C
Hand weed control	C	U
Soil fumigation	C	N
Microbiological inoculation	U	N
Land leveling	C	N
Organic amendment	C	U
Fertilization	C	C
Crop rotation	C	U
Irrigation and drainage	C	U
Animal-damage control	C	C
Precise stocking control	C	U

[1]C = common, U = uncommon, N = rarely or never.

The major native conifer species of the Northwest are uniquely large. Individuals may reach the greatest sizes and standing volumes found in any natural ecosystem [18], and natural stands have the greatest productivities (highest rates of biomass accumulation over time) [7]. Thus, these Northwest conifers (and hardwoods) have diverse genetic potential for

In Duryea, Mary L., and Thomas D. Landis (eds.). 1984. *Forest Nursery Manual: Production of Bareroot Seedlings.* Martinus Nijhoff/Dr W. Junk Publishers, The Hague/Boston/Lancaster, for Forest Research Laboratory, Oregon State University, Corvallis. 386 p.

use in production forestry, although their relatively slow juvenile height-growth rates make nursery production cycles longer than those in the southeastern U.S. If nursery practices can be developed to speed juvenile growth of Douglas-fir [*Pseudotsuga menziesii* (Mirb.) Franco] without sacrificing survival ability, then production cycles in the nursery could be shortened from 2 to 3 years to 1 to 2 years, with attendant lower costs, and faster juvenile growth rates in the field would result in lower relative investments in competition control and higher productivity over a fixed rotation length.

Producing stock types better adapted to site conditions will ultimately affect harvest scheduling and allowable-cut calculations by improving regeneration success. For example, in a 5-county area in southwest Oregon, it is estimated that 270,000 acres are withdrawn from the Bureau of Land Management's (BLM) allowable cut base alone because of inability to regenerate them within 5 years after harvest [9]. Similar acreages probably are withdrawn on other public and private lands within the same region. The BLM withdrawal alone results in a direct current loss of stumpage receipts to the area of at least $25,000,000 annually. Hobbs [10] and other researchers have demonstrated the role of stock type and seedling quality in successful regeneration of southwest Oregon problem sites. Obviously, producing nursery stock types better adapted to these sites could have a large, immediate economic effect. Nor is this opportunity restricted to areas usually regarded as difficult or less productive. The Willamette National Forest currently has 85,000 acres withdrawn from its standard component because of regeneration difficulties [pers. commun., 11]. Though not all regeneration problems are traceable to nursery practices, successful artificial regeneration by planting depends absolutely on the availability of hardy nursery stock adapted to specific site conditions and objectives [4].

Obviously, large gains in current harvest levels depend on our ability to produce suitably adapted nursery stock. If this "allowable cut effect"—the effect on harvest of replacing withdrawn acres—were calculated, with the appropriate multipliers, for the entire Northwest, a very large number would result. Any failure to regenerate stands immediately after harvest can have sizable costs; Brodie and Tedder [3] emphasize that, at today's prices for forest products, efforts at reducing regeneration delay will probably be cost effective. The importance of replacing withdrawn acres and reducing delays on timber supply can scarcely be overstated because of the seriousness of countervailing influences such as loss of forest land to urbanization and wilderness. Thus, it is crucial that nursery management make its maximum contribution to sustaining harvests through prompt, efficient regeneration.

30.3 Nursery Research and Development

Because nurseries present such an obvious opportunity for effectively improving yields, one would suppose them to be a major focus of research and development. However, a recent report [13] indicated that nursery research needs were far from being adequately met; indeed, many nursery managers surveyed felt such needs were quite low on researchers' and specialists' priority lists. Relative to field regeneration and stand management, very little research in Oregon and Washington has been aimed directly at nurseries; moreover, much of the work done has been carried out as a "spare time" activity by people—such as nursery managers—who already have too much to do.

Although no really good figures are available, probably no more than 1% of the total research expenditure on Northwest public forestry is directed at nursery practices and problems [pers. commun., 14]. I believe this scandalous situation springs from two important causes:

- **Poor communication between the forestry research community and nursery operators:** Few forestry researchers, as with foresters generally, are trained in nursery operations. In this instance, *un*familiarity breeds contempt. Researchers tend to view economic payoff and professional recognition as resulting from silvicultural activities outside the nursery. Unfortunately, nursery managers have not spoken with a unified voice, as have, for example, people interested in forest fertilization and genetic tree improvement. The latter case is particularly ironic—because nursery effectiveness is an absolute constraint on genetic tree improvement as it is currently practiced; if "superior" seedlings are not raised and handled effectively, investment in selection and breeding programs is futile.
- **Lack of clearly stated "blue sky" targets:** Nursery needs have failed to excite the imagination of those capable of directing investment in research and development to nursery problems. Everyone's attention is focused when publicity is given to substantial projected improvements in yield, such as the percentage figures often quoted for genetic tree improvement and fertilization [1, 12]. Really imaginative nursery-related concepts have been lacking or have not been widely publicized and examined, partly because research on areas closely related to nursery practice has been relatively slight. For example, several comprehensive models of seedling development have been proposed (e.g., [pers. commun., 15]), but none, to my knowledge, has been pursued to completion.

The responses to the OSU Nursery Survey (see chapter 1, this volume) clearly indicate which problems nursery managers want solved. When asked in what areas current information is sufficient, most answered "none." A few named topics in which existing information was adequate, but only one topic was repeated; storage was given 3 votes as an area about which enough was now known. When respondents were asked to list areas where more information was needed, a great variety of interests emerged. Those most repeated were related to (1) seedling physiology, dormancy, and hardiness induction, including topics such as watering schedules and top pruning, (2) seed-sowing equipment and spacing control, and (3) seedling nutrition and fertilization. Many addressed the need for "nursery-specific" information. Also obvious from the Survey was the desire of managers to continue to rely on written information, with workshops and personal contact as necessary, additional research-communication aids. The strong implication was that more focused, briefer, and better illustrated publications were desired, and that scientific journals were not heavily used. A coordinated effort to address these priority problems on a nursery-specific basis will not only solve them more rapidly and efficiently, it will generally improve communication between nursery personnel and the research community at large. Researchers will become aware both of nursery problems and of the recognition that accompanies their solution.

Providing better targets is a more difficult problem—one that depends on an increased level of fundamental research related to nursery concerns. One Survey respondent suggested a comprehensive study of the cost effectiveness and social impacts of nursery practices; if such a study could be carried out, particularly in relation to timber harvest levels, better quantitative targets for nursery research and practice would be one immediate result. Similarly, basic physiological (mechanistic) models of seedling growth and development could be used to screen new or modified practices for impact on costs and outplanting success before expensive, site-by-site field tests are undertaken. These models could also be used to predict maximum nursery production levels and to calibrate practices, at least roughly, to species and stock-type combinations.

Better organized efforts to attract nursery research and development thus will require (1) improved communication between nursery operators and researchers and (2) more clearly articulated and focused targets. One of the most promising methods for achieving both of these—the research cooperative—has already successfully marshalled opinion and resources behind several forestry research programs (e.g., [17]). A cooperative effort focused on nursery practice is now underway in the Northwest [6].

30.4 Future Considerations

As harvests increase from their current low level, there will inevitably be an increased demand for nursery stock. Although it is possible to project a stable demand for stock in the Northwest as sustained harvest levels are realized and backlog regeneration is reduced, several large potential changes in the quality and kind of nursery stock produced can be expected:

- **Genetically improved seed:** The Northwest has not yet substantially shifted to seed-orchard seed but, given current plans, will do so over the next 2 decades. Reliance on this more costly, higher potential seed will intensify the focus on seed handling and sowing and seedbed survival. Nurseries, as the first custodians of this precious seed-orchard commodity, will receive even more attention from high-level managers and the public. In turn, objective evidence of the efficacy of nursery practices will be in greater demand. Thus, more attention will be given to recordkeeping and research substantiation of "obvious" solutions to problems.
- **Other species:** True firs (*Abies* spp.), spruce (*Picea* spp.), and hardwoods may increase radically in importance, relative to current production levels. As higher elevation sites are managed intensively, true firs will be increasingly prescribed and therefore will require more attention from both nursery managers and researchers. If the tip weevil that infests Sitka spruce [*Picea sitchensis* (Bong.) Carr.] can effectively be controlled, that species may be much more widely planted in Oregon and Washington; such control measures are currently under development [pers. commun., 8]. Hardwoods will gradually increase in regional importance, particularly if conifer supply and demand falter, as forecast. Although hardwoods are unlikely to ever constitute a major fraction of nursery stock produced in the Northwest, the transition from none to some, particularly for red alder (*Alnus rubra* Bong.) and black cottonwood (*Populus trichocarpa* Torr. & Gray), will require future attention. As eastern Oregon, eastern Washington, and Idaho produce relatively more timber [2], conifer species other than Douglas-fir will increase in importance. We should anticipate this change now with comprehensive efforts to better understand and grow pines (*Pinus* spp.), larch (*Larix* spp.), and white fir [*Abies concolor* (Gord. & Glend.) Lindl. ex Hildebr.].
- **New systems:** Radically new approaches to bareroot nursery management will be suggested and tried. Many of these will be useful as "blue sky" targets that stimulate thinking, even if their operational impact is not large. An example is Cooper's [5] concept of a bareroot nursery without soil; he suggests that in some places, nutrient film technique—a simple, low-cost hydroponic system—can be used to produce bareroot woody-plant nursery stock without the disease, drainage, and other problems associated with soil. Although certain barriers stand in the way of widespread application of such systems, the idea of a soilless nursery has many attractive points. The future will bring a steady flux of "radical" new ideas and, with it, the need for a well-developed mechanism for screening them and for developing and adapting the promising ones. Again, research cooperatives can play this role effectively.

In sum, nurseries present a unique opportunity to multiply the effects of technology and research on wood yields. Though neglected in the past, nursery research and development should be infused with new purpose and support, perhaps most effectively through a cooperative approach. Widespread use of genetically improved seed and diverse species in conjunction with new systems and ideas will merit the serious future attention of the entire nursery community.

References

1. Bengtson, G. W. 1979. Forest fertilization in the United States: progress and outlook. J. Forestry 78:222-229.
2. Beuter, J. H., K. N. Johnson, and H. L. Scheurman. 1976. Timber for Oregon's tomorrow: an analysis of reasonably possible occurrences. Oregon State Univ., Forest Res. Lab., Corvallis. Res. Bull. 19. 111 p.
3. Brodie, J. D., and P. Tedder. 1982. Regeneration delays: economic cost and harvest loss. J. Forestry 80(1):26-28.
4. Cleary, B. D., and B. R. Kelpsas. 1981. Five steps to successful regeneration planning. Oregon State Univ., Forest Res. Lab., Corvallis. Special Publ. 1. 31 p.
5. Cooper, A. J. 1975. Crop production in recirculating nutrient solution. Scientia Horticulturae 3(3):251-258.
6. Duryea, M. L. 1982. Prospectus for nursery technology center and cooperative. Dep. of Forest Science, Oregon State Univ., Corvallis. 18 p.
7. Fujimori, T. 1971. Primary productivity of a young *Tsuga heterophylla* stand and some speculations about biomass of forest communities on the Oregon Coast. U.S.D.A. Forest Serv., Pacific NW Forest and Range Exp. Sta., Portland, Oregon. Res. Pap. PNW-123. 11 p.
8. Gara, R., and C. Oliver. 1982. Personal communication, Univ. of Washington, Seattle.
9. Helgerson, O. T., S. D. Hobbs, D. H. Lysne, D. H. McNabb, and S. D. Tesch. 1982. Adaptive FIR annual report. Southwest Oregon Forestry Intensified Research (FIR) program, Medford, Oregon. 70 p.
10. Hobbs, S. D. 1982. Stock type selection and planting techniques for Douglas-fir on skeletal soils in southwest Oregon. Pages 92-96 *in* Proc., Reforestation of skeletal soils workshop (S. D. Hobbs and O. T. Helgerson, eds.). Oregon State Univ., Forest Res. Lab., Corvallis.
11. Kerrick, M. 1982. Personal communication, Willamette National Forest, Eugene, Oregon.
12. Silen, R. R., and J. G. Wheat. 1979. Progressive tree improvement program in coastal Douglas-fir. J. Forestry 77:78-83.
13. Tarrant, R. 1980. Report of the nursery task force. Dep. of Forest Science, Oregon State Univ., Corvallis. 30 p.
14. Tarrant, R. 1982. Personal communication, Dep. of Forest Sci., Oregon State Univ., Corvallis.
15. Timmis, R., and L. Fuchigami. 1982. Personal communication, Weyerhaeuser Co., Tacoma, Washington, and Dep. of Horticulture, Oregon State Univ., Corvallis.
16. U.S.D.A. Forest Service. 1981. Forest planting, seeding, and silvicultural treatments in the United States, 1980 report. Washington, D.C. Publ. FS 368. 15 p.
17. Walstad, J. D., J. C. Gordon, M. Newton, and L. A. Norris. 1982. CRAFTS: comprehensive research program in forest vegetation management. Dep. of Forest Science, Oregon State Univ., Corvallis. 27 p.
18. Waring, R. H., and J. F. Franklin. 1979. Evergreen coniferous forests of the Pacific Northwest. Science 204:1380-1386.

Appendix A: Questionnaire #1

QUESTIONNAIRE FOR

BAREROOT NURSERY TECHNOLOGY WORKSHOP

Nursery **Name** ______________________

Address ______________________

____

Phone # ______________________

Names of Persons Completing Questionnaire:

Name	**Position**
____________	____________
____________	____________
____________	____________

Nursery Questionnaire (continued) 2

Annual Nursery Production (1980)

Please list the major species grown at your nursery. Give the stock types and number produced in thousands of seedlings (M) for each species in 1980.

Species	Stock Type	Number Produced (Harvested)
		M
		M
		M
		M
		M
		M
		M
		M
Other minor species and	Stock types	M
Total Number of Seedlings Produced =		M

Land Usage

Please indicate the total number of acres/or hectares (circle one) your nursery utilizes for each area below.

Fallow or cover-cropped land	____________	Acres or Hectares
Seedbed area	____________	Acres or Hectares
Transplant area	____________	Acres or Hectares
Uncultivated land (buildings, roads, etc)	____________	Acres or Hectares
Unused land	____________	Acres or Hectares
Other ______________	____________	Acres or Hectares
Total -----	____________	Acres or Hectares

Nursery Site Characteristics

From the list below, please indicate the five (5) most important criteria used to select your nursery site, where 1 is the most important, 2, the second most important, 3, the third most important, 4, the fourth most important, and, 5, the fifth most important.

_____ Climate
_____ Elevation
_____ Aesthetics
_____ Proximity to markets
_____ Water supply
_____ Soil depth
_____ Soil workability and drainage
_____ Cost of land
_____ Proximity to work force
_____ Soil fertility (including pH and cation exchange capacity)
_____ Local topography
_____ Politics
_____ Previous land use
_____ Freedom from weeds
_____ Soil texture
_____ Other ______________

Nursery Questionnaire (continued)

Site Problems - Most bareroot nurserymen realize that they do not have the""ideal" nursery site but "make-do" with what they have. The purpose of this portion of the survey is to identify major site problems and determine what can be done about them. On the table below, please indicate the 5 most serious site problems at your nursery, where 1 is the most serious problem; 2, the second most serious; 3, the third most serious; 4, the fourth most serious and, 5, the fifth most serious. Next, for these 5 site problems, list any corrective treatments that you have tried and whether these treatments have alleviated the problem.

Problem	Major Site Problem (Rank 1 through 5)	Corrective Treatments			
		Treatment	Is it effective?	Treatment	Is it effective?
SOIL					
Acidity					
Alkalinity					
Salinity					
Too "heavy"					
Too "light"					
Too much variation					
Compaction					
Poor Drainage					
Rocks					
Organic matter maintenance					
Soil splash					
Workability (tilth)					
Uneven topography					
CLIMATE					
Intense rainfall					
Frost pockets					
Frozen soil					
Excessively high temperatures					
Wind abrasion					
WATER					
Poor quality					
High water table					
Availability					
OTHER					

SOIL

Soil Characteristics

Please fill in the table below discussing the soil characteristics of your major soil types. For each soil type indicate the % of cultivated land that it occupies, its pH, particle size distribution: % sand, silt and clay, the drainage quality: good, fair or poor, the cation exchange capacity (me/100g), the bulk density (g/cm^3) and organic matter (%).

Soil type	% of cultivated land	pH range	Particle size distribution % sand silt clay (range)	Drainage (good, fair, or poor	Cation exchange capacity (me/100g)	Bulk Density (g/cm^3) (range)	Organic Matter (%) (range)
	%		% % %		me/100g	g/cm^3	%
	%		% % %		me/100g	g/cm^3	%
	%		% % %		me/100g	g/cm^3	%

Sowing and Seedbed Density

Please fill in the table below discussing the sowing practice for the stock types produced in your two major species. For each, indicate the sowing date: optimum time (e.g. early May; mid April) and the actual range (e.g. late May to early June); the sowing depth in inches or centimeters; the type of mulch used, if any, and the depth applied; the target growing density (in the 2-0 year) ($seedlings/ft^2$); the seedling inventory ($seedlings/ft^2$) for 1-0's and 2-0's; and the cull % during grading.

Species	Stock Type of Final Product	Sowing Date		Sowing Depth (circle whether inches or cm)	Mulch		Target Growing Density	Seedbed Inventory ($Seedlings/ft^2$)		Cull % During Grading
		Optimum Date (e.g. early May)	Actual Range (mid May to late June)		Type	Depth (in or cm)	$Seedl/ft^2$	1-0's	2-0's	
1.	1-0		to	in cm		in cm	$/ft^2$	$/ft^2$	$/ft^2$	%
	2-0		to	in cm		in cm	$/ft^2$	$/ft^2$	$/ft^2$	%
	2-1		to	in cm		in cm	$/ft^2$	$/ft^2$	$/ft^2$	%
	Other		to	in cm		in cm	$/ft^2$	$/ft^2$	$/ft^2$	%
2.	1-0		to	in cm		in cm	$/ft^2$	$/ft^2$	$/ft^2$	%
	2-0		to	in cm		in cm	$/ft^2$	$/ft^2$	$/ft^2$	%
	2-1		to	in cm		in cm	$/ft^2$	$/ft^2$	$/ft^2$	%
	Other		to	in cm		in cm	$/ft^2$	$/ft^2$	$/ft^2$	%

Fertilization

Please fill in the table below with the general fertilizer schedule for your 2-0 stock starting with the applications before sowing, then the applications made in the first year and, finally, the applications made in the second year up to the time of lifting. Indicate time of year of application (month), what type of fertilizer is applied, the amount applied in lbs/acre or kg/ha and finally the purpose for the application, e.g. to stimulate fall root growth.

This is the entire 2-0 fertilizer schedule for the major species grown at our nursery which is ____________________
Species Name

	Age of Stock at Time of Application	Time of Year of Application (Month)	Type of Fertilizer Applied (e.g. Ammonium Phosphate 11-55-0)	Amount Applied lbs/acre or kg/ha (circle one)	Purpose of Application (e.g. fall root growth)
Applications Before Sowing	0			lbs/acre or kg/ha	
				lbs/acre or kg/ha	
				lbs/acre or kg/ha	
Applications in First Year	1-0			lbs/acre or kg/ha	
				lbs/acre or kg/ha	
				lbs/acre or kg/ha	
Applications in Second Year	2-0			lbs/acre or kg/ha	
				lbs/acre or kg/ha	
				lbs/acre or kg/ha	

Root Culturing

Please fill in the table below discussing your root culturing regime for growing 2-0 seedlings of your major species. For undercutting, wrenching and lateral pruning, list the age of the stock at the time of the operation (e.g. 1-0), the date(s) of the operation, the depth at which you draw the blade(s) (in. or cm.) and the purpose for doing the operation (e.g. to harden-off seedlings, stimulate root growth, etc.).

This is the root culturing regime for 2-0 seedlings of our major species, ______________________
Give Species Name

Root Culturing	Age of Seedlings at Time of Operation (e.g. 1-0)	Date(s) of Operation (month/day)	Depth at which you Draw the Blade (in or cm) (circle one)	Purpose of Operation (e.g. to Harden-off Seedlings)
Under-Cutting		/	in cm	
		/	in cm	
Wrenching		/	in cm	
		/	in cm	
		/	in cm	
		/	in cm	
		/	in cm	
		/	in cm	
Lateral Pruning or Sidecutting		/	in cm	
		/	in cm	
		/	in cm	
		/	in cm	

TRANSPLANTING

In the table below discuss the transplanting regime for your major species and stock types. Indicate the species, stock type of final product (eg. plug-1, 1-1, 2-1), the time of year (month(s)) seedlings are lifted, stored, and transplanted, the root length of seedlings to be transplanted, the density of the bed immediately after transplanting and at the time of lifting (seedlings/ft^2) and the final number of seedlings produced after culling (seedling/ft^2).

Species	Stock Type (final product)	Time of Year ...			Pruned Root length in/cm (circle one)	Density of Transplant Bed		Number of seedlings produced after culling (seedlings/ft^2)
		Lifted month(s)	Stored month(s)	Transplanted month(s)		At time of transplanting	At time of lifting	
					in/cm	/ft^2	/ft^2	/ ft^2
					in/cm	/ft^2	/ft^2	/ ft^2
					in/cm	/ft^2	/ft^2	/ ft^2
					in/cm	/ft^2	/ft^2	/ ft^2
					in/cm	/ft^2	/ft^2	/ ft^2
					in/cm	/ft^2	/ft^2	/ ft^2
					in/cm	/ft^2	/ft^2	/ ft^2

Please, indicate the percentage of seedlings transplanted in the fall versus spring:

Fall = ______ %

Spring = ______ %

Total = 100 %

Which do you prefer, fall or spring transplanting and why? ____________________

____

Nursery Questionnaire (continued)

WEEDS

Please fill in the table below with your weed control schedule. Indicate what conifer species are in the seedbed at the time, the timing of the application (g pre-seeding, spring of 2-0 year, etc.), the machine hours per acre of mechanical weeding and/or the person hours/acre of handweeding and/or the full name of the herbicide applied in lbs ai/acre or kg ai/acre, and finally list the plants which you are attempting to control.

Conifer Species in Seedbed	Application Timing (e.g. pre-seeding)	Mechanical Weeding (machine hours/acre)	Handweeding (person hours/acre)	Full Name of Herbicide and Amount (lbs ai/acre or kg ai/acre) (circle one)	For control of what weeds? (e.g. annual grasses)
		machine hrs/acre	person hrs/acre	lbs ai/acre kg ai/acre	
		machine hrs/acre	person hrs/acre	lbs ai/acre kg ai/acre	
		machine hrs/acre	person hrs/acre	lbs ai/acre kg ai/acre	
		machine hrs/acre	person hrs/acre	lbs ai/acre kg ai/acre	
		machine hrs/acre	person hrs/acre	lbs ai/acre kg ai/acre	
		machine hrs/acre	person hrs/acre	lbs ai/acre kg ai/acre	
		machine hrs/acre	person hrs/acre	lbs ai/acre kg ai/acre	
		machine hrs/acre	person hrs/acre	lbs ai/acre kg ai/acre	

Nursery Questionnaire (continued)

Pests

Please discuss your pest management program in the table below. First rank the major pest groups where 1 is the greatest problem as a pest, 2, the next greatest problem, 3, the next, 4, the next, 5, the next, and, 6, the least important problem. Then discuss the severity of the problem for each specific pest; is it heavy, moderate, slight or non-existent as a problem; list the species and stock-type that the pest affects; answer whether the pest is increasing or decreasing as a problem (check one) and finally discuss the major methods of control that you have used: soil fumigation, seed treatment, drench, sprays (check as many as you use) and cultural controls (list those used).

Major Pest Group (rank from 1 through 6)	Specific Pest	Severity of Problem (check one)				Species and Stock Type which Pest Affects	Is this Pest Increasing or Decreasing as a Problem? (check one)		Methods of Control (check as many as you use)				Cultural Controls (List)
		Heavy	Moderate	Slight	Non-Existent		Increasing	Decreasing	Fumigation	Fungicide/Insecticide Seed Treatment	Drench	Sprays	
___ Diseases of Seeds and Germinants	1. Damping off (Pythiums, Rhizoctonia, etc.)												
	2. Seed Fungi												
___ Diseases of Roots and Root Collars	1. Fusarium Root Rot												
	2. Nematodes												
	3. Phytophthora												
	4. Cortical Rot (Fusarium Roseum)												
	5. Charcoal Root Rot (Macrophomina Phaseoli)												
___ Diseases Affecting Shoots	1. Sirococcus Blight												
	2. Fusarium Top Blight												
	3. Melampsora Foliage Rust												
	4. Western Gall Rust												
	5. Gray Mold (Botrytis)												
___ Molding of Stored Seedlings	1. Fungal Molds												
___ Insects and Allied Pests	1. Springtail Insects												
	2. Cutworms												
	3. Marsh Crane Fly												
	4. Root and Vine Weevils												
	5. Aphids												
	6. Cranberry Girdler												
	7. European Pine Shoot Moth												
	8. Balsam Woolly Aphid												
___ Birds and Mammals	1. Deer												
	2. Rodents												
	3. Birds												
	4. Rabbits												
	5. Moles												

Inventory of Equipment or Method Used

Please indicate the kind of equipment or methods now used at your nursery. If it is commercially available, please list the make, model, or type. Also indicate whether or not the equipment or method fulfills its function well? And, if the answer is no, discuss why the equipment or method is unsatisfactory, for reasons that may be related to the high cost of equipment, maintenance or fuel or is it due to low operational efficiency, i.e., the equipment or method does not do the job well resulting in lower product quality or product loss.

1. CONE STORAGE AND HANDLING

Equipment or Method	Make, Model, or Type	Does Equipment or Method Fulfill its Function Well?	If No: Why is Equipment or Method Unsatisfactory?
______ Forklift	______________	______ Yes ______ No	______________
______ Conveyor	______________	______ Yes ______ No	______________
______ Manual	______________	______ Yes ______ No	______________
______ Tray Storage	______________	______ Yes ______ No	______________
______ In Sack Storage	______________	______ Yes ______ No	______________
______ Rack Storage	______________	______ Yes ______ No	______________
______ Loose Storage	______________	______ Yes ______ No	______________
______ ______________	______________	______ Yes ______ No	______________
______ ______________	______________	______ Yes ______ No	______________

2. SEED PROCESSING

Equipment or Method	Make, Model, or Type	Does Equipment or Method Fulfill its Function Well?	If No: Why is Equipment or Method Unsatisfactory?
____Preheat Bins		____Yes ____No	
____Extractor		____Yes ____No	
____Scalper		____Yes ____No	
____Dewinger		____Yes ____No	
____Fanning Mill/Clipper		____Yes ____No	
____Specific Gravity Separator		____Yes ____No	
____Powered Conveyors		____Yes ____No	
____Cone Grinders		____Yes ____No	
____ ____________		____Yes ____No	
____ ____________		____Yes ____No	

3. SEED STORAGE AND HANDLING

Equipment or Method	Make, Model, or Type	Does Equipment or Method Fulfill its Function Well?	If No: Why is Equipment or Method Unsatisfactory?
____Refrigerated Storage		____Yes ____No	
____Nonrefrigerated Storage		____Yes ____No	
____Freezer Storage		____Yes ____No	
____Cans		____Yes ____No	
____Sacks		____Yes ____No	
____Cartons		____Yes ____No	
____ ____________		____Yes ____No	
____ ____________		____Yes ____No	

Nursery Questionnaire (continued)

14

4. FUMIGATION

Equipment or Method	Make, Model, or Type	Does Equipment or Method Fulfill its Function Well?		If No: Why is Equipment or Method Unsatisfactory?
____ Granular Applicator	____	____ Yes	____ No	____
____ Liquid Applicator	____	____ Yes	____ No	____
____ Shank Injector	____	____ Yes	____ No	____
____ Tarp Layer	____	____ Yes	____ No	____
____ Tarp Remover	____	____ Yes	____ No	____
____	____	____ Yes	____ No	____
____	____	____ Yes	____ No	____

5. GENERAL CULTIVATION & GROUND PREPARATION

Equipment or Method	Make, Model, or Type	Does Equipment or Method Fulfill its Function Well?		If No: Why is Equipment or Method Unsatisfactory?
____ Plow	____	____ Yes	____ No	____
____ Rototiller	____	____ Yes	____ No	____
____ Disc	____	____ Yes	____ No	____
____ Harrow	____	____ Yes	____ No	____
____ Chisel	____	____ Yes	____ No	____
____ Leveler	____	____ Yes	____ No	____
____ Packer	____	____ Yes	____ No	____
____ Rod Weeder	____	____ Yes	____ No	____
____ Rock Picker	____	____ Yes	____ No	____
____ ________	____	____ Yes	____ No	____
____ ________	____	____ Yes	____ No	____

6. SOWING

Equipment or Method	Make, Model, or Type	Does Equipment or Method Fulfill its Function Well?	If No: Why is Equipment or Method Unsatisfactory?
____Bed Marker	____________	____Yes ____No	____________
____Bed Maker	____________	____Yes ____No	____________
____Drill Seeder	____________	____Yes ____No	____________
____Broadcast Seeder	____________	____Yes ____No	____________
____Mulch Spreader	____________	____Yes ____No	____________
____Bed Sander	____________	____Yes ____No	____________
____Bed Roller	____________	____Yes ____No	____________
____ ____________	____________	____Yes ____No	____________
____ ____________	____________	____Yes ____No	____________

7. IRRIGATION

Equipment or Method	Make, Model, or Type	Does Equipment or Method Fulfill its Function Well?	If No: Why is Equipment or Method Unsatisfactory?
____Ditch	____________	____Yes ____No	____________
____Overhead Oscillation	____________	____Yes ____No	____________
____Overhead Impulse (Fixed)	____________	____Yes ____No	____________
____Overhead Impulse (Portable)	____________	____Yes ____No	____________
____ ____________	____________	____Yes ____No	____________
____ ____________	____________	____Yes ____No	____________

8. FERTILIZATION

Equipment or Method	Make, Model, or Type	Does Equipment or Method Fulfill its Function Well?	If No: Why is Equipment or Method Unsatisfactory?
____Granular Applicator	____________	____Yes ____No	____________
____Liquid Gravity Applicator (Drenching)	____________	____Yes ____No	____________
____Liquid Pressure Sprayer	____________	____Yes ____No	____________
____Soil Injector	____________	____Yes ____No	____________
____Irrigation	____________	____Yes ____No	____________
____Manure Spreader	____________	____Yes ____No	____________
____ ____________	____________	____Yes ____No	____________
____ ____________	____________	____Yes ____No	____________

9. SOIL AMENDMENTS

____Manure Spreader	____________	____Yes ____No	____________
____Mulch Spreader	____________	____Yes ____No	____________
____Apricultural Seed Drill	____________	____Yes ____No	____________
____ ____________	____________	____Yes ____No	____________
____ ____________	____________	____Yes ____No	____________

10. SEEDBED CULTIVATION

Equipment or Method	Make, Model, or Type	Does Equipment or Method Fulfill its Function Well?	If No: Why is Equipment or Method Unsatisfactory?
____Path Cultivator (Hand)	____________	____Yes ____No	____________
____Path Cultivator (Mechanized)	____________	____Yes ____No	____________
____Row Cultivator (Hand)	____________	____Yes ____No	____________
____Row Cultivator (Mechanized)	____________	____Yes ____No	____________
____Pipeline Cultivator (Hand)	____________	____Yes ____No	____________
____Pipeline Cultivator (Mechanized)	____________	____Yes ____No	____________
____ ____________	____________	____Yes ____No	____________
____ ____________	____________	____Yes ____No	____________

11. ROOT PRUNING

Equipment or Method	Make, Model, or Type	Does Equipment or Method Fulfill its Function Well?	If No: Why is Equipment or Method Unsatisfactory?
____Side Root Pruner (Disc)	____________	____Yes ____No	____________
____Side Root Pruner (Fixed)	____________	____Yes ____No	____________
____Bottom Root Pruner (Fixed)	____________	____Yes ____No	____________
____Bottom Root Pruner (Reciprocating)	____________	____Yes ____No	____________
____Root Wrencher	____________	____Yes ____No	____________
____ ____________	____________	____Yes ____No	____________
____ ____________	____________	____Yes ____No	____________

12. TOP PRUNING

Equipment or Method	Make, Model, or Type	Does Equipment or Method Fulfill its Function Well?	If No: Why is Equipment or Method Unsatisfactory?
____ Hand Shearing	____	____ Yes ____ No	____
____ Rotary Mower	____	____ Yes ____ No	____
____ Sickle Bar	____	____ Yes ____ No	____
____ ____	____	____ Yes ____ No	____
____ ____	____	____ Yes ____ No	____

13. HERBICIDE AND INSECTICIDE SPRAYING

Equipment or Method	Make, Model, or Type	Does Equipment or Method Fulfill its Function Well?	If No: Why is Equipment or Method Unsatisfactory?
____ Hand Sprayer	____	____ Yes ____ No	____
____ Tractor Mounted Boom Sprayer	____	____ Yes ____ No	____
____ Trailer Mounted Boom Sprayer	____	____ Yes ____ No	____
____ Mist or Dust Blower	____	____ Yes ____ No	____
____ ____	____	____ Yes ____ No	____
____ ____	____	____ Yes ____ No	____

14. FROST PROTECTION

Equipment or Method	Make, Model, or Type	Does Equipment or Method Fulfill its Function Well?	If No: Why is Equipment or Method Unsatisfactory?
____ Irrigation	____	____ Yes ____ No	____
____ Smudge Pots	____	____ Yes ____ No	____
____ Covering (Plastic Sheet)	____	____ Yes ____ No	____
____ ____	____	____ Yes ____ No	____
____ ____	____	____ Yes ____ No	____

15. TRANSPLANTING

Equipment or Method	Make, Model, or Type	Does Equipment or Method Fulfill its Function Well?	If No: Why is Equipment or Method Unsatisfactory?
____Hand Transplanting Board	________	____Yes ____No	________
____Self-Propelled Transplanter	________	____Yes ____No	________
____Tractor Drawn Transplanter	________	____Yes ____No	________
____ ________	________	____Yes ____No	________
____ ________	________	____Yes ____No	________

16. FIELD LIFTING

____Rigid Undercutting Blade	________	____Yes ____No	________
____Rigid Undercutting Blade W/Agitator	________	____Yes ____No	________
____Manual Lifting ___%	-- If seedlings are lifted both manually and	____Yes ____No	________
____Mechanical Harvesting ___%	mechanically, give approximate percentage	____Yes ____No	________
____ ________	of each used.	____Yes ____No	________
____ ________	________	____Yes ____No	________

17. FIELD HANDLING

____Boxes	________	____Yes ____No	________
____Bins	________	____Yes ____No	________
____Tubs	________	____Yes ____No	________
____Fabric Slings	________	____Yes ____No	________
____ ________	________	____Yes ____No	________
____ ________	________	____Yes ____No	________
____ ________	________	____Yes ____No	________

18. FIELD TRANSPORT

Equipment or Method	Make, Model, or Type	Does Equipment or Method Fulfill its Function Well?	If No: Why is Equipment or Method Unsatisfactory?
_____Trailer	________	_____Yes _____No	________
_____Truck	________	_____Yes _____No	________
_____Forklift	________	_____Yes _____No	________
_____ ________	________	_____Yes _____No	________
_____ ________	________	_____Yes _____No	________
_____ ________	________	_____Yes _____No	________

19. GRADING AND COUNTING

_____Stationary Table	________	_____Yes _____No	________
_____Moving Belt	________	_____Yes _____No	________
_____Counter (Mechanical or Electrical)	________	_____Yes _____No	________
_____Counter (Weight)	________	_____Yes _____No	________
_____Counter (Manual)	________	_____Yes _____No	________
_____ ________	________	_____Yes _____No	________
_____ ________	________	_____Yes _____No	________
_____ ________	________	_____Yes _____No	________
_____ ________	________	_____Yes _____No	________

20. SEEDLING PACKAGING

Equipment or Method	Make, Model, or Type	Does Equipment or Method Fulfill its Function Well?	If No: Why is Equipment or Method Unsatisfactory?
____Mechanical Bundling	____	____Yes ____No	____
____Manual Bundling	____	____Yes ____No	____
____Boxes	____	____Yes ____No	____
____Bags	____	____Yes ____No	____
____Bales	____	____Yes ____No	____
____Crates	____	____Yes ____No	____
____Stapler	____	____Yes ____No	____
____Taper	____	____Yes ____No	____
____Stitcher	____	____Yes ____No	____
____Baler	____	____Yes ____No	____
____Packing Medium	____	____Yes ____No	____
____ ____	____	____Yes ____No	____
____ ____	____	____Yes ____No	____

21. SEEDLING STORAGE

Equipment or Method	Make, Model, or Type	Does Equipment or Method Fulfill its Function Well?	If No: Why is Equipment or Method Unsatisfactory?
____Refrigerated Storage	____	____Yes ____No	____
____Non-Refrigerated Storage	____	____Yes ____No	____
____Humidity Control	____	____Yes ____No	____
____Permanent Racks	____	____Yes ____No	____
____Pallet System	____	____Yes ____No	____
____Palleteer System	____	____Yes ____No	____
____ ____	____	____Yes ____No	____
____ ____	____	____Yes ____No	____

22. SEEDLING HANDLING

Equipment or Method	Make, Model, or Type	Does Equipment or Method Fulfill its Function Well?	If No: Why is Equipment or Method Unsatisfactory?
____ Forklift	____	____ Yes ____ No	____
____ Skids	____	____ Yes ____ No	____
____ Carts	____	____ Yes ____ No	____
____ Belt Conveyor	____	____ Yes ____ No	____
____ Roller Conveyor	____	____ Yes ____ No	____
____ ____	____	____ Yes ____ No	____
____ ____	____	____ Yes ____ No	____

23. SHIPPING FOR OUTPLANTING

____ Refrigerated	____	____ Yes ____ No	____
____ Non-Refrigerated	____	____ Yes ____ No	____
____ Common Carrier	____	____ Yes ____ No	____
____ ____	____	____ Yes ____ No	____
____ ____	____	____ Yes ____ No	____

24. SPECIALIZED ON-SITE TRANSPORTATION

____ Buses	____	____ Yes ____ No	____
____ Crew Carriers	____	____ Yes ____ No	____
____ Scooters	____	____ Yes ____ No	____
____ ____	____	____ Yes ____ No	____
____ ____	____	____ Yes ____ No	____

25. TRACTORS

Make, Model, and Horsepower	Crawler	Wheeled	Gas	Diesel	Does Equipment or Method Fulfill its Function Well?
________	____	____	____	____	____ Yes ____ No
________	____	____	____	____	____ Yes ____ No
________	____	____	____	____	____ Yes ____ No
________	____	____	____	____	____ Yes ____ No
________	____	____	____	____	____ Yes ____ No
________	____	____	____	____	____ Yes ____ No
________	____	____	____	____	____ Yes ____ No

What 3 pieces of nursery equipment need the most improvement? List in priority order.

	Equipment	What needs to be improved?
1.	________	________
2.	________	________
3.	________	________

Please estimate the number of seasonal workers employed by your nursery last year in the four functions listed.

	Number of Seasonal Employees Minimum	Maximum
Cone harvesting	________	________
Seed processing	________	________
Nursery seedling production	________	________
Seedling packing shed operation	________	________

How many permanent employees do you have? ______________________________

We realize that this questionnaire has taken a considerable amount of your time and want to sincerely thank you. The information produced by this survey will insure the success of the Bareroot Nursery Conference and make the resulting Proceedings a valuable bareroot nursery manual.

Mary Duryea, Oregon State Univ.

Tom Landis, USDA - Forest Service

Appendix B: Questionnaire #2

ORAL QUESTIONNAIRE

FOR

BAREROOT NURSERY TECHNOLOGY WORKSHOP

Nursery Name ______________________

Address ______________________

____

Phone # ______________________

Names of Persons Present at Interview:

Name	Position

10. Where are they stored?

____

11. Under what environmental conditions are they stored?

Relative humidity______Temperature______Kind of ventilation____________

12. Do you extract and clean your own seed?

1 Yes (skip to 14)
2 No (go to 13)

13. Who extracts and cleans your seed?

______________________ (go to 17)

14. Where do you dry, extract and clean seed?

____

15. How do you transport your seed to cold storage? (In what containers?)

____

16. What is the average purity of the seed of your major species?

Species #1 ______________

Species #2 ______________

17. Do you purchase any of your seed from seed companies?

1 Yes (go to 18)
2 No (skip to 20)

18. What percent (%) of your seed?

______________________%

19. Where is it purchased? (Name and address of seed company)

____

____

Seed Testing

20. What seed tests do you do at your nursery? (e.g. purity, germination, weight of 1,000 seeds, moisture content, x-rays, etc.)

____

____

____

____

21. What seed tests are done at another location?

____

____

____

____

22. Where?

____

23. Do you set minimum purity standards for each species?

    1 Yes (go to 24)
    2 No (skip to 25)

24. What are they for your major species?

    Species #1 ________ ________

    Species #2 ________ ________

25. Under what conditions do you store your seed?

    Temperature ________________

    Humidity ________________

    Packaging ________________

26. Do you retest seed after it has been in cold storage?

    1 Yes (go to 27)
    2 No (skip to 29)

27. What tests do you use?

____

____

____

28. How often do you retest a given batch of seed?

____

Stratification

29. Do you stratify your own seed?

    1 Yes (go to 31)
    2 No (go to 30)

30. Where is your seed stratified?

____________________ (go to 39)

31. How long do you soak your seed in water? And in what container?

____________________ ________________

32. Do you aerate the water?

    1 Yes
    2 No

33. At what temperature do you stratify seed? And in what container?

____________________ ________________

34. How long do you stratify each of your major species?

Species	Length of Stratification
_____	____________
_____	____________
_____	____________

35. What is the maximum time you store your seed before sowing?

__________ weeks

And in what container is it stored?

____

Nursery Workshop Questionnaire (continued) 5

36. Have you experienced premature germination in stratification?

1 Yes (go to 37)
2 No (go to 39)

37. In which species did it occur?

____

____

38. What do you do when it occurs?

____

39. What do you do if you have stratified too much seed?

____

____

____

____

<u>Preparing the Seedbed and Sowing</u>

40. Are your seedbeds raised?

1 Yes (go to 41)
2 No (skip to 43)

41. How high above the original ground line do you raise them?

____________________ in or cm

42. Why do you raise them?

____

43. How do you compute how much seed to sow? (What formula do you use?)

____

____

____

____

Nursery Workshop Questionnaire (continued) 6

44. Describe the layout of your beds.

Bed width (row to row) ________________ in or cm

Path width (outside row to outside row) ________________ in or cm

Rows per bed ________________

Number of beds between irrigation lines ________________

Distance between irrigation lines ________________ ft or m

<u>Fumigation</u>

45. Do you fumigate your soil?

1 Yes Do you contract or do your own fumigation?

Contract ____________ Own ____________

2 No (skip to 55)

46. What time of the year do you fumigate?

____________ Spring

____________ Fall

____________ Other ____________

47. Why do you fumigate at this time?

____

48. a. What fumigant do you use? b. At what rate is it applied?

____________________ ____________________ lbs/acre

c. Why have you selected this fumigant? ________________________

d. Have you had problems with any other fumigants?

1 Yes (go to 48e)
2 No (go to 49)

e. What problems have you had?

____

49. Under what soil conditions do you fumigate?

________________ Temperature

________________ Moisture

50. What is your minimum time length for tarping?

____________________ days

51. How often do you fumigate?

__________ every rotation

__________ other __________

52. a. Do you fumigate your transplant beds?

1 Yes (go to 52b)
2 No (skip to 53)

b. What fumigant? __________

c. At what rate? __________ d. When? __________

53. Why do you fumigate? (rank in order)

_____ Weeds

_____ Pests

_____ Other _____

54. Do you fumigate as a preventative?

1 Yes
2 No

55. Do you consider fumigation an economic practice?

1 Yes
2 No

56. Do you use a bioassay for determining pest populations?

1 Yes
2 No

57. Have you tested to see if fumigants are effective?

1 Yes (go to 58)
2 No (skip to 59)

58. What were the results of your tests?

____

Seedling and Soil Analysis

59. Do you have your soil analyzed?

1 Yes (go to 60)
2 No (skip to 66)

60. Where is it analyzed?

____

61. How often is it analyzed?

____

62. At what time of the year, stage in your seedbed preparation or growing regime?

____

63. How many samples are collected and what area do they represent?

____

64. At what depth are your samples collected?

__________ in or cm

65. For what soil characteristics is your soil analyzed?

Yes	No	
1	2	pH
1	2	Organic Matter
1	2	Cation Exchange Capacity (CEC)
1	2	Soluble Salts
1	2	Lime Requirement
1	2	Total Nitrogen (N)
1	2	Phosphorous (P)
1	2	Potassium (K)
1	2	Calcium (Ca)
1	2	Magnesium (Mg)
1	2	Boron (B)
1	2	Zinc (Zn)
1	2	Other __________

66. Do you have a map of your various soil types?

 1 Yes

 2 No

67. Do you have a soil management plan?

 1 Yes

 2 No

68. Do you have your seedlings analyzed?

 1 Yes (go to 69)

 2 No (skip to 73)

69. Where are they analyzed?

____

70. At what stage in seedling growth?

____

71. How often are they analyzed?

____

72. For what nutrients?

Yes	No	
1	2	N
1	2	P
1	2	K
1	2	S
1	2	Ca
1	2	Mg
1	2	B
1	2	Mn
1	2	Mo
1	2	Zn
1	2	Cu
1	2	Fe
1	2	Other _______________

Fertilization

73. How do you determine what levels of fertilizer to apply?

____

____

74. How do you coordinate fertilizer applications to harden-off seedlings?

____

____

75. Do you have optimum soil nutrient levels established for each soil type?

 1 Yes (go to 76)

 2 No (skip to 77)

76. Describe the optimum nutrient levels for your major soil type.

_________ pH

_________ ppm P

_________ ppm K

_________ me/100 g Ca

_________ me/100 g Mg

_________ total N (%)

<u>Organic Matter</u>

77. What was the percent (%) organic matter in your major soil type at time of your last analysis?

_________________ %

_________________ date

78. What percent (%) would you like it to be?

_________________ %

79. Why?

____

____

____

____

80. Do you add organic amendments to your soil?

1 Yes (skip to 82)
2 No (go to 81)

81. What are your reasons for not adding organic materials? (skip to 89)

____

82. What materials do you add, when are they added and at what rates?

What Material?	When Added?	At What Rate?
_________	_______	_________
_________	_______	_________
_________	_______	_________

83. Do you compost your organic matter sources?

1 Yes (go to 84)
2 No (skip to 86)

84. Do you add fertilizer, lime, fungal inoculant or other additives to the composting material?

1 Yes (go to 85)
2 No (skip to 86)

85. List additives.

_____________________ _____________________

86. Why do you choose these specific organic materials?

____

____

____

____

87. Do you see a shortage of supply or high prohibitive costs for this material in the future?

1 Yes (go to 88)
2 No (skip to 89)

88. Explain.

____

____

____

89. Do you sow a cover crop?

1 Yes (go to 90)
2 No (skip to 95)

90. Is this a:

_____________ summer cover crop or a

_____________ winter cover crop or

_____________ both

91. How often do you sow a cover crop?

____________ every rotation

____________ every other rotation

____________ other ____________

92. Describe when you sow:	Summer	Winter
	____________	____________
When you plow under:		
	____________	____________
93. What plants do you use?		
	____________	____________
94. Why do you cover crop?		
	____________	____________
	____________	____________

Water

95. What equipment or methods do you use to generally determine when irrigation is needed? (First check which of the six methods you use, then give details only for those areas checked.)

_____ 1. Visual and tactile examination of the soil

Check one, stating where this is done:

_____ a. in the root zone

_____ b. in the surface layers

_____ c. both

Explain briefly how soil is examined (example: can't squeeze water out of soil, etc.).

____

____

_____ 2. Soil moisture tensionmeters

Check one, stating where these are located:

_____ a. above the main root zone

_____ b. in the root zone

_____ c. below the root zone

_____ d. two tensionmeters, one above and one below the main root zone area.

_____ e. other ____________

Explain briefly how information is used to irrigate (example: start irrigation at "X" bars, stop at "Y", etc.)

____

____

_____ 3. Electrical (Bouyoucous) Resistance Blocks (soil moisture blocks)

Check one, stating where these are located:

_____ a. above the main root zone

_____ b. in the root zone

_____ c. below the root zone

_____ d. two blocks, one above and one below the main root zone

_____ e. other ____________

_____ 4. <u>Water Budget Method</u> (calculation of evapotranspiration of the crop per day and a running balance sheet of accumulated deficit below field capacity. Irrigation takes place whenever the deficit exceeds a certain predetermined value.

What water deficit guides do you tolerate at:

a. initiation of irrigation: _______________ (in or mm)

b. cessation of irrigation: _______________ (in or mm)

How do you calculate evapotranspiration of your crop?

____

____

_____ 5. <u>Pressure Bomb</u> (direct measurement of internal plant water potential)

What plant moisture stress do you use for:

a. initiation of irrigation:

_______ bars (or other unit _______)

b. Do you employ this method routinely?

1 Yes
2 No

c. How often?

____

_____ 6. Other method or guidelines used to signal need for irrigation (plant wilting, weather, etc.).

96. Do you change irrigation monitoring methods as the crop gets older or bigger?

1 Yes (go to 97)
2 No (skip to 98)

97. Explain how you change your monitoring.

1-0's: ____

2-0's and others: _______________________________________

98. Do you think there is a need for better equipment or guides to monitor nursery irrigation?

1 Yes
2 No

Why? ____

99. Do you have a soil moisture retention curve (percent soil moisture content by weight/soil matric potential) for your major soil types?

1 Yes
2 No

100. Do you irrigate during the day or night or at both times? (circle one or both)

101. Do you irrigate for cooling of recently germinated seedlings?

1 Yes (go to 102)
2 No (skip to 103)

102. At what air and soil temperatures and for how many minutes do you irrigate for cooling?

_____________ air temperature

_____________ soil temperature

_____________ minutes

103. Do you irrigate for frost protection?

1 Yes (go to 104)
2 No (skip to 105)

104. At what air temperature do you irrigate for frost protection?

_____________ temperature

105. Do you reduce watering to harden seedlings in the fall?

1 Yes (go to 106)
2 No (skip to 107)

106. Describe the procedure for restricted watering for each stock type of your major species.

1-0 ____________

2-0 1st year ____________

2nd year ____________

		Spring Transplant	Fall Transplant
2-1	1st year	______	______
	2nd year	______	______
	3rd year	______	______

107. Have you done anything to improve the water drainage of your soil (e.g. tiling)?

1 Yes (go to 108)
2 No (skip to 109)

108. Describe what you have done to improve drainage.

____

____

109. Have you ever had your irrigation water tested

a. for nitrates, nitrites or pathogens?

1 Yes
2 No

b. for cation content and pH?

1 Yes
2 No

110. Describe the results of the test.

____________ ____________

____________ ____________

Cultural Regimes

111. Do you top prune?

1 Yes (go to 112)
2 No (skip to 117)

112. What stock types do you top prune?

______ ______ ______

113. When do you top prune? (When in the seedlings' growing regime?)

____

____

114. How often do you top prune?

____

115. To what height?

____

116. Why do you top prune?

____

____

117. Do you have a target or optimum morphology which you try to achieve for each stock type?

1 Yes (go to 118)
2 No (skip to 119)

118. Describe the target morphology (height, root:shoot ratio, and caliper) for your two (2) major species.

Species	Stock Type	Height	S:R or R:S	Caliper	Other
1.______	______	______ in cm	______	______ mm	______
______	______	______ in cm	______	______ mm	______
______	______	______ in cm	______	______ mm	______
2.______	______	______ in cm	______	______ mm	______
______	______	______ in cm	______	______ mm	______
______	______	______ in cm	______	______ mm	______

119. What stock type are you growing in greater quantity today than in 1975?

____

In lesser quantity?

____

Why?

____

____

120. What stock type do you foresee as being grown in greater quantity by 1985?

____

In lesser quantity?

____

Why?

____

____

Weed Control

121. Is there an herbicide which you would rather not use because of health risk concerns?

1 Yes (go to 122)
2 No (skip to 123)

122. What herbicide and why would you rather not use it?

____

____

123. Have you observed any coniferous seedling damage from herbicides?

1 Yes (go to 124)
2 No (skip to 125)

124. Which herbicides and tree species? What type of damage was observed?

Herbicides	Tree Species	Type of Damage Observed

Disease and Insects

125. Where do you get help on insect and disease problems?

Consultants ______ Government Pest Specialists ______

Chemical Representatives ______ Other Nurseryman ______

In-House Specialists ______

126. What is your preferred way to control insects and diseases?

______ Cultural Means (Varying Nursery Practices)

______ Pesticides

______ Biological Control

______ Other

127. What kind of information would be valuable to you to manage these pests? (Rank in order of importance).

______ Life History

______ Identification

______ Control Methods

______ Storage and Disposal of Pesticides

______ Other ______

128. Do you think your cover crops contribute to weed or other pest problems?

1 Yes Why? ______________________________
2 No

Mycorrhizae

129. Do you notice an abundance of mycorrhizae on your seedling root systems when lifted? Or are there very few?

Abundance ______________________ Few
5 4 3 2 1

130. Have you noticed fungal fruiting bodies in your nursery?

1 Yes (go to 131)
2 No (skip to 132)

131. Are there many types or just one type of mushroom?

__________ Many types __________ One type

Describe these mushrooms.

____

____

Seedling Growth in Relation to Nursery Environment

132. What range of seed zones do you grow? (i.e., coastal, mountain, East side). (List areas for which trees are grown).

__________ __________

__________ __________

__________ __________

133. How many seed zones do you grow for each of your major species? (One seed zone includes one elevation zone).

Species	# of Seed Zones
__________	__________
__________	__________
__________	__________
__________	__________

134. Do you have separate growing regimes for different seed zones and/or elevations? (e.g. sowing dates, irrigation schedules, lifting dates, etc.)

1 Yes (go to 135)
2 No (skip to 136)

135. Describe these regimes.

Sowing dates: ______________________

Irrigation schedules: ______________________

Lifting dates: ______________________

Other: ______________________

Lifting

136. When do you lift your stock? State the normal range of dates for each major species.

Species	Lifting Dates
__________	__________ to __________
__________	__________ to __________
__________	__________ to __________

137. How is your choice of lifting dates arrived at?

____

____

138. What are your lifting and pre-sort handling procedures?

a. Is stock undercut before lifting?

1 Yes (go to 138b)
2 No (skip to 138c)

b. How long in advance is stock undercut? (maximum time)

__________ hours

c. When the soil moisture level is low, do you irrigate before lifting?

1 Yes
2 No

139. Do you cover or water-down seedlings in field containers?

a. Cover?

1 Yes   Type of Cover ____________
2 No

b. Waterdown?

1 Yes
2 No

140. How long is lifted stock held before grading? State normal and maximum period.

Normal = __________ days

Maximum = __________ days

141. How is ungraded stock held?

a. In what container?

____

b. At what temperature and relative humidity?

__________ temperature

__________ relative humidity

c. How is it protected from dessication?

____

142. Do you shut down lifting operations if certain weather conditions arise?

1 Yes (go to 143)
2 No (skip to 144)

143. Under what conditions do you shut down lifting?

__________ temperature

__________ moisture

__________ wind speed

__________ wet soil

__________ high PMS

Grading

144. What culling standards do you use on the grading table?

caliper ____ height ____ root length ____

multiple tops ____ physical damage ____

145. How are your culling standards arrived at?

____

146. What percent of your stock is root pruned?

__________%

a. To what length?

__________ in cm (circle one)

147. Are the environmental conditions controlled in your packing shed?

1 Yes (go to 148)
2 No (skip to 149)

148. At what temperature and humidity is your packing shed controlled?

__________ temperature

__________ humidity

Packaging

149. What type and size of storage/shipping container do you use?

Waxed box ____ Unwaxed box ____ Polybag ____ Bundle(cloth) ____

Size of container ______

150. Are seedlings bundled when packaged?

1 Yes (go to 151)
2 No (skip to 152)

151. What is used to tie the bundles?

____

152. Do you use a moisture-holding medium such as sphagnum moss in your containers?

1 Yes (go to 153)
2 No (skip to 154)

153. What medium do you use?

____

Storage

154. At what ambient temperature and relative humidity are your seedlings stored?

__________ temperature

__________ relative humidity

155. Do you monitor inside your containers?

1 Yes (go to 156)
2 No (skip to 157)

156. What do you monitor?

________ temperature

________ mold development

________ seedling moisture stress

________ root viability

157. What percent of your seedlings are shipped in refrigerated trucks?

__________%

In non-refrigerated trucks?

__________%

158. What are the normal and maximum times stock is in transit?

__________ normal

__________ maximum

159. Are stock temperatures monitored in transit?

1 Yes (go to 160)
2 No (skip to 161)

160. How warm does the stock get in transit?

__________ temperature

161. Are seedlings shipped directly to the planting site?

1 Yes (skip to 163)
2 No (go to 162)

162. If not planted immediately, how long and under what conditions are they held after being shipped from the nursery?

____

____

Seedling Evaluation

163. Which of the following tests do you use to assess seedling vigor or condition:

	Tests (check which tests are used)						
Operational Stage	Dormancy	Cold Hardiness	Water Status	Root Growth Potential	Stress	Foliage Nutrients	Other
During the Growing Regime?	___	___	___	___	___	___	___
Before Lifting?	___	___	___	___	___	___	___
During Lifting and Processing?	___	___	___	___	___	___	___
During Cold Storage?	___	___	___	___	___	___	___
After Cold Storage?	___	___	___	___	___	___	___

164. How often and when do you take size measurements on your 2-0 stock?

__________________ __________________

165. What measurements do you take?

____ Caliper ____ Height ____ R:S Ratio ____ New Root Tips ____ Other ____

166. Do you plot and follow growth curves for your stock each year?

1 Yes
2 No

167. **What experimental trials have you done at your nursery?** Are the results available?

Trial	Are the Results Available?
______________	1 Yes 2 No
______________	1 Yes 2 No
______________	1 Yes 2 No
______________	1 Yes 2 No
______________	1 Yes 2 No
______________	1 Yes 2 No
______________	1 Yes 2 No
______________	1 Yes 2 No

168. Do you have any experimental trials at your nursery at the present time?

1 Yes (go to 169)
2 No (skip to 170)

169. List types of trials.

______________ ______________ ______________

______________ ______________ ______________

170. Do you monitor field growth and survival of your stock?

Yes (go to 171)
No (skip to 172)

171. Describe tests.

____

____

____

____

General Questions

172. In what areas of nursery technology do you feel more information is needed, i.e., further research is necessary? (List in order of importance).

1. ______________________________

2. ______________________________

173. In what areas of nursery technology do you feel the current level of technology is sufficient, i.e., no more research is necessary?

____

____

174. What is the best form in which you would like to receive results of new research?

____

____

175. Do you read:

	Often	Occasionally	Never
Tree Planter's Notes			
American Nurseryman			
Forestry Update			
Journal of Forestry			
Forest Science			
Canadian Journal of Forest Research			
Other ______________			

176. Whom do you contact concerning specialized nursery problems?

Soil? ______________

Insects? ______________

Seedling Quality? ______________

Diseases? ______________

Herbicides? ______________

Other? ______________

____

Appendix C: Conversion Tables

Length

Abbreviations

English	Metric
inches (in.)	millimeter (mm)
feet (ft)	centimeter (cm)
yard (yd)	meter (m)
mile (mi)	kilometer (km)

Conversions

English to English	Metric to Metric
1 ft = 12 in.	1 cm = 10 mm
1 yd = 3 ft	1 m = 100 cm
1 mi = 1,760 yd	1 km = 1,000 m
1 mi = 5,280 ft	

English to Metric	Metric to English
1 in. = 2.54 cm	1 cm = 0.3937 in.
1 ft = 0.3048 m	1 m = 3.281 ft
1 yd = 0.9144 m	1 m = 1.094 yd
1 mi = 1.609 km	1 km = 0.6214 mi

Area

Abbreviations

English	Metric
square inches (in.^{2})	square millimeters (mm^2)
square feet (ft^2)	square centimeters (cm^2)
square yards (yd^2)	square meters (m^2)
acres (ac)	hectares (ha)
square miles (mi^2)	square kilometers (km^2)

Conversions

English to English	Metric to Metric
1 ft^2 = 144 in.^{2}	1 cm^2 = 100 mm^2
1 yd^2 = 9 ft^2	1 m^2 = 10,000 cm^2
1 ac = 4,840 yd^2	1 ha = 10,000 m^2
1 ac = 43,560 ft^2	
1 mi^2 = 640 ac	1 km^2 = 100 ha
1 mi^2 = 27,880,000 ft^2	

English to Metric	Metric to English
1 in.^{2} = 6.452 cm^2	1 cm^2 = 0.1550 in.^{2}
1 ft^2 = 0.09290 m^2	1 m^2 = 10.7639 ft^2
1 yd^2 = 0.8361 m^2	1 m^2 = 1.196 yd^2
1 ac = 0.4047 ha	1 ha = 2.471 ac
1 mi^2 = 259.0 ha	1 ha = 0.003861 mi^2

Volume

Abbreviations

English	Metric
cubic inches (in.^{3})	cubic centimeters (cm^3)
cubic feet (ft^3)	cubic meters (m^3)
cubic yards (yd^3)	milliliters (mL)
teaspoon (tsp)	liters (L)
tablespoon (tbsp)	hectoliters (hL)
fluid ounces (fl oz)	
pint (pt)	
quart (qt)	
U.S. gallon (gal.)	
Imperial gallon (Imp gal.)	
bushel (bu)	
acre feet (ac ft)	

Conversions

English to English	Metric to Metric
1 tbsp = 3 tsp	1 m^3 = 1,000,000 cm^3
2 tbsp = 1 fl oz	1 mL = 1 cm^3
1 cup = 8 fl oz	1 L = 1,000 mL
1 pt = 16 fl oz	1 hL = 100 L
1 qt = 32 fl oz	
1 gal. = 128 fl oz	
1 Imp gal. = 1.201 gal.	
1 bu = 1.244 ft^3	
1 ac ft = 43,560 ft^3	

English to Metric	Metric to English
1 in.^{3} = 16.39 cm^3	1 cm^3 = 0.06102 in.^{3}
1 ft^3 = 0.02832 m^3	1 m^3 = 35.32 ft^3
1 yd^3 = 0.7646 m^3	1 m^3 = 1.308 yd^3
1 bu = 0.3524 hL	1 hL = 2.838 bu
1 ac ft = 1,234 m^3	1 m^3 = 0.0008107 ac ft

Mass (weight)

Abbreviations

English	Metric
avoirdupois ounces (oz)	grams (g)
pounds (lb)	kilograms (kg)
short ton (t)	metric ton (mt)

Conversions

English to English	Metric to Metric
1 lb = 16 oz	1 kg = 1,000 g
1 t = 2,000 lb	1 mt = 1,000 kg

English to Metric	Metric to English
1 oz = 28.35 g	1 g = 0.03527 oz
1 lb = 0.4536 kg	1 kg = 2.205 lb
1 t = 907.2 kg	1 kg = 0.001102 t
1 t = 0.9072 mt	1 mt = 1.102 t

All conversions contain four significant figures or are exact.

Pressure (mass per unit area)

Abbreviations

English	Metric
pounds per square inch (psi)	bar (b)
atmospheres (atm)	megapascal (MPa)

Conversions

English to English

1 psi = 0.06805 atm

Metric to Metric

1 MPa = 10 b

English to Metric

1 psi = 0.06895 b
1 psi = 0.006895 MPa
1 atm = 1.013 b
1 atm = 0.1013 MPa

Metric to English

1 b = 14.50 psi
1 MPa = 145.0 psi
1 b = 0.9869 atm
1 MPa = 9.869 atm

Temperature

Abbreviations

English	Metric
degrees Fahrenheit (°F)	degrees Celsius (°C)

Conversions

English to Metric

°F = 1.800 (°C) + 32

Metric to English

°C = 0.5556 (°F − 32)

Flow (volume per unit time)

Abbreviations

English	Metric
gallons per minute (gpm)	liters per minute (Lpm)
cubic feet per second (cfs)	cubic meters per second (cms)
cubic feet per minute (cfm)	

Conversions

English to English

1 cfs = 448.8 gpm
1 cfm = 0.01667 cfs
1 cfm = 7.480 gpm

Metric to Metric

1 cms = 60,000 Lpm

English to Metric

1 cfs = 0.02832 cms
1 cfs = 1,699 Lpm
1 gpm = 0.00006309 cms
1 gpm = 3.785 Lpm
1 cfm = 0.0004720 cms
1 cfm = 28.32 Lpm

Metric to English

1 cms = 35.32 cfs
1 Lpm = 0.0005886 cfs
1 cms = 15,850 gpm
1 Lpm = 0.2642 gpm
1 cms = 2,119 cfm
1 Lpm = 0.03532 cfm

All conversions contain four significant figures or are exact.

Seedbed density

Abbreviations

English	Metric
lineal bed foot (bed ft)*	lineal bed meter (bed m)*
square foot (ft^2)	square meter (m^2)

Conversions

42-inch usable bed space**

English to English

1 bed ft = 3.5 ft^2

Metric to Metric

1 bed m = 1.067 m^2

English to Metric

1 bed ft = 0.3252 m^2
1 ft^2 = 0.09290 m^2
1 ft^2 = 0.08709 bed m

Metric to English

1 bed m = 11.48 ft^2
1 m^2 = 10.76 ft^2
1 m^2 = 3.075 bed ft

48-inch usable bed space**

English to English

1 bed ft = 4 ft^2

Metric to Metric

1 bed m = 1.219 m^2

English to Metric

1 bed ft = 0.3716 m^2
1 ft^2 = 0.09290 m^2
1 ft^2 = 0.07620 bed m

Metric to English

1 bed m = 13.12 ft^2
1 m^2 = 10.76 ft^2
1 m^2 = 2.691 bed ft

*One lineal bed ft (or 1 lineal bed m) equals an area of seedbed 1 ft (or 1 m) long times the width of the bed.
**Usable bed space is the area of seedbed actually occupied by seedlings.

Fertilizer

Abbreviations

English	Metric
ounces per square foot (oz/ft^2)	grams per square meter (g/m^2)
pounds per acre (lb/ac)	kilograms per hectare (kg/ha)
parts per million (ppm)	

Conversions

English to English

1 oz/ft^2 = 2722 lb/ac

Metric to Metric

1 g/m^2 = 10 kg/ha

English to Metric

1 lb/ac = 1.121 kg/ha

Metric to English

1 kg/ha = 0.8921 lb/ac

Other useful fertilizer conversions

phosphorus (P) = phosphoric acid (P_2O_5) x 0.4364
P_2O_5 = P x 2.291
potassium (K) = potash (K_2O) x 0.8301
K_2O = K x 1.205

pounds per acre to parts per million

If 1 acre-foot of soil weighs approximately 4 million pounds, then 1 plow-slice (9 inches deep) weighs 3 million pounds per acre. Therefore:

1 ppm = 3 lb/ac
1 lb/ac = 0.3333 ppm

Appendix D: Glossary

Compiled by S. K. Omi and T. D. Landis

In addition to the definitions of terms contributed by the chapter authors of the *Manual*, the following sources have been used in compiling this glossary:

Abercrombie, M. C. J. Hickman, and M. L. Johnson. 1973. A dictionary of biology. 6th ed. Penguin Books, New York. 309 p.

Bohmont, B. L. 1983. The new pesticide user's guide. Reston Publishing Co. Inc., Reston, Virginia. 452 p.

Cleary, B. D., R. D. Greaves, and R. K. Hermann (eds.). 1980. Regenerating Oregon's forests. Oregon State Univ. Extension Serv., Corvallis, Oregon. 286 p.

Donahue, R. L., R. W. Miller, and J. C. Shickluna. 1983. Soils. An introduction to soils and plant growth. 5th ed. Prentice-Hall Inc., Englewood Cliffs, New Jersey. 667 p.

Hanan, J. J., W. D. Holley, and K. L. Goldsberry. 1978. Greenhouse management. Springer-Verlag, Berlin, Heidelberg, New York. 530 p.

Miller, A. V., and S. M. Craig (eds.). 1979. Handbook for pesticide applicators and pesticide dispensers. 4th ed. Pesticide Control Branch, Ministry of Environment, Victoria, B.C. 2?3 p.

Soil Science Society of America. 1965. Glossary of soil science terms. Soil Science Society of America Proceedings 29(3):330-351.

Williams, R. D., and S. H. Hanks. 1976. Hardwood nurseryman's guide. U.S.D.A. Forest Serv., Washington, D.C. Agric. Handb. 473. 78 p.

A

Absorption: Movement of a pesticide from the surface into a body of water or of nutrients into a plant (compare with **adsorption**).

Acid soil Soil having a pH value less than 7.0.

Active ingredient (a. i.): Portion of a pesticide formulation that produces the desired toxic, stimulatory, or repelling effect, expressed as a percentage.

Adjusted sodium adsorption ratio (ASAR): Irrigation-water quality index that is the ratio of deleterious ions (i.e., Na^+, CO_3^{--}, and HCO_3^-) to beneficial ions (i.e., Ca^{++} and Mg^{++}). The ASAR value is used to assess the effect of salts on soil permeability compare with **residual sodium carbonate, RSC**).

Adsorption: Attraction or bonding of ions or compounds, usually temporarily, to the surface of a solid (compare with **absorption**).

Advection frost: Frost occurring when cold air flows from a higher location to a lower one, displacing lighter, less dense air and often pooling to form a frost pocket (compare with **radiation frost, wind frost**).

Aeration (soil): Process by which (1) oxygen diffuses through the soil to the root and (2) carbon dioxide and other gases released from the root diffuse to the soil surface.

Aerobic: Occurring or growing in the presence of oxygen.

Alkaline soil: Soil having a pH value greater than 7.0.

Allelopathy: Production of chemical compounds by one plant which are released into the soil environment and are harmful to other nearby plants.

Amendment: Any substance added to a soil to alter its physical or chemical properties and thereby make that soil more useful for plant production.

Anaerobic: Occurring or growing in the absence of oxygen.

Anion: Ion having a negative charge, e.g., Cl^-, NO_3^- (compare with **cation**).

Annual: Plant that completes its entire life cycle from seed germination to seed production and death within a single season (compare with **perennial, biennial**).

Artificial regeneration: Reforestation of a stand by direct seeding or planting (compare with **natural regeneration**).

Available nutrient: Any essential element or compound in the soil that can be readily absorbed and assimilated by plants.

Available water: Portion of soil water that can readily be absorbed by plant roots. Generally considered to be that water held in the soil within a water-potential range of approximately − ⅓ to − 15 bars.

B

Band application: Spreading of a chemical or fertilizer to a restricted area (such as in, on, or along a crop row), rather than over an entire field or area (compare with **broadcast application**).

Bar: Unit of pressure equaling 1,000,000 dynes/cm²; typically used to measure plant and soil water potential.

Base saturation percentage: Proportion of the cation exchange capacity (CEC) occupied by cations other than hydrogen or aluminum expressed as a percentage of the total CEC.

Bed: Elongated strip of soil in which seedlings or transplants are grown.

Bed foot (meter): Area of seedbed 1 lineal foot (or 1 lineal meter) long times the width of the bed.

Biennial: Plant that requires 2 years to complete its entire life cycle, from seed germination to seed production and death. Food is stored in the first season and then used in the second season when flowers and seed are produced (compare with **annual, perennial**).

Biometrics: Use of statistical techniques to evaluate biological problems.

Box pruning: Root-culturing technique that consists of the lateral pruning of roots in a four-sided box shape around a seedling in a seedbed (compare with **lateral pruning**).

Broadcast application: Spreading of a chemical or fertilizer over an entire area or field rather than only on rows, beds, or individual plants (compare with **band application**).

Broadcast seeding: Method of sowing in which seeds are distributed across the seedbed by hand (compare with **drill seeding**).

Broadleaf species: Those plants classified as Dicotyledoneae; characterized by having round and flattened leaves (compare with **narrow leaf species**).

Brushing: Frost-protection technique in which shields are placed near plants to deflect radiation to the plant and soil during the day and reduce heat loss to the sky at night; used primarily for agricultural crops.

Buffer capacity: Ability to resist change in pH. A soil with a high buffer capacity will have stable soil pH.

Bulk density (soil): Mass (weight) of dry soil divided by soil volume, commonly expressed as pounds per cubic foot or grams per cubic centimeter.

C

Caliper: Stem diameter of a seedling, usually measured just above the root collar.

Candle: Terminal shoot of conifers, especially pines, that has elongated but not yet produced needles. Color of bud scales or a waxy secretion gives the terminal shoot the appearance of a white candle.

Capillary water: Water held in the small pores of a soil, usually with attraction forces exceeding the pull of a 60-cm column of water. Most capillary water is available to plants.

Carbon:nitrogen (C:N) ratio: Ratio of carbon content to nitrogen content. The C:N ratio in plant residues is often a convenient predictor of decomposition rates but is not the only determinant.

Catch crops: Crops which are grown and then incorporated into the soil, primarily to capture and retain nutrients on the site but also to increase soil organic matter (compare with **cover crops, green manure crops**).

Cation: Ion having a positive charge, e.g., Ca^{++}, Mg^{++} (compare with **anion**).

Cation exchange capacity (CEC): Sum total of exchangeable cations that a soil can adsorb, expressed in milliequivalents per 100 grams of soil, clay, or organic colloid. Because of the high CEC of organic matter, the buffer capacity of a soil is increased after organic amendments are added, and the soil is better able to hold cationic nutrients.

Chelation: Formation of strong bonds between metals and organic compounds. Those chelates used in fertilizers are soluble and increase and maintain the availability of micronutrients (such as iron, zinc, manganese, and copper) for plant uptake.

Chilling requirement: A certain period (usually expressed in hours) of cold temperatures (0 to 10°C) that seeds or seedlings must experience before growth occurs in periods of warmer temperatures. This adaptive mechanism helps to protect buds from flushing during temporary warm-temperature spells in winter or early spring and being subsequently killed when cold temperatures return.

Chilling sum: Cumulative number of hours that a seedling is exposed to a certain range of temperatures, of either soil or air, known to effectively release dormancy for the species of interest (e.g., for Douglas-fir in the Northwest, the temperature range is 0 to 5°C). Chilling sums are used to assess seedling dormancy status and to determine proper lifting dates.

Chiseling: Breaking or loosening soil, without inverting it, with a cultivator or chisel plow, generally below the normal plow depth (compare with **subsoiling, ripping**).

Chlorosis: Yellowing of normally green plant tissue, due to a lack of chlorophyll. Chlorosis can be a symptom of disease, nutrient deficiency, or inadequate light.

Clay: Soil particle less than 0.002 mm in diameter; soil textural class characterized by a predominance of clay particles.

Claypan: Dense, compact layer in the subsoil which has a much higher clay content than the overlying material, resulting from the downward movement of clay or the synthesis of clay in place during soil formation. Claypans, separated from the soil material above by a sharply defined boundary, are typically hard when dry and plastic and sticky when wet. They usually impede water and air movement and growth of plant roots (compare with **hardpan**).

Coefficient of variation (CV): Relative measure of variation in which the standard deviation is expressed as a percentage or fraction of the treatment mean.

Compaction (soil): Increase in bulk density, hence lower porosity, of a soil, due to the rearrangement of soil aggregates from applied loads, pressure, or vibration. The reduction of pore spaces between particles impedes gas and water exchange and also root penetration.

Complete fertilizer: Fertilizer containing all three of the major fertilizer elements—nitrogen (N), phosphorus (P), and potassium (K); the concentration is expressed as the percentage of N but as the percentage of the *oxide* of phosphorus (P_2O_5) and potassium (K_2O).

Compost: Organic residues or a mixture of organic residues and other materials (e.g., sawdust combined with nitrogen fertilizer or sewage sludge) that have been piled and allowed to undergo biological decomposition.

Concretion (soil): Local concentration of a chemical deposit, such as calcium carbonate or iron oxide, in the form of an aggregate or nodule of varying size, shape, hardness, and color.

Confidence interval: Specified range within which the quantity of interest (e.g., treatment mean) will be contained for a specific level of confidence (e.g., 95%).

Contact herbicide: Herbicide that kills plant tissue by direct contact rather than by translocation or root uptake (compare with **systemic herbicide**).

Control treatment: Zero-level application of a treatment. A control treatment (e.g., no wrenching) is used to judge whether particular treatment levels (e.g., multiple wrenchings) are effective (compare with **standard treatment**).

Cotyledon: First leaf or leaves of the embryo in seed plants. In conifers, the cotyledon stage occurs after the seedling has emerged and until the primary (true) leaves develop.

Cover crops: Crops grown principally to control various forms of erosion but also incorporated into the soil to increase organic matter (compare with **catch crops, green manure crops**).

Cull: Seedling which is not acceptable because it does not meet certain size and quality standards and which is thought to have low survival and growth potential.

Cull factor: Number of seedlings that do not meet shippable standards (e.g., diseased, poor form or size, damaged), expressed as a percentage.

Cultural control: Indirect control measure (as opposed to direct killing) used to prevent pest damage—e.g., use of a previous crop that is antagonistic to the pest.

D

Damping-off: Disease characterized by either seed decay in the soil or seedling wilting and death after germination, usually caused by soil-borne fungi.

Deciduous plants: Plants which shed their leaves at a certain season usually autumn.

Deductive reasoning: Drawing conclusions or making inferences which logically follow from a well-defined base of biological, chemical, or physical principles (compare with **inductive reasoning**).

Designed experiment: Detailed investigation in which the experimenter or researcher requires precise and unbiased (1) conclusions and (2) measures of uncertainty associated with those conclusions. Treatments are usually replicated and applied to plots at random (compare with **operational trial**).

Determinant growth: A 2-year process of shoot growth in which primordia are laid down in buds in the first year but elongate in the second year. Species exhibiting determinant growth may only have one growth flush per year (compare with **indeterminant growth**).

Dicot (Dicotyledoneae): Larger of the two classes of flowering plants (Angiospermae), distinguished from the smaller class (Monocotyledoneae) by the presence of two seed leaves (cotyledons) in the embryo and by other structural features, e.g., net-veined leaves; includes oak, elm, alder. Certain herbicides (e.g., 2,4-D) are effective against dicots but do not harm monocots.

Disking: Breaking up surface layers of soil with a disk implement, to destroy weeds, prepare the soil for planting, or incorporate a pesticide or fertilizer.

Dormancy: Condition in which a tissue predisposed to elongate does not do so even if environmental conditions are suitable for growth. Dormancy, composed of different phases, is a plant adaptation to survival under stress (e.g., frost, drought).

Drench: Saturation of a soil with pesticide, usually to control root diseases.

Drill seeding: Nursery sowing method in which seeds are planted in rows with a seed-drilling implement (compare with **broadcast seeding**).

Dysgenic: Causing harmful changes in the genetic makeup of a population.

E

Ectendomycorrhiza(e): Group of mycorrhizae which have both intercellular and intracellular fungal penetrations of root cortical cells. The branching and Hartig net formation are similar to those in ectomycorrhizae; infection within cortical cells is similar to that in endomycorrhizae.

Ectomycorrhiza(e): Group of mycorrhizae in which the fungal hyphae penetrate between the host root cells, often forming a mantle or sheath over the feeder roots. Ectomycorrhizae are common on members of the Pinaceae, Fagaceae, Betulaceae, and Salicaceae.

Electrical conductivity: Reciprocal of electrical resistance, expressed in "mhos" (reverse of "ohms"). A method for expressing salt concentration in soil or water.

Embryo: Young plant developing after the union of male and female gametes. In seed plants, the embryo is contained in the seed.

Emulsifiable concentrate (EC): Liquid pesticide formulation consisting of an active ingredient, a solvent, and an emulsifier that mixes with water to form an emulsion.

Emulsifier: Material which helps to suspend one liquid in another, such as oil in water.

Emulsion: Mixture of two or more immiscible liquids, such as oil and water, in which one is suspended or dispersed in the other in the form of very minute droplets and remains so through the use of an emulsifier.

Endomycorrhiza(e): Group of mycorrhizae in which the hyphal infections of host roots are intracellular. This group is not as common in conifer species as it is in many angiosperms and herbaceous species, although cedars and redwoods have endomycorrhizae.

Evapotranspiration: Sum of the water transpired by vegetation plus that evaporated from the soil.

Experiment: Planned inquiry designed to obtain new facts or to confirm or deny information from previous results, to aid in making recommendations or decisions.

Experimental plot: Smallest physical unit (e.g., specific length of nursery bed) to which a treatment is applied independent of other treatments.

F

Factorial experiment: Experiment in which each level of a given factor is tested across each level of one of more other factors.

Fallow: Allow cultivated land to lie idle during the entire or greater portion of the growing season.

Fertilizer: Any organic or inorganic substance, either of natural or synthetic origin, which is added to the soil to provide elements essential for plant growth.

Field capacity: Soil water content resulting after the free water has been allowed to drain from a saturated soil for 1 to 2 days; expressed as a percentage on a dry-weight basis.

Field efficiency: Amount of area growing trees divided by the amount of area cultivated in a given field; expressed as a percentage. Field efficiency primarily depends on the distance between irrigation lines and the width of tractor paths.

Frost cracking: Type of cold injury caused by freezing and thawing in which exterior and interior portions of the bark expand and contract at different rates, causing the bark to crack.

Frost hardiness: Ability of plant tissue to survive and resist the stress from freezing temperatures without sustaining irreversible physical damage.

Frost heaving: Lifting of the soil surface due to growth of ice crystals in the underlying soil; when this recurs over a period of time, seedlings can be physically lifted out of the ground.

Frost pocket: Area whose topographic features cause cold air to accumulate, increasing frost hazard to seedlings.

Frost smothering: Condition occurring when a saturated soil freezes, allowing little or no oxygen to reach plant roots. If the condition persists over 48 hours, seedling damage or death can result.

Fumigant: Chemical applied as liquid or powder which volatilizes to gases and kills insects, nematodes, fungi, bacteria, seeds, roots, rhizomes, or entire plants. Fumigants are usually applied beneath a tarp, sheet, or other enclosure.

Fumigation: Use of chemicals in gaseous form to destroy pests, usually applied under a cover or shelter.

Fungicide: Chemical used to kill or inhibit fungi.

Fungistasis: Inhibition of fungal growth, without destroying the fungus, by preventing the germination of conidia and other spore types.

G

Genotype: Genetic constitution of an organism, i.e., the set of genes belonging to an individual. Genotype interacts with environment to produce the phenotype (compare with **phenotype**).

Germination: The beginning of growth of a mature, generally dormant seed.

Germination percent (seed): Percentage of seeds that germinate under standard treatment and after a given time period. This value, considered a principal index of seed quality, is used to calculate seedbed sowing density.

Germinative capacity: Number of seeds in a given sample that actually germinate regardless of time required, expressed as a percentage.

Germinative energy: Number of seeds that have germinated at the time of peak germination, expressed as a percentage.

Gley: Layer of mineral soil developed under conditions of poor drainage, characterized by reduction of iron and other elements, and gray colors and mottles.

Grading: Process of identifying and subsequently separating various classes of acceptable (shippable) and inferior (cull) stock to improve stock quality. This operation occurs after lifting and before packing and storing.

Gravitational potential: Component of soil water potential caused by the force of gravity.

Green manure crops: Crops grown primarily as organic amendments for the soil. Green manure crops are incorporated into the soil while green but before seedset, to benefit succeeding crops (compare with **catch crops, cover crops**).

H

Hardening off: Natural process of adaptation by plants to cold or drought. Hardening off may be induced in the nursery by reducing water or by root culturing, thus preparing the seedling for outplanting or transplanting.

Hardpan: Hardened soil layer caused by cementation of soil particles with materials such as silica, sesquioxides, or calcium carbonate. The hardness does not change appreciably with changes in moisture content (compare with **claypan**).

Hartig net: Hyphal network of ectomycorrhizae which penetrates between root cortical cells of the host. This extensive contact between fungus and plant root facilitates the exchange of nutrients and other substances.

Herbicide: Chemical used to kill or inhibit unwanted plants or weeds.

Hormone: Growth-regulating substance synthesized in one location, usually in small amounts, and then transported to another location within the organism where if affects growth and differentiation.

Humic acid: Mixture of dark-colored organic materials of indefinite composition extracted from soil with dilute alkali and precipitated by acidification.

Humification: Breakdown of organic residues to humus.

Humus: Fraction of soil organic matter remaining after most plant and animal residues have decomposed; usually dark colored. The chemical composition of humus is very different from that of the original parent compound. Humic substances (1) help the soil retain water, (2) increase the cation exchange capacity, and (3) stabilize soil pH.

Hydraulic conductivity (K): Flow-rate constant, expressed in centimeters/second, which indicates the ability of soils to transmit flowing water. Values of K commonly range from 1×10^{-1} cm/sec in sands to 1×10^{-9} cm/sec in tight clays.

Hydroponics: Commercial production of plants in sand or gravel cultures. The sand or gravel is relatively inert providing mechanical support for growth, and nutrients are supplied by liquid solutions.

Hypothesis testing: Statistical technique which uses experimentation to support or reject a (null) hypothesis, formulated before the experiment, with a certain level of confidence.

IJK

Incorporation: Mixing of a fertilizer or chemical into nursery soil before sowing.

Indeterminant growth: Shoot growth resulting from the successive periods of initiation and elongation of apical meristem cells during the growing season without extended periods of rest (compare with **determinant growth**).

Inductive reasoning: Making inferences or drawing conclusions, based on a limited number of observations, about a wider sphere of interest (compare with **deductive reasoning**).

Infiltration rate: Rate at which water can be absorbed into a soil surface. Infiltration rate can be altered by nursery practices influencing the porosity and structure of surface soils. A soil with a poor infiltration rate is subject to surface runoff and erosion.

Inoculation: Process of introducing microorganisms for some beneficial effect, such as the addition of *Rhizobium* bacteria to legume seed or of mycorrhizal fungi to nursery seedbeds.

Inoculum: Portions of a pathogen (e.g., fungal spores) capable of causing infection or initiating mycorrhizae upon contact with the host.

Interception system (drainage): Type of closed drainage system located below the water-table level, intended to remove water before it enters the soil. The system may vary from a simple line of pipes positioned at the source of seepage to a complex grid pattern and is usually used in small land areas (compare with **relief system**).

Ions: Atoms or groups of atoms which are electrically charged, i.e., cations or anions.

L

Lammas shoots: Additional flushes of growth on the terminal shoot which result from bursting of current-year buds, thought to be stimulated by excess fertilization or irrigation. A seedling with lammas shoots may be more susceptible to frost damage or less resistant to the stresses of lifting, storage, and planting.

Lateral pruning: Root-culturing technique in which blades or coulters are passed between drill rows to sever long lateral roots. The purpose of lateral pruning is to facilitate lifting, stimulate root growth and fibrosity, and retard height growth (compare with **box pruning**).

Leaching: Downward movement of materials in the soil solution. Soluble nutrients such as nitrate are often leached out of the seedling root zone.

Lifting window: Time period of the year believed to be the best for harvesting seedlings from the seedbed, i.e., when seedlings are most resistant to handling stresses and when subsequent survival and growth potential upon outplanting are high. The lifting window will vary from year to year depending on seed source, nursery location, and cultural regimes used before lifting.

Lignification: Deposition of lignin (complex aromatic compounds) in the cell walls of sclerenchyma, xylem vessels, and tracheids making them rigid.

Lime (calcium) requirement: Amount of agricultural limestone required per acre to raise the soil pH to an optimal value for seedling growth; usually calculated for a soil depth of 15 cm or per 910,000 kg of soil.

Liming: Addition of calcium, sometimes including magnesium (dolomite), in the form of calcium carbonate, ground limestone, or hydrated lime to furnish elements for plant growth and to neutralize soil acidity.

Loam: Textural class for a soil having moderate amounts of all three soil separates—sand, silt, and clay.

Luxury consumption: Absorption of excess amounts of nutrients beyond those necessary for optimal growth.

M

Macronutrient: Chemical element necessary for plant growth in large amounts (usually greater than 1 part/500 in the plant) often supplied artificially in fertilizers (compare with **micronutrient**).

Maintenance dressing: Fertilizer application which functions to keep soil macronutrient and micronutrient levels adequate during the crop rotation.

Material attributes: Individual measurements of one aspect of seedling quality which can indicate physiological condition, e.g., leaf osmotic potential, root starch concentration, and shoot:root ratio. Performance attributes reflect the sum total of material attributes.

Matric potential: Largest component of total soil-water potential; caused by capillary action and attraction of water by soil particles. Matric potential can be measured with tensiometers and is usually expressed in negative pressure units such as − bars.

Measurement plot: That part of the experimental plot in which the observational units are measured.

Megagametophyte: Haploid generation portion of a seed plant representing the female contribution to the developing embryo.

Metabolism: Chemical processes comprising the synthesis and degradation of constituents within an organism.

Micronutrient: Chemical element necessary for plant growth in very small amounts (less than several parts per million in the plant). Micronutrient fertilizers are not normally needed except under soil conditions such as excessively high or low pH (compare with **macronutrient**).

Milliequivalent (meq): One milligram (mg) of hydrogen or the amount of any other ion that will combine with it. Milliequivalents are units used in cation exchange capacity and fertility calculations. For example, 1 meq of a calcium ion (Ca^{++}) is computed as its atomic weight in grams (40) divided by the valence (2), or 20 mg.

Mineralization: Conversion of organic elements to the inorganic state as a result of microbial decomposition.

Mineral soil: Soil consisting largely of mineral matter, with organic matter usually less than 20%.

Mineral spirits: Derivatives of naphthenic petroleum, containing 10 to 20% aromatic hydrocarbons, used as a solvent for some pesticides; sometimes used in tree nurseries as a contact herbicide.

Monocot (Monocotyledoneae): Smaller of the two classes of flowering plants (Angiospermae), distinguished from the larger class (Dicotyledoneae) by the presence of a single leaf (cotyledon) in the embryo and by other structural features, e.g., parallel-veined leaves; includes grasses, lilies, orchids. Certain herbicides are effective against monocots but do not harm dicots.

Mulch: Layer of plant residues or other material (e.g., plastic film, paper fiber) spread upon the soil surface to protect soil, seeds, or plant roots from the effects of freezing, evaporation, crusting, etc.

Mycorrhiza(e): The biological association, usually symbiotic, between plant roots and particular fungi.

N

Narrow leaf species: Those plants classified as Monocotyledoneae; characterized by having narrow, parallel-veined leaves (compare with **broadleaf species**).

Natural regeneration: Reforestation of a stand by natural seeding (compare with **artificial regeneration**).

Nitrification: Biological process in which (1) ammonium is oxidized to nitrites, and (2) nitrites are further oxidized to nitrates.

Nonselective pesticide: Material that is toxic to a wide range of pests or to more than one plant or animal.

Null hypothesis: Specific hypothesis about a population that is being investigated by analyzing data from a sample of that population. For both statistical and biological reasons, elements under investigation are hypothesized to have *no* effect on the response variable—thus, the use of "null."

O

Observational unit: Item to be measured within an experimental plot (e.g., tree seedling).

Operational trial: Preliminary inquiry in which each treatment is applied to only one plot (i.e., treatments are unreplicated) (compare with **designed experiment**).

Organic matter: The complex interaction of (1) plant, animal, and microbial residues in various stages of decay, (2) humus, and (3) live organisms. Organic matter increases the buffer capacity, cation exchange capacity, and water retention of the soil and provides a substrate for microbial activity.

Organic soil: Soil usually containing 20% or more organic matter.

Ornamentals: Plants, including trees, shrubs, and flowers, which function to beautify homes, gardens, and lawns; refers to stock used for landscaping rather than wildland plantings.

Osmotic potential: Pressure that would have to be applied to a solution to prevent water from moving from the less concentrated solution to the more concentrated one when two solutions are separated by a semipermeable membrane. Soil osmotic potential refers to solute concentration in the soil solution; plant osmotic potential refers to solute concentration of the cell sap.

Outplanting: Planting of seedlings on a forest site.

Oxidation: Chemical process of combining with oxygen; removal of hydrogen or electrons.

P

Parasite: Organism that lives on other living organisms.

Pathogen: Specific agent (usually fungus, bacterium, virus, or nematode) that can cause infectious disease.

Peat: Largely undecomposed or slightly decomposed organic matter accumulated under conditions of excessive moisture and low oxygen availability; soil amendment used to increase soil organic matter and lower soil pH.

Perched water table: Surface of a local zone of water saturation held above the main body of ground water by an impermeable layer, usually clay or rock, and separated from the main body of ground water by an unsaturated zone.

Percolation rate: Downward movement of water through the soil, particularly the downward water flow in saturated or nearly saturated soil. Percolation rate is used to calculate the internal drainage requirements of a soil.

Perennial: Plant that continues growing from year to year. Tops may die back in winter but roots or rhizomes persist (compare with **annual, biennial**).

Performance attributes: Attributes of seedling quality measured by assessing the performance of seedlings subjected to environmentally controlled test conditions, e.g., root-growth potential and frost hardiness. Performance attributes reflect the sum total of material attributes.

Permanent wilting percentage: Water content of a soil when indicator plants growing in that soil wilt and fail to recover when placed in a humid chamber; varies with plant species but is generally considered to occur at approximately −15 bars water potential.

Permeability (soil): Soil attribute that enables water or air to move through it; determined by soil porosity.

pH: Numerical measure (negative logarithm of the hydrogen ion activity) of the acidity or alkalinity in a soil or solution. A pH reading of 7 is neutral for soils measured in water paste.

Phenotype: Morphological appearance of an organism. The phenotype results from the interaction of genotype and environment (compare with **genotype**).

Photoperiodism: Plant response to relative length of day and night. Temperature and photoperiod are the principal environmental factors affecting plant dormancy.

Photosynthesis: Production by plants containing chlorophyll of organic compounds from water and carbon dioxide, using energy absorbed by the chlorophyll from light.

Phytotoxic: Causing injury or death to plants.

Plant moisture stress (PMS): Measure of plant water status; equal to the absolute value of plant water potential. PMS is an integrated index of the current moisture status of a plant, and is influenced by soil moisture status and evaporative demand of the atmosphere.

Plant water potential: Current water status of a plant; consists of two components, turgor potential and osmotic potential, and is measured in negative pressure units (− bars).

Plasmolysis: Shrinkage of cell protoplasm away from the cellulose wall due to osmotic withdrawal of water.

Plug plus one (p+1): Transplanted seedling that was started in a container and then transplanted to the field, usually for 1 year (compare with **stock type**).

Point estimate: Number that estimates a certain quantity in the population of interest (e.g., treatment mean standard deviation).

Pore space: Total space not occupied by soil particles in a bulk volume of soil.

Porosity (soil): Volume of total soil bulk not occupied by solid particles, expressed as a percentage. Percent porosity equals the volume of pores divided by total soil volume.

Postemergence: Time period *after* crop plants or weeds emerge through the soil surface.

Preemergence: Time period *before* crop plants or weeds emerge through the soil surface.

Preplanting treatment: Application of, e.g., herbicide or fertilizer before a crop is planted.

Profile (soil): Vertical section of soil extending through all of its horizons and into the parent material.

Prolepsis shoots: Shoot growth resulting from the expansion of lateral buds at the base of the terminal bud or on lateral shoots. Prolepsis shoots may be more susceptible to winter injury because of inadequate time to harden off.

Propagule: Any part of a plant that is capable of growing into a new organism.

QR

Quiescence (seedling): Period of physiological inactivity preceding true dormancy when plants still can grow if environmental conditions are suitable.

Radiation frost Frost occurring when large amounts of heat in soil and plants are dissipated into the atmosphere, allowing temperatures near the ground to reach the freezing point. This condition occurs on cloudless nights with little or no wind (compare with **advection frost, wind frost**).

Radicle: Root of the embryo in seed plants.

Randomization: Assignment of treatments to a set of plots such that all plots are equally likely to receive any treatment.

Relief system (drainage): Type of closed drainage system located below the water-table level, designed to drain already saturated soils. The system may consist of a simple pipeline or a complex interconnecting network (compare with **interception system**).

Replication: Repetition of a treatment in an experiment.

Residual sodium carbonate (RSC): Irrigation-water quality index that measures the difference between the sum of calcium and magnesium ions and the sum of carbonate and bicarbonate ions. RSC values are used to determine the effect of salts on soil permeability (compare with **adjusted sodium adsorption ratio, ASAR**).

Respiration: Metabolic process of taking oxygen from the environment to produce energy and release carbon dioxide (in organisms); oxidative breakdown of fuel molecules and subsequent release of energy (in cells).

Response variables: Characteristics which an experiment and its treatments are designed to test (e.g., height, caliper).

Ripping: Cultural practice used to ameliorate compacted subsoils by pulling shanks through the soil at a depth of 40 to 80 cm. Usually, the shanks are then pulled at right angles to the first pass to produce a grid pattern (compare with **subsoiling, chiseling**).

Rolling: Cultural practice used before sowing to ensure good contact between seeds and soil particles. A cylindrical roller is passed over the land to firm the soil without causing a great deal of compaction.

Root culturing: General term for those nursery cultural practices designed to modify seedling root growth (e.g., undercutting, wrenching).

Root-growth capacity (root-growth potential): Physiological capability of a plant for producing new roots under optimal environmental conditions for root growth. High root-growth capacity is thought to be one of the best indicators of seedling field survival.

S

Salt: Water-soluble chemical compound that either is found naturally in the soil or is applied as fertilizer. In soil solutions, salts are generally dissociated into cations (e.g., calcium, sodium, potassium) and anions (e.g., sulfates, nitrates, bicarbonates). High salt levels are detrimental to plant growth. Conifer seedlings are especially sensitive to salts.

Sand: Soil particle between 0.05 and 2.00 mm in diameter; soil textural class characterized by a predominance of sand particles.

Saprophyte: Organism that lives on dead or decaying organic matter.

Scarification (land): Type of site preparation in which the duff and litter layers of the forest floor are removed or the mineral soil is mechanically mixed with the organic surface layer.

Scarification (seed): Process of scratching the seedcoat with abrasive material to improve germination of seeds with hard seedcoats which are relatively impervious to water.

Seedbed: Elongated strip of prepared soil in which seeds are sown and seedlings raised.

Seedbed density: Number of seedlings growing in a seedbed, expressed relative to area (e.g., number per square meter or foot) or lineal measure (e.g., number per lineal meter or foot).

Seedling: Young tree propagated from seed.

Seedling quality: Potential of a seedling to survive and grow successfully after outplanting.

Seedling water potential: See **plant water potential**.

Seedlot: Quantity of seeds from a particular location and elevation (seed zone) which are reasonably similar or uniform in quality. The identity and integrity of each seedlot (one of the basic divisions in seedling recordkeeping) are maintained during seed storage and during the nursery production period.

Seed protectant: Pesticide applied to seed before planting to protect seeds and new seedlings from diseases and insects.

Seed purity percent: Proportion of the total seedlot that is seed and not debris, expressed as a percentage of the seedlot weight.

Seed zone: Area of similar environmental conditions. Plants originating from the same seed zone are believed to be similarly adapted to the environment.

Serotinous: Cones which remain closed after maturing and which do not release seeds until several years after reaching maturity or exposure to high temperatures (e.g., some ecotypes of lodgepole pine).

Shale: Flat, layered rock consisting of consolidated clay and silt.

Shippable percent: Percentage of seedlings remaining at the end of the nursery growing period which meet certain size and form specifications (compare with **tree percent, yield percent**).

Silt: Soil particle between 0.05 and 0.002 mm in diameter; soil textural class characterized by a predominance of silt particles.

Sludge: General term for solid wastes, usually collected by sedimentation from water. Sludge is derived from many sources including agricultural wastes, brewery and cannery wastes, and sewage.

Soil horizon: Layer of soil approximately parallel to the land surface and differing from adjacent genetically related layers in physical, chemical, and biological properties or characteristics such as color, structure, texture, consistency, amount of organic matter, and degree of acidity or alkalinity.

Soil-moisture retention curve (soil-water characteristic curve): Curve depicting the relationship between soil matric potential and soil water content; varies with soil textural class and is usually calculated by soil-testing laboratories. This curve aids nursery managers in regulating soil matric potential and therefore irrigation scheduling.

Soil test: Chemical or physical analysis of a soil to determine texture, acidity, total salt concentration, or concentration of nutrient elements.

Soil water potential: Amount of work that a plant must exert to absorb water from the soil, usually expressed in units of negative pressure (– bars). It is composed primarily of matric potential but can also have as components osmotic potential, pressure potential, and gravitational potential.

Sowing: Process of placing seeds in the seedbed at a specific depth and density.

Specific gravity: Ratio of weight of a volume of matter to weight of an equal volume of water at a specific temperature.

Standard treatment: Treatment which simulates the operational procedures of a current practice (compare with **control treatment**).

Statistical inference: Process of making generalizations, based on sample data, about a population or wider sphere of interest.

Statistics: Use of mathematical techniques to draw conclusions from a limited number of observations.

Stock type: Seedling classification, usually by age and location in the nursery—e.g., 1+0, 2+0, etc. The first of the two digits represents the number of growing seasons spent in the seedbed, the second digit the number of growing seasons spent in a transplant bed. "1+0 for 1+1" or "2+0 for 2+1" means seedlings grown for transplanting, often under specific cultural conditions (e.g., high seedbed density) (compare with **plug plus one, p+1**).

Stratification (seed): Treatment applied before germination to overcome seed dormancy. Cold stratification consists of placing seeds in an environment of cold temperature, sufficient moisture, and oxygen for a specified time period.

Subsoiling: Tillage of subsurface soil without inverting it, to break up dense soil layers that restrict water movement and root penetration (compare with **ripping, chiseling**).

Summer annuals: Plants which germinate in the spring, do most of their growing in the summer, produce flowers or seeds, and then die in the fall of the same year (compare with **winter annuals**).

Surfactant: Chemical agents (e.g., spreaders, detergents, wetting agents) added to pesticides to make mixing easier and to assist application of a solution and adherence to the treated surface.

Symbiosis: Association of two dissimilar organisms, usually referring to cases in which the relationship is beneficial to one or both organisms.

Systemic: Entering and then acting within an entire organism; used especially to describe the action of pesticides or diseases within a plant.

Systemic herbicide: Herbicide which is absorbed by and then distributed within a plant, as opposed to one which functions only on contact with the plant's surface (compare with **contact herbicide**).

T

Table pruning: Pruning of seedling roots at the time of grading and packing.

Tensiometer: Instrument for measuring the matric potential of soil water, often used for monitoring irrigation.

Tilth: Physical condition of soil as related to its ease of tillage, fitness as a seedbed, and impedance to seedling emergence and root growth.

Tissue analysis: Chemical analysis of plant tissue to determine the concentrations of essential elements in the plant.

Top dressing: Application of chemical or fertilizer after a crop has been established.

Top pruning: Clipping of seedling terminal leaders with a sharp blade to alter shoot:root ratio, facilitate handling, achieve uniformity in crop size, and control height growth.

Trace element: See **micronutrient**.

Transpiration: Process of water movement through a plant to the atmosphere as a result of evaporation of water from leaves.

Transplant: Cultural practice of moving seedlings from one bed to another to promote additional growth. Also, a seedling after it has been lifted and then replanted one or more times in the nursery.

Transplant shock: Reduced growth rate of a young tree after it has been transplanted or outplanted.

Tree percent: Number of seedlings, irrespective of size or form, in a nursery bed at lifting compared to the number of viable seed sown, expressed as a percentage (compare with **yield percent, shippable percent**).

Turgor potential: One of the main components of plant water potential; reflects a positive force exerted inward by the cell wall.

Type 1 error: Rejection of the null hypothesis when it is true.

Type 2 error: Acceptance of the null hypothesis when it is false.

UV

Undercutting: Root pruning in the nursery bed using a sharp blade drawn parallel to the soil surface at a regulated depth to stimulate root growth and fibrosity (compare with **wrenching**).

Vesicular-arbuscular (VA) mycorrhiza(e): Group of mycorrhizae in which the fungal hyphae form two characteristic structures—vesicles and arbuscules—which are both intercellular and intracellular. Vesicles are saclike storage and reproductive structures, whereas finely branched arbuscules facilitate nutrient exchange. Though not visible to the naked eye, VA mycorrhizae can be seen by staining infected cells and then inspecting by microscope.

Viability: Ability of a seed to germinate and grow under a given set of conditions; usually estimated by germination percent or other tests.

WXYZ

Water content (soil): Index of soil moisture status, calculated as the amount of water lost from the soil upon drying to constant weight at 105°C; usually expressed as the weight of water per unit weight of dry soil.

Waterlogged: Saturated with water. Waterlogged soil, which may result from a high water table caused by overirrigation, seepage, or inadequate drainage, is detrimental to plant growth.

Water potential: See **plant water potential, soil water potential**.

Water table: Upper surface of the ground-water level, below which the soil is saturated with water.

Wettable powder (WP): Powder formulation of a pesticide which contains a wetting agent so that it will readily form a suspension in water.

Wetting agent: Compound added to a pesticide solution causing the spray droplets to spread and more thoroughly wet the leaf surface.

Wind frost: Frost occurring when winds in excess of 4 mph from cold regions displace warmer air. Wind frosts can occur day or night and are not necessarily dependent on topography (compare with **radiation frost, advection frost**).

Winter annuals: Plants which germinate in the fall or early winter, exist in a rosette form over the winter, do most of their growing the following spring, and then die after flowering and producing seeds (compare with **summer annuals**).

Winter burn: Type of cold injury to foliage. Foliage is warmed above freezing by the winter sun during the day (even though air temperature is below freezing), then refreezes after sunset (compare with **winter scald**).

Winter desiccation: Type of foliage injury which occurs on warm days when the ground is frozen; actually a type of physiological drought caused by excessive transpiration when frozen soils prohibit water absorption.

Winter scald: Type of cold injury to tree bark. Bark is warmed above freezing by the winter sun during the day (even though air temperature is below freezing), then refreezes after sunset (compare with **winter burn**).

Wrenching: Passing of an *angled* horizontal blade beneath the soil surface of the nursery bed at a specified depth to cut newly penetrating roots and to loosen and aerate soil. Wrenching is used to stimulate root growth and fibrosity and to regulate seedling growth (compare with **undercutting**).

Yield percent: Number of trees which meet a specific size criterion, regardless of form; expressed as a percentage. These seedlings may have multiple tops or damage from insects, disease, or other agents—characteristics that may make them unacceptable for shipping (compare with **shippable percent, tree percent**).

Index

Page numbers set in **boldface** indicate index entries located in figures or tables; page numbers set in regular type indicate index entries located in the text.

H

I

J

K

L

M

Page numbers set in **boldface** indicate index entries located in figures or tables; page numbers set in regular type indicate index entries located in the text.

Q

Page numbers set in **boldface** indicate index entries located in figures or tables; page numbers set in regular type indicate index entries located in the text.

R

S

Page numbers set in **boldface** indicate index entries located in figures or tables; page numbers set in regular type indicate index entries located in the text.